COMPETITION AND COOPERATION IN ECONOMICS AND BUSINESS

PROCEEDINGS OF THE ASIA-PACIFIC RESEARCH IN SOCIAL SCIENCES AND HUMANITIES, DEPOK, INDONESIA, 7–9 NOVEMBER 2016, TOPICS IN ECONOMICS AND BUSINESS

Competition and Cooperation in Economics and Business

Editors

Lindawati Gani, Beta Y. Gitaharie, Zaäfri A. Husodo & Ari Kuncoro

Faculty of Economics and Business, Universitas Indonesia, Indonesia

Routledge is an imprint of the Taylor & Francis Group, an informa business

Typeset by V Publishing Solutions Pvt Ltd., Chennai, India

Published by: CRC Press/Balkema
Schipholweg 107C, 2316 XC Leiden, The Netherlands
e-mail: Pub.NL@taylorandfrancis.com
www.crcpress.com – www.taylorandfrancis.com

ISBN: 978-1-138-62666-9 (Hbk)
ISBN: 978-1-315-22522-7 (eBook)

Competition and Cooperation in Economics and Business – Gani et al. (Eds)
© 2018 Taylor & Francis Group, London, ISBN 978-1-138-62666-9

Table of contents

Competition and Cooperation in Economics and Business – Gani et al. (Eds)
© 2018 Taylor & Francis Group, London, ISBN 978-1-138-62666-9

Preface

This book is one of the five volumes of proceedings that provides an opportunity for readers to engage with selected papers presented at the 1st Asia-Pacific Forum for Research in Social Sciences and Humanities (1st APRISH) conference held in Depok (Indonesia) on November 7–9, 2016. Quoting the theme of the conference "Competition and Cooperation in the Globalized World," this book would be a valuable contribution to get a better understanding of the future global challenges. Readers will find a broad range of research reports on topics ranging from globalization, poverty, population and some macroprudential perspectives on managing the risk of a more integrated world. One of the unique and valuable dimensions of this conference is the capability for blending the chemistries among educators, students, and researchers from around the world to discuss globalization in a broad spectrum. The reader will discover both common challenges and creative solutions emerge from very diverse settings.

The relevance of the conference's theme, towards an evidence-based society, for a variety of disciplines, is reflected in the diverse range of papers that have been submitted for publication. These serve to demonstrate the natural attraction of the conference for sharing ideas and findings within an international community. Overall, the conference was undoubtedly very successful. Plenary lectures and presentations were successfully bridging the gap between the different fields of social sciences, making it possible for non-experts in a given area to gain insight into new areas. Also, among the speakers there were some young scientists, namely, doctoral, master's and undergraduate students, who bring new perspectives to their fields.

The editors would like to thank all those who have contributed to submit full papers for this conference. We received 101 papers from the conference and after a rigorous review, only 49 papers were accepted to be published in this volume. We would like to express our appreciation and gratitude to the scientific committee and the reviewers who have selected and reviewed the papers, and also to the technical editor's team who helped carry out the page layout and check the consistency of the papers with the publisher's template. It is an honor to publish selected papers in this volume by CRC Press/Balkema (Taylor & Francis Group). Finally, we would like to thank the steering committee, the chairman of the conference, the members of organizing committee involved in preparing and organizing the conference, and the financial support from Universitas Indonesia.

The Editorial Board of the 1st APRISH Proceedings for Topics in Economics and Business

Prof. Dr. Lindawati Gani
Department of Accounting, Faculty of Economics and Business, Universitas Indonesia

Dr. Beta Y. Gitaharie
Department of Economics, Faculty of Economics and Business, Universitas Indonesia

Zaäfri A. Husodo, Ph.D.
Department of Management, Faculty of Economics and Business, Universitas Indonesia

Prof. Ari Kuncoro, Ph.D.
Department of Economics, Faculty of Economics and Business, Universitas Indonesia

Competition and Cooperation in Economics and Business – Gani et al. (Eds)
© 2018 Taylor & Francis Group, London, ISBN 978-1-138-62666-9

Organizing committee

STEERING COMMITTEE

Rosari Saleh, *Vice Rector of Research and Innovation, Universitas Indonesia*
Topo Santoso, *Dean Faculty of Law, Universitas Indonesia*
Ari Kuncoro, *Dean Faculty of Economics and Business, Universitas Indonesia*
Adrianus L.G. Waworuntu, *Dean Faculty of Humanities, Universitas Indonesia*
Arie Setiabudi Soesilo, *Dean Faculty of Social and Political Sciences, Universitas Indonesia*

INTERNATIONAL ADVISORY BOARD

Peter Newcombe, *University of Queensland, Australia*
Fred Piercy, *Virginia Tech University, Australia*
Frieda Mangunsong Siahaan, *Universitas Indonesia, Indonesia*
James Bartle, *University of New South Wales, Australia*
Elvia Sunityo Shauki, *University of South Australia, Australia*

SCIENTIFIC COMMITTEE

Manneke Budiman
Isbandi Rukminto Adi
Beta Yulianita Gitaharie
Surastini Fitriasih
Sri Hartati R. Suradijono
Elizabeth Kristi Poerwandari

CONFERENCE DIRECTOR

Tjut Rifameutia Umar Ali

CONFERENCE VICE-DIRECTOR

Turro Wongkaren

ORGANIZING COMMITTEE

Dewi Maulina, *Faculty of Psychology, Universitas Indonesia*
Intan Wardhani, *Faculty of Psychology, Universitas Indonesia*
Elok D. Malay, *Faculty of Psychology, Universitas Indonesia*
Josephine Rosa Marieta, *Faculty of Psychology, Universitas Indonesia*
Teraya Paramehta, *Faculty of Humanities, Universitas Indonesia*

Nila Ayu Utami, *Faculty of Humanities, Universitas Indonesia*
Priskila Pratita Penasthika, *Faculty of Law, Universitas Indonesia*
Efriyani Djuwita, *Faculty of Psychology, Universitas Indonesia*
Destri Widaya, *Faculty of Economics and Business, Universitas Indonesia*

WORKSHOP COMMITTEE

Corina D.S. Riantoputra, *Faculty of Psychology, Universitas Indonesia*
Fithra Faisal Hastiadi, *Faculty of Economics and Business, Universitas Indonesia*
Mirra Noormilla, *Faculty of Psychology, Universitas Indonesia*

TREASURERS

Robby Oka Yuwansa, *Faculty of Psychology, Universitas Indonesia*
Nurul Husnah, *Faculty of Economics and Business, Universitas Indonesia*

Competition and Cooperation in Economics and Business – Gani et al. (Eds)
© 2018 Taylor & Francis Group, London, ISBN 978-1-138-62666-9

Supporting factors and the effect of an Initial Public Offering (IPO) on an Islamic bank: The first case in Indonesia

D. Siswantoro
Department of Accounting, Faculty of Economics and Business, Universitas Indonesia, Depok, Indonesia

ABSTRACT: An Islamic bank has been established in Indonesia since 1992, but there is only one Islamic bank, Bank Panin Syariah, which has listed its shares on the stock exchange, which happened in 2014. This listing was initiated by a new fully fledged Islamic bank, which was established in 2009. Thus, the objective of this paper is to investigate the supporting factors of an Initial Public Offering (IPO) and its effect on an Islamic bank. This is the first case in Indonesia, and it may offer a few good lessons for others. The qualitative method was used by the author to analyse relevant documents and information regarding the Islamic bank's IPO, such as prospectuses, news, research reports and stock prices. Topics that covered the supporting factors in an IPO include the utilisation of funds and an event study in share issuance by an Islamic bank in Indonesia. The results may indicate that it is quite difficult to sell shares for a new bank, because the utilisation of funds to boost new branches and the performance of the stock prices are affected by specific issues.

1 INTRODUCTION

It is a common phenomenon in Indonesia for a bank to be listed on the stock exchange. A total of 37 banks listed their shares on the stock market in 2013. But this is not true for Islamic banks, which were established in Indonesia in 1992. There is only one Islamic bank, Bank Panin Syariah, which has issued and traded its shares on the Indonesian Stock Market (IDX), which happened in 2014. This issue may be caused by the smaller capital market of the Islamic banks. In addition, listing shares on the stock exchange may be viewed as an unlawful transaction because traders can manipulate or 'short sell' the transaction, which is unlawful in Islamic teaching.

Yet Islamic banks may benefit from getting funds by issuing shares. Most Islamic banks in Indonesia have a large demand from customers, so funding is increased from savings. This has limited time constraints, such as time deposits, so it can only be delivered for a specific period. Until 2013, Islamic banks in Indonesia only used sukuk (Islamic bond) and equity injections from internal or external institutions.

This paper will investigate the supporting factors of an IPO and its effect on an Islamic bank when it lists its shares on the stock exchange; the first listing in Indonesia was implemented by Bank Panin Syariah. This paper begins with a brief explanation of the topic. It then continues with a literature review, which includes relevant theories about IPO; the research methodology (i.e. how the research was conducted); and an analysis, which focuses on evaluating the research. A conclusion, which addresses the objectives of the research, will follow.

2 LITERATURE REVIEW

There are many reasons why companies choose to offer an Initial Public Offering (IPO). The biggest reason may be to obtain a large amount of flexible funds and to strengthen capital. Another reason may be to promote the company, which encourages investors and is more transparent.

In general, some of the motives behind an IPO are as follows: (a) to create an efficient capital structure; (b) to increase bargaining power with the banks because of a better cost of capital; (c) to increase the market price of a company; and (d) to create publicity (Ising, 2013). Most of these requirements must be met in order to get better funding for an IPO. The better the market valuation, the higher the valuation of the share price of an IPO. Specifically, Kim and Heshmati (2010) found that, for an IT company, an IPO has some benefits and adds confidence to the market. Islamic banks in Indonesia still have a low market share and stock valuation; this may affect their stock's activity in the stock market.

Cecchini et al. (2012) state that after an IPO, companies tend to write off their bad debts to increase income. However, Black and Gilson (1998) characterise countries based on stock market and bank funding domination. Thus, not all countries may be suited to capital market instruments. We will next discuss the theory of the paper—namely, the theory of underpricing and performance.

2.1 *Underpricing*

Underpricing is a condition whereby the price of an IPO is lower than it is in a secondary market transaction. This condition may not occur on purpose, but it can attract investors to buy shares because the share price looks cheap and it is seen as producing a bigger capital gain in the future. In fact, a theoretical model of IPOs for underpricing consists of the following: (a) adverse selection or underpricing caused by asymmetric information; (b) a principal agent; (c) signalling or knowing the market better; and (d) a heterodox explanation, which is characterised by the underwriters' reputation and the possible risks involved (Anderson et al., 1995). Underpricing can be an incentive for investors to buy on the IPO date; otherwise the underwriter may incur a bad reputation. Similarly, underpricing an IPO can be caused by market sentiment, dividend forecasts, options (Dimovski & Brooks, 2004), and document completeness (Chen & Guo, 2010).

Borges (2007) found that underpricing IPOs usually leads to better stock performance in the long run. Investors gain the confidence to buy more shares and the price increases consistently. In addition, having an underwriter tends to stabilise the market price after an IPO, especially in situations of underpricing (Mazouz et al., 2013).

In the case of a post-IPO bank, dividends can be a signal of a bank's maturity, which can mean a higher stock value for investors (Cornett et al., 2011). Conducting an IPO and an efficient funds placement can create effective asset valuation and development.

2.2 *Performance*

Some investors would like to keep their shares for a longer period. However, in the long run, the IPO stock price underperforms due to various factors, such as poor performance (Ecker, 2005). However, most companies underperform in their first year (Dimovski & Brooks, 2004). In fact, in China the IPO process can be manipulated, but we can see the stock performance in the long run (Cai et al., 2008). In the case of a bank, an IPO does not affect the bank's performance if it has acquired large debt loans before the IPO (Houge & Loughran, 1999).

The underwriter may play an important role in the IPO. Dunbar (2000) states that the market share of underwriters is determined by the success of IPOs. In addition, to reduce asymmetric information and signal issues, some companies choose a qualified underwriter for an IPO (Baginski et al., 1999). However, Fields and Fraser (2004) found that there is no difference between commercial and traditional banks in terms of underwriting. In other cases, Fargher et al. (2000) stated that a new underwriter charges higher fees for an IPO and assumes a bigger risk because of uncertainty. However, Balkovskaya and Stullova (2012) found that, following an IPO by a Russian bank, stock prices were affected by macroeconomic factors. In addition, companies that maintain good relationships with banks perform better after an IPO (Gonzalez & James, 2007). Finally, Ghosh (2012) found that the banking sector in developing countries performs better after an IPO.

Kim (1999) found that, after an IPO, smaller companies face more constraints than do bigger companies. In addition, an asset quality factor affects banks that conduct IPOs (Ahmad & Kashian, 2010). Allen et al. (2012) show that China's model involves offering IPOs in small share portions to foreign institutions.

3 RESEARCH METHODOLOGY

To conduct the research on the first IPO for an Islamic bank, the author analysed the prospectuses of IPOs, research reports and stock news, and stock price movement after an IPO. The data is analysed based on the specific issues for this research:

1. Prospectus: A prospectus contains information about the company, the purpose of an IPO and other related information.
2. Research report and stock news for stock price movement analysis: A company's securities give a view of the company, including the corporate actions, issues, and other important matters. News from the Indonesian Stock Exchange can also be used.
3. Benchmarking analysis: In the benchmarking analysis, the stock price would be compared with that of the parent company and with related information that could affect the price. The period was from 15th January 2014 to 9th June 2014.

The research is limited to 2014, both before and after the IPO, and is based on stock prices and other information and news. From the above, we can see what news or information could affect the fluctuation of the stock prices.

4 ANALYSIS

4.1 *The prospectus*

On the prospectus cover, we can see the number of shares that were issued: 5 billion at Rp 100 (equivalent to 50%) with 1 billion warrant at Rp 100 on 17th January 2017 (equivalent to 20%). The underwriters were PT Evergreen Capital and PT RHB OSK Securities Indonesia. However, the largest market share for underwriting was held by Mandiri Sekuritas in 2013. As shown in Table 1, holding the market share of the underwriting can affect the IPO process (Dunbar, 2000). In Table 2, we can see the fund proposal before and after the IPO. The public would own 50% of the planned shares. In addition, PNBS (Bank Panin Syariah) also has a warrant, which can add to the number of shares and the Employee Stock Allocation (ESA) that is taken from the society (see Table 3). Eighty per cent of the IPO was used for working capital, whereas 20% was used for developing networks. In fact, the total funds from the society were only Rp 121.7 billion out of a total of Rp 500 billion. This is below expectations, then the parent company, Bank Panin buys the rest of shares (see Table 4). The IPO did not

Table 1. Underwriter market share in 2009.

Underwriter	%
Mandiri Sekuritas	28.4%
PT CIMB Securities Indonesia	24.5%
PT Bahana Securities	9.7%
PT BNP Paribas	8.6%
PT Danareksa Sekuritas	8.6%
Others	20.2%

Source: http://www.ibpa.co.id/News/ArsipBerita/tabid/126/EntryId/701/Danareksa-Mandiri-Kuasai-Pasar-Underwriting.aspx

Table 2. Composition before and after an IPO (in Rupiah).

	Before		After	
Owner	Amount	%	Amount	%
PT Bank Panin Tbk	499,995,179,000	99.9	499,995,179,000	49.9
H. Ahmad Hidayat	4,821,000	0	4,821,000	0
Society	0	0	500,000,000,000	50
Total	500,000,000,000	100	1,000,000,000,000	100

Source: Panin Bank Syariah (2013), (2014).

Table 3. Composition after warrant and ESA (in Rupiah).

	After Warrant		After ESA	
Owner	Amount	%	Amount	%
PT Bank Panin Tbk	499,995,179,000	45.4	499,995,179,000	45.4
H. Ahmad Hidayat	4,821,000	0	4,821,000	0
Society	500,000,000,000	45.4	450,000,000,000	40.9
Warrant Stock	100,000,000,000	9.0	100,000,000,000	9.0
ESA			50,000,000,000	4.5
Total	1,100,000,000,000		1,100,000,000,000	100

Source: Panin Bank Syariah (2013), (2014).

Table 4. Composition after and IPO realisation (in Rupiah).

	After		Realisation (31st March 2014)	
Owner	Amount	%	Amount	%
PT Bank Panin Tbk	499,995,179,000	49.9	853,245,179,000	87.51
H. Ahmad Hidayat	4,821,000	0	0	
Society	500,000,000,000	50	121,754,821,000	12.49
Total	1,000,000,000,000	100	975,000,000,000	100

Source: Panin Bank Syariah (2013), (2014).

significantly affect the fixed assets; this may be caused by large withdrawals from time deposits (Rp 508.7 billion). So the IPO may not have been conducted well on this occasion. However, as a plan, 80% of IPO (Rp 369.6 billion) funds are used for working capital and 20% are used for networking and infrastructure development (Rp 92.4 billion).

4.2 *Research report and stock news*

Not many securities covered PNBS news but we found some highlights from the RHB Research Institute Sdn. Bhd. On 20th November 2013 (40 pages):

i. Promising growth (asset, revenue) for the Islamic bank in Indonesia, which was ranked eighth in total assets, and
ii. Low valuation compared to other banks; its ROE (Return on Equity) was 7.4% (2013), which is below the cost of equity (14.5%).

Other information is from Reliance Securities (2013), which only covered the company profile in terms of strategy, risk and activities. Analyses were on lower Non Performing Fund (NPF) and good Capital Adequacy Ratio (CAR). It also covered decreasing efficiency costs and cheap IPO stock prices.

We could not find much research coverage or information about securities after an IPO. Some of the research and news items are as follows:

i. 7th March 2014 (Indonesia Finance Today—news). Bank Panin Syariah would like to sell 25% of its shares, which are owned by Bank Panin. The potential investor, Affin Holdings Berhad, is from Malaysia.
ii. 26th May 2014 (Paramitra—research). The following news is that Bank Panin Syariah would like to sell 24.9% of its shares at an amount of Rp 120. The investor would like to buy 40%; they are from Dubai Islamic Bank.
iii. 4th June 2014 (Myanmar Eleven—news). Dubai Islamic Bank bought Bank Panin Syariah from Bank Panin, so its shares decreased to 64.01% from 87.51%.
iv. 5th June 2014 (Trust Securities—research). Dubai Islamic Bank bought 24.9% of Bank Panin Syariah for Rp 115, at a total amount of Rp 2.29 billion. Ising (2013) stated that by listing on the stock exchange, a company could become well known by investors.

Interestingly, the share price increased when there was news of corporate action (i.e. that Bank Panin Syariah would be acquired by foreign investors, see Figure 1). The price then tended to decrease when the parent company sold its shares on 5th March 2014, selling at Rp 103, and on 10th March 2014. The stock price then fluctuated until Dubai Islamic Bank (DIB) planned to buy about 24.9% of the shares. After the transaction was completed, the

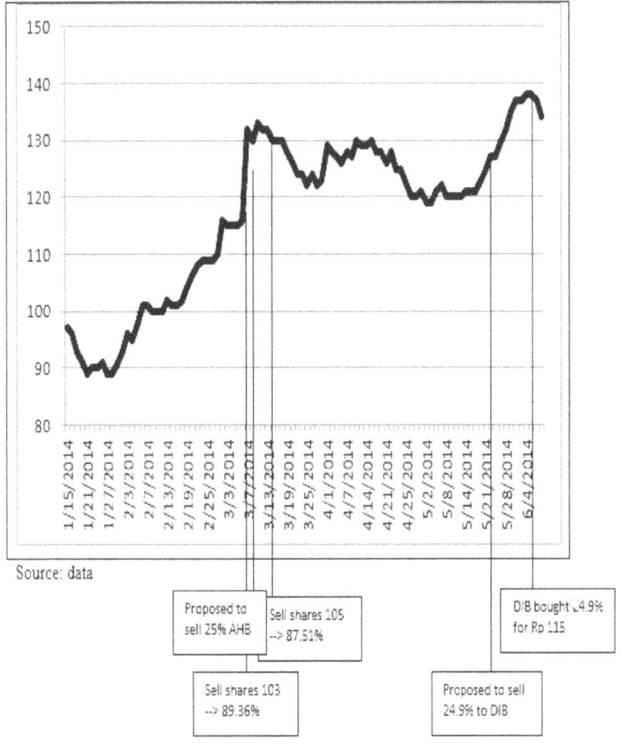

Figure 1. Stock price movement analysis.

share price went down; this decrease may have been caused by the selling price being below the proposed offer price (down from Rp 120 to Rp 115).

4.3 *Benchmarking analysis*

In the benchmarking analysis, we add Bank Panin (PNBN), as a shareholder, to the analysis. To measure the stabilisation of the price movement from 15th January to 9th June 2014, the formula is high price minus low price for each day's transaction. This formula is used to see the gap in each day's transaction—the bigger the gap, the more volatile the stock. Then the mean and standard deviation are each determined for the stock. PNBS has a mean of 4.64 and a standard deviation of 2.76, whereas PNBN has a mean of 24.79 and a standard deviation of 13.8. The result is a volatility of 1.68 for PNBS and 1.79 for PNBN.

For robustness, we used the stock price movements for the volatility analysis. We found that the mean of PNBS is 117.08 and the standard deviation is 14.35; PNBN has a mean of 798.33 and a standard deviation of 75.63. The volatility is still greater for PNBN than PNBS: 10.55 for PNBN compared to 8.15 for PNBS. The first approach may be stronger for the purposes of analysis because it uses a gap analysis of the high-low price for each day. One may assume that the buyer of PNBS is more cautious because Islamic teaching prohibits speculation.

The next analysis is the correlation between PNBS and PNBN, the result is insignificant. From this, it would appear that the stock movement between the two stocks might not be different.

From Figure 2, we can see that PNBS did not experience underpricing because the PNBN and Jakarta Islamic Index (JII) stocks were decreasing. In some cases, even though the market did not support an IPO, a company's IPO stock may increase significantly on the first day of the offering, which may indicate that the stock is supported by strong investors. However, in the case of PNBS, it reached 100 again at the beginning of February 2014. Then, it increased significantly at the beginning of March 2014, reaching 133 in the middle of March. Borges (2007) would have supported the condition if it were in a normal IPO situation. In addition, an underwriter's reputation can also be a factor in underpricing conditions (Anderson et al., 1995). A good underwriter can be seen from the market share of an IPO (Dunbar, 2000). Again, although concrete references to this specific case would be helpful, a qualified underwriter can reduce the asymmetric information for an IPO (Baginski et al., 1999).

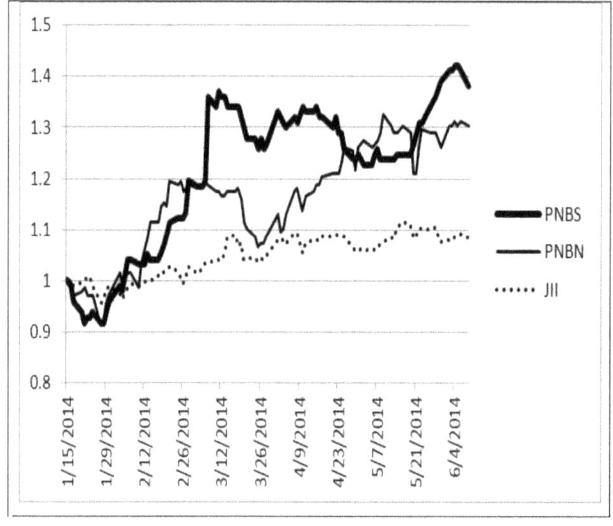

Figure 2. Benchmarking analysis.
Source: Data.

From mid-March until the end of May 2014, the stock price fluctuated. It reached its peak of 138 on 4th June 2014. From Figure 2, we may infer that the correlation between both stocks is different. Besides, the objective of each trader is different, but not significantly so. A similar opinion was held by Houge and Loughran (1999), who found that banks have different characteristics pre- and post-IPO.

5 CONCLUSION

Factors contributing to the success of an IPO may be (a) the reputation and market share of the underwriter, (b) the economic conditions of the market and (c) the proposed type of shares, such as a warrant or an Employee Stock Allocation (ESA). The more sweeteners that there are, the more reluctant long-term investors are to buy the stock. After an IPO, stock price is affected by (a) rumours, (b) news from securities and the media and (c) potential foreign information. Bank Panin Syariah may enjoy the benefits from the IPO as it becomes better known in the market, both locally and internationally.

This research may be useful for other Islamic banks that would like to offer an IPO. An IPO can be one possible financial instrument for expanding the bank network. The limitation of the paper is that it used qualitative analysis. Quantitative analysis, such as econometrics, should be applied to test the significant factors.

REFERENCES

Ahmad, Y. & Kashian, R. (2010). Modeling the time to an Initial Public Offering: When does the fruit ripen? *Journal of Economics and Finance, 34*(4), 391–414. doi:10.1007/s12197-008-9073-z.

Allen, F., Shan, S.C., Qian, J. & Zhao, M. (2012). The IPO of industrial and commercial bank of China and the 'Chinese Model' of privatizing large financial institutions. *The European Journal of Finance, 20*(7–9). doi:10.1080/1351847X.2012.671780.

Anderson, S., Beard, T.R. & Born, J. (1995). *Initial Public Offerings: Findings and theories*. New York: Kluwer Academic Publishers.

Baginski, S.P., Hassell, J.M. & Neill, J.D. (1999). Predicting subsequent management forecasting behavior at the date of an Initial Public Offering. *Review of Quantitative Finance and Accounting, 12*(1), 5–20. doi:10.1023/A:1008304206249.

Balkovskaya, D. & Stullova, N. (2012). IPO of Russian banks: The crisis effect on the market value of the stocks. *International Journal of Business and Social Science, 3*(1), 59–65.

Black, B.S. & Gilson, R.J. (1998). Venture capital and the structure of capital markets: Banks versus stock markets. *Journal of Financial Economics, 47*(3), 243–77. doi:10.1016/S0304-405X(97)00045-7.

Borges, M.R. (2007). Underpricing of Initial Public Offerings: The case of Portugal. *International Advances in Economic Research, 13*(1), 65–80.

Cai, X. Liu, G.S. & Mase, B. (2008). The long-run performance of initial public offerings and its determinants: The case of China. *Review of Quantitative Finance and Accounting, 30*(4), 419–432.

Cecchini, M., Jackson, S.B. & Liu, X. (2012). Do Initial Public Offering firms manage accruals? Evidence from individual accounts. *Review of Accounting Studies, 17*(1), 22–40.

Chen, H.C. & Guo, W.C. (2010). Divergence of opinion and Initial Public Offerings. *Review of Quantitative Finance and Accounting, 34*(1), 59–79. doi:10.1007/s11156-009-0125-z.

Cornett, M.M., Fayman, A., Marcus, A.J. & Tehranian, H. (2011). Dividends, maturity, and acquisitions: Evidence from a sample of bank IPOs. *Review of Financial Economics, 20*(1), 11–21. doi:10.1016/j.rfe.2010.11.001.

Dimovski, W. & Brooks, R. (2004). Initial Public Offerings in Australia 1994 to 1999, recent evidence of underpricing and underperformance. *Review of Quantitative Finance and Accounting, 22*(3), 179–198. doi:10.1023/B:REQU.0000025759.39918.89.

Dunbar, C.G. (2000). Factors affecting investment bank Initial Public Offering market share. *Journal of Financial Economics, 55*(1), 3–41. doi:10.1016/S0304-405X(99)00043-4.

Ecker, F. (2005). *Information risk and long-run performance of Initial Public Offerings*. Wiesbaden: Gabler.

Fargher, N., Fields, L.P. & Wilkins, M.S. (2000). The Impact on IPO assurance fees of commercial bank entry into the equity underwriting market. *Auditing, 19* (SUPPL.), 23–35. doi:10.2308/aud.2000.19.s-1.23.

Fields, L.P. & Fraser, D.R. (2004). Effects of IPO mispricing on the risk and reputational capital of commercial banks. *Review of Financial Economics, 13*(1–2), 65–77.
Ghosh, S. (2012). *The post offering performance of IPOs from the banking industry*. Retrieved from *oii.igidr.ac.in:8080/jspui/bitstream/2275/212/1/saurabh.pdf*.
Gonzalez, L. & James, C. (2007). Banks and bubbles: How good are bankers at spotting winners?" *Journal of Financial Economics, 86*(1), 40–70. doi:10.1016/j.jfineco.2006.07.004.
Houge, T. & Loughran, T. (1999). Growth fixation and the performance of bank Initial Public Offerings, 1983–1991. *Journal of Banking & Finance, 23*(8), 1277–1301.
Indonesia Finance Today. (2014). *Panin Bank Siap Lepas 25% Saham Bank Panin Syariah*. 7th March 2014, Jakarta: Indonesia Finance Today.
Ising, P. (2013). *Earnings accruals and real activities management around Initial Public Offerings*. Wiesbaden: Springer Gabler.
Kim, J. (1999). The relaxation of financing constraints by the Initial Public Offering of small manufacturing firms. *Small Business Economics, 12*(3), 191–202. doi:10.1023/A:1008090609649.
Kim, Y. & Heshmati, A. (2010). Analysis of Korean IT startups' Initial Public Offering and their post-IPO performance. *Journal of Productivity Analysis, 34*(2), 133–49. doi:10.1007/s11123-010-0176-0.
Mazouz, K., Ampomah, S.A., Saadouni, B. & Yin, S. (2013). Stabilization and the aftermarket prices of Initial Public Offerings. *Review of Quantitative Finance and Accounting, 41*, 417–439.
Myanmar Eleven. (2014). *Dubai Islamic bank acquires shares in Indonesian bank*. 4th June 2014, Yangoon: Myanmar Eleven.
Panin Bank Syariah. (2013). *Prospektus awal*. Jakarta: Bank Panin Syariah.
Panin Bank Syariah. (2014). *Financial statement 31 March 2014 and 31 December 2013*. Jakarta: Bank Panin Syariah.
Paramita Alfa Sekuritas. (2014). *Market review*. 26th May 2014, Jakarta: Paramitra Alfa Sekuritas.
Reliance Securities. (2013). *Equity research Panin bank syariah*. 4th December 2013, Jakarta: Reliance Securities.
RHB Research Institute Sdn Bhd. (2013). *Panin bank syariah IPO Report*. 20th November 2013, Kuala Lumpur: RHB Research Institute Sdn Bhd.
Trust Securities. (2014). *Stock news*. 5th June 2014, Jakarta: Trust Securities.

Competition and Cooperation in Economics and Business – Gani et al. (Eds)
© 2018 Taylor & Francis Group, London, ISBN 978-1-138-62666-9

Analysis of behavior and determinants of cost stickiness in manufacturing companies in Indonesia

Farizy Yunaz & Catur Sasongko
Department of Accounting, Faculty of Economics and Business, Universitas Indonesia, Depok, Indonesia

ABSTRACT: This research aims to provide the empirical evidence regarding cost stickiness behavior and its determinants in listed manufacturing companies. Hypothesis testing is performed using the pooled least square method. The result concludes that there is cost stickiness behavior in selling, general, and administrative costs. In terms of determinants, firm specific adjustment costs measured by asset intensity and employee intensity have a positive significant impact on the level of cost stickiness. Meanwhile, earnings target and leverage have a negative significant impact on the level of cost stickiness. However, the management empire building incentives measured by free cash flow has no positive significant impact.

1 INTRODUCTION

In a traditional cost behavior concept, costs are classified into fixed cost and variable cost, which are assumed to have a proportional relationship with the level of activities. These costs may either increase or decrease according to the change in activities (Noreen, 1991). In fact, previous studies have suggested that the traditional cost behavior assumption does not fully apply. Cooper and Kaplan (1998) find that there are differences in the cost behavior response from its activities since management has the desire to increase the cost changes more when activities are rising than when activities are declining. Anderson *et al.* (2003) find that selling, general, and administrative costs respond to upward or downward changes in activity. This condition is referred to as cost stickiness behavior. The increase of these costs when revenues increase is higher compared to the decrease caused by the decline in revenues by an equivalent amount. This cost stickiness behavior is influenced by several factors. Firm-specific adjustment costs measured by asset intensity and employee intensity as suggested by Anderson *et al.* (2003) have a positive impact on cost stickiness. These adjustment costs happen when the firm increases or decreases its committed resources. Kama and Weiss (2010) also mention that an earning target could be a trigger for firms' cost saving. As a result, the degree of cost stickiness will decrease. According to Chen *et al.* (2012), the degree of cost stickiness could be influenced by the agency problem considering the amount of incentive to the management measured by free cash flow (FCF). Higher FCF could increase the management's opportunity to overinvest in operating costs in response to an increase in output demand and to delay the cutting of operating costs in response to a decrease in output demand. This condition is called a management empire building phenomenon, which could distort the cost behavior. On the other hand, Calleja *et al.* (2006) find that leverage increases the creditor's control of management in administering the firm's costs. Therefore, leverage affects the stickiness of cost behavior.

This study replicates the model in the previous studies. The objectives of the study are to investigate the cost stickiness behavior in operational costs, especially Sales and General Administrative (SandGA) Costs, and to find the determinants of this cost stickiness.

2 LITERATURE REVIEW

One of the basic concepts in cost accounting is cost behavior. In general, cost behavior relates to how cost responds to change in activities and volume. Many types of costs could respond proportionally, but some are independent with a change in activities and volume. Cost behavior is classified into fixed cost, variable cost, and semi-variable cost (Anderson and Raiborn, 1977; Carter and Usry, 2002).

1. Fixed Cost is the cost which remains constant within a relevant range of activities although there is fluctuation in business operations.
2. Variable Cost is the cost which will change in direct proportion to the changes in volume or activity.
3. Semi-variable Cost is the combination of variable and fixed costs in terms of characteristics.

Related to cost behavior, existing empirical studies find that some of the costs in a company might not respond proportionally to the change in volume and activity. It means that those costs will increase when the volume and activity increase, either larger or smaller than the decrease in costs when the volume and activity decreases. This cost behavior is called cost stickiness. Prior research by Anderson *et al.* (2003), Chen *et al.* (2012), Bruggen and Zehnder (2014), and Venieris *et al.* (2015) find cost stickiness behavior on sales, general, and administrative costs. On the other hand, Calleja *et al.* (2012) state that operating costs in companies have sticky behavior. In Indonesia, research done by Hidayatullah *et al.* (2011) and by Windyastuti (2013) demonstrates that sticky cost behavior on sales, general, and administrative costs is in tune with change in sales.

Cost stickiness is defined as asymmetric cost behavior because the relative magnitude of an increase in costs for an increase in sales is greater than the relative magnitude of decrease in costs for a decrease in sales (Anderson *et al.*, 2003). Bruggen and Zehnder state that cost stickiness is cost behavior which depends on a change in managerial decision making, especially when sales revenue decreases. The sticky cost behavior tends to rise in the short term but will not automatically decline along with a decrease in the activity (Mak and Roush, 1994 in Baumgarten, 2012). Instead of the cost stickiness phenomenon, the cost behavior could lead to cost anti-stickiness which occurs when the increase in costs caused by the increase in the activity is smaller than the decrease in costs resulting from the decrease in the activity. To explain more about this cost behavior, Figure 1 shows asymmetric cost behavior according to Balakrishnan *et al.*, (2004).

One of the main reasons for cost stickiness behavior is that there is committed resources cost adjustment in a company, which is not based on a degree of change in business activities

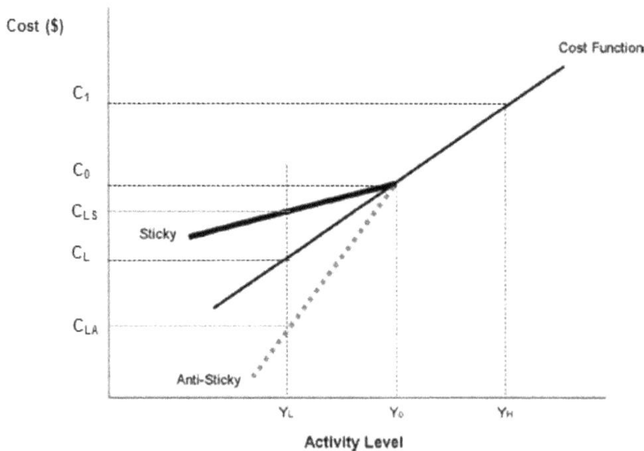

Figure 1. Asymmetric cost behavior function.
Source: Balakrishnan (2004).

(Guenther *et al.*, 2014). This happens when the management is involved in the resources adjustment process, especially through their decisions and intentions for the company. Anderson *et al.* (2003) state that there are two main arguments why the management would not decrease the company's costs when the business activities are declining. Firstly, there could be adjustment costs which are the potential cause of cost stickiness because when demand declines, managers are faced with a trade-off between the costs of retaining redundant resources on the one hand, and resource-adjustment costs on the other hand. Secondly, the management has their individual considerations as well as personal interests. Another reason stated by Guenther *et al.* (2014) is that cost stickiness could occur from several factors, such as:

1. Laws and regulations
2. Corporate social and personnel policy
3. Firm and operating policy
4. Psychological and agency-related reasons.

After explaining the reasons for sticky cost behavior, Baumgarten (2012) mentions the characteristics relating to cost stickiness:

1. General characteristics of cost stickiness.
2. Firm-specific characteristics of cost stickiness.
3. Industry-specific characteristics of cost stickiness.
4. Country-specific characteristics of cost stickiness.

Based on previous studies, there are several factors affecting the level of cost stickiness. These factors include firm-specific adjustment costs, earnings target, management empire-building state incentives, agency problem, and other factors. The firm-specific adjustment cost is measured by asset intensity and employee intensity. Earnings target is measured by return on equity (ROE). Management empire-building states incentives could be indicated by free cash flow (FCF). Agency problem is measured by the company's leverage. Finally, other factors are measured by GDP growth. Figure 2 shows the conceptual framework for this research.

Related to the cost stickiness phenomenon, there are several hypotheses developed in this research. First, the cost that will be tested includes the sales, general, and administrative (SG&A) costs (Anderson *et al.*, 2003).

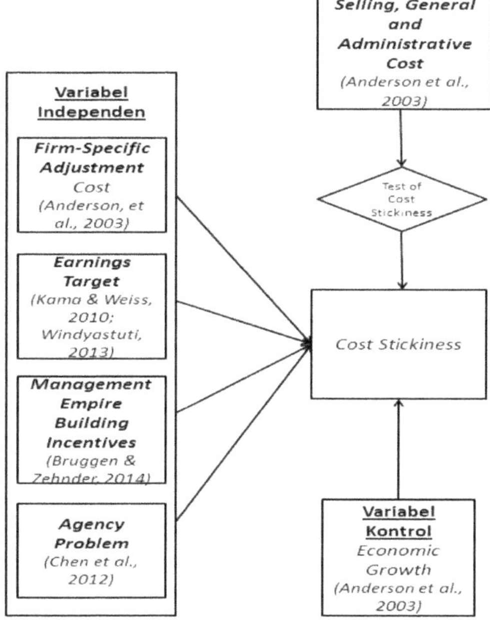

Figure 2. Conceptual framework.

H1a: The magnitude of an increase in SG&A costs for an increase in net sales is greater than the magnitude of a decrease in SG&A costs for a decrease in net sales.

Secondly, the company's resources related to assets and employees will be tested as one of the factors affecting the degree of cost stickiness. Those variables are indicated by asset intensity and employee intensity. Thus, the hypothesis could be derived as follows:

H2a: Asset intensity has a positive impact on the level of cost stickiness.

H3a: Employee intensity has a positive impact on the level of cost stickiness.

In addition, the level of cost stickiness is influenced by earnings target. Calleja *et al.* (2006) show that the earning target through return on equity (ROE) causes a decrease in the degree of cost stickiness in many companies in Germany. Kama and Weiss (2010) find that the profit target could motivate the management to achieve cost saving when sales are declining. Based on those studies, the hypothesis is:

H4a: Earnings target has a negative impact on the level of cost stickiness.

Chen *et al.* (2012) express that when the level of free cash flow (FCF) is high, the management has the opportunity to get empire-building incentives causing over-investment in operating costs. As a result, free cash flow (FCF) as an indicator of management empire-building incentives is related to the degree of cost stickiness.

H5a: Free cash flow has a positive impact on the level of cost stickiness.

The last main factor in this research is the agency problem. Calleja *et al.* (2006) discover that higher debt intensity in companies could lead to asymmetric cost behavior. Furthermore, Xue and Hong (2015) state that leverage in companies could improve the companies' performance as well as corporate governance which will minimize the distortion of cost behavior coming from discretionary management.

H6a: Leverage has a negative impact on the level of cost stickiness.

3 RESEARCH METHOD

The sample used for this research is manufacturing companies listed in Indonesia Stock Exchange (IDX) ten years starting from 2005 to 2014. Manufacturing companies are chosen because they have various types of costs compared to other industries. This research constitutes quantitative research using multiple regressions through the pool least square method. The research model is developed from previous studies. Variables to be tested are the ratio of sales, the general and administrative expenses as a proxy to measure the degree of cost stickiness, and the ratio of net sales as the main factor which determines the sticky cost behavior (Anderson *et al.,* 2003). Other variables are asset intensity, employee intensity, return on equity, free cash flow, and leverage. There is one control variable used, namely the GDP growth. The research models are presented below:

Model 1.

$$\log\left[\frac{SG\&A_{i,t}}{SG\&A_{i,t-1}}\right] = \beta_0 + \beta_1 \log\left[\frac{Sales_{i,t}}{Sales_{i,t-1}}\right] + \beta_2 * DecDummy_{i,t} * \log\left[\frac{Sales_{i,t}}{Sales_{i,t-1}}\right] + \varepsilon_{i,t}$$

Model 2.

$$log\left[\frac{SG\&A_{i,t}}{SG\&A_{i,t-1}}\right] = \beta_0 + \beta_1 log\left[\frac{Sales_{i,t}}{Sales_{i,t-1}}\right] + \beta_2 * Decrease\ Dummy_{i,t} * log\left[\frac{Sales_{i,t}}{Sales_{i,t-1}}\right] + \sum_{m=3}^{Z} \beta_m * Decrease\ Dummy_{i,t} * log\left[\frac{Sales_{i,t}}{Sales_{i,t-1}}\right] * Factors_{m,i,t} + \varepsilon_{i,1}$$

Factors in model 2 are asset intensity, employee intensity, return on equity, free cash flow, leverage, and GDP growth. Asset Intensity is measured by the logarithm of the ratio of total assets to net sales. Employee Intensity is measured by the logarithm of the ratio of the number of employees to net sales. Return on Equity is derived from the ratio of EBIT to equity. Free Cash Flow is calculated from cash flow from operation minus dividend and divided by total assets. GDP growth is measured by the percentage of yearly GDP growth. Finally, leverage is simply the ratio of total debt to total assets.

In addition to the research, the robustness of the test will be achieved by changing SG&A expenses with total operating costs. Finally, to test the factors, the interaction or moderation effects between variables will be eliminated.

Model 3.

$$log\left[\frac{TC_{i,t}}{TC_{i,t-1}}\right] = \beta_0 + \beta_1 log\left[\frac{Sales_{i,t}}{Sales_{i,t-1}}\right] + \beta_2 * DecDummy_{i,t} * log\left[\frac{Sales_{i,t}}{Sales_{i,t-1}}\right] + \varepsilon_{i,1}$$

Model 4.

$$STICKINESSit = \beta 0 + \beta 1SALESCHGit + \beta 2DECDUMit*SALESCHGit + \beta 3ASSETit + \beta 4EMPLOYEEit + \beta 5ROEit + \beta 6FCFit + \beta 7LEVit + \beta 8GDPit + \varepsilon it$$

This research has also been tested for any possible multicollinearity, heteroscedasticity, and auto-correlation for all models. Based on the test, this research is free from those classical assumptions.

3.1 *Empirical tests*

The first empirical testing is about descriptive statistics as shown in Tables 1 and 2,

Table 1. Descriptive statistics of *net sales* and *SG&A costs*.

	Mean	Median	Max	Min	Std. Dev.
Sales (in billion Rupiah)	5,346.07	1,196.65	201,701.00	29.23	15,656.69
SG&A Costs (in billion Rupiah)	526.15	85.98	17,885.00	1.05	1,490.20
SG&A Costs/Sales (%)	11.30%	7.85%	96.92%	0.48%	11.20%

Table 2. Descriptive statistics of variables.

Variable	Mean	Median	Max	Min	Std. Dev.
STICKINESS	0.03399	0.04232	0.96310	−0.75358	0.18306
SALESCHG	0.05017	0.05222	1.19132	−0.59127	0.11004
DECDUM	0.21748	0	1	0	0.41273
*DECDUM*SALESCHG*	−0.01814	0	0	−0.59127	0.05365
ASSET	−0.03033	−0.03395	0.72397	−1.20104	0.23321
*DECDUM*SALESCHG*ASSET*	−0.00129	0	0.17198	−0.16920	0.01694
EMPLOYEE	−2.98511	−2.98458	−1.80370	−4.86838	0.43187
*DECDUM*SALESCHG*EMPLOYEE*	0.05325	0	1.67481	0	0.15930
ROE	0.15957	0.12828	4.45089	−5.95062	0.47583
*DECDUM*SALESCHG*ROE*	−0.00003	0	0.79512	−0.20728	0.03101
FCF	0.00598	0.00315	2.55886	−1.07237	0.15481
*DECDUM*SALESCHG*FCF*	−0.00040	0	0.06304	−0.07790	0.00628

Table 3 describes the result of the level of cost stickiness in SG&A costs. The coefficient and expected signs indicate that there is cost stickiness in SG&A costs.

Table 4 shows the result of testing factors that influence the level of cost stickiness as well as the answer to the previous hypothesis. In general, the value of coefficients for all variables is the same as the expected sign except for the GDP variable. These results indicate that each factor of cost stickiness has either a positive or negative impact on the level of cost stickiness.

To analyze the result of the robustness testing, Tables 5 and 6 exhibit the summary. Table 5 shows that the variable of SG&A costs is replaced by operating costs. In Table 6, the factors of cost stickiness are re-tested by eliminating the moderating impact of each variable.

Table 3. Regression testing results of the first model.

Variable		Expected sign	Coeff.	Std. error	t-stat	Prob.
C	β0		−0.0086	0.0049	−1.7494	0.0805*
SALESCHG	β1	+	0.7259	0.0448	16.1992	0.0000***
DECDUM*SALESCHG	β2	–	−0.1567	0.0968	−1.6564	0.0979*
R-squared		0.2867				
Adjusted R-squared		0.2853				
F-statistic		206.423				
Prob (F-Statistic)		0.0000				

Table 4. Regression testing results of the second model.

Variable		Expected sign	Coeff.	Std. Error	t-stat	Prob.
C	β0		−0.0137	0.0039	−3.4789	0.0005***
SALESCHG	β1	+	0.7541	0.0361	20.8660	0.0000***
*DECDUM*SALESCHG*	β2	–	−0.2128	0.1308	−1.6267	0.1042
*DECDUM*SALESCHG *ASSET*	β3	–	−0.4135	0.1696	−2.4382	0.015**
*DECDUM*SALESCHG *EMPLOYEE*	β4	–	−0.0742	0.0176	−4.2062	0.000***
*DECDUM*SALESCHG*ROE*	β5	+	0.1170	0.0704	1.6611	0.0971*
*DECDUM*SALESCHG*FCF*	β6	–	−0.6076	0.7313	−0.8309	0.4063
*DECDUM*SALESCHG*LEV*	β7	+	0.6439	0.2686	2.3973	0.0167**
*DECDUM*SALESCHG*GDP*	β8	–	0.5468	0.3887	1.4068	0.1599
R-square		0.5159				
Adjusted R-square		0.5111				
F-statistic		108.563				
Prob (F-Statistic)		0.0000				

Table 5. Result of robustness testing for the third model.

Variable		Expected sign	Coeff.	Std. error	t-stat	Prob.
C	β0		0.0011	0.0006	1.7293	0.0841
SALESCHG	β1	+	0.9898	0.0071	139.6679	0.0000***
DECDUM*SALESCHG	β2	–	0.0055	0.0176	0.3143	0.7534
R-squared		0.9683				
Adjusted R-squared		0.9683				
F-statistic		15702.08				
Prob (F-Statistic)		0.0000				

Table 6. Result of robustness testing for the fourth model.

Variable		Expected sign	Coeff.	Std. error	t-stat	Prob.
C	β0		0.0560	0.0282	1.9861	0.0473
SALESCHG	β1	+	0.7273	0.0455	15.9724	0.0000***
DECDUM*SALESCHG	β2	–	–0.2001	0.0979	–2.0438	0.0412**
ASSET	β3	+	0.0313	0.0175	–1.7883	0.074*
EMPLOYEE	β4	+	0.0221	0.0090	2.4618	0.014**
ROE	β5	–	–0.0030	0.0070	–0.4307	0.6668
FCF	β6	+	–0.0303	0.0277	–1.0936	0.2744
LEV	β7	–	0.0008	0.0157	0.0482	0.9616
GDP	β8	+	0.0203	0.0246	0.8277	0.4080
R-squared		0.2940				
Adjusted R-squared		0.2884				
F-statistic		53.13627				
Prob (F-Statistic)		0.0000				

4 RESULT AND DISCUSSION

According to Table 3, the coefficient of *saleschg* (*β1*) is 0.7259 and *decdum*saleschg* (*β2*) has the value –0.1567. Each of the variables has a significant result in terms of *t-stat*. Previous studies state that the hypothesis will be accepted if $\beta 1 > \beta 1 + \beta 2$ or if $\beta 1 > 0$ and $\beta 2 < 0$. Anderson *et al.* (2003) express that the value of *β1* indicates an increase in SG&A costs when sales are high. Meanwhile, *β1+β2* denote a decrease in SG&A costs when sales are declining. Therefore, based on the regression testing result, when net sales increase by 1%, SG&A costs will rise by 0.73%. On the other hand, when net sales decrease by 1%, SG&A costs will decrease by 0.57%. Based on this condition, H_0 is rejected. The accepted hypothesis is that the magnitude of an increase in SG&A costs for an increase in net sales is greater than the magnitude of a decrease in SG&A costs for a decrease in net sales. The result of this study is consistent with the previous research by Anderson *et al.* (2003), Windyastuti (2010), Hidayatullah *et al.* (2011), Chen *et al.* (2012), Bruggen and Zehnder (2014), and also Venireis *et al.* (2015).

To determine the factors affecting the level of cost stickiness, Table 4 presents the result of regressions. The first factor is asset intensity. Based on Table 4, the coefficient of *β3* is -0.4135 and is significant at the 95% confidence level. It could be inferred that when net sales increase by 1%, SG&A costs will rise by 0.75%, and also when net sales decrease by 1% related to asset intensity, SG&A costs will decrease by 0.34%. The hypothesis that states that asset intensity has positive impact on the level of cost stickiness could be accepted. This result supports the previous studies by Anderson *et al.* (2003), Calleja *et al.* (2006), Anderson and Lenan (2007), Chen *et al.* (2012).

The second factor is employee intensity. In Table 4, the value of *β4* is -0.0742 and is significant at the 99% confidence level. This condition shows that when net sales increase by 1%, SG&A costs will rise by 0.75%, and when net sales decrease by 1% related to employee intensity, SG&A costs will decrease by 0.68%. As the number of employees in a company grows, the cost stickiness will rise. The hypothesis that states that employee intensity has a positive impact on the level of cost stickiness could be accepted. This result is consistent with Anderson *et al.* (2003), Calleja *et al.* (2006), and also Venieris *et al.* (2015).

The third factor is the earning target measured by return on equity. Table 4 shows that coefficient of *β5* has a value of 0.1170 and is significant at the 90% confidence level. This result denotes that when net sales increase by 1%, SG&A costs will rise by 0.75%, and when net sales decrease by 1% related to return on equity, SG&A costs will decrease by 0.87%. The earning target from return on equity causes a decrease in the level of cost stickiness.

SG&A costs have anti-sticky behavior since the decrease in costs when net sales are declining is greater than the increase in costs when net sales increase. As a result, the hypothesis that states that the earning target has a negative impact on the level of cost stickiness could be accepted.

The next factor is management empire-building state incentive, which is measured by free cash flow (FCF). Table 4 shows that *β6* has a value of -0.6076. It means that when net sales increase by 1%, SG&A costs will rise by 0.75% and also when net sales decrease by 1% in relation with free cash flow, SG&A costs will decrease by 0.15% but not significant. As a result, the hypothesis that states that free cash flow has a positive impact on the level of cost stickiness could not be accepted. This result is not consistent with studies done by Chen *et al.* (2012) and Venieris *et al.* (2015).

The last main factor is leverage. Based on Table 4, the value of *β7* is 0.6439 and is significant at the 95% confidence level. Therefore, when net sales increase by 1%, SG&A costs will rise by 0.75%, and when net sales decrease by 1% in relation with leverage, SG&A costs will decrease by 1.34%. The level of cost stickiness decreases, and SG&A costs have anti-sticky behavior when the company has higher leverage. Hence, the hypothesis that states that leverage has a negative impact on the level of cost stickiness could be accepted. This result is in line with Calleja *et al.* (2006).

In relation with control variables, according to Table 4, the coefficient of *β8* is 0.5468 with t-statistic 0.1599. It means that GDP growth does not have significant influence on the level of cost stickiness.

As part of robustness testing, there are two models to be tested. The first testing is summarized in Table 5. Meanwhile, the second testing is shown in Table 6. Based on Table 5, the value of *saleschg* (*β*1) is 0.9898 and the coefficient variable *decdum*saleschg* (*β*2) is 0.0055. It can be concluded that when net sales increase by 1%, operating costs will rise by 0.99%, but when net sales decrease by 1%, operating costs will decrease by 1% and not in a significant level. This condition demonstrates that operating costs have proportional cost behavior to the change of sales. This is the same with the variable of cost behavior in general. As a result, operating cost does not have any cost stickiness behavior.

Then, referring to Table 6, there are some results regarding factors of cost stickiness which are without a moderation variable to the change of net sales. The cost stickiness of SG&A in this model is still consistent with the first model. In addition, only two variables have consistent outcome with the main research model, namely asset intensity and employee intensity. Both variables have a positive impact on the level of cost stickiness. Return on the equity variable has the same expected sign but not significant statistically. The other variables, namely free cash flow, leverage, and GDP growth, do not have any impact on the degree of cost stickiness.

Overall, the value of coefficients according to Table 6 is less than 1% for the factors of cost stickiness. This shows that determinants with no interaction with a change in sales have less relation and influence to SG&A cost stickiness behavior. The existence of moderation or interaction variables could substantiate the influence of cost stickiness factors.

5 CONCLUSION

The result of this research can be concluded in two main points. First, cost stickiness behavior could be found on sales, general, and administrative costs. However, the operating cost does not have sticky cost behavior. Second, factors that affect the level of cost stickiness are asset intensity, employee intensity, earnings target, and leverage, and this is consistent with the previous studies.

REFERENCES

Anderson, Henry R., and Mitchell H. Raiborn. 1977. Basic Cost Accounting Concept. Boston: Houghton Mifflin Company.

Anderson, Mark C., Rajiv D. Banker, and Surya N. Janakiraman. 2003. Are Selling, General and Administrative Costs "Sticky"?. Journal of Accounting Research 41: 47–63.
Anderson, Shannon W., and William N. Lanen. 2007. "Understanding Cost Management: What Can We Learn from the Empirical Evidence on "Sticky Costs"?." Working Paper. http://papers.ssrn.com/sol3/papers.cfm?abstract_id = 975135
Balakrishnan, Ramji, Michael J. Petersen, and Naomi S. Soderstrom. 2004. "Does Capacity Utilization Affect the 'Stickiness' of Cost?." Journal of Accounting, Auditing, and Finance 19(3): 283–299.
Baumgarten, Daniel. 2012. "The Cost Stickiness Phenomenon: Causes, Characteristics, and Implications for Fundamental Analysis and Financial Analysts' Forecasts." PhD Dissertation, University of Cologne. Germany: Springer Fachmedien Wiesbaden.
Baumgarten, Daniel. 2012. The Cost Stickiness Phenomenon. Gabler Verlag: Springer Fachmedien Wiesbaden.
Bruggen, Alexander and J.O. Zehnder. 2014. SG&A Cost Stickiness and Equity-Based Executive Compensation: Does Empire Building Matter?. Journal of Management Control 25: 169–192.
Calleja, K.enneth, Michael Steliaros and Dylan C. Thomas. 2006. A Note of Cost Stickiness: Some International Comparison. Management Accounting Research 22: 1–22.
Carter, William K., and Milton F. Usry. 2002. Cost Accounting (13th edition). Ohio: Thomson Learning Custom Publishing.
Chen, Clara X., Hai Lu, and Theodore Sougiannis. 2012. The Agency Problem, Corporate Governance, and the Asymmetrical Behavior of Selling, General, and Administrative Costs. Contemporary Accounting Research 29: 252–282.
Cooper, Robin, and Robert S. Kaplan. 1998. The Design of Costs Management Systems: Text, Cases, and Readings. Upper Saddle River, NJ: Prentice Hall.
Guenther, Thomas W., Anja Riehl, and Richard Robler. 2014. Cost Stickiness: State of the Art of Research and Implications. Journal Management Control 24: 301–318.
Hidayatullah, Idi J., Wiwik Utami, and Yudhi Herliansyah. 2011. Analisis Perilaku Sticky Cost and Pengaruhnya terhadap Prediksi Laba Menggunakan Model Cost Variability and Cost Stickiness (CVCS) pada Emiten di BEI untuk Industri Manufaktur. Banjarmasin: Simposium Nasional Akuntansi 15.
Kama, Itay and Dan Weiss. 2010. "Do managers' Deliberate Decisions Induce Sticky Cost?" Working Paper. http://ssrn.com/abstract = 2148888.
Nachrowi, Nachrowi D. and Hardius Usman. 2006. Pendekatan Populer and Praktis Ekonometrika untuk Analisis Ekonomi and Keuangan. Jakarta: Lembaga Penerbit FEUI.
Noreen, Eric. 1991. Conditions under which Activity-Based Cost Systems Provide Relevant Costs. Journal of Management Accounting Research 3: 159–68.
Pervan, Maja, and Pervan, Ivica. 2012. "Analysis of Sticky Cost: Croatian Evidence." Working Paper. http://www.wseas.us/e-library/conferences/2012/Porto/AEBD/AEBD-23.pdf.
Venieris, George, Naoum, Vasilos Christos, and Orestes Vlismas. 2015. Organization Capital and Sticky Cost Behavior of Selling, General, and Administrative Expenses. Management Accounting Research 26: 54–82.
Windyastuti. 2013. Penetapan Target terhadap Stickiness Cost. Jurnal Keuangan and Perbankan 17: 71–77.
Xue, Shuang and Yun Hong. 2015. Earnings Management, Corporate Governance, and Expense Stickiness. China Journal of Accounting Research 9 (1): 41–58.

Anderson, Mark C., Rajiv D. Banker, and Surya N. Janakiraman. 2003. Are Selling, General and Administrative Costs "Sticky"?. Journal of Accounting Research 41: 47–63.

Anderson, Shannon W., and William N. Lanen. 2007. "Understanding Cost Management: What Can We Learn from the Empirical Evidence on "Sticky Costs"?." Working Paper. http://papers.ssrn.com/sol3/papers.cfm?abstract_id = 975135

Balakrishnan, Ramji, Michael J. Petersen, and Naomi S. Soderstrom. 2004. "Does Capacity Utilization Affect the 'Stickiness' of Cost?." Journal of Accounting, Auditing, and Finance 19(3): 283–299.

Baumgarten, Daniel. 2012. "The Cost Stickiness Phenomenon: Causes, Characteristics, and Implications for Fundamental Analysis and Financial Analysts' Forecasts." PhD Dissertation, University of Cologne. Germany: Springer Fachmedien Wiesbaden.

Baumgarten, Daniel. 2012. The Cost Stickiness Phenomenon. Gabler Verlag: Springer Fachmedien Wiesbaden.

Bruggen, Alexander and J.O. Zehnder. 2014. SG&A Cost Stickiness and Equity-Based Executive Compensation: Does Empire Building Matter?. Journal of Management Control 25: 169–192.

Calleja, K.enneth, Michael Steliaros and Dylan C. Thomas. 2006. A Note of Cost Stickiness: Some International Comparison. Management Accounting Research 22: 1–22.

Carter, William K., and Milton F. Usry. 2002. Cost Accounting (13th edition). Ohio: Thomson Learning Custom Publishing.

Chen, Clara X., Hai Lu, and Theodore Sougiannis. 2012. The Agency Problem, Corporate Governance, and the Asymmetrical Behavior of Selling, General, and Administrative Costs. Contemporary Accounting Research 29: 252–282.

Cooper, Robin, and Robert S. Kaplan. 1998. The Design of Costs Management Systems: Text, Cases, and Readings. Upper Saddle River, NJ: Prentice Hall.

Guenther, Thomas W., Anja Riehl, and Richard Robler. 2014. Cost Stickiness: State of the Art of Research and Implications. Journal Management Control 24: 301–318.

Hidayatullah, Idi J., Wiwik Utami, and Yudhi Herliansyah. 2011. Analisis Perilaku Sticky Cost and Pengaruhnya terhadap Prediksi Laba Menggunakan Model Cost Variability and Cost Stickiness (CVCS) pada Emiten di BEI untuk Industri Manufaktur. Banjarmasin: Simposium Nasional Akuntansi 15.

Kama, Itay and Dan Weiss. 2010. "Do managers' Deliberate Decisions Induce Sticky Cost?" Working Paper. http://ssrn.com/abstract = 2148888.

Nachrowi, Nachrowi D. and Hardius Usman. 2006. Pendekatan Populer and Praktis Ekonometrika untuk Analisis Ekonomi and Keuangan. Jakarta: Lembaga Penerbit FEUI.

Noreen, Eric. 1991. Conditions under which Activity-Based Cost Systems Provide Relevant Costs. Journal of Management Accounting Research 3: 159–68.

Pervan, Maja, and Pervan, Ivica. 2012. "Analysis of Sticky Cost: Croatian Evidence." Working Paper. http://www.wseas.us/e-library/conferences/2012/Porto/AEBD/AEBD-23.pdf.

Venieris, George, Naoum, Vasilos Christos, and Orestes Vlismas. 2015. Organization Capital and Sticky Cost Behavior of Selling, General, and Administrative Expenses. Management Accounting Research 26: 54–82.

Windyastuti. 2013. Penetapan Target terhadap Stickiness Cost. Jurnal Keuangan and Perbankan 17: 71–77.

Xue, Shuang and Yun Hong. 2015. Earnings Management, Corporate Governance, and Expense Stickiness. China Journal of Accounting Research 9 (1): 41–58.

Competition and Cooperation in Economics and Business – Gani et al. (Eds)
© 2018 Taylor & Francis Group, London, ISBN 978-1-138-62666-9

The effect of ownership structure, board of commissioners' effectiveness and audit committee effectiveness on firm value: An empirical study of listed companies in the Indonesia stock exchange 2014

F. Yosi & R. Yuniasih
Department of Accounting, Faculty of Economics and Business, Universitas Indonesia, Depok, Indonesia

ABSTRACT: The objective of this research is to examine the effect of the ownership structure, board of commissioners and audit committee effectiveness on the value of a firm. Institutional and managerial ownerships were used to measure ownership structure. The board of commissioners' effectiveness was measured by its independence, activity, size and competence, based on Hermawan (2009). Audit committee effectiveness was measured by the educational background of its members and the number of meetings it held. The value of the firm was measured using Tobin's Q. Involving 412 companies listed in the Indonesia Stock Exchange (BEI) in 2014, this empirical study with multiple regression shows that institutional and managerial ownerships, board of commissioners' effectiveness and audit committee effectiveness have no significant effect on the value of a firm.

1 INTRODUCTION

A company is an organisation founded by a group of people working to achieve the common goal of gaining profits and increasing prosperity for its shareholders. In the long term, the objective is to increase the value of the company (Sudana, 2001). In the process of maximising value, sometimes a conflict of interests between the agent and the employer (principal), often called an agency problem, may arise. The management are more concerned with their own interests, which are not in accordance with the company's goal of increasing the shareholders' value. Without good corporate governance in place, this would lead to increased costs, which later on would affect the value of the company.

In companies with weak corporate governance, the actions of managers tend to be selfish (Darmawati, 2004). Managers, as the agents of the shareholders, are entrusted with the company. However, without good corporate governance, shareholders may lose confidence. Therefore, it is important to protect the interests of the shareholders from management expropriation. A corporate governance mechanism can be in the form of institutional ownership, because the presence of an institution will empower the monitoring of the management performance so that the management would be more cautious when making decisions.

Another corporate governance mechanism that can be put in place to discourage the selfish actions of managers is providing managers with ownership of the company. Management ownership will encourage the management to enhance shareholder value. The results of a study conducted by Jensen and Meckling (1976) prove that the ownership of shares by the management has a positive effect on a firm's value. The greater the proportion of management ownership, the more likely it is for the management to try harder for the benefit of the shareholders, who are none other than themselves.

Public companies in Indonesia are also subject to the appointment of independent directors who come from outside the issuer or public company. The number of independent directors shall be at least 30% of the total members of the board of commissioners. The higher the percentage of independent board members, the better the oversight and co-ordination of the company; this will increase the value of the company. Other characteristics of the board of commissioners' effectiveness in monitoring the company's performance are the activity, size and competence of the board of commissioners.

It is also mandatory for public companies to have an audit committee to help the board of commissioners to exercise their duty of oversight, especially related to financial information and reports. The existence of the audit committee will contribute to the quality of financial reporting, which can enhance a firm's value. The purpose of establishing an audit committee is to assist the commissioners or board of supervisors with ensuring the effectiveness of the implementation of the tasks of the external auditor and the internal auditor. The characteristics of the audit committee that can help the monitoring process are the educational background of its members and the number of meetings.

This research analysed factors hypothesised to be affecting a firm's value, namely (i) institutional ownership, (ii) managerial ownership, (iii) effectiveness of the board of commissioners, and (iv) effectiveness of the audit committee.

2 LITERATURE REVIEW

Jensen and Meckling (1976) describe an agency relationship as a contract in which one or more principal assign another person (the agent) to perform some service for them by delegating decision-making authority to the agent. According to Jensen and Meckling (1976), the ownership structure is believed to have the ability to influence the running of the company, which in turn can affect the performance of the company. Institutional ownership is a resource that can be used to promote optimal control of management performance in order to anticipate managerial actions that do not conform to shareholders' wishes. Institutional ownership will encourage supervision, which will ensure the increased wealth of shareholders. The concentration of institutional ownership increases public confidence in the company, which increases trading volume and stock price, and this will increase the firm's value (Tarjo, 2008). Amba (2014) states that institutional owners are more concerned about the return on their investment, so their contribution to the effective corporate governance of the company can enhance financial performance. Abukosim et al. (2014) state that the higher the level of institutional ownership in the company, the greater the interest and control, which can improve the function for overseeing the management; this will boost the performance and value of the company.

H1: Institutional ownership significantly affects a firm's value

Managerial stock ownership means that managers have a stake in the company, or that the managers are also corporate shareholders (Rustiarini, 2006). Share ownership by the management will lead to better control over the policies taken by the management itself. Fauzi and Locke (2012) found positive relations between managerial ownership and a firm's value. This is based on the agency theory that high managerial ownership can decrease the agency costs, which later on increases the firm's performance and its value. Other studies by Debby et al. (2014) and Abukosim et al. (2014) found that managerial ownership does not have a significant relationship to the firm's value; this happens because the management are under the control of shareholders, so that even though they have shares in the company, they do not try to maximise the shareholders' wealth.

H2: Management ownership significantly affects a firm's value

Corporate governance mechanisms are geared up to ensure and oversee the governance system at work in an organisation (Sudiyatno & Puspitasari, 2010). The monitoring mechanisms in corporate governance are divided into two groups, internal and external mechanisms (Lastanti, 2004). An internal mechanism is a way to control the company by using the structure and internal processes, such as the general meeting of shareholders, the composition of the board of directors and audit committee, and meetings with the board of directors.

Previous studies concluded that the larger the size of the board, the greater the monitoring of the company and the higher the availability of independent commissioners who can mitigate management wrongdoings, which will certainly increase the firm's value (Barnhart & Rosenstein, 1998; Fauzi & Locke, 2012). Other studies by Debby et al. (2014) found no significant relationship between the proportion of independent commissioners and the firm's value. However, an active board of commissioners reflects actual and updated control over the management, which enhances the possibility of achieving better results and increasing the firm's value.

H3: Board of commissioners' effectiveness significantly affects a firm's value

McMullen (1996) found that investors, analysts and regulators consider that an audit committee provides a positive contribution to the quality of financial statements. This implies that the existence of an audit committee positively and significantly affects the value of the company. Amer et al. (2014) found that audit committee members with an accounting and finance background have a positive effect on a firm's value.

H4: Audit committee effectiveness significantly affects a firm's value

3 RESEARCH METHODS

The analysis in this study used multiple linear regressions, with the following model:

$$Q_i = b_0 + b_1KI + b_2KM + b_3EDEKOM + b_4EKA + b_5lnMA + b_6lnTA + \varepsilon_i$$

Qi is the Tobin's Q for the firm's value. Tobin's Q is calculated by (ME+D)/TA; where ME is the number of ordinary shares outstanding multiplied by the closing share price. D is calculated as total liabilities + inventories—current assets. TA is the book value of total assets. KI is the institutional ownership, calculated as the percentage of total shares outstanding. KM is the managerial ownership, calculated as the percentage of total shares outstanding. EDEKOM was measured using a checklist developed by Hermawan (2009). This checklist measures board independence, activities, size, expertise and competence. Table 1 provides the detailed questions in the checklist. EKA is audit committee effectiveness, calculated as the

Table 1. List of questions.

	Question item
Board independence	The percentage of independent commissioners on the Board of Commissioners
	The chairman of the Board of Commissioners is an independent commissioner
	There is a clear definition of independent commissioner in the annual report
	The percentage of commissioners working for shareholders or any affiliation
	Company has a nomination and remuneration committee
	The average assignment period for the Board of Commissioners
Board activities	There is a clear statement of the job descriptions for the Board of Commissioners
	The number of meetings of the Board of Commissioners in a year
	The average attendance of the Board of Commissioners in a year
	The Board of Commissioners evaluate the financial statement
	The Board of Commissioners evaluate the annual management performance
	The Board of Commissioners evaluate business projects proposed by management
Board size	The number of people on the Board of Commissioners
board expertise and competence	The percentage of commissioners having a finance and accounting background
	The percentage of commissioners having business and managerial experience
	The percentage of commissioners having an understanding of the company's business process
	The average age of the Board of Commissioners

percentage of audit committee members with accounting and financial backgrounds. LnMA is the natural logarithm of the number of audit committee meetings in one year. LnTA is the natural logarithm of total assets. The samples used in this study were the companies listed in the Indonesia Stock Exchange in 2014, other than those from the financial sector.

4 EMPIRICAL RESULTS

The final samples were 412 companies. Table 2 shows the descriptive statistics, and Table 3 shows the correlations among variables.

Table 3 shows that there was only a significant correlation with the dependent variable with the control variable of LnTA and that other independent variables did not have a significant correlation with a firm's value. There was no potential multicollinearity expected from the data. Table 4 provides the regression results.

Table 2. Descriptive statistics.

Variable	Mean	Median	Maximum	Minimum	Std. Dev.
Q	1.20	0.88	7.94	–9.76	1.28
KI	0.64	0.67	0.99	0.00	0.24
KM	0.05	0.00	0.85	0.00	0.13
EDEKOM	41.15	42.00	51.00	0.00	5.43
EKA	0.58	0.67	1.00	0.17	0.23
MA	5.9	4.00	45.00	1.00	5.28
TA	6,948,670	2,079,180	85,938,885	7,000	12,846,942

Table 3. Correlation results.

Variable	Q	KI	KM	EDEKOM	EKA	LnMA	LnTA
Q	1.000000						
KI	–0.026076	1.000000					
KM	0.004404	–0.368381*	1.000000				
EDEKOM	0.095547	0.038488	–0.166319	1.000000			
EKA	–0.018634	0.078317	–0.028510	–0.009367	1.000000		
LnMA	–0.033221	–0.088078	–0.007276	0.040863	0.018918	1.000000	
LnTA	0.286686*	–0.125177*	0.056815	0.080965	–0.101541	–0.017966	1.000000

*significant at 5%.

Table 4. Regression results.

Variable	Coefficient	t-Statistic	Prob.
C	–0.355337	–1.011657	0.3123
KI	0.018956	0.069609	0.9445
KM	0.021282	0.044672	0.9644
EDEKOM	0.190604	1.512809	0.1312
EKA	0.061643	0.240807	0.8098
LnMA	–0.053791	–0.917916	0.3592
LnTA	0.112324***	8.317736	0.0000
	R-squared	0.088564	
	Adjusted R-squared	0.074248	
	F-statistic	6.186441	
	Prob(F-statistic)	0.000003***	

5 DISCUSSION

5.1 *Ownership structure and firm value*

The regression analysis did not provide any evidence of a relationship between institutional ownership and a firm's value. This result is consistent with Rusiti and Welim (2014), who stated that institutional share ownership does not affect a firm's value. The regression analysis also did not provide any evidence of a relationship between managerial ownership and a firm's value. Indonesian stock market is still considered as having a weak form market efficiency, so the flow of information towards those undertaking transactions in the market is still not in the best condition. The rise and flow of stock prices are still affected by the reputation of the company; the ownership structure is not a factor in determining the price. The market only focuses on the results and performance of the company, which therefore means that control by an institution or by the management is an insignificant factor in affecting a firm's value.

5.2 *Board of commissioners' and audit committee effectiveness and firm value*

The regression analysis did not provide any evidence of a relationship between the effectiveness of the board of commissioners and a firm's value. The regression analysis also did not provide any evidence of a relationship between the audit committee and a firm's value. This shows that neither the monitoring by the board of commissioners nor the help of the audit committee is very effective in affecting the performance of the company. This is probably due to the fact that most of the companies listed in the study samples appointed only one accounting expert to meet the regulatory requirements, namely 'one independent director should be the accounting expert on the audit committee'. Moreover, the audit committee member is usually derived from academic institutions and lacks experience in the industry.

6 CONCLUSION

The regression results show that only the variable of total asset that has a relationship with firm's value. All other variables do not have empirical support to having any impact to firm value. The hypothesis stating that a firm's value is affected by the ownership structure and corporate governance mechanism is not proven by the 2014 data in Indonesia. This shows that the model itself should be tested for further research.

REFERENCES

Abukosim, M., Ferina, I.S. & Nurcahaya, C. (2014). Ownership structure and firm values: Empirical study on Indonesia manufacturing listed companies. *International Refereed Research Journal, 5*(4), 41.

Afza, T. & Nazir, M.S. (2014). Audit quality and firm value: A case of Pakistan. *Research Journal of Applied Sciences, Engineering and Technology, 7*(9), 1803–1810. http://maxwellsci.com/print/rjaset/v7-1803-1810.pdf.

Al-Najjar, B. (2012). The determinants of board meetings: Evidence from categorical analysis. *Journal of Applied Accounting Research, 13*(2), 178–190. http://dx.doi.org/10.1108/09675421211254867.

Amba, S.M. (2014). Corporate governance and firm's financial performance. *Journal of Academic and Business Ethics, Vol. 8.*

Amer, M., Ragap, A.A. & Shehata, S.E. (2014). *Audit committee characteristics and firm performance: Evidence from Egyptian listed companies.* Paper presented at the Proceedings of 6th Annual American Business Research Conference, Sheraton LaGuardia East Hotel, New York, USA, June 9-1-, 2014. http://wbiworldconpro.com/uploads/new-york-conference-2014/accounting/1401859517_116-Mrwan.pdf.

Azharia, A. (2007). *The effect of institutional ownership, managerial ownership, sales, and age to firm value.* (Unpublished thesis). Universitas Indonesia.

Barnhart, S.W. & Rosenstein, S. (1998). Board composition, managerial ownership, and firm performance: An empirical analysis. *The Financial Review, 33*(303), 1–16. doi:10.1111/j.1540–6288.1998.tb01393.x.

Bonn, I. (2004). Board structure and firm performance: Evidence from Australia. *Journal of Management and Organization,10*(1), 14–24. doi:10.5172/jmo.2004.10.1.14.

Brigham, E.F. & Houston, J.F. (2006). *Fundamentals of financial management* (10thed., translated version). Jakarta: Salemba Empat.

Daniri, M.A. & Simatupang, A.I. (2009). Internal audit transformation to create good corporate governance. *National Committee for Governance Policy.* 1–3. http://www.kadin-indonesia.or.id/enm/images/dokumen/KADIN-167-3751-15042009.pdf.

Darmawati, D. (2004). *The relation of corporate governance and firm performance*. Paper presented at Simposium Nasional Akuntansi VII, Denpasar, December 2–3, 2004.

Debby, J.F., Mukhtaruddin, Yuniarti, E., Saputra, D.& Abukosim. (2014). Good corporate governance's characteristics and firm's value: Empirical study of listed banking on Indonesia Stock Exchange. *GSTF Journal on Business Review (GBR), 3*(4), 81–88. doi: 10.5176/2010–4804_3.4.346.

Fauzi, F. & Locke, S. (2012). Board structure, ownership structure and firm performance: A study of New Zealand listed-firms. *Asian Academy of Management Journal of Accounting and Finance, 8*(2), 43–67.

Finance Ministry of the Republic of Indonesia. (2012). Decree No: KEP-643/BL/2012 on the Creation and Operational Guidance of Audit Committee.

Ghozali, I. (2005). *The application of multivariate analysis using SPSS*. Semarang: Badan Penerbit Universitas Diponegoro.

Giri, M.I.A. (2008). *The effect of independent commissioner, managerial ownership, institutional ownership, and concentrated ownership to performance and firm value in Indonesia Stock Exchange.* (Unpublished Thesis). Universitas Indonesia.

Haruman, T. (2008). *Ownership structure, financial decision, and firm value*. Paper presented at Simposium Nasional Akuntansi XI, University of Tanjungpura, Pontianak, Juli 23–26, 2008.

Hasibuan, M.S.P. (2002). *Human Resource Management*. Jakarta: BumiAksara.

Herawati, V. (2008). *The role of corporate governance practices as a moderating variable to the effect of earning management to firm value*. Paper presented at Simposium Nasional Akuntansi XI, University of Tanjungpura, Pontianak, Juli 23–26, 2008.

Hermawan, A.A. (2009). *The effect of board of commissioners' and audit committee's effectiveness, family ownership, and bank monitoring role to earnings information content*. (Unpublished doctoral thesis). Universitas Indonesia.

Isshaq, Z., Bokpin, G.A. & Onumah, J.M. (2009). Corporate governance, ownership structure, cash holdings and firm value on the Ghana Stock Exchange. *The Journal of Risk Finance, 10*(5), 448–499. http://dx.doi.org/10.1108/15265940911001394.

Jensen, M.C. & Meckling, W.H. (1976). Theory of the firm: Managerial behaviour, agency costs and ownership structure. *Journal of Financial Economics, 3*(4), 305–60. doi:10.1016/0304-405X(76)90026-X.

Kumar, J. (2004). *Agency theory and firm value in India*. Goregaon (East), Mumbai: Indira Gandhi Institute of Development Research. http://unpan1.un.org/intradoc/groups/public/documents/APCITY/UNPAN023822.pdf.

Lastanti, H.S. (2004). *The Relation of Corporate Govenance Structure with Firm Performacne and Market Reaction*. Paper Presented at National Conference of Accounting: Accountant Role in Building Good Corporate Governance.

Li, J., Pike, R. & Haniffa, R. (2008). Intellectual capital disclosure and corporate governance structure in UK firms. *Accounting and Business Research, 38*(2), 137–59. doi:10.1080/00014788.2008.9663326.

McMullen, D. (1996). Audit committee performance: An investigation of the consequences associated with audit committees. *Auditing: A Journal of Practice and Theory*, 15 (1), 87–103.

Nurlela, R. & Islahuddin. (2008). *The effect of corporate social responsibility to firm value using managerialownership as moderating variabel*. Paper presented at Simposium Nasional Akuntansi XI, University of Tanjungpura, Pontianak, Juli 23–26, 2008.

Ojucari, O. (2012). *Corporate governance: The relationship between audit committees and firm values*. The Department of Management Sciences, Kwara State University, Malete.

Otoritas, J.K. (2014). *Peraturan Otoritas Jasa Keuangan* No: 33/POJK.04/2014 on Directors and Board of Commissioners of Publicly Listed Companies. Jakarta: Sekretariat Negara.

Rachmawati, A. & Triatmoko, H. (2007). *Analysis on the factors affecting earnings quality and firm value.* Paper presented at Simposium Nasional Akuntansi X, Unhas Makassar, Juli 26–28, 2007.

Rusiti & Welim, M.A. (2014). *The effect of managerial ownership and institutional ownership to firm value (Empirical study on listed banks in Indonesia Stock Exchange)*. Unpublished Tesis. Universitas Atmajaya.

Rustiarini, N.W. (2006). *The effect of stock ownership structure on the disclosureof corporate social responsibility*. Paper presented at Simposium Nasional Akuntansi 9, Padang, August 23–26, 2006.

Sartono, A. (2001). *Theories and Application of Financial Management*. Yogyakarta BPFE.

Siallagan, H. & Machfoedz, M. (2006). *Corporate governance mechanism, earnings quality, and firm value*. Paper presented at Simposium Nasional Akuntansi 9, Padang, August 23–26, 2006.

Sudana, I.M. (2001). *Corporate Financial Management*. Jakarta: Erlangga.

Sudiyatno, B. & Puspitasari, E. (2010). Tobin's Q and Altman Z-Score as indicators of company performance measurement. *JurnalIlmiahKajianAkutansi, 2*(1), 9–21.

Sugiarto. (2009). *Capital structure, firms ownership structure, agency problem, and information asymmetry*. Yogyakarta: GrahaIlmu.

Sugiyono.(2009). *Quantitative and Qualitative Research Methods*. Bandung: CV. Alfabeta.

Sukamulja, S. (2004). Good corporate governance in finance sector: The impact of GCG to firms performance (A case at the Jakarta Stock Exchange).*BENEFIT JurnalManajemendanBisnis*. http://journals.ums.ac.id/index.php/benefit/article/view/1193.

Swastika, S. (2013). *The effect ownership structure and good corporate governance practice on thevalue of non financial companies listed in Indonesia Stock Exchangeduring 2006–2011*. (Unpublished thesis). Universitas Gajah Mada.

Tarjo. (2008). *The effect of institutional ownership concentration and leverage to earnings management, stockholders values, and cost of equity capital*. Paper presented at Simposium Nasional Akuntansi 9, Padang, August 23–26, 2006.

Vafeas, N. (1999). Board meeting frequency and firm performance. *Journal of Financial Economics, 53*(1), 113–42. doi:10.1016/S0304-405X(99)00018–5.

Vintilă, G. & Gherghina, S.C. (2015). Does ownership structure influence firm value? An empirical research towards the Bucharest Stock Exchange listed companies. *International Journal of Economics and Financial Issues, 5*(2), 501–14. doi:10.1016/S2212-5671(14)00500-0.

Wahyudi, U. & Pawestri, H.P. (2006). *The implication of ownership structure to firm value with financial decisions as intervening variable*. Paper presented at Simposium Nasional Akuntansi 9, Padang, August 23–26, 2006.

Yin, F., Gao, S.,Li, W. & Lv, H. (2012). Determinants of audit committee meeting frequency: Evidence from Chinese listed companies. *Managerial Auditing Journal, 27*(4), 425–444. http://dx.doi.org/10.1108/02686901211218003.

Competition and Cooperation in Economics and Business – Gani et al. (Eds)
© 2018 Taylor & Francis Group, London, ISBN 978-1-138-62666-9

Perception of accounting students and alumni of the influence of internships on skill development

E.Y. Favorieta & R. Yuniasih
Department of Accounting, Faculty of Economics and Business, Universitas Indonesia, Depok, Indonesia

ABSTRACT: This research aims to investigate the perception of accounting students and alumni of the Faculty of Economic and Business Universitas Indonesia (FEB UI) of the influence of internships on skill development. Online questionnaires were distributed to accounting students of FEB UI batch 2012 and alumni batches 2010 and 2011 who are working in public accounting firms. Using an independent sample t-test and the Mann-Whitney test, 308 sets of data were analysed. The result indicates that both accounting students and alumni have positive perceptions on the influence of internships on skill development.

1 INTRODUCTION

The accounting profession is one of the few professions that can compete freely within the scope of the ASEAN Economic Community (AEC) in Indonesia (ASEAN, 2014). This means that all levels of Indonesian society, including students–especially students of accounting–need to be prepared. Under the AEC, employment opportunities are possible for citizens of ASEAN. However, for Indonesians, this can be an opportunity and a challenge.

Although there is strength in terms of quantity, the quality of accountants in Indonesia still needs to be improved (Tarkosunaryo, 2015). According to M. Jusuf Wibisana, the level of competence of public accountants in Indonesia is still low and inadequate. Some companies also complained that even the best graduates from universities in Indonesia still could not answer basic accounting questions during job interviews (Subari, 2016). Moreover, cultural constraints and English skills also become an obstacle for Indonesians when they are competing with accountants from Malaysia, Singapore, the Philippines, and Brunei Darussalam (Budilaksono, 2014).

Departing from the notion that experience is the best teacher, one approach to learning is by a system of practice and real experience, as well as by discovery and exploration, which is best known as experiential learning. One of the methods of experiential learning is through an internship programme (Northern Illinois University, n.d). The internship programme is considered to be one way to improve the readiness of accounting students to face work (Cheong et al., 2014; Cord et al., 2010; Crisostomo, 2015).

Various studies have been conducted to determine the benefits of the programme and the internship experiences of the students (Beard & Morton, 1999; Hymon-Parker, 1998; McCosh, 1957; Pasewark et al., 1989; Rigsby et al., 2013; Scott, 1992; Swift & Russel, 1999; Gault et al., 2000; Tovey, 2001; Yu et al., 2013). Internships can be a learning tool that helps students to better understand and apply the theory acquired in the classroom and put it in to practice at work, as well as a means to explore career options and understand the industrial world. With internship experience, students will possess networks and obtain an advantage over other job applicants. Rigsby et al. (2013) also argue that one of the main benefits of an internship is the increase in job offer opportunities. In addition, the internship programme can also provide benefits by improving communication skills (Beard, 1998; Beck & Halim, 2008; Burnett, 2003; Maelah et al., 2012; Milhail, 2006; Paisey & Paisey, 2010; Wesley & Bickle, 2005), problem solving characteristics (Burnett, 2003), confidence (Mohd Jaffri et al., 2011), and the networking capabilities of students (Gerken et al., 2012; Mohd Jaffri et al., 2011). It can be concluded that the internship programme provides many benefits for students.

Based on data from the Department of Accounting, the Faculty of Economics and Business, Universitas Indonesia, the number of students who attend an internship programme in public accounting firms increases from year to year. Students have expectations and a positive perception of the benefits of internships, thus helping to provide even more students with an internship programme. But do students and alumni have a positive perception of internship as being beneficial for a career in public accounting firms? There is an absence of evidence showing that alumni have a positive perception of the benefits of internship on the development of their capabilities and their careers in public accounting firms in Indonesia, which prompted this study to determine how the perceptions of students and alumni of the Department of Accounting, the Faculty of Economics and Business, Universitas Indonesia influence internships with regards to the development of competence and capabilities. The study uses research instruments used by Muhamad et al. (2009) and Yu et al. (2013). This study focuses on the education sector, the result of this study is expected to provide feedback to universities especially on the importance of internship program as part of the undergraduate curriculum.

2 LITERATURE REVIEW

2.1 *Perceptions on internship and competence development*

The research conducted by Martin and Wilkerson (2006) on a sample of 132 students of the Master of Science in Administration programme at Wake Forest University shows an increase in knowledge and changes in the behaviour of students after the internship. Students found that the internship experience they had gone through was useful for their academic activities; this, in turn, helps to strengthen their academic understanding of issues they have learned previously, provides additional insight into any new courses they take, as well as improving their understanding of business concepts and accounting. In terms of behaviour, student interns feel the positive effects on their maturity, both personally and as a student, and an increase in confidence in their academic abilities.

Furthermore, the research conducted by Muhamad et al. (2009) on a sample of 243 student interns at Universiti Malaya shows that students' expectations before the internship programme are not achieved after they have completed the internship programme. Students have higher expectations of the effects of an apprenticeship, particularly in improving their understanding of external audit, while their expectations of the influence of an apprenticeship in enhancing their understanding of public sector accounting were met. Similarly, the results of research conducted by Warinda (2013) on a sample of 112 student interns at the University of Zimbabwe shows that the expectations that the students had before undertaking the internship, of its influence in improving their understanding of information systems, had been achieved after they had completed an apprenticeship programme.

The research by Yu et al. (2013) towards alumni who had graduated from college the previous year and who were students at a university in the Midwest from 2000 to 2008 shows that alumni (as compared to students) has a higher perception on the ability of internship in increasing competence in financial reporting, accounting, auditing, management fees, tax and accounting information systems. The research conducted by Crisostomo (2015) shows that a student has a neutral perception on the influence of an internship experience in improving their understanding of tax accounting, but provided increased understanding of accounting and auditing and reporting in aspects that are considered most keenly felt by the students after the apprenticeship.

2.2 *Perceptions on internship and skills/capabilities development*

Muhamad et al. (2009) conclude that the expectations of students before the internship programme, particularly in terms of the development of problem solving ability, communication, interpersonal relations, and also confidence, is higher than their perception of the actual level of skills acquired after the completion of the internship programme. However, the research conducted by Mohd Jaffri et al. (2011) on 33 business students and 51 students of accounting of Universiti Sultan Zainal Abidin shows that business students perceive that the internship experience that they undergo will help to develop their interpersonal skills, while accounting students perceive that their internship experience will help to boost their confidence. Overall, both groups of students feel that going through an internship experience will help to improve their communication skills.

The research conducted by Warinda (2013) shows that, overall, students of the University of Zimbabwe agreed about the influence of an internship on the development of their soft skills, with communication capability being regarded as the most developed capability after a student internship programme. Research conducted by Crisostomo (2015) is also in line with the research by Mohd Jaffri et al. (2011) and Warinda (2013), with results showing that the students perceive that their internship experience has helped to develop their skills of problem solving, communication and interpersonal relations, and that the development of their communication skill is the aspect considered most keenly felt by the students after internship.

Research on the perceptions of students and alumni on the influence of an apprenticeship towards the development of capabilities is also specifically made by Yu et al. (2013). The results show that students perceived that the internship experience would help to greatly develop their overall ability, except for the ability to master the accounting system and also the use of databases, while the alumni perceived that the internship experience had helped them to develop skills in spread sheets, problem solving and written communication, but had not worked so well with the development of oral communication skills and effectiveness, although it had helped to develop their ability to use their database. It was concluded that the alumni have a perception of skill development that is lower than that of the students, especially for the development of effective oral communication skills, the ability to master the accounting system and the use of the database.

3 RESEARCH METHOD

This study was conducted on all accounting students of the Faculty of Economics and Business, Universitas Indonesia in the fourth year (class of 2012) and alumni of FEB UI accounting students with work experience of one to two years at a public accounting firm in Indonesia (classes of 2011 and 2010). Data collection was conducted by distributing online questionnaires using google forms. The questionnaire was adapted from an instrument of research studies conducted by Muhamad et al. (2009), Yu et al. (2013) and Rigsby et al. (2013) with modifications to adjust to the competencies and skills expected from internship conducted by students of the Accounting Department FEB UI. The questionnaire uses a Likert scale and is processed through a simple descriptive analysis (i.e. by calculating mean of the perception scale).

4 RESULTS AND DISCUSSION

Data was collected from 308 respondents, consisting of 205 students and 103 alumni; 106 students attended an internship in an accounting firm ("KAP"), 73 alumni attended an internship in KAP. Almost 60% of the respondents are female and 40% are male; they ranged in age from 18 to 25, with the majority (41.56%) being 22 years of age. Of the student respondents, 7.32% have a GPA below 3, while all the alumni respondents have a GPA of more than 3.

Table 1 shows that the students found that an internship experience provides the greatest impact in helping to improve their understanding of interpreting and evaluating financial statements. But according to alumni, an internship experience provides the greatest impact in helping to improve their knowledge and understanding of auditing, whether externally or internally, and also helps to improve their knowledge and understanding of financial accounting and reporting. This is consistent with the results of the research conducted by Crisostomo (2015), which showed that the increased understanding of accounting, auditing and reporting are the aspects that are considered to be most influential during an internship programme.

Both students and alumni perceive that an internship experience has very little effect in helping them to improve their knowledge and understanding of taxation. This could have been caused by the fact that the majority of the respondents do not work or have internship experience in the tax department. Of the 148 students, only 31 students have an internship focusing on taxation. Also, out of 82 alumni, there are only nine alumni who have an internship focusing on taxation.

Table 2 shows that the students found that an internship experience helps them to develop effective communication skills. This is in line with the results of Mohd Jaffri et al. (2011), Warinda (2013) and Crisostomo (2015), which show that the development of communication skills is considered to be the most perceived influence by students after the internship programme.

Table 1. Results from questionnaire for internship experience and development of competence (knowledge and competence).

Question item	Detailed question	Response rate	
		Students	Alumni
comp5	My internship experience helped me to improve my knowledge and understanding of auditing (internal or external)	3.80	4.04
comp6	My internship experience helped me to improve my knowledge and understanding of financial accounting and reporting	3.89	4.04
comp7	My internship experience helped me to improve my knowledge and understanding of taxation	3.17	3.41
comp8	My internship experience helped me to improve my knowledge and understanding of information systems	3.51	3.44
comp9	My internship experience helped me to improve my skills in composing a financial statement	3.72	3.83
comp10	My internship experience helped me to improve my skill in understanding and analysing financial statements	3.93	3.82

Table 2. Results from questionnaire for internship experience and the development of ability (skill).

Question item	Detailed question	Response rate	
		Students	Alumni
skill11	My internship experience helped me to develop analytical skills and judgement	4.02	3.96
skill12	My internship experience helped me to develop effective communication skills	4.26	4.30
skill13	My internship experience helped me to develop oral communication skills	4.22	4.24
skill14	My internship experience helped me to develop written communication skills	4.04	3.98
skill15	My internship experience helped me to develop technical spreadsheet skills	4.22	4.40
skill16	My internship experience helped me to develop problemsolving skills	3.93	3.94
skill17	My internship experience helped me to develop interpersonal skills	4.16	4.21
skill18	My internship experience helped me to develop self-confidence	4.14	4.27

On the one hand, in line with the results of Yu et al. (2013), the alumni found that an internship experience helps them to develop technical spreadsheet capabilities. This is because the nature of auditors or public accountants' work is greatly connected to using spreadsheet programmes. Therefore, technical spreadsheet capability has become important. In addition, both students and alumni alike perceived that an internship experience had very little effect in helping them to develop their problem solving ability. This is in contrast with the results of Yu et al. (2013), who state that the alumni found that their internship experience had helped their problem solving abilities. It is also contrary to the research results of Warinda (2013), which state that problem solving ability is the ability most developed after an internship.

5 CONCLUSION

From the results of the descriptive analysis (i.e. through on the perception of students and alumni on the influence of an internship on the development of competencies, capabilities and career opportunities in public accounting firms, it can be concluded that students and alumni both have a positive perception on the effect of an internship towards the development of competence and capability development. A further study can make additional comparisons on the differences in the perception between student respondents and alumni respondents or between those having experience in KAP and those in other types of firm. So what? What is it for? What do you want to achive in conducting this study?

REFERENCES

ASEAN. (2014). *ASEAN mutual recognition arrangement on accountancy services*. Nay Pyi Taw, Myanmar. http://www.fap.or.th/images/column_1401267099/03%20Accountancy%2014%20MRA%20(Final%20 Draft%2027May2014)%20clean.pdf.

Beard, D.F. (1998). The status of internships/cooperative education experiences in accounting education. *Journal of Accounting Education, 16*(3–4), 507–16. doi:10.1016/S0748–5751(98)00021-9.

Beard, F. & Morton, L. (1999). Effects of internship predictors on successful field experience. *Journalism and Mass Communication Educator, 53*, 42–53. doi: 10.1177/107769589805300404.

Beck, J.E. & Halim, H. (2008). Undergraduate internships in accounting: What and how do Singapore interns learn from experience? *Accounting Education: An International Journal, 17*(2), 151–72. doi:10.1080/09639280701220277.

Budilaksono, I. (2014). Indonesian accountants still have problems with English language. *Antara News,* Mei 6. http://www.antaranews.com/berita/432881/akuntan-indonesia-masih-terkendala-bahasa-inggris.

Burke, E. & Gibbs, T. (2014). *Driving new success strategies in graduate recruitment—Talent report 2014*. UK: CEB SHL Talent Measurement.

Burnett, S. (2003). The future of accounting education: A regional perspective. *Journal of Education for Business, 78*(3), 129–34. doi:10.1080/08832320309599709.

Cheong, A.L.H., Yahya, N.B., Shen, Q.L. & Yen, A. (2014). Internship experience: An in-depth interview among interns at a business school of a Malaysian private higher learning institution. *Procedia—Social and Behavioral Sciences,123*(1995), 333–43. doi:10.1016/j.sbspro.2014.01.1431.

Cord, B., Bowrey, G. & Clements, M. (2010). Accounting students' reflections on a regional internship program. *Australasian Accounting Business and Finance Journal, 4*(3), 47.

Crisostomo, D.T. (2015). Students' perception of the accounting internship program: Ebscohost. *Academy of Educational Leadership Journal, 19*(1), 167–74. http://nebulosa.icesi.edu.co:2517/ehost/detail/detail?vid=3&sid=8c35cd79–27c8-46a8-b460-640d945ab3e0%40sessionmgr120&hid=128&bdata=Jmxhbmc9ZXMmc2l0ZT1laG9zdC1saXZl#AN=108525859&db=bth.

Gault, J., Redington, J. & Schlager, T. (2000). Undergraduate business internships and career success: Are they related? *Journal of Marketing Education, 22*(1), 45–53. doi:10.1177/0273475300221006.

Gerken, M., Reinties, B.,Giesbers, B. & Konings, K.D. (2012). Enhancing the academic internship experience for business education: A critical review and future directions. *Advances in Business Education and Training, 4*, 7–22. doi: 10.1007/978-94-007-2846-2_2.

Hymon-Parker, S. (1998). Benefits and limitations of internships as viewed by educators and retailers/commentary. *Journal of Family and Consumer Sciences, 90*, 76–81.

Maelah, R., Aman, A., Mohamed, Z.M. & Ramli, R. (2012). Enhancing soft skills of accounting undergraduates through industrial training. *Procedia—Social and Behavioral Sciences, 59*, 541–49. doi:10.1016/j.sbspro.2012.09.312.

Martin, D.R. & Wilkerson, J.E. (2006). An examination of the impact of accounting internships on student attitudes and perceptions. *The Accounting Educators' Journal, 16*, 129–138. http://www.aejournal.com/ojs/index.php/aej/article/viewFile/70/49.

McCosh, R.B. (1957). Internship programs in accounting. *The Accounting Review, 32*(2), 306–308. http://www.jstor.org/stable/241490.

Meyer, A. (2011). GPA importance. *Aurorameyer,* April 20. https://aurorameyer.com/tag/screening-applicants-based-on-their-gpa/.

Milhail, D.M. (2006). Internship at Greek universities: An exploratory study. *Journal of Workplace Learning,18*, 28–41. http://dx.doi.org/10.1108/13665620610641292.

Mohd Jaffri, A.B., Harun, R J.,Yusof, K.N.C.K. & Tahir, I.M. (2011). Business and accounting students' perceptions on industrial internship program. *Journal of Education and Vocational Research, 1*(3), 72–79. http://ifrnd.org/Research%20 Papers/V1(3)2.pdf.

Muhamad, R., Yahya, Y., Shahimi, S. & Mahzan, N. (2009). Undergraduate internship attachment in accounting: The interns perspective. *International Education Studies, 2*(4), 49–54. http://dx.doi.org/10.5539/ies.v2n4p49.

Northern Illinois University. n.d. *Experiential learning.* DeKalb, Illinois, US: Northern Illinois University. http://www.niu.edu/facdev/_pdf/guide/strategies/experiential_learning.pdf.

Paisey, C. & Paisey, N.J. (2010). Developing skills via work placements in accounting: Student and employer views. *Accounting Forum, 34*(2), 89–108. doi:10.1016/j.accfor.2009.06.001.

Pasewark, W.R., Strawser, J.R. & Wilkerson, J.E. (1989). An empirical examination of the effect of previous internship experience on interviewing success. *Journal of Accounting Education, 7*(1), 25–39. doi:10.1016/0748-5751(89)90020-1.

Rigsby, J.T., Addy, N., Herring, C. & Polledo, D. (2013). An examination of internships and job opportunities. *Journal of Applied Business Research, 29*(4), 1131–44. doi:10.19030/jabr.v29i4.7921.

Scott, M.E. (1992). Internships add value to college recruitments. *Personnel Journal, 71*, 59–62.

Subari, W.A. (2016). Public accountants' quality is not enough. *Media Indonesia,* April 7. http://mediaindonesia.com/news/read/38995/kualitas-akuntan-publik-masih-belum-memadai/2016-04-07#.

Swift, C.O. & Russel, K. (1999). Business school internships: Legal concerns. *Journal of Education for Business, 75*, 23–26. http://dx.doi.org/10.1080/08832329909598985.

Tarkosunaryo. (2015). We can export Indonesian CPAs. In CPA Indonesia 2015: ASEAN Economic Community is coming 14–15.

Tovey, J. (2001). Building connections between industry and university: Implementing an internship program at a regional university. *Technical Communication Quarterly, 10*(2), 225–38. doi:10.1207/s15427625tcq1002_7.

Warinda, T. (2013). Accounting students' evaluation of internship experience from a skills perspective. *International Journal of Asian Social Science, 3*(3), 783–799.

Wesley, S. & Bickle, M.C. (2005). Examination of a paradigm for preparing undergraduates for a career in the retailing industries: Mentors, curriculum and internship. *College Student Journal, 39*(4), 680–691.

Yu, S., Churyk, N.T. & Chang, A. (2013). Are students ready for their future accounting careers? Insight from observed perception gaps among employers, interns, and alumni. *Global Perspectives on Accounting Education, 10*, 1–15.

Competition and Cooperation in Economics and Business – Gani et al. (Eds)
© 2018 Taylor & Francis Group, London, ISBN 978-1-138-62666-9

Characteristics of local government as Zakat (tithe) collector

D. Siswantoro
Department of Accounting, Faculty of Economics and Business, Universitas Indonesia, Depok, Indonesia

ABSTRACT: This paper aims to analyse the characteristics of local government as a zakat collector in Aceh, Indonesia. In Aceh province, zakat can be collected by the government as original regional income in the local budget. This study uses primary and secondary data gathered from in-depth interviews and the audit board's database. The interviews were conducted in June 2013 in Aceh at Baitul Maal and local administrative offices. Only 6 out of 23 local governments collect zakat. Data is analysed based on descriptive statistics and crosstab analysis. The paper provides evidence that most local governments did not collect zakat as original regional income and present the expenditure accordingly and explicitly. It is also found that the special allocation fund has a significant negative correlation with zakat collection, while significant differences are found between local governments that collect zakat in original regional income, budget surplus, assets, fixed assets and debt. Local governments that have not yet collected zakat as original regional income in their local budget can learn from those regions that have collected it. Some specific characteristics may influence regions that have collected zakat in their regional budgets.

1 INTRODUCTION

Zakat, or tithe, has been collected by private institutions in Indonesia professionally since 1994 by Dompet Dhuafa. It was started by a newspaper, Republika, which collected funds from the public. The institution gained the trust of the public and began to be managed professionally. Currently, they have many branches in some provinces in Indonesia, including Aceh, which practices Islamic law. Meanwhile, 18 other zakat institutions have been established. They are recognised by the Ministry of Finance for taxable income deduction. Muslims can deduct their taxable income only if they pay zakat to those institutions.

Historically, zakat was collected by the government in the era of Prophet Muhammad PBUH in the 7th century. As the leader of the Arab territory, he also supervised Baitul Maal (House of Wealth) as the central zakat collection. Currently, the issue of whether zakat must be collected by the government has become an interesting topic. In the 21st century, in countries based on Islamic law, such Saudi Arabia, Pakistan, Sudan, Libya and Yemen, zakat is collected and managed by the government by establishing the Ministry of Zakat. Meanwhile in Iran, Iraq and Brunei, the zakat payer can voluntarily pay zakat through a specific institution (Hassan, 2010). In countries that do not explicitly state that Islam is their basic system, such as Qatar, Bahrain, Kuwait, Jordan and Malaysia, zakat is also collected by the government. In Indonesia, the private sector is permitted to collect zakat in addition to the government, which is similar to the system in Egypt. However, there is also a separate government institution that collects zakat, namely BAZNAS (National Zakat Institution, Indonesia), which is also similar to Egypt. In other countries where Muslims are not the majority, zakat is collected by the private sector under Muslim associations and unions, such as in the US, the UK, Australia, Hong Kong, Japan, the Netherlands and Canada. However, in Singapore, where Muslims are the minority, zakat is collected under an organisation registered by the government.

The potential of zakat collection in Indonesia is still below expectations. This causes zakat collectors in the private sector to develop more quickly compared to government organisations, which face bureaucratic and systematic problems. In addition, in 2012 Dompet Dhuafa was targeted to collect zakat funds to a total of Rp205 billion, Rumah Zakat Rp324 billion, PKPU Rp120 billion and Al Azhar Rp100 billion. This significant amount encouraged the government to centralise zakat collection. This also caused the government to try to establish a zakat organisation to conduct the collection at the local government level. In Aceh, a province that applies Islamic law, zakat must be collected by the government, as it was historically in the Islamic state. Interestingly, they have separated the zakat institution, Baitul Mal, which does not have to report to BAZNAS as the national zakat organisation under the government in Indonesia.

In fact, zakat is a source of funds for the national budget and historically and conceptually must be managed by the government in the Islamic state. If we look at it on an historical and conceptual basis, this is the reason why zakat must only be collected by the government. However, some scholars stated that zakat can also be disbursed directly to recipients if it is based on non-physical assets, which cannot be seen by the government. This argument may support the opinion that non-government institutions can also collect and manage the zakat funds.

The issue of zakat collection in Indonesia is on the trust factor. The success factor of private institutions in increasing zakat collection in Indonesia is due to the fact that they can convince stakeholders that they disburse zakat funds accordingly and transparently, which can be seen from financial reports. Most private zakat institutions provide their financial reports easily to customers through their website. On the other hand, local government zakat institutions do not provide similar services. In Aceh, if zakat is collected as part of the original regional income of the local budget, it is reported in the financial report and audited by the audit board. However, the disbursement side is not stated clearly, as it is assumed in social expenditure. In addition, only 6 out of 23 local regions have collected zakat, even though it has been required since 2006 (Act 11/2006 and Qanun (Islamic regulation) No. 10/2007).

2 LITERATURE REVIEW

2.1 *Zakat in historical practices*

Bakar and Ghani (2011) give the example that the Prophet Muhammad PBUH directly distributed zakat funds to the correct recipients, especially the poor and needy. He sent people to other regions to collect zakat. Then in the period of Omar, the second caliph, with the Islamic territory expanding, he established a zakat organisation to manage zakat funds (Helal, 2012).

Qardawi (1999) states that the government must be responsible for managing zakat. This is because zakat is not an individual activity, but more a systemic and community-based activity. This is based on the Quran (holy book), which states that the wealth of Muslims should be purified through zakat. In addition, in hadith (Prophet Muhammad's sayings), it is stated that the government must be responsible for managing zakat. This was also practised during the Islamic state initiated by Prophet Muhammad PBUH, which his followers continued for some years. Currently, the issue is when Muslim countries neglect zakat due to colonisation and the adoption of Western economic concepts, such as the tax system. However, the government is still concerned with providing for the needy and the poor, but does so from tax resources, which is actually different from zakat. Zakat is deducted from a Muslim's wealth, while tax is deducted from income.

2.2 *Jurisprudence of zakat collection institution*

Bakar (2007) highlights that, in Malaysia, zakat is operated as a fiscal system where it is deducted as corporate tax. Corporate zakat is calculated based on the wealth of the company and is then multiplied by 2.5%. She states that zakat is part of the fiscal system for that reason, which highlights an interesting issue as to whether zakat is part of the government budget or otherwise. On the other hand, Toor and Nasar (2004) state that zakat is not part of the government budget as it is under central bank supervision in Pakistan. Zakat should be independent and the government should only monitor the institution.

In Indonesia, zakat can be collected by independent organisations, which receive funding from the government budget. Zakat is not included in either the national or local budget (Beik & Arsyianti, 2013), but in Aceh zakat is included in the local budget. Buehler (2008) states that Indonesia is unique for zakat collection as each region has different regulations for collecting it. This may be caused by the decentralisation system adopted by the Indonesian government. In Pakistan, a similar institution has been formed by local committees. They, however, spent almost 62.5% of zakat collection, with the rest sent to central zakat (Adebayo, 2011). This causes zakat fund management to be ineffective and inefficient.

Helal (2012) states that it is important to have a government zakat organisation in order to collect zakat legally and optimally. However, there would be a problem if the country was in financial crisis, such as in Egypt. He gives the example of Saudi Arabia as having the best government involvement in zakat management by establishing the Ministry of Zakat and Income Tax. Similarly, this also occurs in Malaysia and Sudan. In Jordan, zakat collection is centralised by the government (Benthall, 2008). However, the amount of zakat and other social funds are lower than social assistance (almost by half). In fact, zakat has not been optimally collected from Muslims (Wartonick, 2011).

The poverty alleviation programme in Indonesia uses the social safety net budget. Each year the budget almost doubled, but it only lowered the poverty rate by a small percentage. It would be interesting to see if zakat could be implemented in such a programme. Zakat is recognised to have successfully lowered the poverty rate. However, there should be clear regulations to manage zakat collection for social activities, which uses social activities expenditure in the budget (Yaumidin, 2011). This, however, is to avoid double activities in one budget. Sulaiman (2008) claims that the government must be able to collect zakat optimally and to eradicate poverty. This is because, compared to the tax rate, the zakat rate is lower. Therefore, the collection method should be better than tax collection.

Hassan and Khan (2007) claim that zakat funds in Bangladesh can replace social expenditure in the government budget, especially for foreign aid (Hassan, 2010), but not for disaster management (Nurzaman, 2011). For this to be the case, the government must collect and distribute zakat. This is because a legal system can support a zakat mechanism appropriately and with more effective supervision (Hassan, 2010) and also support the collection of zakat (Ahmed, 2008). However, this should not decrease other sources of budget revenue and it must also be autonomous.

The objective of zakat in Bangladesh is to alleviate poverty. It has the same objective in Malaysia (Shariff, 2011) as well as in Pakistan, where zakat has been collected optimally (Quraishi, 1999).

2.3 *Zakat and government budget*

The issue of zakat realisation (actual) vs. budget is interesting. Rahman (2007) states that the performance indicator of zakat institutions is the actual result of the zakat budget. This is also compared to other factors, which relates to the actual result and budget. This shows that the current zakat management may be different from what it was in the past, when it was included in the government budget.

The important thing is how zakat can be disbursed to the right people so that it benefits the community and society. Each country has a different system and background and cannot be treated equally.

In addition, the countries that charge zakat as a deductible tax are Jordan, Bangladesh, Malaysia, the United Arab Emirates and Bahrain (Powell, 2010). This can encourage Muslims to pay more zakat, but, in fact, it may not be successful. In Indonesia, zakat can be deducted from taxable income.

3 METHODS

The research is based on secondary data, comprising financial statements, which were taken by the audit board between 2007 and 2010. The problem is that not all the financial

Table 1. Filter sample.

Description	Data
Total regional government	23
Total 4 years	92
Not available	(19)
Available data	73
Do not collect zakat	(56)
Collect zakat	17
Zero collection	(3)
Final sample	14

Variables used in the research are:
Zakat = amount of zakat collection by local governments
Social = social expenditure of local governments
PAD = Pendapatan Asli Daerah (original regional income)
DAK = Dana Alokasi Khusus (special allocation funds)
DAU = Dana Alokasi Umum (general allocation funds)
Operational = operational expenditure of local governments
Capital = capital expenditure of local governments
Asset = asset of local governments
SILPA = Sisa Lebih Perhitungan Anggaran (budget surplus)
Fixed = fixed asset of local governments
Debt = debt of local governments
Variables are based on factors that may affect other indicators in local government (Harianto & Hadi, 2007; Putro & Pamudji, 2011; Ardhini, 2011; Darwanto & Yustikasari, 2007, Solikin, 2007; Abdullah & Halim, 2006).

statements can be collected easily and some are not fully up to date. Until 2010, only 6 out of 23 local regions in Aceh collected zakat as original regional income, which was only 26%. In detail, from 92 financial reports over 4 years, there are only 14 available financial statements from the period, which was only 15.2% (see Table 1). The primary data is taken from the interviews with Baitul Mal and local administrative offices in June 2013 in Aceh.

4 ANALYSIS

From the discussion above, the critical issues are who should manage the zakat funds and whether they should be included in the government budget, which depends on several factors.

1. If the government has a good infrastructure and system, zakat may be effectively and efficiently managed by the government. If not, private zakat institutions can participate to manage the zakat funds. This must also be co-ordinated by the government.
2. To ease the management and zakat disbursement, zakat should not be included in the government budget, as it may cause difficulties in disbursement and bureaucratic problems. Zakat may just be reported to the government for accountability and monitoring.

In Indonesia, the management of zakat should be flexible in order to enable the involvement of those private institutions who would like to participate, as the conditions are very complex in this country. On the other hand, the government must prepare the infrastructure and system for the collection and disbursement of the zakat funds so that it is effective and efficient. The government's involvement in the zakat budget can be grouped into three types:

1. Zakat is collected by the government
   - Zakat is part of the government budget, for example Aceh, Saudi Arabia, Sudan, Jordan and Kuwait
   - Zakat is excluded from the government budget, such as Iran and Pakistan
2. Zakat is collected by an institution appointed by the government (Powell, 2010) and is not part of the government budget, such as Malaysia
3. Zakat is collected by a registered zakat institution (mixed, private and state), such as Indonesia and Egypt

The important thing is how zakat can be collected as efficiently as possible and distributed to the correct recipients. However, the tax system may also be different between countries. This is also based on the beliefs and principles of each country.

Zakat began to be collected in Aceh in 2007, based on the Act No. 11/2006 dated 1st August 2006 and supported by Qanun No. 10/2007 dated 17th January 2008. Total zakat collection in Aceh was Rp5.4 billion in 2007, then Rp15.8 billion in 2008, Rp18.7 billion in 2009 and only Rp16.4 billion in 2010. Not all of the data for 2010 was available, so the amount decreased in 2010. However, zakat was collected from only 6 local regions out of a total of 23 local regions. Therefore, the potential is still promising, as the amount could be fivefold if all the local regions collect zakat.

Between 2007 and 2010, the minimum zakat collection for local governments in Aceh was Rp1 billion and the biggest was Rp8.3 billion (see Table 2). This was quite a big contribution to the original regional income (PAD). However, zakat disbursement is restricted to 8 groups of recipients. The problem was that there were differences in presenting zakat disbursement in the financial reports. In fact, some local governments presented zakat disbursement under social expenditure, showing the same amount as the zakat collected, some showed differences between collection and disbursement, and others did not state any collection and disbursement figures. Similar cases occurred in Pakistan (Adebayo, 2011). This is different from private zakat institutions, which stated the disbursement (for 8 groups) clearly and in detail. However, the amount of social expenditure may be similar to zakat collection. This means that social expenditure has a specific target for each year. Zakat recipients are part of the object of social expenditure in some local regions.

Original regional income (PAD) had a larger average than did zakat collection and also a higher spread. This could mean that actually regional governments may not be dependent on zakat collection, which had a small contribution to PAD. On the other hand, DAU and DAK had specific and determined amounts, which relied on specific regional budget deficits and activities. Furthermore, the amount of operational expenditure was larger than the capital expenditure; this is common in Indonesia where the budget is heavily focused on operational activities.

A higher and more significant correlation of zakat is seen only with DAK (see Table 3). The correlation is negative, meaning that the larger the DAK (special allocation fund), the smaller the zakat collection. However, there is no strong theory that supports the phenomenon. Zakat

Table 2. Statistical description (in billion IDR).

Variable	Min (in billion IDR)	Max (in billion IDR)	Mean (in billion IDR)	SD (in billion IDR)	Skewness	Kurtosis
Zakat	1.03	8.33	3.21	2.06	1.46	1.86
Social	1.12	20.1	8.45	5.59	0.73	–0.28
PAD	6.66	2,830	923.88	995.28	0.51	–1.21
DAU	39.9	395	286.16	83.5	–2.12	5.89
DAK	21.4	48.6	40.3	6.84	–1.69	3.85
Operational	1.12	507	322.97	150.17	–1.43	1.76
Capital	4.64	149	74.47	37.56	–0.32	0.93
SILPA	0	3.740	836.88	1,202.83	1.5	1.43
Asset	848	302,000	78,629	104,660	1.49	1.36
Fixed	576	71,700	6,716.9	18,734.8	3.72	13.92
Debt	0	21,200	396.33	796.85	1.72	1.33

Table 3. Correlation variables.

	Social	DAU	DAK	Operational	Capital	SILPA	Asset	Fixed	Debt
Zakat	0.32	0.42	−0.58	0.48	0.13	−0.12	−0.47	−0.10	−0.22
Sig.	0.25	0.12	0.02**	0.08*	0.65	0.68	0.87	0.72	0.43

*sig 10%.
**sig 5%.

Table 4. Test of mean difference.

	Social	PAD	DAU	DAK	Operational	Capital	SILPA	Asset	Fixed	Debt
Sig.	0.19	0.00***	0.75	0.27	0.19	0.42	0.00***	0.00***	0.00***	0.00***

***sig 1%.

is collected based on a Muslim's wealth, while DAK is based on specific activities in the local government. Zakat should have a higher correlation with social expenditure, whereby zakat collection is disbursed. However, zakat has a positive correlation with operational expenditure with a 10% significance. This could mean that operational expenditure consists of employee salaries, subsidies and grants. In this case, zakat is collected from employees' salaries.

Some zakat officers claimed that zakat was not budgeted for in the local budget, it was just recorded for collection and disbursement. However, it was stated as budgeted and realised in the financial report. If zakat is included in DAU calculation, the amount would be larger than in normal calculations; however, there is no supporting regulation for this issue.

It is interesting to identify which factors drive local regions to collect zakat in Aceh. The Mann-Whitney test, with a non-parametric test, was used in this study. The significant differences were on PAD, SILPA, Asset, Fixed and Debt. PAD can mean that the local regions can collect local resources optimally (see Table 4). So it would not be a problem for a local region to collect zakat. Similar issues also occur with SILPA, asset, fixed and debt. A larger amount means that local governments can be effective and have the economic scale and tools to collect zakat. However, this may also relate to the wealth of the zakat payer.

5 CONCLUSION

The issue of who must collect and manage zakat in any one country is actually based on the readiness of the government. Conceptually and historically speaking, the government must be responsible for collecting and disbursing zakat accordingly. The government has the authority to conduct that activity. However, for reasons of effectiveness and efficiency, private zakat institutions can participate in cases where the government is not well prepared to manage zakat. However, it is an irony in Muslim countries that they can collect taxes with a good system and infrastructure, but not zakat. In the case of Aceh, Indonesia, which applies Islamic law, the government would like to collect zakat directly and disburse it to 8 groups. However, the government has a strong authority to collect zakat optimally with the support of regulations. There should be an independent institution that monitors and reviews zakat management in Aceh, Indonesia.

Not many local regions can collect zakat in Aceh, which shows that the government is not well prepared for that task. This paper shows that only local regions with bigger assets, fixed assets, debt, PAD, and SILPA can collect zakat. This may be caused by those local regions already having the system and infrastructure to collect zakat. In addition, the accountability of the local government must be enhanced by reporting zakat disbursement in financial reports

in detail. In addition, supporting regulations to include zakat in the local budget should be provided by lobbying the central government to accommodate that activity. Otherwise, this can cause legal problems in the implementation.

REFERENCES

Abdullah, S. & Halim, A. (2006). Studi atas Belanja Modal pada Anggaran Pemerintah Daerah dalam Hubungannya dengan Belanja Pemeliharaan dan Sumber Pendapatan. *Jurnal Akuntansi Pemerintah, 2*(2), 17–32.

Adebayo, R.I. (2011). Zakat and poverty alleviation: A lesson for the fiscal policy makers in Nigeria. *Journal of Islamic Economics, Banking and Finance, 7*(26), 25–42.

Ahmed, H. (2008). Zakah, macroeconomic policies, and poverty alleviation: Lessons from simulations on Bangladesh. *Journal of Islamic Economics, Banking and Finance, 4*(2003), 81–105.

Ardhini. (2011). *Pengaruh Rasio Keuangan Daerah terhadap Belanja Modal untuk Pelayanan Publik Dalam Perspektif Teori Keagenan (Studi pada Kabupaten dan Kota di Jawa Tengah)* (Master's thesis). Universitas Diponegoro.

Bakar, M.H.A. & Ghani, A.H.A. (2011). Towards achieving the quality of life in the management of zakat distribution to the rightful recipients (the poor and needy). *International Journal of Business and Social Science, 2*(4), 237–245.

Bakar, N.B.A. (2007). A zakat accounting standard (ZAS) for Malaysian companies. *The American Journal of Islamic Social Sciences, 24*(4), 74–92.

Beik, I.S. & Arsyianti, L.D. (2013). Optimization of zakat instrument in Indonesia's poverty alleviation programme. *Poverty Alleviation—Academic Conference 2013*. Retrieved from http://www.seadiproject.com/post/poverty-alleviation—academic-conference-2013.

Benthall, J. (2008). *The Palestinian zakat committees 1993–2007 and their contested interpretations*. Geneva: Graduate Institute.

Buehler, M. (2008). The rise of Shari'a by-laws in Indonesian districts: An indication for changing patterns of power accumulation and political corruption. *South East Asia Research, 16*(2), 255–85. doi:10.5367/000000008785260473.

Darwanto, & Yustikasari, Y. (2007). *Pengaruh Pertumbuhan Ekonomi, Pendapatan Asli Daerah, dan Dana Alokasi Umum terhadap Pengalokasian Anggaran Belanja Modal*. Simposium Nasional Akuntansi X, Makassar.

Harianto, D. & Hadi, A.P. (2007). *Hubungan antara dana alokasi umum, belanja modal, pendapatan asli daerah dan pendapatan per kapita*. Simposium Nasional Akuntansi X, Makassar.

Hassan, K.M. & Khan, J.M. (2007). Zakat, external debt and poverty reduction strategy in Bangladesh. *Journal of Economic Cooperation, 28*(4), 1–38.

Hassan, M.K. (2010). An integrated poverty alleviation model combining zakat, awqaf and micro-finance. *The Tawhidi Epistemology: Zakat and Waqf Economy, 1*, 261–81. doi:10.1109/INCOS.2010.100.

Helal, H.E.A.A. (2012). *What is the required role of the state in Egypt concerning zakat?* Research Paper, International Political Economy and Development (IPED), The Netherlands.

Nurzaman, M.S. (2011). Zakat and human development: An empirical analysis on poverty alleviation in Jakarta, Indonesia 1. *8th International Conference on Islamic Economics and Finance* (pp. 1–26).

Powell, R. (2010). Zakat: Drawing insights for legal theory and economic policy from Islamic jurisprudence. *University of Pittsburgh Tax Review 7(43), Seattle University School of Law Research Paper: 10–17*. Retrieved from http://digitalcommons.law.seattleu.edu/faculty.

Putro, N.S. & Pamudji, S. (2011). Pengaruh Pertumbuhan Ekonomi, Pendapatan Asli Daerah dan Dana Alokasi Umum terhadap Pengalokasian Anggaran Belanja Modal. http://eprints.undip.ac.id/26411/2/jurnal.pdf.

Qardawi, Y. 1999. Hukum Zakat. Lentera, Jakarta.

Quraishi, M.A. (1999). The institution of zakat and its economic impact on society. *Proceedings of the Second Harvard University Forum on Islamic Finance: Islamic Finance into the 21st Century* (pp. 77–81). *Cambridge, Massachusetts: Center for Middle Eastern Studies, Harvard University*.

Rahman, A.R.A. (2007). Pre-requisites for effective integration of zakah into mainstream Islamic financial system in Malaysia. *Islamic Economic Studies, 1*(2), 91–107.

Shariff, A.M., Jusoh, W.N.H.W., Mansor, N. & Jusoff, K. (2011). A robust *zakah* system: Towards a progressive socio-economic development in Malaysia. *Middle-East Journal of Scientific Research, 7*(4), 550–554.

Solikin, I. (2007). Hubungan Pendapatan Asli Daerah dan Dana Alokasi Umum dengan Belanja Modal di Jawa Barat. Retrieved from http://file.upi.edu/Direktori/FPEB/PRODI.AKUNTANSI/196510122001121-IKIN_SOLIKIN/Jurnal_PAD.pdf.

Sulaiman, W. (2008). Modern approach of zakat as an economic and social instrument for poverty alleviation and stability of ummah. *Jurnal Ekonomi dan Studi Pembangunan, 9*(1), 105–118.

Toor, I.A. & Nasar, A. (2004). Zakat as a social safety net: Exploring on household welfare in Pakistan. *Pakistan Economic and Social Review, 42*(1/2), 87–102.

Wartonick, D. (2011). *Jordan fiscal reform II: Social protection public expenditures perspectives*. USAID/Jordan Economic Growth Office.

Yaumidin, U.K. (2011). Assessing the role of zakat for poverty alleviation in Indonesia. *The Review of Indonesian Economic and Business Studies, 2*(1), 89–112.

Competition and Cooperation in Economics and Business – Gani et al. (Eds)
© 2018 Taylor & Francis Group, London, ISBN 978-1-138-62666-9

Internal audit's role as a coordinator of combined assurance implementation

Rama Kurnia & Lufti Yulian
Department of Accounting, Faculty of Economics and Business, Universitas Indonesia, Depok, Indonesia

ABSTRACT: This research focuses on the role of internal audit as a coordinator in the combined assurance approach from the planning, implementation, and reporting points of view. This qualitative research study on SKK Migas Financial Reporting Assurance analyzes data from the literature and interviews to reach conclusions. The results of this study reveal that organizations can perform independent measurements for readiness-combined assurance implementation. Business processes which become the assurance objects must be mapped to the embedded risks. Coordination among assurance providers can be implemented using an assurance map. A combined assurance planning report states the role of each assurance provider and assists the auditee in performing resource planning for facilitating the assurance providers. The value added for the organization which implements combined assurance comes from reducing duplication and focusing internal audit activities on significant areas where assurance has not been performed. A thorough implementation of these activities reduces assurance gap and assurance fatigue.

1 INTRODUCTION

1.1 *Background*

Organizations through their will use their resources to achieve their objectives. By performing activities, companies have to manage related risks to an acceptable level (Kaplan and Mikes, 2012). To ensure that risks are properly managed and the organization can achieve their objectives, the management performs assurance process to monitor the process and implement good governance (Decaux, 2015).

There are many ways to treat risks. One of them is to mitigate the risks by performing controls. Control implementation by the management involves a risk owner, a risk manager, and assurance providers in the three lines of defense (Alijoyo, 2011). To implement the three lines of defense concept, the organization has to create the design, implementation and coordination (Daugherty and Anderson, 2012). The assurance process can occur in every line of defense. Each assurance provider has its own objectives and also reports the result to its stakeholders. This situation can create (1) assurance fatigue, due to the management being too focused on facilitating the assurance process in certain areas and (2) assurance gap, due to unclear and inefficient risk response and assurance (Sarens *et al.*, 2012).

To make the assurance process efficient and help the company achieve its objectives, the assurance process needs to be coordinated. Assurance coordination is the main concept of combined assurance (Decaux and Sarens, 2015). Coordinated assurance will reduce assurance redundancy and provide benefits such as: (1) effective and efficient assurance, (2) reduction of assurance gap and assurance fatigue (Daugherty and Anderson, 2012).

Good assurance quality in financial statements will reduce the likelihood of restatement due to material misstatement (Veronica and Bachtiar, 2005). From that statement, good assurance planning will create good assurance quality. Financial statement is important for management. It can be used as a governance process for the stakeholders. Generally, the assurance of financial statement will involve several assurance providers, either internal or external.

Research done by Decaux and Sarens (2015) reveal that combined assurance is one of many ways to improve an organization's governance process. In addition, Decaux (2015) states that combined assurance is a new phenomenon related to implementation and benefits.

Previous research reveal that: (1) an internal auditor is the most legitimate candidate to lead the combined assurance activities from facilitating, coordinating, and reporting (Decaux, 2015), (2) the implementation and coordination of combined assurance performed by an internal auditor will ensure the continuous improvement of the organization (Huibers, 2015), (3) combined assurance methods will provide benefits for the internal auditor in improving the effectiveness of activities and increase the role of internal audit in the organization's governance process (Decaux and Sarens, 2015). Based on previous research, there are still many unanswered questions about combined assurance, especially from the perspectives of the internal auditors as coordinators.

1.2 *Research questions*

1. Does the organization have the tools and ability to fully implement combined assurance?
2. Is there any sufficiency on the significant risk identification regarding financial statement process?
3. What are the forms of the assurance map that will describe relations between significant risks and the assurance providers involved?
4. What is the form of the combined assurance report plan that will be proposed to the audit committee?
5. What is the coordinator's role in the combined assurance implementation?

1.3 *Research goals*

1. Analyze the organization's ability to fully implement combined assurance.
2. Identify and analyze the significant risks related to financial statement process.
3. Create assurance maps based on the determined significant risks and assurance providers involved.
4. Create a combined assurance plan report.
5. State the roles to be performed by the combined assurance coordinator.

2 LITERATURE REVIEW AND RESEARCH METHODOLOGY

In theory, the Combined Assurance concept aims to help the board ensure that the key stakeholder needs are met in strategic direction and monitoring direction (Decaux, 2015). The monitoring roles of the board consist of assurance and risk management activities. These activities are performed by the risk management department and the internal audit department. The separation can cause redundancy in activities. Decaux (2015) confirms that there should be coordination between the risk management process and the internal audit process. Sarens *et al.*, (2012) describe this concept as Combined Assurance. Assurance providers coordinate on delivering the result to the board and reporting the assurance about the management of risks that impact the company's objectives.

The topic of this paper is the financial reporting assurance process at SKK Migas. SKK Migas is a special task force for upstream oil and gas activities in Indonesia. There are several reasons why the research was performed there.

First, SKK Migas is a public service agency; therefore, it has to prepare a financial statement based on the Indonesian Accounting Standards (IAS). Starting from the 2015 fiscal year, SKK Migas' budget has to be included in the Indonesian State Budget or Anggaran Pendapatan dan Belanja Negara (APBN), and so the financial statement needs to be prepared in accordance with the Indonesian Government Accounting Standards or Standar Akuntansi Pemerintahan (SAP). The management decided to continue preparing one financial statement based on the IAS in order to maintain the bookkeeping balance, and to prepare another financial statement based on the SAP to comply with the government rules.

Second, because the organization decided to prepare two kinds of financial statements, additional efforts will be required from the internal accounting department. In addition, the internal accounting department as the auditee has to facilitate several assurance providers internally and externally.

Third, from a risk management perspective, the risk register related to financial statement preparation has not been updated for several years, and the business process risks in preparing two kinds of financial statements have not been identified and measured.

The methods used in this research are descriptive qualitative. The purpose of using this approach is to gain understanding on the characteristics of a group and to create the systematic descriptive related to the research aspect. In the end, the method will help to make a simple decision (Sekaran and Bougie, 2013) and gain understanding from the interviewee's interpretation (Cooper and Schindler, 2014).

Our research is performed in four steps. The first step involves literature review from journals, books, and the organization's procedures to ensure that this research will be useful and that it has not been done before (Sekaran and Bougie, 2013).

The second step involves observation of the involved parties regarding the financial reporting assurance from both the auditor's and the auditee's perspective. The purpose is to reach a deep understanding which will lead to better analysis and interpretation.

The third step consists of a combination of a structured and unstructured interview. The purpose is to get a broad background and interviewee analysis.

The fourth and final step will involve analyzing the data gathered to interpret the literature about the impact and implementation within the research object (Shauki, 2015).

3 RESULTS

3.1 *Measurement of readiness for the combined assurance implementation*

Decaux and Sarens (2015) use six components to measure the readiness of an organization for the combined assurance implementation: (1) establishment of a mature risk management framework, (2) awareness of combined assurance, (3) identification of the champion, (4) development of the assurance strategy, (5) creation of the assurance provider map of the assurance activities, and (6) reporting of the combined assurance findings.

Our finding is that the organization is ready to implement combined assurance but it lacks a formalized audit charter and support from the management.

3.2 *Identification of significant risks*

The identification of significant risks has revealed that risks related to the organization's financial statement are under confidence. The causes are: (1) risks aren't linked to the business processes, (2) risks haven't been updated since the changes of business processes related to the preparation of financial statement related to Indonesian Government Accounting Standards.

3.3 *Design of assurance map*

Currently, SKK Migas does not have an assurance map. Based on the assessment, there are several concerns.

First, based on EY (2013) the organization has a clear understanding of their risk policy and goals. In contrast, risk event, risk cascading and risk mapping with roles and responsibilities of the stakeholders need to be updated.

Second, based on the auditee's workload measurement, the activities for facilitating assurance providers can impact the normal working hours and there is a potential for assurance fatigue due to the pressure to maintain work fulfilment and facilitate multiple assurance providers.

Third, the auditee may be experiencing assurance gap, due to the fact that the assurance activities are still uncoordinated and narrow-sighted.

To prepare an assurance map, the organization has to assess based on PWC (2010) all of the assurance providers involved in the area and then place them into each line of defense.

Based on each identified significant risk, there will be an assignment of expected inherent risk rating and level of assurance extensiveness. The assurance map will be created based on significant risks and consist of the assurance objective which will be performed by each assurance provider throughout the line of defense.

3.4 *Combined assurance plan report*

The purposes of this report are to document all activities related to combined assurance implementation. This report helps the coordination among assurance providers, and facilitates resource preparation by the auditee. In addition, the report contains the schedule and frequency of the activities that will be performed by the assurance providers. Internal assurance providers cannot communicate directly with external assurance providers. By using this report the external assurance providers will be able to use the work already done by the internal assurance providers and thus reduce the assignment duration. The report can also be used as a communication tool in reporting the assurance activities to the audit committee.

3.5 *Internal audit activity as an implementation coordinator of combined assurance*

Combined assurance has to be coordinated by an internal auditor. The objective is to perform efficient and non-redundant assurance activities. A combined audit program links all the activities together. The successful implementation of combined assurance can also help reduce assurance fatigue by allowing auditee resources to be used by the assurance providers. To reduce assurance gap, joint usage of assurance report can be an efficient way to reduce the redundancy of assurance activities. Each assurance provider focuses on their strengths. To cover the remaining activities, the internal auditor will take care of the areas that have not been addressed by other assurance providers.

A summary of the combined assurance process from the resource identification, planning, implementation, and reporting point of view is represented in Table 1:

Table 1. A summary of the combined assurance process.

Process	Activities	Objective
1. Identification	1.1 Identification of readiness to implement combined assurance	To measure the readiness of the organization to implement combined assurance
	1.2 Identification of ability to create an assurance map	To gather information about the resource availability in order to create an assurance map
2. Assessment	2.1 Initial identification of assurance fatigue	To measure the level of fatigue of the auditee
	2.2 Initial identification of assurance gap	To measure the magnitude of the current assurance gap of the auditee
	2.3 Assessment of the existing risk register	To measure the relevance of the existing risk register under the current conditions
	2.4 Measurement of the auditee's capacity to assist assurance providers	To form the basis of the resource availability for assisting assurance providers
	2.5 Assessment of all the assurance providers involved	To get the description of the reliability of assurance providers

(*Continued*)

Table 1. *Continued.*

Process	Activities	Objective
	2.6 Assessment of the control rating	To assess whether the control has effectively managed the risk. The result will useful as the basis of the assurance objective and rating
3. Implementation	3.1 Updating the risk register	To align the significant risks to the organization's current business processes
	3.2 Mapping the assurance providers to the lines of defense	To map the assurance providers to existing lines of defense; forms the basis of the assurance map
	3.3 Defining assurance objective	To state the objective of the assurance process that will be done by assurance providers, focusing on reducing assurance gap
	3.4 Defining assurance rating	To state the level of assurance that will be performed by the assurance providers. This activity focuses on reducing assurance fatigue.
4. Reporting	4.1 Making a combined assurance report plan	To summarize the entire assessment implementation plans of the combined assurance. This document is used as a communication channel between the assurance providers and the audit committee.

4 CONCLUSION

The organization has the ability to implement combined assurance based on the Institute of Directors Southern Africa (2009). Several things that need to be encouraged within the organization are: (1) There should be a formal appointment from the management, ensuring that the internal audit department will be the coordinator of the implementation, (2) the implementation of combined assurance can also be performed on other areas by other departments. The concept is that some of the departments have to perform assurance for the stakeholders, so they also can coordinate with other assurance providers within the scope.

There should be regular updates for the identification and measurement of current risk register. The point is that the risk embedded to each business process has to be identified properly and accurately. This is an important thing for the assurance providers; since combined assurance planning will use a risk-based approach, failure to identify significant risks will cause the inefficiency of assurance resources.

The usage of assurance map will give clarity about the assurance providers' profile, their assignment to each line of defense, and state the reliability level of the assurance to be performed. The assurance map will also contribute to the preparation process, leading to the efficient usage of the assurance resources. In order to decrease the assurance gap, the assurance map also has to be attached to the coordinated audit or review program; as a result, the assurance will be performed effectively.

Documentation of the combined assurance on the report will give valuable information to both sides. For the assurance providers, documentation ensures that the assurance process covers all significant risks. Coordination prepared by internal assurance providers can help the external assurance providers plan the activities.

The benefits of the combined assurance process are:

1. Assurance planning reduces duplication and clarifies the assurance providers' positions and roles related to significant risks.
2. Assurance planning identifies and measures all risks related to the organization's objectives and communicates them to the risk management department.
3. Internal audit activities will be focused on significant areas where assurance has not been performed.

REFERENCES

Alijoyo, Antonius. 2011. "Knowledge: Pertahanan 3 Lapis (The 3 Lines of Defense) - Konteks ERM Perusahaan Publik di Indonesia." CRMS Indonesia website. Accessed February 26, 2016 http://crmsindonesia.org/knowledge/crms-articles/pertahanan-3-lapis-3-lines-defense-konteks-erm-perusahaan-publik-di-indonesia.

Cooper, Donald, and Pamela Schindler. 2014. Business Research Method 12th Edition. New York: McGraw - Hill Education (Asia).

Daugherty, Brian, and Urton Anderson. 2012. "The Third Line of Defense: Internal Audit's Role in the Governance Process." Internal Auditing July/August 38–41.

Decaux, Loic. 2015. Internal Auditing and Organizational Governance: the Combined Assurance Approach. Doctoral's thesis, Université catholique de Louvain.

Decaux, Loïc and Gerrit Sarens. 2015. "Implementing Combined Assurance: Insights from Multiple Case Studies." Managerial Auditing Journal 30 (1): 156–79. doi:10.1108/maj-08-2014-1074.

EY. 2013. "Maximizing Value from Your Lines of Defense." Insight on Governance, Risk, and Compliance. http://www.ey.com/Publication/vwLUAssets/EY-Maximizing-value-from-your-lines-of-defense/$FILE/EY-Maximizing-value-from-your-lines-of-defense.pdf.

Huibers, Sam C.J. 2015. "Combined Assurance: One Language, One Voice, One View." Common Body of Knowledge. https://www.iia.nl/SiteFiles/Downloads/2015-1481_Combined%20 Assurance_CBOK_IIARF_S.Huibers.pdf.

Institute of Directors Southern Africa. 2009. King Code of Governance for South Africa 2009. Cape Town: Institute of Directors Southern Africa.

Kaplan, Robert S, and Anette Mikes. 2012. Strategic Planing: Managing Risks A New Framework. Accessed March 7, 2016 https://hbr.org/2012/06/managing-risks-a-new-framework/ar/1.

PWC. 2010. "Preparation, Preserverance, Payoff: Implementing a combined assurance approach in the era of King III." Business School Risk Assurance.

Sarens, Gerrit, Loic Decaux, and Rainer Lenz. 2012. Combined Assurance: Case Studies on a Holistic Approach to Organizational Governance. Florida: The Institute of Internal Auditors Research Foundation.

Sekaran, Uma, and Roger Bougie. 2013. Research Method for Business: a skill building approach 6th edition. Chichester: John Wiley and Sons Ltd.

Shauki, Elvia. 2015. "Research Design in Accounting: Topic 6 Analysing Data Apporaches to Quantitative and Qualitative Data Analysis." Surabaya: Airlangga University.

Veronica, Sylvia, and Yanivi Bachtiar. 2005. "The Role of Corporate Governance in Preventing Misstated Financial Statement." Jurnal Akuntansi dan Keuangan Indoncesia 2(1): 159–173.

Competition and Cooperation in Economics and Business – Gani et al. (Eds)
© 2018 Taylor & Francis Group, London, ISBN 978-1-138-62666-9

Islamic accountability index of cash waqf institution in Indonesia

Dodik Siswantoro
Department of Accounting, Faculty of Economics and Business, Universitas Indonesia, Depok, Indonesia

Haula Rosdiana
Department of Public Administration, Faculty of Administrative Sciences, Universitas Indonesia, Depok, Indonesia

Heri Fathurahman
Department of Business Administration, Faculty of Administrative Sciences, Universitas Indonesia, Depok, Indonesia

ABSTRACT: The objective of paper is to create an Islamic accountability index of cash waqf institution in Indonesia. Cash waqf is a new phenomenon in Indonesia since the Act was just issued in 2004. Some new cash waqf institutions were established but unfortunately without tight monitoring from the government. An accountability index can show the reported activity of cash waqf institution. This may help society to see the accountability condition of the institution. Research method conducted is based on qualitative method with literature review. In-depth interview is added to confirm and to enrich the index analysis. The index is tested to cash waqf institution based on web basis. The result may show that accountability index may describe some activities and financial report of cash waqf institutions. Most cash waqf institutions have not high index for the accountability. This may be caused by limited support of the cash waqf institutions to show their accountability. Cash waqf institutions should prepare the supported information system to show the accountability especially in the online system.

1 INTRODUCTION

Cash waqf was formally regulated under Act No. 41/2004, which permits non fixed asset for waqf. Then, 82 registered cash waqf institutions were established until 2014. So far, those institutions should disclose their financial statements to public through Badan Wakaf Indonesia (BWI). In fact, BWI do not disclose them to public. The issue accountability may occur here as stakeholders need to know how institutions manage and report their activities properly. Accountability can be measured in terms of several important aspects.

Accountability measurement is also an important issue for not-for-profit organizations, especially for cash waqf institution. Many not-for-profit organizations utilize the Internet to publish their reports, which is easier and more efficient. In addition, a wider target of accountability can be reached by web-based reporting (Gandia, 2011; Saxton and Guo, 2011; Tremblay-Boire and Prakash, 2014). Dainelli *et al.* (2013) state that the number of stakeholders and funds received can be an indicator of accountability measurement for not-for-profit. Good accountability can lead to bigger donation amounts, as people will trust not-for-profit organizations with accountability more (Waters, 2011).

Accountability measurement for not-for-profit organizations has been studied by Dhanani and Connolly (2012); Dainelli *et al.* (2013); Dumont (2013); and Tremblay-Boire and Prakash (2014), revealing a conventional theme. The cash waqf institution based on Islamic teaching has different characteristics that may cause specific treatment for accountability measurement.

2 LITERATURE REVIEW

Coy and Dixon (2004) have completed studies on accountability measurements in the public sector. They created the Modified Accountability Index (MAD). Further research was conducted by Ryan *et al.* (2002) for regional government; Nelson *et al.* (2003) for a Canadian university; and Ismail and Barizah (2010) in Malaysia. Lima and Pereira (2011) analyze other education sectors. These measurements show the accountability of public institutions. Each type of public institution has specific characteristic; therefore, the weighted components may be different (Carnegie and Wolnizer, 1996).

In the not-for-profit organization sector, some scholars' studies on the accountability index are based on Global Reporting Initiatives (GRI) with seven indicators, such as Tremblay-Boire and Prakash (2014). Dainelli *et al.* (2013) study examines museums with 18 indicators, while Dhanani and Connolly (2012) identify four main indicators, such as strategic accountability, fiduciary accountability, financial accountability, and procedural accountability.

Kang and Norton (2004) state that disclosing information on the Internet is useful and helps to increase credibility (Kenix 2008), but it is not very effective for communication. Communication with donors is also important; it can increase the amount of donations (Waters, 2010). Gandia (2011) states that the use of websites for disclosure is important to increase accountability as well as to increase donations. Sloan (2008) also states that a higher rating of accountability leads to bigger donations being received. In addition, Rodríguez *et al.* (2012) state that the age of an organization, as well as its experiences, can affect the transparent quality. Dumont (2013) shows that the accountability index on websites is based on five characteristics: accessibility, engagement, performance, governance, and mission.

3 RESEARCH METHODOLOGY

The research is based on qualitative method with constructivist paradigm. Research objects in this study were four persons who are waqf expert and shariah scholars. It was conducted by asking them to fill the weight of proposed themes and components of accountability measurement. Then, those experts reconfirmed the result of the weighted component. To check the accountability index of the cash waqf institution, each institution was given a 'one' if it fulfilled the component and 'zero' if it was not available. The others checked the component list by reviewing online resources, such as websites and social media (YouTube, Facebook, and Twitter).

The result then was discoursed by two experts and commented by two Islamic teaching experts. After that, the result was suggested to cash waqf institutions as the consequence of constructivist paradigm by authenticity parameter (fairness, ontology, educative, catalytic, and tactical authenticity) (Lincoln and Guba, 1985).

4 ANALYSIS

Research of the accountability index for not-for-profit organizations was studied by Dhanani and Connolly (2012); Dainelli *et al.* (2013); and Tremblay-Boire and Prakash (2014). Dumont (2013) has created the accountability index based on an organization's website. The Islamic perspective refers to Iqbal and Lewis (2009), while the measurement method refers to Mohammed and Razak (2008). The Islamic accountability index for the cash waqf organization consists of five themes taken from previous research. The theme that was most appropriate in the cash waqf context was selected. Each theme consists of various components that are also appropriate in the cash waqf context (see Table 1). The themes with the greatest support are finance and performance. Four studies support these themes are Dhanani and Connolly (2012); Dainelli *et al.* (2013); Dumont (2013), and Tremblay-Boire and Prakash (2014). Other themes, such as public engagement and the Islamic aspect, are only supported by one or two research studies, but are still suitable for this case.

Table 1. Islamic accountability index component.

Theme	Source
Finance	Dhanani and Connolly (2012); Dainelli, Manetti, and Sibilio (2013); Dumont (2013); Tremblay-Boire and Prakash (2014)
Audited financial statements	Dhanani and Connolly (2012); Dumont (2013); Tremblay-Boire and Prakash (2014); Dainelli, Manetti, and Sibilio (2013); Iqbal and Lewis (2009)
Expenditure information based on category	Tremblay-Boire and Prakash (2014); Mohammed and Razak (2008);
List of donors	Tremblay-Boire and Prakash (2014); Dainelli, Manetti, and Sibilio (2013)
Composition and management structure	Tremblay-Boire and Prakash (2014)
Activity and shariah board information	Iqbal and Lewis (2009)
Management note	Tremblay-Boire and Prakash (2014); Dainelli, Manetti, and Sibilio (2013); Mohammed and Razak (2008);
Performance	Dhanani and Connolly (2012); Dainelli, Manetti, and Sibilio (2013); Dumont (2013); Tremblay-Boire and Prakash (2014)
Income on investment (net income/average total asset)	Dhanani and Connolly (2012); Tremblay-Boire and Prakash (2014); Mohammed and Razak (2008)
Suggestions from beneficiaries	Tremblay-Boire and Prakash (2014); Mohammed and Razak (2008)
Program effectiveness and effects of evaluation	Dhanani and Connolly (2012); Tremblay-Boire and Prakash (2014)
Employee competence upgrading program	Dhanani and Connolly (2012); Dumont (2013); Tremblay-Boire and Prakash (2014); Mohammed and Razak (2008)
Employee code of ethics and regulations	Tremblay-Boire and Prakash (2014)
Good governance information	Dhanani and Connolly (2012); Dumont (2013); Dainelli, Manetti, and Sibilio (2013)
Performance indicators	Dumont (2013); Dainelli, Manetti, and Sibilio (2013)
Public	Tremblay-Boire and Prakash (2014)
Benefits of waqf for beneficiaries information	Tremblay-Boire and Prakash (2014); Dainelli, Manetti, and Sibilio (2013); Mohammed and Razak (2008);
Information of percentage of funds used for social empowerment	Tremblay-Boire and Prakash (2014); Mohammed and Razak (2008)
Environment report	Tremblay-Boire and Prakash (2014)
Response to suggestions	Dumont (2013); Tremblay-Boire and Prakash (2014); Dainelli, Manetti, and Sibilio (2013); Mohammed and Razak (2008)
Program description and activity	Tremblay-Boire and Prakash (2014)
Involvement	Dumont (2013); Tremblay-Boire and Prakash (2014)
Contact access and availability	Dumont (2013); Tremblay-Boire and Prakash (2014); Mohammed and Razak (2008)
Waqf online participation	Dumont (2013)
Updates and organizational news	Dumont (2013); Tremblay-Boire and Prakash (2014); Mohammed and Razak (2008)
FAQ	Dumont (2013); Mohammed and Razak (2008)
Islamic aspect	Mohammed and Razak (2008); Iqbal and Lewis (2009)
Shariah compliance	Mohammed and Razak (2008); Iqbal and Lewis (2009)
Regulation compliance	Dumont (2013); Tremblay-Boire and Prakash (2014); Iqbal and Lewis (2009)

Sources: Various.

The Islamic aspect is an important issue to help hold the waqf institution accountable. This may differ from similar conventional endowment institutions which are not based on Islamic teaching. The research that resulted from informants, verified by reconfirmation, can be seen at Table 2.

The highest weight is audited financial statements (10.56%), followed by shariah compliance (8%) and expenditure information based on category (7%). Audited financial statements were the important issue in the previous research by Dainelli *et al.* (2013), but not for Dumont (2013). Dhanani and Connolly (2012) state that financial statements are an important aspect. Audited financial statements show that one institution has already had the administrative transaction checked. The shariah compliance aspect has 8% weight. This shows that shariah compliance is an important aspect that must be fulfilled by the cash waqf institution. Expenditure information based on category has a 7% weight, which means that expenditure allocation is an important component to disclose.

The fulfillment theme shows that the average cash waqf institutions report and disclose the accountability component. The highest fulfillment is the public theme at 79.1%, while the engagement theme is at 74.86%. The highest fulfillment theme may be caused by a lower weight component in each theme with 3.5% for public and 3.31% for engagement (see Table 3). The lowest fulfillment is the accountability theme, which has the highest average weight component

Table 2. Islamic accountability index.

Theme	Weight (%)
Finance	
Audited financial statements	10.56
Expenditure information based on category	7.00
List of donors	3.90
Composition and management structure	3.35
Activity and shariah board information	2.50
Management note	2.69
Performance	
Income on investment (net income/average total asset)	5.99
Suggestions from beneficiaries	3.31
Program effectiveness and effects of evaluation	5.73
Employee competence upgrading program	3.25
Employee code of ethics and regulations	2.26
Good governance information	2.55
Performance indicators	2.42
Public	
Benefits of the waqf for beneficiaries information	5.25
Information of percentage of funds used for social empowerment	4.05
Environment report	2.08
Response to suggestions	2.38
Program description and activity	3.75
Engagement	
Contact access and availability	4.76
Waqf online participation	3.05
Updates and organizational news	3.35
FAQ	2.09
Islamic aspect	
Shariah compliance	8.00
Regulation compliance	5.75
Total	100.00

Source: Data processed.

ratio (Islamic aspect 31.36%, performance 33.98%, and finance 38.39%). The average weight component rank is 6.78% for Islamic aspect, 3.64% for performance, and 5% for finance. It can be stated that the higher weight of component accountability in one theme usually corresponds to the lowest accountability fulfillment of the accountability index.

The implementation of the accountability index is shown by giving a 'one' if the cash waqf institution provides component requirements, otherwise it is given a 'zero.' The highest index score is DD, which may fulfill all components for each theme. The lowest is DT. The biggest gap is seen in the finance theme (10.56), with a high score of 17.56 and a low of 7. This shows that one cash waqf institution did not show accountability in the finance aspect (see Table 4). The lowest gap is in the public theme (4.05, 15.43-11.38). This may imply that each cash waqf institution in this theme can fulfill the requirements. Only one cash waqf institution has zero for the Islamic aspect theme. However, the cash waqf institution based on Islamic teaching should be more proactive in disclosing any Islamic teaching compliance activity.

If we look at each component in detail, not all cash waqf institutions can fulfill the requirements, for example, activity and shariah board information, management note, good governance information, and shariah compliance. All cash waqf institutions have a score of zero for this component. Most of the component weights are small (from 2.5% to 2.69%), except for shariah compliance (8%). This may occur due to ignorance and unpreparedness of the cash waqf institutions (see Table 5).

Authenticity is used to check the constructivist paradigm. Some requirements discussed are as follows:

1. Fairness–this study claims to give better suggestions for the institution to be more accountable. The institution would be concerned with components that create a better accountability index.
2. Ontological authenticity–cash waqf institutions realize that reporting to the public is important, and they should support a good reporting system and resources. Good accountability leads to better support and response from society to participate in the cash waqf program in the future.
3. Educative authenticity–some institutions try to improve their reporting on their website and social media. They try to understand the necessity of having a good activity report

Table 3. Islamic accountability index and fulfillment (%).

Theme	Weight	Fulfillment	Ratio
Finance	30	11.52	38.39
Performance	25.5	8.67	33.98
Public	17.5	13.84	79.1
Engagement	13.25	9.92	74.86
Islamic aspect	13.75	4.31	31.36
Total	100	48.26	

Source: Data processed.

Table 4. Islamic accountability index score (%).

Theme	DD	Water	BWI	DT
Finance	17.56	14.25	7.5	7
Performance	11.71	9.04	8.25	5.67
Public	13.45	15.43	11.38	15.13
Engagement	8.11	13.25	10.20	8.11
Islamic aspect	5.75	0	5.75	5.75
Total	56.59	51.96	42.83	41.65

Source: Data processed.

Table 5. Islamic accountability index and fulfillment.

Theme	Weight (%)	Average fulfillment (%)
Finance		
Audited financial statements	10.56	25
Expenditure information based on category	7.00	75
List of donors	3.90	50
Composition and management structure	3.35	50
Activity and shariah board information	2.50	0
Management note	2.69	0
Performance		
Income on investment (net income/average total asset)	5.99	50
Suggestions from beneficiaries	3.31	25
Program effectiveness and effects of evaluation	5.73	50
Employee competence upgrading program	3.25	25
Employee code of ethics and regulations	2.26	25
Good governance information	2.55	0
Performance indicators	2.42	25
Public		
Benefits of the waqf for beneficiaries information	5.25	100
Information of percentage of funds used for social empowerment	4.05	50
Environment report	2.08	50
Response to suggestions	2.38	75
Program description and activity	3.75	100
Engagement		
Contact access and availability	4.76	100
Waqf online participation	3.05	25
Updates and organizational news	3.35	100
FAQ	2.09	50
Islamic aspect		
Shariah compliance	8.00	0
Regulation compliance	5.75	75
Total	100.00	

Source: Data processed.

and financial statement. Institutions can report the simpler component first in order to learn for the future.

4. Catalytic authenticity–some cash waqf institutions want to fulfill the component requirement of the accountability index. This depends on the institution's capability of preparing it. Support from related systems and resources are significant roles for greater accountability.
5. Tactical authenticity–participation from other staff in the institution should be conducted. This takes time and effort for the organization. One institution assigned staff to prepare the system and better reports, especially for information systems based on the Internet.

5 CONCLUSION

Constructing an accountability index for the cash waqf institutions in Indonesia can compare and increase awareness of accountability components within institutions. An institution needs a competent resource that is aware of this area to prepare a good accountability index.

The most important issues are auditing financial statements and compliance with Islamic teachings. Institutions should not only be able to conduct the activity, but also report it to the public. However, the institution can utilize social media and the Internet to increase its accountability to the public by disclosing information.

REFERENCES

Carnegie, Garry D, and Peter W. Wolnizer. 1996. "Enabling Accountability in Museums." *Museum Management and Curatorship* 15 (4): 371–86. doi:10.1016/S0260-4779(96)00053-2.

Coy, David and Keith Dixon. 2004. "The Public Accountability Index: Crafting a Parametric Disclosure Index for Annual Reports." *The British Accounting Review* 36, 79–106.

Dainelli, Francesco, Giacomo Manetti, and Barbara Sibilio. 2013. "Web-Based Accountability Practices in Non-Profit Organizations: The Case of National Museums." *Voluntas* 24 (3): 649–65. doi:10.1007/s11266-012-9278-9.

Dhanani, Alpa, and Ciaran Connolly. 2012. "Discharging Not-for-profit Accountability: UK Charities and Public Discourse." *Accounting, Auditing & Accountability Journal* 25 (7): 1140–69. doi:10.1108/09513571211263220.

Dumont, Georgette E. 2013. "Nonprofit Virtual Accountability: An Index and Its Application." *Nonprofit and Voluntary Sector Quarterly* 42 (5): 1049–67. doi:10.1177/0899764013481285.

Gálvez Rodríguez, María del Mar, María del Carmen Caba Pérez, and Manuel López Godoy. 2012. "Determining Factors in Online Transparency of NGOs: A Spanish Case Study." *Voluntas* 23 (3): 661–83. doi:10.1007/s11266-011-9229-x.

Gandia, Juan L. 2011. "Internet Disclosure by Nonprofit Organizations: Empirical Evidence of Nongovernmental Organizations for Development in Spain." *Nonprofit and Voluntary Sector Quarterly* 40 (1): 57–78. doi:10.1177/0899764009343782.

Iqbal, Zafar, and Mervyn K Lewis. 2009. *An Islamic Perspective on Governance*. Cheltenham: Edward Elgar Publishing Limited.

Ismail, Suhaiza, and Nur Barizah Abu Bakar. 2010. "Annual Report of Malaysian Public Universities: The Extent of Compliance and Accountability." Presented at the *Asia Pacific Interdisciplinary Research in Accounting,* University of Sydney.

Kang, Seok, and Hanna E. Norton. 2004. "Nonprofit Organizations' Use of the World Wide Web: Are They Sufficiently Fulfilling Organizational Goals?" *Public Relations Review* 30 (3): 279–84. doi:10.1016/j.pubrev.2004.04.002.

Kenix, Linda Jean. 2008. "Nonprofit Organizations' Perceptions and Uses of the Internet." *Television & New Media* 9 (5), 407–428.

Lima, Emanoel Marcos, and Carlos Alberto Pereira. 2011. "Comparative Analysis of the Index of Disclosure and the Importance given by Stakeholders to Information Considered Relevant for Accountability in Philanthropic Institutions of Higher Education of Brazil (PIHEB)." *African Journal of Business Management* 5 (34): 13116–29. doi:10.5897/AJBM11.167.

Lincoln, Yvonna S, and Egon Guba. 1985. *Naturalistic Inquiry.* California: Sage Publications.

Mohammed, Mustafa Omar, and Dzuljastri Abdul Razak. 2008. "The Performance Measures of Islamic Banking Based on The Maqasid Framework." *The IIUM International Accounting Conference* (INTAC IV), Kuala Lumpur.

Nelson, Morton, William Banks, and James Fisher. 2003. "Improved Accountability Disclosures by Canadian Universities." *Canadian Accounting Perspectives* 2 (1): 77–107. doi:10.1506/1H5W-R5DC-U15T-KJPH.

Ryan, Christine, Trevor Stanley, and Morton Nelson. 2002. "Accountability Disclosures by Queensland Local Government Councils: 1997–1999." *Financial Accountability & Management* 18 (August): 261–89. doi:10.1111/1468-0408.00153.

Saxton, Gregory D, and Chao Guo. 2011. "Accountability Online: Understanding the Web-Based Accountability Practices of Nonprofit Organizations." *Nonprofit and Voluntary Sector Quarterly* 40 (2): 270–95. doi:10.1177/0899764009341086.

Sloan, Margaret F. 2008. "The Effects of Nonprofit Accountability Ratings on Donor Behavior." *Nonprofit and Voluntary Sector Quarterly* 38 (2): 220–36. doi:10.1177/0899764008316470.

Tremblay-BoireEmail, Joannie, and Aseem Prakash. 2014. Accountability.org: Online disclosure by U.S. nonprofits. *Voluntas* 26: 692–719. doi:10.1007/s11266-014-9452-3.

Waters, Richard D. 2011. "Increasing Fundraising Efficiency Through Evaluation: Applying Communication Theory to the Nonprofit Organization- Donor Relationship." *Nonprofit and Voluntary Sector Quarterly* 40 (3): 458–75. doi:10.1177/0899764009354322.

Competition and Cooperation in Economics and Business – Gani et al. (Eds)
© 2018 Taylor & Francis Group, London, ISBN 978-1-138-62666-9

Sustainability of the productive cash waqf institutions in Indonesia from an N-Helix perspective

D. Siswantoro
Department of Accounting, Faculty of Economics and Business, Universitas Indonesia, Depok, Indonesia

H. Rosdiana
Department of Public Administration, Faculty of Administrative Sciences, Universitas Indonesia, Depok, Indonesia

H. Fathurahman
Department of Business Administration, Faculty of Administrative Sciences, Universitas Indonesia, Depok, Indonesia

ABSTRACT: The aim of this research is to identify the supporting factors for the sustainable development of productive cash waqf institutions in Indonesia. Currently, cash waqf in Indonesia is still mainly in the social sector, such as mosque development and social hospitals. Therefore, productive cash waqf must be developed further, especially for the supporting factors. The research method used is based on a qualitative case study with four productive cash waqf institutions. Some supporting factors are identified using main nodes in each aspect. The result shows that each aspect has an important issue, such as education (need early education on waqf), economics (supporting system to pay waqf), media culture (habitual activity to pay waqf), environment (not so familiar) and politics (supporting regulations and promotion). The additional helix is the belief of Muslims to participate in waqf which is the main factors.

1 INTRODUCTION

Act No. 41/2004 on waqf has been issued in Indonesia, and so far 82 cash waqf institutions have been registered by Badan Wakaf Indonesia (BWI)—Indonesian Waqf Board as legal government authorities in Indonesia. However, the total fund collection before September 2014 was still about 200 billion IDR (Republika, 2014). This fund is still far from the minimum amount of fund management needed in order to achieve the economics of scale. The institutions may only use the income generated from the waqf fund (the maximum allocation is 10% from income) if it is fully invested (according to BWI regulation No. 1/2009). Therefore, the sustainability of cash waqf will be a problem if the total cash management is still small.

Therefore, the sustainability of the cash waqf institutions is also an important issue. The cash waqf institutions will become more sustainable if they have the minimum amount of cash fund management that can generate sufficient income for their operating expenses. To analyse some supporting factors for the sustainability of cash waqf institutions, this report uses the N-Helix analysis. Previously, it started from a triple helix that used three aspects, such as education, economy and politics. The quadruple helix uses education, economy, politics and media culture. The quintuple helix adds environment as the fifth helix (Carayannis & Campbell, 2011; Carayannis et al., 2012; Carayannis & Rakhmatullin, 2014). Waqf is a social fund that must be maintained and organised in order to allocate income to recipients in need. In this case, it is appropriate to use the quintuple helix approach to analyse the sustainability of the cash waqf institutions.

Siswantoro and Rosdiana (2016) identified in previous research that cash waqf should be productive in terms of current or fixed asset types. Knowledge in creation and innovation is needed in this area to create sustainability. In fact, this is still based on conceptual and theoretical analysis. However, the main issue of cash waqf is actually that of having sufficient funds of cash waqf in order to productively generate income for operational and social activities.

2 LITERATURE REVIEW

To be sustainable, a cash waqf institution needs the minimum amount of capital necessary to achieve an economic scale that can generate enough income to cover operational expenses and targeted allocations for needy. For one cash waqf institution, the minimum amount of capital to be covered is 50 billion IDR. For example, operating costs per year are 400 million IDR. An institution can only take a 10% fee from investment income, so they must generate a minimum income of four billion per year. If the expected return is 8%, the minimum total investment is 50 billion IDR.

Furthermore, Nasution (2005) created a simulation with ten billion Muslims who give 5,000 to 100,000 IDR per month. The total amount of funds collected is three trillion IDR. In fact, the total amount of funds received in more than five years was only 200 billion IDR. The waqf fundraising programme is also not big enough in other countries, such as Bangladesh. In 2013 they only collected 436,010,591 Taka (52 billion IDR) through Islami Bank Bangladesh Limited (IBBL), while Social Investment Bank Limited SIBL raised 84,577,685 Taka (ten billion IDR). In 2012, Perbadanan Wakaf Selangor in Malaysia, where waqf payments can count as a taxable income deduction, collected 126 billion IDR in one year. In Singapore, they use sukuk (Islamic obligation) to raise waqf funds. In one issuance they can collect almost 400 billion IDR. In 2015, the total amount of waqf assets in Singapore was almost 6.5 trillion IDR, which generates an income of 111 billion IDR per year. Only 22.5 billion IDR is allocated to the needy and the social sector. The rest is reinvested in the waqf fund. From this evidence, we can see how waqf funds should be managed and developed in Indonesia. Bigger waqf funds will better sustain the cash waqf institutions.

To analyse the sustainability of cash waqf institutions in Indonesia, the quintuple helix approach can be used to evaluate the institutions from the aspects of education, economy, media culture, environment and politics. Carayannis et al. (2012) analysed the cycle of the quintuple helix in sequence from those five aspects. If each aspect can strengthen the sustainability of the others, it will create an innovation that will lead to sustainable development. Siswantoro and Rosdiana (2016) found that the important aspects in the sustainability of cash waqf development are educational, economic and political systems. In fact, Indonesia still suffers from the existence of cash waqf institutions, so not many people see this as a new issue.

Additional helix may be possible, since each issue has specific characteristics that may be different from previous studies. Carayannis and Campbell (2012) state that the development of technology can cause sustainable development. Leydesdorff (2012) emphasises the need for additional helix to prove the sustainability of specific issues and ease the analysis. Park (2014) explains that economic development can cause changes in factors to adapt for sustainability.

3 RESEARCH METHODOLOGY

The post positivist paradigm is used in this research along with the qualitative method. Research objects in this study were Badan Wakaf Indonesia (BWI), Tabung Wakaf Indonesia (TWI), Wakaf Center (Water) and Pusat Pengembangan Wakaf Daarut Tauhiid (DT). These four institutions have similar characteristics to the cash waqf, as they focus on productive assets.

Unit analysis was classified based on the research objective of six people who have custody of and manage the waqf institutions, government, and Islamic banks. Then, the data analysts applied the qualitative method, content analysis and narrative. The objective of analysis used content analysis by coding the interview transcript with the main theme. The helix aspect required each informant to determine the sustainable development aspect of the cash waqf institutions in Indonesia.

The credibility of the research was achieved by checking other responses from the Directorate of Waqf, Ministry of Religious Affairs and observing research objects. The research may have strong transferability as the unit analysis was taken from qualified and competent resources (Lincoln & Guba, 1985).

4 ANALYSIS

The analysis of the sustainable development of the cash waqf institutions was based on the quintuple helix approach. Additional helixes were added since one issue, such as the iman system or belief, may not fit in a particular helix (see Figure 1). Issues with the cash waqf are different from those of the fixed asset waqf, which requires large assets, such as land or buildings. In the cash waqf, Muslims can pay any amount of money, as there is no minimum amount. The important issue is whether Muslims believe in the cash waqf institution. If they do not believe in the institution, they will not pay the waqf. This could cause less sustainable development.

Society, especially Muslims, needs to be educated to create awareness of the waqf. Therefore, the first effort should be made by the education system. Once they are well educated, the economic system must improve so that Muslims can pay the cash waqf. This must also be supported by the payment system. In addition, Iman or belief in the cash waqf institution is also an important aspect, because a cash waqf donation does not have to be a large amount, a small amount is also possible, so paying the cash waqf is not a burden.

The media must support the cash waqf, and paying it should be part of a society's culture. If this phenomenon occurs, the cash waqf is likely to be sustainable. The media can also educate people to participate in this programme. The cash waqf can also be related to the environmental system if it is allocated to an environmental cash waqf programme. In Indonesia, the environment is an interesting issue that has the potential for further development, as Indonesia is covered with forests and farmland.

The political aspect also supports the sustainable development of the cash waqf institutions in Indonesia. The government should support the development of the cash waqf institutions by easing regulations and providing incentives to the institutions. The political aspect would affect the education aspect by increasing educational awareness and policies that support cash waqf development.

The analysis of the sustainable development for the cash waqf institutions in Indonesia is based on the sextuple helix as follows:

4.1 *Education system*

Socialisation is an important issue for this aspect, for example, giving practice and training on the cash waqf. Other issues that arise are supporting technology in education, research, sponsorship for waqf development, early childhood education, testament and incentives in education (see Figure 2).

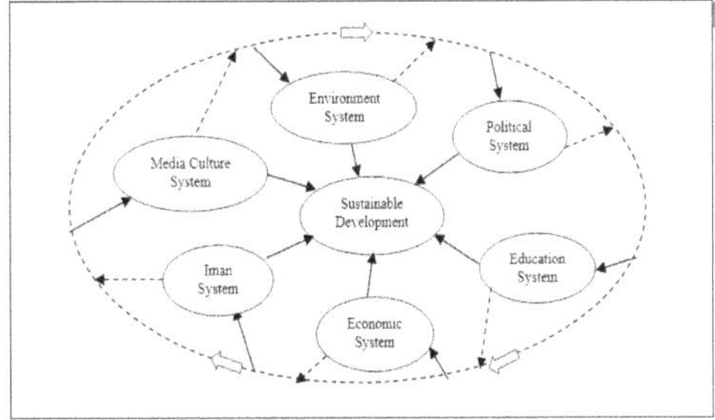

Figure 1. Cycle dimension of the sextuple helix.
Source: Data processed.

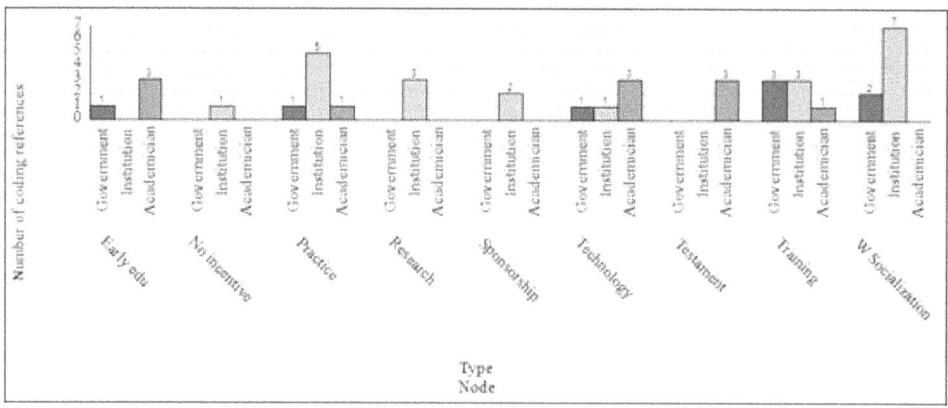

Figure 2. Education aspect of the cash waqf institution.
Source: Data processed.

The biggest issue in education for the waqf is that there has not been very much information dissemination, even though the Act was issued in 2004. This includes training and courses in schools and universities. The image of the cash waqf is not well known, so Muslims have not yet participated in the programme. Society needs to be aggressively educated and involved in marketing programme participation, like agency systems in insurance. Early childhood education should also be geared towards the cash waqf. This would help to create a strong image of the cash waqf for children before they become adults. Course programmes could be offered in senior high schools, since they may already be familiar with the cash waqf programme. In elementary schools, subject material on charity should be introduced to create awareness. The Islamic religion, of course, must focus on the practice of praying and having empathy for the poor and needy.

The cash waqf institutions must utilise technology that is capable of bringing socialisation to the cash waqf. Social media is a cheap and effective tool for marketing. Society needs to see the benefits of a productive waqf asset so that they will have a sense of belonging to this programme. In addition, sponsorship on waqf research and incentives by the government for further study can create triggers to develop the cash waqf.

4.2 *Economic system*

The economic aspect of the sustainable development of the cash waqf institutions in Indonesia allows society to participate in the cash waqf and makes it easy to pay for it. Important issues raised in this aspect are niche markets, trust, priority and unknowns from society. Other interesting issues are limited socialisation and the utilisation of a temporary cash waqf in banks (see Figure 3).

Understanding of the cash waqf becomes an interesting issue. Many people are reluctant to pay the waqf. In fact, it has a specific niche market, but not many people feel comfortable with all of the cash waqf institutions. People have a tendency to use specific institutions that they know well. Trust can also be an important issue for donors. Trust can be built if the institution is transparent and reliable. Society must know that the cash waqf institutions are reliable and meaningful, and that they are also of benefit to them.

A temporary deposit for the cash waqf can be an effective tool for boosting waqf income. The cash waqf institutions should co-operate with Islamic banks to promote this plan, so that they do not feel threatened. Another important issue is that society must know the real physical assets of the cash waqf, for example, hotels, rental houses, shops and office buildings. The poor and needy can also rent the store to sell their products. In other words, this will empower those to improve their life. This is a good example whereby the cash waqf can empower the poor and needy, as the generated income is also beneficial to them. Muslims would have strong incentives to participate in the cash waqf programme if they could see this process and its result.

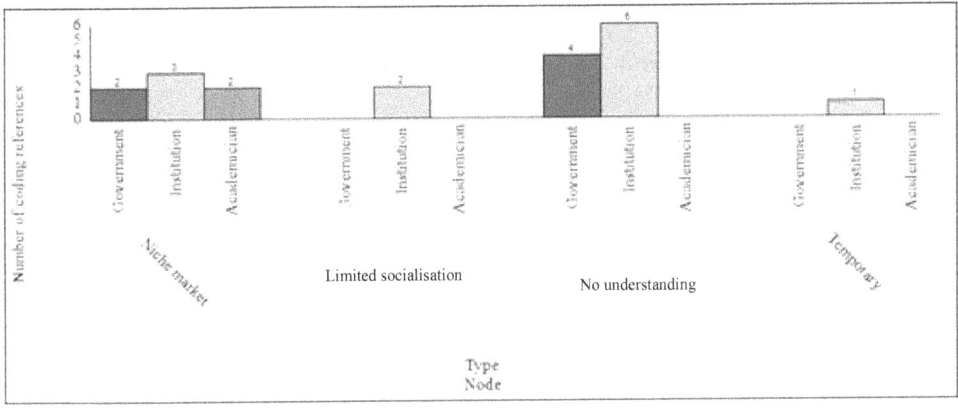

Figure 3. Economic aspect of the cash waqf institution.
Source: Data processed.

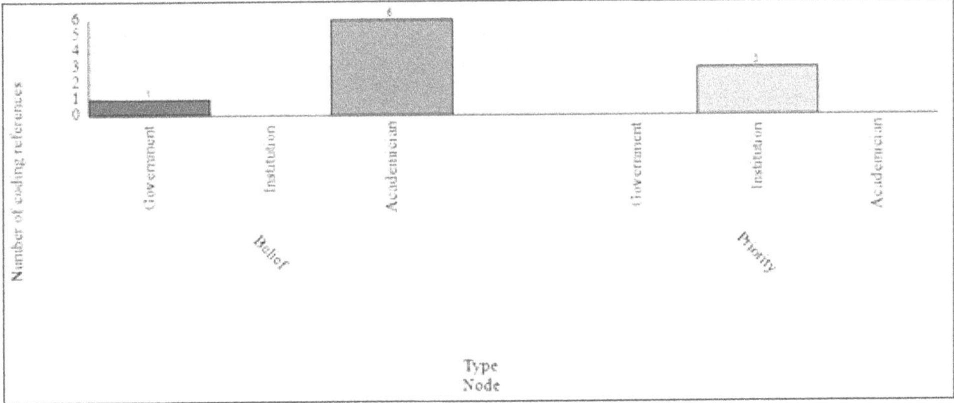

Figure 4. Iman aspect of the cash waqf institution.
Source: Data processed.

4.3 *Iman system*

The waqf is not obligatory for all Muslims. Only those who have strong beliefs will participate in the cash waqf programme. It is necessary that the cash waqf programme be of a larger amount in order to generate a larger income and so increase the economic scale. Even if the participant only gives a small amount to the cash waqf, the total amount will be large and able to generate greater income if it is collected effectively. Therefore, the problem with the cash waqf is not the amount of money or its capability, but the willingness to pay. This stems from the belief or Iman of Muslims, as it is not an obligation, not many Muslims participate in the waqf (see Figure 4).

People are not familiar with the cash waqf as there are other types of charity, such as infaq (donation), shadaqah (charity) or zakat (tithe). In fact, the cash waqf is like waqf in general, but it focuses on productive assets. The cash waqf has lifetime benefits for wakif (people who pay the cash waqf). It does not require large amounts to be paid, but the collective total will be large.

4.4 *Media culture system*

Media has quite a significant role in the sustainability of waqf development. Currently, media is not only based on common media, such as television, radio and newspapers, but on social media, based on the Internet. Some cash waqf institutions have utilised the media, such as radio and television, to promote their programmes. In addition, they have also used social media, such as Facebook, Twitter and YouTube. If the institution promotes the cash waqf aggressively, this can be beneficial to the marketing of the cash waqf. Media promotion also needs a good example of cash waqf practices; therefore, a professional cash waqf institution is needed (see Figure 5).

Developing supporting systems to ease the process of waqf payments can create a culture that supports the sustainable development of the cash waqf institutions. This can be in the form of automatic payment schedule of waqf. The Islamic banks must also support the cash waqf programme by promoting it to their customers. Culture can be shaped if it becomes a habit in society. The government should also give incentives and support to make it easier for people to give cash waqf systematically. The cash waqf can be in the form of productive fixed assets that generate income (productive assets).

4.5 *Environment system*

This aspect is suitable in an Indonesian context. Feasible programmes, such as plantation waqf or agriculture waqf, should support it. Agriculture waqf can be in the form of profit sharing with the farmer who cultivates the land. This would benefit to farmers (see Figure 6).

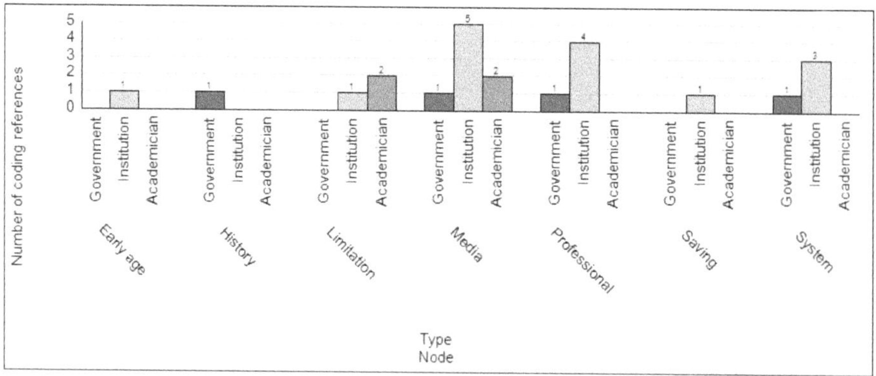

Figure 5. Media & culture aspect of the cash waqf institution.
Source: Data processed.

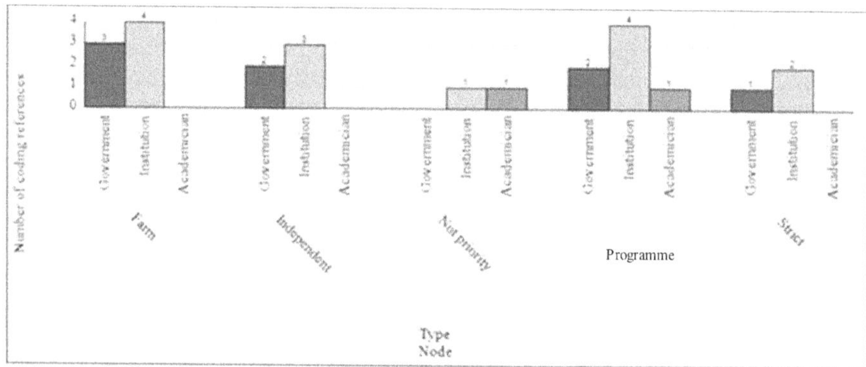

Figure 6. Environment aspect of the cash waqf institutions.
Source: Data processed.

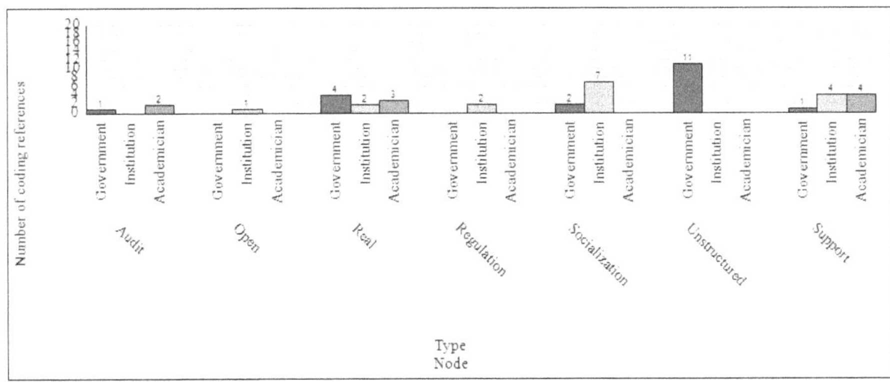

Figure 7. Political aspect of the cash waqf institution.
Source: Data processed.

Waqf in this area can help to preserve the environment and protect the earth from further damage. The cash waqf can also be allocated to a natural electricity power supply. This can benefit institutions by saving energy, which means that the cash waqf supports the environment. The impact may be indirect, but the benefits are long term. Nevertheless, this programme is still not likely to be seen as a top priority by society, as the effect is not direct and it takes a long time to see the results. However, this may affect the life and sustainability of human beings.

4.6 *Political system*

The political system in Indonesia may also be able to support the sustainable development of the cash waqf institutions. Indonesia is a developing country that still requires alignment from the government. Issues that occur in this aspect are unstructured waqf mechanisms, lack of government socialisation, good practice and support from others. BWI and the Ministry of Religious Affairs should have clear job descriptions for reporting, including the monitoring and control of the cash waqf institutions. The report must be disclosed to the public by an online system so that the performance of the cash waqf institution can be easily accessed (see Figure 7).

The government should have promotion programmes for cash waqf, similar to those for taxation. There should be targets for each city, district and province. The amount of the cash waqf would be increased by region and time period. Currently, the government has prepared cash waqf information by publishing an e-book that can be downloaded on the website, but this has not been promoted to the public.

The problem with the socialisation of cash waqf is that there is no interesting cash waqf programme that the government can promote. A good pilot project is needed from a productive cash waqf programme that would encourage the public to participate in it. At this time, this only occurs in the private sector. However, corporations can ask to be involved in the cash waqf programme by offering programmes from Corporate Social Responsibility (CSR). Companies would benefit from this, as they will gain a positive image in society.

5 CONCLUSION

The sustainable development of the cash waqf institutions can be seen from the N-Helix perspective, based on the quintuple helix. From this, we can see which aspects are important and what institutions need to be concerned about. This covers education, economics, media culture, the environment and political systems. For additional consideration, a sixth helix is added, which is the Iman system or belief of the benefits of cash waqf. The cash waqf can be paid in small amounts so that any Muslim can join, as long as they have these beliefs.

This research may add to the study of Siswantoro and Rosdiana (2016), which is still at the preliminary stage at this time.

REFERENCES

Badan Wakaf Indonesia. (2009). Regulation of Badan Wakaf Indonesia No. 1/2009 on Pedoman Pengelolaan dan Pengembangan Harta Benda Wakaf Bergerak Berupa Uang (Guide to management and development of moveable cash waqf).

Carayannis, E.G. & Campbell, D.F.J. (2011). Open innovation diplomacy and the 21st century fractal research, education and innovation (FREIE) ecosystem: Building on the quadruple and quintuple helix innovation concepts and the "Mode 3" knowledge production system. *Journal of Knowledge Economics, 2*, 327–372.

Carayannis, E.G. & Campbell, D.F.J. (2012). *Mode 3 knowledge production in quadruple helix innovation systems.* New York: Springer.

Carayannis, E.G., Barth, T.D. & Campbell, D.F.J. (2012). The quintuple helix innovation model: Global warming as a challenge and driver for innovation. *Journal of Innovation and Entrepreneurship, 1*(1), 2. doi:10.1186/2192-5372-1-2.

Carayannis, E.G. & Rakhmatullin, R. (2014). The quadruple/quintuple innovation helixes and smart specialisation strategies for sustainable and inclusive growth in Europe and beyond. *Journal of the Knowledge Economy, 5*(2), 212–39. doi:10.1007/s13132-014-0185-8.

Leydesdorff, L. (2012). The triple helix, quadruple helix, ..., and an N-tuple of helices: Explanatory models for analyzing the knowledge-based economy? *Journal of the Knowledge Economy, 3*(1), 25–35. doi:10.1007/s13132-011-0049-4.

Lincoln, Y.S. & Guba, E. (1985). *Naturalistic inquiry.* California: Sage Publications.

Nasution, M.E. (2005). Wakaf tunai dan sektor volunteer. In M.E. Nasution & U. Hasanah (Eds.), *Wakaf tunai inovasi financial Islam peluang dan tantangan dalam mewujudkan kesejahteraan umat* (pp. 28–50). Jakarta: Pusat Kajian Timur Tengah dan Islam, Universitas Indonesia.

Park, H.W. (2014). Transition from the triple helix to N-tuple helices? An interview with Elias G. Carayannis and David F.J. Campbell. *Scientometrics*. doi:10.1007/s11192-013-1124-3.

Republika. (2014). *Dana Wakaf Capai Rp 200 Miliar.* http://www.republika.co.id/berita/koran/khazanah-koran/14/09/18/nc2z881-dana-wakaf-capai-rp-200-miliar.

Siswantoro, D. & Rosdiana, H. (2016). Sustainability of cash waqf development in Indonesia: A quintuple helix perspective. *Sains Humanika, 8*(1–2), 111–116.

Competition and Cooperation in Economics and Business – Gani et al. (Eds)
© 2018 Taylor & Francis Group, London, ISBN 978-1-138-62666-9

Do accrual accounting practice and the supreme audit institution role improve government financial disclosure? Cross-country empirical studies

S.W. Kartiko & S. Utama
Department of Accounting, Faculty of Economics and Business, Universitas Indonesia, Depok, Indonesia

ABSTRACT: The study aims to examine accrual choice practices and the role of the supreme audit institution at the level of government financial statements disclosure in different countries. By using the data from a range of 65 observations in 2012, this study finds that the accrual level and independence of the supreme audit institution affects the level of government disclosure. The disclosure level increased sequentially based on the choice of accounting practices, as follows: cash basis, modified cash, modified accrual and full accrual. Nevertheless, the role of the supreme audit institution is only visible in strengthening disclosure levels in countries that implement full accrual, as found in previous studies.

1 INTRODUCTION

The improvement of transparency and accountability in government accounting practice is an important agenda of public sector management reform. It has encouraged the management and accounting practices that are used in the business sector to be adopted in the government sector as a New Public Management (NPM) concept (Hood, 1995; Osborne, 1993). Efficiency, effectiveness and economical principles in the NPM require the disclosure of accounting information, such as government financial performance indicators, recognition of rights and obligations, cost of public services and the availability of resources in the future (IFAC, 2014). Financial statements based on the accrual accounting system have provided one full set of managerial actions (Kwon, 1989) to be adopted by public sector entities (Pallot, 1992; Pallot, 1999).

There is a wide spectrum of accrual practices reflecting such meeting points as political compromise, cultural background (Hyndman & Connolly, 2011; Lapsley et al., 2009), structure of the economy and characteristics of business infrastructure (Pina et al., 2009). The latest survey conducted by PwC (Sturesson et al., 2013) shows the adoption of various accrual levels in each country. Government accounting practices are still dominated by cash-based accounting methods (Table 1). However, it is predicted that five years from now, over 50% of the countries in the world, especially the developing countries, will have converted their accounting system to a full accrual basis (Sturesson et al., 2013).

Table 1. Survey of accounting basis choice and commitment towards full accrual in 140 countries (Sturesson et al., 2013).

Basis of accounting	March 2013 (%)	2018 (%) (Estimated)
Cash	31	14
Modified Cash	23	9
Modified Accruals	20	14
Full Accruals	26	63

It is believed that the application of the accrual basis of accounting will improve the quality of government financial reporting. The benefits obtained from using the accrual basis are to improve the cost of services measurement efficiently and effectively and to complete service performance (value for money) information that can be compared between agencies and business partners. In addition, accrual practice would increase government functional and administrative accountability (IFAC, 2014), as well as improving the transparency of fiscal sustainability (Evans III & Patton, 1987; Mellor, 1996; Robinson, 1998).

Many studies advocate various benefits and advantages of adopting the accrual basis rather than a traditional or cash basis, such as increased transparency (Lapsley & Oldfield, 2001; Lye et al., 2005), enhanced performance information (Sturesson et al., 2013) and improved accountability (IFAC, 2014). However, there are a limited number of empirical studies that explain that the accrual level encourages transparency, especially in central government.

The growing adoption of international accounting standards, such as the International Public Sector Accounting Standards (IPSAS), in a number of central governments creates an opportunity to explore accounting disclosure studies (Pina & Torres, 2003). A previous study explains the spectrum and variability of accounting information items that are publicly disclosed when trying to implement IPSAS (Pina et al. 2009; Torres, 2004). However, the depth of the disclosure lists in a financial statement, such as cash flow and changes in the equity report, was not compared thoroughly. Moreover, a comparative accounting disclosure list used in these studies has not accommodated a cash basis practice.

IPSAS adoption studies at central government level have been surveyed for developed countries (Christiaens et al., 2010) and also extended to developing countries (Christiaens et al., 2014). These studies grouped accounting practices such as cash basis and accrual adoption (IPSAS, IPSAS accrual GAAP-like and business-like) descriptively in each country, including the reasons and commitment in the next IPSAS adoption. However, the discussion that links the adoption of accrual choice and the disclosure level has not been tested explicitly and empirically in these studies.

The determinant factor of the disclosure level has been discovered by various researchers, but is limited to the local government level, such as the United States context of adopting US-GAAP (i.e. Copley, 1991; Gore, 2004; Malone, 2006). A previous study found that external auditors have an important role in enhancing the disclosure of accounting information in local government (Copley, 1991). Studies at the central government level generally indicate a new audit role as part of the element of governance that contributes to improving the level of transparency (Benito & Bastida, 2009; De Renzio & Masud, 2011). The role of the Supreme Audit Institution (SAI) was assessed based on the perception of auditor independency (Seifert et al., 2013). Therefore, the SAI role in improving the disclosure levels is a gap that will be filled in this study.

This study aimed to test the viability of accrual on accounting information disclosure in central government. This test requires a disclosure index that gives an opportunity at various accrual levels, including cash basis reporting, such as cash receipts and disbursement statements. Lastly, the role of SAIs in strengthening transparency and accountability is also tested. Hence, the research question posed is "Do accrual accounting practice and the SAI role improve government financial disclosure levels"? The general hypothesis is "There is a positive association of accrual accounting choice and strength of the SAI role on disclosure levels".

There are at least three things that makes this study unique: first, it empirically models the effects of the various accrual level options on the level of disclosure of accounting information. Second, it creates a list of disclosures based on the depth level of the accounting information items and accommodates various options, including cash basis disclosure. The disclosure index is rebuilt based on the accrual basis of disclosure checklist (Pina et al., 2009; Sturesson et al., 2013) and according to the statement of cash IPSAS standards (IFAC, 2004). The disclosure index also accommodates central governments that do not provide a complete financial information report (i.e. budget realisation reports and annual reports). Third, it empirically tests the role of the SAI, which is associated directly and moderately (strengthened) with the disclosure level.

2 THEORY AND HYPOTHESIS DEVELOPMENT

2.1 *Level of accruals adoption and financial disclosure*

The accrual accounting basis has many advantages over the cash basis. Information related to government rights and obligations in the statement of financial position reflects the fiscal conditions that will be encountered in the future. The government's fiscal strategy is more precise when considering accrual information (such as taxes receivable, long-term assets and long-term debt). From the theory of stewardship standpoint, the recording of accruals avoids misappropriate disclosure of assets (Copley, 1991; Monsen & Näsi, 2001) and takes into account provision and contingent liability.

INTOSAI's (1995) Accounting Standards Framework mentions four characteristics of the accounting system: cash basis, modified cash, modified accrual and full accrual. Cash basis recognises transactions when cash is received or paid. Financial statements prepared on this basis would show cash received and paid during the reporting period (INTOSAI, 1995). Disclosure by this method produces a balance sheet, but it is limited to current assets and current liabilities (Gross et al., 2005).

Modified cash basis recognises what is beyond the cash receipt and disbursement in the reporting period, as well as cash received and paid during the short period after the end of the reporting period (INTOSAI, 1995). Although there are many variations of this method, adjustments at the end of the year resulted in some of the information being disclosed (i.e. receivables and short-term debt) (Van der Hoek, 2005). In other cases, it acknowledges the receipt of cash from liabilities, such as long-term debt, and its reclassification into short-term debt as a current portion of long-term debt (Bremser & Hiltebeitel, 1992).

Modified accruals basis presents economic events when a transaction occurs regardless cash received/paid with several conditions. For example, in the US, tax revenue are recognised when the cash flow are both: 1) reasonably estimated, and 2) available for current expenditure payment no later than 60 days. However, disclosures on the balance sheet still include the accounts that are accrued as both assets and liabilities, just like full accrual. In addition, amortisation/depreciation is commonly ignored (Voorhees & Kravchuk, 2001).

Full accrual basis uses an extensive practice of accounting in the business sector that presents complete government performance and financial position statements. The recognition of revenues and expenses has been done during an economic event (INTOSAI, 1995). Presentation of assets shows the various categories of fixed assets and also intangible assets (Chan, 2003).

Implementation of accrual encourages the completeness of future rights and obligations, sustainability of resources (i.e. account receivables, account payables and contingent liabilities), revenues, expenses (Chan, 2008) and the performance of public entities (Pallot, 1992; Pallot, 1999). Thus, the choice of the accrual accounting level determines the managerial accountability, transparency and completeness level of government financial information.

H1: The accruals level is associated positively with the disclosure level of government financial statements.

2.2 *Role of supreme audit institution on financial disclosure*

The purpose of an audit is to minimise the agency problems that arise in the form of asymmetric information between principal and agent (Watts & Zimmerman, 1990). Multi-agency problems that occurred in the government led to the crucial role of the Supreme Audit Institution (SAI). SAI establishes pillars of good governance to promote the transparency of public sector administration. SAI has committed to attest the information of government policy independently and profesionally for stakeholders (INTOSAI, 2013).

However, the success in achieving the objectives of the audit depends on the level of audit quality. According to DeAngelo (1981) the auditor's competency is related to their ability to find material errors in the client's accounting system. In-depth understanding and application of audit skills on accounting and audit standards in the process of the public entity examination reflects auditor professionalism and credibility.

The auditor's independence relies on a mental attitude to disclose material errors freely from any interventions (DeAngelo, 1981). Independent government auditors try to maintain their reputation at the highest level in order to preserve an adequate level of government financial disclosure (Copley, 1991)

In country level, the audit quality measurement in Europe has been constructed by English & Guthrie (2000), which consists of two dimensions: the accountability of the SAI and the strength of the SAI. Accountability criteria can be undertaken by parliament in order to examine the professionalism of the SAI in performing their duties and oversight functions. SAI strength dimensions are the level of mandate and the independence of the SAI in exercising authority (Clark et al., 2007).

The survey assesses the strength of the SAI in exercising the supervision and inspection at state level that has been conducted by the International Budget Partnership (IBP). IBP publishes a report on the degree of openness of the budgeting process of a country, the so-called Open Budget Index (OBI). The components that determine the effectiveness of budget transparency include the role of the SAI in a country (Seifert et al., 2013).

In summary, the SAI has the role of testing the suitability of the use of public funds. To meet the audit objectives, the SAI has committed to ensure transparency in public financial management on behalf of the community.

H2: The role of the SAI is associated positively with the disclosure level of government financial statements.

The IBP survey shows that institutional strengthening of the SAI is also determined by the level of transparency of the budget (Seifert et al., 2013). Previous studies also confirmed that strengthening the role of the audit as part of the element of governance can improve a country's transparency (Benito & Bastida, 2009; De Renzio & Masud, 2011). It supports the argument that enforcing the level of accountability and transparency by strengthening the SAI can improve the accounting disclosure in a country.

However, studies of the accruals practice in developed countries are caused by the institutional pressure of the SAI. Studies conducted by Baker and Rennie (2006) in Canada and Lye et al. (2005) in New Zealand have demonstrated that the motivation of accrual practice is to improve effectiveness and efficiency. For developing countries, the institutional pressure came from international donors, such as the World Bank, ADB and IMF, to support transparency assistance programmes (Oulasvirta, 2014).

H3: The role of the SAI has a moderating effect to strengthen the relationship.

3 RESEARCH METHOD

3.1 *Sample selection*

This study used the 2012 or 2012/2013 fiscal year as observations related to the survey conducted by PwC. The total samples obtained are from 82 countries, with 56 financial reports obtained, and the rest are only in the form of financial information, such as budget realisation reports and financial and statistical reports (see Table 2). Data were collected from each country's website that presents financial and statistical reports.

3.2 *Research model and measurement*

The hypothesis test of the equation is estimated using Ordinary Least Squares (OLS). Equation 1 is a model of accrual accounting choice and the effect of SAI roles on the level of disclosure of financial statements.

$$\begin{aligned}\text{DISCLOSURE}_i = \beta_0 &+ \beta_1\text{D\_MODIF\_CASH}_i + \beta_2\text{D\_MODIF\_ACCRUAL}_i \\ &+ \beta_3\text{D\_FULL\_ACCRUAL}_i + \beta_4\text{SAI}_i + \beta_5\text{GOV\_EFFECTIVE}_i \\ &+ \beta_6\text{POLITICAL\_RIGHTS}_i + \beta_7\text{LN(GDP)}_i + \beta_8\text{GOV\_DEBT}_i \\ &+ \beta_9\text{D\_FIN\_STATEMENTS}_i + \varepsilon_i\end{aligned} \tag{1}$$

This model aims to test H1 and H2 hypothesis. Accrual choices variables (D_MODIF_CASH, D_MODIF_ACCRUAL, and D_FULL_ACCRUAL) are expected to have positive associations with the disclosure level. The increase of the role of the SAI is also expected to have a positive effect on higher disclosure levels.

Previous studies found prominent determinant factors of the disclosure level on local government, such as governance level (García & García-García, 2010; Yilmaz et al., 2008), size of government (Copley, 1991; Gore, 2004; Malone, 2006) and fiscal conditions (Copley, 1991; Gore, 2004; Malone, 2006). Therefore, GOV_EFFECTIVE, POLITICAL_RIGHTS, LN(GDP), GOV_DEBT, and D_FIN_STATEMTS are control variables that supposedly have a positive effect on the disclosure level.

Table 2. Sample selection.

All countries with governance indicator data from the World Bank	201
Countries with insufficient accounting basis from Sturesson et al. (2013) and Christiaens et al. (2014)	(61)
	140
Countries with insufficient strength of SAI data from IBP (2012)	(40)
	100
Unpublished annual report on the website of each country	(18)
Final sample	82
Annual report consists of:	
Financial statements	56
Budget realisation report and/or the financial statistics	26

Table 3. Operational variables.

Variables	Definitions
$DISCLOSURE_i$	The financial disclosure index in this study is rebuilt from Pina et al. (2009) and Torres (2004) checklists, Ernst and Young (EY, 2012) disclosure checklist and cash-based IPSAS (IFAC, 2004). This disclosure index includes fair value and hedge accounting practices that have not been considered in previous studies.
The category accounting system is based on a study conducted by Sturesson et al. (2013) and Christiaens et al. (2014):	
$D\_MODIF\_CASH_i$	dummy variable, 1 = modified cash basis of accounting and 0 = other.
$D\_MODIF\_ACCRUAL_i$	dummy variable, 1 = modified accrual basis of accounting and 0 = other.
$D\_FULL\_ACCRUAL_i$	dummy variable, 1 = full accrual accounting basis and 0 = other.
SAI_i	score of strength of Supreme Audit Institutions (SAI) role surveyed by The International Budget Partnership (Seifert et al., 2013).
$GOV\_EFFECTIVE_i$	index of government effectiveness obtained from the World Bank.
$POLITICAL\_RIGHTS_i$	proxy of the economic size of the country that is the natural logarithm of the constant dollar GDP obtained from the World Bank.
GOV_DEBT_i	ratio of government debt to GDP.
$D\_FIN\_STATEMTS_i$	dummy variable, 1 = countries that publish financial statements and 0 = no information available yet to publish the financial statements. and 0 = no information available yet to publish the financial statements of the budget realisation reports and/or economic statistics.
SAI_i x $D\_MODIF\_CASH_i$	interaction variable SAI's role in the country with the cash accounting system modifications.
SAI_i x $D\_MODIF\_ACCRUAL_i$	interaction variable role of SAI in the country with a modified accrual accounting system.
SAI_i x $D\_FULL\_ACCRUAL_i$	interaction variable role of SAI in the country with full accrual accounting system.

$$
\begin{aligned}
\text{DISCLOSURE}_i = {} & \beta_0 + \beta_1\text{D\_MODIF\_CASH}_i + \beta_2\text{D\_MODIF\_ACCRUAL}_i \\
& + \beta_3\text{D\_FULL\_ACCRUAL}_i + \beta_4\text{SAI}_i + \beta_5\text{SAI}_i\text{xD\_MODIF\_CASH}_i \\
& + \beta_6\text{SAI}_i\text{xD\_MODIF\_ACCRUAL}_i + \beta_7\text{SAI}_i\text{xD\_FULL\_ACCRUAL}_i \\
& + \beta_8\text{GOV\_EFFECTIVE}_i + \beta_9\text{POLITICAL\_RIGHTS}_i + \beta_{10}\text{LN}(\text{GDP})_i \\
& + \beta_{11}\text{GOV\_DEBT}_i + \beta_{12}\text{D\_FIN\_STATEMENTS}_i + \varepsilon_i
\end{aligned} \tag{2}
$$

Equation 2 is a moderating model that represents the enforcement effect of the SAI role on the positive relationship between accounting accrual choice and the level of disclosure. To test the H3 hypothesis, a strong SAI role is expected to impose the accrual choice level on enhancing the disclosure level. Table 3 indicates the operational variables of models 1 and 2.

4 RESULTS AND DISCUSSION

Table 4 presents the descriptive statistics of the variables to be tested by Equation 1. The level of disclosure of government financial reports has an average of one third of the maximum score and a distribution of value width that is a quarter of the maximum score. Most of the samples have a level of disclosure score that exceeds the average. The categorical basis of accounting in Table 4 has a similar percentage to the PwC survey results (Sturesson et al., 2013). In general, the accounting practice samples in this study are still dominated by the cash basis. Based on a survey, the SAI role index on the average country has, in general, strengthened the independence of the SAI's role, as shown by the results of the study conducted by the IBP (Seifert et al., 2013). The number of countries that have published financial reports consists of over 60 per cent of the samples.

Table 4. Descriptive statistics.

	Mean	Dev.	Min	Median	Max
DISCLOSURE_i	34.678	24.477	3.125	29.688	100
D_CASH_i	0.388	0.491	0	0	1
D_MODIF_CASH_i	0.224	0.420	0	0	1
D_MODIF_ACCRUAL_i	0.239	0.430	0	0	1
D_FULL_ACCRUAL_i	0.149	0.359	0	0	1
SAI_i	76.358	25.426	0	83	100
GOV_EFFECTIVE_i	0.21	0.845	−1.4	0.01	1.94
$\text{POLITICAL\_RIGHTS}_i$	2.955	1.894	1	3	7
LN(GDP)_i	19.345	1.861	15.703	19.425	23.488
GOV_DEBT_i	23.434	35.559	0	0	127.206
D_FIN_STATMTS_i	0.612	0.491	0	1	1
N	67				

DISCLOSURE_i: score of the disclosure level of financial statements based on the studies of Pina et al. (2009) and Torres (2004), checklists by Ernst and Young (EY, 2012) and statement of cash-based IPSAS (IFAC, 2004). **D_CASH_i**: dummy variable, 1 = cash accounting basis and 0 = other. **D_MODIF_CASH_i**: dummy variable, 1 = modified cash basis of accounting and 0 = other. **D_MODIF_ACCRUAL_i**: dummy variable, 1 = modified accrual basis of accounting and 0 = other. **D_FULL_ACCRUAL_i**: dummy variable, 1 = full accrual accounting basis and 0 = other. **SAI_i**: score of the role of Supreme Audit Institutions (SAI) surveyed by the International Budget Partnership (IBP, 2012). **GOV_EFFECTIVE_i**: government effectiveness index obtained from the World Bank. **$\text{POLITICAL\_RIGHTS}_i$**: index of political rights were obtained from the World Bank. **LN (GDP)_i**: a proxy of the economic size of the country that is the natural logarithm of the constant dollar GDP obtained from the World Bank. **GOV_DEBT_i**: the ratio of government debt to GDP. **D_FIN_STATEMTS_i**: dummy variable, 1 = countries that publish financial statements and 0 = no information available yet to publish the financial statements of the budget realisation reports and/or economic statistics.

Table 5 represents the correlation results of the variables to be estimated in the regression model. Generally, the univariate test shows that all the explanatory variables have positive significant associations on the disclosure level. Cash and modified cash basis are significantly associated with low accounting disclosure. On the other hand, modified and full accrual basis are significantly associated with high accounting disclosure.

The regression results from 65 countries are shown in Table 6, both before and after the moderated effect caused by strengthening the level of the SAI role. Under the moderating effect model, generally all independent variables affect the level of disclosure. Based on the coefficient test (Table 7), this shows that the disclosure level is increasing on accrual choices sequentially as follows: cash, modified cash, modified accrual and full accrual. Cash basis financial statement disclose less information according to limited demonstrate items (accept H1).

The level of independence of the SAI, based on the regression results, has prompted the government to disclose more financial information (accept H2). External oversight function with the attributes of professionalism and the independence of auditors supports the government in improving the transparency of public finances. This is in line with the study by Copley (1991), which states that audit quality also determines the level of public sector financial statement disclosures.

Under the after moderating effect model, enforcement on the disclosure level is only demonstrated by the countries applying full accrual (accept H3 partially). These findings support previous studies, which show that accrual adoption, which is mostly done in developed countries, has been done due to SAI pressure, as in Canada (Baker & Rennie, 2006) and New Zealand (Lye et al., 2005). However, in developing countries, disclosing more financial information has been required by donor agencies, such as the World Bank, ADB and IMF. They advocate the importance of the adoption of the accounting standard IPSAS to facilitate accountability and transparency in public programmes (Oulasvirta, 2014).

The effectiveness of the government in improving their administration results in improved financial transparency. Studies conducted by Benito and Bastida (2009) and De Renzio and Masud (2011) show that good governance is associated with the level of transparency in a

Table 5. Correlation analysis.

	1	2	3	4	5	6	7	8	9	10	11
DISCLOSURE	1										
D_ CASH	−0.67***	1									
D_MODIF_CASH	−0.03	−0.43***	1								
D_MODIF_ACCRUAL	0.27**	−0.45***	−0.30**	1							
D_FULL_ACCRUAL	0.63***	−0.33**	−0.22*	−0.23*	1						
SAI	0.39***	−0.29**	−0.06	0.21*	0.21*	1					
GOV_EFFECTIVE	0.58***	−0.45***	−0.08	0.13	0.55***	0.38**	1				
POLITICAL_RIGHTS	−0.47***	0.44***	−0.01	−0.15	−0.41***	−0.51***	−0.65***	1			
LN(GDP)$_i$	0.37**	−0.30**	−0.01	0.11	0.29**	0.21*	0.41***	−0.14	1		
GOV_DEBT$_i$	0.35**	−0.34**	−0.1	0.05	0.53***	0.14	0.48***	−0.34**	0.40***	1	
D_FIN_STATEMTS	0.44***	−0.12	−0.09	−0.06	0.33**	0.13	0.11	−0.2	0.03	0.23*	1

DISCLOSURE$_i$: score of the disclosure level of financial statements based on the studies of Pina et al. (2009) and Torres (2004), checklists by Ernst and Young (EY, 2012), and statement of cash-based IPSAS (IFAC, 2004). **D_CASH$_i$**: dummy variable, 1 = cash accounting basis and 0 = other. **D_MODIF_CASH$_i$**: dummy variable, 1 = modified cash basis of accounting and 0 = other. **D_MODIF_ACCRUAL$_i$**: dummy variable, 1 = modified accrual basis of accounting and 0 = other. **D_FULL_ACCRUAL$_i$**: dummy variable, 1 = full accrual accounting basis and 0 = other. **SAI$_i$**: score of the role of Supreme Audit Institutions (SAI) surveyed by the International Budget Partnership (IBP, 2012). **GOV_EFFECTIVE$_i$**: government effectiveness index obtained from the World Bank. **POLITICAL_RIGHTS$_i$**: index of political rights were obtained from the World Bank. **LN (GDP)$_i$**: a proxy of the economic size of the country that is the natural logarithm of the constant dollar GDP obtained from the World Bank. **GOV_DEBT$_i$**: the ratio of government debt to GDP. **D_FIN_STATEMTS$_i$**: dummy variable, 1 = countries that publish financial statements and 0 = no information available yet to publish the financial statements of the budget realisation reports and/or economic statistics.

Table 6. Regression results.

		Model		Model	
		Before moderating effect		After moderating effect	
		Coefficient	t-stat	Coefficient	t-stat
D_MODIF_CASH	+	17.954	6.153***	25.520	2.999***
D_MODIF_ACCRUAL	+	28.595	9.374***	25.621	4.869***
D_FULL_ACCRUAL	+	48.091	4.986***	−88.311	−1.437*
SAI	+	0.076	1.698**	0.026	0.771
SAI x D_ MODIF_CASH	+			−0.097	−0.751
SAI x D_ MODIF_ACCRUAL	+			0.049	0.554
SAI x D_ FULL_ACCRUAL	+			1.573	2.561***
GOV_EFFECTIVE	+	3.370	1.444*	0.115	0.053
POLITICAL_RIGHTS	+	0.531	0.632	−0.307	−0.453
LN(GDP)	+	1.430	1.791**	1.122	1.469*
GOV_DEBT	+	−0.135	−1.893**	−0.097	−1.228
D_FIN_STATEMTS	+	13.387	5.706***	13.191	5.523***
C		−24.721	−1.677**	−13.342	−0.980
N		65		65	
F		23.08***		35.85***	
R2		0.758			0.841
adj R2		0.718			0.804

DISCLOSURE$_i$: score of the disclosure level of financial statements based on the studies of Pina et al. (2009) and Torres (2004), checklists by Ernst and Young (EY, 2012) and statement of cash-based IPSAS (IFAC, 2004). **D_ CASH**$_i$: dummy variable, 1 = cash accounting basis and 0 = other. **D_MODIF_CASH**$_i$: dummy variable, 1 = modified cash basis of accounting and 0 = other. **D_MODIF_ACCRUAL**$_i$: dummy variable, 1 = modified accrual basis of accounting and 0 = other. **D_FULL_ACCRUAL**$_i$: dummy variable, 1 = full accrual accounting basis and 0 = other. **SAI**$_i$: score of the role of Supreme Audit Institutions (SAI) surveyed by the International Budget Partnership (IBP, 2012). **GOV_EFFECTIVE**$_i$: government effectiveness index obtained from the World Bank. **POLITICAL_RIGHTS**$_i$: index of political rights were obtained from the World Bank. **LN (GDP)**$_i$: a proxy of the economic size of the country that is the natural logarithm of the constant dollar GDP obtained from the World Bank. **GOV_DEBT**$_i$: the ratio of government debt to GDP. **D_FIN_STATEMTS**$_i$: dummy variable, 1 = countries that publish financial statements and 0 = no information available yet to publish the financial statements of the budget realisation reports and/or economic statistics.

Table 7. Coefficient test from Equation 1.

Coefficient test	Prob.
$\beta_1 = \beta_2, \beta_2 = \beta_3, \beta_2 = \beta_4$	0.0016
$\beta_1 = \beta_2$	0.0051
$\beta_2 = \beta_3$	0.0444
$\beta_2 = \beta_4$	0.0030

country. Other control variables, such as the size of the government and the availability of financial reports, bring positive effects on the level of disclosure.

However, the ratio of debt to GDP shows the opposite, which is that it is negative on the level of disclosure. This is in line with previous research between transparency and the level of debt or deficits, which show that negative associations have happened in studies in developed countries (Alt et al., 2006; Hameed, 2005). It is stated that the reduction in debt occurred when fiscal transparency increased.

5 CONCLUSION

This study shows that the choice of accrual accounting in central government determines the level of transparency of public financial information. The degree of accrual is related to the amount of accrual financial information disclosed. The level of financial statement disclosures followed by accrual basis, modified accrual basis, modified cash, and cash, respectively. Report types that are generated from accrual are more detailed than those generated from the cash method, which generally only present a statement of receipt and disbursement.

The role of the SAI is also associated with more disclosure of government transactions. The independence and professional level of the SAI as an external auditor imposes government accountability and transparency. However, strengthening the role of the SAI on the accrual method, which enhances the disclosure of financial statements, is more pronounced in developed countries. Meanwhile, in developing countries, strengthening the level of transparency of public finances is probably more influenced by outside factors, such as international donors (Oulasvirta, 2014).

This study has its limitations, such as the small sample size. Not many countries publish their full financial report on their website. A survey conducted by the IBP to assess the strength of the role of the SAI does not cover entire countries, thereby reducing the number of observations in this study. The next study is expected to expand the sample by using more than one year for observation.

REFERENCES

Alt, J.E., Lassen, D.D. & Rose, S. (2006). The causes of fiscal transparency: Evidence from the US states. *IMF Staff Papers, 53*(Special Issue), 30–57. doi:0908-7745.

Baker, R. & Rennie, M.D. (2006). Forces leading to the adoption of accrual accounting by the Canadian Federal Government: An institutional perspective. *Canadian Accounting Perspectives, 5*(1), 83–112. doi:10.1506/206K-RV7L-2JMN-W3D3.

Benito, B. & Bastida, F. (2009). Budget transparency, fiscal performance, and political turnout: An international approach. *Public Administration Review, 69*(3), 403–17. doi:10.1111/j.1540-6210.2009.01988.x.

Bremser, W.G. & Hiltebeitel, K.M. (1992). A look at the modified cash basis. *The CPA Journal, 62*(2), 49.

Chan, J.L. (2003). Government accounting: An assessment of theory, purposes and standards. *Public Money & Management, 23*(1), 13–20. doi:10.1111/1467-9302.00336.

Chan, J.L. (2008). International public sector accounting standards: Conceptual and institutional issues. Working Paper. http://www.jameslchan.com/papers/Chan2008IPSAS3.pdf.

Christiaens, J., Reyniers, B. & Rolle, C. (2010). Impact of IPSAS on reforming governmental financial information systems: A comparative study. *International Review of Administrative Sciences, 76*(3), 537–54. doi:10.1177/0020852310372449.

Christiaens, J., Vanhee, C., Manes-Rossi, F., Aversano, N. & van Cauwenberge, P. (2014). The effect of IPSAS on reforming governmental financial reporting: An international comparison. *International Review of Administrative Sciences, 81*(1), 158–177.

Clark, C., De Martinis, M. & Krambia-Kapardis, M. (2007). Audit quality attributes of European Union supreme audit institutions. *European Business Review, 19*, 40–71. doi:10.1108/09555340710714144.

Copley, P.A. (1991). The association between municipal disclosure practices and audit quality. *Journal of Accounting and Public Policy, 10*(4), 245–66. doi:10.1016/0278-4254(91)90001-Z.

DeAngelo, L.E. (1981). Auditor size and audit quality. *Journal of Accounting and Economics, 3*(3), 183–99. doi:10.1016/0165-4101(81)90002-1.

De Renzio, P. & Masud, H. (2011). Measuring and promoting budget transparency: The open budget index as a research and advocacy tool. *Governance, 24*(3), 607–16. doi:10.1111/j.1468-0491.2011.01539.x.

English, L., & Guthrie, J. (2000). Mandate, Independence and Funding: Resolution of a Protracted Struggle Between Parliament and the Executive Over the Powers of the Australian Auditor-General. *Australian Journal of Public Administration*, 59 (1): 98–114.

Ernst and Young (EY). (2012). *International public sector accounting standard's (IPSAS's) disclosure checklist.* EYGM Limited. http://www.ey.com/Publication/vwLUAssets/EY_-_International_Public_Sector_Accounting_Standards_Disclosure_Checklist_-_September_2012_Edition/$FILE/EY-IPSAS-Ctools-%20Disclosurechecklist-Sept2012.pdf.

Evans III, J.H. & Patton, J.M. (1987). Signaling and monitoring in public-sector accounting. *Journal of Accounting Research, 25*(1987), 130–58. doi:10.2307/2491084.

García, A.C. & García-García, J. (2010). Determinants of online reporting of accounting information by Spanish local government authorities. *Local Government Studies, 36*(5), 679–95. doi:10.1080/03003930.2010.506980.

Gore, A.K. (2004). The effects of GAAP regulation and bond market interaction on local government disclosure. *Journal of Accounting and Public Policy, 23*(1), 23–52. doi: 10.1016/j.jaccpubpol.2003.11.002.

Gross, M.J., McCarthy, J.H. & Shelmon, N.E. (2005). *Financial and accounting guide for not-for-profit organizations.* New York: Wiley.

Hameed, F. (2005). *Fiscal transparency and economic outcomes*. IMF Working Paper.

Hood, C. (1995). The 'new public management' in the 1980s: Variations on a theme. *Accounting, Organizations and Society, 20*(2–3), 93–109. doi:10.1016/0361-3682(93)E0001-W.

Hyndman, N. & Connolly, C. (2011). Accruals accounting in the public sector: A road not always taken. *Management Accounting Research, 22*(1), 36–45. doi:10.1016/j.mar.2010.10.008.

International Budget Partnership (IBP). (2012). Open Budget Survey – 2012. Washington, D.C. www.internationalbudget.org/wp-content/uploads/OBI2012-Report-English.pdf.

International Federation of Accountants (IFAC). (2004). *Cash basis IPSAS: Financial reporting under the cash basis of accounting*. New York, NY: International Federation of Accountants. .//www.ifac.org/system/files/publications/files/financial-reporting-under-t.pdf.

International Federation of Accountants (IFAC). (2014). *The importance of accrual-based financial reporting in the public sector*. New York, NY: International Federation of Accountants. https://www.ifac.org/system/files/uploads/IPSASB/IPSASB-The-Importance-of-Accrual-based-Financial-Reporting.pdf.

International Organization of Supreme Audit Institutions (INTOSAI). (1995). *Accounting standards framework implementation guide for SAIs: Departmental and government-wide financial reporting*. Vienna.

International Organization of Supreme Audit Institutions (INTOSAI). (2013). *The International Standards of Supreme Audit Institutions (ISSAIs) 100: Fundamental principles of public-sector auditing*. Vienna.

Kwon, Y.K. (1989). Accrual versus cash-basis accounting methods: An agency-theoretic comparison. *Journal of Accounting and Public Policy, 8*(4), 267–81. doi:10.1016/0278-4254(89)90015-X.

Lapsley, I., Mussari, R. & Paulsson, G. (2009). On the adoption of accrual accounting in the public sector: A self-evident and problematic reform. *European Accounting Review, 18*(4), 719–23. doi:10.1080/09638180903334960.

Lapsley, I. & Oldfield, R. (2001). Transforming the public sector: Management consultants as agents of change. *European Accounting Review, 10*(3), 523–43. doi:10.1080/713764628.

Lye, J., Perera, H. & Rahman, A. (2005). The evolution of accruals-based crown (government) financial statements in New Zealand. *Accounting, Auditing and Accountability Journal, 18*(6), 784–815. doi:10.1108/09513570510627711.

Malone, D. (2006). An exploration of municipal financial disclosure and certain dimensions of political culture. *Academy of Accounting and Financial Studies Journal, 10*(1), 7–23.

Mellor, T. (1996). Why governments should produce balance sheets. *Australian Journal of Public Administration, 55*(1), 78–81.

Monsen, N. & Näsi, S. (2001). Comparing cameral and accrual accounting in local governments. In A.D. Bac (ed.) *International Comparative Issues in Government Accounting* (pp.135–157). Boston, MA: Springer US.

Osborne, D.E. (1993). Reinventing government. *Public Productivity & Management Review, 16*(4), 349.

Oulasvirta, L. (2014). The reluctance of a developed country to choose international public sector accounting standards of the IFAC. A critical case study. *Critical Perspectives on Accounting, 25*(3), 272–85. doi:10.1016/j.cpa.2012.12.001.

Pallot, J. (1992). Elements of a theoretical framework for public sector accounting. *Accounting, Auditing & Accountability Journal, 5*(1), 38–59.

Pallot, J. (1999). Beyond NPM: Developing strategic capacity. *Financial Accountability and Management, 15*(3&4), 419–426.

Pina, V. & Torres, L. (2003). Reshaping public sector accounting: An international comparative view. *Canadian Journal of Administrative Sciences, 20*(4), 334–50.

Pina, V., Torres, L. & Yetano, A. (2009). Accrual accounting in EU local governments: One method, several approaches. *European Accounting Review, 18*(4), 765–807.

Robinson, M. (1998). Accrual accounting and the efficiency of the core public sector. *Financial Accountability and Management, 14*(1), 21–37. doi:10.1111/1468-0408.00048.

Seifert, J., Carlitz, R. & Mondo, E. (2013). The Open Budget Index (OBI) as a comparative statistical tool. *Journal of Comparative Policy Analysis: Research and Practice, 15*(1), 87–101.
Sturesson, J., Rouvet, J.L. Schumesch, P. & Duval, J.P. (2013). *Towards a new era in government accounting and reporting*. PwC Global survey on accounting and reporting by central governments. Brussels.
Torres, L. (2004). Accounting and accountability: Recent developments in government financial information systems. *Public Administration and Development, 24*(5), 447–56. doi:10.1002/pad.332.
Van der Hoek, P.M. (2005). From cash to accrual budgeting and accounting in the public sector: The Dutch experience. *Public Budgeting & Finance, 25*(1), 32–45. doi:10.1111/j.0275-1100.2005.00353.x.
Voorhees, W.R. & Kravchuk, R.S. (2001). The new governmental financial reporting model under GASB Statement No. 34: An emphasis on accountability. *Public Budgeting & Finance, 21*(3), 1–30. doi:10.1111/0275-1100.00040.
Watts, R.L. & Zimmerman, J.L. (1990). Positive accounting theory: A ten year perspective. *The Accounting Review, 65*(1), 131–56. doi:10.2307/247880.
Yilmaz, S., Beris, Y. & Serrano-Berthet, R. (2008). Local government discretion and accountability: A diagnostic framework for local governance. Washington, DC: World Bank.

Competition and Cooperation in Economics and Business – Gani et al. (Eds)
© 2018 Taylor & Francis Group, London, ISBN 978-1-138-62666-9

Player's value, field performance, financial performance and stock performance of European football clubs

M.R.W. Putra & G.H. Wasistha
Department of Accounting, Faculty of Economics and Business, Universitas Indonesia, Depok, Indonesia

ABSTRACT: The purpose of this research is to analyse the relationship between a player's value, field performance, financial performance and stock performance of European football clubs. The samples used in this research consist of 17 European football clubs listed in the stock exchange. The method used to analyse the data is the Generalized Structural Component Analysis (GSCA). The results show that, firstly, a player's value has a positive impact on a club's financial performance, but its impact on field performance is insignificant. Secondly, field performance and financial performance have a significant impact on each other. Finally, both field performance and financial performance do not have a significant impact on a football club's stock performance.

1 INTRODUCTION

In this century, football has entered a new industrial level, especially in Europe. The revenue of the 237 clubs participating in European competition has been rising by 9.9% season by season, standing at EUR 8.1 billion. For the biggest 25 clubs, the growth is even more staggering, showing a 41.9% rise from season 2009/2010 to season 2013/2014 (UEFA, 2015a). The driving force behind this revenue growth is the increase in television rights, sponsorship money and merchandise sales (UEFA, 2015b).

However, along with the increase in revenue, players' transfer prices and wages are also inflated. Tomkins (2015) states that the inflation rate of players' transfer prices exceeds the average inflation level in the United Kingdom. Furthermore, Rikardsson and Rikardsson (2013) state that the wage bill level is now at 60–80% of club revenue. Football clubs are more susceptible to bankruptcy as they react more to negative shocks (Szymanski, 2012).

Aside from being susceptible to bankruptcy, owning a football club is perceived as a losing investment, particularly with regards to its stock. Evidence for this is the continual decrease of the stock value of Italian clubs on the Milan Stock Exchange (Dragoni, 2014).

This research aims to evaluate these phenomena by explaining the impact of a player's value on the financial performance and field performance of the club, as well as the effect of such performances on the club's stock performance.

The results will be useful for club managers and investors. For Indonesia, the results of this paper will be useful to plan the national football landscape. After a series of bad experiences from 2013 to 2016, Indonesian football is still looking for a way to recover. The rising costs, the bankruptcy phenomenon and the losing investment paradigm among European football clubs can be avoided in Indonesia if a proper plan is applied.

2 LITERATURE REVIEW

The indicators for players' values are the wages and players' registration rights. Szymanski and Smith (1997) use players' wages as the indicator of the value of such players. Clubs pay their players according to their ability and, therefore, players' wages reflect their value.

The value of a player's registration rights is derived from the player's transfer value. The latter is widely used in European football since the adoption of the Financial Reporting Standard (FRS) 10 – Goodwill and Intangible asset by the Union of European Football Associations (UEFA). FRS 10 requires that a player's transfer value is capitalised as an intangible asset.

Specifically, football clubs' performance measurement can be seen from two aspects, namely field and financial performances (Samagaio et al., 2009). Field performance can be measured by a simple measure, such as their league position at the end of the season, or by more complex indicators, such as the one developed by Barajas et al. (2005). Barajas et al. (2005) developed a new coefficient formula to incorporate clubs' performances in all competitions. Financial performance can also be measured by a simple indicator, such as revenue, or more complex indicators, such as the financial ratio.

Early research on a player's value in relation to the club's field performance and financial performance was done by Szymanski and Smith (1997). The result shows that a player's value has a positive impact on field performance and, in turn, field performance has a positive impact on financial performance. Kase et al. (2006) concluded that financial performance has a positive impact on long-term field performance. Samagaio et al. (2009) found that a player's value has no impact on field performance, but confirm that there is an impact on financial performance.

It is expected that a player's value will give a boost to both the field and financial performance of the club. Star players with high wages, such as Cristiano Ronaldo and Lionel Messi, can change the course of the game. Clubs can also benefit from an increase in television rights, gate receipts and merchandise sales if they have star players.

Szymanski and Smith (1997) and Samagaio et al. (2009) support the claim that field performance has a positive impact on financial performance. It is expected that clubs with a better field performance will make higher revenue from competition prizes, television rights and gate receipts.

H1. Player's value has a positive impact on field performance.
H2. Player's value has a positive impact on financial performance.

Both Kase et al. (2006) and Samagaio et al. (2009) support the claim that financial performance has a positive impact on field performance. A club needs to have a healthy financial position in order to build a strong squad and pay their expenses. A club with an unhealthy financial position will find it difficult to survive among the competition and ultimately may end up bankrupt if they fail to settle their liabilities.

H3. Financial performance has a positive impact on field performance.
H4. Field performance has a positive impact on financial performance.

Regarding football clubs' stock, Renneboog and Vanbrabant (2000) state that field performance has a positive impact on stock performance. This result is further supported by Floros (2014). Renneboog and Vanbrabant (2000) also add that financial performance has a positive impact on stock performance. Only Samagaio et al. (2009) conclude that both field performance and financial performance have no significant impact on stock performance.

H5. Financial performance has a positive impact on stock performance.
H6. Field performance has a positive impact on stock performance.

3 RESEARCH METHOD

Quant/qual/Mixed-method? Pimary/Secondary Data? Population? Sampling Technique? This research uses a sample of 17 European football clubs that are actively listed on the stock exchange. The data were collected using the Eikon Database for financial report data, Yahoo! Finance for stock related data and www.statto.com for field performance related data. The model used in this research is based on the model proposed by Samagaio et al. (2009) with some modifications. The player's value is added into a separate, observed variable, while field performance (FIELD) indicators are modified into indicators developed by Barajas et al. (2005).

Figure 1 gives an overview of the research model.

Table 1 gives a short explanation of the latent variables used in the model:

Data for variables W/S, INT, TNV, OOC, NPP and NI are taken directly from the clubs' financial reports. The FIELD variable is measured by indicators developed by Barajas et al. (2005). This variable is measured by the total number of points achieved in the domestic league (DOM), domestic cup (CUP) and UEFA Champions League (CL) or EURO Cup/ Europa League (EUR). For the cup format tournament (DOM, CL and EUR), the coefficient is calculated as shown in Table 2:

To combine all the variables into a single field performance indicator, Barajas et al. (2005) assign a weight to each variable as shown in Equation 1:

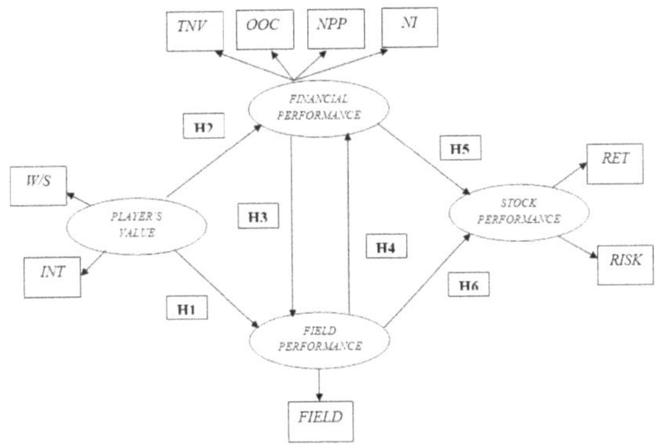

Figure 1. Initial research model.

Table 1. Short explanation of latent variables.

Variable	Short explanation
W/S	Wages and Salaries
INT	Player's Contract Rights Registration
TNV	Turnover
OOC	Other Operating Expenses
NPP	Net Profit from Sales of Player's Rights Registration
NI	Net Income
FIELD	Field Performance Indicator as developed by Barajas et al. (2005)
RET	Stock Return
RISK	Stock Risk

Table 2. Cup format coefficient according to Barajas et al. (2005).

Achievement	Coefficient
Champion	39
Runner-up	33
Semi-final	27
Quarter-final	21
Round 16	15
Round 32	9
Round 64	6
Round 128	3
< Round 128	0

$$FIELD = 3.CL + 2.EUR + 2.DOM + CUP \tag{1}$$

Next, *RISK* is measured by the daily stock return (*RET*) standard deviation, where *RET* is defined as shown in Equation 2:

$$Daily\,Stock\,Return(RET) = \ln\left(\frac{Stock\,Value\,at\,t}{Stock\,Value\,at\,t-1}\right) \times 100 \tag{2}$$

The data will be analysed using the Generalized Structured Component Analysis (GSCA) with the Alternate Least Square (ALS) method. GSCA is used because of its advantage in analysing data with a few samples (Hwang & Takane, 2004).

The structural model used in this research is:

$$\gamma = B'\gamma + \xi$$

The measurement model used in this research is:

$$\begin{aligned} Z &= C'\gamma + \varepsilon \\ Z\xi 1 &= \gamma 1 Ci + i \text{ if } i \leq 2 \\ Z\xi 1 &= \gamma 3 Ci + i \text{ if } i = 3 \\ Z\xi 1 &= \gamma 3 Ci + i \text{ if } i = 4 \\ Z\xi 1 &= \gamma 4 Ci + i \text{ if } i = 5 \end{aligned}$$

Z: Standardised Indicators Variable's Matrix
C: Loading Matrix between Latent Variables and Indicators
B: Path Coefficient Matrix that connects Latent Variables
ε: Residual Vector
γ: Latent Variable Vector
ξ: Residual Vector for γ
i: Amount of Indicators Variable

4 RESULTS

The initial model was modified in order to pass the data reliability test. The modified model was developed by selecting only a single indicator to measure financial performance and stock performance. Based on Szymanski and Smith (1997), financial performance is measured solely by revenue, while according to Renneboog and Vanbrabant (2000), stock performance is measured solely by stock returns.

The results as presented in Table 3 show that 3 out of the 6 hypotheses are accepted. Consistent with the hypotheses, a player's value has a positive impact on the club's financial performance. Financial performance has a positive impact on field performance. Field performance also has a positive impact on financial performance. The analysis will be provided in the next section.

Table 3. Hypothesis testing results.

Hypothesis testing variable	Estimate	SE	CR
Player's Value→Financial Performance	0.858	0.035	24.38**
Player's Value→Field Performance	0.149	0.133	1.12
Financial Performance→Field Performance	0.350	0.124	2.83**
Financial Performance→Stock Performance	0.049	0.058	0.85
Field Performance→Financial Performance	0.086	0.039	2.23**
Field Performance→Stock Performance	0.057	0.083	0.68

**: Significant at 0.05 level.

5 DISCUSSION

A player's value has no significant effect on the club's field performance. This implies that clubs cannot rely solely on the value of their players to achieve good results on the field. This is consistent with Samagaio et al. (2009). For example, in the 2015/2016 season of the English Premier League, out of the 7 clubs with the highest wage bills, 3 of them finished at the bottom of the table. One of them, Newcastle United, was relegated. On the other hand, Leicester City, ranked only 17th in wage expenditure (Horsefield, 2015), came out on top at the end of the 2015/2016 season.

Clubs can use their scouting network and academy to enhance their field performance without relying on a player's value. Di Minin et al. (2014) found that the good use of both their scouting network and the club's academy had a significant impact on Udinese's success. Bigger clubs can also benefit from this. For example, Barcelona got their best player, Lionel Andres Messi, from their scouting network in Argentina, and nurtured him in their reputable La Masia academy.

Other big clubs are following in the footsteps of FC Barcelona, for example, Manchester City. Until now, the club, which is also known as "The Citizens", has spent much more than most European clubs. However, they have also equipped themselves with a better club academy. Manchester City hired former FC Barcelona Sporting Director Txiki Begiristain and began the project to build their own "La Masia". Now promising young stars, such as Denis Suarez and Kelechi Iheanacho, have emerged after completing several years training in Manchester City's academy.

The next result shows that a player's value has a positive effect on the club's financial performance. This implies that clubs can get additional revenue by acquiring highly paid star players. The revenue comes from sales of merchandise, sponsors and additional match coverage on television. However, due to the proneness of football clubs to negative shocks (Szymanski, 2012), they have to consider their financial condition when acquiring highly paid players.

This result also marks the shift in the business model of football clubs. Clubs are gradually shifting from the SSSL (Spectator, Sponsorship, Subsidies and Local) model to the MCMM (Media, Corporate, Merchandise and Market) model. The SSSL model relies heavily on ticket sales, sponsorship and funds from the government and local communities as sources of revenue. On the other hand, the MCMM model relies more on television rights, sales of image rights and sales of merchandise. The example of this is Real Madrid. Under the stewardship of Florentino Perez, Los Blancos transformed itself into a Galactico, a club filled with superstar players. The result has been successful. For 12 consecutive years since 2004, Real Madrid topped the Deloitte Money League Chart as the club with the biggest revenue.

For clubs that are smaller than the likes of Real Madrid or Manchester United, acquiring highly paid star players can still be useful. However, they need to be mindful of their financial condition in order to avoid financial distress.

Financial performance has a positive impact on a club's field performance. This is consistent with Kase et al. (2006) and Samagaio et al. (2009). Kase et al. (2006) emphasises that a club's financial health has a significant impact on their long-term field performance. This implies that clubs cannot sacrifice their financial health for the sake of short-term success if they want to survive in the long term.

Manchester United and Real Madrid set examples for this. Their current success is because of their careful financial management. Both Real Madrid and Manchester United had set themselves up to be a global brand as early as the late 1990 s under the leadership of the clubs' respective directors and presidents.

Establishing the club's financial security and winning trophies should be done simultaneously. Under spending can lead to underperformance, as the club will not be able to fortify its squad with strong players. This problem can be solved by careful transfer management and by using an extensive scouting network and having an excellent academy.

On the other hand, overspending can lead to financial distress, despite winning trophies. The Milan duo, AC Milan and Internazionale Milan, are prime examples of clubs whose financial condition declined significantly shortly after winning trophies (Bandini, 2015). Both clubs failed to qualify for the UEFA Champions League from season 2013/2014 to season 2015/2016, less than 5 years after they won the Italian Serie A.

The next result shows that a club's field performance has a positive impact on their financial performance. This implies that a club's good performance brings them more revenue. This is consistent with the results of Szymanski and Smith (1997) and Samagaio et al. (2009). Successful clubs can benefit from more gate receipts, television rights and merchandise sales revenue. For example, Juventus received EUR 80 million from television rights in the 2014/2015 season, EUR 50 million of which came from their participation in the 2014/2015 UEFA Champions League final. Clubs play more home matches as they progress into the later stages of the competition. Therefore, it is logical that they get more revenue from gate receipts.

It is natural that clubs want the best results on the field. However, as they cannot rely on their players' value, alternatives have to be made. For example, clubs can use the human resource management approach by motivating players through other means than money. Mathis and Jackson (2011) state that employees' satisfaction comes through psychological factors rather than monetary factors. In football, players' job satisfaction is very important because they are the core of the business.

UEFA attempted to solve the problems related to the financial condition of the clubs. They came up with the Financial Fair Play (FFP) rules. However, the implementation of these rules has not been entirely successful, as clubs can still manage to overspend (Gibson, 2015). Two cases in point are Manchester City and Galatasaray. Manchester City's squad players were reduced to 21 men, and the club had to pay a GBP 42 million fine in 2015. One year later, Galatasaray had to serve a one-year ban from European competition due to their violation of the Financial Fair Play rules. So far, the FFP rules have not stopped clubs from overspending.

An alternative to the FFP rules can be found in American sports. According to Totty and Owens (2011), the salary cap implemented in American sports has successfully increased the leagues' competitiveness and ensured the clubs' financial health. For example, in the NBA and NFL, due to the salary cap, no clubs are in danger of insolvency, and the league remains competitive. However, the implementation of a salary cap in football may find opposition from the big clubs.

The last result shows that a club's stock performance is not affected by either their financial or field performance. This is consistent with the findings of Samagaio et al. (2009). This result implies that investors cannot rely solely on clubs' financial or field performances to make an investment decision., and therefore the effect is not significant. Another explanation is that investors are irrational. That is, their investment decision is based on their fanatical support towards the club. Such investors do not care whether their club is winning or losing. They also do not care whether the club's financial report is healthy or not, or whether it records any profit or not. They are just proud that they own a part of their favourite football club.

Other indicators, such as squad harmony, clubs' transfer and management policies and clubs' investment decisions, may affect clubs' stock performance. Therefore, investors are advised to consider not only the field and financial performances in their investment decisions but also these other factors.

6 CONCLUSION AND RECOMMENDATIONS

This paper concludes that a player's value has a positive impact on the club's financial performance but has no significant impact on field performance. Next, field performance has a positive impact on financial performance, and financial performance also has a positive impact on field performance. Lastly, neither financial performance nor field performance has a significant effect on a football club's stock performance. So? What are the implications then?

For the football club managers, it is recommended that they use an extensive scouting network and maximise the club's academy to enhance field performance. The signing of highly paid star players is encouraged as a means to increase revenue, as long as it does not affect the club's financial balance. In addition, managers should balance their field performance target with the club's financial health. To enhance field performance without further burdening the club's financial condition, managers can use a psychological approach instead of a monetary approach with the players.

For investors, as field performance and financial performance are not enough to make investment decisions on a football club's stock, they should also consider other indicators. Those indicators are squad harmony and the club's transfer and managerial policies.

7 LIMITATIONS

This paper has several limitations, which are:

- It only has a small number of samples. Unfortunately, this limitation cannot soon be overcome due to the small number of listed football clubs.
- The measurement for a player's value still excludes the fair value of players. The measurement of academy graduate players, which has no value on current accounting standards and the fair value of transfer, should be part of the next research.

REFERENCES

Bandini, P. (2015, January 17). Milan and Internazionale face up to new realities of life in Serie A. *The Guardian*. https://www.theguardian.com/football/2015/jan/17/milan-internazionale-serie-a-italian-football-city-rivals.

Barajas, A., Fernandez-Jardon, C. & Crolley, L. (2005). *Does sports performance influence revenues and economic results in Spanish football?* MPRA Munich Personal RePEc Archive, no. July. http://mpra.ub.uni-muenchen.de/3234/%5Cnhttp://scholar.google.com.br/citations?view_op=view_citation&continue=/scholar?hl=pt-BR&as_sdt=0,5&scilib=1&citilm=1&citation_for_view=X7l8wt4 AAAAJ:AvfA0Oy_GE0C&hl=pt-BR&oi=p.

Di Minin, A., Frattini, F., Bianchi, M., Bortoluzzi, G. & Piccaluga, A. (2014). Udinese Calcio soccer club as a talents factory: Strategic agility, diverging objectives, and resource constraints. *European Management Journal, 32*(2), 319–36. doi:10.1016/j.emj.2013.04.001.

Dragoni, G. (2014, December 31). Investing in a football team is not a good deal: In 2014 all three clubs Listed in Milan lost value. *Italy Europe 24:Il Sole 24 Ore Digital Edition*. http://www.italy24.ilsole24ore.com/art/panorama/2014-12-30/investing-a-football-team-is-not-good-deal-2014-all-three-clubs-listed-milan-lost-value-202426.php?uuid=ABj8h9WC.

Floros, C. (2014). Football and stock returns: New evidence. *Procedia Economics and Finance,14*, 201–209. doi:10.1016/S2212-5671(14)00703-5.

Gibson, O. (2015). Financial fair play: How can Manchester City still spend £130 m in one window? *The Guardian*. https://www.theguardian.com/football/blog/2015/sep/02/financial-fair-play-manchester-city.

Horsefield, I. (2015). Every Premier League club's wage bill revealed: So who spends the most on their players? *Express*. http://www.express.co.uk/sport/football/656624/Pictures-Premier-League-club-wage-bill-revealed.

Hwang, H. & Takane, Y. (2004). Generalized Structured Component Analysis. *Psychometrika, 69*(1), 81–99. doi: 10.1007/BF02295841.

Kase, K., Gómez, S., Urrutia, I., Opazo, M. & Martí, C. (2006). Real Madrid–Barcelona: Business strategy vs. sports strategy, 2000–2006. *Occasional Paper,* 06/12. http://www.iese.edu/research/pdfs/op-06-12-e.pdf?q=la-liga-results.

Mathis, R.L. & Jackson, J.H. (2011). *Human resource management: Essential perspectives.* Mason, OH, USA: Cengage Learning.

Renneboog, L. & Vanbrabant, P. (2000). Share price reactions to sporty performances of soccer clubs listed on the London Stock Exchange and the AIM. *Center Discussion Paper Tilburg University, 2000-19,* 1–31. https://pure.uvt.nl/portal/files/534671/19.pdf.

Rikardsson, H. & Rikardsson, L. (2013). Strategic management in football. MA Thesis, Linköping University.
Samagaio, A., Couto, E. & Caiado, J. (2009). Sporting, financial, and stock market performance in English football: An empirical analysis of structural relationships. *Centre for Applied Mathematics and Economics (CEMAPRE) Working Papers.* http://cemapre.iseg.utl.pt/archive/preprints/395.pdf.
Szymanski, S. & Smith, R. (1997). The English football industry: Profit, performance and industrial structure. *International Review of Applied Economics, 11*(1), 135–53. doi:10.1080/02692179700000008.
Szymanski, S. (2012). Insolvency in English professional football: Irrational exuberance or negative shocks? *International Association of Sports Economists Working Paper* 1202. http://web.holycross.edu/RePEc/spe/Szymanski_Insolvency.pdf.
Tomkins, P. (2015, August 5). Football inflation and the transfer window. *Vice Sports*. https://sports.vice.com/en_uk/article/football-inflation-and-the-transfer-window.
Totty, E.S. & Owens, M.F. (2011). Salary caps and competitive balance in professional sports leagues. *Journal for Economic Educators, 11*(2), 46–56. http://frank.mtsu.edu/~jee/2011/fall/5_MS1011-pp45to56.pdf.
UEFA (Union of European Football Associations). (2015a). *UEFA Club Benchmarking Report 2013-2014 Season.* Switzerland: UEFA. http://www.uefa.org/MultimediaFiles/Download/Tech/uefaorg/General/01/99/91/07/1999107_DOWNLOAD.pdf.
UEFA (Union of European Football Associations). (2015b). *UEFA Club Licensing Report 2014: The European Club Footballing Landscape*. Switzerland: UEFA. http://www.uefa.org/MultimediaFiles/Download/Tech/uefaorg/General/02/29/65/84/2296584_DOWNLOAD.pdf.

Competition and Cooperation in Economics and Business – Gani et al. (Eds)
© 2018 Taylor & Francis Group, London, ISBN 978-1-138-62666-9

Portfolio formation using the Fama-French five-factor model with modification of a profitability variable: An empirical study on the Indonesian stock exchange

C. Hapsari & G.H. Wasistha
Department of Accounting, Faculty of Economics and Business, Universitas Indonesia, Depok, Indonesia

ABSTRACT: This study aims to analyse portfolio formations using the Fama-French five-factor model with a modification on the profitability variable. Different portfolio formations are performed for three kinds of profitability variables, which are annual operating profit per total equity (RMW), monthly operating profit per total equity (ROE) and annual operating profit per total assets (ROA). The method used in this study is based on the Fama and French (2015) five-factor model. The result shows that the portfolio formation for the RMW variable has the highest impact on stock return. This result is consistent with the results of Fama and French (2015). This result means that it would be better to use annual operating profit per total equity as the proxy for profitability.

1 INTRODUCTION

Profitability is one of the main aspects for measuring the effectiveness of company performance. Sharpe (1964) states that, besides profitability, there are other factors that affect the return on an investment, such as market risk, the book-value-to-market-value ratio, or the investment itself. Capital Asset Pricing Model (CAPM) theory explains that the only relevant risk from stock return is the variance from the share itself. This means that an individual risk can be explained by the contributed risk on a portfolio as a whole. However, factors that affect stock return do not simply come from one variable.

Fama and French (1995) introduced the Fama-French three-factor model, which analyses the relationship between stock return and portfolio formation based on the market size and the price-to-book ratio. Continuing on from the 1995 model, Fama and French developed their five-factor model in 2015 (Fama & French, 2015). The variables that they added were profitability and investment variables. The portfolio formation for the profitability variable used in this model is based on the yearly operating profit divided by total equity (RMW variable).

Other than the yearly operating profit, there are several ways to measure profitability, such as dividing the monthly operating profit by the total equity or the yearly operating profit per total asset. Hou, et al. (2014), on their factor-q model, use the monthly operating profit per total equity as the base for their formation of the profitability variable (ROE variable). On the other hand, Muhammad and Scrimgeour (2014) study stock return using several types of factors, one of which is profitability. Portfolio formation for the profitability variable in Muhammad and Scrimgeour's (2014) research uses the yearly operating profit per total asset (the ROA variable).

The objective of this research is to assess whether the portfolio formation using the RMW variable's influence on stock return is better than using the ROE and the ROA. The previous studies concerning this issue mostly focused on testing the model on their own country or use the same model on the different time period, whereas this research focuses more on the relationship between profitability proxy and portfolio formation.

This research will use three models, each with a different profitability proxy, i.e. RMW, ROE and ROA. Then each model will be analysed through comparisons (Model I-Model II) and (Model I-Model III).

2 LITERATURE REVIEW

If the Fama-French three-factor model is designed only to explain the relationship between return, size and price ratios, the five-factor model tries to explain this relationship by adding two more variables, which are profitability and investment. The Fama and French (2015) five-factor model shows that a high return could be gained from the stock of a company that is small in size, by having a high price-to-book capitalisation ratio, being profitable and being conventional in their investment strategy.

There are five variables that are used in this model. The first variable is SMB (Small Minus Big). Positive SMB means that a company with a small size will give a better return compared to a company with a bigger size, and vice versa. The next variable is HML (High Minus Low). Positive HML means that a stock with a lower book-to-price ratio will give a better return compared to a stock with a bigger book-to-price ratio, and vice versa. The third variable is market premium or Rm-Rf. Market premium is an additional return received by investors because they invest in stocks and not in risk-free assets, such as short-term treasury bills. The fourth variable is CMA (Conservative Minus Aggressive). Positive CMA means that the stock of a company that is more aggressive in undertaking investments will get less return compared to the stock of a more conservative company. The reason why the stocks of conservative companies will give a better return is because these companies only make investments when they actually have the funds to do so, and they choose their options carefully in order not to make a mistake that will result in a less stable financial condition (Rankine, 2001). The last variable is RMW (Robust Minus Weak). Positive RMW means that the stock of a company with a higher profitability will give greater returns compared to the stock of a company with lower profitability (Fama & French, 2015).

Fama and French (2015) also find that when they use the five-factor model, size is not a factor that influences stock return. The Fama-French five-factor model shows that a high return can be obtained from the stock of a company that has small capital, a high book-to-market capitalisation ratio, is profitable and is conventional in developing its investment strategy (Fama & French, 2015).

3 HYPOTHESIS BUILDING

Profitability is an indicator of company's performance in carrying out its activities in order to achieve the company's objectives, namely to optimise profits for stockholders. High profitability reflects the company's financial situation, in which the company has good financial conditions that allow them to develop their business when needed.

RMW variable is the new profitability variable that was introduced by Fama and French in 2015 in their new research model, the Fama-French Five-Factor Model (FF5F Model). *RMW* variable is obtained from the difference between the stock return for a portfolio with a high yearly operating profit per total equity and the stock return for a portfolio with a low yearly operating profit per total equity. In their research, Fama and French found a positive relationship between a *RMW* variable and return. This result means that a company with higher profitability will give a better return on their stock.

Hou et al. (2014) use the *ROE* variable on their model to identify the relationship between certain factors and stock return. *ROE* variable is computed from the difference between the stock return of a portfolio with a higher ratio of the monthly operating profit per total equity and the stock return of a portfolio with a lower ratio of the monthly operating profit per total equity. They found that the portfolio formation for the *ROE* variable that is calculated from the monthly operating profit per total equity will give a better stock return based on neoclassical economic theory.

*H*1: Portfolio formation using the *RMW* variable is better than portfolio formation using the *ROE* variable as a profitability proxy to stock return.

Muhammad and Scrimgeour (2014) used the *ROA* variable in their research to analyse the effect of the financial ratio on stock return in the Australian Stock Exchange. In their research, the *ROA* variable is defined as the difference between stock return on a portfolio with a high ratio of the yearly operating profit per total asset and stock return on a portfolio with a low ratio of the yearly operating profit per total asset.

*H*2: Portfolio formation for the *RMW* variable is better than portfolio formation for the *ROA* variable as a profitability proxy to stock return.

4 EMPIRICAL MODEL

This research uses the model based on the FF5F Model (2015):

$$R_{it} - R_{Ft} = a_i + b_i(R_{Mt} - R_{Ft}) + s_i SMB_t + h_i HML_t + r_i RMW_t + c_i CMA_t + e_{it} \quad (1)$$

where:
R_{it} = Stock return for portfolio *i* at period *t*.
R_{Ft} = Risk-free return at period *t*.
R_{Mt} = Market average portfolio return at period *t*.
SMB_t = Portfolio return that is diversified on small size stock minus portfolio return that is diversified on a big size stock at period *t*.
HML_t = Portfolio return that is diversified on a company with a big price-to-book ratio minus portfolio returnthat is diversified on a company with a small price-to-book ratio at period *t*.
RMW_t = The difference between portfolio return that is diversified on stock and high profitability and low profitability at period *t*.
CMA_t = The difference between portfolio return that is diversified on stock of a conservative company and an aggressive company in investment strategy at period *t*.

There are two other models that have the same variables as the model above, except for the profitability variables. The other two models replace the profitability variable with a different type of profitability variable, which are the ROE variable and the ROA variable.

The samples for this research are the listed companies whose stocks are actively traded in the Indonesian Stock Exchange, excluding financial companies, from January 2012 to December 2014. The total number of companies that are used in this research is 316.

The steps of the portfolio formation for the three models are as follows:

1. Data of stock return portfolio from 316 companies are arranged based on capitalisation value.
2. Stock portfolio is divided into two sizes based on capitalisation: small and big.
3. The small size stock portfolio will be rearranged later into three parts based on the BE/ME ratio, 30% lowest, 40% medium and 30% highest. The big size stock portfolio will also be arranged into three parts based on the BE/ME ratio, 30% lowest, 40% medium and 30% highest.
4. The small size stock portfolio will be arranged based on market premium, investment and profitability into three parts, 30% lowest, 40% medium and 30% highest. The same treatment also applies to the big size stock portfolio.
5. Profitability factor will be arranged based on three variable types used in this research, which are *RMW*, *ROE*, and *ROA*.

To determine which portfolio formation is better as the profitability proxy, the significance level will be used as the basis. Significance level is a way to determine whether the dependent variable has any relation to the independent variable. Thus, a smaller significant level means that the dependent level has a better relation to the independent level.

5 RESULTS AND ANALYSIS

A normality test, multicollinearity test and heteroscedasticity test are performed, and no irregularity is found. The results for multiple regressions are as shown in Table 1, Table 2 and Table 3:

The results of the multiple regression show a significant value for the RMW variable. Positive RMW means that the stock of a company with higher profitability will give more return compared to the stock of a company with lower profitability. One of the reasons why RMW has positive effects on stock return is because the stock of a company with lower profitability

Table 1. Result of regression model (with the *RMW* variable).

	Unstandardised coefficients		Standardised coefficients		
Model	*B*	*Std. Error*	*Beta*	*T*	*Prob. Val*
(Constant)	−0.027	0.007		−4.020	0.000
SMB	0.004	0.002	0.238	2.061	0.048
CMA	0.004	0.002	0.328	2.415	0.022
RMW	0.006	0.001	0.532	4.326	0.000
HML	0.001	0.002	0.053	0.358	0.723
RmRf	0.520	0.106	0.545	4.897	0.000

Table 2. Result of regression model (with the *ROE* variable).

	Unstandardised coefficients		Standardised coefficients		
Model	*B*	*Std. Error*	*Beta*	*T*	*Prob. Val*
(Constant)	−0.018	0.010		−2.450	0.020
SMB	0.005	0.000	0.260	1.930	0.060
CMA	0.002	0.000	0.140	0.930	0.360
ROE	0.004	0.000	0.280	2.330	0.030
HML	0.000	0.000	−0.030	−0.150	0.880
RmRf	0.641	0.120	0.670	5.470	0.000

Table 3. Result of regression model (with the *ROA* variable).

	Unstandardised coefficients		Standardised coefficients		
Model	*B*	*Std. Error*	*Beta*	*T*	*Prob. Val*
(Constant)	−0.027	0.008		−3.546	0.001
SMB	0.004	0.002	0.229	1.864	0.072
CMA	0.005	0.002	0.443	2.767	0.010
HML	0.002	0.002	0.125	0.754	0.457
RmRf	0.522	0.116	0.546	4.482	0.000
ROA	0.007	0.002	0.599	3.558	0.001

is more vulnerable to macroeconomic and industry conditions (Hao et al., 2011). It will be difficult for companies that have low profitability to compete when the general economic situation is not good. These companies usually lack the resources, including the adequacy of funds, which makes it harder for them to compete with companies with a high profitability.

Companies with small RMW mean that companies with same range of book value are relatively less profitable than companies with a high RMW. This also means that they are less than optimal in managing equity to achieve the goals of the company. Equity here includes common shares and dividends. As a result, the company's profit achieved during this period will also be low, and this will have an impact on the amount of dividends received by the stockholders. Furthermore, this condition will be reflected in the company's stock price and ultimately in its stock return. Equally, companies with a high RMW will give a higher return. These companies can use the funds from the undistributed profit to stockholders (retained earnings) to increase the company's performance, so it can reach the optimum capacity. The result of this research is in accordance with the findings proposed by Fama and French (2015).

ROE also shows a significant value to stock return. It means that the portfolio formation for ROE has an influence on stock return, even if the significance level is lower than that of the RMW variable. This result is consistent with the result of Hou et al. (2014).

Hence, RMW gives a better explanation for the relationship between profitability and stock return. The difference between these two variables is the period used on these variables, which are monthly and yearly. According to Fama and French (2015), the Fama-French five-factor model is formed or derived from the model formulation of the Dividend Discount Model (DDM). Thus, this model has already included the information on dividends. This means that, in addition to capital gain, dividend is also a factor that drives the investors in Indonesia to invest in stocks. Indonesian investors are attentive to a company's annual financial statement data in deciding whether or not to buy a company's stock.

The implication of this result is that investors who choose to focus on stocks from companies with low profitability will get a smaller return than those who buy stocks from companies with high profitability.

The multiple regressions also show a significant result for the ROA variable. The ROA variable is the variable that is used in many studies. One of these studies is done by Muhammad and Scrimgeour (2014). The ROA variable in this research is acquired based on portfolio formation of the yearly operating profit divided by total assets. This result shows a company's profitability on an accounting term divided by total assets also related to return. A positive ROA coefficient also means that a higher value of ROA will give higher returns in the future. This result is in accordance with Muhammad and Scrimgeour's (2014) result, which also finds the relationship between the ROA variable and return.

The ROA variable tries to find out the effectiveness of the company's management in managing its asset in order to achieve their planned operating profit for the year. A high value ROA means that the company gets high operating profit for every asset that is used.

The relationship between ROA and stock returns is that a higher ROA will give a better return for investors who invest in those companies. Companies with a high ROA will have advantages in optimising the use of their assets to increase their operating profit. Effective performance will be reflected in the company's annual financial report, so that ultimately this information will be reflected in the company's stock price.

The difference between the RMW variable and the ROA variable in this research is the denominator, which is total equity for the RMW and total asset for the ROA.

Although the ROA variable has a positive effect on the return, RMW still has a better significance level. Also, the RMW variable is a better variable in explaining the dependent variable, so it can be said that generally RMW is a better variable for explaining the relationship between profitability and return. RMW links with the denominator of each variable, which is total equity and total assets. It can also be linked with DuPont analysis, which states that ROE is ROA that has a company's leverage information. Leverage information is a factor that needs to be considered in order to understand a company's profitability. Thus, the RMW variable has more influence on returns compared to the ROA variable.

The implication of this research is that investors who choose to invest based on the yearly operating profit divided by total equity will get greater returns compared to those who choose to invest based on the yearly operating profit divided by total assets.

6 CONCLUSION

This research concludes that all of the three profitability variables used in this research model have a positive effect on stock return. That means that the portfolio formation for RMW, ROE and ROA has a positive coefficient, as can be seen in the multiple regression results.

From the models used in this research, the RMW variable is better than the ROE variable in representing profitability. This is related to the Fama and French (2015) FF5F Model, which is based on the Dividend Discount Model. Fama and French (2015) use the data of the yearly operating profit because the data already includes dividend information that is not included in the monthly operating profit data.

This research also concludes that the RMW variable is better than the ROE variable in representing profitability. This is related to the leverage factor. The ratio that is used in portfolio formation for the RMW variable already contains the leverage factor, which is not applicable in ROA, so the RMW variable has a greater effect on stock return.

To sum up the results of this research, the portfolio formation for the profitability factor based on the yearly operating profit per total equity will give a higher return compared with the monthly operating profit per total equity or yearly operating profit per total asset.

In addition to the profitability factor, other variables that have an effect on stock return, based on the results of this research, are market premium, portfolio formation for the SMB variable, and the CMA variable. This research also finds that market premium, portfolio formation for SMB and CMA variables have a positive effect on stock return.

For further advancement on this topic, more types of ratios, such as EPS (Earning per Share) or DPR (Dividend Payout Ratio), can be used as a base for portfolio formation.

REFERENCES

Fama, E.F. & French, K.R. (1995). Size and book-to-market factors in earnings and returns. *The Journal of Finance, 50*(1), 131–55. doi:10.1111/j.1540–6261.1995.tb05169.x.

Fama, E.F. & French, K.R. (2015). A five-factor asset pricing model. *Journal of Financial Economics, 116*(1), 1–22. doi:10.1016/j.jfineco.2014.10.010.

Hao, S., Jin, Q. & Zhang, G. (2011). Relative firm profitability and stock return sensitivity to industry-level news. *Accounting Review, 86*(4), 1321–47. doi:10.2308/accr-10042.

Hou, K., Chen, X. & Zhang, L. (2014). *A comparison of new factor models*. NBER Working Paper 20682, National Bureau of Economic Research, Cambridge, MA.

Muhammad, N. & Scrimgeour, F. (2014). Stock returns and fundamentals in the Australian market. *Asian Journal of Finance & Accounting, 6*(1), 271–90. doi:10.5296/ajfa.v6i1.5486.

Rankine, Denzil. Why *Acquisitions Fail: Practical Advice for Making Acquisitions Succeed*. London: Pearson Education, 2001.

Sharpe, W.F. (1964). Capital asset prices: A theory of market equilibrium under conditions of risk. *The Journal of Finance, 19*(3), 425–42. doi:10.1111/j.1540–6261.1964.tb02865.

Competition and Cooperation in Economics and Business – Gani et al. (Eds)
© 2018 Taylor & Francis Group, London, ISBN 978-1-138-62666-9

Analysing the impact of the double taxation treaty on foreign direct investment in Indonesia

A. Rizky & C. Tjen
Department of Accounting, Faculty of Economics and Business, Universitas Indonesia, Depok, Indonesia

ABSTRACT: Foreign Direct Investment (FDI) is an important element in the era of globalisation and links between economies. Agreement between partner countries, particularly on the avoidance of double taxation, is believed to contribute to increasing flows of FDI. This study aims to examine whether agreements on the avoidance of double taxation have an effect on FDI and to discover what other factors have an influence on FDI in Indonesia. This study uses a number of data samples of double taxation treaties in Indonesia with partners from developed countries within the period from 1990 to 2014. Data processing uses the OLS regression analysis with time series data structure. The result of this study shows that an avoidance of double taxation treaty increases the flow of FDI into Indonesia. In addition, GDP per capita, resources rent and political conditions also affect FDI in Indonesia. Among these influencing factors, the treaty on the avoidance of double taxation has been the main factor that has had the most influence on FDI in Indonesia.

1 INTRODUCTION

Developing countries often use bilateral agreements as a signal of their commitment to foreign investors. Their aim is to increase the flow of FDI into developing countries. By signing the double taxation agreement, developing countries will ensure stability and security for foreign investors on the issue of double taxation. In addition to signing a double taxation treaty, developing countries have also signed a bilateral investment treaty. This has been done to demonstrate the commitment from developing countries to give equal treatment to both local and foreign investors in the regulations and standards (Neumayer, 2007).

Global FDI in 2014 declined by 16% compared to 2013, and in 2013 the value reached 1.47 billion USD to 1.23 billion USD in 2014. Several factors led to a decline in 2014, including global economic conditions that were less stable, policy uncertainty for investors and the emergence of geopolitical risks in some places in the world. FDI flow to developed countries fell by 28% to 499 billion USD and the flow of FDI to economic transition countries fell by 52% to 48 billion USD. While FDI flow to developing countries increased by 2% to 681 billion USD, see Figure 1, this was driven by the flow of FDI to developing countries in Asia (UNCTAD, 2015).

FDI flow to the continent of Asia increased by 9% in 2014 to 465 billion USD, as East Asia, Southeast Asia and South Asia experienced an increase in receiving FDI flow. Only the West Asia region has continued to show a downward trend over a period of 6 years. In 2014, the West Asia region only received FDI flow amounting to 43 billion USD, and this is because of the unstable security situation in the region. China received FDI flow amounting to 129 billion USD in 2014, which increased by 4% compared to 2013. The service sector became the main support for China's growth. An increase in FDI flow also occurred in Hong Kong, Singapore, India and Indonesia, see Figure 2 (UNCTAD, 2015).

In the period from the 1960 s to the 2000 s, Indonesia made many agreements with developed countries, both bilateral investment treaties and double taxation treaties. This was done by the Indonesian government in order to improve the attractiveness of investment in Indo-

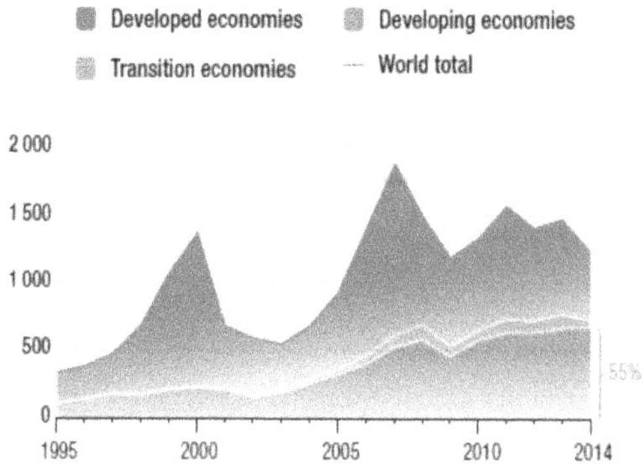

Figure 1. Trend of FDI flow from the years 1995–2014 (in billion USD).
Source: UNCTAD (2015).

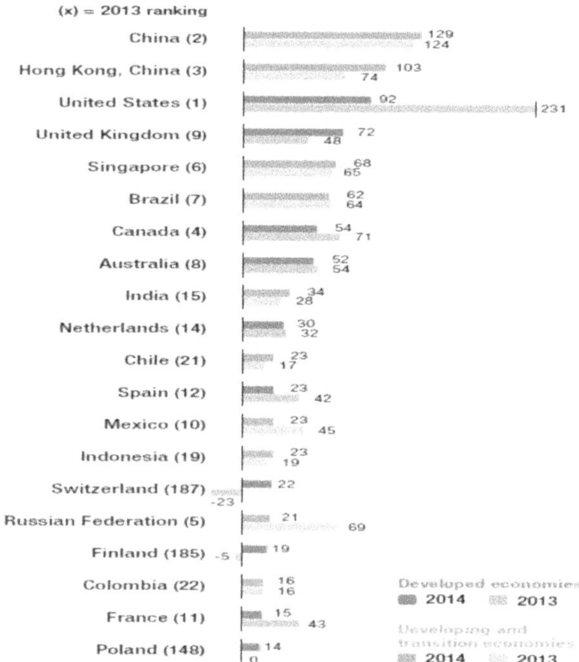

Figure 2. The 20 countries with the largest FDI flow in 2013–2014 (in billion USD).
Source: UNCTAD (2015).

nesia, as well as providing certainty in the field of law and taxation. With many foreign investors investing directly in Indonesia, this will generally result in increasing the overall Indonesian economy. In 2015 President Joko Widodo revised the bilateral investment treaty, which is not considered relevant in the current economic conditions. Things that would be revised in bilateral investment treaty agreements include an expiration date, automatic renewal of the agreement and equal conditions for both countries (Ariyanti, 2015).

The double taxation treaty in Indonesia is called P3B or an agreement on the avoidance of double taxation. P3B allegedly has the effect of raising the flow of FDI, as the results of several previous studies support this contention, however, several other studies have concluded that there is no effect between FDI and P3B (Badan Kebijakan Fiskal, 2012). According to the Badan Kebijakan Fiskal (2012), "P3B is the agreement made by two/more countries/tax jurisdictions governing the tax treatment of income tax on income earned by the domestic taxpayer". This agreement relates to the passive income of the beneficial owner. Tax problems arise when the two countries want to impose a tax on the same kind of income. Some of the desired goals of the Indonesian government in making P3B with partners from other developing countries, as well as with developed countries, are namely: the loss of double taxation, the increase in foreign investment, human resources development, the exchange of information between the tax authorities to fight tax evasion and fair conditions on taxation between P3B partners (Badan Kebijakan Fiskal, 2012).

2 LITERATURE REVIEW

According to the research from Rahayu (2008), FDI is one indicator of the economic system in the era of globalisation today. FDI comes from a company in one country investing over a long period of time in companies from other countries. The company may have either partial or full control over the company that is in another country. The majority of the forms of FDI lead to full control over a company. Easson (2000) argues that the forms of foreign investment may be the transfer of assets to the state's investment objectives, investing back in the country of investment, buying assets in an investment destination country, as well as the acquisition of the majority of shares or providing loans to subsidiaries or affiliated companies (Rahayu, 2008).

The FDI flow to Indonesia can come in two ways, according to Gunadi (1997). First, to establish a subsidiary, we could create a new body called a PT. PMA (Foreign Investment Company), or we could buy a majority company's shares in Indonesia that is already in operation. These subsidiaries have a legal standing as being separate from their parent company, although the sources of capital and the control of the business are conducted by the parent company. From a fiscal standpoint, transactions can be made between the two for the value used in fair transactions because they have a special relationship. The second way is by setting up a branch company. A foreign branch company is, in fact, a division that is operated in a separate geographical area. From a legal point of view, as well as for taxes, the branch company and headquarters are a single entity. The establishment of a branch company will create BUT (Permanent Establishment), while the establishment of a subsidiary will be treated as a domestic taxpayer (Rahayu, 2008).

The conventional view of tax practitioners and tax authorities stated that tax treaties will attract foreign investment. Economists also argue that tax treaties will increase investment because tax treaties indicate the co-operation towards taxation between treaty partners. How can it happen? First, the harmonisation of the definition of taxes and tax jurisdiction between treaty partners would reduce double taxation on investment. For example, the income is taxed by the state as the source of income is derived from permanent establishments (BUT), and the definition of the BUT is set up very clearly in the tax treaties. If there is no agreement between the two countries, then there is a possibility that the definition of the BUT will be different and this will lead to double taxation on that income. Second, tax treaties have an effect on taxation carried out by a multinational company. This may occur because of the mechanisms of double taxation system either the credit system or the exemption and the withholding taxes on income derived from foreign affiliated company such as dividends, interest, and royalties.. With the tax treaties, the tax rate on such income will drop, which will cause the tax burden of companies to also decline, and this is expected to increase the company's investment abroad. Developing countries often act as an importer of capital. Therefore, in their tax treaties with partners from developed countries, developing countries act as a source country, while the developed countries act as the state of domicile. The application of tax treaties, when the tax can be levied by the source state, is reduced, which is actually reducing tax revenues from developing countries. Discussion of tax treaties in developed countries as exporters of capital, largely driven by multinational companies originating from these countries. Like, the motivation to create a new treaty or modify an existing treaty. For example, tax treaties between Mexico and Indonesia were encouraged by the national oil company Pemex, which allowed the Mexico oil industry to invest in Indonesia. Furthermore, the tax treaties between the United States and the Netherlands are triggered by the investment company Shell and Philips (Loncarevic, 2005)

From an economic standpoint, the presence of double taxation would provide a greater burden on the business activity, investment and other activities, which would cause a disruption of the flow of resources between countries that work together. To overcome this, most of the tax authorities in the world work together and make agreements to avoid double taxation. Surrey (1980) provides a definition of double taxation treaties, namely the agreement made by two or more countries, which aims to find a way out of double taxation (Sophia, 2013).

Until 2015, Indonesia had had double taxation agreements with 65 partner countries. The agreements focused only on income tax. The position of Indonesia's double taxation treaty also embraced aspects of international law. So if the agreement on the avoidance of double taxation (P3B) has already been effective, then this may change the tax subject or tax object that has been determined in the domestic tax laws of Indonesia (Sophia, 2013). Indonesia's legal basis enables it to make a double taxation treaty (Sophia, 2013):

1. Article 11 paragraph 1 of the Constitution of 1945.
In this chapter, it is mandated that the President can make treaties with other states with the consent of the House of Representatives.
2. Article 4 paragraph 1 of Law No. 24 of 2000.
This law regulates an international agreement by Indonesia with other countries. This article states that the conclusion of a treaty should be based on the agreement, and the parties involved are obliged to implement it in good faith.
3. Section 32 A of the Income Tax Act No. 7 of 1983, as amended into the Law No. 36 of 2008.
This Article provides that, in order to avoid the incidence of double taxation and prevent tax evasion, the government has the authority to make agreements with other countries.

Research conducted by Neumayer (2007) aimed to discover whether the United States' double taxation treaty with partners from developing countries would increase FDI to developing countries. His research shows that developing countries that have a tax treaty with the United States have benefited by getting a higher flow of FDI, compared to the developing countries that do not have a double taxation treaty with the United States. Several other previous studies that discuss the relationship between the tax treaty and FDI can be seen in Table 1.

Table 1. Previous studies on treaty and FDI.

Research information	Samples	Type	Result
Davies, 2003	US treaty network, 1966–2000	Dyadic studies	No effect on FDI
Bloningen & Davies, 2004	US inward and outward investment stocks in term of 88 partners, 1980–1999	Dyadic studies	No effect on FDI
Bloningen & Davies, 2005	OCED data on bilateral FDI stocks and flows, 1982–1992	Dyadic studies	No effect on FDI
Egger *et al.*, 2006	Bilateral outward FDI from OECD source countries, 1985–2000	Dyadic studies	No effect on FDI
Giovanni, 2005	193 countries, 1990–1999	Monadic studies	Increase FDI
Barthel et al., 2010	30 FDI source countries and 105 FDI host countries, 1978–2004	Monadic studies	Increase FDI
Coupe, Orlova and Skiba, 2008	Investment from OECD countries to 9 transition economies, 1999–2001	Monadic studies	No consistent findings

Source: www.dannydarussalam.com.

Table 2. Variable data sources.

Symbol	Data source
FDI	World Bank, BKPM
P3B	www.dannydarussalam.com
GDP	World Bank, BI
Population	World Bank, BPS
Economic Growth	World Bank, BI
Inflation	World Bank, BI, BPS
Resource Rents	World Bank
Trade	World Bank
Politic Condition	www.systemicpeace.org

3 RESEARCH METHOD

The research approach used in this study is a quantitative research approach, which uses multiple linear analysis regression to analyse the effect of double taxation treaties in Indonesia on FDI into Indonesia. The measurement results of this study will be used as an evaluation of the Indonesian government in making decisions related to agreements with partner countries on the avoidance of double taxation in Indonesia.

The data used in this research are the data from the P3B in Indonesia with partners from the developed countries within the period of 1990 to 2014. The definition and grouping of developed countries was taken from the 2016 World Bank Development Indicator Data. The data used in this study are time series and secondary data. Data were collected from the literature and from studies. See Table 2 for the data source variables used in this study.

This study uses two types of variables, an independent variable and a dependent variable. An independent variable is a variable whose movement is not affected by the dependent variable. A dependent variable is a variable whose motion is affected by the independent variable. The independent variables are divided into two: explanatory variables and control variables (Neumayer, 2007).

The model of this study is:

$$Y = \beta_o + \beta_1 X_1 + \beta_2 X_2 + \beta_3 X_3 + \beta_4 X_4 + \beta_5 X_5 + \beta_6 X_6 + \beta_7 X_7 + \beta_8 X_8 + \varepsilon\tau \quad (1)$$

where

Y = Log of Total Foreign Direct Investment Flows into Indonesia

X_1 = Cumulative Number of P3B Signed with Developed Countries, Weighted by the Number of Inward FDI Flow to Indonesia Relative to Total World FDI Flow.

X_2 = Log of GDP per Capita Indonesia

X_3 = Log of Total Indonesia Population

X_4 = Economic Growth

X_5 = Inflation

X_6 = Resources Rent (% of GDP)

X_7 = Trade (% of GDP)

X_8 = Politic Condition

β_o = Constant

β_{1-} *to* β_8 = Regression Coefficients

ε_τ = Error

4 RESULTS

The R square (adjusted R square) is very useful for measuring the affinity between the predicted value and the actual value of the dependent variable. The larger the R square, the greater the relationship between a dependent variable and one or more independent variables.

In order to meet the criteria for an optimal fashion, the criteria in Table 3 must meet the criteria of optimal models, namely:

By looking at the output above, the regression model of this study suggests optimal results, because it:

- Has a high value of adjusted R^2, which means that the variations of the five independent variables of this model, namely P3B, GDP per capita, Resources Rent, Politics and Trade, are able to explain the FDI as the dependent variable of 87.80%, while the rest of the 12.20% is explained by other factors that are not included in the research model.
- Has a value of F-statistics less than 5%.
- Has a value of Akaike Information Criterion (AIC) that is relatively low at –0.198242.
- Carries Schwarz Information Criterion (SIC) that is relatively low at 0.100477. This shows the simplicity of the model.

Looking at the results from Table 4, we can conclude which variables have a level of significance below 0.05 and can be used in the research model to describe the factors that influence FDI in Indonesia. From the results of the regression that has been made to the existing variables, the equation model is obtained as follows:

$$\text{FDI} = 14\ \text{P3B} + 2.18\ \text{GDPpc} + 7.872\ \text{Resources Rent} - 0.04\ \text{Politics} \tag{2}$$

The overall independent variables provide a positive effect on the increase in the flow of FDI, except for the political variables, which have a negative correlation. From the above equation model, we can interpret that the addition of double taxation treaties (P3B) will make an impact of 14% for every 1% increase in FDI.

Indonesia has double taxation treaties with developed countries, which could allow Indonesia to receive a greater direct investment flow when compared with other developing countries that do not have a tax treaty agreement with these countries.

From Figure 3 it can be seen that when there is the addition of double taxation treaties, the FDI into Indonesia will increase, except during a period of economic crisis. This study shows that there is a strong correlation between the addition of double taxation treaties and the increase in FDI into Indonesia. Indonesia, which serves as a source country, will keep investors from their partners in developed countries interested in investing in Indonesia. In terms of the signing of the double taxation treaty, Indonesia made a deal with many countries in the period from the 1990 s to the 2000 s. This is also consistent with the results of this study, which show that the addition of double taxation treaties, especially with partners in developed countries, will increase the flow of investment from foreign investors. These

Table 3. Coefficient of determination results.

Criteria	Value
R squared	91.01%
Adjusted R squared	87.80%
Prob (F-statistic)	0.000001
Akaike info criterion	–0.198242
Schwarz criterion	0.100477

Table 4. Hypothesis testing results.

Independent variable	Coefficient	Significance level	Decision
P3B	14	0.0172	Reject Ho
GDP per Capita	2.182	0.0021	Reject Ho
Resources Rent	7.872	0.0029	Reject Ho
Politics	–0.04	0.0252	Reject Ho
Trade	0.454	0.7569	Accept Ho

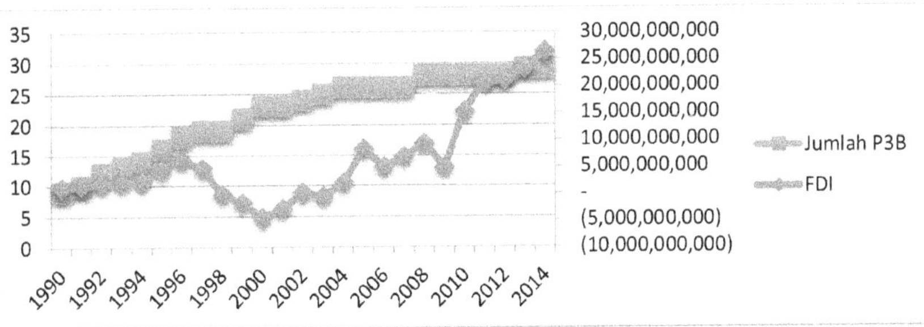

Figure 3. FDI trend & number of P3B with developed countries.

results are also consistent with the results from Neumayer (2007), which show that developing countries that have double taxation treaties with more developed countries are benefited by getting a higher flow of FDI. In addition, the results of this study are also in tune with economists, who argue that tax treaties will increase investment because tax treaties indicate the co-operation in taxation between treaty partners. Thus, it can be concluded that there is a correlation between P3B and FDI.

On the control variables, gross domestic products per capita will make an impact of 2.18% for every 1% increase in FDI, and the resources rent will have an impact of 7.87% for every 1% increase in FDI. As for the political situation, it will have an impact of 0.04% for every 1% decrease in FDI. Indonesia is a capital importer country that has a large GDP, reserves of natural resources in oil, gas and mining, and has stable political conditions, so it will get an opportunity to have a higher inflow of FDI because investors also look at these three aspects in determining their decision on whether or not to invest in Indonesia.

If the total number of mineral resources rents in Indonesia is higher, then the FDI into Indonesia will also increase. However, it is also influenced by the technical factors of the fields or wells of mineral resources that Indonesia has. Foreign investment flow to Indonesia has continued to increase during the last fifteen years. This is supported because of the improved condition of Indonesia's economy and more stable political conditions. During Indonesia's economic crisis of 1997–1998, which was followed by political and security instability, much foreign investment flow went out of Indonesia.

Thus, when the GDP per capita of Indonesia continues to increase, it is expected that this will give a signal to investors that the economic situation in Indonesia is very good, which will create a conducive business climate. This will encourage foreign investors to invest in Indonesia, which, in turn, will increase the flow of FDI into Indonesia. Thus, it can be concluded that in terms of both mineral resources and political and economic conditions, they affect the increase in FDI in Indonesia, which is in line with the results of this study.

5 CONCLUSION

The avoidance of double taxation agreement (P3B) significantly affects the flow of FDI into Indonesia. This is consistent with the conventional view of tax practitioners and tax authorities, who state that tax treaties will attract foreign investment.

The addition of double taxation treaties between Indonesia and partners from other countries will significantly affect the increase in FDI into Indonesia. These results are also consistent with the results of the research of Neumayer (2007), which shows that developing countries that have double taxation treaties with more developed countries are benefited by getting a higher flow of FDI. In addition to P3B, other factors that have an effect on FDI in Indonesia are GDP per capita, political conditions and the availability of natural resources, mining, oil and gas fields. P3B is a major factor in increasing FDI in Indonesia compared to the GDP factor, resources rent and political conditions. Although these three factors also significantly affect FDI in Indonesia, the presence or absence of P3B with certain countries is a major consideration for investors of other countries when deciding on their capital investment in Indonesia.

In order to attract more foreign investors from developed countries and others, governments should renegotiate or amend the P3B that exists today. Its goal is to provide optimum benefits for the Indonesian economic condition. The Indonesian government and the other parties concerned should start to consider adding a new partner of P3B, especially from countries that have substantial investments around the world, and particularly in Indonesia, such as Brazil, Ireland and Chile.

The government, in this case the executive, is expected to maintain a stable political condition by maintaining a good relationship with the legislature, either with the DPR or with the MPR agency. In addition, the government should improve the condition of the national economy, by creating policies that can create a good business climate and provide incentives in accordance with current conditions in the mining, oil and gas industries. Hence, it will give a sense of security and stability to existing foreign investors and will attract other investors to invest in Indonesia. The government is expected to provide legal certainty in the field of taxation, by making new regulations that support it, and also to build awareness by the tax authorities in Indonesia to obey the rules in a treaty agreement.

REFERENCES

A. Easson, (2000). *Do We Still Need Tax Treaties?*, 54 Bull. Intl. Fiscal Docn. 12, pp. 619–625, Journals IBFD.

Ariyanti. (2015). *Sering Digugat Pemerintah Jokowi Revisi Perjanjian Investasi (Often Sued Jokowi Government Revised Investment Agreement)*. http://bisnis.liputan6.com/read/2229991/sering-digugat-pemerintah-jokowi-revisi-perjanjian-investasi.html.

Badan Kebijakan Fiskal Kementrian Keuangan. (2012). Tax Treaty dan Pengaruhnya Terhadap Arus Investasi antara Indonesia dengan Negara-negara Mitra (Tax Treaty and its Influence on Investment Flows between Indonesia and Partner Countries). Pusat Kebijakan Regional dan Bilateral Jakarta.

Barthel, F., Busse, M. & Neumayer, E. (2010). The impact of double taxation treaties on foreign direct investment: Evidence from large dyadic panel data. *Contemporary Economic Policy, 28*(3), 366–77. doi:10.1111/j.1465–7287.2009.00185.x.

Blonigen, B., and R, Davies. (2004). The Effect of Bilateral Tax Treaties on U.S FDI Activity. *International Tax and Public Finance*.

Blonigen, B.A. and Davies, R.B. (2005), "Do Bilateral Tax Treaties Promote Foreign Direct Investment?", in Choi/Hartigan (eds.) Handbook of International Trade, Volume II (London: Blackwell), pp. 526–546.

Coupé, Tom, Irina Orlova, and Alexandre Skiba, 2008, The Effect of Tax and Investment Treaties on Bilateral FDI Flows to Transition Countries, mimeo.

Danny Darussalam Tax Consulting. (2016). *www.dannydarussalam.com*. Retrieved from www.dannydarussalam.com.

Davies, R. (2003), *Tax Treaties, Renegotiations, and Foreign Direct Investment*, Economic Analysis & Policy 33(2), pp. 251–273.

Di Giovanni, J. (2005). "What Drives Capital Flows? The Case of Cross-border M&A Activity and Financial Deepening", 65 Journal of International Economics 1, pp. 127–149.

Egger, P., Larch, M., Pfaffermayr, M., and Winner, H. (2006), "The Impact of Endogenous Tax Treaties on Foreign Direct Investment: Theory and Evidence", 39 Canadian Journal of Economics 3, pp. 901–931.

Gunadi. (1997). *Pajak Internasional*. Jakarta: Lembaga Penerbit UI.

Loncarevic, I. (2005). Economic relevance of double taxation conventions. In M. Stefaner & M. Zuger (Eds.), *Tax treaty policy in development*. Austria: Linde Verlag.

Neumayer, E. (2007). Do double taxation treaties increase foreign direct investment to developing countries? *Journal of Development Studies, 43*(8), 1501–1519.

Rahayu, N. (2008). Praktik Penghindaran Pajak (tax avoidance) pada foreign direct investment Yang Berbentuk subsidiary company (PT. PMA) Di Indonesia (Suatu Kajian Tentang Kebijakan anti tax avoidance) (Tax Avoidance Practices in the Foreign Direct Investment of a Subsidized Company (PT PMA) In Indonesia (An Anti Tax Avoidance Policy)) (Published Dissertation). Universitas Indonesia. http://lib.ui.ac.id/opac/themes/libri2/detail.jsp?id = 121296&lokasi = local.

Sophia. (2013). *Konsep Dasar Pajak Internasional (Basic Concepts of International Tax)*. http://sophiaririnkali.blogspot.com/2013/05/konsep-dasar-pajak-internasional.html.

Surrey, Stanley. (1980), United Nations model convention for tax treaties between developed and developing countries: a description and analysis, Amsterdam: IBFD.

UNCTAD. (2015). *World investment report*. http://unctad.org/en/PublicationsLibrary/wir2015_en.pdf

World Bank. (2016). *World development indicators*. http://data.worldbank.org/data-catalog/world-development-indicators.

Competition and Cooperation in Economics and Business – Gani et al. (Eds)
© 2018 Taylor & Francis Group, London, ISBN 978-1-138-62666-9

The determinants of working capital requirements of manufacturing firms in Indonesia

S. Wiguna & G.H. Wasistha
Department of Accounting, Faculty of Economics and Business, Universitas Indonesia, Depok, Indonesia

ABSTRACT: The purpose of this study is to examine the effect of the cash conversion cycle, operating cash flow, firm size, profitability, leverage, growth opportunities and the real GDP growth rate on working capital requirements. This research uses 85 manufacturing companies that were listed on the Indonesian Stock Exchange from 2010 to 2014. Multiple linear regression and panel data are used as the analysis tools. The study finds that the cash conversion cycle and profitability have a positive effect on working capital, whereas leverage has a negative effect on working capital. However, operating cash flow, firm size, growth opportunities and the real GDP growth rate have no significant effect on working capital.

1 INTRODUCTION

Almost every financial management textbook starts by describing the three major decisions to be taken by a financial manager, which are investment decisions, financing decisions and working capital related decisions. Hanafi (2014) states that approximately 60% of the time spent by financial managers is related to working capital activities. This is because working capital has a direct relationship to a company's activities, so that when there is a disruption to working capital decisions, then the activities will also get delayed.

A lack of working capital can lead to a company becoming bankrupt. Sudarmadi (2013) explains that in 2007 the aircraft manufacturer PT Dirgantara Indonesia (Persero) went bankrupt because it was sued by its former employees. At that time, President Susilo Bambang Yudhoyono tried to save the company and appointed Budi Santoso as the Chief Executive Officer. One of the reasons for the company's deterioration was its lack of working capital, so the company could not accept any new projects.

Abbadi and Abbadi (2013) conducted research into the working capital requirements in Palestine. The research was performed on 11 industrial companies listed on the Palestine Securities Exchange and showed that the cash conversion cycle, return on assets and operating cash flow have positive effects on working capital requirements. Meanwhile, leverage and firm size have negative effects on working capital requirements. Other variables, such as the interest rate and the real GDP growth rate, have no significant effects on working capital requirements.

Another study was conducted by Onaolapo and Kayjola (2015). The results show that the leverage, size, industry classification, return on assets and the operating cycle have a significant effect on working capital requirements. This research was conducted on non-financial companies listed in the Nigerian Stock Exchange during the period of 2004 to 2011. In addition to these variables, other variables tested in the study, namely the level of economic activities and growth opportunities, have been found to have no effects on working capital requirements.

The difference between the research of Abbadi and Abbadi (2013) and Onaolapo and Kayjola (2015) is in the size variable. Abbadi and Abbadi (2013) show that size is a factor that does not have any significant effect in their research, while Onaolapo and Kayjola (2015) state that size has a significant effect on working capital requirements. Onaolapo and Kayjola

(2015) also give a recommendation that further research should add an operating cash flow variable, as well as macroeconomic components such as inflation. This research uses size and operating cash flow as variables, and uses the real GDP growth rate as a variable of macroeconomic factors.

This study is different from previous research. First, the factors that are used by Abbadi and Abbadi (2013), such as the cash conversion cycle, return on assets, operating cash flow, leverage, firm size, interest rate and the real GDP growth rate, are used as a whole in this study; meanwhile, the variables in Onaolapo and Kayjola's research (2015), such as leverage, size, industry classification, return on assets, operating cycle, growth opportunities and the level of an economic activity, are used only for additional testing. Nun (1981) in Onaolapo and Kayjola (2015) states that sales are a very important component and that their growth must be measurable. This component is highly influenced by the amount of inventory available, which will indirectly affect the size of the working capital. The sales growth will be measured by the growth opportunities in this study.

Second, Abbadi and Abbadi (2013) use the logarithm of total assets as a proxy measurement of firm size. Meanwhile, Onaolapo and Kayjola (2015) use the logarithm of sales for this measurement. Other tests proposed by Kwenda and Holden (2014) use the natural logarithm of the market value of equity or market capitalisation. This study uses the proxy used by Onaolapo and Kayjola (2015), because sales have a more direct relationship with working capital requirements compared to other factors.

2 LITERATURE REVIEW

Abbadi and Abbadi (2013) published their research on working capital using an econometric model in the form of panel data in 11 industries during the years 2004 to 2011. The result shows that the Cash Conversion Cycle (CCC), Return on Assets (ROA) and Operating Cash Flow (OCF) have significant and positive effects on working capital requirements, while leverage and firm size have negative effects on working capital requirements. The interest rate and the GDP growth rate do not have significant effects on working capital requirements.

Subsequent research conducted by Onaolapo and Kayjola (2015) uses the same research methods as Abbadi and Abbadi (2013), with a panel data model and Ordinary Least Squares (OLS) as an estimation technique. The regression results indicate that five variables, such as leverage, size, industrial classification, return on assets and the operating cycle, have significant effects on working capital requirements.

Figure 1, below, shows the relationship between the tested variables: independent variables and dependent variables:

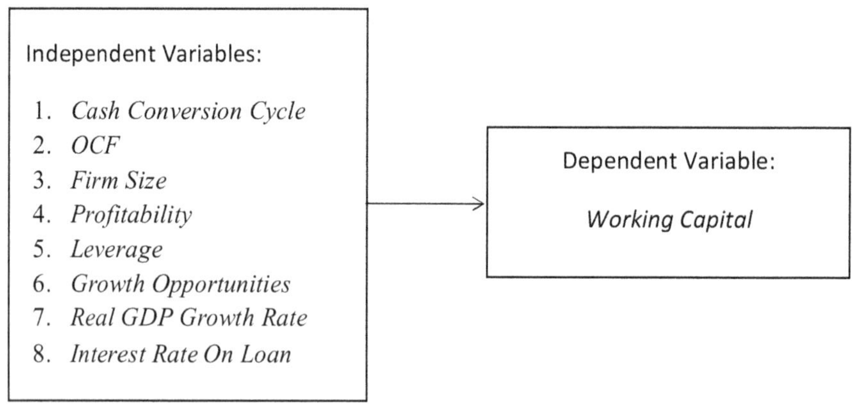

Figure 1. Research model.

3 HYPOTHESES

The relationship between the CCC and working capital is that the shorter the CCC level, the less working capital is needed. The CCC will move in the same direction as the working capital. The longer the period of the CCC, the larger the amount of working capital that is needed to be invested by a company in order to undertake its operational activities. If the CCC has a shorter period of time, the investment that a company needs to make on its working capital will also be less.

H1 = The shorter the period of the CCC, the lower the amount needed for working capital.

When the OCF is low, the working capital needed will be higher. If the cash flow resulting from the company's daily activities is halted, this means that the company needs to invest a larger amount of working capital. If the OCF is high, the cash needed for operational activities is already sufficient so the company does not need more investment on working capital.

H2 = The higher the OCF, the lower the amount needed for working capital.

Big companies usually have more involvement in the money markets, so they do not need a large amount of working capital. These companies also have the means to get a discount on their inventory purchases and a longer credit term, so they have less need for working capital.

H3 = The bigger the size of the firm, the less need for working capital.

The demand for a company's inventory from potential customers will increase their revenue, which will indirectly increase their expected profit. These continuous demands from potential customers will make the company's profitability higher. Companies with high profitability have bargaining power over their suppliers and customers, which will lessen their need for working capital.

H4 = The higher the company's profitability, the less need for working capital.

If the company has a high level of leverage, the investment needed for working capital will be less. Working capital can be funded by two means: internal and external. If the investor who funds the company is from an external party, then the company's information will be shared with creditors and shareholders. The impact of this is that the leverage cost or the cost of financing will be higher. If this happens, the company will be able to reduce their level of working capital.

H5 = The higher the leverage level, the less need for working capital.

The growth of a company will be followed by an increasing need for working capital. A company will need more working capital if it continuously grows. If the company is stuck on the same level and does not grow, then the level of working capital needed will be less.

H6 = The higher the growth opportunities, the higher the need for working capital.

If economic conditions improve, then the working capital needed will be larger, because better economic conditions mean higher purchasing power. This purchasing power will be followed by more demand for the company's inventory (product) and in larger amounts. The result is that the company will need to produce more products, so they will need more working capital.

H7 = The higher the real GDP growth rate, the larger the amount of working capital that is needed.

The cost that a company has to pay for a loan is interest. The interest expense will be higher if the company has higher debt or loan. If the company funds its working capital by getting a loan, the investment needed for working capital will be smaller, but the interest rate will be higher.

H8 = The higher the interest rate on a loan, the less working capital is needed.

4 RESEARCH METHODOLOGY

This research examines 85 manufacturing companies that were listed in the Indonesian Stock Exchange from 2010 to 2014. The model is adapted from research conducted by Abbadi and Abbadi (2013) and Onaolapo and Kayjola (2015), as follows:

$$WCR_{it} = a + ACC_{it} + OCF_{it} + SIZE_{it} + ROA_{it} + LEV_{it} + GROWTH_{it} + GDPR_{it} + R_{it} + e_t \tag{1}$$

where:

WCR_{it} = Working capital ratio
ACC_{it} = Average cash conversion cycle
OCF_{it} = Operating cash flow
$SIZE_{it}$ = Firm size
ROA_i = Profitability
LEV_{it} = Leverage
$GROWTH_{it}$ = Growth opportunities
$GDPR_{it}$ = Real GDP growth rate
R_{it} = Interest rate on loan
Ross et al. (2015) give the formula:

$$\text{Current assets} - \text{Current liabilities} = \text{Net working capital} \tag{2}$$

Abbadi and Abbadi (2013) use a comparison of total assets and net working capital, so that the ratio of working capital is formulated as follows:

$$\text{Inventory turnover} = \frac{\text{Cost of goods sold}}{\text{Average inventory}} \tag{3}$$

$$\text{Inventory period} = \frac{365 \text{ days}}{\text{Inventory turnover}} \tag{4}$$

$$\text{Receivable turnover} = \frac{\text{Credit sales}}{\text{Average account receivable}} \tag{5}$$

$$\text{Receivable period} = \frac{365 \text{ days}}{\text{Receivables turnover}} \tag{6}$$

$$\text{Payable turnover} = \frac{\text{Cost of goods sold}}{\text{Average payable}} \tag{7}$$

$$\text{Payables period} = \frac{365 \text{ days}}{\text{Payable turnover}} \tag{8}$$

$$\text{Cash Cycle} = \text{Operating cycle} - \text{Account payable period} \tag{9}$$

Working capital is needed every day, so a period of inventory turnover, accounts receivable and accounts payable are specified in a general need per day. The proxy used in the research of Abbadi and Abbadi (2013) is also used in this study. The proxies are:

$$\mathrm{ACCC} = \frac{\mathrm{CCC}}{365} \tag{10}$$

Hill et al. (2010) in Abbadi and Abbadi (2013) give a proxy:

$$\mathrm{OCF} = \frac{\mathrm{EBIT} + \mathrm{Depreciation} - \mathrm{Taxes}}{\mathrm{Total\ Assets}} \tag{11}$$

Onaolapo and Kayjola (2015) as follows:

$$\text{Firm size} = \text{Logarithm of Sales} \tag{12}$$

Abbadi and Abbadi (2013) also give a proxy to measure profitability as follows:

$$\mathrm{ROA} = \frac{\text{Net Income}}{\text{Total Assets}} \tag{13}$$

Abbadi and Abbadi (2013) as follows:

$$\text{Leverage} = \frac{\text{Short Term Loans} + \text{Long Term Loans}}{\text{Total Assets}} \tag{14}$$

Onaolapo and Kayjola (2015) give a proxy measurement of the growth opportunities of the company as follows:

$$\text{Growth opportunities} = \text{Change in the natural log of sales} \tag{15}$$

Onaolapo and Kayjola (2015) give a proxy measurement of the real GDP growth rate of the company as follows:

$$\text{Real GDP growth rate} = \text{Change in the natural log of GDP} \tag{16}$$

Interest rates used in this research are taken from the Bank Indonesia website.

5 RESULTS AND ANALYSIS

Based on the above data, there is a multicollinearity for the interest rate on loans and firm size variables, so one of those variables needs to be dropped. The variable chosen to be dropped is the interest rate on loans, because this has the highest multicollinearity level. This research also has a problem with heteroscedasticity. To correct this problem, the model uses Generalized Least Square. The Chow test, Breusch-Godfrey test, Hausman test and classic assumption tests have been applied to the model (see the results in Table 1).

The multiple linear regression model is as follows:

$$\begin{aligned} \mathrm{WCR}_i = {} & 0{,}1050036 + 0{,}4557108CCC - 0{,}0128045OCF \\ & + 0{,}0017265SIZE + 0{,}5676952ROA - 0{,}3321341LEV \\ & + 0{,}2415028GROWTH + 0{,}102873GDPR + e_i. \end{aligned} \tag{17}$$

The relationship of the CCC to working capital is in accordance with the previous research of Abbadi and Abbadi (2013). The financial manager, as the one who plans the related

Table 1. Results of regression model.

Variable	Coefficient	Positive/Negative	Prob.
CCC	0.4557108	Positive	0.000***
OCF	–0.0128045	Negative	0.913
SIZE	0.0017265	Positive	0.911
ROA	0.5676952	Positive	0.000***
LEV	–0.3321341	Negative	0.000***
GROWTH	0.2415028	Positive	0.629
GDPR	0.102873	Positive	0.400

***significant at 1% level.

financial activities, should follow the flow of the working capital closely, as suggested by Halim and Sarwoko (2013). They should do this so that the working capital can be efficiently used to finance the company's activities.

Damodaran (2001) in Karina (2012) states that companies who generate cash from operations may use these funds for various motives. One motive is for working capital needs. However, sometimes companies hold a fund that has been created for other reasons, such as for future profitable projects, providing funds to anticipate recession, etc. Thus, companies that have particular funds can split the funds for working capital and for other investments that provide a more favourable return.

Firm size has no significant effects on working capital. Ross et al. (2015) state that the policies for giving credit are based on the Five Cs of credit, consisting of a desire to pay credit (credit), the ability to pay credit (capacity), financial capability (capital), corporate guarantees (collateral) and the company's business cycle (condition). If a small company has the same potential consumers as large enterprises based on the 5Cs criteria, then the small companies will have the opportunity to shorten the billing period of receivables.

Peel and Wilson (1996) in Karina (2012) state that the big companies with great liquidity can pay their customers in a timely manner. Damodaran (2001) in Karina (2012) also explains that the volume of inventory can be determined by product availability, cost and the production lines of the company. Typically, large companies have more production lines, which will have an effect on the availability of inventory in bulk. If this continues, then the big companies would have the potential to acquire working capital in large amounts.

Chiou et al. (2006) in Karina (2012) find that mature companies will most likely not require a large amount of working capital. The company that needs a large amount of working capital is a growing company. Thus, it can also be said that company size has no effects on working capital, because a mature company does not require large amounts of working capital.

This study also supports the ideas of Wu (2001) in Karina (2012),who states that ROA is a tool used to measure the performance of companies. If the company has a good performance, then the company can manage its working capital at the lowest level. In addition, Chiou et al. (2006) also reveal that companies with a good performance will act more carefully, so the value of their inventory and accounts receivable will increase. Gill et al. (2015) also state that companies that have a good performance tend to pay less attention to working capital management; thus, working capital needs to be increased.

This study supports the research by Abbadi and Abbadi (2013), which states that leverage has a negative effect on working capital. This result suggests that an increase in debt would cause a decrease in working capital requirements. In addition, this study also supports the result of Mansoori and Muhammad (2012), who explain that a company needs to be aware of keeping its working capital balanced. Borrowing costs are expensive, so a company needs to reduce its working capital to the minimum level.

According to Chiou et al. (2006) in Karina (2012), companies that use funds that are borrowed from other parties are experiencing a recession period. That means that, as the manufacturing firms in Indonesia used in this study are in a recession period, so those companies need to borrow funds from other parties, which will be followed by a reduction in their working capital requirements.

Different types of companies will have a different response to growth. According to Bodie et al. (2009) in Karina (2012), each company has a different strategy towards investing. The growth strategy is based on three types of companies, which are risk averse, risk neutral and risk lover. Each company will respond to a decrease or an increase in the company's growth in different ways. For example, a risk lover company will continue to develop its business, although it is experiencing a decline in growth. In such conditions, the company will require a large amount of working capital. In contrast with the previous example is a company in the risk averse category; if this type of company experiences decreased growth it will do nothing and live according to its existing circumstances. If so, the company will not require a large amount of working capital.

Both Abbadi and Abbadi (2013) and Onaolapo and Kayjola (2015) state that the real GDP growth rate has no effects on working capital.

6 CONCLUSION

This research tries to analyse the variables that determine the amount of working capital that manufacturing companies hold. The results show that the cash conversion cycle has positive effects on working capital, and that operating cash flow has no effects on working capital. Meanwhile, firm size has no effects on working capital, and profitability has positive effects on working capital. In addition, leverage has negative effects on working capital, and growth opportunities have no effects on working capital. Finally, it is found that the real GDP growth rate has no effects on working capital.

7 RECOMMENDATION

Recommendations for management are as follow: the cash conversion cycle must be maintained at the lowest level, while management should have a policy on the type and the term of payment that keeps the balance at a minimum. Management should also control their operating cash flow. Profitability is an important thing and management should keep the minimum possible level of working capital. In addition, management should decide the source of funds for the working capital, and management has two choices for the source of funding, internal and external. The costs of financing should be kept to a minimum. Growth is one of the company's goals, but management should consider the risk of investment in the expansion of their business. Finally, management should not only consider internal factors but also external factors, such as the real GDP growth rate.

Some suggestions for future research are a separation of cash sales and credit sales, considering other macroeconomic factors, such as inflation, and extending the period of observation.

REFERENCES

Abbadi, S. M. & Abbadi, R. T. (2013). The determinants of working capital requirements in Palestinian industrial corporations. *International Journal of Economics and Finance, 5*(1), 65–75. doi:10.5539/ijef.v5n1p65

Adekunle. (2015). What are the determinants of working capital requirements of Nigerian firms? *Research Journal Finance and Accounting, 6*(6), 118–127.

Bodie, Z., Kane, A., & Marcus, A. J. (2009). *Investments*. New York: The McGraw-Hill.

Chiou, J., Cheng, L., & Wu, H. W. (2006). The Determinants of Working Capital Management. *Journal of American Academy of Business*, 10(1), 149–155.

Damodaran, A. (2001). Corporate Finance: Theory and Practice. 2nd ed. United State of America: Jhon Wilet & Sons.

Gill, A., Biger, N. & Obradovich, J. (2015). The impact of independent directors on the cash conversion cycle of American manufacturing firms. (2011): https://www.google.co.id/url?sa=t&rct=j&q=&esrc=s&source=web&cd=3&cad=rja&uact=8&ved=0CDEQFjACahUKEwjVuduP453JAhUBH5QKHe4HBtc&url=http%3A%2F%2Fdigitalcommons.liberty.edu%2Fcgi%2Fviewcontent.cgi%3Farticle%3D1031%26context%3Dbusl_fac_pubs&usg=AFQjCNHz3k59Uw11Dyt9ZbjMnh47cF3Orw (Accessed Sept 21, 2015).

Halim, A. & Sarwoko. (2013). *Corporate Finance (Principles of Investnment) Book 1: Management and Analysis of Asset 2nd Ed.* Yogyakarta: BPFE-Yogyakarta.

Hanafi, M. M. (2014). *Corporate Finance 1st Ed.* Yogyakarta: PFE-Yogyakarta.

Hill, M. D., Kelly, G. W. & Highfield, M. J. (2010). Net operating working capital behavior: A first look. *Financial Management, 39*(2), 783–805. doi:10.1111/j.1755-053X.2010.01092.x.

Karina, A. D. (2012). The effect of firm characteristic on the working capital requirements. https://www.google.co.id/url?sa=t&rct=j&q=&esrc=s&source=web&cd=1&cad=rja&uact=8&ved=0ahUKEwjym-rMvf3JAhUSC44KHfQ6DEkQFggZMAA&url=http%3A%2F%2Flib.ui.ac.id%2Ffile%3Ffile%3Ddigital%2F20298570-T29947-Pengaruh%2520karakteristik.pdf&usg=AFQjCNFo8Z14RQLbiiC6Eammw7U91b7y2Q&bvm=bv.110151844,d.c2E (*Accessed on Dec 21,* 2015).

Kwenda, F. & Holden, M. (2014). Determinants of working capital investment in South Africa : Evidence from selected JSE—listed firms. *Journal of Economics and Behavioral Studies, 6*(7), 569–80.

Mansoori, E. & Muhammad, D. J. (2012). Determinants of working capital management: Case of Singapore firms. *Research Journal of Finance and Accounting,3*(11), 15–24.

Onaolapo, A. A. & Kayjola, S. O. (2015). What are the determinants of working capital requirements of Nigerian firms? *Research Journal of Finance and Accounting, 6*(6), 118–28.

Peel, M., & Wilson, N. (1996). Working Capital and Financial Management Practices in the Small Firm Sector. *International Small Business Journal* Vol. 14 No. 2, 52–68.

Ross, S. A., Westerfield, R. W. & Jaffe, J. (2015). *Corporate finance Asia global edition.* New York: McGraw-Hill Education.

Sudarmadi. (2013). The awakening after long drought. http://www.indonesian-aerospace.com/view.php?m=news&t=news-detil&id=142 (*Accessed on April 1,* 2015).

Competition and Cooperation in Economics and Business – Gani et al. (Eds)
© 2018 Taylor & Francis Group, London, ISBN 978-1-138-62666-9

Audit opinion, internal control system, bureaucracy reform, political background and the level of corruption in government institutions

D.P.A. Rahayuningtyas & F. Yulia
Department of Accounting, Faculty of Economics and Business, Universitas Indonesia, Depok, Indonesia

ABSTRACT: This study empirically examines the impact of audit opinions, internal control system weaknesses, bureaucracy reform and the political background of ministries and government institution leaders on the level of corruption in Indonesia. This study, involving a sample of 86 ministries in 2011 and 2012, was processed by categorical and non-parametric statistical methods, and revealed that weaknesses of the internal control system and the political background of the leaders were factors that may affect the probability of corruption, while audit opinions and bureaucracy reform were proved to have no significant impact on the probability of corruption. Therefore, the Indonesian government must increase its control in those ministries that have leaders with a political background and must also make improvements to the internal control system of government agencies.

1 INTRODUCTION

Indonesia's Corruption Eradication Commission (*Komisi Pemberantasan Korupsi or KPK*) (CNN Indonesia, 2015) reported that corruption is most prevalent in ministries and government institutions, commonly referred to as "K/L" (*Kementerian dan Lembaga*). Corruption cases in K/Ls significantly increased from 2004 to 2015, reaching a peak in 2013 with a total of 416 cases.

An examination of government financial statements that only addresses limited administrative and compliance procedures leads to corruption in the public sector. Often highlighted are the small irregularities, while the systematic abuse that causes material losses remains undetected (Dye, 2007). The Indonesian government has enacted the Law Number 17 of 2003, concerning public finance management. The quality of financial reporting accountability is interpreted in the opinions of the Supreme Audit Institution (*Badan Pemeriksa Keuangan or BPK*). The results are expected to have a strong influence on mitigating errors in the recording of financial data and deviant actions, eventually mitigating readers' misinterpretation of financial reports (Ismiyati & Widiyanto, 2015).

One way to mitigate the level of corruption is through the mechanism of bureaucracy reform (Reformasi Birokrasi or RB). RB is expected to improve public services and information quality by reinforcing incentives and disincentives, which in turn will inhibit corruption (Klitgaard et al., 2000). Government institutions that implement RB will receive incentives in the form of benefits for the employees who perform their job well.

Factors such as no rotation for leaders with political backgrounds in government institutions, poor quality of public services and unfinished processes of democratic transitions have caused a high level of corruption (Zafarullah & Siddiquee, 2001). Democratic processes requires a high cost, which eventually leads to the connection between politicians and business people who provide funding for their political campaigns (Neu et al., 2012). A KPK survey in 2013 revealed that 71% of Indonesian people acknowledged that money politics generally occurs during elections., Money politic occur when voters are given some money to vote the candidates who participate in election. Approximately 92% of the respondents stated that the relationship between government officials and politicians in a corruption case is commonplace (Tempo, 2014).

Heriningsih and Marita (2013), Milal (2013) and Liu and Lin (2012) studied the effects of government institutions' accountability on corruption. But, there has been no research on the impact of bureaucracy reform reflected by the provision of performance benefits for government officials on corruption in K/Ls. Research on benefits from the implementation of RB becomes important because the income of government officials could influence the level of corruption (Dong & Torgler, 2013; Huang & Snell, 2003). De Graaf (2007) states that, besides the accountability and economic factors, corruption cannot be separated from political factors. This study aims to fill the research gap by further examining the effect of factors that have been previously studied and also by examining the effect of RB and the political background of K/L leaders on the level of corruption in Indonesia.

This study makes several contributions to the literature. Theoretically, it provides empirical evidence on the effect of audit opinion, internal control systems, bureaucracy reform and leaders' political background on corruption in government institutions. Practically, it provides information regarding factors that must be highly considered in order to tackle corruption in government institutions.

2 THEORY AND HYPOTHESIS DEVELOPMENT

2.1 *Agency theory*

Hopkin (1997) states that the community (principal) entrusts authority to government officials (agents). The high costs of monitoring all government officials gives freedom to the agents in carrying out their functions, causing different levels (asymmetry) of information between the agents and principal, which raises a moral hazard problem that may lead to several deviant actions by agents who take as much advantage as possible, which will eventually sacrifice public interests.

Incentives can motivate people and control their performance (Verbeeten, 2008). Theoretically, agents would only be able to maximise their interests based on the incentives they receive. In practice, the implementation of incentives in the public sector brings about a complex impact, due to diversities of organisational objectives that weaken the impact of incentives. Difficulties in measuring the level of performance in every element of the public sector and the lack of professional management also weaken that impact.

2.2 *Accountability of financial statements*

Greiling and Spraul (2010) found a close relationship between accountability and agency theory. The role of accountability in reducing agency conflict focuses on improving supervision, reducing information asymmetry and increasing transparency. Besides measuring the accountability level of K/L financial reports, the BPK also provides a list of the findings from their examinations. These results are reported to the authorities. The finding become a whistle-blower in case of divergence (Dye, 2007). Four types of opinion are given by the BPK: Qualified (WTP), Unqualified (WDP), Disclaimer (TMP) and Adverse (TW) (BPK, 2014).

2.3 *Internal control*

K/L reports should be reliable, and this is a strong reason for government institutions to implement internal control systems (SPI) in compliance with laws and regulations. Good SPI is significantly determined by the top management of the government institutions that are responsible for creating a conducive framework in order to achieve an effective internal control (Harbord, 1994).

2.3.1 *Leadership*

Many individuals might become leaders, but it is not easy to find people who have the capabilities to act as a leader (Idakwoji, 2010). There are similarities and differences between managers in the public and private sectors. Managers in the private sector need a longer time for problem solving. Government officials or authorities could shorten the decision-making process, but the monitoring system is weaker, which eventually increases the level of corruption (Rainey & Bozeman, 2000).

The character and behaviour of leaders affects the moral atmosphere of the organisation. Leadership is closely associated with politics, especially in democratic countries, where in some cases the leaders of the countries are responsible for the acts of corruption (Charron, 2011).

2.3.2 *Corruption*

Corruption is not always a systematic and planned act, but it may include various acts such as bribery in the licensing process or offering gifts to government officials in order to expedite the bureaucratic process (Widjajabrata & Zacchea, 2004).

Several causes of corruption are the government monopoly system in the procurement of goods and services (Klitgaard et al., 2000); the limitation of goods and services available, lack of transparency, prolonged bureaucracy and lack of law enforcement (Barthwal, 2003); the greediness of government officials, asymmetry information, the excessive lifestyle of corruptors and minor punishment for corruptors (Bologna, 1993).

2.4 *Audit opinion and corruption*

Audit opinions can prevent, detect and investigate white collar criminal activities (Hopwood, 2009 in Neu et al., 2012). The BPK examination results could drive K/Ls to increase the reliability of their financial reports in order to reflect the real financial situation in K/Ls. Accountability promotes transparency and a level of compliance with accounting rules, so acts of fraud and irregularities in K/Ls can be decreased.

The BPK gives a better opinion if a K/Ls' financial reports are presented in accordance with the government financial reporting standards. Auditors in some corruption cases are able to detect corruption, but they do not have the authority to investigate further (Kayrak, 2008). Heriningsih and Marita (2013) indicate that the BPK's audit opinions cannot significantly affect the level of corruption, while Masyitoh et al. (2014) found that audit opinions play an important role in mitigating the level of corruption.

The different inferences from previous research cannot give a definite conclusion regarding the relationship between audit opinions and corruption; therefore, this study's first hypothesis is:

H1: The probability of corruption during the following year for K/Ls that obtain Qualified (WTP) or Unqualified (WDP) opinions is different from the probability of corruption for K/Ls that do not obtain those opinions (see Figure 1).

2.5 *Weaknesses of internal control systems and the level of corruption*

A weak internal control system also contributes to the lack of transparency and accountability in K/Ls (Reginato et al., 2011). The BPK audit opinions and findings of irregularities could be indicators of corruption. The more irregularities that are reported, the weaker the internal control system in K/Ls. The number of auditor's findings could significantly lower the corruption level in subsequent years (Liu & Lin, 2012; Ekasani, 2016).

The second hypothesis proposed is:

H2: The probability of corruption in K/Ls that have a weak internal control system, based on BPK findings in one year, tends to increase in the following year (see Figure 1).

2.6 *Bureaucracy reform and corruption*

China's government employees are poorly paid in contrast to those working in the private sector, while in Singapore, corruption levels are very low because the remuneration received by a person is equivalent to his/her contribution (Huang & Snell, 2003). Poor management in the public sector administrative system and low salaries are triggers for corruption in Iran (Ahmadi & Houmani, 2011). Government officials' income can significantly increase the level of corruption (Dong & Torgler, 2013).

Incentives provided for K/Ls that can reach the RB targets are in the form of a relatively large amount of performance benefits. The use of incentives has a positive impact on the quantity of, but not the quality of, K/L performance, whereas clear and measurable targets have a significant impact on both (Verbeeten, 2008).

Klitgaard et al. (2000) found a close relationship between incentives and corruption. An individual carries out an act of corruption because the risks and sanctions are low, while the increase to his welfare could be high. Bureaucracy reform should be carried out by providing performance benefits as an incentive for government employees to increase their total remuneration and to reduce corruption.

The third hypothesis proposed is:

H3: The probability of corruption in the following year tends to decrease in K/Ls whose employees receive more income in the form of performance benefits, due to their success in implementing bureaucracy reform (RB) (see Figure 1).

2.7 *Political background of K/L leaders and the level of corruption*

Corruption is the abuse of government positions in the fulfilment of a person's or a group's interests (Ahmadi & Houmani, 2011). Leaders of government institutions who are initially from political parties might commit acts of corruption in order to increase their political interests (Zafarullah & Siddiquee, 2001). Even in developed countries, such as Japan, many political leaders in the cabinet have committed acts of corruption (Choi, 2007).

Fitzsimons (2009) states that the political situation might significantly affect the performance of government institutions. Politics in K/Ls often occurs during a leadership election. The composition of K/L leaders in Indonesia is dominated by certain parties, either professional or political, that have supported the President in order to increase government efficiency.

Leaders elected through a general election mechanism are proved to have lower levels of performance compared to those from a professional background who are directly appointed (Garmann, 2015). The weak financial performance of the elected leaders might increase the level of local government spending, while leaders from professional backgrounds do things the other way around. This situation happens because the elected leaders are from a political party; their duties are supervised by members of the local council who are also from political parties, either from the same party as the leaders or from others, resulting in weaker supervision from the council members.

Hopkin (1997) claims that the decline in political quality has forced political actors to accept the salary of public officials, pushing them even harder to do anything in order to increase their own welfare and interests, which eventually makes corruption inevitable.

The last hypothesis proposed is:

H4: The probability of corruption in the K/Ls tends to increase in the following year if their leaders are from political parties (see Figure 1).

3 RESEARCH METHODOLOGY

3.1 *Population and samples*

The population used in this study was the overall K/Ls in Indonesia from 2011 to 2014. However, there were some K/Ls that had not been established and some audit reports were still incomplete in the sampling years; therefore, we ended up with 250 K/Ls selected for the three year sample period.

3.2 *Types and sources of data*

This study used secondary data, i.e. corruption cases determined by the court regarding K/Ls that were reported in the KPK Performance Reports and published on the KPK official website. The sample period was from 2012 until 2014. Audit opinions were gained from the Executive Summary, which was published by the KPK in 2011 and 2012. Data regarding bureaucracy reforms were gained from the official website of the Ministry of Empowerment of State Apparatus and Bureaucracy Reform (Kemenpan RB), including the publications of all K/Ls that have gained performance benefits under the RB Acts of 2011, 2012, and 2013.

Information regarding the weaknesses of K/L internal control systems (SPI) was sourced from the official website of the BPK, under the Internal Control System Report section. The data included the findings of BPK auditors from years 2011 to 2013. The political backgrounds of K/L leaders were obtained from the official websites of the K/Ls or other sources from 2011 to 2013. The information that was used concerned whether or not the leaders were from a political party.

3.3 *Measurement of variables*

1. Dependent Variable

This study aims to determine the probability level of corruption in K/Ls and the measurement selected was based on the occurrence of corruption in K/Ls. The corruption data retrieved are corruption cases that were already decided by the court from 2012 to 2014. It takes the value of 1 if corruption did occur in K/Ls and 0 if otherwise. For dummy variables, the measurement scale is nominal. The use of a dichotomous dependent variable is the basis for selecting the regression method, which is the Logistics or Logit Method (Salih, 2013).

2. Independent Variables

– BPK Audit Opinions

Ekasani (2016) used a dummy variable that takes the value of 1 if the BPK gives qualified (WTP) or unqualified (WDP) opinions, and the value of 0 if the BPK gives adverse (TW) or disclaimer (TMP) opinions. The objective of this research is to examine the categorical data, so the data should be transformed into nominal (Agung, 2002). This study used a dummy variable measurement with the value of 1 if the BPK gave WTP or WDP opinions, and the value of 0 if the BPK gave TW or TMP opinions.

– Bureaucracy Reforms

Kotera et al. (2012) and Dong and Torgler (2013) used per capita income as well as the salaries of government employees as a proxy for income. This study aims to relate income with good governance as a reflection of the implementation of the bureaucracy reform (RB), so we used ordinal scale measurement that is 0 if the K/L did not receive performance benefits because they had not met the requirements of RB, and 1 if the K/L received the benefits.

– Internal Control System Weaknesses

Liu and Lin (2012) used the weaknesses of internal control systems (SPI) as a proxy in the form of the number of government auditors' findings. Ekasani (2016) used the number of findings as the proxy. This study aims to examine the probability of SPI weaknesses in audit findings by using categorical data, so it used the proxy in the form of categorical number 1 if there were findings and 0 if there were not any (Agung, 2002).

– Political Background of K/L Leaders

This study used a dummy variable; it takes the value of 1 if the leaders came from political parties and 0 if they did not. This measurement is based on the research by Garmann (2015) regarding the election of local government leaders.

4 RESEARCH FRAMEWORK

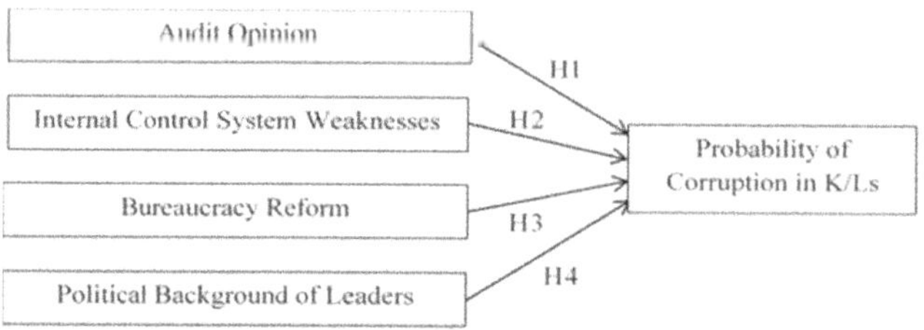

Figure 1. The mechanism of audit opinion, internal control system (SPI) weaknesses, bureaucracy reform (RB) and political background of K/Ls' leaders on the probability of corruption in K/Ls.

5 RESEARCH MODELS

The model applied in this study is:

$$logitkor_{it+1} = ln\left[\frac{Pit+1}{1-Pit+1}\right] = \beta_o + \beta_1 Opini_t + \beta_2 SPI\ Weak_t + \beta_3 RB_t + \beta_4 Political_t + \varepsilon$$

Variable	Definition
$\text{logitkor}_{\text{it+1}}$	Use dummy variable, the value of 1 if corruption did occur in K/Ls and 0 if otherwise at year $t + 1$
β_0	Constanta
$\text{WTP\_WTPDPP}_{\text{it}}$	Audit opinion at year t
$\text{SPI Weak}_{\text{t}}$	categorical number 1 if there were findings and 0 if there were not in t
RB_{it}	ordinal scale measurement that is 0 if the K/L did not receive performance benefits because they had not met the requirements of RB, and 1 if the K/L received the benefits in year t
$\text{Political}_{\text{it}}$	dummy variable; it takes the value of 1 if the leaders came from political parties and 0 if they did not at year t
ε	Error

6 RESULTS

6.1 *Analysis of descriptive statistics*

Table 1. Descriptive statistics.

Variables	Obs	Mean
Logitkor	250	0.192
Opini	250	0.752
SPI Weak	250	0.275
RB	250	0.540
Political	250	0.428

The average value (see Table 1) of dependent variable logitkor is 0.192, which means that in three years there were 19.2% of corruption cases in all K/Ls, or approximately 48 out of

250 samples. Independent variable opini has an average value of 0.752, which means that most of the K/Ls (75.2%) obtained qualified (WTP) or unqualified (WDP) opinions. SPI weak has an average value of 0.275, meaning that internal control system weaknesses were found only 27.5% of all K/Ls. The RB variable has an average value of 0.540; this number indicates that 54% of all K/Ls gained performance benefits for the achievement of the RB in their institutions. Political has an average value of 0.428, which means that 42.8% of leaders of the K/Ls came from political parties in Indonesia.

6.2 *Analysis of regression results*

This study used a non-parametric test because not all of the data could meet the assumption of normality. The abnormal data reflected in the results of skewness and kurtosis with a probability value > Chi2 of 0, smaller than $\alpha = 0.05$. In order to determine the level of association between categorical variables, this study used a Kendall Tau test (Agung, 2002). The correlation coefficient between variables is less than 0.8, which can be interpreted as showing that there was no multicollinearity problem. There was a positive, significant association between the internal control system weaknesses findings and the probability of corruption. There was a negative, significant relationship between audit opinion and corruption, which can be interpreted as the weaker the internal control system (SPI) in K/Ls, the bigger the probability of corruption, while the more that K/Ls can obtain qualified (WTP) or unqualified (WDP) opinions, the smaller the probability of corruption will be. Surprisingly, there was no significant relationship between the implementation of RB and the probability of corruption, while it is no surprise that there was a significant positive association between K/L leaders that came from political parties and the likelihood of corruption. Table 2 shows the results of the test using Kendall's Tau, while Table 3 shows the results of regression tests using Stata 12.

Table 2. The results from Kendall's Tau tests.

	Logitkor	Opini	SPI weak	RB	Political
Logitkor	1				
Opini	–0.0493	1			
	0.4376				
SPI weak	0.1988*	–0.2049*	1		
	0.0017	0.0012			
RB	0.0016	0.1018	–0.0765	1	
	0.9805	0.1083	0.2279		
Political	0.2146*	–0.1210	0.0446	–0.0938	1
	0.0007	0.0564	0.4820	0.1393	

Table 3. The regression results.

Dependent Variable: logitkor

Variable	Prediction signs	Odds ratio	Std. error	z-statistic	Prob > z
Opini	+	1.0746	0.4153	0.19	0.852
SPI weak	+	2.8793	1.0177	2.99	0.003
RB	+	1.2275	0.4198	0.60	0.549
Political	+	3.1054	1.0747	3.27	0.001
C	+	0.0788	0.0386	–5.19	0.000

LR $chi^2(4) = 20.56$; Log likelihood = –111.99738; Prob > $chi^2 = 0.0004$; Pseudo $R^2 = 0.0841$.

The findings regarding the weaknesses of internal control systems in K/Ls and the political background of K/L leaders show that they might significantly affect the probability of corruption in K/Ls. This conclusion is derived from the probability values, which are smaller than $\alpha = 0.05$. If there were findings regarding SPI weaknesses, the number of corruption cases in the next year increased by 2.88% and the probability value (0.003) is less than 0.05. If K/L leaders came from political parties, the probability of corruption in the following year increased by 3.11%. Audit opinion from the BPK and the implementation of bureaucracy reform (RB) in K/Ls did not significantly affect the corruption level in K/Ls in the following year. The coefficient of the dependent variable (0.08) indicates that if there were no findings of SPI weaknesses, K/Ls received qualified and unqualified audit opinions. If K/L leaders were not from political parties, the probability of corruption in the next year was 0.08%. The value of Pseudo R2 indicates that the independent variables are able to explain 8.41% of the variability of the dependent variable, while the rest is explained by other factors.

7 CONCLUSION

Audit opinions from the BPK (Supreme Audit Institution) cannot significantly reduce the level of corruption in ministries and government institutions (K/Ls). This suggests that the opinions of the BPK cannot be a signal for the presence of corruption in K/Ls.

K/Ls' internal control system weaknesses significantly affect the probability of corruption in K/Ls. This can be interpreted as showing that the findings might function as a signal for the presence of corruption in K/Ls.

Bureaucracy reform in K/Ls cannot make a significant impact on the reduction of corruption in K/Ls. The reform reflected by the performance benefits received by the employees of K/Ls that already implement RB is not yet able to realise its objective to fight corruption in government institutions.

The political background of K/L leaders negatively affects the probability of corruption in K/Ls, which suggests that the probability of corruption in a K/L increases if their leaders come from political parties.

8 RECOMMENDATIONS

This study used non-parametric and logistic/categorical methods to show significant factors affecting the level of corruption in ministries and government institutions. Future research could further examine the theme of corruption by using other factors and also strengthen the results of this study by using numerical data.

Bureaucracy reform policy in Indonesia might need further reassessment. Even though the policy has already been in place for almost five years, it has not made a significant impact on the reduction of corruption, especially in K/Ls. Government officials should participate in the efforts to oversee the institutions and prohibit any acts of corruption, as well as to enhance the role of bureaucracy reform implementation, not merely aiming to obtain bigger incentives but also actively participating in decreasing the level of corruption.

This study shows that the increasing likelihood of corruption in government institutions occurs partly because of many findings regarding the weaknesses of internal control systems in K/Ls and also because some K/L leaders come from a political background.

REFERENCES

Agung, I.G.N. (2002). *Statistika Analisis Hubuangan Kausal Berdasarkan Data Kategorik*. Jakarta: Rajawali Pers.

Ahmadi, F. & Houmani, G. (2011). The role of good governance in fighting against corruption and financial crimes. *Interdisciplinary Journal of Contemporary Research in Business*.

Barthwal, C.P. (2003). E-governance for good governance. *The Indian Journal of Political Science, 4*(3/4), 285–308.
Bologna, J. (1993). *Handbook of corporate fraud.* Boston: Butterworth Heinemann.
BPK. (2012). *Laporan Hasil Pemeriksaan BPK RI atas Laporan Keuangan Pemerintah Pusat Tahun 2011 – Laporan Hasil Pemeriksaan atas Sistem Pengendalian Intern.* BPK. Jakarta.
BPK.. (2013). *Laporan Hasil Pemeriksaan BPK RI atas Laporan Keuangan Pemerintah Pusat Tahun 2012 – Laporan Hasil Pemeriksaan atas Sistem Pengendalian Intern.* BPK. Jakarta
BPK. (2014). *Laporan Hasil Pemeriksaan BPK RI atas Laporan Keuangan Pemerintah Pusat Tahun 2013 – Laporan Hasil Pemeriksaan atas Sistem Pengendalian Intern.* BPK. Jakarta
BPK. (2014). *Laporan Hasil Pemeriksaan BPK RI atas Laporan Keuangan Pemerintah Pusat Tahun 2013 – Ringkasan Eksekutif.* BPK. Jakarta.
Charron, N. (2011). Party systems, electoral systems and constraints on corruption. *Electorial Studies, 30*(4), 595–606.
Choi, J.W. (2007). Governance structure and administrative corruption in Japan: An organizational network approach. *Public Administration Review, 67*(5), 930–942.
CNN Indonesia. (2015, September 9). www.cnbindonesia.com. Retrieved September 9, 2015, from CNN Indonesia: http://www.cnnindonesia.com/nasional.
De Graaf, G. (2007). Causes of corruption: Towards a contextual theory of corruption. *Public Administration Quarterly, 31*(1), 39–86. doi:10.1017/CBO9781107415324.004
Dong, B. & Torgler, B. (2013). Causes corruption: Evidence from China. *China Economic Review, 26*, 152–169. doi: 10.1016/j.chieco.2012.09.005*
Dye, K.M. (2007). Corruption and fraud detection by public sector auditors. *Edpacs, 36*(5–6), 6–15. doi:10.1080/07366980701805026
Ekasani, P.W. (2016). *Pengaruh Desentralisasi Fiskal dan Akuntabilitas Laporan Keuangan terhadap Tingkat Korupsi pada Pemerintah Daerah di Indonesia* (Master's thesis). Universitas Indonesia.
Fitzsimons, V.G. (2009). A troubled relationship: Corruption and reform of the public sector in development. *Journal of Management Development, 28*(6), 513–521. doi http://dx.doi.org/10.1108/02621710910959675
Garmann, S. (2015). Elected or appointed? How the nomination scheme of the city manager influences the effects of government fragmentation. *Journal of Urban Economics, 86*, 26–42. doi:10.1016/j.jue.2014.12.004
Greiling, D. & Spraul, K. (2010). Accountability and the challenges of information disclosure. *Public Administration Quarterly, 34*(3), 338–377.
Harbord, J. (1994). Competition and effective internal control in local authorities. *Management Accounting, 1994.*
Heriningsih, S. & Merita. (2013). Pengaruh Opini Audit dan Kinerja Keuangan Pemerintah Daerah terhadap Tingkat Korupsi Pemerintah Daerah. *Jurnal Manajemen, Akuntansi dan Ekonomi Pembangunan—Volume 11.*
Hopkin, J. (1997). *Political Parties, Political Corruption, and the Economic Theory of Democracy. Crime, Law and Social Change. Kluwer Academic Publisher.*
Huang, L.J. & Snell, R.S. (2003). Turnaround, corruption and mediocrity: Leadership and governance in three state owned enterprises in mainland China. *Journal of Business Ethics, 43*, 111–24. doi:10.1023/A:1022919418838
Idakwoji, S P. (2010). Leadership, corruption and development/leadership, corruption et development. *Canadian Social Science, 6*(6), 173–179. doi: 10.3968/j.css.1923669720100606.020
Ismiyati, & Widiyanto, A. (2015). *Potret Akuntabilitas Keuangan Daerah di Jawa Tengah.* http://www.bpkp.go.id.
Kayrak, M. (2008). Evolving challenges for supreme audit institutions in struggling with corruption. *Journal of Financial Crime, 15*(1), 60–70. doi:10.1108/13590790810841707
Kemenpan. (2015). Pokok-Pokok Kebijakan Reformasi Birokrasi. *Kemenpan.* http://www.menpan.go.id.
Klitgaard, R., Maclean-Abaroa, R. & Parris, H. L. (2000). *Penuntun Pemberantasan Korupsi dalam Pemerintahan.* Jakarta: Yayasan Obor Indonesia.
Kotera, G., Okada, K. & Samreth, S. (2012). Government size, democracy, and corruption: An empirical investigation. *Economic Modelling, 29*(6), 2340–48. doi:10.1016/j.econmod.2012.06.022
Liu, J. & Lin, B. (2012). Government auditing and corruption control: Evidence from China's provincial panel data. *China Journal of Accounting Research, 5*(2), 163–86. doi:10.1016/j.cjar.2012.01.002
Masyitoh, R.D., Wardhani, R. & Setyaningrum, D. (2014). *Pengaruh Opini, Temuan Audit, dan Tindak Lanjut Audit terhadap Persepsi Korupsi pada Pemerintah Daerah Tingkat II di Indonesia* (Masters's Thesis). Universitas Indonesia.

Milal, Z.A. (2013). Makna Opini Audit WTP Bagi Kementerian/Lembaga (Studi Kasus pada Kementerian Sosial). *Jurnal Ilmiah Mahasiswa FEB. Vol 2No. 2*, Universitas Brawijaya.

Neu, D., Everett, J., Rahaman, A.S. & Martinez, D. (2012). Accounting and network of corruption. *Accounting, Organizations and Society.*

Rainey, H.G. & Bozeman, B. (2000). Comparing public and private organizations: Empirical research and the power of the A Priori. *Journal of Public Administration Research & Theory, 10*(2), 447.

Reginato, E., Paglietti, P. & Fadda, I. (2011). Formal or substantial innovation: Enquiring the internal control system reform in the Italian local government. *International Journal of Business and Management, 6*(6), 3–16. doi:10.5539/ijbm.v6n6p3

Salih, A.R.M. (2013). The determinants of economic corruption: A probabilistic approach. *Advances in Management & Applied Economics, 3*(3), 155–169.

Tempo. (2014). *tempo.co*. Accessed December 04, 2014 http://www.tempo.com

Verbeeten, F.H.M. (2008). Performance management practices in public sector organizations. *Accounting, Auditing & Accountability Journal, 21*(3), 427–54. doi:10.1108/09513570810863996

Widjajabrata, S. & Zacchea, N.M. (2004). International corruption: The Republic of Indonesia is strengthening the ability of its auditors to battle corruption. *The Journal of Government Financial Management, 53*(3), 34–43. http://qut.summon.serialssolutions.com/2.0.0/link/0/eLvHCXMwY2BQSDNIA4IkC2PjFGBlk2iUDJqMAbY00oDVhbkxeCUhYrANqTR3E2JgSs0TZZB1cw1 × 9tCFFY3xKTk58 WaWFsCOiwUwa4ox8CaCFn_nlYA3iaUAAKptG-A.

Zafarullah, H. & Siddiquee, N.A. (2001). Dissecting public sector corruption in Bangladesh: Issues and problems of control. *Public Organization Review, 1*(4), 465.

Competition and Cooperation in Economics and Business – Gani et al. (Eds)
© 2018 Taylor & Francis Group, London, ISBN 978-1-138-62666-9

The effect of corporate governance mechanisms and managerial incentives on corporate tax avoidance in Indonesia

R.Y.F. Perdana & R. Yuniasih
Department of Accounting, Faculty of Economics and Business, Universitas Indonesia, Depok, Indonesia

ABSTRACT: The aim of the research is to examine the link between corporate governance, managerial incentives and corporate tax avoidance. Agency problems may lead managers to make different decisions regarding tax compliance. Managers may see tax avoidance as an investment, along with other similar opportunities. To see a broadened view of the level of tax avoidance, this research uses quantile regression to exhibit the link between corporate governance and tax avoidance. This research finds a positive relationship between corporate governance mechanisms and a low level of tax avoidance, instead of a negative relationship for high levels of tax avoidance. These results indicate that the governance mechanism has a stronger relationship with more extreme levels of tax avoidance. Meanwhile, this research finds a negative influence between managerial incentives and tax avoidance. The implication of this research is that tax avoidance, at both high and low levels, can be explained by corporate governance.

1 INTRODUCTION

This study examined the relationship of corporate governance and management incentives to tax management. Management reports on the company's performance in the form of regular financial statements. The manager sometimes has different interests in a firm's performance than those of the shareholders, which can cause conflicts of interest. Therefore, a system of corporate governance is needed to supervise the performance of management in carrying out their duties, in order to protect the interests of stakeholders and to create value for the company (Desai & Dharmapala, 2006).

The implementation of good corporate governance requires management to work harder, and requires the company to provide incentives for the management. Management incentives can be in the form of share compensation to the directors. Armstrong et al. (2015) and Rego and Wilson (2012) stated that management expects a higher personal gain when the company experiences an increase in the level of tax avoidance. Management will try to find loopholes in order to increase the level of compensation they receive. Furthermore, management will choose investments that have a level of risk that is higher than the level of risk that can be accepted by the shareholders.

Armstrong et al. (2015) found that there is a positive relationship between the attributes of good corporate governance mechanisms, i.e. the independence of the board of commissioners and the educational backgrounds or financial careers of the commissioners, and a low level of tax avoidance. Conversely, a negative relationship was found between the independence and educational backgrounds of commissioners and a high level of tax avoidance. In summary, the commissioners who have sufficient knowledge in financial management will try to simplify tax decisions when the tax risk is low. The opposite happens if the tax risk is high, when the commissioner will provide a disincentive to tax management (Armstrong et al., 2015). It appears that corporate governance mechanisms play an important role in management decisions when faced with a difficult tax position. Friese et al. (2006) say that the corporate governance structure affects the way that companies meet their tax obligations, so that tax planning depends on the company's corporate governance dynamics.

This study analyses the influence of corporate governance mechanisms and management incentives on the tax avoidance measures of non-financial companies listed in the Indonesia

Stock Exchange in the years 2010 to 2014. This study is an adaptation of the research conducted by Armstrong et al. (2015) and Rego and Wilson (2012), with some adjustments. In this study, the proxy variables related to management incentives are measured by both the cash and stock compensations obtained by the directors, in contrast to Armstrong et al. (2015), who use the property portfolio of directors. Scores of corporate governance uses those developed by Hermawan (2009), but only on the part of commissioner.

This research is the first research to undertake an observation of tax avoidance in Indonesia. Data is analysed using quantile regression, a method that has never previously been applied in Indonesia for tax avoidance research. The remainder of this paper includes the literature review, methodology, results and analysis, and conclusions and suggestions.

2 THEORETICAL BACKGROUND

Indonesia adopts a two-tier top level management in which the directors are responsible to the commissioner as the representative of the shareholders. This structure has several advantages, such as better control over the management and increased independence of the directors (Brandle & Noll, 2004). However, such systems can bring agency problems between the management and the owners of the company (Jensen & Meckling, 1976). Similar to other investment opportunities that have a cash flow risk, agency problems can make the manager engage in tax avoidance that does not align to the selected policy of the shareholders (Armstrong et al., 2015).

Good corporate governance restrains management from engaging in more aggressive tax avoidance behaviour (Desai & Dharmapala, 2006). The application of the principles of corporate governance led to increased monitoring from the board of commissioners, including monitoring on tax management. Companies with good governance have an internal control mechanism that prevents the transfer of interest (diversion) and supports a negative correlation between equity incentives and tax avoidance. Armstrong et al. (2015) found a positive relationship between the independence, ability and financial background of the board of commissioners and a low level of tax avoidance, but found a negative relationship with a high level of tax evasion. These results indicate that governance factors have a strong relationship with tax evasion, especially at extreme levels of tax avoidance.

Agency problems that arise from a conflict of interest between management and shareholders may trigger opportunistic actions. Therefore, in order to decrease the conflict, management compensation is needed. Compensation received by the directors can be in the form of cash or shares. Equity risk incentives support managers in their willingness to take more risks in investment and financial planning, including in tax management, in the hope of getting higher stock prices and market portfolios. Aggressive tax strategies may lead to high uncertainty, but also to higher savings in the future, so Rego and Wilson (2012) predict that the equity risk incentives positively influence the aggressiveness of tax.

3 METHODOLOGY

3.1 *Hypothesis building*

This study assumed that good corporate governance mechanisms would negatively affect tax avoidance. The mechanism, in the form of oversight by the board of commissioners, can mitigate the agency problems, assuming that management, i.e. the company's directors, are a group of risk takers and that the risk taken by the management exceeds the shareholders' risk preferences (Desai & Dharmapala, 2006). Meanwhile, assuming that tax avoidance is an effort of investment, the corporate governance mechanism can be a positive influence on tax avoidance and encourage more risk management in order to optimise the chances of getting better returns. The hypotheses developed in this study are:

H1A: The mechanism of corporate governance negatively affects tax avoidance at a high level of tax avoidance.

H1B: The mechanism of corporate governance positively affects tax avoidance at a low level of tax avoidance.

Desai and Dharmapala (2006) stated that in companies with good corporate governance, the granting of equity compensation would increase the level of tax avoidance. This statement is supported by Armstrong et al. (2015) and Minnick and Noga (2010), who state that equity incentives encourage managers to take risks in investing over the long term, so that tax evasion will be higher, and there is a tendency that managers will seek to take advantage of the results of the tax evasion. This is in line with the view expressed by Armstrong et al. (2015) that there is a risk that makes directors more opportunistic and want to take more risks in terms of tax avoidance.

H2: Management incentives positively affect tax avoidance.

3.2 *Sample selection*

This study uses data obtained from Thomson Reuters EIKON. The sample uses companies listed in the Indonesia Stock Exchange (BEI) by Jakarta Stock Industrial Classification (JASICA), consisting of nine major sectors, except the financial sector. The range of the observation period is 2010 to 2014. Companies with a negative Effective Tax Rate (ETR) and those who do not disclose the compensation to directors and commissioners separately are excluded from this study. Based on these criteria, a sample of 279 companies was obtained.

3.3 *Research methodology*

Koenker and Hallock (2001) say that the quantile regression technique is useful in order to see the difference in the behaviour of the independent variables at various levels of the distribution of the dependent variable. Furthermore, this technique provides more in-depth statistical analysis of the stochastic relationship between random variables. Here is a model built in this study.

$$\begin{aligned} TAETR_{i,t} = \beta_{0,i} &+ \beta_1 BOARDCOM_{i,t} + \beta_2 COMPDIR_{i,t} + EQUITYD \\ &+ \beta_3 PERFORMANCE_{i,t} + \beta_4 MVE_{i,t-1} + \beta_5 SIZE_{i,t} - 1 \\ &+ \beta_6 LEVERAGE_{i,t} + \beta_7 PROFITABILITY_{i,t} + \varepsilon_{i,t} \end{aligned}$$

TAETR is a 3-year average ETR industry reduced by 3-year average company. BOARDCOM is the value of corporate governance for BOC using a checklist by Hermawan (2009). CASHDIR is cash compensation given to the Board of Directors. EQUITYD is a dummy variable for the company that provides compensation to directors in the form of shares. PERFORMANCE comes from operating cash flow divided by total assets in the previous year. MVE is the natural logarithm of the market value of equity in the previous year. SIZE is the natural logarithm of total assets. LEVERAGE is debt to equity ratio. PROFITABILITY is the return on assets.

4 RESULTS AND ANALYSIS

The negative numbers shown by TAETR in Table 1 represent that the samples used have high levels of tax avoidance, which is smaller than the industry average of similar companies in the sector [1]. The selected sample is of companies that reveal their complete data and that generally have good transparency and accountability. Corporate governance is represented by the effectiveness of the BOC [2] in Indonesia has a corporate governance at the level of the leading Fair to Good.

The first hypothesis in this study looked at the relationship between corporate governance mechanisms and tax avoidance in Indonesia. According to Table 2, the BOARDCOM variable as a proxy for corporate governance has a negative correlation when using OLS. The coefficient indicates the value –0.0078173 with a p-value of 0.000. Quantile regression obtained stronger results at extreme points, namely the first and eighth quantile.

The first quantile shows a value of 0.2033, while all eight quantiles show a value of 0.2071. When compared with 0.1464 then the OLS only the second quantile shows a stronger

Table 1. Descriptive statistics.

Variables	Obs	Average	Std. Dev.	Minimum	Maximum
TAETR	279	−0.00964	0.11804	−0.1985	0.1975
BOARDCOM	279	40.65591	3.19789	34	48
CASHDIR*	279	17,999.34	17,642.51	360.22	101.408
EQUITYD	279	0.42366	0.49948	0	1
Control variables	Obs	Average	Std. Dev.	Minimum	Maximum
PERFORMANCE	279	0.17483	0.08038	−0.1533	0.951
MVE**	279	14,100,000	31,900,000	4,695,555	252,000,000
SIZE*	279	9,535,042	16,900,000	137,750.4	128,000,000
LEVERAGE	279	0.55636	0.42358	0.0627	1.3389
PROFITABILITY	279	0.07759	0.07065	−0.0172	0.2017

relationship. The first quintile shows a BOARDCOM coefficient of 0.0054711 with a p-value of 0.001. This value shows a low level of tax avoidance and that corporate governance has a significant and positive relationship. The condition is different from that of the quintile distribution of all eight. The BOARDCOM coefficient has a value of −0.0163241 with a p-value of 0.000. This shows a high level of tax avoidance, a mechanism significantly negatively related to corporate governance. Armstrong et al. (2015) state that corporate governance also affects the company's decisions on tax avoidance, even though it has a level of good or bad.

The results are consistent with the hypothesis, which found that the corporate governance mechanism had a different influence at each distribution of tax avoidance. The corporate governance mechanism positively affects the level of tax avoidance when it is at a low level, which is in line with the state of management underinvestment against tax avoidance in the absence of monitoring. As has previously been said, if tax avoidance is one of the investments, then those commissioners who have financial expertise and independence will seek to protect the interests of shareholders by encouraging management to take risks in tax avoidance, since there is still an opportunity for the management to optimise investment in accordance with the interests of shareholders.

Incentive management in this study is proxies into two variables: cash compensation (COMPDIR) and stock compensation (EQUITYD). The use of OLS shows that cash compensation has a coefficient of −0.11380 and a p-value of 0.063. These results indicate that cash compensation has a significant negative relationship to tax avoidance at a significance level of 10%. When referring to the quantile regression, almost all quantiles indicate a negative value against tax avoidance.

Using OLS, the stock compensation coefficient shows a value of 0.01166 and a p-value of 0.408. Whereas if you look at the quantile regression, the distribution of compensation for existing shares is different. Quantile of the distribution provided none p-value compensation for stocks that have p-value $<\alpha$, which means a lack of influence between the stock compensation tax avoidance. At high levels of tax avoidance which shares granted compensation coefficient tends to be negative, while at low levels tend to be positive. However, because there is no significant value under alpha (α) it can be concluded that the stock compensation tested in this study had no effect on tax avoidance. The results of this study are different from that performed by Armstrong et al. (2015). So the second hypothesis developed in this study is not accepted.

These results are consistent with the research conducted by Crocker and Slemrod (2005). According to the management, a company that has good corporate governance will act as a shareholder by aiming to reduce tax obligations sacrifice anything without seeing a cost of compensation received. In line with Desai and Dharmapala (2006), the incentives that are received by directors can negatively affect tax avoidance in companies with good internal control. Desai and Dharmapala (2006) also found a random relationship between incentives and tax avoidance in companies with poor corporate governance. It can be concluded that the cash compensation received by directors in companies with good corporate governance has a negative relationship with tax avoidance.

Table 2. Ordinary least square and quintile regression results.

Value/Variable		OLS	Quintile regression 1st	2nd	3rd	4th	5th	6th	7th	8th	9th
R2/Pseudo R2		0.1464	0.2033	0,1538	0.0948	0.0764	0.0896	0.1308	0.1816	0.2071	0.1477
Prob > F		0.0000***									
BOARDCOM	C	–0.0078173	0.0054711	0.0043698	0.00111187	–0.0046414	–0.0096704	–0.0130114	–0.015044	–0.0163241	–0.0128627
	S	0.000***	0.001***	0.058*	0.746	0.189	0.002***	0.000***	0.000***	0.000***	0.001***
CASHDIR	C	–0.11380	–0.0007758	–0.0005661	–0.0004402	–0.0062561	–0.0133243	–0.142099	–0.177711	–0.0141394	–0.0287741
	S	0.063*	0.882	0.949	0.965	0.463	0.049**	0.055*	0.059*	0.138	0.009***
EQUITYD	C	0.0116558	0.0110341	0.0206738	0.010509	0.0005395	–0.0030684	–0.0115263	–0.0143037	–0.0306248	0.0007429
	S	0.408	0.305	0.160	0.634	0.981	0.879	0.517	0.390	0.112	0.975
PERFORMANCE	C	0.1262581	0.0580785	0.0426076	0.0014006	0.1102339	0.1360795	0,0761285	–0.0112034	0.0106188	0.3404002
	S	0.154	0.390	0.644	0.992	0.436	0.284	0.496	0.915	0.930	0.0204**
MVE	C	0.0257732	0.0093785	0.207107	0.0244896	0.0226305	0.0219462	0.0248042	0.0145823	0.0046535	0.0112702
	S	0.007***	0.198	0.038**	0.102	0.138	0.109	0.040**	0.195	0.720	0.486
SIZE	C	–0.0013079	–0.0157671	–0.025022197	–0.0185467	–0.0053395	0.0100396	0.138131	0.0277846	0.394886	0.035525
	S	0.910	0.077*	0.326	0.310	0.774	0.548	0.348	0.044**	0.013**	0.074*
LEVERAGE	C	0.0374852	0.0044318	0.0222197	0.0634756	0.049419	0.35717	0.0295379	0.196478	0.0337279	–0.0094441
	S	0.084*	0.789	0.001***	0.062*	0.154	0.251	0.281	0.443	0.254	0.797
PROFITABILITY	C	0.3112057	0.5207711	0.4833301	0.5511532	0.3662408	0.2708094	0.0867898	0.1014351	0.0804559	–0.1014565
	S	0.019**	0.000***	0.305	0.008***	0.083*	0.153	0.603	0.515	0.655	0.651

C: Coefficient.
S: Significance level: * is at 10%, ** is at 5%, *** is at 1%.

5 CONCLUSION

At a low level of tax avoidance, the mechanism of corporate governance positively affects tax avoidance activities. This relationship means that if tax avoidance is low, the company's board of commissioners will call on companies to perform tax planning or tax avoidance. Companies with low tax avoidance have a low risk level, so undertaking income tax planning can optimise the company's final performance.

At high levels of tax avoidance, the result is reversed. The mechanism of corporate governance negatively affects tax avoidance activities, and the relationship is stronger compared to their relationship when the tax avoidance level is low. This result shows that, when the level of tax avoidance is high, the company's board of commissioners will call on companies to be more cautious and reduce tax avoidance activities. At this level, the company is facing a higher risk, so that the company's directors would be more prudent, taking care not to violate any tax laws, and tend to lower tax avoidance activities.

Furthermore, compensation in the form of cash given to the directors of companies in Indonesia was negatively related to tax avoidance. The cash compensation will influence directors to not take the risks associated with tax avoidance. Unlike the cash compensation, stock compensation has no influence on tax avoidance. If the directors receive compensation in the form of shares they will not be motivated either way with regards to tax avoidance. However, ownership of a company by directors will make the directors try to undertake tax planning so that net income rises. The aim is to increase the value of the company, which, in turn, can increase the value of shares.

The mechanism of corporate governance is a check and balance procedure against a company's policy on tax planning. The better the mechanism, the better the tax planning done by the company.

This research can be expanded from a regional aspect, such as in Southeast Asia, to determine the factors that affect tax avoidance, which can be used as input to the regulator in Indonesia.

This study only observed the effects of incentives on all of the directors on the board, not specifically those directors who are in charge of financial or tax management, so bias arising from this data is expected. This study is only looking at the motivation for tax avoidance on the internal side of the company, so the external factors of tax avoidance motivation need to be observed. Further studies can explore the use of quadratic equations to analyse the relationship between the mechanism of corporate governance and tax avoidance.

REFERENCES

Armstrong, C.S., Blouin, J.L., Jagolinzer, A.D. & Larcker, D.A. (2015). Corporate governance, incentives, and tax avoidance. *Journal of Accounting and Economics, 60*, 1–17.

Brandle, U. & Noll, J. (2004). Power of monitoring. *The German LJ, 5*, 1349.

Crocker, K.J. & Slemrod, J. (2005). Corporate tax evasion with agency costs. *Journal of Public Economics, 89*, 1593–1610.

Desai, M. & Dharmapala, D. (2006). Corporate tax avoidance and high powered incentives. *Journal of Financial Economics, 79*, 145–179.

Friese, A., Link, S. & Mayer, S. (2006). *Taxation and corporate governance*. Berlin Heidelberg: Springer.

Hermawan, A.A. (2009). *Pengaruh Efektivitas Dewan Komisaris dan Komite Audit, Kepemilikan Oleh Keluarga, dan Peran Monitoring Bank Terhadap Kandungan Informasi Laba* (Dissertation). FEB Universitas Indonesia, Jakarta.

Jensen, M.C. & Meckling, W.H. (1976). Theory of the firm: Managerial behavior, agency costs and ownership structure. *Journal of Financial Economics, 3*(4), 305–360.

Koenker, R. & Hallock, K. (2001). Quantile regression. *Journal of Economics Perspective, 15*(4), 143–156.

Minnick, K. & Noga, T. (2010). Do corporate governance characteristics influences tax management. *Journal of Corporate Finance*, 703–718.

Rego, S.O. & Wilson, R. (2012). Equity risk incentives and corporate tax aggressiveness. *Journal of Accounting Research*, 775–810.

Weber, M. (2004). *Executive equity incentives, earning management and corporate governance* (Dissertation). University of Texas, Texas.

Competition and Cooperation in Economics and Business – Gani et al. (Eds)
© 2018 Taylor & Francis Group, London, ISBN 978-1-138-62666-9

CEO tenure period and earnings management in the banking industry in Indonesia

M.R. Khasandy & D. Adhariani
Department of Accounting, Faculty of Economics and Business, Universitas Indonesia, Depok, Indonesia

ABSTRACT: The purpose of this research is to determine the effect of the Chief Executive Officer (CEO)'s tenure period on earnings management behaviour. CEOs will gain a good reputation if they can generate higher profits for the company during their tenure. This study analyses CEOs' behaviour in managing earnings by using their discretion in the early years of their tenure. The research samples were taken from banks listed in the Indonesia Stock Exchange. The findings show that, in the banking industry, the CEOs' early years of tenure affect their earnings management behaviour. In addition, an analysis was done to understand the characteristics of the samples that might strengthen the results. The additional analysis shows that the competition in the CEO job market in the Indonesian banking industry is very low, so they might not face external pressures to maintain a good reputation, which may affect their performance. In other words, the performance of Indonesian CEOs in the banking industry is not affected by their career concerns, but by their reputation with the shareholders internally.

1 INTRODUCTION

Agency problems are very common in a company. These problems usually arise when an agent acts on behalf of the principal to make a specific decision. This research focuses on the negative impact of agency problems on CEOs through the earnings management behaviour that will enhance the financial reports and will increase the CEO's reputation indirectly in the presence of shareholders. A CEO's reputation is usually associated with the success of the company during their tenure, for example, the increase of market capitalisation, financial ratios and increased profits. CEOs have a strong incentive to maintain their reputation, especially with the shareholders (Zhang, 2009). Individually, a good reputation will lead to long-term benefits for them, for example, reappointment as the CEO in the next election.

This research used "*CEO Tenure and Earnings Management*" by Ali and Zhang (2015) as the main reference. Earnings management is usually undertaken by a CEO in an effort to maintain his or her reputation. CEOs usually manage earnings in the early years of their tenure (during the first three years of the CEO's tenure) by overstating the expenses, so that in the following year the company's profit will increase and they can take the credit for it (Ali & Zhang, 2015). Unfortunately, Ali and Zhang (2015) state that the association between earnings management and CEO tenure is weak for companies with a strong supervision and monitoring system. Hence, this research used banks listed on the Indonesia Stock Exchange as samples, with the aim of understanding the earnings management behaviour of CEOs in a highly regulated industry (not only supervised by the Indonesia Stock Exchange but also specifically by the Central Bank (Bank Indonesia) and the Financial Services Authority of Indonesia).

Using a study that was conducted based on a US context as the main reference for this research may be considered rather inappropriate. However, despite the fact that US companies use a one-tier system for their Corporate Governance (CG), which might lead to agency problems, it is still possible for agency problems to occur in the Indonesian two-tier CG system. This is caused by a conflict of interest that might arise between the principal and the agent, where the agent does not act on behalf of the principal, and hence will affect the company's performance (Jensen & Meckling, 1976).

The research questions are:

Can CEO tenure affect earnings management?
Does institutional ownership weaken the effect of CEO tenure on earnings management?

The results of this research show that, for the banking industry, a CEOs' early years of tenure affect the company's earnings management behaviour. An additional analysis was also done in order to understand the CEOs' motives for undertaking earnings management in their early tenure. The additional analysis showed that competition in the CEO job market in the Indonesian banking industry is very low; therefore, CEOs might not face external pressures to maintain a good reputation in the job market, which will affect their performance. In other words, the performance of CEOs in the banking industry in Indonesia is not affected by their career concerns, but rather by the need to gain a good reputation with the shareholders.

1.1 *Theoretical review*

Discussions on earnings management cannot be separated from earnings quality. Earnings management behaviour shows that earnings quality is low. Agency cost is needed to maintain earnings quality and also to control moral hazards caused by the conflict of interest between the principal and the agent (Scott, 2012). Sometimes, agency conflict is needed to motivate CEOs, so that they can work optimally for the shareholders' welfare. However, when moral hazard behaviour (by CEOs) appears, a good supervision system by the principal is needed to control the agent. Supervision and monitoring from shareholders is one indication of good earnings quality (Ujiyantho & Pramuka, 2007).

It is very important for a company to have a high earnings quality. Earnings quality is how a company's earnings information can predict its future performance. Schipper and Vincent (2003) state that earnings quality will be high if it can accurately reflect the company's long-term performance. Good earnings quality can be used to predict a company's future condition; otherwise, low earnings quality shows that the current financial report information could not be used to predict a company's future performance (Scott, 2012).

Earnings quality can be measured by using accrual quality (Dechow & Dichev, 2002). Net profit should be the sum of the cash flow from operations added to (or reduced by) net accrual. Accrual itself may be raised by operational activity and management discretion (conducted with the aim to enrich them). If most of the accrual is based on management discretion, there is an indication that earnings management behaviour has been undertaken.

Earnings management itself is one indicator of poor earnings quality (Schipper & Vincent, 2003). Accrual based on management discretion is usually called discretionary accrual. The bigger the discretionary accrual, the lower the earnings quality; this is because the earnings are not in accordance with the company's performance (Dechow & Dichev, 2002). Management discretion can also be done through discretionary expenses by decreasing certain costs so that the profits will increase during the tenure (Pan et al., 1999, as cited in Ali & Zhang, 2015).

Discretionary accrual was also used by Ali and Zhang (2015) to measure earnings management. Earnings management occurs when the management use their judgement and discretion in financial reporting and in financial transactions with the aim of modifying the financial report, not only to mislead stakeholders regarding the company's economic performance but also to influence contractual outcomes related to the accounting numbers reported (Healy & Wahlen, 1999 as cited in Adiasih & Kusuma, 2011).

For the banking industry, management discretion can be undertaken by using special tools, such as LLP (*Loan Loss Provisioning*) and RSGLS (*Realised Gain or Loss on Available-for-sale Securities*). LLP itself is stated by Cornett et al. (2009) as the main tool used by banks to manage their earnings. Increasing LLP will decrease the profits, and vice versa (Grougiou et al.,

2014). RSGLS is also considered as one of the management tools that can be used to meet certain goals, since its fair value is estimated by management using their discretion. Cornett et al. (2009) mention that RSGLS is also one way of earnings smoothing, since there are no regulations about this and auditors cannot interfere with any decision related to RSGLS sales.

1.2 *Hypothetical review*

The market will evaluate a CEO's ability during their tenure (Zhang, 2009). This can motivate them to maintain their good reputation with specific expectations, such as being re-elected as the CEO when their tenure ends. The market or shareholders can reappoint the current CEO through the General Meeting of Shareholders, by considering his or her experience and good reputation during their tenure (Francis et al., 2008).

Information related to the CEOs' experience and reputation is not yet available when they are newly appointed. The market can only use the available information to assess their ability, which can be seen through their performance during the early years of their tenure (Ali & Zhang, 2015). This matter pushes them to work hard during the early years of their tenure to deliver a good performance (Zhang, 2009). However, this can also end badly if they decide to overstate earnings with the aim of increasing their performance in the eyes of shareholders or the market (Ali & Zhang, 2015). Nevertheless, Francis et al. (2008) found that CEOs with a good reputation will tend to manage earnings because they deal with greater pressure, especially related to analyst prospects, compared to disreputable CEOs or CEOs who do not have a reputation yet.

H1 = CEO tenure could affect earnings management.

Earnings management, as stated above, can be handled by supervision and monitoring by the shareholders. By having good corporate governance, any manipulation or fraud can be easily detected. In Indonesia, the monitoring system has been regulated in Limited Liability Company Constitution number 40 year 2007. A good monitoring system can be assessed and seen from certain components, such as the General Meeting of Shareholders, Independency of Board of Commissioners and the Audit Committee.

Having high institutional ownership is one indication of a strong monitoring system in a company. If an institution is one of the large holding shareholders, they will tend to supervise the company as a way to protect their investment. Institutional ownership shows that companies have a better monitoring system than others that do not have it. Compared to the individual shareholders, an institutional ownership monitoring system not only consists of the minimum regulation required by government, but also by that of the institutions that have a significant shareholding. This statement is consistent with Koh (2003), who states that companies with low institutional ownership have a tendency to undertake discretionary accrual to increase the earnings, and vice versa. A strong monitoring system could mitigate the opportunistic behaviour of the management (Ali & Zhang, 2015).

H2 = Institutional ownership weakens the influence of CEO tenure on earnings management.

To empirically test these hypotheses, a different research model was used for each hypothesis. This research used institutional ownership as the moderation variable and some of the control variables used by Ali and Zhang (2015). The writer also added one variable, capital, as the indicator of the banks' health (according to Bank Indonesia Regulation number 13/1/PBI/2011). Capital is a proxy for Capital Adequacy Ratio (CAR) (Astuti, 2003).

1.3 *Data*

This research used a purposive sampling method by using banks listed on the Indonesia Stock Exchange in the years from 2008–2014 (see Table 1). The base year used was 2008, because based on ICMD data, most of the CEO turnover occurred from 2007–2010 (Wijaya &

Table 1. Sample descriptions.

Number of banking companies listed in 2008	29
Number of years (2008–2014)	7
Initial samples	203
Less:	
Number of incomplete data	–83
Final samples	120

Ardiana, 2014), and a period of 7 years was selected in order to be able to capture the differences in CEO tenure within a span of 5 years. Thomson Routers, ICMD and Annual Report were used as databases for financial and non-financial data.

2 METHOD

Research model:

$$H1 = EM = \alpha + \beta 1 Early\,Years + \beta 2 Leverage + \beta 3 ROA + \beta 4 CAR + \beta 5\ InstOwn \qquad (1)$$

Notes:
EM = Earnings Management
Early Years = Early years of tenure
Leverage = Ratio of total debt to total assets in the first year
ROA = Ratio of earnings before tax to total assets before tax in the first year
CAR = Capital Adequacy Ratio
InstOwn = Institutional Ownership

$$H2 = EM = \alpha + \beta 1\ Early\,Years + \beta 2\ Leverage + \beta 3\ ROA + \beta 4\ CAR + \beta 5\ InstOwn\ Early\,Years * InstOwn + \beta 5\ InstOwn \qquad (2)$$

Note:
*Early Years*INSTOWN* = Earnings management within high institutional ownership in the early years of tenure.

2.1 *Operationalisation of variables*

The data were analysed using the Ordinary Least Square (OLS) method. Previously, discretionary accrual was measured using a regression model from a Cornett et al. (2009) study, which was also adopted by Grougiou et al. (2014). This model refers to the Cook (1977) model by measuring earnings management through discretionary management with LLP and RSGLS. This discretionary level was used as the dependent variable in Equation 1 and in Equation 2 as the EM variable.

This research used only one independent variable, EarlyYears, as the indicator of the early years of a CEO's tenure. If the tenure was in the first or second year, 1 was used as the indicator and 0 otherwise. The 2 years assumption was determined from the median of the samples, which was 2. Early Years was predicted negative, consistent with Hypothesis 1. This research also used one moderation variable, *Early Years*INSTOWN*. This variable was predicted negative; the higher the institutional ownership, the lower the possibilities of earnings management in the early years of a CEO's tenure. This variable refers to Ali and Zhang (2015).

Four control variables were used. Three of them, referring to Ali and Zhang (2015), were ROA, Leverage and Institutional Ownership, and the other one was from Astuti (2003). Three financial ratios were chosen as control variables in order to enhance more explanation about CEOs' motives in earnings management during their early years of tenure.

Leverage was measured by dividing Total Debt by Total Assets at the beginning of the year. Leverage was predicted negative; the higher the leverage level, the lower the possibilities of earnings management occurring. ROA is one of the variables usually used to measure the effectiveness of management in the use of assets and resources to obtain greater profits. ROA is measured by dividing Earnings before Extraordinary Items by Total Assets. Data of Capital Adequacy Ratio were obtained from the database. The higher the CAR, the higher the bank profitability (Astuti, 2003); therefore, the CAR coefficient was predicted positive.

3 RESULTS

Table 2 shows that 41.67% of the samples were CEOs in the early years of their tenure (the first year or second year of tenure). In other words, the CEOs only held the position for two years before being replaced by new CEOs. The table also shows the average of institutional ownership of the samples (25.92%).

Table 3 displays the regression results of Equation 1. The results indicate that the variable EarlyYears was not related to earnings management. However, after the regression using Equation 2 was done, where one moderation variable, *EARLYYEARS*INSTOWN,* was added, the variable EarlyYears was shown as being positively related to earnings management. This shows that Hypothesis 1, that CEO tenure affects earnings management, was accepted.

This result is consistent with Ali and Zhang (2015), who also found that CEOs are assumed to have a specific goal, which is to affect the market perception of their ability to manage the company. The results for Leverage and ROA are also consistent with Ali and Zhang (2015). Companies with a high leverage ratio will tend to meet financial problems and need to renegotiate their debt contracts. Companies with a high ROA also tend to have high accrual levels; therefore, they have higher possibilities to encounter earnings management than companies with a low ROA.

Table 2. Descriptive statistics.

Descriptive statistics	Mean	St. Dev.	Median
EM	0.00023	0.00614	0.04147
EarlyYears	0.4167	0.4951	1.0000
InstOwn	0.2592	0.2684	0.9800
Leverage	0.0597	0.0431	0.1800
ROA	0.0251	0.0144	0.0530
CAR	0.2983	1.4462	16.0000

Table 3. Regression results (H1).

Equation	Variable	Coefficient	Prob
1	*INSTOWN*	–0.000524	0.7997
1	*LEVERAGE*	–0.040964	0.0020**
1	*ROA*	0.077182	0.0717*
2	*EARLYYEARS*	0.003183	0.0521**

Table 4. Regression results (H2).

Variable	Coefficient	Prob
EARLYYEARS	0.003183	0.0521**
INSTOWN	0.002045	0.4476
*EARLYYEARS*INSTOWN*	–0.006117	0.1410
LEVERAGE	–0.042680	0.0013***
ROA	0.080751	0.0589*

Table 5. Samples of CEOs who once served in more than one bank.

Number of banks	Number of CEOs
One	42
More than 1	1
Total	43

The regression results of Equation 2, displayed in Table 4, show that Hypothesis 2, that CEOs do not manage the earnings if the institutional ownership is high, was not accepted. This can be seen from the insignificance of the variable *EARLYYEARS*INSTOWN*. This result is not consistent with Ali and Zhang's findings (2015). This inconsistency can be caused by the different characteristics of the samples; the samples in this study were taken from Indonesia, while Ali and Zhang's (2015) study used samples from the USA.

In this study, the mean was 26% and the median was 16%, as shown in Table 2. These characteristics are quite different compared to the US samples, where the average institutional ownership was 55% (Ali & Zhang, 2015) and the median was 0.6824. In other words, the inconsistency of Hypothesis 2 can be attributed to the low institutional ownership in Indonesia, hence the no significance effect, indicating that institutional ownership can be used as one way to control and monitor management.

3.1 *Additional analysis*

As seen in Table 3, Hypothesis 1 was accepted after seeing the results of Equation 2. In order to be able to explain the inconsistency occurring in the regression results of Equation 1, an additional analysis was done to better understand the sample characteristics, which might be different from the sample characteristics in the previous study (Ali & Zhang, 2015).

The tabulation of the above data (see Table 5) was done to see the labour market conditions of CEOs in the Indonesian banking industry. From the sample data analysis, from 2008 to 2014, only one CEO served in more than one bank. This shows that the competition among CEOs in the Indonesian banking industry is quite low. If the competition was high, a CEO in Bank X could be recruited by Bank Y, provided that the CEO showed a good performance in Bank X. CEOs will also have individual pressure to have a good performance that will lead to a good reputation. This reputation is proportional to the possibility of earnings management behaviour (Francis et al., 2008). This will lead to a conclusion that a CEOs' performance in the Indonesian banking industry is not affected by their need to have a good reputation for career purposes; therefore, they tend not to manage their earnings.

4 CONCLUSION

The objective of this research is to understand the tendency of earnings management behaviour in the early years of CEOs' tenure using discretionary management. This behaviour will usually lead to an increase in company profits, which unfortunately is not followed by an increase in company performance.

This research attempted to prove two hypotheses. The first hypothesis was accepted, indicating that CEO tenure could affect earnings management, even in highly regulated industries. This result is consistent with Ali and Zhang (2015), who state that CEOs will tend to manage their earnings in the early years of their tenure in order to maintain their reputation in the market by increasing company profits. Nevertheless, CEOs do not maintain their reputation due to their career concerns. There is a possibility that they maintain their reputation in order to be reappointed at the next election or to keep their current position.

The second hypothesis, regarding how institutional ownership could weaken the influence of CEO tenure on earnings management, was rejected. This result is inconsistent with Ali and Zhang (2015). This may be caused by the magnitude of differences in the sample characteristics of institutional ownership between Indonesia and the US.

It is suggested that during the CEO turnover phase, especially in the early years of the CEO's tenure, investors should be careful in assessing the company performance. Investors should also not base their valuation on institutional ownership, because it is still uncertain whether high institutional ownership will lead to high earnings quality.

There are some limitations in this research. One is that the sample characteristics between the samples used in this research and those used in the main reference are quite different. These differences occur because Ali and Zhang (2015) used all types of industries in their research. This may also lead to the inconsistency in the results.

Future research might observe the differences between CEO behaviour in the early years and the final year of tenure, specifically related to future career concerns (CEOs have a tendency to manage their earnings in the final year of their tenure in order to earn bigger bonuses and higher compensation). Academics and regulators should start to collaborate to evaluate and research other factors that can affect earnings management in the early years, the results of which can be used by regulators to draft new regulations in Indonesia. This is very important in order to minimise earnings management behaviour in Indonesia.

REFERENCES

Adiasih, P. & Kusuma, I.W. (2011). Manajemen Laba Pada Saat Pergantian CEO (Dirut) Di Indonesia. [Earnings Management during CEO Changes in Indonesia]. *Jurnal Akuntansi Dan Keuangan, 13*(2), 67–79. http://puslit2.petra.ac.id/ejournal/index.php/aku/article/view/18458.

Ali, A. & Zhang, W. (2015). CEO tenure and earnings management. *Journal of Accounting and Economics, 59*(1), 60–79. doi:10.1016/j.jacceco.2014.11.004.

Astuti, F. (2003). *Pengaruh Tingkat Kecukupan Modal (CAR) dan Likuiditas (LDR) terhadap Profitabilitas (ROA) pada Bank.* [The Effect of Capital Adequacy Level (CAR) and Liquidity (LDR) to Bank's Profitability (ROA)]. (Undergraduate Thesis), Faculty of Economics, Universitas Widyatama.

Bank Indonesia. (2011). *Peraturan Bank Indonesia Nomor: 13/ 1 /PBI/2011 tentang Penilaian Tingkat Kesehatan Bank Umum.* [Bank Indonesia Regulation Number: 13/1/PBI/2011 towards Commercial Banking Health Measurement]. Jakarta: Bank Indonesia.

Cook, R.D. (1977). Detection of influential observations in linear regression. *Technometrics, 19*(1), 15–18.

Cornett, M.M., McNutt, J.J. & Tehranian, H. (2009). Corporate governance and earnings management at large U.S. bank holding companies. *Journal of Corporate Finance, 15*(4), 412–30. doi:10.1016/j.jcorpfin.2009.04.003.

Dechow, P.M. & Dichev, I.D. (2002). The quality of accruals and earnings: The role of accrual estimation errors. *Accounting Review, 77*, 35–59. doi:10.2308/accr.2002.77.s-1.61.

Francis, J., Huang, A.H., Rajgopal, S. & Zang, A.Y. (2008). CEO reputation and earnings quality*. *Contemporary Accounting Research, 25*(1), 109–47. doi:10.1506/car.25.1.4.

Grougiou, V., Leventis, S., Dedoulis, E. & Owusu-Ansah, S. (2014). Corporate social responsibility and earnings management in U.S. banks. *Accounting Forum, 38*(3), 155–69. doi:10.1016/j.accfor.2014.05.003.

Jensen, M.C. & Meckling, W.H. (1976). Theory of the firm: Managerial behavior, agency costs and ownership structure. *Journal of financial economics, 3*(4), 305–360.

Koh, P.S. (2003). On the association between institutional ownership and aggressive corporate earnings management in Australia. *British Accounting Review, Volume 35, Issue 2; 102–128.* doi:10.1016/S0890-8389(03)00014-3.

Schipper, K. & Vincent, L. (2003). Earnings quality. *Accounting Horizons, 17*, 97–110.

Scott, W.R. (2012). *Financial accounting theory* (6th ed.). Toronto, Ontario, Canada: Pearson Prentice Hall.

Ujiyantho, M.A. & Pramuka, B.A. (2007). *Mekanisme Corporate Governance, Manajemen Laba dan Kinerja Keuangan.* [Corporate Governance, Earnings Management and Financial Performance Mechanism]. Simposium Nasional Akuntansi X, Makassar, 26–28 Juli 2007.

Wijaya, B.A. & Ardiana, P.A. (2014). Manajemen Laba pada Peristiswa Pergantian Chief Executive Officer. [Earnings Management during CEO Turnover Event]. *E-Jurnal Akuntansi Udayana, 8*(2), 263–278.

Zhang, W. (2009). CEO Tenure and earnings quality. Working Paper, School of Management University of Texas, Dallas.

Competition and Cooperation in Economics and Business – Gani et al. (Eds)
© 2018 Taylor & Francis Group, London, ISBN 978-1-138-62666-9

Factors affecting business lending of regional development banks in Indonesia

R.C. Hapsari & D.A. Chalid
Department of Management, Faculty of Economics and Business, Universitas Indonesia, Depok, Indonesia

ABSTRACT: Despite the main role of the Regional Development Bank (RDB), which is to increase the economic activity in each area, it is found that most Regional Development Banks in Indonesia distribute more consumer loans than business ones. This study aims to analyse several factors affecting both the business lending activity in general and the productive loans of the RDB. Using the panel random effect model, this research analysed 26 RDBs from 2001 until 2014. The result shows the differences in the variables affecting business lending activity between general and productive loans. The most prominent result obtained from this study is that regional productivity influences the lending of productive loans but not general loans. This means that, even though the role of the RDB in Indonesia as an agent of regional development is still limited, regional productivity can be used as a key factor to increase productive loans.

1 INTRODUCTION

Local government banks (BPD) perform the same function as private banks, which is to connect between the surplus units that need more money for their investment activity and the deficit units that have excess money. Law No. 13 of 1963 about the Basic Provisions of Local Government Banks presents another BPD function, which is to assist regional development by providing financing that is in line with the Pembangunan Nasional Semesta Berencana (National Development Planning) framework.

An overview comparison between the total loans disbursed by the BPD and by the other types of commercial banks that were operating in Indonesia at the end of 2014 confirm that the BPDs have low lending performances (see Table 1).

One reason why the BPDs' lending volumes are small compared to the other types of banks may be due to the placement of local government funds as the major resource of BPD financing, which are mostly deposited in the Bank of Indonesia (BI) in the form of SBI. The liquidity and profitability issues become the main reason considered by the local government. On the other hand, BPDs have limited access to household consumers, due to a market share that is smaller than that of private and state banks, even though they have greater access to local people.

Table 1. Total BPD and other bank lending in 2014 (in millions rupiah).

Bank type	Total lending
Private Bank	1,481,808
State-Owned Bank	1,325,087
Local Government-Owned Bank	301,456
Foreign Bank	244,031
Mixed Bank	195,925
Others	126,001

Source: SPI Data 2010–2014.

Pratama (2010) and Oka et al. (2015) found that the increasing DPK owned by the BPDs affected the increase in lending amounts that are distributed to deficit units. Both studies concluded that DPK had a positive influence on bank lending. In other words, the greater the DPK compiled, the more credit that can be disbursed by banks.

In addition, there are other variables that influence the level of bank lending. According to Agustine (2009), credit offers made by commercial banks are influenced by internal and external factors. In the research, the internal factors are represented by the Capital Adequacy Ratio (CAR), Third Party Fund (DPK), Non-Performing Loan (NPL) and Return on Assets (ROA), while external factors affecting the loan portfolio are represented by the interest rate of Bank Indonesia Certificates (SBI).

The BPD lending issues addressed by Indonesia Financial Services Authority (IFSA) are dominated by the lending for the consumption sector rather than for the productive sector, such as working capital and investment. It is a fact that the increase in the BPD's lending from 2010 to 2014 was caused by increasing amounts of consumer lending (see Table 2).

Indonesian Banking Booklet (IBB) (2014) mentioned that the BPDs' lending in the productive sector is still not optimal in terms of lending. The BPDs' lending activity is considered to not be in line with the BPDs' function as an agent of regional development.

As one type of bank in Indonesia, BPDs should have similarities with the other banks concerning lending. However, researchers believe that there are other factors that can affect the level of lending in BPDs. Factors that describe regional economic conditions are considered as the other factors affecting lending, especially productive credit, which is presented by non-oil imports, Gross Domestic Product (GDP) and revenue (PAD) variables.

Total non-oil imports that can be produced by a region describe the size of the level of consumption by the community in an area. When consumption activities are carried out in large

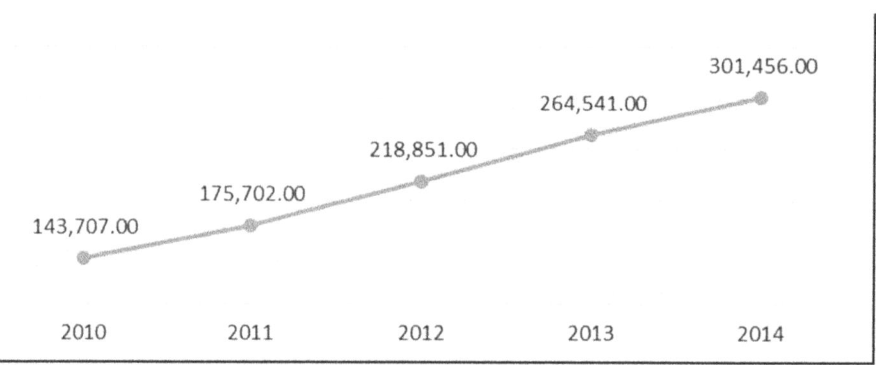

Graph 1. Total BPD lending in 2010–2014 (in million rupiah).
Source: SPI Data 2010–2014.

Table 2. Productive and consumptive lending in 2010–2014.

		Loan proportion (%)	
Year	Total loan	Productive	Consumption
2010	140,391.78	36.61	65.54
2011	170,816.54	34.48	67.80
2012	212,296.07	20.54	67.67
2013	234,680.75	31.15	68.85
2014	269,836.63	30.70	69.30

Source: SPI Data 2010–2014.

numbers, the possibility of filing bank credit as capital for the purchase of imported goods will increase. This is what causes the BPD credit channel, especially productive credit, not to run optimally.

The amount of the Gross Regional Domestic Product (GRDP) generated in a region indicates the level of productivity of the people in the area. If a higher amount of GRDP is generated, this will have an increasingly positive impact on the economic situation in a region. BPDs, which are located in every province in Indonesia, should be more aware of the economic potential of their area. Therefore, it can be said that the BPD will extend more credit, especially productive credit, in the local area, which is shown by the dramatic increase in the GRDP.

Local Revenue (PAD) is revenue that is generated by the production sectors in every province. The total PAD generated by the local government will largely be placed in the banking sector. These funds can be used to help finance the productive sectors. Therefore, the increase in PAD owned by a region is expected to increase lending, especially productive credit, by the BPD.

In addition, bank size, which is proxied by total banking assets, is also one of the factors that can determine the BPDs' lending. According to Indiapsari (2012), the greater the amount of total assets generated by the bank will lead to an increase in the level of loans extended. BPD, as one type of bank in Indonesia, should have similar characteristics to other types of banking. Therefore, if a greater amount of total assets is produced by the BPD, this will increase the level of credit lines from the BPD.

Therefore, based on the description above, this study will examine the effect of some variables, such as non-oil imports, GRDP, PAD, ROA, CAR, NPL, DPK, size and lending rates, in order to investigate total lending and productive lending from the BPD. This is due to BPD's lending are still not productive yet There are also problems that make the BPD unproductive in improving local development by consumption lending.

2 LITERATURE REVIEW

2.1 *Previous research*

Several studies have found some related factors that affect bank lending. Agustine (2009) conducted research on internal and external factors that could affect the bank's SME credit supply in four types of commercial banks operating in Indonesia, such as the state-owned banks, national private banks, BPD and foreign mix banks. The variables used in the study were CAR, DPK, ROA and SBI. The study asserted that each variable has different levels of significance in each of the banks studied. In the state-owned banks, the variables that significantly influenced the MSME lending were CAR, DPK and SBI. In the national private banks, SME lending was influenced significantly by the DPK and ROA. MSME lending in the BPD was influenced significantly by the CAR and ROA. Meanwhile, in the foreign banks and mix, SME lending was influenced significantly by the CAR, NPL and SBI.

Another research was conducted by Giovanny (2014) to observe the effects of DPK and CAR during the period 2008–2012. The results of this study suggested that DPK and LDR had a positive and significant impact on lending by the BPD, while CAR, NPL and ROA left a negative and significant impact on lending by the BPD.

Meydianawathi (2007) also conducted research on the effects of DPK, CAR, ROA and the NPL on investment credit and working capital at commercial banks, especially in the SME sector. The results showed that the DPK, ROA and CAR had a positive and significant impact on the supply of credit and working capital investment by commercial banks to the SME sector. Meanwhile, the NPL had a negative and significant effect on the supply of credit to the SME sector.

In Bali, Yoga and Yuliarmi (2013) conducted a study on the factors that influence their loan portfolio, as one source of financing for the SME sector. These factors were DPK, GRDP, interest rates on loans and NPL. The results of the study suggested that DPK had a positive effect, while loan interest rates had a negative influence, and the two variables significantly affected credit. Variable GRDP and NPL did not affect their loan portfolio significantly in the province of Bali.

Tomak (2013) conducted a similar study on public and private banks in Turkey. The study aimed to analyse the effect of bank level (size and access to funds) and the market (interest rates, inflation and GDP) on the behaviour of lending in 2003 to 2012. The study concluded that bank lending in Turkey was very dependent on the size of the bank, total liabilities, NPL and the rate of inflation.

Guizani (2014) suggested in his research that bank which has a good health assets capitalization will increase its loans. In addition, the influence of macroeconomic factors could also affect bank lending. However, both of these were temporary because they were influenced by the regional regulations in Tunisia.

In this study, the research is aimed at the BPDs in Indonesia. The financial ratios of banks and lending rates are used as an option to find the factors affecting the level of lending in BPDs. In addition, there are other variables used in this study, such as non-oil imports, GRDP, PAD and the bank size (size). These variables are used in order to see the potential effects of the local economic and financial conditions on the BPDs' lending by considering the BPDs' function as regional development agents. This study will also discuss how these factors may affect the distribution of productive credit by determining the factors that are likely to increase productive lending by the BPDs. Several of the literature reviews used to construct the model are attached in Appendix A.

3 RESEARCH METHODS

This research is an empirical study using the BPDs in Indonesia as research objects. This study uses a panel data model and the data was obtained from various sources—secondary data—which provides the information required for the research. The website of each BPD, BI, BPS, IFSA and several other sources are used to obtain the data that are considered relevant to this study. The data used comprises the period from 2001 to 2014.

3.1 *Research model*

Multiple linear regression is a method that will be used to process the data in this study (see Figure 1). This method is used to test the influence of imports, GRDP, PAD, ROA, CAR, NPL, DPK, bank size and lending rates proxy variables to lending, as a whole, and productive lending in Indonesian BPDs. The multiple regression models are shown in the following equations:

$$\text{Kredit}_{it} = \alpha + \beta_1\text{Impor}_{it} + \beta_2\text{GRDP}_{it\,+}\ \beta_3\text{PAD}_{it\,+}\ \beta_4\text{ROA}_{it} + \beta_5\text{CAR}_{it} + \beta_6\text{NPL}_{it} + \beta_7\text{DPK}_{it} + \beta_8\text{Size}_{it\,+}\ \beta_9\text{Sbkredit}_{it\,+}\ e \ldots. \quad (1)$$

$$\text{KProd}_{it} = \alpha + \beta_1\text{Impor}_{it} + \beta_2\text{GRDP}_{it\,+}\ \beta_3\text{PAD}_{it\,+}\ \beta_4\text{ROA}_{it} + \beta_5\text{CAR}_{it} + \beta_6\ \text{NPL}_{it} + \beta_7\text{DPK}_{it} + \beta_8\text{Size}_{it\,+}\ \beta_9\text{Sbkredit}_{it\,+}\ e \ldots.. \quad (2)$$

where:

Kredit_{it}	:	Total Loan	NPL_{it}	:	*Non-Performing Loan*
KProd_{it}	:	Total Productive Loan	DPK_{it}	:	Third Party Funds
A	:	Constanta	Size_{it}	:	Total Asset
Impor_{it}	:	Non-Oil and Gas Import	Sbkredit_{it}	:	Interest Rate
GRDP_{it}	:	Gross Regional Domestic Product	$\beta_1 \ldots \beta_9$	:	Coef.Var. $X_1 \ldots X_9$
PAD_{it}	:	Regional Income	e	:	*Error*
ROA_{it}	:	Return on Asset	i	:	Cross Section Term
CAR_{it}	:	Capital Adequacy Ratio	t	:	Observation Period

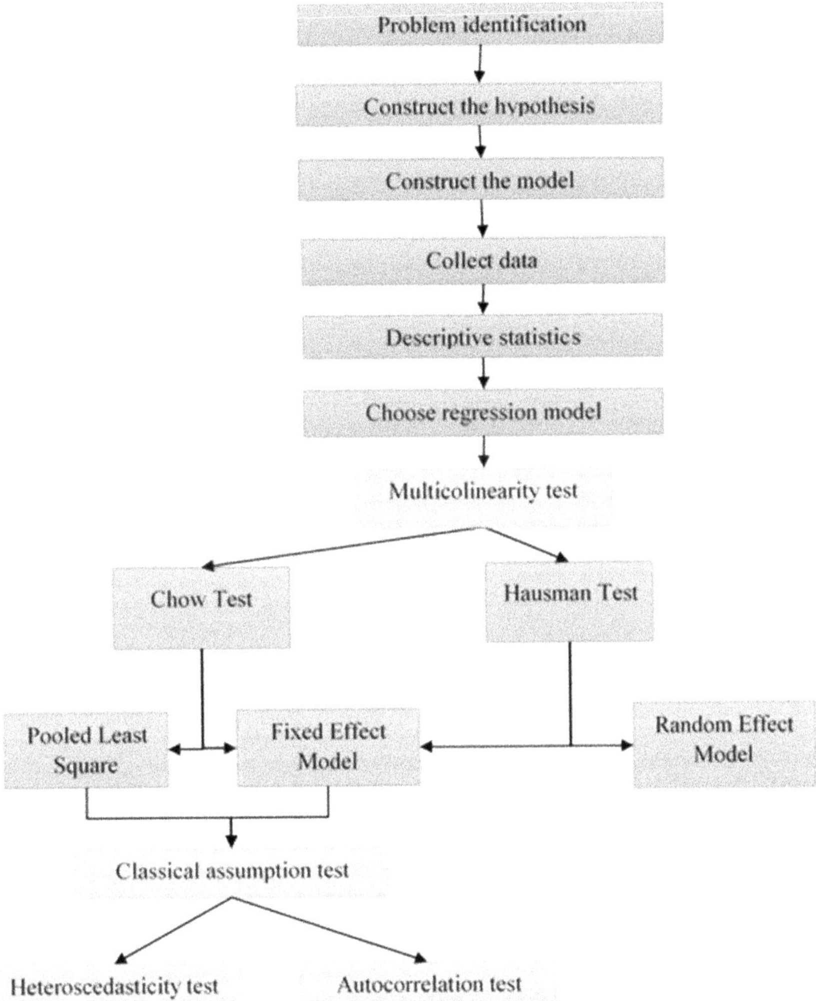

Figure 1. Research scheme.
Source: Author (2016).

The regression estimation uses two different methods, the Fixed Effects Model (FEM) and the Random Effects Model (REM). Furthermore, the regression coefficients obtained are tested with v regression coefficient (t-test) and model regression coefficient (F test). In addition, a classic assumption test is conducted to ensure that the model fits the Best Linear Unbiased Estimator (BLUE) criteria in linear regression.

4 RESULT AND DISCUSSION

Before regressing the model, the researcher carried out the model selection test to select the correct model that can be used between the fixed effect model, random effect model or pooled least square. The result is shown in Table 3. Thereafter, the researcher did the classical assumption tests, namely the heteroscedasticity test, autocorrelation test and multicollinearity test, the results of which are shown in Table 3 below.

From the table below, the researcher used the fixed effect model for Equation 1 and the random effect model for Equation 2. Since there were some classical assumption problems, the researcher also used robust coefficient correction. The final result of the regression is shown in Table 5.

Table 3. Model selection test.

		(1)		(2)	
		Total loan		Total productive loan	
Test	Hypothesis	Result	Model selection	Result	Model selection
Chow *test*	H_0: PLS H_1: FE	Reject H_0	FE	Reject H_0	FE
Hausman *test*	H_0: RE H_1: FE	Reject H_0	FE	Not Reject H_0	RE

Source: Author (2016).

Table 4. The result of BLUE test.

		Blue test		
Equation	Model selected	Multicollinearity	Heteroscedasticity	Autocorrelation
(1)	FE	v	v	v
(2)	RE	v	–	–

Source: Author (2016).

Table 5. Regression results in Equation 1 and Equation 2.

Variable	Equation 1		Equation 2	
Dependent \ Independent	Total loan		Total productive loan	
	Coefficient	Probability value	Coefficient	Probability value
Non-Oil and Gas Import	–0.010	0.071*	0.021	0.447
GRDP	0.023	0.626	0.103	0.035**
PAD	0.308	0.000***	–0.098	0.718
ROA	0.003	0.879	0.079	0.457
CAR	–0.012	0.002***	–0.032	0.008***
NPL	–0.003	0.069*	0.020	0.595
DPK	–0.072	0.226	0.294	0.093*
SIZE	0.608	0.000***	0.987	0.002***
Interest Rate	–0.169	0.000***	0.066	0.473
Constant	4.507	0.003	–7.011	0.043
R^2	0.9598		0.7178	
Prob > F / chi2	0.0000		0.0000	

Source: Author (2016).

4.1 *Equation model (1)*

In the equation model (1), the BPDs' loan distribution depends on some variables, such as:

Non-oil and gas imports have a negative impact on the BPDs' lending distribution, suggesting that high non-oil and gas imports imply a smaller lending distribution. Credit reduction in the consumption sector is supported by the BRC programme (BPD Regional Champion), which serves to push the BPD to be more active in increasing lending to the productive sectors (Indonesian Banking Booklet 2014). Therefore, any increase in the imports of non-oil and gas in an area will reduce the BPD's lending rate as well as the BPDs' commitment to promote regional development.

The PAD variable is positively associated with BPD lending. The increase in the number of PAD will increase the amount of lending of the BPD in the region. According to Yuwono & Warsito (2001) in Pangastuti (2015), PAD is local revenue collected by local governments, which is normally placed in the BPD by the local governments (Parwito, 2015). This is because the governments fear that the BPD will fail to repay their money which is funded for projects or potential business in the region. That condition ensures that almost 60% of the deposits owned by the BPD come from funds belonging to the local government. Greater lending to the public, especially in the productive sector, is expected to stimulate regional economic growth (Rokhim, 2007).

However, considering that PAD funds placed at the BPD are only temporary, even though the nominal amount is large enough, the BPD is reluctant to channel PAD in the form of lending allowance. Therefore, many BPDs in Indonesia wait for PAD from the local government revenue rather than maximising their own capital and deposits to conduct banking activities. This then makes the placement of funds in the BPD very important, since most of the BPD capital for lending to the public comes from the placement of government revenue or the PAD.

The CAR variable is negatively associated with the lending of the BPD. This indicates that an increasing CAR ratio will reduce the amount of BPD lending in a region. The results are consistent with the research conducted by Anindita (2011), Giovanny (2014) and Pratama (2010), which state that the CAR negatively and significantly influences bank lending.

A greater CAR ratio owned by a bank is a signal of its efforts to strengthen capital (Anindita, 2011). The BPD has the same character as other commercial banks in Indonesia, in that when CAR is produced the amount of credit will be reduced by the BPD. Thus, the determination of the size of the CAR must be adapted to the conditions of the BPD because it relates to the capital owned by the BPD.

The NPL variable is negatively associated with BPD lending. This indicates that an increase in the NPL ratio will reduce the amount of BPD lending in the region. The results are consistent with the results of research conducted by Mukhlis (2011) and Pratama (2010), which state that the NPL negatively and significantly influences bank lending. This result does not correspond with the research of Yoga and Yuliarmi (2013) and Trimulyanti (2013), which state that the NPL has no significant effect on bank lending.

Variable size is positively related to BPD lending. This indicates that increasing the size of a BPD will increase the amount of BPD lending in the region. This is consistent with the results of research conducted by Tomak (2013), which states that size has a positive effect on the bank lending rate. The study mentioned that government-owned banks located in developing countries, which, in general, are larger in size, have a better ability to access long-term financing.

The lending rate variable is negatively related to the BPD lending. This indicates that an increase in the lending rates of the BPD will reduce the amount of BPD lending in the region. This is consistent with the results of research conducted by Anindita (2011) and Tyastika (2013), which state that the lending rate has a significant negative effect on bank lending. However, it is not compatible with Yoga and Yuliarmi's findings (2013), which state that the lending rates do not significantly affect the increase in the bank credit channel.

4.2 *Equation model (2)*

In the equation model (2), the distribution of productive credit is influenced by variables such as the following:

The GRDP variable is positively related to BPD productive lending. The increase of GRDP is a picture of a region's success in managing its economic resources. This then stimulates the BPD to lend in regions that have good economic conditions. Therefore, this implies that the larger the GRDP generated by a region, the greater the potential for the BPD to distribute productive lending.

The CAR variable is negatively related to BPD productive lending. This indicates that an increase in CAR ratio will reduce the amount of BPD productive lending in the region. Thus, the determination of the size of the CAR must be adapted to the BPD conditions because it relates to capital owned by the BPD, given that the primary function of the BPD, similar to that of other banks, is to channel the funds that have been collected, primarily for the development of the growth of the productive sector.

The DPK variable is positively related to BPD productive lending. This indicates that an increase in the deposits of the BPD will increase the amount of BPD productive lending in the region. The results are consistent with the research conducted by Pratama (2010) and Oka et al. (2015), which state that the DPK significantly and positively influences bank lending.

In the productive lending portfolio, the BPDs also use funds from DPK. The funds raised from the public are used as the operating capital of the bank (Oka et al., 2015). The availability of these funds will affect the level of the productive lending portfolio of the BPD, since productive lending is the priority for the BPD in order to help local development.

The productive sector in Indonesia, especially SMEs, have still not been funded because of the uncertainty of future financial conditions. Therefore, the BPD needs to raise more capital for productive lending and bear all the risks related to the loan portfolio.

Variable size is positively related to BPD productive lending. This indicates that increasing the size of a BPD will increase the amount of BPD productive lending in the region. Tomak (2013) states that size has positive effects on the bank lending rate. The study mentions that the government-owned banks that are located in developing countries are, in general, of a larger size because of a better ability to access long-term financing.

5 CONCLUSION

According to the results, BPD lending is significantly influenced by non-oil and gas imports, PAD, CAR, NPL, size and lending rates. PAD and size positively influence BPD lending, which indicates that an increase of PAD and size will affect the increase in BPD lending. On the other hand, non-oil and gas imports, CAR, NPL and lending rates have a negative effect on BPD lending. Increasing one, or all, of them will lead to a decrease in BPD lending.

In the next research, we found that BPD productive lending is significantly influenced by two factors, namely the CAR and size. The availability of capital and the size of banks are the main factors that causes BPD productive lending to fluctuate. The establishment of the BPD to move funds in the form of credit, especially for the productive sector, is only visible if the BPD has a strong amount of capital.

In order to increase financing in the productive sector, the BPD is expected to have invested sufficient capital in the certain level of risk taking in the existing productive sector. Poor credit, which could be experienced due to the cessation of production processes, disturbs BPD fund distribution in the form of credit.

In addition, the BPD could also be reluctant to distribute lending because of limited amounts of savings. In other words, the BPD still relies on deposits from the local government. The savings will be reflected in the size of each BPD. This is due to the fact that the ability of the BPD to access long-term funding is affected by the amount placed by the local governments through local revenue.

REFERENCES

Agustine, A.M. (2009). Analysis the influence of internal and external factors of banking on lending of small and medium business (Case of Bank in period 2007–2008) (Master's thesis). University of Indonesia.

Anindita, I. (2011). Analysis the influence of interest rate, car, npl, and ldr on lending of small and medium enterprises (Study case of private bank in the period of 2003–2010) (Master's thesis). University of Diponegoro.

Giovanny, A. (2014). Analysis the impact of deposits, *capital adequacy ratio, nonperforming loan, loan to deposit ratio, and return on asset lending of regional development bank in Indonesia* (Master's thesis). University of Sumatera Utara.

Guizani, B. (2014). Capital requirements, banking supervision and lending behavior: Evidence from Tunisia. *Munich Personal RePEc Archive*. https://mpra.ub.uni-muenchen.de/54234/1/MPRA_paper_54234.pdf.

Indiapsari, P. (2012). The effect of liquidity, Assets Quality, Market Sensitivity, Efficiency and Profitability to Capital Adequacy Ratio (CAR) of Listed Banks in Indonesia (thesis). STIE Perbanas.

Indonesian Banking Booklet (2014). "Development and Banking Policy" Jakarta: Indonesia financial services authority. http://www.bi.go.id/id/publikasi/perbankan-dan-stabilitas/booklet-bi/Default.aspx.

Meydianawathi, L.G. (2007). Analysis the lending behavior of banking to small and medium business in Indonesia in the period 2002–2006. *Buletin Studi Ekonomi, 12*(2).

Mukhlis, I. (2011). Lending behaviour: the role of third-party funding and non-performing loans. *Jurnal Keuangan dan Perbankan, 15*(1), 130–38.

Oka, K.W.L., Purnamawati, G.A. & Sinarwati, N.K. (2015). The impact of deposits, 5Cs valuation, and lending quality in lending decision of regional development bank in Bali, branch of Singaraja,. *e-Journal S1 Ak Universitas Pendidikan Ganesha, Jurusan Akuntansi Program S1, 3*(1).

Olusanya, S.O., Oluwatosin, O.A. & Chukwuemeka, O.E. (2012). Determinants of lending behavior of commercial banks: Evidence from Nigeria, a co-integration analysis (1975–2010). *IOSR Journal of Humanities and Social Science, 5*(5), 71–80.

Pangastuti, Y. (2015). Analysis determinants of employment in central java in 2008–2012.. *Economics Development Analysis Journal, 4*(2), 224–34.

Parwito (2015). *Fiscal system make regional fund become idle.* http://www.unisosdem.org/article_detail.php?aid=6560&coid=2&caid=30.

Pratama, B.A. (2010). *Analysis determinants affecting lending decision of banking (study case of Indonesian commercial bank in year 2005–2009)* (Master's thesis). University of Diponegoro.

Rokhim, R. (2007). Regional fund optimalization. *Bisnis Indonesia*. Retrieved May 20, 2016, from http://www.unisosdem.org/article_fullversion.php?aid = 7456&coid =1&cai d= 26&gid =2.

Tomak, S. (2013). Determinants of commercial banks' lending behavior: Evidence from Turkey. *Asian Journal of Empirical Research, 3*(8), 933–43.

Triandaru, S. & Budisantoso, T. (2008). Bank and non-bank financial institutions. Jakarta: Salemba Empat.

Trimulyanti, I. (2013). *Analysis internal factors on lending growth (case study of rural bank in Semarang city in year 2009–2012)* (Master's thesis). Universitas Dian Nuswantoro.

Tyastika, M. (2013). *Analysis he impact of microeconomic performance and macroeconomic factors on farming lending (case study of regional development bank in Indonesia)* (Master's thesis). Bogor Agricultural University.

Wicaksono, B. (2004). Analysis the behaviour of banking lending in terms of intermediation function (case study of Jakarta, west java, east java, and west sumatera) (Master's thesis). University of Indonesia.

Yoga, G.A.D.M. & Yuliarmi, N.N. (2013). Determinants affecting banking lending in Rural bank in Bali province. *E-Jurnal Ekonomi Pembangunan Universitas Udayana, 2*(6), 284–93.

Yuwono, T. & Warsito. (2001). Regional autonomy management. Semarang: CLoGAPPS, Universitas Diponegoro.

APPENDIX A.

Author	Paper title	Variable: Dependent	Variable: Independent	Result: Positive	Result: Negative	Result: Significant	Result: Not significant
Budi Wicak-	Analysis of Bank-		Credit Capacity (CAP)	v		v	
Budi Wicak-sono (2004)	Analysis of Bank Credit Distribution in Region as Intermediation Function (Case Study in DKI Jakarta, West Java, East Java and West Sumatera)	Loan	Equity to Asset Ratio (CA)	v		v	
			Real GDRP	v		v	
			Interest Rate	v		v	
			Crysis (dummy)		v	v	
Luh Gede Meydianawathi (2007)	Analysis of Credit Offering for SMEs in Indonesia (2002–2006)	Investment and Working Capital Loan	Third Party Fund (DPK)	v		v	
			Return on Asset (ROA)	v		v	
			Capital Adequacy Ratio (CAR)	v		v	
			Non-Performing Loan (NPL)		v	v	
Irma Anindita (2011)	Analysis of Interest Rate, CAR, NPL and LDR Impact on SMEs Loan (Case Study in National Private Bank in 2003–2010)	SMEs Loan	Capital Adequacy Ratio (CAR)		v	v	
			Non-Performing Loan (NPL)				
			Loan to Deposit Ratio (LDR)		v	v	v
			Interest Rate		v	v	
Samuel Olumuyiwa Olusanya-	Determinants of Lending Behaviour of Commerc-	Loan and Advances	Volume of Deposits	v		v	

Samuel Olumuyiwa Olusanya, Oyebo Afees Oluwatosin & Ohadebere Emmanuel Chukwuemeka (2012)	Determinants of Lending Behaviour of Commercial Banks: Evidence From Nigeria, A Co-Integration Analysis (1975–2010)	Loan and Advances	Annual Average Exchange Rate	v		v	
			Investment Portfolio		v		v
			Interest Rate		v		v
			Gross Domestic Product	v		v	
			Cash Reserve Requirement Ratio	v		v	
Yoga and Yuliarmi (2013)	Factors Which Affect Loan Distribution in Bali Province	Credit	Third Party Fund (DPK)	v		v	
			Gross Domestic Regional Product (GRDP)				v
			Interest Rate		v	v	
			Non-Performing Loan (NPL)				v
Serpil Tomak (2013)	Determinants of Commercial Banks' Lending Behaviour: Evidence From Turkey	Total Business Loan	Size	v		v	
			Total Liability	v		v	
			Non-Performing Loan (NPL)		v	v	
			Inflation Rate	v		v	
			GDP Growth Rate	v			v
			Interest Rate		v		v
Aditya Giovanny (2014)	Analysis of Third Party Fund, CAR, NPL, LDR and ROA Impact on BPD's Credit Allocation in Indonesia	Credit	Third Party Fund (DPK)	v		v	
			Capital Adequacy Ratio (CAR)				

(Continued)

APPENDIX A. *(Continued)*

Author	Paper Title	Variable Dependent	Independent	Result Positive	Negative	Significant	Not significant
			Non-Performing Loan (NPL)		v		v
			Loan to Deposit Ratio (LDR)	v		v	
			Return on Asset (ROA)		v		v
Brahim Guizani (2014)	Capital Requirements, Banking Supervision and Lending Behaviour: Evidence from Tunisia	Lending Change	Lending Percentage Changes	v		v	
			Interest Rate Differential	v			v
			Capital Adequacy Ratio (CAR)	v		v	
			GDP Growth Rate	v		v	
			Percentage of NPL		v	v	

Competition and Cooperation in Economics and Business – Gani et al. (Eds)
© 2018 Taylor & Francis Group, London, ISBN 978-1-138-62666-9

The impact of competition on bank stability in ASEAN-5

H. Lindawati & D.A. Chalid
Department of Management, Faculty of Economics and Business, Universitas Indonesia, Depok, Indonesia

ABSTRACT: Southeast Asian Nations Economic Community (AEC) initiatives are expected to increase financial integration, which potentially increases banking competition and eventually impacts the stability of the banking industry. From a theoretical point of view, there are two contradictory perspectives regarding the relationship between banking competition and stability: competition-fragility and competition-stability. This study use samples of commercial banks listed in the stock exchange in the ASEAN countries of Indonesia, Philippines, Malaysia, Singapore and Thailand, from 2005 to 2014. This research supports the 'competition-stability' hypothesis in the banking industry in the scope of the ASEAN region.

1 INTRODUCTION

A bank is one of the institutions that has an intermediary function to gather and distribute funds to the community so that it can increase the country's economic growth. When banks do not act cautiously when taking decisions, this intermediation function of the bank can be disrupted. This can cause instability in the banking industry itself, which will make a great impact on the stability of the country's financial system (Kocabay, 2009).

One of the triggers of the financial markets' instability over the last few years is the fact that the banking sector takes excessive risks to be able to maintain profitability in a highly competitive market (Soedarmono et al., 2011). The Asian financial crisis, occurring in 1997–1998, made many banks in some ASEAN countries, such as Indonesia, Thailand, Malaysia and the Philippines, become bankrupt. Responding to the phenomena, there were changes in the supervision structure and regulation of the banking sector of ASEAN countries (Soedarmono et al., 2011).

Up to now, ASEAN countries have continued to make an effort to improve and strengthen the performance of the banks, so that they can keep the banks stable. The effort now is implementing ABIF (ASEAN Banking Integration Framework), which was approved by the Central Bank Governors in December 2014. The integrated banking sector contributes to economic growth and financial inclusion (Wihardja, 2013). On the other hand, having an integrated banking sector can also mean that the risk of financial instability that occurs in a country can spread rapidly to other member countries, especially due to the lack of an adequate regulatory framework (Wihardja, 2013). Thus, implementing ABIF will increase banks' competition in each country, which will ultimately affect the behaviour of the bank in maintaining its stability.

However, the relationship between competition and the stability of the bank has not found a clear consensus. Theoretically, there are two views regarding the relationship between competition and stability: competition-fragility and competition-stability. Some research findings also vary. Shaeck et al. (2009), analysing banks in 45 countries for a research period from 1980 to 2005, showed that the higher the competition level of the banking system, the greater will be the tendency to experience a systemic crisis and with a longer duration to the crisis. The research conducted by Berger et al. (2009) reviewed a sample of banks in 23 developed countries and observed that an increase in the market power of the banks will direct the banks to a risky portfolio and that the effects on stability can be balanced with franchise value. Banks in ASEAN are expected to be able to compete in the future since the existence of liberalisation of the banking sector has become an important agenda in the Southeast Asian Nations Economic Community (AEC). The study provides an empirical evidence on the competition in the ASEAN banking sector. The research contribute to the creation of policies that support the stability of the banking sector in ASEAN.

2 LITERATURE REVIEW

Theoretically, there are two conflicting views to the relationship between competition and financial stability: competition-fragility and competition-stability. The theory of competition-fragility stated that competition has an adverse impact on a bank's stability. The impact is defined in the form of the failure or bankruptcy of the bank (bank failures). In other words, this theory wants to explain that the more competitive the bank industry is, the greater the chance that it will cause the market to become more vulnerable and unstable. In contrast, the competition-stability theory states that the greater the competition level, the lower the bankruptcy risk, which in turn increases financial stability. In other words, this theory supports the existence of a positive relationship between competition and stability (when the competition level is low it will improve financial instability).

There are several theories which explain the competition fragility. One of the theory is franchise value hypothesis (Marcus, 1984; Chan et al., 1986). The decrease in franchise value will tend to make the banks less cautious and increase their incentive to take excessive risks (excessive risk-taking) in order to achieve a higher profit. The bank may choose to allocate funds on assets and credit that have a higher risk, but give a higher return, as a form of risk-taking behaviour. As a result, this can increase the risk of the bank's bankruptcy, which enhances the risk towards financial instability. Therefore, a banking system that has a high level of competition is more vulnerable to crisis (Beck, 2008). In addition, a more concentrated banking system has a smaller number of large-sized banks, making it easier to supervise and monitor than a competitive banking system with many banks (Beck, 2008). The concentration and the consolidation of the banking industry's activities will form large-sized banks, and, in general, these have a more diversified portfolio The more diversified the bank's portfolio, the lower are the bank's profitability of risk and failure (Beck, 2008).

The 'competition of stability' can be explained through the mechanisms of risk shifting paradigm and policies of 'too big to fail' or 'too important to fail'. According to the risk shifting paradigm, as the interest rate rises, the borrowing costs also rises, and the borrower will tend to make riskier project, so that the possibility of default will be higher. This condition results in a high number of bad debts or non-performing loans and increases the risk to the bank, which then leads to financial instability (Berger et al., 2009). So on the basis of this paradigm, a low level of competition thus encourages financial instability. In addition, large-sized banks are believed to be one of the triggers of the occurrence of systemic risk. Policymakers or regulators would not be likely to let the big banks suffer solvency problems. Therefore, the regulator and the government guarantee the business continuity of large-sized banks in order to avoid a crisis in the country. Based on the assumption that banks who have a higher market power will receive larger subsidies, this would cause the bank to intensify their risky activities. This indicates that a lack of competition in the system will harm the bank's stability (Beck, 2008). The study conducted by Allen and Gale (2004), using data from as many as 23 developed countries, found that a financial crisis is more likely to occur in a bank in a less concentrated market.

The theory of bank risk-taking and competition suggests that the relationship between competition and stability is negative. However, a research implemented by Boyd and De Nicolo (2005), found a different conclusion. This research proved that higher market power will tend to increase the interest rate, which then causes the possibility of facing a default risk caused by the increasing ratio of default by the borrower, known as the risk shifting paradigm. One study that supports this theory is the research conducted by Yeyati and Micco (2007), who state that in the banking sector of 8 Latin countries there was evidence of a positive relationship between competition and financial stability. In addition, the study by Arogaki et al. (2011) also found evidence in the countries of Central and Eastern Europe that supports the idea that competition in the banking industry can improve stability. The research conducted by Boyd et al. (2006) also came to the same conclusion, using samples from the United States and 134 other countries. Using the measurement of structural competition, HHI, they found a negative and significant relationship between concentration and stability. This study was further improved by De Nicolo and Loukoianova (2007) using data from 133 non-industrial countries in the period from 1993 to 2004, and they found that the negative relationship between concentration and stability is stronger when it includes the structure of the ownership of the bank.

3 RESEARCH METHODS

The main purpose of this research is to analyse the effect of bank competition on bank stability in the ASEAN-5 banking industry (Indonesia, Malaysia, Thailand, Philippines and Singapore) during the period from 2005 to 2014. Based on the criteria of sample selection, we obtained 63 listed banks located in 5 countries in ASEAN—Indonesia, Thailand, Malaysia, Philippines and Singapore. The secondary data is obtained from the banks financial statements in each of the ASEAN-5 countries. The banks financial statements are obtained from the data source in Reuters Eikon.

The dependent variable in the research is bank stability and the independent variable is bank competition. Bank stability is measured using a proxy Z-Score. Meanwhile, bank competition uses the size of the market power (Lerner Index). This research also uses several control variables, such as bank concentration on the credit markets, herding, liquidity risk, composition of assets and credit risk, and some variables interaction, such as bank competition with the bank concentration on the credit markets, bank competition with the herding bank and bank competition with the period of the financial crisis. The indicators of dependent variables and independent variables will be explained below.

3.1 *The bank competition*

The level of banking competition is measured by using the proxy Lerner Index. This research uses the Lerner Index as the indicator for measuring bank competition indirectly, for several reasons. The Lerner Index is more accurate to measure the banking competition with a large number of market participants than the Panzar and Rosse measurement (H-Statistics) or Concentration Ratio (Brissimis et al., 2008). In addition, the Lerner Index is the appropriate size to determine the market strength for each bank based on the type of ownership, size and different specialisation (Claessens & Laeven, 2004; Brissimis, 2011), which is in line with the sample banks that have different types of ownership and size in each country.

The Lerner Index in this research is calculated by using the following formula:

$$LI_{i,t} = \frac{P_{i,t} - MC_{i,t}}{P_{i,t}} \tag{1}$$

where P is the total income (interest income and non-interest income) to total asset ratio of the bank in that year. MC is the marginal cost of each bank, which is calculated by using the translog cost function.

Marginal cost in this research is measured by using a translog cost function which is in line with the previous research conducted by Berger et al. (2009), which used the Lerner Index as variables. The translog cost function is described by the function in Equation 2 below:

$$\ln TC_{i,t} = \alpha_0 + \alpha_1 \ln Q + \frac{\alpha_2}{2} \ln Q^2 + \sum_{j=1}^{3} \beta_j \ln P_j + \frac{1}{2} \sum_{j=1}^{3} \sum_{k=1}^{3} \delta_{jk} \ln P_j \ln P_k + \frac{1}{2} \sum_{j=1}^{3} \gamma_j \ln Q \ln P_j + \tau_1 t + \frac{\tau_2}{2} t^2 + \tau_3 t \times \ln Q + \sum_{k=1}^{3} \omega_j t \ln P_j + \varepsilon_{i,t} \quad (2)$$

The Lerner Index is estimated for each country using the robust fixed effect, where each of the elements will be explained as shown in Table 1:

Furthermore, the marginal cost for each bank can be calculated as follows:

$$MC_{i,t} = \frac{TC_{i,t}}{Q_{i,t}} \left[\hat{\alpha}_1 + \hat{\alpha}_2 \ln Q + \sum_{j=1}^{3} \hat{\gamma}_j \ln P_j + \hat{\tau}_3 t \right] \quad (3)$$

By getting the value of the marginal cost for each bank, the Lerner Index for each bank can be calculated. The Lerner Index is the measurement of the market power, so that a lower value shows that the bank does not have market power in determining the price, which further indicates that the bank has a high level of competition.

3.2 *The bank stability*

In measuring the stability of the bank, this research uses a proxy Z-Score, as used in previous empirical studies conducted by Fiordelisi and Mare (2014), Boyd et al. (2006), Iannotta et al. (2007) and Laeven and Levine (2009). Z-Score is used as the indicator that describes the level of health of the financial system of individual banks, so that it represents the stability of the bank. This research uses the calculation of Z-Score in accordance with the research conducted by Fiordelisi and Mare (2014), as follows:

$$Z - score_{i,t} = \frac{ROA_{i,t} + CAR_{i,t}}{\sigma(ROA_{c,t})} \quad (4)$$

where $ROA_{i,t}$ is the *return on assets* ratio for each bank in that year. $CAR_{i,t}$ is the capital to assets ratio for each bank in that year, calculated by comparing the bank equity to the total assets of the bank. Then $\sigma(ROA_{c,t})$ is the standard deviation from the *return on assets* ratio for each country in that year. To reduce heteroscedasticity in the data, this research uses natural

Table 1. The definition of independent variables.

The variables	Description
TC (Total cost)	As a proxy of the total cost of the bank measured using the interest expense and the non-interest expense.
Q (Assets)	As a proxy of bank output measured using the total assets of the bank.
P1 (Input prices of labour)	As a proxy of bank input that is described by the level of wages and is calculated as the ratio of personnel expense to total assets of the bank.
P2 (Input price of physical capital)	As a proxy of bank input that describes the level of interest rate and is calculated as the ratio of interest expenses to total deposits of the bank.
P3 (Input prices of borrowed funds)	As a proxy of bank input that describes the level of other expenses and is calculated as the ratio of the operational expenses to the total assets of the bank.

logarithms to calculate the *Z-Score*. In accordance with the method of counting used to get the value of the *Z-Score,* the banks with high capitalisation and low volatility income will be more stable. This means that the higher the value of the *Z-Score*, the better the stability of the bank.

One of several control variables to be considered that might affect this research is herding; this indicator shows the possibility of the bank expanding its operational activities, apart from its core business, in response to an increase in bank industry competition (Fiordelisi & Mare, 2014). The measurement used as a proxy from the bank concentration is the Herfindahl-Hirschman concen-

Table 2. Research variables.

The variables	The indicator	The calculation
The dependent variables		
Bank stability	*Z-Score*	$\frac{\mu(ROA_{i,t})+CAR_{i,t}}{\sigma(ROA_{c,t})}$
Independent variables		
Bank competition	*LI*	$\frac{P_{i,t}-MC_{i,t}}{P_{i,t}}$
Control variables		
Bank concentration in the credit markets	*HHIL*	$\sum_{i=1}^{N} s_i^2$
Herding	*HERD*	$\sigma_{i,t}\left(\frac{Non\ Interest\ Income_{i,t}}{Total\ Assets_{i,t}}\right)$
Liquidity risk	*LIQUIDITY cushion*	$\frac{Cash_{i,t}+Due\ from\ other\ banks_{i,t}}{Total\ Assets_{i,t}}$
Credit risk	*CREDITRISK*	$\frac{Loan\ loss\ provisions_{i,t}}{Interest\ Income\ Margin_{i,t}}$
The composition of assets	*ASSETCOMPOSITION*	$\frac{Total\ Loans_{i,t}}{Total\ Assets_{i,t}}$
Interaction variable		
Herd and Lerner Index	*HERD_LI*	The interaction between the dummy variable *Herd* and *Lerner Index*. *Dummy variable* value herd 1 when 3 is the lowest value between the sample countries in the year t.
HHI loan and Lerner Index	*HHIL_LI*	The interaction between the dummy variable *HHI loan* and *Lerner Index*. Dummy variable *HHI loan* worth 1 when 3 is the highest value in between the sample countries in the year t.
Financial crisis and Lerner Index	*FINCIR_LI*	The interaction between the dummy variable crisis period with *Lerner Index*. *Dummy variable* value crisis period 1 in the year 2008–2009.

tration Index (HHI). In this study the HHI used is the HHI on credit markets. In addition, there are liquidity risk variables described by the liquidity bank, credit risks described by the proportion of loan loss provisions, and the composition of credit on assets owned by the bank.

The explanation for each of the variables used in this research will be described in Table 2:

This research model refers to the model used in the previous research on the same topic conducted by Fiordelisi and Mare (2014). This research analyses the effect of banking industry competition on the bank's stability using samples of listed banks in ASEAN-5 countries during the period from 2005 to 2014. The model used in this research is as follows:

$$Z-score_{i,t} = \alpha + \left(\beta_1 + \beta_2 X_{i,t-1}\right) \times LI_{i,t-1} + \gamma X_{i,t-1} + \vartheta K_{i,t-1} + \varepsilon_{i,t} \quad (5)$$

where $Z-score_{i,t}$ is $Z-score$ for each bank i at period t, $LER_{i,t-1}$ is the *Lerner Index* for each bank *i* at period *t–1*, $X_{i,t-1}$ is the control variable as *herd*, bank concentration factor on the market *loan* period t, and period of global crisis (2008–2009), $K_{i,t-1}$ is a variable control on *the bank fundamental level* that includes liquidity risk, credit risk and the composition of the assets of the bank *i* period *t*.

4 RESEARCH RESULTS

Table 3 is a summary of the descriptive statistics from all of the variables used in the research, both dependent variables (Z-Score), independent variables (Lerner Index) and bank control variables, such as herd, liquidity, credit risk and asset composition.

Table 4, below, shows a summary of the results of the regression using the *Generalized Least Square Method* to solve the heteroscedasticity problems and autocorrelation.

The Lerner Index coefficient of –0.6070 means that there is a negative effect of the variable to the stability of the bank. In other words, the lower the Lerner Index, the higher the banks stability. A lower Lerner Index indicates less market power, so the market competition level will be increasing greatly. Then, a higher level of competition in the banking industry will improve the stability of the banks. The partial test results indicate that the level of bank competition in ASEAN-5 from 2005 to 2014 had a positive effect on the stability of the banks. This result is in line with the research done by Fiordelisi and Mare (2014) and Schaeck et al. (2009), in which they also conducted research with a cross-countries scope. These researches support the hypothesis that banks will face a higher risk of bankruptcy in a less competitive industry. In addition, the regression result shows that in both a crisis and non-crisis period, the market power still has a negative impact on the stability of the banks.

The value of the herding variable coefficient from the regression is positive, so it can be concluded that the herding variables in ASEAN-5 banks have a positive significant effect on the banks. Herding variables in this research are measured using the deviation from the ratio of total non-interest income to total assets. The results of this research showed that the more heterogeneous the income of a bank, the better the stability of the bank. This means that ASEAN-5 banks that extend their business to non-traditional lines (non-interest income activities) will have better stability.

Table 3. Descriptive statistics.

The variables	Obs.	Mean	Max	Min	Std. Dev
Z_SCORE	595	2.4644	5.5968	–2.5501	0.8069
LI	595	0.2604	0.7784	–0.5393	0.1596
HERD	595	0.0084	0.0252	0.0016	0.0042
HHI_LOAN	595	0.1507	0.3409	0.1076	0.0507
ASSET_COMPOSITION	595	0.6324	2.6203	0.0237	0.1981
LIQUIDITY cushion	595	0.0863	0.3501	0.0009	0.0619
CREDIT_RISK	595	0.1568	1.0990	–0.1423	0.1799

Table 4. The results of the regression using the generalized least square method.

Dependent variables: *Z-Score*		
The variables	Drag coefficient	Prob.
Planck	4.6902	0.0000***
LI	−0.6070	0.0209**
HERD	12.4186	0.0030***
HHI_LOAN	−10.9932	0.0000***
LIQUIDITY cushion	0.8981	0.0000***
CREDIT_RISK	0.1158	0.6942
ASSET_COMPOSITION	−0.5173	0.0003***
HERD_LI	−1.1177	0.0090***
HHIL_LI	−0.7419	0.0121**
FINCIR_LI	−0.7682	0.0313**
R-squared	0.7179	
Adj. R-squared	0.6802	
Prob. (F-statistic)	0.0000	
DW Stat	1.4317	
The observation	595	

***Significant at the level of $\alpha = 1\%$.
**Significant at the level of $\alpha = 5\%$.
*Significant at the level of $\alpha = 10\%$.

The effect of the concentration variables of the loan market is negative, so that the concentration of the loan market in ASEAN-5 has a significant negative impact on the stability of the banks. This means that the more concentrated the loan markets are in ASEAN-5 banks, in the case of credit distribution, the more likely that it will disrupt the bank's stability.

Liquidity risk is negatively associated with bank stability, so the liquidity of banks in ASEAN-5 will increase the instability of the banks. This is in line with Wagner's (2007) opinion, that greater liquidity makes a crisis become 'less costly' for the bank. As a result, the bank has an incentive to take a number of new risks that cannot offset the positive impact to improve stability, and thus increases the risk of failure.

This research endorses the fact that more loan composition in ASEAN-5 bank assets will reduce the stability of the banks. The results are also in line with the research conducted by Hesse and Cihak (2007), which states that banks with a higher ratio of composition of the loan on the assets will have higher risks due to the possibility of experiencing default.

If the effect of credit risk on the stability of the bank is negative, this means that a greater credit risk in ASEAN-5 banks IT will disrupt the stability of the banks. According to Liu et al. (2012), loan loss provisions reflect the actual amount that is issued by the bank to cover credit losses, so that the bigger the value, the bigger the credit losses, which is bad for the stability of the banks.

5 CONCLUSION

Competition level is measured using the proxy market power Lerner Index, which has a negative and significant effect on the stability of the banks in ASEAN-5. The conclusion supports the theories of competition-stability or competition-fragility. By supporting the competition-stability theory, the conclusion of this research indicates that the more competitive the banking sector in ASEAN-5, the lower the risk of experiencing bankruptcy, and this will eventually improve financial stability. This also remained the same during the crisis period used in the research.

In the context of the liberalisation implementation of the banking sector as one of the important agendas in the AEC, ASEAN Banking Integrated Framework (ABIF), then this research provides an overview that a high competition level in financial integration is proven to be able to improve the stability of the banks in ASEAN-5. This becomes empirical evidence for the banks to continue to improve their performance and competitiveness as a form of maintaining the stability of the bank and for the regulator to be able to make policies and implement supervision of the banks in order to support financial stability when ABIF is applied.

REFERENCES

Allen, F. & Gale, D. (2004). Competition and financial stability. *Journal of Money, Credit and Banking, 36*(3), 453–80.

Arogaki, M., Delis, M.D. & Pasiouras, F. (2011). Regulations, competition and bank risk-taking in transition countries. *Journal of Financial Stability, 7*(1), 38–48.

Association of Southeast Asian Nations. (2015). *ASEAN Economic Community 2015: Progress and key achievements.*

Beck, T. (2008). Bank competition and financial stability: Friends or foes? *World Bank Policy Research Working Paper* 4656, Washington DC, USA. doi: http: 10.1596/1813-9450-4656.

Beck, T., De Jonghe, O. & Schepens, G. (2013). Bank competition and stability: Cross-country heterogeneity. *The Journal of Financial Intermediation, 22*(2), 218–244.

Berger, A.N., Klapper, L. & Turk-Ariss, R. (2009). Bank competition and financial stability. *Journal of Financial Services Research, 35*(2), 99–118.

Boyd J.H. & De Nicoló, G. (2005). The theory of bank risk taking and competition revisited. *Journal of Finance, 60*(3), 1329–1343.

Boyd J.H., De Nicoló, G. & Jalal, A.M. (2006). *Bank risk taking and competition revisited: New theory and new evidence* (IMF Working Paper 06/297), Washington DC, USA.

Brissimis, S. & Delis, M.D. (2011). Bank-level estimates of market power. *European Journal of Operational Research, 212*(3), 508–517.

Brissimis, S., Delis, M.D. & Athanasoglou, P.P. (2008). Bank-specific, industry-specific and macroeconomic determinants of bank profitability. *Journal of International Financial Markets, Institutions and Money, 18*(2), 121–136.

Chan, Y.S., Greenbaum, S.I., & Thakor, A.V. (1986). Information reusability, competition and bank asset quality. *Journal of Banking & Finance, 10*(2), 243–253.

Claessens, S. & Laeven, L. (2004). What drives bank competition? Some international evidence. *Journal of Money, Credit, and Banking, 36*(3), 563–583.

De Nicolò, G. & Loukoianova, E. (2007). *Bank ownership, market structure and risk: A general equilibrium exposition* (IMF Working Paper 09/105), Washington DC, USA.

Fiordelisi, F. & Mare, D.S. (2014). Competition and financial stability in European cooperative banks. *Journal of International Money and Finance, 45*, 1–16. doi: 10.1016/j.jimonfin.2014.02.008.

Hesse, H. & Čihák, M. (2007). *Cooperative banks and financial stability* (IMF Working Papers 07/2), Washington DC, USA.

Iannotta, G., Nocera, G. & Sironi, A. (2007). Ownership structure, risk, and performance in European banking industry. *Journal of Banking & Finance, 31*(7), 2127–2149.

Jiménez, G., Lopez, J.A. & Saurina, J. (2013). How does competition affect bank risk-taking? *Journal of Financial Stability, 9*(2), 185–195.

Kocabay, S.A. (2009). *Bank competition and banking system stability: Evidence from Turkey* (Thesis). Middle East Technical University.

Laeven, L. & Levine, R. (2009). Bank governance, regulation and risk taking. *Journal of Financial Economics,* 93(2), 259–275.

Liu, H., Molyneux, P. & Nguyen, L.H. (2012). Competition and risk in South East Asian commercial banking. *Applied Economics, 44*(28), 3627–3644.

Marcus, A.J. (1984). Deregulation and bank financial policy. *Journal of Banking & Finance, 8*(4), 557–565.

Schaeck, K., Čihák, M. & Wolfe, S. (2009). Are competitive banking system more stable? *Journal of Money Credit and Banking, 41*(4), 711–734.

Soedarmono, W., Machrouh, F. & Tarazi, A. (2011). Bank market power, economic growth and financial stability: Evidence from Asian banks. *Journal of Asian Economics, 22*(6), 460–470.

Turk Ariss, R. (2010). On the implications of market power in banking: Evidence from developing countries. *Journal of Banking and Finance, 34*(4), 765–775.

Wagner, W. (2007). The liquidity of bank assets and banking stability. *Journal of Banking, 31*(1), 121–139.

Wihardja, M.M. (2013). *Financial integration challenges in ASEAN beyond 2015* (ERIA Discussion Paper Series 2013/27), Jakarta, Indonesia.

Yeyati, E.L., & Micco, A. (2007). Concentration and foreign penetration in Latin American banking sectors: Impact on competition and risk. *Journal of Banking & Finance, 31*(6), 1633–1647.

Competition and Cooperation in Economics and Business – Gani et al. (Eds)
© 2018 Taylor & Francis Group, London, ISBN 978-1-138-62666-9

Making sense of an airline's logo makeover: The case of garuda Indonesia

A. Andana & H. Mahardika
Department of Management, Faculty of Economics and Business, Universitas Indonesia, Depok, Indonesia

ABSTRACT: This study aims to explain the relationship between a logo makeover and its company's performance in the airline industry. Through the semiotic method, the authors propose that a logo makeover plays an important role in determining the ability of an airline to improve its service performance. Garuda Indonesia Airline has been selected as the sole case for the study. The company made significant changes to the shape, colour schemes, and typography of its logo in 2009. The Garuda's logo makeover was analysed using the Peirce's semiotic approach. The results indicate that a logo makeover has a positive influence, as much as 19.6% on the customer's brand attitude and 16.5% on the customer's brand commitment, while the rest is determined by other factors that are not included in the scope of this research, such as service quality and marketing strategy.

1 INTRODUCTION

Today, the airline industry has become more important and continues to be one of the fastest-growing service industries. The implications have been profound for the airlines, for competition intensifies and consumers expect a higher quality of services. This led to a significant shift in the industry, where safety and operational excellence are no longer the most important factors to win the competition.

As many airlines have changed their course into one that is more service-oriented, several paths have been taken in a number of different cases. Garuda Indonesia is one airline company that offers a unique perspective in this regard. The company emerged as a five-star airline and held the best cabin crew award, in a matter of three years after commencing its Quantum Leap programme in 2009 (Zerlina et al., 2016). Part of Garuda's Quantum Leap strategy is its logo makeover.

In the airline industry, corporate identity plays a crucial role as a medium of communication to build corporate image in the eyes of the public and competitors (Seo & Park, 2016). Prior research found that corporate identity is related to various issues, such as credibility and philosophy (e.g. Melewar, 2003). As experienced by Garuda Indonesia, to showcase its commitment to a better and new spirit of quality services, the company also needs the presence of an improved corporate image as an airline. As a flag carrier airline for Indonesia, the new image formulation has to reflect the sense of a distinctive Indonesia, professionalism, adherence to international standard, and premium service.

One way to find out the meaning of logo makeover is by using the study of semiotics. Pateda in Sobur (2013) contends that the Peirce semiotic approach can help to understand how a 'signal' creates a unique state of mind, and therefore a strong brand name or logo (which creates a strong 'signal') is needed to achieve a competitive advantage. Hence, the appropriate logo redesign can help increase brand awareness in consumers' minds. Keller (1998) found strong evidence that the clarity of logos creates certain meanings and associations which can shape consumers' perceptions of the company. Like brand names, logos can acquire associations through their inherent meaning as well as through the supporting marketing programme.

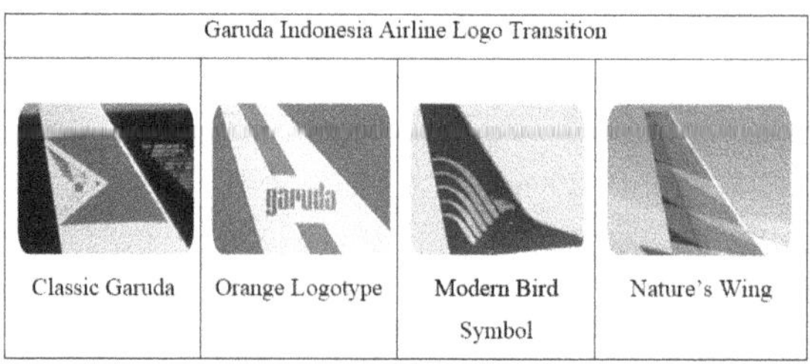

Figure 1. Garuda indonesia airline logo transition.

Zerlina et al. (2016) comments that the urgency of Garuda's management to present a more representative symbol in the form of a revised logo, is needed to signal the business restructuring that they were working on. The evolution of Garuda Indonesia's logo can be summarised as follows (see Figure 1).

As soon as the new logo was proposed in 2009, several big questions arose: 'Will the new logo play an important part in guiding Garuda on its road towards service excellence? If yes, can Garuda's path to excellence be replicated by other airlines?'

Answering these questions will be the main objective of this study. We would like to understand how the logo makeover affects service improvement in the airline industry. Logo redesign can be very expensive for some companies, and there is a high risk related to it. For example, a new logo sometimes has negative effects on the company's brand equity because the new design of the logo is too complex, unclear and hard to remember, unpleasant, or totally fails to create meanings for the consumers (Walsh et al., 2010).

2 LITERATURE REVIEW

Brand has always played a vital role as a corporate or product identity. It differentiates one from the rest and facilitates the introduction of the interaction between consumers and companies/products. On the brink of information overload, such as what is happening today, brand is a necessary cue for competition.

While undoubtedly important, there is no single effective approach to developing a good brand. According to Kotler et al. (2009), a brand name must be memorable, meaningful, likeable, transferable, adaptable, and protectable. The name, slogan, and logo of a product have also been used extensively to represent a brand. They function as a brief explanation or definition of what the brand is and what makes it special.

Among the different pillars of a brand, the logo plays a particularly key role in determining consumers' awareness towards the brand. Therefore, a logo must be special, recognisable, and unique. As difficult as it gets, many companies continuously change their logos in their search for perfection. However, changing the corporate logo is not a simple task to do. It is associated with resources, costs, and risks. The company will not only be burdened by the cost of changing the logo, but also by the risk of dispositioning the corporate image. A logo also contains the vision and missions of the company (most of the time, its founders) and also the value of the company.

As a widely-used terminology, 'logo' is often used to refer to a variety of graphs and typeface elements ranging from word-driven, conceptually simple logotypes and word marks, to image-driven, conceptually complex brand marks (Pittard et al., 2007). It is a vision of delivering a positive impression through a simple display in the form of a symbol. However, marketers encounter dilemmas as a logo should be able to convey its message over a prolonged period of time, and it must be able to adapt to cultural changes, while most logos do not. It might be exciting for the company to design a logo that is influenced by a trendy typeface, but before long it will become outdated and need to be replaced in later years.

In responding to a brand, consumers convey their attitude by using brand elements that they have encountered. Attitude reflects a person's overall evaluation of a concept, and it is an inner feeling (affections, emotions, and moods) towards a product or a service offering (Mahardika et al., 2009). Consumers' attitude is always towards some concepts, including towards a certain brand. Recalling from the aforementioned discussion, brand attitude functions as the consumers' overall evaluation of various physical and social objects, including products, brands, models, stores, and people, as well as aspects of marketing strategy. Consumers can also have attitudes towards intangible objects, such as concepts and ideas.

In any circumstances, a brand must consistently stimulate a certain image in consumers' minds. Consistency is very important for a brand in today's competitive situation. Many companies have to deal with problems after making a very simple mistake, following a long period of excellent reputation. It is not easy for consumers to forgive mistakes, and they tend to punish any mistake as they rely on three factors in processing information regarding a brand, namely perception, emotions, and behaviour.

Consistency in the long run will create commitment. Commitment is the final phase in branding. Happy consumers will have a higher tendency to be committed to the brand, and therefore will prioritise it as their primary preference. In the case of the Garuda logo makeover, this brand commitment is the ultimate goal to achieve. However, it is also a risky venture since many companies fail in doing so. In order to induce a strong brand commitment, airlines should change their service delivery in accordance to the brand image. This is why understanding brand makeover is very important in the service industry.

3 METHODOLOGY

This research uses both qualitative and quantitative approaches. The qualitative approach was performed using a semiotic analysis on the logo, design, and other elements of the brand. It was also broadened into a more detailed semiotic analysis related to shape, colour, typography, branding, logo, design, and other graphic elements. In addition, we also discuss the role of each element of the logo related to the process of building brand attitude and brand commitment. The quantitative approach, on the other hand, was performed using a correlation analysis. We distributed questionnaires to respondents who had been frequent passengers (frequent flyer members of Garuda Indonesia Airline).

Overall, this research uses the mix of qualitative and quantitative approaches in order to complement each other. The result from exploration through the qualitative approach is explained further by a quantitative approach.

3.1 *Semiotic analysis*

As stated previously, the qualitative approach in this research mainly focuses on semiotics. According to Peirce and Welby (1977), three elements in semiotics are: icon, index, and symbol. Icon is the relationship between the sign and the object, or a reference in the form of two or three dimensions that are similar. Index is a sign that indicates natural relationships between signs and markers that are causal or have a causal relationship; it is a sign that directly refers to the fact. Symbol is a relationship that is arbitrary and based only on conventions. Symbols also include all languages in general (i.e. the language of a particular tribe or nation, alphabetical characters, punctuation, vocabulary, phrases, and sentences). As for numbers, the Morse code, traffic signs, and flag state need to be studied further to avoid mistakes. It is fundamentally a process of giving meaning to the signs consisting of three interconnected elements, namely: Representamen [R], Object [O] and Interpretant [I]. [R] is a sign that refers to something represented by [O], or the object as a reference, which is something referred to by the sign (denotatum). On the other hand, [I] is part of the process of interpreting the relationship between [R] and [O].

To measure the validity of this qualitative study, the process of interpretation of icon, index, and symbol, supported by the use of secondary data sources (literature from books, journals, and

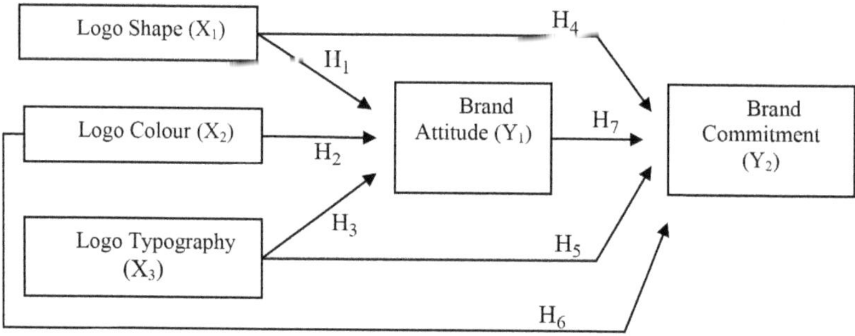

Figure 2. Research model.

online psychological articles) is also conducted. They have been obtained in order to engage with the meaning and message (verbal and non-verbal) contained in the new logo of Garuda Indonesia Airline through narrowing the logo elements into shape, colour, and typography. These main elements (shape, colour, and typography) were then put into the Peirce's semiotic instrument.

3.2 *Correlation analysis*

In order to strengthen the result of this study, we perform a follow-up examination by conducting a correlation analysis. It is aimed at examining the relationships between the logo makeover and the consumers' attitude. It is also aimed at providing an assessment of the new Garuda Indonesia Airline logo, as to whether consumers like or dislike it. We recruited 100 frequent flyer passengers of Garuda Indonesia Airline. The sampling technique chosen for this research is convenience sampling, using a non-random questionnaire distribution. We distributed the questionnaire both online and offline. The online questionnaire was made using the Google Docs service.

3.3 *Hypotheses*

Ultimately, this study aims at examining the influence of logo (shapes, colours, and typography) as a key brand element towards Garuda Indonesia's consumers' judgment (brand attitude and brand commitment), after the logo makeover. The conceptual framework can be seen as follows (see Figure 2).

In summary, based on the aforementioned discussion, we formulated each hypothesis as follows:

H1: Logo shape makeover is positively related to consumers' brand attitude.
H2: Logo colour makeover is positively related to consumers' brand attitude.
H3: Logo typography makeover is positively related to consumers' brand attitude.
H4: Logo shape makeover is positively related to consumers' brand commitment.
H5: Logo colour makeover is positively related to consumers' brand commitment.
H6: Logo typography makeover is positively related to consumers' brand commitment.
H7: Logo shape, logo colour, logo typography makeover is positively related to consumers' brand commitment with brand attitude as a mediating variable.

Descriptive and inferential statistics were analysed using SPSS.

4 RESULTS

4.1 *Semiotics analysis*

Based on the findings as a result of the survey of various literatures, it is revealed that most airline logos are associated with flying objects, which are dominated by symbols of a bird.

This can be explained easily since the bird is an animal that has the ability to fly high in the sky. By employing the bird symbol, airline companies want their fleets to have a bird-like quality and agility (i.e. being capable of flying away around the world). A variety of birds have been traditionally selected by airline companies to be used as their logos, such as pigeon, eagle, and mythical birds like Garuda.

Several airlines also combine the bird symbol with other objects. Flag carriers (the national carriers) are commonly the ones with historic logos that are the most eye-catching. Most of the ornate ones have been refreshed for the modern era and hence most of their shapes are decorated with curved lines and pastel colours. For example, some of the Middle East airlines infuse Arabic calligraphy or typography into their logos, and it works to convey brand distinctiveness. Low-cost airlines, on the other hand, have been at the forefront in making their airline colourful. Colour has played a key role for them since they need to get as much attention as they can for the lowest possible cost. Low-cost airlines also use celestial objects, such as suns, globes, and stars, similarly to signify a sense of distance or remoteness in space/air, and arguably with an added potential for global or world-wide reach.

Through a careful review of most airlines logos, we conclude that there are a significant number of airlines that use bird symbols in their logos. They mostly combine the bird illustration with other objects in order to show the national identity and airline classification, as well as to strengthen its existence. Usually, the most specific airline logo can be found on the tail fin of the aeroplanes.

In the case of Garuda Indonesia Airline, in the process of the logo makeover, they have added an element of graphical effects to the logo. If we look more carefully at the logo located on the tail fin of Garuda Indonesia's aeroplanes, it contains some graphical effects, such as brush stroke, diagonalising, streamlining, and gradation. These effects represent the aspects of motion and speed as the manifestation of energy passing through substance, such as when the wind blows a flag or when something is painted clearly with the stroke of a paintbrush. In addition, it also works through the association to the direct effect of velocity. Streamlining is related to the technological and physical reduction of friction for faster speed, just like the speed of a flying arrow. The final category is gradation, which is the gradual thinning of lines. That element of the logo is found on Garuda Indonesia's new logo, which is called the 'Nature's Wings' logo. It seems to work slightly differently through a geometric illusion. Specifically, it is called the kinetic effect. It represents the increasing level of abstraction in the representation of speed/velocity/motion. The spiky shapes (streamlining) in the 'Nature's Wings' logo of Garuda Indonesia Airline refer to stability. This shows that Garuda Indonesia Airline is a stable airline company that stands unwavering through any threats. The 'spiky' shapes look like a solid triangle, implying a stable and dynamic condition (Munger, 2006).

With regard to this new logo, it can be inferred that the streamlining style combines aesthetics and technology. In addition, it has a high symbolic value, and this made Garuda Indonesia Airline decide to use this style on its tail fin logo. It is expected to show the public that Garuda Indonesia is a reputable, sophisticated, fast, and professional airline. It is expected to enhance the product contact, so that it can attract many passengers and give a positive contribution to the 'Quantum Leap' programme of the airline.

In the semiotic analysis, colours also have an important role. Colours have psychological meanings that can be delivered to the viewer. These meanings can support the branding for the airlines. Each airline company has its own consideration in implementing certain colours on its tail fin. The colours on Garuda Indonesia's tail fin are blue, turquoise, white, tosca green and grey. Blue is a colour associated with business because it evokes a sense of balance as well as calm intelligence. Turquoise and tosca green are basically the same but different in the colour element. These two colours are associated with finance, safety, and nature. They also represent harmony, freshness, ambition, and greed. White reflects innocence, purity, and cleanliness. Meanwhile, grey is closely related with authority, maturity, security, and stability.

To summarise, Garuda Indonesia Airline changes its logo by incorporating some elements. Those elements have philosophical messages whose purpose is to strengthen the logo in order to express corporate identity to potential passengers.

4.2 *Results from the correlation analysis*

Overall, the results of the analysis and discussion of the study have shown that all hypotheses have been proven. The following are important points that need to be highlighted:

1. The Garuda logo makeover has a positive influence, with as much as 19.6% on the consumers' brand attitude. Each element of the Garuda logo is positively related to the consumers' brand attitude, and typography is the most crucial element.
2. The Garuda logo makeover has a positive influence, with as much as 16.5% on the consumers' brand commitment. Each element of the Garuda logo is positively related to the consumers' brand commitment, and colour is the most crucial element.
3. Considering the consumers' brand attitude and commitment, 19.6% and 16.5% of determination is considered a small portion of contribution (the maximum is of course 100%), but even though it is only a small portion, the Garuda logo makeover has proven to be aligned with service improvement and the company's profitability.
4. Other factors, such as safety, ground handling, service level of front desk, air customer service, punctuality and having a pleasant flight, are also important for respondents. These factors also induce a positive influence towards the consumers' brand attitude and brand commitment.
5. Brand attitude is partially mediated by the relationship between the Garuda logo makeover and brand commitment.

5 CONCLUSION

The current Garuda Indonesia Airline logo makeover has a distinctive design in shape, colour, and typography. These changes can be analysed using the Peirce's semiotic approach.

The old Garuda bird symbol is still used, but the colour composition has changed. The new logo, which is called the 'Nature's Wings', has a spiky shape. From the design perspective, it is known as the kinetic effect. From the semiotic perspective, the spiky shape represents the increasing level of abstraction of speed/velocity/motion. By this semiotic perspective, as a proud flag carrier of Indonesia, Garuda Indonesia Airline intends to offer a fast airline service to bring passengers to their destinations.

The colours on the Nature's Wings logo are blue, turquoise, tosca green, white, and grey. From the semiotic and psychological perspectives, blue means professionalism, tosca green means freshness, white means cleanliness, grey means reputability and integrity. These colours deliver the message that Garuda Indonesia Airline is the proud flag carrier of Indonesia; it offers prime service quality and always brings the Indonesian atmosphere in its flight. The ground handling and aircraft cabin crew provide a warm and friendly service to the passengers. Through the colours, Garuda Indonesia Airline emphasises itself as a leading airline in Indonesia and always presents the Indonesian spirit in its service.

The results of the research on the Garuda logo makeover in relation to the brand attitude and brand commitment indicate several points as follows:

1. The Garuda logo makeover overall and partially has a positive influence, as much as 19.6% on the consumers' brand attitude. The rest is determined by other factors that are not included in the scope of this research. Based on the magnitude of the t value, the typography element of the Garuda logo is the most crucial factor that affects the customers' brand attitude.
2. The Garuda logo makeover overall and partially has a positive influence, as much as 16.5% on the consumers' brand commitment. The rest is determined by other factors that are not included in the scope of this research. Based on the magnitude of the t value, the colour element of the Garuda logo is the most crucial factor that affects the customers' brand commitment.
3. Brand attitude is partially mediated by the relationship between the Garuda logo makeover and brand commitment.

4. From those findings, even though the portions (19.6% and 16.5%) are considered small, the Garuda logo makeover has a positive influence on the consumers' brand attitude and brand commitment.

Overall, the results from this study provide insight into the role of the Peirce's semiotic approach in the logo design. It has validated the relationship between the logo makeover and the consumers' brand attitude and commitment. Future research needs to be undertaken in general with regards to the effect of the logo makeover on the overall image of a company, on both quantitative and qualitative levels.

There are several limitations inherent in this study. The limited object of the study and sample size may restrict generalisation of the study findings. The sampling technique adapted in the study may also limit the generalisation of the study findings. There are also other factors that are not included in the scope of this research, such as service quality, marketing strategy, safety, competition from other airlines, and organisation culture; however, those factors, in fact, have a bigger influence on determining the brand image.

REFERENCES

Baron, R. M., & Kenny, D. A. (1986). The moderator-mediator variable distinction in social the moderator-mediator variable distinction in social psychological research: Conceptual, strategic, and statistical considerations. *Journal of Personality and Social Psychology, 51*(6): 1173–1182. doi:10.1037/0022-3514.51.6.1173.

Carlson, N. R., and Buskist, W. (1997). *Psychology: the science of behavior*. Allyn and Bacon.

Gujarati, D. (2003). *Basic Econometrics*. Sixth Edition. Jakarta: Erlangga.

Hasan, A. (2013). *Marketing dan Case Study*. Yogyakarta: Center for Academic Publishing Service (CAPS).

Kotler, P., Keller, K. L., Brady, M., Goodman, M. & Hansen, T. (2009). *Marketing Management*. Essex, UK: Pearson Education Limited.

Mahardika, H., Ewing, M. & Thomas, D. (2009). Comparing the temporal stability of behavioral intention, behavioral expectation and implementation intention. In P*roceedings of the 2009 Australian and New Zealand Marketing Academy Conference, Melbourne, Australia.*

Malhotra, N. K. (2010). *Marketing research: An applied orientation*. New Jersey: Pearson.

Melewar, T. C. (2003). Determinants of the corporate identity construct: a review of the literature. *Journal of Marketing Communications, 9*(4), 195–220. doi:10.1080/1352726032000119161.

Munger, D. (2006). The Emotion of Shapes. *ScienceBlogs*, 27 March. http://scienceblogs.com/cognitivedaily/2006/03/27/the-emotion-of-shapes/#comments.

Peirce, C. S., & Welby, V. (1977). *Semiotic and Significs the Correspondence between Charles S. Peirce and Lady Victoria Welby*. Indiana University Press.

Peter, J. P. & Olson, J. C. (2010). *Consumer behavior & marketing strategy*. New York: McGraw Hill.

Pittard, N., Ewing, M. Jevons, C. (2007). Aesthetic theory and logo design: examining consumer response to proportion across cultures. *International Marketing Review, 24*(4), 457–473. doi:10.1108/02651330710761026.

Seo, E. J. & Park, J. W. (2016). A study on the impact of airline corporate reputation on brand loyalty. *International Business Research, 10*(1), 59. http://dx.doi.org/10.5539/ibr.v10n1p59.

Sobur, A. (2013). *Semiotics for Communication* Bandung: Remaja Rosdakarya.

Supranto, J. & Limakrisna, N. (2012). Guidance for Research. Jakarta: Mitra Wacana Media.

Turlow, C. & Aiello, G. (2007). National pride, global capital: A social semiotic analysis of transnational visual branding in the airline industry. *Visual Communication, 6*(3), 305–344. https://doi.org/10.1177/1470357207081002.

Walsh, M. F. (2005). Consumer response to logo shape redesign: The influence of brand commitment. Order No. 3206844, University of Pittsburgh. https://search.proquest.com/docview/305449396?accountid=17242.

Walsh, M. F., Winterich, K. P. & Mittal, V. (2010). Do logo redesigns help or hurt your brand? The role of brand commitment. *Journal of Product & Brand Management, 19*(2), 76–84. doi:10.1108/10610421011033421.

Widagdo. (2000). *Design and Culture*. Bandung: Dirjen Dikti Depdiknas.

Zerlina, A., Haryono, E. & Elvianti, W. (2016). Optimizing the implementation of quantum leap program to face the challenges of ASEAN open sky policy: Lessons from Garuda Indonesia. Aegis: J*ournal of International Relations, 1*(1). http://e-journal.president.ac.id/presunivojs/index.php/AEGIS/article/view/75/588.

Competition and Cooperation in Economics and Business – Gani et al. (Eds)
© 2018 Taylor & Francis Group, London, ISBN 978-1-138-62666-9

Reminiscent bundling and product type influence on intention to purchase: A case in music industry

C.A.E. Hindarto, A.Z. Afiff, T.E. Balqiah & H. Mahardika
Department of Management, Faculty of Economics and Business, Universitas Indonesia, Depok, Indonesia

ABSTRACT: Bundling is one of the options for marketers in providing benefits for consumers in a form of an offering package consisting of two or more products with a cheaper price than the total price of those products when they are sold separately. Experience products are also offered as a bundle by marketers. However, studies on bundling of experience products remain limited despite the potential to explore their influence on intention to purchase. This study examines how experience product combination in a bundle might influence intention to purchase. There are differences in intention to purchase between physical and experience product bundles. Reminiscent bundling is used as the term for the bundling of two experience products in this study. The main finding of this study indicates that intention to purchase is significantly higher in the reminiscent bundling condition. Another important finding is that intention to purchase depends on the product type.

1 INTRODUCTION

Bundling is an alternative for marketers in adding an offering value to the customers. The combinations of products can take several forms: two or more tangible products, two or more intangible products, or a combination of tangible and intangible products. This approach is common for introducing a novel product to the market in order to raise awareness or intention to purchase. In some conditions, the new product is bundled with the established product. The discussion about bundling can be referred to as typologies. Three related bundling issues are the offering form, the strategy, and the product combination. Marketers have options to choose one of those alternatives or to combine them. Bundling enables profit when marketers encounter price competition (Mantovani, 2013). Nevertheless, bundling profitability depends on product demand, bundling costs, and the relationship between the products in the bundling offering (McCardle et al., 2007).

Most bundling discussions have centred around tangible products, both consumer and industrial goods. Since in practice marketers also apply bundling for intangible products, there are possible research avenues to further explore bundling of this type of product. Specifically, with the rising of experiential marketing, which creates consumers' engagement, there is a fertile ground to discuss bundling efficacy. Experiences can be classified into five levels, which are sense, feel, think, act, and relate (Schmitt, 1999). These aspects facilitate consumers and product attachment in the form of memorable experience (Pine & Gilmore, 1999).

Music as one of the experience products (Lacher, 1989) can be offered as a bundled product as well. Marketers in the music industry might apply bundling as one of the alternatives in selling their products. The forms of bundling in the music industry can be relevant to selling recording products, ticket concerts, merchandise, or any other products that relate to the artists. Marketers can combine music articles into a package consisting of tangible and intangible products.

Research on music product bundling remains limited although other industries have successfully applied it. Music as an experience product has uniqueness when it is offered as in a bundle. A specific term for the experience product in this study is called reminiscent bundling, where each product raises memory of another product (Ekananda, 2014). It is an addition to the current terminology of bundling (Chung et al., 2013). The term is necessary since both products cannot be identified as independent, substitutes, or complementary. Although there is a connection between them, the products do not fit the existing categories.

The research gap in this type of bundling is also related to inconclusive results from the previous studies. One study explored intention to purchase physical music, in a CD format, which was bundled with an experience product, a live music concert. For a comparison, a bundle of CD and merchandise was also used in order to find out which offer was more preferred by the research participants. This study discovered that the research participants preferred a bundle of music physical product and live concert in solo artist, but not in a band (Ekananda, 2013a).

Previous studies discussed product bundling which consisted of physical and experience products (Ekananda, 2013a; 2013b). However, there is still a possibility to explore this issue further. Regarding the intention to purchase music products, there have been studies investigating the general condition (Flynn et al., 1995; Lacher & Mizerski, 1994) and digital music (Chiang & Assane, 2007; Chu & Lu, 2007). No research has examined intention to purchase a music bundle that consists of physical and experiential products.

Venkatesh and Mahajan (1993) discuss intention to purchase between single and season tickets. They discussed identical experiential products with different quantities, but their research did not discuss the performance types. There is a possibility that consumers prefer one product type to its alternative offer, and product type can increase or decrease the intention to purchase. It is worth examining the impact of product type in a bundling offer on the intention to purchase.

Based on these arguments, there are two main questions that guided this research:

1. Which bundle, independent or reminiscent, is more influential in intention to purchase music products?
2. Which type of product is more influential in intention to purchase in music products?

The main aim of this research is in exploring the influence of reminiscent bundling as an experiential offer on the intention to purchase, which is novel in bundling literature. The second aim is in recognising the effect of product type on intention to purchase a bundling offer. The findings serve as evidence for marketers that product type has a different impact, although the bundling offer is the same.

This research is expected to contribute to an improved understanding of product bundling that consists of experiential products. Academicians and practitioners need to acknowledge that the kind of product bundling has a different character and effect on consumers' intention to purchase. Another contribution is to emphasise the importance of experiential product type, which can have an effect on intention to purchase.

2 THEORETICAL BACKGROUND

2.1 *Bundling*

Bundling is a marketing effort to offer two or more products in one package at a cheaper price than when those products are sold separately. Two definitions of bundling are 'marketing practice for two or more products and/or service in a single package with a special price' (Guiltinan, 1987) and 'selling two or more separate products in one package' (Stremersch & Tellis, 2002).

Bundling can be discussed in three categorisations: kind of offer, strategy, and component. As a kind of offer, bundling can be categorised as price bundling and product bundling. The former one is defined as selling two or more separate products with discount, without any integration between them (Stremersch & Tellis, 2002). Implicitly, the separate products can be inferred as being individual products, which can be sold independently (Stremersch & Tellis, 2002). While price bundling is a tool for setting the price and promotion, the product package is more about strategic element that relates to sustaining value-added creation with a longer time-frame.

As a strategy, there are options in applying bundling, which are unbundling, pure-bundling, and mixed-bundling. Marketers sell the products independently in the first strategy (Adams & Yellen, 1976). Pure-bundling is defined as a strategy where marketers sell the products only in a bundling format or the products are not sold separately (Stremersch & Tellis, 2002). There is solely a single price for the bundle. In the mixed-bundling strategy, marketers offer separate products and a bundle of the products (Adams & Yellen, 1976; Stremersch & Tellis, 2002). The latter strategy is called mixed-price bundling (Guiltinan, 1987). Mixed-bundling has the highest profitability among the other two strategies (Chen & Riordan, 2013; Venkatesh & Kamakura, 2003; Yang & Ng, 2010). Furthermore, research participants preferred mixed-bundling over pure-bundling (Hamilton & Koukova, 2008; Venkatesh & Mahajan, 1993).

According to the literature, there are three categories of bundling component (Chung et al., 2013). The first category is independent package, where there is no relation between products. The second category is complementary bundling, where products are accompanying each other. There is integration between products, which is a characteristic of product bundling. The third category is substitution package, where the products are able to replace each other.

Most of the previous research on bundling discusses tangible products, services, and information (Adilov, 2011; Arora, 2011; Chang & Yang, 2012; Chung et al., 2013; Hamilton & Koukova, 2008; Koukova et al., 2008; Sheng & Pan, 2009). Although some bundles include experience products, there is no particular term for this specific kind of bundling. Chung et al. (2013) do not mention a category that includes experience products. A specific category for experience products is reminiscent bundling, where each product triggers a positive memory (Ekananda, 2014). Specifically, this terminology is used for bundling of experience products in the music industry, which are a CD and a concert ticket. These products potentially trigger positive memory of each other when customers experience one of them.

In order to understand the effect of reminiscent bundling offering on intention to purchase, the following hypothesis was proposed:

H1: Intention to purchase a reminiscent bundle is higher than that to purchase an independent bundle.

2.2 *Music as an experience product*

The essence of music products is an audial consumption. However, music is not only a product for listening experience since music can be in the form of a product that engages multiple senses. Consumers enjoy music individually through personal devices or collectively by attending a concert. Music as an audial product develops from a physical recording artefact to digital format, which is intangible. Consumers' engagement and the format of music products have influence on the selection and purchase of the products.

A model of music is a product based on individual characteristics (personality type, affection intensity, and music training) that respond to music emotionally, experientially, and analytically (Lacher, 1989). Furthermore, these responses affect music preference and intention to purchase. Sensory, emotional, imagery, and analytical responses towards music have a direct influence on affection and experiential (Lacher & Mizerski, 1994). The strongest indicator of intention to purchase music is to re-experience. A response towards imagery and congruency between music and self is potentially creating a positive emotion (Kwortnik & Ross, 2007).

Three psychological aspects important for consumers to collect music recording products are the product as a sacred object, collecting as a self-facet, and music as a sensual experience (Giles et al., 2007). The first and second aspects are related to tangible music products, while the third one refers to both digital download and physical format. Physical music format shapes a high engagement, where music is positively correlated with subjective knowledge, tangibility preference and portable device usage (Styvén, 2010). The engagement with the physical music product is more valuable than the digital one (Citrin et al., 2003; Styvén, 2010).

Hedonic products through the experiential process have a strong influence on satisfaction and loyalty (Bigné et al., 2008). Consumers' satisfaction, price fairness perception, intention to pay, customer value, and product quality are significantly higher among a concert audience than CD listeners (Rondán-Cataluña et al., 2010). An audience's activities in co-creating and co-producing during a concert are the possible reasons (Chaney, 2012). Research findings have shown that discovered experience product purchase tended to facilitate happiness more compared to material-based purchase (Nicolao et al., 2009). In conclusion, music consumers enjoy music more in the concert setting than when listening to CD. They also perceive concert ticket price as being fairer than CD price.

The types of artist and music genre speculatively influence consumers along with the product format. Results from qualitative research indicated that identity expression and family background where consumers grew up influenced music listener categories (Nuttall, 2008). The three categories were experiential, chameleon, and defender listeners. Individuals symbolically consume recorded music as a self-representation, related to self-concept, symbolic tools and consumption situation (Larsen et al., 2009; 2010). Personal self-identity is related to music through music genre, artist, and favourite CD (Berger & Heath, 2007). It was also found that individuals' differences variables were better predictors compared to demographic variables in predicting buyer behaviour in rock music (Flynn et al., 1995). It can be summarised that although family background influences music taste, individual characters play a dominant role in shaping behaviour in purchasing music products.

Individuals' music preferences reflect information about their personal quality and membership in a particular social group (Rentfrow & Gosling, 2007). Four dimensions of music preference were identified, and they are reflective and complex (such as blues, jazz, classical, folk); intense and rebellious (such as rock, alternative, heavy metal); upbeat and conventional (such as country music, soundtrack, religious, popular); and energetic-rhythmic (such as hip-hop, soul/funk, electronic/dance) (Delsing et al., 2008). The majority of music consumers purchase popular music (Ballard et al., 1999; Schellenberg & von Scheve, 2012). Popular music positively pertains to extraversion, agreeableness, and conscientiousness, while it is negatively associated with openness to new experience (Rentfrow & Gosling, 2003). Using the Thayer model, it was identified that popular music falls into a quadrant of positive valence and low arousal, which creates relaxed, peaceful and calm feelings (Bang et al., 2013). Rebellious artists, on the contrary, have an intense music dimension, which is suitable for audiences who are looking for a strong sensation-seeking scale (Higdon & Stephens, 2008). This kind of artist is positioned at a quadrant of negative valence and high arousal, which creates annoying, angry, and nervous feelings (Bang et al., 2013).

Artists are triggering music product creation, and marketing roles are ensuring the music to be heard (Ogden et al., 2011). Consumer behaviour in purchasing music products deals with self-identity (Berger & Heath, 2007). Ten types of male artists and six types of female artists were identified. Anti-heroes, which in this study is called rebellious, have aggressive, boisterous, brash, angry, and angst-ridden characteristics (Donze, 2011), while romantic pop artists were identified as tender, relaxed, soothing, and intimate. These two specific artist characters were chosen as product types in this research.

Based on the arguments, this research differentiated between romantic and rebellious as product types. Hence, the hypothesis was proposed:

H2: Intention to purchase a reminiscent bundle with a romantic artist is higher than the one with a rebellious artist.

3 DESIGN AND METHODS

This research was conducted using the experimental approach. This method enabled the researchers to know the causal inference of behaviour, and it had high validity (Myers &

Hansen, 2006). Two ways to answer the research questions were in doing observation without any intervention, or by manipulating environmental aspects, and observing the influence of the independent variables on the dependent variable in order to discover their effects (Field & Hole, 2003).

Research on reminiscent bundling is relatively new in marketing. The reason for featuring music products was the consumers' familiarity with the forms, in this case recording and performing products. Since the featured artists had a very strong influence on consumers' purchase decision, it was necessary to disguise the artist. Therefore, a fictional artist was created for this purpose. The survey was conducted to find out the preferences for artists' gender, music product prices (CD, concert, and merchandise), concert venue, merchandise type, and bundling price. A questionnaire was developed based on the findings from the survey.

Prior to the main study, a pilot study was conducted to ensure whether the types of artists were identified correctly by the respondents. Another purpose of the pilot study was to obtain findings regarding the influence on intention to purchase for each of the music products in every artist type.

3.1 *Survey and pilot study*

The questionnaire in this study was developed through preliminary surveys, which were conducted to obtain the preference for artists' gender, product price, concert venue, merchandise type, and bundling price. The participants in this survey and the research study were in a similar age range. The participants' age range for this research was between 18 and 24 years old, chosen based on the previous research (Hemming, 2013; Holbrook & Schindler, 1989). CD price was used as the reference point in setting the product price. The majority of respondents preferred a male solo singer, IDR50,000 for the product price, a concert venue at Rolling Stone Cafe Jakarta, T-shirt as merchandise, and IDR80,000 for bundling price. The CD price was set equal to that of concert ticket or merchandise. Instead of paying IDR100,000 for both products, the bundling price was IDR80,000 for CD/ticket and CD/merchandise options.

The next step was conducting the pilot study. The objectives of the pilot study were to find information about the intention to purchase individual products as control conditions and to confirm the difference between the two artist types. There were 172 participants (69 females and 103 males) who were divided into six groups with two independent variables, product offering (CD, concert ticket, or T-shirt) and artist type (rebellious or romantic). Intention to purchase as a dependent variable was measured using a 5-point Likert scale ranging from 1 (very disagree) to 5 (very agree). The scenarios were presented using a new unknown solo male artist to mitigate participants' subjective musical taste. However, there was music influence information for each artist type to give reference for the participants. Music genres were not explicitly mentioned in order to avoid bias towards a specific music genre. The only exposure was the scenario and questionnaire. There was neither music demo played nor artists' pictures to mitigate the participants' subjective consideration.

A manipulation check was performed in order to confirm the difference between romantic and rebellious artists. Using three statements with a score from 1 (very rebellious) to 5 (very romantic), the three groups were assigned for each artist type. The groups in romantic artist condition had a mean of 4.36, while the ones in rebellious condition had a mean of 2.13. The mean difference between the two conditions was significant ($p < 0.00$). It indicates that the participants could classify the artist types correctly. Another manipulation check was used to differentiate focal from companion products for study 2. The difference between the roles of products in bundling was significant ($p < 0.05$). It indicates that the participants could identify which product was positioned as the focal product.

Table 1 displays the pilot study descriptive statistics. Intention to purchase the CD and concert ticket was higher in romantic artist condition than in the rebellious one. For both artist types, intention to purchase the T-shirt was relatively similar. Intention to purchase the concert ticket was the highest compared to the other two products for both conditions. This finding is consistent with prior research that claimed intention to purchase a concert ticket was higher than that to purchase a CD (Rondán-Cataluña et al., 2010).

Using one-way ANOVA for product offering (Table 2), a significant difference between groups was found. The difference between groups for artist type (Table 3) was insignificant. This result could be attributed to the effect of the new artist being unknown to the participants.

A t-test was performed to understand the difference of intention to purchase different product offerings. Intention to purchase the CD was significantly higher than that to purchase the T-shirt ($M_{CD} = 2.83 > M_{T\text{-shirt}} = 2.47$, $p < 0.05$). There was also a significant difference between intention to purchase the concert ticket and the T-shirt ($M_{concert} = 3.09 > M_{T\text{-shirt}} = 2.47$, $p < 0.01$). However, there was an insignificant difference between artist types.

3.2 *Research study*

Scenarios for research study used mixed-bundling strategy, where the participants had options to purchase the products either as individual products or in a bundle. The distinct features in this research were bundling components, product type, and industry context. There were two component-based bundles in the study, which were independent (experience and functional products) and reminiscent (both experience products). The objective of the study was to find the differences between the intention to purchase bundling component and the artist type as independent variables. The research design was 2 (bundling component: reminiscent vs independent) × 2 (artist type: rebellious vs romantic) between-subject. The reminiscent

Table 1. Descriptive statistics in the pilot study.

Product offering	Artist type	Mean	Standard deviation	N
CD	Rebellious	2.72	0.751	29
	Romantic	2.93	0.961	29
	Total	2.83	0.861	58
Concert ticket	Rebellious	2.93	0.874	27
	Romantic	3.24	0.830	29
	Total	3.09	0.859	56
T-shirt	Rebellious	2.48	1.056	29
	Romantic	2.45	0.948	29
	Total	2.47	0.995	58
Total	Rebellious	2.71	0.911	85
	Romantic	2.87	0.962	87
	Total	2.79	0.938	172

Table 2. Product offering in the pilot study.

	Sum of squares	df	Mean square	F	Sig.
Between groups	11.205	2	5.602	6.799	0.001
Within groups	139.26	169	0.824		
Total	150.465	171			

Table 3. Artist type in the pilot study.

	Sum of squares	df	Mean square	F	Sig.
Between groups	1.209	1	1.209	1.377	0.242
Within groups	149.256	170	0.878		
Total	150.465	171			

bundle consisted of two experience products, a CD and a concert ticket, while the independent bundle included a CD and a T-shirt. In both conditions, each product did not complement and substitute each other. However, there was a memorable relation between the CD and the concert. That was the reason to call it a reminiscent bundle. Although the T-shirt had a connection with the music, its combination with the CD was categorised as independent.

163 participants (61% males) from a private university in Jakarta were involved in this study. Table 4 displays the study descriptive analysis. For both artist types and in total, intention to purchase the reminiscent bundle was higher than that to purchase the independent bundle. Intention to purchase the romantic artist in both bundling conditions was higher than that of the rebellious artist. Using one-way ANOVA for each bundling offer in the inferential analysis (Table 5), a significant difference between them was identified. It indicates that the reminiscent bundling significantly differs from independent bundling. Intention to purchase in reminiscent bundling condition was higher than in independent bundling condition ($M_{reminiscent} = 3.05 > M_{independent} = 2.73$, $p < 0.10$). For the artist types (Table 6), there was a strong indication of higher intention to purchase bundling offers in romantic artist condition, compared to the rebellious artist ($M_{romantic} = 3.098 > M_{rebellious} = 2.683$, $p < 0.05$).

Next, an inferential analysis for bundling offers in rebellious artist was conducted. In reminiscent bundling condition (Table 7), the purchase intention was higher than in independent bundling ($M_{reminiscent} = 2.9024 > M_{independent} = 2.4634$, $p < 0.1$). However, the difference between bundling offers in romantic artist condition was insignificant.

Since there were options to purchase the products individually in mixed-bundling offers, intention to purchase between component conditions in the pilot study was compared with

Table 4. Descriptive statistics in the research study.

Artist	Bundling	Mean	Standard deviation	N
Rebellious	Reminiscent	2.9024	1.09098	41
	Independent	2.4634	0.9246	41
	Total	2.6829	1.02894	82
Romantic	Reminiscent	3.2	1.26491	40
	Independent	3	1.04881	41
	Total	3.0988	1.15764	81
Total	Reminiscent	3.0494	1.18217	81
	Independent	2.7317	1.01894	82
	Total	2.8896	1.11115	163

Table 5. One-way ANOVA for bundling offers in the research study.

	Sum of squares	df	Mean square	F	Sig.
Between groups	4.112	1	4.112	3.38	0.068
Within groups	195.9	161	1.217		
Total	200.012	162			

Table 6. One-way ANOVA for artist types in the research study.

	Sum of squares	df	Mean square	F	Sig.
Between groups	7.046	1	7.046	5.879	0.016
Within groups	192.966	161	1.199		
Total	200.012	162			

Table 7. One-way ANOVA for bundling offer in rebellious artist condition in the research study.

	Sum of squares	df	Mean square	F	Sig.
Between groups	3.951	1	3.951	3.864	0.053
Within groups	81.805	80	1.023		
Total	85.756	81			

that in bundling condition in the research study. Intention to purchase the CD was higher in the bundling than in the control condition, while intention to purchase the concert ticket or merchandise was higher when they were sold in unbundling condition.

4 CONCLUSION

The findings of this study present several contributions to bundling literature. First, reminiscent bundling is preferable to independent bundling. The participants had higher intention to purchase a bundle consisting of two experience products that reminded them of each other, compared to a bundle of tangible and experience products. Second, this work also highlights the importance of product type as bundling element. Although the reminiscent bundle consisted of similar items, the participants' intention to purchase was influenced by the product type.

The reminiscent bundle was perceived by the participants to give higher satisfaction than the independent bundle, based on transaction utilisation (Chang & Yang, 2012). In the music industry, a combination of a CD and a concert ticket as a reminiscent bundle is interesting. The findings in this research have confirmed that the participants valued a concert ticket more than a T-shirt (Ekananda, 2013a). The consumers' engagement in a concert setting deals with multiple senses, similar to the definition of experiential marketing. In this case, consumers get more than just material benefits. They are involved in the experience mentally and imagery (Schmitt, 1999). From the marketers' view, the integration of recording and concert selling provides better profitability (Grönroos & Voima, 2013).

The research also supported the influence of artist type on the marketing (Dewenter et al., 2012) regarding purchase intention in the bundling offers. The positive image of a romantic artist influenced the participant's intention to purchase more than the rebellious one who was perceived negatively. Previous literature did not explicitly discuss the influence of artist type on consumers' preference (Ogden et al., 2011). This research found that there was an impact of artist type on the marketing context. Negative perception of the rebellious artist could be the participants' way in protecting their personal qualities and membership in a certain social group (Rentfrow & Gosling, 2007). This research also discussed the relationship between individual identity and music (Berger & Heath, 2007). Specifically, they mentioned that music genre, artist, and favourite CD were three of four stronger issues in influencing self-identity.

Another finding in this study was a preference for attending the concert of the rebellious artist over purchasing a T-shirt although the artist was unknown. Involvement in this artist type's concert raised enthusiasm, like in the escapist condition where active participation and immersion occurred (Pine & Gilmore, 1999). Since the rebellious dimension is connected with openness to a new experience (Rentfrow & Gosling, 2003), it can be assumed that consumers will possibly get a novel experience by participating in the concert of this artist type.

REFERENCES

Adams, W.J. & Yellen, J.L. (1976), Commodity bundling and the burden of monopoly. *Quarterly Journal of Economics*, *90*(3), 475–98. doi:10.2307/1886045.

Adilov, N. (2011). Bundling information goods under endogenous quality choice. *Journal of Media Economics*, *24*(1), 6–23. http://dx.doi.org/10.1080/08997764.2011.549425.

Arora, R. (2011). Bundling or unbundling frequently purchased products: A mixed method approach. *Journal of Consumer Marketing, 28*(1), 67–75. doi:10.1108/07363761111101967.
Ballard, M.E., Dodson, A.R. & Bazzini, D.G. (1999). Genre of music and lyrical content: Expectation effects. *The Journal of Genetic Psychology, 160*, 476–87. doi:10.1080/00221329909595560.
Bang, S.W., Kim, J. & Lee, J.H. (2013). An approach of genetic programming for music emotion classification. *International Journal of Control, Automation and Systems, 11*(6), 1290–1299. doi:10.1007/s12555-012-9407-7.
Berger, J. & Heath, C. (2007). Where consumers diverge from others: Identity signaling and product domains. *Journal of Consumer Research, 34*(2), 121–134. doi:10.1086/519142.
Chaney, D. (2012). The music industry in the digital age: Consumer participation in value creation. *International Journal of Arts Management, 15*(1), 42–52.
Chang, K.F. & Yang, H.W. (2012). An empirical study to construct a systematic model for product bundles. *Information Technology Journal, 11*(6), 699–706. http://docsdrive.com/pdfs/ansinet/itj/2012/699–706.pdf.
Chen, Y. Riordan, M.H. (2013). Profitability of product bundling. *International Economic Review, 54*,(1), 35–57. doi:10.1111/j.1468–2354.2012.00725.x.
Chiang, E.P. Assane, D. (2007). Determinants of music copyright violations on the university campus. *Journal of Cultural Economics, 31*(3), 187–204. doi:10.1007/s10824-007-9042-y.
Chu, C.W. & Lu, H.P. (2007). Factors influencing online music purchase intention in Taiwan: An Empirical Study Based on the Value-Intention Framework. *Internet Research, 17*(2), 139–155. doi:10.1108/10662240710737004.
Chung, H.L., Lin, Y.S. & Hu, J.L. (2013). Bundling strategy and product differentiation. *Journal of Economics/Zeitschrift Fur Nationalokonomie, 108*(3), 207–229. doi:10.1007/s00712–012–0265–9.
Citrin, A.V., Stem, D.E., Spangenberg, E.R. & Clark, M.J. (2003). Consumer need for tactile input: An Internet retailing challenge. *Journal of Business Research, 56*(11), 915–922. doi:10.1016/S0148-2963(01)00278-8.
Delsing, M.J.M.H., Ter Bogt, T.F.M., Engels, R.C.M.E. & Meeus, W.H.J. (2008). Adolescents' music preferences and personality characteristics. *European Journal of Personality, 22*(2), 109–130. doi:10.1002/per.665.
Dewenter, R., Haucap, J. & Wenzel, T. (2012). On file sharing with indirect network effects between concert ticket sales and music recordings. *Journal of Media Economics, 25*,(3), 168–178. doi:10.1080/08997764.2012.700974.
Donze, P.L. (2011). Popular music, identity, and sexualization: A latent class analysis of artist types. *Poetics, 39*(1), 44–63. doi:10.1016/j.poetic.2010.11.002.
Ekananda, A.B. (2013a). Does the product bundling influence consumers to purchase compact disc? In *Proceeding of the 10th EBES Conference, Eurasia Business and Economics Society, Istanbul, May 23–25, 2013.* Istambul: Taksim Nippon Hotel.
———. (2013b). Product bundling in music business: Free cd or free concert. In *Proceeding of 12th International Colloquium on Nonprofit, Arts, Heritage, and Social Marketing, Edinburgh, 2013.* Edinburgh: Heriot-Watt University.
———. (2014). Experiential product bundling category: A case in music industry. *In Proceeding of Spring 2014 MMA Conference, Marketing Management Association, Once Retro Now Novel Again, Chicago, March 26–28, 2014*, 46–50. Chicago, IL.
Enrique Bigné, J., Mattila, A.S. & Andreu, L. (2008). The impact of experiential consumption cognitions and emotions on behavioral intentions. *Journal of Services Marketing, 22*,(4), 303–315. doi:10.1108/08876040810881704.
Field, A. & Hole, G.J. (2003). *How to design and report experiments*. London: Sage Publication.
Flynn, L.R., Eastman, J K. & Newell, S.J. (1995). An exploratory study of the application of neural networks to marketing: Predicting rock music shopping behavior. *Journal of Marketing Theory & Practice, 3*(2), 75. http://search.ebscohost.com/login.aspx?direct=true&db=buh&AN=5382022&site=ehost-live.
Giles, D.C., Pietrzykowski, S. & Clark, K.E. (2007). The psychological meaning of personal record collections and the impact of changing technological forms. *Journal of Economic Psychology*, 28(4), 429–443. doi:10.1016/j.joep.2006.08.002.
Grönroos, C. & Voima, P. (2013). Critical service logic: Making Sense of value creation and co-creation. *Journal of the Academy of Marketing Science, 41*(2), 133–150. doi:10.1007/s11747–012–0308–3.
Guiltinan, J.P. (1987). The price bundling of services: A normative framework. *Journal of Marketing, 51*(2), 74–85. doi:10.2307/1251130.
Hamilton, R.W. & Koukova, N.T. (2008). Choosing options for products: The effects of mixed bundling on consumers' inferences and choices. *Journal of the Academy of Marketing Science, 36*(3), 423–433. doi:10.1007/s11747–007–0083–8.

Hemming, J. (2013). Is there a peak in popular music preference at a certain song-specific age? A replication of Holbrook & Schindler's 1989 study. *Musicae Scientiae*, *17*(3), 293–304. https://doi.org/10.1177/1029864913493800.

Higdon, L.G. & Stephens, E.G. (2008). Preferred music genre: The influence of major personality factors. *Psi Chi Journal of Undergraduate Research*, *13*(3), 140–147.

Holbrook, M.B. & Schindler, R.M. (1989). Some exploratory findings on the development of musical tastes. *Journal of Consumer Research*, *16*(1), 119–124. doi:10.1086/209200.

Koukova, N.T., Kannan, P.K. & Ratchford, B.T. (2008). Product form bundling: Implications for marketing digital products. *Journal of Retailing*, *84*(2006), 181–194. doi:10.1016/j.jretai.2008.04.001.

Kwortnik, R.J. & Ross, W.T. (2007). The role of positive emotions in experiential decisions. *International Journal of Research in Marketing*, *24*(4), 324–335. doi:10.1016/j.ijresmar.2007.09.002.

Lacher, K.T. (1989). Hedonic consumption: music as a product. *Advances in Consumer Research*, *16*(1), 367–373. http://search.ebscohost.com/login.aspx?direct=true&db=buh&AN=6487732&site=ehost-live.

Lacher, K.T. & Mizerski, R. (1994). An exploratory study of the responses and relationships involved in the evaluation of, and in the intention to purchase New Rock music. *Journal of Consumer Research*, *21*(2), 366–380. http://www.jstor.org/stable/2489827.

Larsen, G., Lawson, R. & Todd, S. (2009). The consumption of music as self-representation in social interaction. *Australasian Marketing Journal*, *17*, (1), 16–26. doi:10.1016/j.ausmj.2009.01.006.

———. (2007). The content and validity of music-genre stereotypes among college students. *Psychology of Music*, *35*(2), 306–326. doi:10.1177/0305735607070382.

Mantovani, A. (2013). The strategic effect of bundling: A new perspective. *Review of Industrial Organization*, *42*(1), 25–43. doi:10.1007/s11151-012-9361-9.

McCardle, K.F., Rajaram, K. & Tang, C.S. (2007). Bundling retail products: Models and analysis. *European Journal of Operational Research*, *177*(2), 1197–1217. doi:10.1016/j.ejor.2005.11.009.

Myers, A. & Hansen, C.H. (2006). *Experimental Psychology* (6th ed.). Belmont: Thomson Wadsworth.

Nicolao, L., Irwin, J.R. Goodman, J.K. (2009). Happiness for sale: Do experiential purchases make consumers happier than material purchases? *Journal of Consumer Research*, *36*(2), 188–198. doi:10.1086/597049.

Nuttall, P. (2008). For those about to rock: A new understanding of adolescent music consumption. *Advances in Consumer Research*, *35*, 401–408. http://www.acrwebsite.org/volumes/v35/naacr_vol35_5.pdf.

Ogden, J.R., Ogden, D.T. & Long, K. (2011). Music marketing: A history and landscape. *Journal of Retailing and Consumer Services*, *18*(2), 120–125. doi:10.1016/j.jretconser.2010.12.002.

Pine, B.J. & Gilmore, J.H. (1999). *The Experience economy: Work is theatre and every business is a stage*. Boston: Harvard Business School Press.

Rentfrow, P.J. & Gosling, S.D. (2003). The Do Re Mi's of everyday life: The structure and personality correlates of music preferences. *Journal of Personality and Social Psychology*, *84*(6), 1236–1256. doi:10.1037/0022-3514.84.6.1236.

———. (2010). The symbolic consumption of music. *Journal of Marketing Management*, *26*(7–8), 671–685. doi:10.1080/0267257X.2010.481865.

Rondán-Cataluña, F.J. & Martín-Ruiz, D. (2010). Customers' perceptions about concerts and CDs. *Management Decision*, *48*(4), 1410–1421. doi:10.1108/00251741011082152.

Schellenberg, E.G. & von Scheve, C. (2012). Emotional cues in American popular music: Five decades of the Top 40. *Psychology of Aesthetics, Creativity, and the Arts*, *6*(3), 196–203. doi:10.1037/a0028024.

Schmitt, B.H. (1999). *Experiential marketing: How to get consumers to sense, feel, think, act, relate to your company and brands*. New York: The Free Press.

Sheng, S. & Pan, Y. (2009). Bundling as a new product introduction strategy: The role of brand image and bundle features. *Journal of Retailing and Consumer Services*, *16*(5), 367–376. doi:10.1016/j.jretconser.2009.04.003.

Stremersch, S. & Tellis, G.J. (2002). Strategic bundling of products and prices: A new synthesis for marketing. *Journal of Marketing*, *66*(1), 55–72. doi:10.1509/jmkg.66.1.55.18455.

Styvén, M.E. (2010). The need to touch: Exploring the link between music involvement and tangibility preference. *Journal of Business Research*, *63*(9–10), 1088–1094. doi:10.1016/j.jbusres.2008.11.010.

Venkatesh, R. & Kamakura, W.A. 2003. Optimal bundling and pricing under a monopoly: Contrasting complements and substitutes from independently valued products. *The Journal of Business*, *76*(2), 211–231. doi:10.1086/367748.

Venkatesh, R. & Mahajan, V. (1993). A probabilistic approach to pricing a bundle of products or services. *Journal of Marketing Research*, *30*(4), 494–508. doi:10.2307/3172693.

Yang, B. & Ng, C.T. (2010). Pricing problem in wireless telecommunication product and service bundling. *European Journal of Operational Research*, *207*(1), 473–480. doi:10.1016/j.ejor.2010.04.004.

Competition and Cooperation in Economics and Business – Gani et al. (Eds)

The measurement of operational risk capital costs with an advanced measurement approach through the loss distribution approach (A case study in one of the Indonesia's state-owned banks)

R. Hartini, S. Hartoyo & H. Sasongko
School of Business, Bogor Agricultural University, Bogor, Indonesia

ABSTRACT: The rapid growth of the banking business requires banks to adapt quickly and to be supported by reliable risk management. In contrast to the market and credit risks, an operational risk is the first risk type known by the banks, but the least understood compared to market and credit risks. Basel II (International Committee for setting up bank risk management) defines an operational risk as the arising risk from the failure of internal processes, people, systems, or external events. Basel II also sets the standard and internal calculation modelling that must be applied by the banks. This research discusses the method for a bank to measure the operational risk capital cost accurately with the Advanced Measurement Approach (AMA), that requires historical data (Loss Event Database) regarding operational loss events. This advanced approach uses mathematics and probabilistic calculation, that highly likely provides an accurate result. This research found that the Loss Distribution Approach has high accuracy for calculating operational risk on every event of the eight bank business lines. It is known that the largest fraud is derived from internal bank operation.

1 INTRODUCTION

Indonesia is one of the countries that has a banking industry with good performance. This can be seen from the level of ROA (Return on Assets) of 3.03% (Bank Indonesia, 2011). In South East Asia, the average banking ROA level has reached 1.14% (Firmanzah, 2011). This is a positive performance indicator of the banking industry in Indonesia, but there is still a lot of work to improve competitiveness regionally and internationally.

The ratio of the Indonesian banking BOPO level (Operational Cost of Operational Income) per December 2011 reached 85.42% (Bank Indonesia, 2011a), or larger than the average BOPO in ASEAN (Assocoation of South East Asia Nations) by 40%–60% (Firmanzah, 2011). According to Firmanzah (2011), another indication discovered is the slow response of Bank Indonesia (BI) to decrease its benchmark interest rate, which points to 6%. This rate is expected to improve the competitiveness of national banks against other neighbouring countries, whose own benchmark interest rates are relatively lower, such as Malaysia (3.25%), Thailand (3.50%), the Philippines (4.50%), or Korea (3.25%). Therefore, in order to increase the national banking competitiveness, the government needs to make interventions in improving the banks' performance, as well as in improving the efficiency of their business operations.

The government needs to develop bank mechanisms for supervision with high concentration. This is important in order to control the banks' operational activities so that they become more efficient, leading towards high competitiveness and avoiding practices that potentially lead to moral hazard. The central bank will implement a policy that will evaluate the ownership of shares through the value of the level of health and governance implicated against the level of health or bank performance.

Operational risks become the most dominant type of risk compared to other risks. Different from various other risk types, an operational risk is the earliest known type of risk in the banking world, but little understood compared to other risks. Operational risks also have unique characteristics, because they are not associated with the expectations of the rate of return, but they occur naturally and appear as a result of business activities. Basel II (International Committee for setting up bank risk management) defines an operational risk as the risk that arises because of the failure of the internal process, man, system, or from external events. Basel II also sets up the model of the calculation and internal standards that must be applied by the banks.

The calculation of the operational risk burden, that might be the source of the potential losses to the bank, is yet to be measured accurately in the calculation of weighed assets according to the risk (ATMR). The bank must calculate ATMR to operational risks in the calculation of the minimum capital participation obligation (or Minimum Capital Requirement with various approaches:

1. Basic Indicator Approach;
2. Standardised Approach; and/or
3. AMAs (Advanced Measurement Approaches).

One of the internal approaches that is used and will be the focus of discussion in this research is the AMA, that is believed to have a high level of accuracy because it uses the mathematical and probabilistic approach. The advanced method requires historical data (Loss Event Database) on operational loss events. With the database, the bank can make an operational risk quantification model so that the projection of the capital charge can describe the estimated losses.

The understanding of the operational risks concept, along with the mathematical and probabilistic approach, becomes very important to be understood by practitioners in the business environment, especially for bankers and academics. The problem that arises is how the bank can measure operational risks and then implement mitigation (Operational Value at Risk/OpVar). This is the main focus in this research that was conducted in one of the Indonesian state-owned banks.

This research was implemented in one of Indonesia's state-owned banks, by considering its status as one of the largest banks in Indonesia, that will inevitably have systemic impacts should a crisis happen. This Indonesian state-owned bank is also considered as a representative bank as a reference regarding the number of assets managed owned by the customer and also the number of branch office, a cash office, and other business units reaching out to the remote rural areas.

2 LITERATURE REVIEW

2.1 *Risk definition*

Risk is associated with uncertainty or irregularities. The general understanding of a risk is also stated by Lee et al. (2001), explaining that risk can be defined as the potential for events or trends in progress that cause losses in the future or cause future income fluctuations. Hardanto (2006) defines risk as the likelihood of a bad result and the greatness of the opportunity that can be estimated.

The risk in the bank's context can be interpreted as a potential event that can be expected but which cannot be expected to have a negative impact on earnings and capital expenditure of the bank (Yulianti, 2009).

2.2 *Operational risks*

Operational risk is defined as the risk of direct or indirect losses resulting from the internal process and system that are inadequate or have failed, but can also be influenced by the external process (Esch et al., 2005). Buchelt and Unteregger (2004) define operational risks as one unity of risks of very diverse and interconnected risks with a different origin.

Djohanputro (2008) defines operational risks as a potential deviation from the results expected due to system malfunctioning, human resources, technology, or other factors. Fahmi (2010) states that the operational risk is the risk that generally originates from internal company issues, and in this case the risk of this happening is caused by a weak management control system that is implemented by the company's internal party. Operational risks according to Crouhy et al. (2001) are the operational risks as the external event risks or weaknesses in the internal control system to the detriment of the company.

2.3 *Operational risk capital burden*

According to Article 31 of the Bank Indonesia Regulation No. 10/15/PBI/2008 24 September 2008 regarding the obligation of providing general Bank Minimum Capital (or Minimum capital requirement), it states that the bank must calculate ATMR operational risks in the calculation or Minimum capital requirement using the following approaches:

1. Basic Indicator Approach (BIA)
2. Standardised Approach (SA)
3. Advanced Measurement Approach (AMA)

However, in the implementation of the initial phase, the calculation of ATMR required is done by using the basic indicator approach (PID), where ATMR is determined by the following formula:

$$\boxed{\text{ATMR} = 12.5 \times \text{Operational Risk Capital Burden}}$$

where

$$K_{PID} = \frac{\left[\sum (GI_{1...n} \times \alpha)\right]}{n}$$

K_{PID}: operational risk capital burden using the basic indicator approach (PID)
GI: annual positive gross income in the last three years
n: the number of years of positive gross income
α: 15%
Some of the things that need to be noted are:

- Gross revenue is net interest income plus non-operational income of certain other interests, calculated by a cumulative net from the early period of January to the end of December each year.
- Calculation of gross income uses data that is delivered through the monthly bank report (LBU). In this case, the system changes LBU, and the bank uses the gross income according to the old LBU of the corresponding year.
- When the bank owns a sharia business unit, the gross income calculation from the sharia business unit is then converted in accordance with the characteristics of the bank business and sharia principles.
- The bank should make corrections if, based on the financial reports that have been audited by a Public Accountant, there are corrections regarding the amount of gross revenues.
- When calculating the average gross income for three years, and there are one or two years where the bank's gross income is negative or zero, the calculation of the average annual gross income of the bank should reveal the value of gross revenues from the negative quantifiers, and both should be operable when calculating the average gross income.
- When in a span of three years, and the bank's gross income is negative or zero, then the average gross income of the bank must be added to the operational risk capital burden using the latest annual positive gross revenue.

2.4 *The model of the loss distribution approach*

The methods are most often used in the AMA and the Loss Distribution Approach (LDA). Using LDA, the bank can measure the frequency distribution and how great the loss of operational risks for each point risk (business line/event type) is within a period of one year (Chernobai et al., 2007).

2.5 *Value at risk*

The bank uses customer costs for risks, so that (hopefully) average losses in one market segment are compensated under Pakistani with other benefits. Other risks, especially the risks of the market and increased credit risks, are protected value (insured) through the market derivatives. Unexpected losses are not diversified or protected value (hedged), which is covered by the capital equity bank and how much bank capital is necessary to cover it. The risk is determined by what is called as Value at Risk (VaR).

Formally, VaR measures the lack of q-quintile as a result of the distribution of loss (losses distribution) that exceeds the expected loss (loss), EL in the time period, T discounted on the level of r risk-free for time t = 0:

$$VaRq, T = (QqL(T)] - EL) e - rT,$$

where q – quintile is the case of worst loss (loss), Qq[L(T)] defined on the level of trust q through

$$Prob (L(T) > Qq[L(T))) = 1 - q.$$

2.6 *The process of back testing*

According to Cruz (2002), the operational back testing is done by comparing VaR prediction based on historical data with the actual loss occurring. The model is acceptable when the number of deviations from the value of VaR with the actual loss does not exceed the requirements. The procedure for testing the validity of the model with back testing is to compare the value of the VaR operational risks with the realisation of operational losses in a certain period (Muslich, 2007).

3 RESEARCH METHOD

3.1 *Data and variables used*

The data used in this research is secondary data. The data used comes from the incident Management in Operational Risk Assessor Application (OPRA) and is combined with data collected from a variety of sources, including: the data loss that is managed by the division of the central operation of the data to the Internal Audit findings (AIN) which has not been put into OPRA applications, and other data which has definitely not been put into the incident management application. The data used in this research is loss frequency data and severity data, which is the monthly data from January 2008 to December 2012.

The variables used in this research are based on the assessment of the risk profile, consisting of the assessment of Inherent Risk (the risk inherent in the activity of the bank) and the assessment of the Risk Control System (the control of risk inherent) to eight business lines, namely corporate finance, trading and sales, retail banking, commercial banking, payment and settlement, agency services, asset management and retail brokerage, with seven categories of Genesis, namely internal fraud, external fraud, employment practices and for student placements safety, clients products and business practices, damage to physical assets, business disruption and system failures and execution delivery, and process management.

3.2 *Data analysis*

3.2.1 *Descriptive analysis*

Descriptive statistics provides an explanation about the collection of quantitative size as risk indicators to describe data. Some numbers in the descriptive statistics give an overview of the characteristics of the data in which used in this paper.

3.3 *Testing the distribution of the frequency and severity*

The intended test is to prove the hypothesis as to whether the spread of the frequency of some specific distribution opportunities already meets the criteria of an opportunity spread. The distribution test uses the Chi-Square test and the Anderson Darling test.

3.4 *LDA analysis*

The test was intended to prove whether the hypothesis regarding the spread of the frequency of some specific distribution opportunities meets the criteria of a spread. Testing used the Chi-Square test and Anderson Darling test.

The LDA model used for the calculation of OpVar is a model that combines the frequency distribution and the distribution of the severity of the sample data loss. Genesis frequency distribution data is operational as the distribution of discrete graphics, while data operational loss severity distribution is a continuous distribution. In this LDA approach, total loss is the number of operations (S) from variable random (N) on individual operational loss (X1, X2,..., XN) so that the number of total operational losses can be stated as follows:

$$S = X_1, X_2 <....., X_N \text{ where } N = 0,1,2,.....$$

The distribution of the total loss is then used to project the potential operational loss. One of the methods used for the conjunction is the Monte Carlo simulation. (Figure 1).

3.5 *The calculation of OpVar with Monte Carlo simulation*

The calculation of the OpVar estimation can be conducted with the help of a Monte Carlo simulation. The Monte Carlo simulation is the best way to produce a loss distribution or the distribution of losses. The Monte Carlo simulation is done through the process of combining the distribution of the frequency of a loss of data with the distribution of the severity of the lost data. This simulation is intended to produce a probability distribution of some possible results of experiments using random number data. The number of data points determines whether the number is big enough to ensure that the quality of the results does not provide

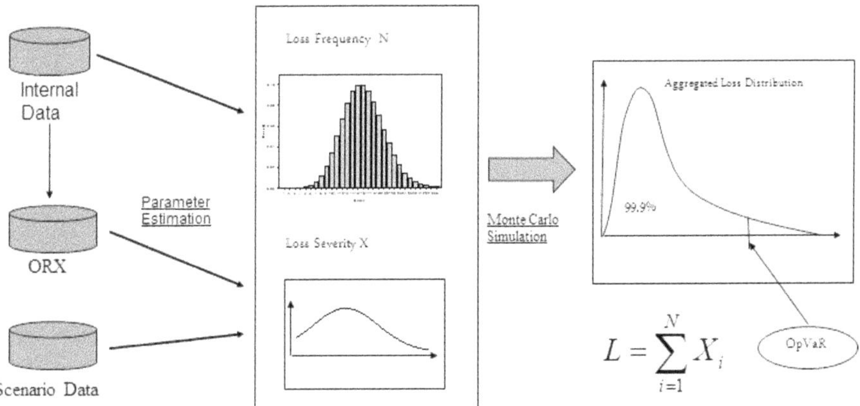

Figure 1. Aggregating severity and frequency models.

a large error deviation. Therefore, the more the number of data simulation points (>10,000), the more accurate the results and the more stable the data (Cruz, 2002).

3.6 *The process of back testing*

Cruz (2002) explains that operational back testing is conducted by comparing the prediction of VaR based on historical data with the actual loss occurring. The model can be accepted when the number of deviations from the value of VaR with the actual loss does not exceed the limit.

Testing the validity of the model of the operational risks utilises the testing hypothesis with the aim to see whether or not the model can be accepted.

H0: correct model in the projected potential operational loss
H1: the incorrect model in the projected potential operational loss

Kupiec uses the formulation of LR to test the validity of the model as follows:

$$LR = -2\ln\left[(1-p)^{T-N} p^N\right] + 2\ln\left[\left(1-\frac{N}{T}\right)^{T-N}\left(\frac{N}{T}\right)^N\right]$$

where

T: number of samples of observation
N: the number of losses that exceeds the value of VaR (failure rate)
p: VaR confidence level

Then the value of the LR is compared with the Chi-Square critical value with 1 degree of freedom. If LR > 3.84 then H0 is rejected, which means that the model is not correct/ not valid. Likewise, if the result of the validation model revealed is true/valid, then this will strengthen the policy of using the model. However, if the model stated is not true/not valid then the model used needs to be reviewed or replaced with another measurement model.

4 RESULTS AND DISCUSSION

To perform the modelling and measurement of potential operational risk losses, the distribution characteristics of operational risk losses must be known in advance. The distribution of operational risk loss data can be grouped into the distribution of data frequency and loss severity distribution of data losses. The data in Table 1 can be used for the process of the modelling and measurement of potential operational risk losses.

Table 1. Based on the Genesis data loss frequency and severity data.

	Genesis category						
Business Line	Internal fraud	External fraud	Employment practices	Product clients	Physical assets damage	Business interruption	Delivery execution
Asset management	Available	Available	Available	Available	Available	Unavailable	Available
Retail banking	Available	Available	Available	Available	Available	Available	Available
Retail brokerage	Available	Available	Available	Available	Available	Available	Available
Commercial banking	Available	Unavailable	Unavailable	Unavailable	Unavailable	Unavailable	Unavailable
Corporate finance	Available	Unavailable	Unavailable	Unavailable	Unavailable	Unavailable	Unavailable
Payment and settlement	Available	Available	Unavailable	Available	Unavailable	Available	Unavailable
Trading and sales	Unavailable	Unavailable	Unavailable	Unavailable	Unavailable	Available	Available

To conduct the calculation of the large Operational VaR, the initial step that must be taken is to perform a test on the type of distribution used in this research. Testing the most suitable distribution type (Goodness of Fit Test) is based on the existing data. Distribution testing is done by using the software Easyfit and function of the Excel spreadsheet. If the operational loss distribution type testing is conducted in a timely manner and is true, then a model will be obtained to calculate the potential losses that will arise.

Incident management data was downloaded from the application of the OPRA on 26 June 2012. Furthermore, data collection from the AIN 2011 and data losses are managed by the division of the Operational Centre (STO). After verification and cleansing, experts were invited as speakers, and the data obtained is regarded as being clean data (believed to be good data), with descriptive statistics as shown in Table 2.

The spread of the data frequency and severity data for each business line per Genesis is determined based on the review of the spread and the form of the spread of the graph. Based on the review of Anderson Darling, it can be deduced that H0 can be rejected or accepted, where H0 is the data spread with certain spread (Table 3).

4.1 *The results of the calculation of the OpVar 99.9% and back testing*

After conducting the process sounding form of the spread of the data frequency and severity, futhermore we also test the spread of the data frequency and severity, the data combined is then implemented to determine the OpVar value with 99.9% degrees of trust of each business line per Genesis completed. After calculating the value of OpVar, back testing must be done to show that the alleged model is valid. The results of the calculation can be explained in Table 4.

Monte Carlo simulation results show that the value of OpVar is 99.9%, produced from business line asset management with the internal event category fraud of 499,977,111,001,062. This means that the bank must provide capital to cover the risk of internal fraud of Rp499,977,111,001,062 (million) with the possibility of the worst incidents 0.1 out of 100.

For the category of external fraud, the OpVar value that is produced with a reliability degree of 99.9% is 523,843,356,045,642. This means that the bank must provide capital to cover the risk of external fraud amounting to Rp523,843,356,045,642 (million) with the possibility of the worst incidents 0.1 from 100.

Table 2. Grouping of the incident operational risks and operational credit field 2007–2011 based on data sources.

Data resource	The number of incidents	Total loss (Rp.)
Fraud Audit TW1-2011	5	15.661.006.090
Fraud Audit TW2-2011	9	9.699.773.871
Fraud Audit TW3-2011	5	935.422.379
Fraud Audit TW4-2011	3	4.093089.054
OPRA-MI	1059	3.725.860.395.758
Special Audit TW1-2011	12	79.340.015.125
Special Audit TW2-2011	2	9.137.051.709
Special Audit TW3-2011	4	17.563.304.145
Special Audit TW4-2011	6	14.183.821.778
STO	22	6.854.550.000
STO—Illegal Card	23	1.138.860.851
STO—Account Book Counterfeiting	9	5.385.100.000
Evidence AIN Major-TW1-2011	37	65.033.608.700
Evidence AIN Major-TW2-2011	58	1.015.796.911.986
Evidence AIN Major-TW3-2011	68	164.605.682.063
Evidence AIN Major-TW4-2011	88	459.473.565.205
Total	1410	5.594.762.158.713

Table 3. The distribution of the spread of frequency data opportunities and operational loss severity.

LINI BISNIS	Data	Kejadian						
		Internal Fraud	Eksternal Fraud	Praktek Ketenagakerjaan	Klien Produk	Kerusakan Aset Fisik	Gangguan Bisnis	Eksekusi Pengiriman
ASSET MANAGEMENT	Frekuensi	Poisson (1,4833)	Poisson (0,25)	Poisson (0,13333)	Uniform (0;1)	Poisson (0,2333)		Geometric (0,18181)
	Severity	Gamma (0,01773;236640;0)	PhasedBi-Weibull (0,18048;1;0;0,18952;1,272;155)	Exponential (0,01378;0)	Gamma (0,01667;2000;0)	Exponential (0,11517;0)		Gamma (0,0291;27041;0)
RETAIL BANKING	Frekuensi	Geometric (0,08683)	Geometric (0,12448)	Poisson (0,3)	Geometric (0,20548)	Poisson (0,48333)	Poisson (1,4167)	Geometric (0,02431)
	Severity	PhasedBi-Exponential (0,00083199;0;0,000028586;1905)	Gamma (0,33252;4103,8;0)	Weibull (0,50153;1;0)	Pareto2 (0,27262;0,00000049557)	PhasedBi-Exponential (0,18786;0;0,00638;10)	PhasedBI-Exponential (1,713;0;0,00144;1,1)	PhasedBi-Exponential (0,00010332;0;0,0000050942;14641)
RETAIL BROKAGE	Frekuensi	Poisson (0,6)	Poisson (0,6)	Poisson (0,6)	Poisson (0,6)	Poisson (0,6)	Poisson (0,6)	Poisson (0,6)
	Severity	Rayleigh (0,00479;0)	Rayleigh (0,00479;0)	Rayleigh (0,00479;0)	Rayleigh (0,00479;0)	Rayleigh (0,00479;0)	Rayleigh (0,00479;0)	Rayleigh (0,00479;0)
COMMERCIAL BANKING	Frekuensi	PoissonRand(0,03333)						
	Severity	ExpRand(0,30415)						
CORPORATE FINANCE	Frekuensi	Poisson(0,14286)						
	Severity	Chi-Squared (12)						
PAYMENT AND SETTLEMENT	Frekuensi	Poisson (0.91667)	Poisson (0.21667)		poisson (0.13333)		poisson (0,05)	
	Severity	Pareto2 (0.26295;7.8646E-10)	Exponential(0.00892)		Burr(1.2 ;0.99999;1.0)		Burr(1,2;0,99999;1,0)	
TRADING AND SALES	Frekuensi						Poission (0.05)	Poisson (0.83333)
	Severity						Burr (1.2;0.99999 ;1.0)	INV.Gausian (0.00299;0.17917)

For the category of employment practices, the OpVar value that is produced with a reliability degree of 99.9% is 44,524,489,999,575. This means that the bank must provide capital to cover the risk of employment practices amounting to Rp44,524,489,999,575 (million) with the possibility of the worst incidents 0.1 from 100.

For the category of the product client, the OpVar value that is produced with a reliability degree of 99.9% is 133,018,844. This means that the bank must provide capital to cover the risk of a Rp133,018,844 (millions) product client with the possibility of the worst incidents 0.1 from 100.

For the category of physical assets of the damage, the OpVar value that is produced with a reliability degree of 99.9% is 483,332, meaning that the bank must provide capital to cover the risk of damage to the physical assets of Rp483,332 (million) with the possibility of the worst incidents 0.1 from 100.

For the category of delivery execution, the OpVar value that is produced with a reliability degree of 99.9% is 1,084,685,483, meaning that the bank must provide capital to cover the risk of damage to the physical assets of Rp1,084,685,483 (million) with the possibility of the worst incidents 0.1 from 100.

4.2 *The process of back testing*

Basically back testing is done by comparing the results of the estimation of the maximum loss (OpVar) with a certain confidence level with the actual operational losses. Based on Table 4, all earnings obtained of OpVar have LR value smaller than chi-sq cv, meaning that H0 accepts the risk calculation model and it is considered fit or valid.

4.3 *Fraud prevention efforts*

Based on the value of the OpVar table with a reliability level at 99.9%, for each category of Genesis losses—internal fraud, external cheating (external fraud), employment practices and the safety of the workplace, clients, products and business practices—the destruction of the physical assets, business interruption and the failure of the system and the execution and delivery, and the management of the process from the ABC seven business line such as asset management, retail banking, retail brokerage, commercial banking, corporate finance, payment and

Table 4. The results of the calculation.

Business line		Internal fraud	External fraud	Employment practices	Product clients	Physical assets damage	Business interruption	Delivery execution
Asset management	Percentile 99.9%	4.9997.11	5238.43	445.24	133.02	48.33		108468.55
	Back Testing	Accept H_0	Accept H_0	Accept H_0	Accept H_0	Accept H_0	Accept H_0	Accept H_0
Retail banking	Percentile 99.9%	432694.17	85548.77	41.73	5254876.79	747.01	4559.52	14001714.51
	Back Testing	Accept H_0	Accept H_0	Accept H_0	Accept H_0	Accept H_0	Accept H_0	Accept H_0
Retail brokerage	Percentile 99.9%	0.03	0.03	0.03	0.03	0.03	0.03	0.03
	Back Testing	Accept H_0	Accept H_0	Accept H_0	Accept H_0	Accept H_0	Accept H_0	Accept H_0
Commercial banking	Percentile 99.9%	12.50						
	Back Testing	Accept H_0						
Corporate finance	Percentile 99.9%	25.88						
	Back Testing	Accept H_0						
Payment and settlement	Percentile 99.9%	236.08	671.74		94.64		35.91	
	Back Testing	Accept H_0	Accept H_0		Accept H_0		Accept H_0	
Trading and sales	Percentile 99.9%						20.78	13.30
	Back Testing						Accept H_0	Accept H_0

settlement and trading and sales, internal fraud contributes most of the loss to the company. This is shown by the value of the largest OpVar located in the category of internal deviation.

In the operational practices of banking, various types of internal fraud often occur, such as embezzlement, procedures, and violations of the law or the existence of a policy issued by each leader at each structural level.

Fraud prevention in the practice of banking operations is brought to the attention of the local and international world. Various prevention efforts, from the restricting supervision system to the prevention through corporate culture, are implemented and include ABC.

ABC has a number of fraud prevention programmes that have been inaugurated in a culture of anti-fraud. All employees to the board of directors and ABC commissioners must be involved in the drafting and socialisation statement of anti-fraud. ABC also has published a book titled *Top 50 Risk Issues* as a reference to employees to make them aware of the critical points which are prone to corruption. Regularly ABC sent letters of increased control to the entire work unit. Any risk management forum will be held to analyse the current problems faced by the company. Employees are also required to fill out an annual statement (annual disclosure) at the end of each year related to the clash of interests in the transaction activities implemented.

4.4 *Managerial implications*

The status of ABC as a public company (Tbk) should allow zero tolerance against fraud that will have a direct impact on the financial losses. In view of this research, the researchers see that ABC needs to evaluate the effectiveness of the anti-fraud system which is currently being implemented.

Basically, risk management banks have international standards that are under the supervision of an institution called the Bank for International Settlement (BIS). BIS oversees the banking risk management committee Basel II. This committee makes the standardisation of risk handling for all the banks in the world.

The board of directors must be aware of the major aspect of the operational risks of the bank that should be managed and approve and periodically review the framework of operational risk management bank. The framework must be trustworthy and refer to the definition of the correct operational risk.

An internal audit must be conducted on a regular basis against the framework of operational risk management and should be conducted by an independent and competent internal team. The board of directors plays an important role in this case to ensure that the audit process is run independently.

5 CONCLUSION

Based on the above, a conclusion can be made that the methods of the LDA has a high accuracy to calculate the number of OpVar on every event for eight business lines and seven categories of Genesis.

Based on calculations results from OpVar with reliability level of 99% also show that the largest deviations are in the field of Internal Fraud. In that case, the board of directors and also the entire line of senior management have the responsibility to establish a corporate culture that is anti-fraudulent and prioritises the systemisation of effective operational risk management, which is subject to a healthy operational control.

6 SUGGESTIONS

According to the discussion and conclusion, the authors suggest that bank ABC should implement an advanced methods in measuring the operational risks, not only using the LDA, but can also make use of other advanced methods in order to see the comparison between the results with the level of increase.

The importance of data management and data input validity will be utilised to calculate the value of the risks, and thus produce accurate risk value sounding so that it can find the correct policy in accordance with the principles of data: garbage in is equal to garbage out.

REFERENCES

Bank Indonesia. (2003). Peraturan Bank Indonesia Nomor 5/8/PBI/2003 tentang Penerapan Manajemen Risiko Bagi Bank Umum. *BI.* Accessed on May 19, 2013, http://www.bi.go.id/biweb/utama/peraturan/pbi-5-8-2003-eng.pdf

Buchelt, R. & Unteregger, S. (2004). Cultural Risk and Risk Culture: Operational Risk After Basel II. *Financial Stability Report 6.* Accessed on June 29, 2013, http://www.oenb.at/en/img/fsr_06_cultural_risk_tcm16-9495.pdf

Chernobai, A.S., Rachev, S.T. & Fabozzi, F.J. (2007). *Operational risk: A guide to Basel II capital requirements, models, and analysis.* New Jersey: John Willey & Sons.

Crouhy, M., Galai, D. & Mark, R. (2001). *Risk management.* New York: McGraw-Hill.

Cruz, M.G. (2002). *Modeling, measuring and hedging operational risk.* West Sussex: John Wiley & Sons.

Djohanputro, B. (2008). *Manajemen risiko korporasi terintegrasi.* Jakarta: PPM.

Djojosoedarso, S. (2003). *Prinsip-Prinsip manajemen risiko asuransi.* Jakarta: Salemba Empat.

Esch, L., Kieffer, R. & Lopez, T. (2005). *Asset and risk management: Risk oriented finance.* West Sussex: John Wiley & Sons.

Fahmi, I. (2010). *Manajemen risiko, teori, kasus dan solusi.* Bandung: Alfabeta.

Firmanzah. (2011). *Daya saing perbankan.* Neraca.

Hardanto, S.S. (2006). *Manajemen risiko bagi bank umum: Kisi-Kisi ujian sertifikasi manajemen risiko perbankan tingkat i.* Jakarta: Elex Media Komputindo.

Lee, F.M., Marshall, A., Szto, Y.K. & Tang, J. (2001). The practice of financial risk management: An international comparison. *Thunderbird International Business Review, 43*(3), 365–375.

Muslich, M. (2007). *Manajemen risiko operasional: Teori dan praktik.* Jakarta: Bumi Aksara.

Yulianti, R.T. (2009). Manajemen risiko perbankan syariah. In *Jurnal Ekonomi Islam, 3,* 151–165.

Competition and Cooperation in Economics and Business – Gani et al. (Eds)
© 2018 Taylor & Francis Group, London, ISBN 978-1-138-62666-9

Hijab phenomenon in Indonesia: Does religiosity matter?

L. Arifah, N. Sobari & H. Usman
Department of Management, Faculty of Economics and Business, Universitas Indonesia, Depok, Indonesia

ABSTRACT: As stated in the holy Quran, a Muslim woman is obligated to wear a jilbab (a veil). In the 1980s, hijab was commonly adopted in the majority in the Islamic boarding schools. In recent years, the number of Muslim women wearing hijab has experienced a significant increase. We suspect the increasing numbers are due to various modern fashion influences in the hijab design. Hijab has been transformed from the traditional designs into the more modern and fashionable ones. Therefore, this research aims to observe the influence of religiosity, the Subjective Norm (SN), and the Perceived Behavioural Control (PBC) against the decision of Muslim women to wear hijab according to the Theory of Planned Behaviour (TPB). This research is undertaken using questionnaires with closed questions about religiosity, SN, and PBC to 270 Muslim women in three provinces in Indonesia. The data analysis used factor analysis and logistic regression. The outcome of this research illustrates that religiosity is indirectly connected to the decision to wear hijab. Religiosity has a positive correlation with SN and PBS. Furthermore, SN and PBC produce a significant impact on the decision of Muslim women to use hijab. This study also attempts to give information for hijab producers to understand their consumers.

1 INTRODUCTION

The hijab obligation to Muslim women can be seen in holy Qur'an, including Surah (Q.S.) Al-Ahzab (33) verse 59:

> *'O Prophet, tell your wives and your daughters and the women of the believers to bring down over themselves [part] of their outer garments. That is more suitable that they will be known and not be abused. And ever is Allah Forgiving and Merciful.'*

Moreover, it is also contained in Q.S. An-Nur (24) verse 31:

> *'And tell the believing women to reduce [some] of their vision and guard their private parts and not expose their adornment except that which [necessarily] appears thereof and to wrap [a portion of] their headcovers over their chests and not expose their adornment except to their husbands, their fathers, their husbands' fathers, their sons, their husbands' sons, their brothers, their brothers' sons, their sisters' sons, their women, that which their right hands possess, or those male attendants having no physical desire, or children who are not yet aware of the private aspects of women. And let them not stamp their feet to make known what they conceal of their adornment. And turn to Allah in repentance, all of you, O believers that you might succeed.'*

The *aurat* (parts of the body that must be covered) clothing cover for Muslim women has several terms, such as *kerudung*, jilbab, and hijab. *Kerudung*, or khimar in Arabic, is defined as a headgear shawl. Meanwhile, jilbab is not only a cover for the head but also the loose apparels which cover the body shape, whereas hijab (headscarf) in the Arabic language means veils[1].

1. Daud, Jilbab, Hijab dan Aurat Perempuan, 5.

In the 1970s and 1980s, the issues of jilbab in Indonesia were something haram (strictly forbidden) for its presence in public, especially at schools[2]. At that time, the veil was still something unpopular in society. The use of jilbab was still limited to those Muslim women who had a high level of religiosity, like in the *pesantren.* The jilbab wearers in the past were not only seen as old-fashioned but also dangerous. Therefore, jilbab at that time did not only show the religious order and the symbol of godliness but also had a political power to be reckoned.

In later years, the awareness of wearing hijab was no longer dominated by the older women or those who come from the *pesantren* but also by young women. In conjunction with that, the producers of fashion started to release clothing designs and the hijab accessories for all ages. Indeed, there has been a significant rise in numbers of people wearing hijab, including in the circles of artists, presenters, and even public officials. Because of that phenomenon, there have also been religious *sinetron* (soap operas/TV series) produced, and those have not only just aired during Ramadhan.

Indonesia, as a country with the highest amount of Muslim population in the world, has been supporting the development of the hijab fashion trend. The rampant phenomenon of hijab trend fashion is also supported by the hijab community like the Hijaber's Community. This is also escalating the interest of Muslim women to wear hijab as well as inspiring women to select the way they dress, thus adding to the treasures of Muslim women's world fashion[3]. The trend develops wider on the international stage. It is even performed in the international fashion shows, as has been done by Dian Pelangi, one of the brands of hijab fashion in Indonesia. Dian Pelangi performed her works in New York Fashion Week (NYFW) in 2014[4].

The growth does not only happen in the fashion world, but also in other products, such as shampoo. In 2004 and 2015, Unilever, by means of the Sunsilk shampoo brand, launched the shampoo product for hijabers (women who wear hijab). In 2015 Sunsilk even held a talent show named 'Sun Silk Hijab Hunt' which was followed by more than 3,000 hijabers. Not only did Sunsilk take profits from the phenomenon, but other brands creating the hijab shampoo also exploited it and endorsed it as the commercial stars.

In the modelling world, the model agency 'Zaura Model Muslimah' is also well-known. This agency specialises in a hijab model. The founding of Zaura Model cannot be separated from the large market opportunities because of the development of the recent hijab products. Moreover, it is also supported by the reasons for wanting to accommodate the model talents of Muslim women that have probably been hidden since the decision of wearing hijab was made.

Most Muslims believe that hijab is one of Allah's written orders in the Al-Qur'an. Consequently, wearing hijab to cover the hair constitutes one of the observance forms. Muslim women who use hijab would better understand about the underlying religious commitment. In fact, the hijab wearing which has been increasing is not based on the commitment to Islam but to a greater desire to express beauty, maturity, and individualism[5].

This condition makes hijab not only a matter of religiosity but also of such matters as fashion and identity. The motivation of someone to use hijab is not only because of the adherence to the rule of sharia, but also is influenced by other things. Therefore, the question arises as to whether or not the religious motivation is the dominant motivation underlying their decision. As stated in the Theory of Planned Behaviour (TPB), intentions to perform behaviours of different kinds can be predicted with high accuracy from attitudes towards the behaviour, Subjective Norms (SN), and Perceived Behavioural Control (PBC); and these intentions, together with perceptions of behavioural control, account for considerable variance in actual behaviour[6].

---

2. Suhendra, Kontestasi Identitas Melalui Pergeseran Interpretasi Hijab dan Jilbab in Alqur'an, 15.
3. Agustina, Hijabers: Fashion Trend for Moslem Women in Indonesia, 1.
4. Ainun Muftiarini. 'Dian Pelangi Siap Tampil di New York Fashion Week'. http://lifestyle.okezone.com/read/2014/03/28/29/962172/dian-pelangi-siap-tampil-di-new-york-fashion-week).
5. Ismail, Muhajababes—meet the new fashionable, attractive, and extrovert Muslim woman. A study of the hijāb-practice among individualized young Muslim women in Denmark, 1.
6. Ajzen, The Theory of Planned Behavior, 1.

According to the explanation above, the purpose of this study is to discover whether religiosity is the substantial factor influencing Muslim women to wear hijab, or if any other surrounding motivations are encouraging the behaviour.

2 LITERATURE REVIEW

2.1 *Religiosity*

Religiosity, or the religious commitment, is defined as a degree to which a person adheres to his or her religious values, beliefs, and practices and uses them in daily life[7].

In accordance with Islamic teachings, the religious commitment cannot be separated from the life of every Muslim. The teachings of Islam affect each aspect of Muslim life, including in terms of food and attire. Muslims are required to be excited to seek for knowledge, to donate part of their treasures, to work with high spirit, and to care about neighbouring conditions or other Muslim's conditions.

Worthington has developed a tool to measure the religious commitment, which is Religious Commitment Inventory (RCI). RCI quantifies two things: interpersonal religiosity and intrapersonal religiosity.

Several studies related to religiosity have shown that religiosity really influences a person's behaviour. Safiek Mokhlis (2009) states that religiosity, both interpersonally and intrapersonally, significantly affects consumers in evaluating a shopping mall.

Dehyadegari et al. (2016) has reviewed the relationship between Islamic veil involvement with SNs and religiosity. The result shows that there is a positive connection between those variables and the eagerness of buying hijab. The study does not illustrate the direct correlation between the variables of religiosity, SNs, and purchase intention.

The link between hijab and religiosity is presented by Chen (2014), who undertook the qualitative research by conducting in-depth interview with the Muslim women in America. The outcome shows that hijab has diverse meanings among Muslim women, such as religious practices, in-group diversity, political symbols, women's rights, fashion icons, and market power.

2.2 *TPB*

The intention of doing something can be predicted by viewing other symptoms associated with such actions. For instance, attitude, SNs (regulations or norms adopted by the environment around the individuals), and PBC (the individual control to an act). Although it is believed that all three are interrelated, the quality of their relationship cannot be ascertained.

Not many studies have related the religiosity element to the aspects which influence behaviour that is compatible to the planned behaviour theory. One study is by Dehyadegari et al. (2016) who investigated the relationship between religiosity and SNs in deciding to wear hijab connected to the desire of making a purchase (Purchase Intention). The result of the study has shown the correlation between the three variables, but there is an indirect correlation between religiosity and SNs towards purchase intention.

3 RESEARCH FRAMEWORK AND HYPOTHESIS

This research aims to find out whether or not religiosity is a main factor of the growing hijab phenomenon in Indonesia nowadays.

7. Worthington, The Religious Commitment Inventory—10: Development, Refinement, and Validation of a Brief Scale for Research and Counseling, 2.

3.1 *Religiosity and intention in wearing hijab*

Indonesia is the country with the largest Muslim population in the world. However, Indonesia has no such compulsory regulation for its female citizens to wear hijab. Thus, the decision on using hijab is based on individual freedom.

Many studies on religiosity affecting various aspects of life have been conducted. One of them is the research on religiosity influencing the purchasing intention (Dehyadegari et al., 2016) and religiosity towards academic achievement (Logan, 2013). Meanwhile, the research about the way women dress, particularly for hijab, has not been undertaken very often. One of the studies is the research from Bachleda et al (2014) that examines religiosity from the point of view of the way Maroko Muslim women dress. The study could support the first hypothesis, which is:

H1: There is a significant relationship between religiosity and the use of hijab.

3.2 *SNs and intention in wearing hijab*

SN is social encouragement around individuals that could affect them in doing something (Ajzen, 1991). The connection between a SN and behaviour has been studied by many researchers (Wiener, 1982; Mahon et al., 2006). According to Mahon et al. (2006), who observed SNs against the purchase of fast food, they have gained the result that SNs have a significant effect on fast food but not on takeaway food. Various studies have supported the second hypothesis:

H2: There is a substantial relationship between SNs and the use of hijab.

3.3 *PBC and intention in wearing hijab*

PBC is the individual perception regarding whether or not a behaviour is easy to do (Ajzen, 1991). PBC is a better predictor of behavioural intention than attitude[8]. Therefore, the third hypothesis is:

H3: There is a substantial connection between PBC and the use of hijab.

Based on the three hypotheses above, the research model can be described as shown in Figure 1:

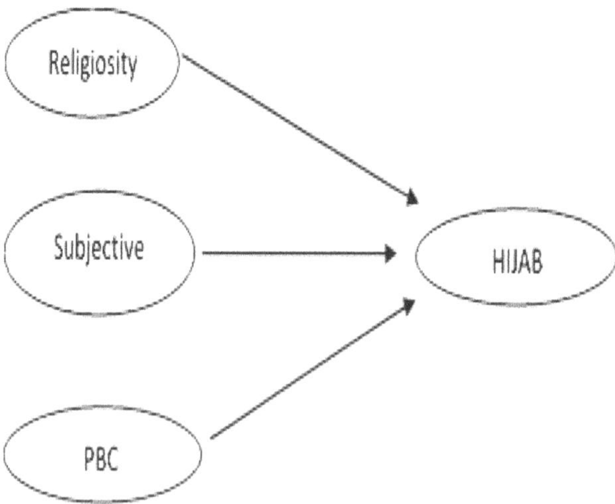

Figure 1. Research model.

8. Chang, Predicting Unethical Behavior: A Comparison of the Theory of Reasoned Action and the Theory of Planned Behavior, 1825.

4 METHODS

4.1 *Samples and data*

The population of the research is productive-aged Muslim women in Indonesia. To represent the population, the respondents were selected from three provinces that are densely populated, specifically West Java, DKI Jakarta, and Banten. Afterwards, the data collection was conducted by giving the questionnaires to 270 Muslim women using nonprobability sampling. Data from 215 of the above samples was used.

4.2 *Measures*

The measure of all variables will be undertaken using the Likert scale of 5 points. Point 1 refers to 'strongly agree' and Point 5 refers to 'strongly disagree'. The questions about religiosity were adopted from the Religious Commitment Inventory (RCI-10) of Worthington et al (2003). Moreover, the questions about the variables of SNs and PBC were adjusted to the questionnaire of the manual direction for TPB. The variables of SNs are measured by four questions, while the variables of PBC are measured by three questions. The questionnaire questions are written in the Indonesian language, so as to be easily understood.

4.3 *Data analysis method*

The analysis methods used are factor analysis and logistic regression analysis. The factor analysis aims to define the matrix data structure and analyse the correlation among various variables by the way of defining one set of similar variables, as mentioned by Demson (Ghozali, 2011). The logistic regression model was used to find the probability of related qualitative variables influenced by independent variables.

5 RESULTS

The collected data of respondents were processed with the SPSS application. The respondents consisted of Muslim women of various ages and education levels. The demography of respondent data can be seen in Table 1.

The results of the factor analysis can be viewed in Table 2.

The purpose of the research is to discover the dominant factor for Muslim women in deciding to wear hijab. To the other questions of how variables of religiosity, SNs, and perceived behaviour would be controlled by the factor analysis, the aim is to get the weight from each variable.

According to the tKaiser-Meyer-Olkin (KMO) test, it is indicated that the Religiosity Data has a score of 0.857, which means that the data is considered to be good enough and that the analysis can be continued.

Table 1. Demography of respondent data.

Age	Total	%	Education	Total	%
17–25 y.o.	109	50.70	Secondary Education	48	22.32
26–30 y.o.	60	27.90	Diploma (D3)	13	6.05
31–35 y.o.	25	11.63	Undergraduate (S1)	129	60.00
36–40 y.o.	6	2.79	Postgraduate	20	9.30
41–45 y.o.	5	2.33	No response	5	2.33
46–50 y.o.	3	1.40			
50 y.o.	1	0.47			

Table 2. Results of the factor analysis.

Variable	KMO	Barlett Test
Religiosity	0.857	806,297
SN	0.789	498,361
PBC	0.588	273,41

Table 3. Outcome of regression model 1.

Variables		B	S.E.	Wald	df	Sig.	Exp(B)
Step 1[a]	SN	0.546	0.226	5.837	1	0.016	1.726
	PBC	0.601	0.216	7.757	1	0.005	1.823
	Constant	2.819	0.321	77.271	1	0.000	16.753

Table 4. Correlation test of religiosity variables.

		Hijab1	Religiosity	SN	PBC
Religiosity	Pearson Correlation	0.068	1	0.375**	0.412**
	Sig. (2-tailed)	0.319		0.000	0.000
	N	215	215	215	215

The correlation table illustrates that religiosity is uncorrelated with hijab, but it is correlated with SN and PBC.

The Barlett Test aims to see whether or not the correlation matrix that has been formed is in the shape of an identity matrix. If the matrix formed is the identity matrix, then this means there is no correlation between variables, and therefore the analysis could not be continued. It is also because the correlations of inter-variables are an important thing in the factor analysis.

The Barlett Test result shows that t the amount of Sig as 0.000 therefore we can reject H0, which means the analysis could be continued.

After the factor analysis was continued, and it was discovered that the analysis for the three variables (religiosity, SNs, and PBC) could be continued, the logit regression was undertaken. The proposed empirical model can be written as:

$$\text{Hijab/Not}=\beta_1 + \beta_2\ \text{Religiosity} + \beta_3\text{SN} + \beta_4\text{PCB} + \varepsilon \tag{1}$$

where:

Hijab/Not = 1 if the respondent is wearing hijab, 0 if otherwise
SN = Subjective Norms
PCB = Perceived Behavioural Control

Regarding the problem of multicollinearity, religiosity must be removed from the model. The outcome can be seen in Table 3.

After religiosity had been taken out from the model, the result of SN and PCB that was obtained significantly affected hijab. It makes hypotheses 2 and 3, which claim a positive connection between SN and PBC with the decision of wearing hijab to become acceptable. Referring to Table 4, we can conclude that the variable of a SN is in a significance level at 5% in order to obtain a decision to refuse H0. This means the SN variable has a positive and significant impact on the choice of using hijab. It also occurs with the variable PBC, which is also at 5% significance level and H0 is rejected. Thus, this supports the third hypothesis that there is a positive and significant relationship between PBC and the decision of wearing

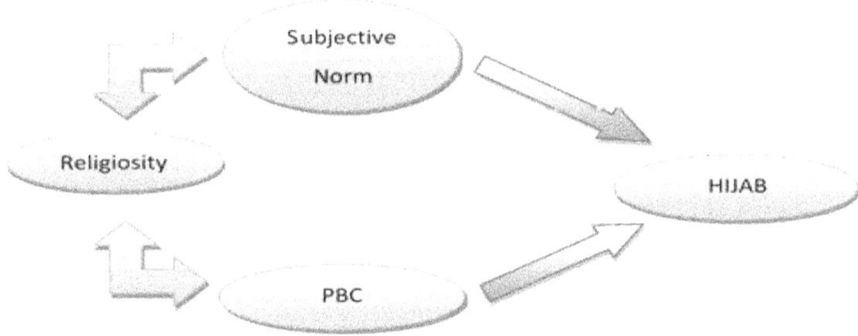

Figure 2. Research model.

hijab. Based on the correlation table below, it shows that the relationship between religiosity and other independent variables (SN and PBC) is strong enough. The correlation test can be seen in Table 4.

6 DISCUSSION AND IMPLICATION

Of the two regressions conducted and the correlation test of the three variables (i.e. religiosity, SN, and PBC) it has been found that religiosity does not significantly affect the decision of using hijab. However, religiosity owns a positive correlation with SN and PBC. Furthermore, SN and PBS have the significant influence on the decision of wearing hijab. Therefore, the research model can be described as shown in Figure 2.

Nevertheless, it can be stated that religiosity still affects the decision of wearing hijab through SNs and PBC. The result differs from the result proposed by Dehyadegari et al. (2016) that there is a positive correlation between religiosity and SNs towards the use of hijab amidst the Muslim women in Iran.

High religiosity indicates high religious awareness. According to Mokhlis (2009), religion is an important cultural factor since religion is the most universal matter that affects the behaviour of an individual and a community. This study indicates that the high religiosity does not necessarily make someone decide to wear hijab if it is not supported by the surroundings, like her school, office, and family. Yet the religiosity of an individual is able to influence the surroundings, and the individual control on behaviour eventually will affect other individuals to take a decision. Therefore, the conclusion of the research is that religiosity affects the decision-making of Muslim women to wear hijab.

This research is expected to provide information for the hijab fashion producers to better understand their consumers and be useful in determining market segmentation.

7 LIMITATION AND FUTURE RESEARCH

The results of the research cannot be taken as being a universal conclusion. This is because the research was only organised in three provinces in Indonesia. The next study will probably be conducted in provinces which have social and cultural differences, like Nanggroe Aceh Darussalam and Bali.

8 CONCLUSION

The outcome of the research puts more emphasis on the idea that TPB is a valid model in predicting a person's behaviour. Meanwhile, religiosity plays a vital role in influencing a person's decision to do or not to do something. Although it indirectly occurs, religiosity is competent to develop the SN and to control the behaviour.

REFERENCES

Agustina, H. N. (2015). Hijabers: Fashion trend for Moslem women in Indonesia. *International Conference on Trends in Social Sciences and Humanities (TSSH 2015)*. Bali.

Ajzen, I. (1991). The theory of planned behavior. *Organizational Behavior and Human Decision Processes*, *50*(2), 179–211. doi:10.1016/0749-5978(91)90020-T

Bachleda, C., Hamelin, N. & Benachour, O. (2014). Does religiosity impact Moroccan Muslim women's clothing choice? *Journal of Islamic Marketing*, *5*(2), 210–226. doi:10.1108/JIMA-05-2013-0038

Chang, M. K. (1998). Predicting unethical behavior: A comparison of the theory of reasoned action and the theory of planned behavior. *Journal of Business Ethics*, *17*, 1825–1834. doi:10.1023/A:1005721401993

Chen, L., Akat, H. D., Xin, C. & Song, S. W. (2014). Rethinking hijab: Multiple themes in Muslim women's perception of the hijab fashion. *International Conference on Communication, Media, Technology and Design, 24–26 April 2014, Istanbul—Turkey.*

Daud, F. K. (2013). Jilbab, hijab dan aurat perempuan (Antara tafsir klasik, tafsir kontemporer dan pandangan muslim feminis). *Al-Hikmah Jurnal Studi Keislaman*, *3*(1), 1–24.

Dehyadegari, S., Esfahani, A. M., Kordnaiej, A. & Ahmadi, P. (2016). Study the relationship between religiosity, subjective norm, Islamic veil involvement and purchase intention of veil clothing among Iranian Muslim women. *International Business Management*, *10*(14), 2624–2631.

Ghozali, I. (2011). *Aplikasi analisis multivariate dengan program IBM SPSS 19.* Semarang: Badan Penerbit Universitas Diponegoro.

Hanzaee, K. H. & Chitsaz, S. (2011). A review of influencing factors and constructs on the Iranian women's Islamic fashion market. *Interdisciplinary Journal of Research in Business*, *1*(4), 94–100.

Ismail, A. M. (2015). Muḥajababes—Meet the new fashionable, attractive and extrovert Muslim woman. A study of the ḥijāb-practice among individualized young Muslim women in Denmark. *Journal of Islamic Research*, *9*(2), 106–129.

Logan, B. C. (2013). The impact of religious commitment and motivation on African American male academic achievement. Electronic Theses & Dissertations Paper 824, Georgia Southern University.

Mahon, D., Cowan, C. & McCarthy, M. (2006). The role of attitudes, subjective norm, perceived control and habit in the consumption of ready meals and takeaways in Great Britain. *Food Quality and Preference*, *17*(6), 474–481. doi:10.1016/j.foodqual.2005.06.001

Mokhlis, S. (2009). Relevancy and measurement of religiosity in consumer behavior research. *International Business Research*, *2*(3), 75–84. doi:10.5539/ibr.v2n3p75

Muftiarini, A. (2014). *Dian Pelangi siap tampil di New York Fashion Week.* http://lifestyle.okezone.com.

Razzaque, M.A. & Chaudhry, S.N. (2013). Religiosity and Muslim consumers' decision-making process in a non-Muslim society. *Journal of Islamic Marketing*, *4*(2), 198–217. doi:10.1108/17590831311329313.

Suhendra, A. (2013). Kontestasi identitas melalui pergeseran interpretasi hijab dan jilbab dalam Al Qur'an. *Jurnal Studi Gender*, 6(1). doi: http://dx.doi.org/10.21043/palastren.v6i1.976

Usman, H. & Sobari, N. (2013). *Aplikasi teknik multivariate untuk riset pemasaran.* Jakarta: Rajawali Pers.

Wiener, Y. (1982). Commitment in organizations: A normative view. *Academy of Management Review*, *7*(3), 418–428. doi:10.2307/257334

Worthington Jr., E. L., Wade, N. G., Hight, T. L., Ripley, J. S., McCullough, M. E., Berry, J. T., Schmitt, M. M., Berry, J. T., Bursley, K. H. & O'Connor, L. (2003). The religious commitment inventory – 10: Development, refinement, and validation of a brief scale for research and counseling. *Journal of Counseling Psychology*, *50*(1), 84–96. doi:10.1037/0022-0167.50.1.84

Competition and Cooperation in Economics and Business – Gani et al. (Eds)
© 2018 Taylor & Francis Group, London, ISBN 978-1-138-62666-9

Role of health corporate social responsibility in enhancing quality of life and loyalty: Customer and society perspectives

T.E. Balqiah, N. Sobari, E. Yuliati & R.D. Astuti
Department of Management, Faculty of Economics and Business, Universitas Indonesia, Depok, Indonesia

ABSTRACT: This paper examines the role of Corporate Social Responsibility (CSR) activities, as perceived by customers, in promoting children's Quality of Life (QOL) and Customer Loyalty. This paper also compares two indicators of social performances from the customers' and society's perspectives. The survey was conducted in five cities in Indonesia: Jakarta, Padang, Surabaya, Makassar and Kupang. The subjects of this research are the customers of three companies/brands operating in Indonesia (Pertamina, Danone-Aqua, and Frisian Flag) and also the society in the communities that had been exposed to the companies' Health CSR activities. The data was collected from 600 respondents (450 respondents are customers, 150 respondents are member of society, and were analysed using multiple regressions to test nine research hypotheses. The result shows that different motivations will be generated from different CSR activities by three brands, and have different paths to influence loyalty.

1 INTRODUCTION

Companies are economic entities that provide goods and services to society. In running their businesses, companies comply with rules and regulations. The implication of ethical responsibilities is that companies must operate according to the high standards, norms, and expectation of their stakeholders. Companies are engaging in Corporate Social Responsibility (CSR) activities as 'window dressing', to appeal to the most influential stakeholder groups (Fatma & Rahman, 2015). They respond by doing philanthropic activities to show that businesses are, in fact, a form of good corporate citizenship. Therefore, companies have the task to take the lead in bringing business and society back together (Erdiaw-Kwasie et al., 2015).

CSR activities are expected to create a strong bond between companies and their stakeholders, including customers and society (Peloza & Shang, 2011). Barnett (2007) stated that the ability of CSR activities to create values for companies depends on their ability to create a positive relationship with the companies' stakeholders. CSR activities can not only create profits, but they can also have impacts on social and environmental issues (Peloza & Shang, 2011). CSR and the issue of sustainability are the two common themes repeatedly used in the discussion of economic contributions, society, the environment, and the consequences of business activities (Torugsa et al., 2013).

The different forms of CSR activities will result in different values for all interested parties that are involved. These CSR activities have different impacts on the companies' profits, and even on the value of those companies (Malik, 2014). Liu et al. (2014) have stated that three different forms of CSR activities will improve perceptions on the quality of a brand, which in turn will make the brand more preferable. Some researchers have also shown that CSR activities have an impact on the consumers' behaviour (Maignan & Ferrel, 2004; Luo & Bhattacharya, 2006; Du et al., 2007). CSR will create an image of a company, which in turn will improve the customers' loyalty (Plewa et al., 2015). CSR activities are not only social activities which will burden the companies' budget, but they could enhance Customer Loyalty.

Satisfied and loyal consumers are intangible assets that will produce income in the future. Therefore, CSR activities are social activities which can also be an investment for the company.

The main objective of CSR activities is to maximise benefits for the social welfare of societies, which also shows the companies' responsibility to the effects of business activities on consumers, employees, shareholders, and the rest of the society in which the companies operate (Narwal & Singh, 2013). Companies must engage in CSR activities to avoid negative image and mistrust from stakeholders.

Companies use CSR activities to build and strengthen the relationship with their multiple stakeholders, such as consumers, suppliers, competitors, and investors (Raghubir et al., 2010). CSR activities are directed to improve education, health, welfare, happiness, and stability of the society. The success of such activities can be measured by using the Quality of Life (QOL) indicators, namely economy, social, health, subjective assessment on happiness, and life satisfaction (Sirgy et al., 2012).

Our previous research in Indonesia showed that CSR activities of Pertamina (a state-owned oil and gas enterprise), Toyota (a global brand of Astra International), Sampoerna (a big local brand of cigarettes), Lifebuoy (a global brand of Unilever), and Aqua-Andone (a big local brand of drinking water, owned by Andone) influence directly and indirectly the perceived QOL in the communities exposed to the companies' CSR activities (Balqiah et al., 2010; Balqiah et al., 2011). Meanwhile, regarding the types and CSR motives, this study had been conducted to examine the more specific relationship of CSR to the QOL of underprivileged children. The results showed a positive influence between CSR motives towards perceived QOL of underprivileged children where the CSR activities were conducted (Balqiah et al., 2012). These studies showed that CSR activities not only have impacts on customers but also on communities or disadvantaged children, who are also stakeholders of the company.

To continue previous studies, this research is conducted by choosing companies which have some forms of CSR activities and which have been shown to make continuous efforts in building a better QOL for society. With children—as our future generation—in mind, we choose CSR activities directed to children's health.

In the next section, this paper will explain the constructs and literatures review as a foundation to develop a conceptual framework. Further, it discusses the methodology adopted in this study, discussion, and conclusion.

2 LITERATURE REVIEW AND HYPOTHESES

2.1 *CSR and stakeholders*

Organisational behaviour reflects, and can be predicted by, the nature of its diverse stakeholders, the norms that they adopt to define right and wrong, and their relative influence on organisational decisions (Helmig et al., 2016). With regard to the social aspect of behaviour, the corporation feels under pressure by its stakeholders. Social responsible corporate behaviour may mean different things in different times (Campbell, 2007).

CSR is one type of activity that can show how they are concerned with social issues. CSR can be defined as the commitment to improve the community well-being through the chosen business activities which contribute the companies' resources (Kotler & Lee, 2005). These social initiatives can be seen through their main activities to support social causes and to fulfil the commitment to do CSR. There are various forms of CSR activities that companies can implement. The types of this social initiative, according to these authors, are cause promotions, cause-related marketing, corporate social marketing, philanthropy, community volunteering, and socially responsible business practices. Cause promotions means companies provide funds or other resources to increase awareness and concerns on a social issue or to support fundraisings, participations and mobilisation of volunteers for that particular cause. Cause-related marketing is the company's commitment to contribute or donate a portion of its revenues to a particular cause based on product sales. Corporate social marketing is when companies support fundraisings or start a campaign to promote changes in behaviour to improve the public health, safety, environment, or well-being in general. Corporate philanthropy is when companies make direct contributions to charities or special causes in the forms of grants, donations or services. Community volunteering is the situation when companies support and encourage their employees, intermediaries or other partners to voluntarily donate their time to support social or community organisations and activities. Finally, socially responsible business practices refers to when companies adopt business practices and make special investments to support social causes in order to improve community well-being and to protect the environment.

Companies try to increase the society welfare through CSR activities. CSR supports the belief that businesses can work together with the government and other stakeholders to promote a better life, since CSR itself is basically the commitment of a business company to contribute to sustainable economic developments by working with employees and local communities to improve their overall QOL (World Business Council for Sustainability Development, 2004). Although many companies have conducted CSR activities, it must be recognised that such activities are driven directly or indirectly by stakeholders (Kiessling et al., 2015). This is the result of globalisation of the market, which makes companies face more diverse situations through interaction with various stakeholders. CSR and sustainability issues are the two common themes in discussions related to the contribution of business activities to economic, social, and environmental aspects (Torugsa et al., 2013).

Although many companies have conducted CSR activities, it must be recognised that such activities are driven directly or indirectly by stakeholders (Kiessling et al., 2015). This is the result of globalisation of the market, which makes companies face more diverse situations through interaction with various stakeholders. CSR and sustainability issues are the two common themes in discussions related to the contribution of business activities to economic, social, and environmental aspects (Torugsa et al., 2013). Therefore, in addition to the provision and delivery of impact-minimising products to consumers directly, companies can also do their CSR activities with the goal of improving QOL for the community at large.

2.2 *The effect of CSR motives on brand attitude*

Marin et al. (2015) explain the possibility that CSR activities will create positive or negative customer perception towards the company. Further, Becker-Olsen et al. (2006) also have some arguments that there are various motivations of consumer responses to CSR actions. When motivations are considered to be profit-related, attitudes towards the firm are likely to be negative; however, when motivations are considered to be socially motivated or society/community focused, positive attitudes are likely to be enhanced.

Ellen et al. (2006) also identify four different motivations that have a different impact on customer response. Egoistic motives related to exploiting the cause rather than helping it, strategic motives that support the attainment of business while benefiting the cause, stakeholder-driven motives related to supporting social causes solely because of pressure from stakeholders, and values-driven motives related to benevolence-motivated giving. When motivations are considered to be firm-serving or profit-related, attitudes towards firms are likely to diminish; when motivations are considered to be socially motivated, attitudes towards firms are likely to be enhanced (Becker-Olsen et al., 2006).

In this paper, the authors investigate specifically CSR activities that are involved with children's health. When customers perceived the CSR activities, which were implemented to focus on society and to give to society (especially children), it enhanced the positive attitude to the brand. This attitude is significant in marketing because it will form a perception in consumer behaviour.

H_1: CSR motives will influence Brand Attitude.

2.3 *The effect of CSR motives on QOL*

According to Sheehy (2014), perception of political participants use CSR as a means to advance particular preferences with respect to corporate and economic policy of distinct political philosophies. On the other hand, customers increasingly expect business to consider human rights in their employment practices and demonstrate stewardship towards the natural environment; therefore, companies around the globe suggest that managers no longer see social engagement as ancillary to economic performance but rather as an integral component of corporate strategy on which they will be judged by their constituents (Bronn & Vidaver-Cohen, 2009).

Social initiative in the business context is defined here as any programme, practice, or policy undertaken by a business firm to benefit society. One type of social activity is to support children's QOL. Jozefiak et al. (2008) define QOL as a subjective assessment related to well-being in terms of physical and mental health, self-esteem, perception of personal activities (playing or hobby), and perceived connection with friends and family as well as school.

However, customers distinguish between other centred, self-centred, and win-win motives, and most customers assume companies have mixed motives for their CSR activities (Öberseder et al., 2011). If CSR motives are sincere in giving back to society by providing social projects for people in need (e.g. children as vulnerable stakeholder), it could increase society's QOL, however, if it was perceived not sincere, it could decrease QOL.

H_2: CSR motives will influence Quality of Life.

2.4 *The moderating role of Subjective Well-Being*

Subjective Well-Being (SWB) is conceptualised as a multidimensional construct. The survey's questions pertain to overall life evaluations, self-assessed financial and health states, and personal views about the future, community, and society (Lin et al., 2014), which are individual characteristics and community level characteristics that will have an impact on SWB, meaning that well-being depends not only on individual abilities, or social position, but on the context, and on the 'goodness of others' (Hooghe & Vanhoutte, 2011). Therefore, a different socio-demographic environment reflects a different welfare condition. This situation induces the effectiveness of CSR activities on customer perception. Referring to the concept of SWB, when the target CSR activities are aimed at people who are in a particular environment, then its impact on society, customer perception, and attitude would be different if carried on in other communities.

H_3: Subjective Well-Being moderates the influence of CSR motives to Brand Attitude.
H_4: Subjective Well-Being moderates the influence of CSR motives to Quality of Life.

2.5 *The effect of QOL on brand attitude*

Concerning the four types of CSR activities of the firm (Carroll 1991, 1999), there must be social activities that focus on supporting core business activities in relation to all stakeholders. There are two sides to a coin regarding CSR activities; one is the social activities that 'give' to society, and the other is to create competitive advantage (Gupta, 2002; Hult, 2011; Hunt, 2011; Huang & Rust, 2011).

Creating a good QOL for children who are the target of CSR activities will surely build a positive attitude in customers towards the brand. Based on this reasoning, we propose the following:

H_5: Quality of Life positively influences Brand Attitude.

2.6 *The effect of QOL and brand attitude on loyalty*

CSR creates strong relationships with the stakeholder, enhances profit, and also has an impact on social and environment issues (Peloza & Shang, 2011). Social exchange is based on the expectation of trust and reciprocation, as the exact nature of the return and the time frame is left unspecified (Lii & Lee, 2012). When a company engages in a CSR activity, consumers may perceive the company to be altruistic, which leads to a more favourable attitudinal and behavioural evaluation of the same company, such as with loyalty. Loyalty is a commitment to buy back or re-subscribe to the preferred product or service in the future, although it may be influenced by situational factors and marketing efforts which lead to switching behaviour (Oliver in Kotler & Keller, 2011).

This paper propose here that these results shed some light on this issue that both a Brand Attitude reflecting the business capabilities and child QOL reflecting social initiative have a joint impact on customer perception and loyalty. It is expected that the perception about a value of company offering (brand) and the QOL will positively influence customers' loyalty to the company's products. Based on this reasoning, propose the following:

H_6: Quality of Life positively influences Customer Loyalty.
H_7: Brand Attitude positively influences Customer Loyalty.

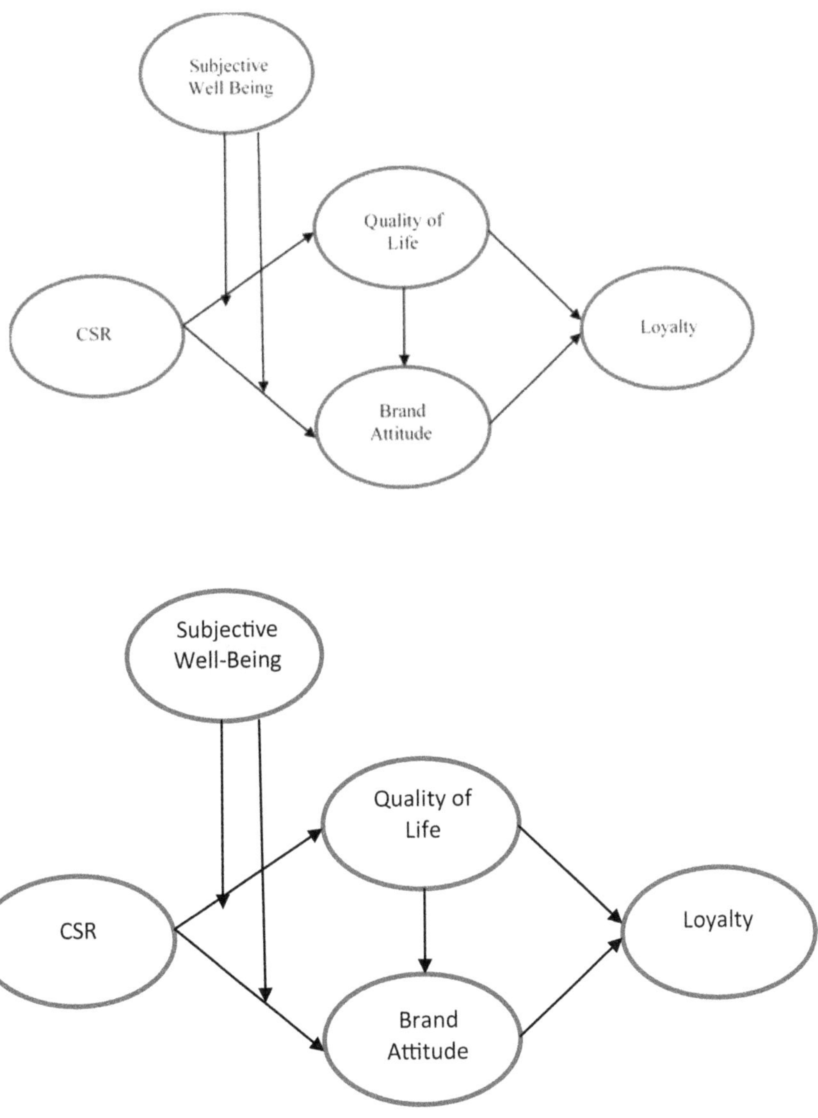

Figure 1. Research model.

2.7 *The difference of customer and community perception on SWB and QOL*

Implementing CSR can create different reactions from different stakeholders. Cantrell et al. (2014) explain about how stakeholder theory aids in the understanding of the influences and influencers on the organisation, and conversely it assists in understanding how the actions of the organisation can affect different stakeholders. When CSR is implemented as social activities of the firm, it could be perceived differently by the stakeholders, because their involvement and attention to those activities are different as the result of their perception of the firm's motivation to conduct the CSR. Subjective Well-Being and Quality of Life are two indicators of community welfare which could be influenced by social or business activities. Based on this reasoning, we propose the following:

H_8: Subjective Well-Being will be perceived differently between customers and community.
H_9: Quality of Life will be perceived differently between customers and community.

The theoretical model for our research, in Figure 1, contains nine hypotheses that the authors derive most dominantly from stakeholder theory and institutional theory.

3 METHOD

3.1 *Sample and data source*

To collect the data, the authors distributed a survey questionnaire in five area/cities: Makassar and Kupang (representing eastern part of Indonesia), and Jabodetabek, Padang, and Surabaya (representing the western part of Indonesia), using self-administered questionnaires from 600 respondents (450 respondents are customers, 150 respondents are member of societies), that were selected by purposive sampling. The objects are AQUA-ANDONE (mineral water), PERTAMINA (lubricant product), and FRISIAN FLAG INDONESIA (milk). These firms conducted different CSR activities that are concerned with children (e.g. AQUA-ANDONE supplies clean water for children; PERTAMINA helps children's and mothers' healthy lifestyle; and FRISIAN FLAG induces children to drink milk and lead a healthy lifestyle). Furthermore, these firms are big and well-known companies, and have done continuous CSR activities in past years.

3.2 *Measurement development*

Questionnaire development is done by conducting a study of the literature and previous studies associated with each construct. To increase the validity of the instrument, the authors conducted preliminary research through a focus group discussion and in-depth interview with several sources (e.g. teachers and housewives) to construct operational definitions of the children's QOL. The questionnaire consists of 43 questions with six points of Likert scale regarding five research constructs: 16 items for Health CSR, 10 items for Subjective Well-Being, 8 items for Quality of Life, 3 items for Brand Attitude, and 6 items for Loyalty.

The pre-test was done to refine the questionnaires by reducing response error. Furthermore, after 600 questionnaires were collected, a factor analysis and multiple regression analysis with SPSS 22 were used to test the hypotheses at $\alpha = 5\%$.

4 RESULTS AND DISCUSSION

The sample included 54.9% females, 55.1% of customers in the sample were under 25 years, 29.6% between 25 and 39 years, and 13.3% over 40 years. Regarding education, 68.3% did not have a bachelor degree, 31.7% have a bachelor degree and higher.

4.1 *CSR's motives*

The result of exploratory factor analysis shows that there are three dimensions of Health CSR which represent the three motives to conduct CSR activities, as perceived by their customers. This result is summarised in Table 1.

Table 1. CSR's motives.

CSR's Motives		
"Moral"	"Stakeholders"	"Business"
– Moral responsibilities	– Meet the customer's expectation	– Get benefit through collaboration
– Care to society	– Respond to employee's expectation	– Support business activities
– Positive benefit beliefs	– Respond to shareholder's expectation	– Minimise income tax
– Easy for customers involved in conservation activities	– Respond to society's expectation	– Publicity
– Balancing take and give to society		– Get more customers – Retain customers – Increase profit

According to Becker-Olsen et al. (2006), when motivations are considered as firm-serving or profit-related, attitudes towards firms are likely to diminish, but when motivations are considered as being socially motivated, attitudes towards firms are likely to be positive. Ellen et al. (2006) divided these driving motives into two groups: positive and negative motives. Positive motives are values-driven and strategic, and negative motives are egotistic and stakeholder-driven. Graafland & Mazereeuw-Van der Duijn Schouten (2012) proposed that the two main motives of the companies' executives to involve in CSR are extrinsic and intrinsic motives. Extrinsic motives are concerned with financial motives, where CSR should contribute to the (long-term) financial performance of the company (Story & Neves, 2014). Intrinsic motives are concerned with the non-financial aspects, such as the managers' personal values and beliefs that might be an important motivating factor for CSR, particularly in SMEs and even in larger companies. The other intrinsic motive is altruism. Executives may contribute to CSR because they enjoy helping others or want to contribute to others' prosperity.

This research shows that business and stakeholder orientations, which concern the economic performance, could harm companies' images. These motives represent strategic and egoistic motives that are perceived by customers as negative motives. This is similar to the perception of internal stakeholders who assume that companies only execute CSR activities because they are forced by their employees and shareholders. This is not perceived as sincere and are considered as being negative motives. This motive might be perceived as institutional motives for social initiatives, suggesting that companies engage in social initiatives primarily due to institutional pressures (Bronn & Vidaver-Cohen, 2009). On the other hand, moral obligation is perceived as a positive motive because this motive shows that companies have concerns about social aspects of their businesses.

4.2 *SWB*

There are four dimensions of SWB that represent a community's welfare concerning health, economic, social, and future conditions. These four dimensions of SWB similar with two dimensions of Hooghe and Vanhoutte (2011), reflect individual and social well-being. The dimensions of SWB are summarised in Table 2.

4.3 *Descriptive statistics*

In Table 3, mean scores for all variables and dimensions included in the analysis are summarised. The result shows that on average, CSR motives are the highest for Moral Motives of CSR, and the lowest for the business motives of CSR. Pertamina was strong in Moral Motives, while Aqua-Andone was strong in Stakeholder and Business Motives. This result shows that different CSR activities will be perceived differently by customers. The highest

Table 2. Dimensions of SWB.

SWB			
"Health"	"Economic"	"Social"	"Hope"
– Life satisfaction	– Support the family	– Happiness	– expectation of better living conditions
– Physical Health	– Sufficient economic conditions	– Live in a social environment kinship	
– Psychological Health	– Financial condition is able to support the needs of life		
	– Decent living conditions		

Table 3. Descriptive statistics.

Dimension/ Motives	Firm	Location					TOTAL
		Jabodetabek	Padang	Surabaya	Makasar	Kupang	
CSR_Moral	Pertamina	4.75	4.69	4.72	4.53	5.25	4.79
	Frisian Flag	4.45	4.81	4.77	4.46	4.82	4.66
	Aqua	4.95	4.83	4.59	4.56	4.89	4.77
CSR_Stakeholder	Pertamina	4.21	3.67	4.16	4.41	4.08	4.11
	Frisian Flag	3.66	3.87	4.06	4.12	3.87	3.91
	Aqua	4.64	3.78	3.88	4.03	4.48	4.17
CSR_Business	Pertamina	4.06	3.46	3.98	3.79	3.53	3.76
	Frisian Flag	4.38	3.75	4.32	2.77	3.57	3.76
	Aqua	4.43	3.72	4.11	2.79	4.06	3.82
SWB_Health	Pertamina	3.81	3.70	3.62	3.17	3.67	3.59
	Frisian Flag	3.71	3.90	3.76	4.21	3.73	3.86
	Aqua	3.82	3.74	3.43	3.77	4.44	3.84
SWB_Economic	Pertamina	3.64	3.69	3.72	3.68	3.60	3.67
	Frisian Flag	4.06	3.79	4.04	4.57	3.94	4.08
	Aqua	3.68	3.62	3.42	4.44	4.26	3.88
SWB_Social	Pertamina	4.20	4.17	4.07	3.90	4.33	4.13
	Frisian Flag	3.85	4.35	4.13	4.60	4.27	4.24
	Aqua	4.37	4.22	3.95	3.95	4.72	4.24
SWB_Hope	Pertamina	4.50	4.47	4.57	5.20	4.93	4.73
	Frisian Flag	4.60	4.70	4.67	4.30	4.87	4.63
	Aqua	5.17	4.87	4.60	3.53	5.10	4.65

score of Moral Motives for Pertamina might be caused by the characteristic of its CSR activities which were not related directly to the core business of Pertamina as an oil and gas company, while Frisian Flag Indonesia and Aqua-Andone's CSR activities are more related to their core business activities. The similarity degree to the core business might be perceived as reflections of extrinsic motives in doing CSR.

For SWB there are, as expected, different results for each company. For Pertamina, the highest score (as perceived by its customers) is for Hope Well-Being in Makassar. Frisian Flag Indonesia has the highest score (as perceived by the customers) for both Economic and Social Well-Being, also in Makassar. Aqua-Andone has the highest score of Hope Well-Being in Padang and Kupang.

Children's QOL (according to customer perception) is perceived highest in Kupang for Pertamina and Aqua-Andone, and highest in Makassar for Frisian Flag Indonesia. The highest mean score for Loyalty is Aqua-Andone in Kupang, followed by Jakarta's mean score. Brand Attitude of Aqua is the highest mean score (according to customer perception). It is perceived highest at almost all locations.

The mean scores of Stakeholder Motive are different among locations. For Pertamina, the highest score is for Makassar, and the lowest score is for Padang. For Frisian Flag Indonesia, the highest score is for Makassar (but still lower than Pertamina's), and the lowest score is for Jabodetabek. For Aqua-Andone, the highest score is for Jabodetabek, and the lowest score is for Padang. As summarised in Table 3, in Padang and Jabodetabek (both represent the western part of Indonesia), the mean score for the Stakeholder Motive of both Pertamina and Frisian Flag Indonesia are relatively lower than for Surabaya, Makassar, and Kupang (which represent the eastern part of Indonesia). This result shows that CSR activities of Pertamina and Frisian Flag Indonesia are perceived as having extrinsic motives in the western part of Indonesia. On the contrary, for Kupang (eastern part of Indonesia), the mean score of the Moral Motive of Pertamina is higher than that of Jabodetabek and Padang, even though it is lower in Makassar. There are no differences in Moral Motive among companies, but the differences occur among locations.

4.4 *The effect of CSR on performance*

This section discusses the results of the multiple regression analysis for each relationship of CSR motive, QOL (represented social performance) and Brand Attitude (represented business performance); relationship of QOL (social performance), Brand Attitude (business performance) and Customer Loyalty (business performance); and the role of SWB as moderating variable. The significance of the path coefficients was evaluated by analysing the value of the parameters.

From the result of regression analysis, as summarised in Table 4, it can be seen that the Moral Motive of CSR positively influences Brand Attitude and Quality of Life. Stakeholder motives only positively influence Quality of Life. Meanwhile, Business motive negatively influence Brand Attitude and Quality of Life. Finally, Quality of Life could increase Brand Attitude. Furthermore, both Brand Attitude and Quality of Life positively influence Customer Loyalty. Thus, this result shows that H_1, H_2, H_5, H_6, and H_7 are supported.

The result show that the respondents perceived the CSR activities were driven by mixed motives. The act of supporting a social initiative may seem to be a public-serving action and sel-serving. The customers' perceptions toward the motivations of act will influence their evaluations about the firm and will impact on their beliefs, attitudes, and intentions (Becker-Olsen et al., 2006).

When presented with evidence of a company's social involvement, consumers are likely to elaborate on the message and assign one of two primary types of motives to the company's self-serving or public-serving intention. In this study, self-serving intention is to increase brand attitude, and public-serving intentions is to increase children's QOL. These results

Table 4. Direct and moderating effect.

Independent variables	Model 1 Brand attitude as dependent	Model 2 Brand attitude as dependent	Model 3 Child-QOL as dependent	Model 4 Child-QOL as dependent	Model 5 Loyalty as dependent
CSR Moral	**0.378***	0.233	**0.351***	**–0.275***	
CSR Stakeholder	0.056	0.361	**0.111***	–0.019	
CSR Business	**–0.115***	**–0.354***	–0.053	**–0.278***	
Child-QOL	**0.165***	0.129			0.648*
Brand Attitude					0.262*
CSR Moral *SWB		0.032		**0.115***	
CSR Stakeholder *SWB		–0.078		–0.008	
CSR Business *SWB		0.059		0.048	
Rsquare	**0.185**	**0.185**	**0.108**	**0.471**	**0.395**

**) sig at $\alpha = 10\%$.

show the two-sided coin of CSR activities, one as social activities that 'give' to society, and the other as to create competitive advantage (Gupta, 2002; Hult, 2011; Hunt, 2011; Huang & Rust, 2011).

This result confirms that instrumental motives revolve fundamentally around managerial beliefs that engaging in social initiatives can have a direct impact on profitability—improving revenue or protecting existing profit levels (Bronn & Vidaver-Cohen, 2009). In this research, Brand Attitude and Customer Loyalty, as a representation of business performance, can enhance revenue and profitability. Different paths in evaluating the CSR activities' effectiveness indicate that different CSR motives will have different impacts on attitude and consumer behaviour (Becker-Olsen et al., 2006; Marın et al., 2015). Being socially responsible is important, but firms must also make a conscious decision about the ratio of doing good things to strategic benefits in their CSR activities. The sustainability concepts direct the need to harmonise the social aspect and the business aspect of the company in creating value.

Vlachos et al. (2009) find that egoistic-driven and strategic-driven attributions have negative effects on word-of-mouth and purchase intentions when consumers are suspicious of the business and the non-profit world. The stakeholders might have different perceptions of the reasons why a company is engaged in CSR activities (Ellen et al., 2006; Bronn & Vidaver-Cohen, 2009; Feldman & Vasquez-Parraga, 2013; Cantrell et al., 2014). In terms of positive motives in the attribution process of the consumers' reactions to CSR action, consumers are likely to accept attributions because they consider companies to be acting with sincere and benevolent intentions, care to society, and tend to view CSR activities as being derived from the companies' moral behaviour (Vlachos et al., 2009).

In this research, although customers perceived companies to have egoistic and strategic-driven attributions, such as Business Motives, the CSR activities still could increase the companies' social performances. This might be caused by the customers' perceptions of the companies' sincerity in doing their social initiatives. The financial motive is not the only reason to contribute to CSR activities. Many companies have business cultures that commit to certain business principles and moral duties. This result confirms that instrumental motives revolve fundamentally around managerial beliefs that engaging in social initiatives can have a direct impact on profitability—improving revenue or protecting existing profit levels (Bronn & Vidaver-Cohen, 2009). In this research, Customer Loyalty, as a representation of business performance, can enhance the companies' revenue and profitability. Interestingly, when considering the SWB of local communities, Moral Motive decreases the perception of the children's QOL as recipient.

4.5 *Moderating effect of SWB*

There is no moderating effect of SWB in building Brand Attitude, but this constructs moderation on relationship between Moral Motive and Quality of Life. Thus, this result shows that H_3 is not supported, and H_4 is supported.

SWB is able to increase the positive impact of CSR on the child's QOL. CSR activities that support social initiatives have an impact on society. The perception that a company participated in the public service will be able to influence customer evaluation, beliefs, attitudes, and intentions (Becker-Olsen et al., 2006). In this study, the perception of SWB is able to increase the positive influence of Moral Motive on Children QOL (social performance), but does not impact the relationship to Brand Attitude (business performance). When customers viewed a company's social activities and concern to society, they tended to infer that these activities are on two main motives (i.e. firm self-serving in this study-related Brand Attitude, or public-serving—on this study of children's QOL). Customers are simplistic in their judgments about CSR initiatives and view them as either serving economic ends or reflecting sincere social concerns (Ellen et al., 2006). When the outcome is Brand Attitude, the customer might see that CSR has business attribution and business objectives; therefore, SWB of society cannot increase the positive impact of CSR to Brand Attitude. The customer perceive Brand Attitude as the company's objective that reflecting insincere motives of CSR activities.

Table 5. Mean difference of SWB and QOL.

Dimension	Firm	Respondents		Sig
		Customers	Societies	
SWB_Health	Pertamina	3.59	5.08	0.000
	Frisian Flag	3.86	4.79	
	Aqua	3.84	4.90	
SWB_Economic	Pertamina	3.67	4.54	0.000
	Frisian Flag	4.08	4.21	
	Aqua	3.88	4.81	
SWB_Social	Pertamina	4.13	4.95	0.000
	Frisian Flag	4.24	4.83	
	Aqua	4.24	4.98	
SWB_Hope	Pertamina	4.73	5.30	0.000
	Frisian Flag	4.63	5.47	
	Aqua	4.65	5.42	
QOL	Pertamina	4.25	5.17	0.000
	Frisian Flag	4.55	5.09	
	Aqua	4.39	5.15	

4.6 *The differences of customers' and society's perceptions on social performances*

From the perspective of Stakeholder Theory, the customers' perceptions of SWB and children's QOL are lower than society's perception. Park et al. (2014) state that ethical responsibilities require that businesses must abide by the moral rules defining appropriate behaviours in society, and that the law prescribes actions that a company must avoid. Ethical responsibilities cover the companies' activities which society expects them to undertake.

Table 5 shows that there are differences on perceptions of social performances between customers and societies, but not among all companies. The difference of SWB is only found in economic well-being. This result concludes that H_8 and H_8 are supported.

5 CONCLUSION

This research has argued that CSR activities related to children's health will create social and business performance. Kiessling et al. (2015) explain that although many companies conducted CSR activities, it must be recognised that such activities are driven directly or indirectly by the stakeholders. The stakeholders might have different perceptions towards the reasons why a company is engaged in CSR activities (Ellen et al., 2006; Feldman & Vasquez-Parraga, 2013; Cantrell et al., 2014). The results of this study are consistent with the findings that Health CSR activities will have positive and negative motives as perceived by customers.

The stakeholders might have different perceptions of the reasons why a company is engaged in CSR activities (Ellen et al., 2006; Bronn & Vidaver-Cohen, 2009; Feldman & Vasquez-Parraga, 2013; Cantrell et al., 2014). The results of this study are consistent with the findings that Health CSR activities have positive and negative motives as perceived by customers, namely Moral, Stakeholder, and Business Motives. Each motive has different impacts on social performance and business performance, indicating that there is still a strategic objective obtained, despite doing social initiatives. Companies that fulfil their responsibilities by carrying out Health CSR activities (social consideration) will be able to improve customer perceptions on the QOL in places where the company conducts its CSR activity (people consideration). This in turn will impact positively on Brand Attitude and Customer Loyalty (business considerations).

This research only focuses on philanthropic activities concerning the health of children. It is hoped that future research could investigate the effectiveness of other types of CSR activities or the combination of some activities. Regarding the QOL, future research should consider the difference of QOL in a different location and refine the measurement of child-QOL.

REFERENCES

Balqiah, T. E., Setyowardhani, H. & Khairani, K. (2010). Study of corporate social responsibility activities influence toward customer loyalty through increasing quality of life in urban area. Unpublished Final Report of Research Grant. Universitas Indonesia.

Balqiah, T. E., Setyowardhani, H. & Khairani, K. (2011). The influence of corporate social responsibility activity toward customer loyalty through improvement of quality of life in urban area. *The South East Asian Journal of Management*, *5*(1), 73–90.

Balqiah, T. E., Setyowardhani, H., Daryanti, S., Mukhtar, S. & Khairani, K. (2012). *Contribution of corporate social responsibility in developing poor's child well-being: Consumer's perspectives.* Unpublished final report of research grant report. Universitas Indonesia.

Barnett, M. L. (2007). *Stakeholder influence capacity and the variability of financial returns to corporate social responsibility.* Academy of Management review. doi:10.5465/AMR.2007.25275520

Becker-Olsen, K. L., Cudmore, B. A. & Hill, R. P. (2006). The impact of perceived corporate social responsibility on consumer behavior. *Journal of Business Research, 59*(1), 46–53. doi:10.1016/j.jbusres.2005.01.001

Bronn, P. S. & Vidaver-Cohen, D. (2009). Corporate motives for social initiative: Legitimacy, sustainability, or the bottom line? In *Journal of Business Ethics*, *87*, 91–109. doi:10.1007/s10551-008-9795-z

Campbell, J. L. (2007). Why would corporations behave in socially responsible ways? An institutional theory of corporate social responsibility. *Academy of Management Review*. doi:10.2307/20159343

Cantrell, J. E., Kyriazis, E. & Noble, G. (2014). Developing CSR giving as a dynamic capability for salient stakeholder management. *Journal of Business Ethics*, *130*(2), 403–421. doi:10.1007/s10551-014-2229-1

Carroll, A. B. (1991). The pyramid of corporate social responsibility: Toward the moral management of organizational stakeholders. *Business Horizons*, *34*(4), 39–48. doi:10.1016/0007-6813(91)90005-G

Carrol, A. B. (1999). Corporate social responsibility. *Business and Society*, *38*(3), 268–295. doi:10.1007/978-3-642-25399-7

Du, S., Bhattacharya, C. B. & Sen, S. (2007). Reaping relational rewards from corporate social responsibility: The role of competitive positioning. *International Journal of Research in Marketing*, *24*(3), 224–241. doi:10.1016/j.ijresmar.2007.01.001

Ellen, P. S., Webb, D. J. & Mohr, L. A. (2006). Building corporate association: Consumer attributions for corporate socially responsible program. *Journal of the Academy of Marketing Science, 34*(2), 147–157.

Erdiaw-Kwasie, M. O., Alam, K. & Shahiduzzaman, M. (2015). Towards understanding stakeholder salience transition and relational approach to 'better' corporate social responsibility: A Case for a proposed model in practice. *Journal of Business Ethics*, 1–17. doi:10.1007/s10551-015-2805-z

Fatma, M. & Rahman, Z. (2015). Consumer perspective on CSR literature review and future research agenda. *Management Research Review*, *38*(2), 195–216. doi:10.1108/MRR-09-2013-0223

Feldman, P. M. & Vasquez-Parraga, A. Z. (2013). Consumer social responses to CSR initiatives versus corporate abilities. *Journal of Consumer Marketing, 30*(2), 100–111. doi:10.1108/07363761311304915

Graafland, J. & Mazereeuw-Van der Duijn Schouten, C. (2012). Motives for corporate social responsibility. *De Economist*, *160*(4), 377–396. doi:10.1007/s10645-012-9198-5

Gupta, S. (2002). Strategic dimensions of corporate image: corporate ability and corporate social responsibility as sources of competitive advantage via differentiation. *ProQuest Dissertations and Theses*, 94–94. http://search.proquest.com/docview/275888617?accountid=14549%5Cnhttp://hl5yy6xn2p.search.serialssoltions.com/?genre=article&sid=ProQ:&atitle=Strategic+dimensions+of+corporate+image:+Corporate+ability+and+corporate+social+responsibility+as+sources+of+compet

Helmig, S., Spraul, K. & Ingenhoff, D. (2016). Under positive pressure: How stakeholder pressure affects corporate social responsibility implementation. *Business & Society*, *55*(2), 151–187.

Hooghe, M. & Vanhoutte, B. (2011). Subjective well-being and social capital in Belgian communities. The impact of community characteristics on subjective well-being indicators in Belgium. *Social Indicators Research*, *100*(1), 17–36. doi:10.1007/s11205-010-9600-0

Huang, M. H. & Rust, R. T. (2011). Sustainability and Consumption. *Journal of the Academy of Marketing Science*, *39*(1) 40–54. doi:10.1007/s11747-010-0193-6

Hult, G. T. M. (2011). Market-Focused sustainability: Market orientation plus! *Journal of the Academy of Marketing Science*. doi:10.1007/s11747-010-0223-4

Hunt, S. D. (2011). Sustainable marketing, equity, and economic growth: A resource-advantage, economic freedom approach. *Journal of the Academy of Marketing Science*, 39(1), 7–20. doi:10.1007/s11747-010-0196-3

Jozefiak, T., Larsson, B. Wichstrøm, L., Mattejat F. & Ravens-Sieberer, U. (2008). Quality of life as reported by school children and their parents: A cross-sectional survey. *Health and Quality of Life Outcomes*, 6(1), 1–11. doi:10.1186/1477-7525-6-34

Kotler, P. & Lee, N. (2005). *Corporate social responsibility: Doing the most good for your company and your cause*. Hoboken: John Wiley and Sons.

Kotler, P. & Keller, K. L. (2011). *Marketing management* (14th ed.). New Jersey: Pearson Education.

Kiessling, T., Isaksson, L. & Yasar, B. (2015). Market orientation and CSR: Performance implications. *Journal of Business Ethics*, *102*(1), 47–55.

Lii, Y. S. & Lee, M. (2012). Doing right leads to doing well: When the type of CSR and reputation interact to affect consumer evaluations of the firm. *Journal of Business Ethics*, *105*(1), 69–81. doi:10.1007/s10551-011-0948-0

Lin, C. C., Cheng, T. C. & Wang, S. C. (2014). Measuring subjective well-being in Taiwan. *Social Indicators Research*, *116*(1), 17–45. doi:10.1007/s11205-013-0269-z

Liu, M. T., Wong, I. A., Shi, G., Chu, R., Brock, J. L. & Introduction Corporate. (2014). The impact of Corporate Social Responsibility (CSR) performance and perceived brand quality on customer-based brand preference. *The Journal of Services Marketing*, *28*(3), 181–194. doi:10.1108/JSM-09-2012-0171

Low, G. S. & Lamb, C. W. (2000). The measurement and dimensionality of brand associations. *Journal of Product & Brand Management*, *9*(6), 350–370. doi:10.1108/10610420010356966

Luo, X. & Bhattacharya, C. B. (2006). Corporate social responsibility, customer satisfaction, and market value. *Journal of Marketing*, *70*(4), 1–18. doi:10.1509/jmkg.70.4.1

Maignan, I. & Ferrell, O. C. (2004). Corporate social responsibility and marketing: An integrative framework. *Journal of the Academy of Marketing Science*, *32*(1), 3–19. doi:10.1177/0092070303258971

Malik, M. (2014). Value-Enhancing capabilities of CSR: A brief review of contemporary literature. *Journal of Business Ethics*, *127*, 419–439. doi:10.1007/s10551-014-2051-9

Marın, L., Cuestas, P. J. & Roman, S. (2015). Determinants of Consumer Attributions of Corporate Social Responsibility. *Journal of Business Ethics*, *138*(2), 247–260. doi:10.1007/s10551-015-2578-4

Narwal, M. & Singh, R. (2013). Corporate social responsibility practices in India: A comparative study of MNCs and Indian companies. *Social Responsibility Journal*, *9*(3), 465–478. doi:http://dx.doi.org/10.1108/SRJ-11-2011-0100

Öberseder, M., Schlegelmilch, B. B. & Gruber, V. (2011). 'Why don't consumers care about CSR?': A qualitative study exploring the role of CSR in consumption decisions. *Journal of Business Ethics*, *104*(4), 449–460. doi:10.1007/s10551-011-0925-7

Oliver, R. L. (1999). Whence consumer loyalty? *The Journal of Marketing*, *63*, 33–44. doi:10.2307/1252099

Park, J., Lee, H. & Kim, C. (2014). Corporate social responsibilities, consumer trust and corporate reputation: South Korean consumers' perspectives. *Journal of Business Research*, *67*(3), 295–302. doi:10.1016/j.jbusres.2013.05.016

Peloza, J. & Shang, J. (2011). How can corporate social responsibility activities create value for stakeholders? A systematic review. *Journal of the Academy of Marketing Science*, *39*(1), 117–135. doi:10.1007/s11747-010-0213-6

Plewa, C., Conduit, J., Quester, P. G. & Johnson, C. (2015). The impact of corporate volunteering on CSR image: A consumer perspective. *Journal of Business Ethics*, *127*, 643–659. doi:10.1007/s10551-014-2066-2

Raghubir, P., Roberts, J., Lemon, K. N. & Winer, R. S. (2010). Why, when, and how should the effect of marketing be measured? A stakeholder perspective for corporate social responsibility metrics. *Journal of Public Policy & Marketing*, *29*(1), 66–77. doi:10.1509/jppm.29.1.66

Sánchez, J. L. F. & Sotorrío, L. L. (2007). The creation of value through corporate reputation. *Journal of Business Ethics*, *76*(3), 335–346. doi:10.1007/s10551-006-9285-0

Sheehy, B. (2014). Defining CSR: Problems and solutions. *Journal of Business Ethics*, *131*(3), 625–648. doi:10.1007/s10551-014-2281-x

Sirgy, M. J., Yu, G. B., Lee, D. J., Wei, S. & Huang, M. W. (2012). Does marketing activity contribute to a society's well-being? The role of economic efficiency. *Journal of Business Ethics*, *107*(2), 91–102. doi:10.1007/s10551-011-1030-7

Story, J. & Neves, P. (2014). When Corporate Social Responsibility (CSR) increases performance: Exploring the role of intrinsic and extrinsic CSR attribution. *Business Ethics: A European Review*, *24*(2), 111–124. doi:10.1111/beer.12084

Torugsa, N. A., O'Donohue, W. & Hecker, R. (2013). Proactive CSR: An empirical analysis of the role of its economic, social and environmental dimensions on the association between capabilities and performance. *Journal of Business Ethics*, *115*(2), 383–402. doi:10.1007/s10551-012-1405-4

Vidaver-Cohen, D. & Brønn, P. S. (2013). Reputation, responsibility, and stakeholder support in Scandinavian firms: A comparative analysis. *Journal of Business Ethics*, *127*(1): 49–64. doi:10.1007/s10551-013-1673-7

Vlachos, P. A., Tsamakos, A., Vrechopoulos, A. P. & Avramidis, P. K. (2009). Corporate social responsibility: Attributions, loyalty, and the mediating role of trust. *Journal of the Academy of Marketing Science*, *37*(2), 170–80. doi:10.1007/s11747-008-0117-x

World Business Council for Sustainability Development. (2004). Doing business with the poor. New York.

Competition and Cooperation in Economics and Business – Gani et al. (Eds)
© 2018 Taylor & Francis Group, London, ISBN 978-1-138-62666-9

Rebalancing strategy analysis on three types of stock portfolios (LQ45, construction, and consumption) in Indonesia for the 2006–2015 period

B. Jesslyn & D.A. Chalid
Department of Management, Faculty of Economics and Business, Universitas Indonesia, Depok, Indonesia

ABSTRACT: This study aims to analyse the rebalancing strategy towards three types of stock portfolios (i.e. LQ45 index, Property, Housing, and Construction sectoral index, and Consumption sectoral index) over a ten-year period. This study used a simulation approach to compare the performance between rebalanced portfolios and non-rebalanced portfolios. It is found that the rebalanced portfolio resulted in a lower risk in the 2008 crisis and in 2013, as well as higher investment returns, than the non-rebalanced portfolio during the investment period of ten years. However, it is found that statistical testing concludes that there is no significant difference between the results of the non-rebalanced and the rebalanced portfolio. Further, it is also found that rebalancing strategy is more suitable for a progressive industry rather than a defensive industry. This simulation study also involves the transaction cost variable.

1 INTRODUCTION

According to *PT Kustodian Sentral Efek Indonesia*, the company that provides securities depository and settlement services, the total number of Indonesia's stock investors has been growing by 20% since 2015 (KSEI Press Release, 2016). It is important for an investor to fully comprehend portfolio management in order to support their investment activities in the capital market. Portfolio management activities are consisting of activities form choosing what asset classes will be invested in, synchronising investment goals, selecting asset allocation, and deciding which strategies to implement in order to achieve a balance between the portfolio's risk and return.

The study of Ibbotson and Kaplan (2000) concludes that around 90% of asset performance and around 40% of a portfolio performance can be explained by the investment strategy used. Then the strategy balance between risks and returns can be explained through accurate asset distribution according to the investor's investment goals, risk profile, and investment period.

One of the most important strategies to which investors should pay attention is portfolio rebalancing, which is a process of reallocating the portfolio's composition that has been changed because of the stock growth rate, to its initial allocation. It is difficult for investors to avoid these changes as each stock grows at a different rate. For example, when one stock grows at a higher rate than others within the portfolio, the portfolio composition will have a tendency to shift towards that stock. This deviation will eventually create a greater risk to the investor. In this case, rebalancing will help to realign the investment back to the initial risk profile and investment goals.

Studies about portfolio rebalancing were initiated by Bernstein and Wilkinson in 1997, where they found that the effective rate of return of a rebalanced portfolio is usually greater than the sum of the weighted geometric mean of each asset in the portfolio. Furthermore, Almadi et al. (2014) analysed the rebalancing strategy performance using forecasted data on stocks, bonds, and T-bills out-of-sample in the United States, and concluded that the best result comes out of the monthly rebalancing strategy. Also in the same year, Kohler and Wittig (2014) did an analysis using the risk-based rebalancing approach through standard deviation and correlation among stocks within a portfolio, and found that the yield is almost the same as the basic value-based rebalancing approach, which results in enhanced performance.

Meanwhile in Indonesia, Putri (2010) did some research regarding a rebalancing strategy on mixed hedge funds in 2003–2009 and found that using the rebalancing strategy did not produce a better yield than not using one.

However, previous research has not really focused on the transaction cost variable. Thus, a question arises as to whether the rebalancing bonus or the return yielded from the rebalancing strategy can still cover the transaction cost each time portfolio rebalancing occurs. Is rebalanced portfolio versus unrebalanced portfolio statistically different?

This study aims to address such gaps using a simulation approach to analyse the performance of the rebalancing strategy towards different types of industries in Indonesia throughout 2006 to 2015 (which are: progressive industry that is represented by the Construction sectoral index, and defensive industry that is represented by the Consumption sectoral index), to analyse the benefit of portfolio rebalancing during crisis, and to see the impact of the transaction cost variable on the study of rebalancing.

2 THEORETICAL BACKGROUND

2.1 *Portfolio rebalancing*

Fabozzi et al. (2007) explain that there are several methods that can be used by investors, and the one that is used in this research is called calendar rebalancing. In the calendar approach, investors determine the rebalancing schedule. For instance, every year (annually), every six months (semi-annually), or every three months (quarterly), and portfolio adjustments are then carried out according to the schedule. For example, an investor schedules rebalancing every year, so in the next one year the investor will sell outperforming assets and buy underperforming assets to return the portfolio composition as the initial allocation. This process will help investors to control the portfolio risk resulting from asset growth. However, we must not forget to consider transaction costs prior to rebalancing since the more often we do rebalancing, the bigger the transaction cost will be.

According to Jaconetti et al. (2010), the process of rebalancing involves the sale of certain assets and the purchase of other assets, and this transaction incurs some costs including taxes, transaction costs to execute and process the trades (i.e. brokerage commissions and bid-ask spreads), and management fees if a professional is hired. All of these costs should be considered in determining how often we will do rebalancing, so that the benefits are not depleted due to higher transaction costs or taxes.

2.2 *Progressive and defensive industry*

Husnan (2005) explained that a progressive industry is an industry that is highly sensitive to changes in economic conditions, while a defensive industry is an industry that is less affected by changes in economic conditions. For example, when the economy is tough, consumer spending will reduce the intensity of buying luxury goods, goods that can still be delayed, as well as postponing purchases in large quantities. Retail sectors, financial services, automotive, technology, and construction are all included in the progressive industry because their stock price will usually fall during recessions and rise again when the economy improves.

In contrast, progressive industry includes food or human consumption needs that cannot be postponed as those are all required for everyday human life. Stock prices in this industry tend to be more stable than in the defensive industry, as they are less affected by the change in economic conditions.

3 RESEARCH METHODOLOGY

3.1 *Data and samples*

Secondary data was used in this study because all of the data sources were obtained from indirect sources, namely documentation from the Indonesia Stock Exchange, such as stocks historical prices and publications from BAPPEPAM, such as prospectuses and the Fund Fact Sheet.

The objects of this research include stocks from the LQ45 index, stocks from the Property, Housing, and Construction sectoral index, and stocks from the Consumption sectoral index, in the period from 1 January 2006 to 31 December 2015.

Sample selection is completed based on several conditions and criteria, and hence, all of the selected stocks will be grouped into three different portfolios according to each sector and index:

1. Portofolio 1: stocks that are always included in the list of the LQ45 index in the period of 1 January 2006 to 31 December 2015.
2. Portfolio 2: 25% of the total population of the sector with the largest market capitalisation and trading volume, as well as daily stock prices during the period being accessible.
3. Portfolio 3: 25% of the total population of the sector with the largest market capitalisation and trading volume, as well as daily stock prices during the period being accessible.

All of the research objects were executed using historical daily prices, and finally, based on the sample selection process, 10 stocks for portfolio 1, 15 stocks for portfolio 2, and 10 stocks for 3 were obtained.

The reason of choosing the LQ45 index is to represent the market performance and it acts as a benchmark to the other two indices, while the Property, Housing, and Construction sectors were chosen to represent the progressive industry, and the Consumption sector was chosen to represent the defensive industry. However, the reason for choosing the period of 1 January 2006 to 31 December 2015 is to see whether there is a difference between portfolio performance during the US subprime mortgage crisis in 2008 that impacted Indonesia and portfolio performance during the trade balance deficit that resulted from plunging commodity prices in 2013.

3.2 *Simulation model*

The subsequent steps of data processing mostly refer to the Bodie et al. (2009) explanation on how to develop an optimal risky portfolio. The simulation model for portfolios 1, 2, and 3 is as follows:

1. Calculate daily Holding Period Yield (HPY) and the arithmetic mean of the rate of the return for each stock:

$$HPY = \frac{p_t - p_{t-1}}{p_{t-1}} \times 100\%$$

where p_t is the stock price in period t,

$$\bar{R} = \sum HPY / (n-1)$$

where $\bar{R}$ is the average rate of return and n is the sum of trading days during the period of 1 January 2006 to 31 December 2015, that is 2,609 days.

2. Calculate the variance and standard deviation for each stock:

$$variance = \sigma^2 = \sum [R_i - \bar{R}_i]^2 / (n-1)$$

where R_i and $\bar{R}_i$ are HPY and the average rate of the return for stock i respectively.

$$standard\ deviation = \sigma = \sqrt{\sigma^2}$$

3. Calculate the reward to the variability ratio for each stock.

   Risk-adjusted returns between stocks in a portfolio are now equivalent as this ratio uses the standard deviation as the denominator:

$$\frac{R}{V} = \frac{\bar{R}_i}{\sigma_i}.$$

4. Calculate the covariance and correlation between each asset for each portfolio:

$$Cov_{ij} = \sum (R_i - \bar{R}_i)(R_j - \bar{R}_j) / (n-1)$$

where Cov_{ij} is the covariance between stock i and stock j,

$$\rho_{ij} = \frac{Cov_{ij}}{\sigma_i \sigma_j}.$$

and ρ_{ij} is the correlation between stock i and stock j.

5. Calculate the optimal portfolio using Solver add-ins in the Microsoft Excel program for each portfolio.

   The conditions incorporated into the Solver calculation include:

- Filling in the 'set objectives' column with the standard deviation cell that has been formulated with the following portfolio standard deviation formula:

$$\sigma_p = (W_i^2 \times \sigma^2_{\ i}) + (W_j^2 \times \sigma^2_{\ j}) + \left(2 \times W_i \times W_j \times Cov_{ij}\right)$$

- Filling in the 'changing variable' cell with all stock weight cells.
- All stock weight values must be greater than or equal (> =) to zero.
- The sum of all stock weights must be equal (=) to one.
- The rate of the return value is the interpolation result from the lowest rate of the return to the highest rate of the return from all stocks in the portfolio.

After the list of optimal portfolio combinations is generated, then the best combination is selected based on the asset allocation that gives a minimum risk or the smallest standard deviation, and this combination will serve as the initial portfolio asset allocation and as a reference in rebalancing simulation. That is, whenever rebalancing is conducted, the composition of each portfolio will be back to this initial allocation. If there are stocks that do not get value from the Solver calculation, they will be dispensed with, starting from the next step.

6. Calculate the rate of returns for each stock as per the rebalancing period, which are the quarterly return, semi-annual return, and annual return.

   The rebalancing period is divided into three, which are quarterly rebalancing where rebalancing is conducted every three months, semi-annual rebalancing where rebalancing is conducted every six months, and annual rebalancing where rebalancing is conducted once a year.

7. Execute the rebalancing simulation for each portfolio, based on the ten year rebalancing period from 2006 to 2015, with the following formula:

$$E(r_p) = w_1 E(r_1) + w_2 E(r_2) + w_3 E(r_3) + \ldots + w_n E(r_n).$$

where $E(r_p)$ is the expected return of the portfolio, w is the stock's weight, $E(r)$ is the expected return of stock, and n is the number of stock.

The stock's weight is a constant value and is obtained from step (5), while the expected return of stock is obtained from step (6) (i.e. quarterly return for quarter rebalancing, semi-annual return for semi-annual rebalancing, and annual return for annual rebalancing).

This step is calculated with the assumption of IDR100 million initial investment, and later, the best rebalancing period will be chosen based on the period that gives the biggest investment result at the end of 2015. Thus, we will have the quarterly rebalanced result using what?

8. Test the significance level of the investment result using the T-test (independent Sample T-Test) and the Mann-Whitney test for each portfolio.

The Mann-Whitney test will only be used for portfolio 2 because the data in portfolio 2 does not qualify the T-test homogeneity term.

The hypothesis for the first test is as follows:

H0: there is no significant difference of the investment result between the best rebalancing period's investment result and unrebalanced portfolio.
H1: there is a significant difference of the investment result between the best rebalancing period's investment result and unrebalanced portfolio.

4 RESULTS

The following results are expected to shed some light on the performance of the rebalancing strategy through a comparison between the rebalancing period (that is, quarterly, semi-annual, and annual), as well as a comparison between three types of portfolios that represent the performance of the market (the progressive industry, and the defensive industry). Figures 1–3 show that the investment that has resulted from each portfolio over the ten-year period using the rebalancing simulation approach.

4.1 *Rebalanced vs unrebalanced portfolio*

Figure 1 shows that investors should implement the annual rebalancing strategy to a market portfolio, and that rebalancing strategies have outperformed the unrebalanced portfolio since 2012 – that is, after six years of investing.

However, Figure 2 shows that investors should implement the quarterly rebalancing strategy to a progressive industry portfolio, and rebalancing strategies have outperformed the unrebalanced portfolio since 2007 – that is, after one year investing.

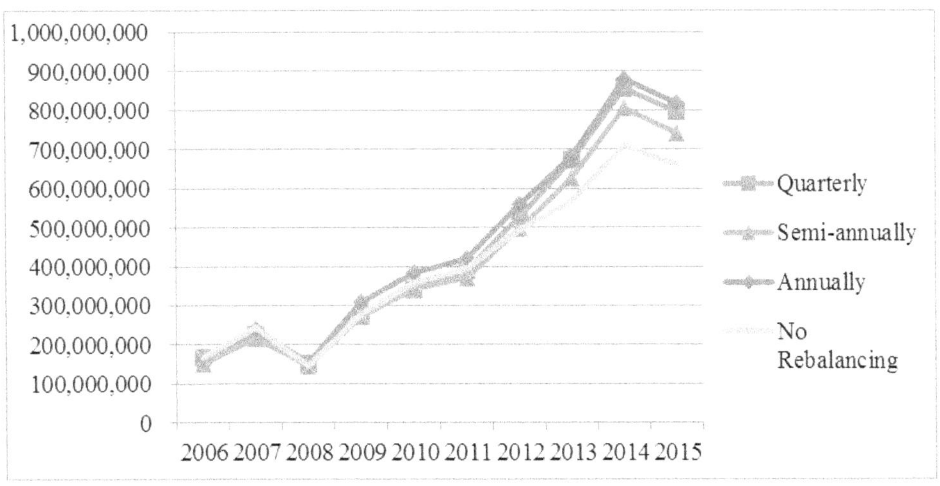

Figure 1. Rebalanced vs unrebalanced portfolio 1.

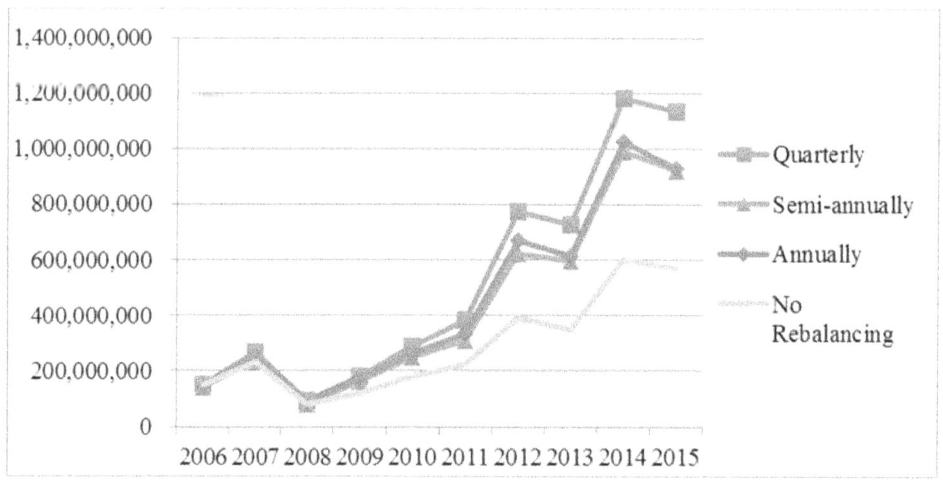

Figure 2. Rebalanced vs unrebalanced portfolio 2.

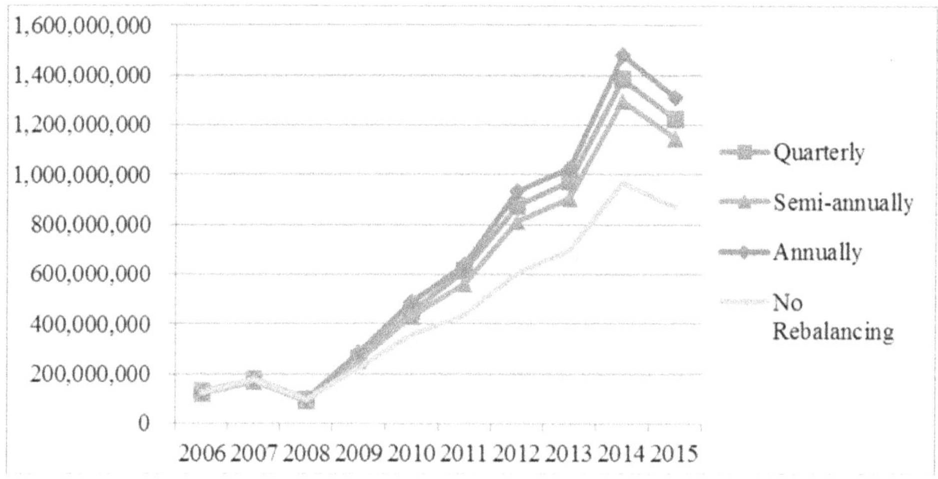

Figure 3. Rebalanced vs unrebalanced portfolio 3.

Similar to a market portfolio, Figure 3 shows that investors should implement annual rebalancing to a defensive industry portfolio, and rebalancing strategies have outperformed the unrebalanced portfolio since 2009 – that is, after three years of investing.

It can be concluded from the above results that for the ten year period of investing, the investment returns of the rebalanced portfolio have exceeded the investment returns of any unrebalanced portfolios. However, in a narrower time period this conclusion does not apply to portfolio 1. The result of the unrebalanced portfolio has exceeded the return of any rebalancing strategies in the first two years. This could happen because stock volatility in portfolio 1 is not as high as other stocks in portfolio 2 and portfolio 3, so it might take a little longer for portfolio 1 to grow and realise the higher rate of the return. Moreover, the graphs also show that during the simulation period, especially from 2009 to 2015, the progressive industry increasingly requires the rebalancing strategy, as the gap between the unrebalanced portfolio and the best rebalancing strategy in portfolio 2 is the largest compared to portfolio

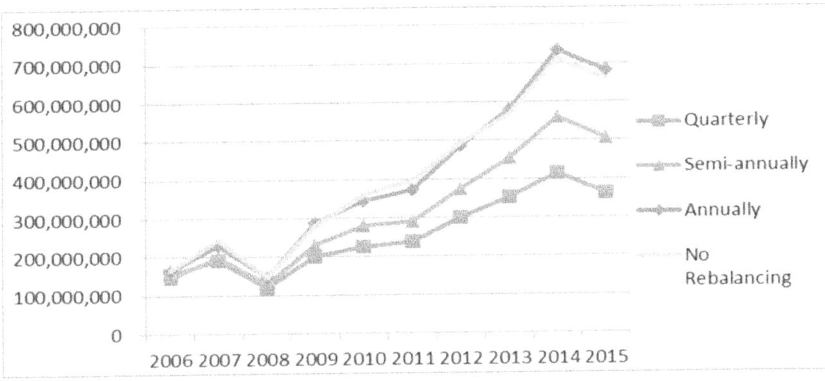

Figure 4. Rebalanced vs unrebalanced portfolio 1 with TC.

1 and portfolio 3. This could happen as the stocks in portfolio 2 have the highest volatility compared to stocks in portfolio 1 and portfolio 3.

Pliska and Suzuki (2004), as well as Tokat and Wicas (2007), conclude that stock volatility is one of asset characteristics that can affect the rebalancing strategy performance. The higher the volatility, the greater the risk of fluctuations as well as deviations from the initial allocation, and that creates greater needs to undertake rebalancing.

However, the transaction cost variable has not been included in this calculation, and therefore the subsequent calculation is to compare the performance of the rebalancing strategy after adding the transaction cost variable to the simulation.

4.2 *Additional Transaction Cost variable*

The Transaction Cost (TC) variable that is used in this simulation at 2%, was obtained from Panin Dana Maksima Mutual Fund Fact Sheet, which is owned by Panin Asset Management and published by BAPPEPAM in January 2016. This 2% fee includes the cost of purchase and resale fees. While the cost of the investment manager's services, custodian services, and transfer fee are not included, investors are assumed not to use professional services in this simulation.

Figure 4 shows that when transaction costs are added, the annual rebalancing strategy still contributes the best performance to a market portfolio, and only annual rebalancing outperforms the no rebalancing strategy in portfolio 1. However, throughout the ten year period, the return of the annually rebalanced portfolio becomes comparable with the return of the unrebalanced portfolio, which is far less complex to be implemented.

Unlike the previous result, Figure 5 shows that when transaction costs are added, investors who hold the progressive industry portfolio should change their rebalancing strategy into annual rebalancing, while semi-annual and quarter rebalancing outperforms the no rebalancing strategy in portfolio 2.

Meanwhile, Figure 6 shows that when transaction costs are added, annual rebalancing still offers the best performance to a defensive industry portfolio, and only annual rebalancing outperforms the no rebalancing strategy in portfolio 3.

Thus, it can be concluded from the results above that when the transaction cost is added to the rebalancing process, the annual rebalancing strategy provides the best offer to the three portfolios, while changes of a strategy in portfolio 2 occur as rebalancing bonus generated from portfolio 2 which is not enough to cover the transaction fee of 2%. Besides, by implementing the rebalancing strategy, investors can manage their portfolio passively to stocks in Property, Housing, and Construction sectors as well as the Consumption sector.

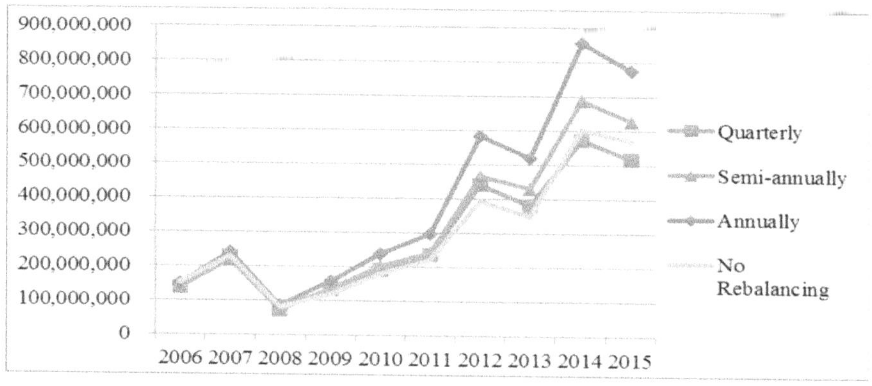

Figure 5. Rebalanced vs unrebalanced portfolio 2 with TC.

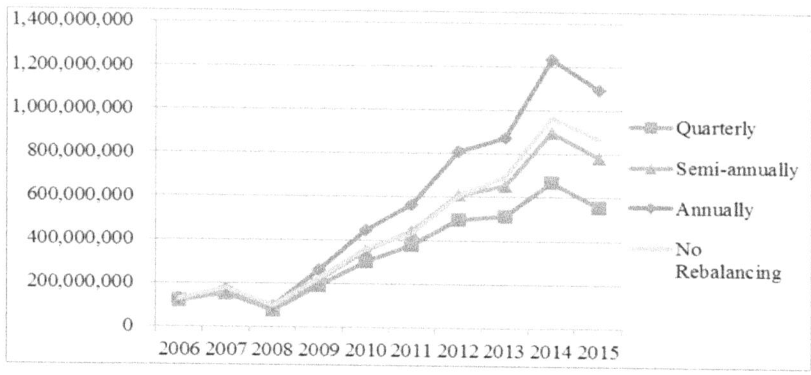

Figure 6. Rebalanced vs unrebalanced portfolio 3 with TC.

Table 1. Summary of the significance test between the average of the rebalanced portfolio vs the unrebalanced portfolio.

	Portofolio 1	Portofolio 2	Portofolio 3
Significance	0.910	0.290	0.472

Table 2. Summary of the significance test between the best rebalancing period vs the unrebalanced portfolio.

	Portofolio 1	Portofolio 2	Portofolio 3
Significance	0.581	0.226	0.299

4.3 *Tests of statistical significance*

Tests of statistical significance were computed using rebalancing simulation results of each portfolio before adding the transaction cost variable so that the significance results would be obtained based on the rebalancing strategy performance itself, without any influences of the transaction cost variable. The results of two statistical significance tests using the SPSS program are as shown in Table 1.

Meanwhile, the second test shown in Table 2 implies that we fail to reject H0 as well, as the levels of significance or p-values are greater than 0.05. Thus, there is no significant difference between the best rebalancing period vs the unrebalanced portfolio in portfolios 1, 2, and 3.

In conclusion, although the rebalancing strategy generates a rebalancing bonus or a positive return through the simulation period of ten years, there might be other factors or variables other than rebalancing that have greater influences on a portfolio performance.

5 CONCLUSION

We acknowledge that portfolio rebalancing has been a common practice in risk management, especially for investment managers. This is a practical approach that does not make any improvement in the science of financial and risk management. On the other hand, the purpose of this study is to maintain, not to reduce, the portfolio risk.

This study concludes that the characteristics of stocks from different types of industries and sectors have an influence towards the rebalancing strategy performance. Throughout the ten year period of investing from 2006 to 2015, the best rebalancing strategy for the market and the defensive industry is annual rebalancing, while for the progressive industry quarter rebalancing gives the best result. Nevertheless, with the addition of the transaction cost variable, this result for the progressive industry shifts into annual rebalancing. It means that the rebalancing strategy is the most suitable to be implemented by investors who manage their portfolio passively.

Finally, during the US subprime mortgage crisis in 2008 and Indonesia's trade balance deficit in 2013, it was proven from the simulations that the rebalancing strategy could help investors to control their portfolio risk because of the prevailing unknown.

Therefore, we need to reassess stock behaviour or characteristics, as different industries require different treatments. This conclusion might be useful for future studies that are interested in analysing how much is the influence of a stock's characteristics (e.g. volatility) in a portfolio towards the benefits of the rebalancing strategy, or how much rebalancing benefits a portfolio consisting of stocks from several industries, or several asset classes (e.g. stocks, bonds, and commodities). The key point is to make sure that the rebalancing bonus is greater than the costs.

REFERENCES

Almadi, H., Rapach, D. E. & Suri, A. (2014). Return predictability and dynamic asset allocation: How often should investors rebalance? *The Journal of Portfolio Management Portfolio Management*, *40*(4), 16–27.

Bernstein, W. J. (1996). *The rebalancing bonus: Theory and practice*. http://www.efficientfrontier.com/ef/996/rebal.htm.

Bernstein, William J. and Wilkinson, David J., Diversification, Rebalancing, and the Geometric Mean Frontier (November 24, 1997). Available at SSRN: https://ssrn.com/abstract=53503 or http://dx.doi.org/10.2139/ssrn.53503.

Bodie, Z., Kane, A. & Marcus, A. J. (2009). *Investments* (8th ed.). Boston, MA: McGraw-Hill Irwin.

Fabozzi, F. J., Kolm, P. N., Pachamanova, D. A. & Focardi, S. M. (2007). *Robust portfolio optimization and management*. Hoboken, NJ: John Wiley.

Husnan, S. (2005). *The Fundamentals of Portfolio Theory and Security Analysis* (5th ed.). Yogyakarta: BPFE.

Ibbotson, R. G. & Kaplan, P. D. (2000). Does asset allocation policy explain 40, 90, or 100 percent of performance? *Financial Analysts Journal*, *56*(1), 26–33. doi:10.2469/faj.v56.n3.2356.

Jaconetti, C. M., Kinniry Jr, F. M. & Zilbering, Y. (2010). *Best practices for portfolio rebalancing*. Vanguard. http://www.vanguard.com/pdf/icrpr.pdf.

KSEI Press Release (2016, December 30). Retrieved from http://www.ksei.co.id/files/uploads/press_releases/press_file/id-id/130_siaran_pers_kinerja_ksei_tahun_2016_raih_penghargaan_sebagai_kustodian_sentral_terbaik_di_asia_tenggara_20161230150627.pdf.

Kohler, A. & Wittig, H. (2014). Rethinking portfolio rebalancing: introducing risk contribution rebalancing as an alternative approach to traditional value-based rebalancing strategies. *The Journal of Portfolio Management*, *40*(3), 34–46. doi:10.3905/jpm.2014.40.3.034.
Perold, A. F. & Sharpe, W. F. (1988). Dynamic strategies for asset allocation. *Financial Analysts Journal*, *44*(1), 16–27. doi:10.2469/faj.v44.n1.16.
Pliska, S. R. & Suzuki, K. (2004). Optimal tracking for asset allocation with fixed and proportional transaction costs. *Quantitative Finance*, *4*(2), 233–243. doi:10.1080/14697680400000027.
Putri, P. R. K. (2010). *Analysis of rebalancing strategy on the combination of equity funds and fixed income mutual funds for the December 2003 – December 2009 period.* Master's thesis, Universitas Indonesia.
Reilly, F. K. & Brown, K. C. (2011). *Investment analysis & portfolio management* (10th ed.). Ohio: South-Western Cengage Learning.
Tokat, Y. & Wicas, N. W. (2007). Portfolio rebalancing in theory and practice. *The Journal of Investing*, *16*(2), 52–59. doi:10.3905/joi.2007.686411.

Competition and Cooperation in Economics and Business – Gani et al. (Eds)
 ISBN 978-1-138-62666-9

Impact of foreign institutional and individual ownership on stock return volatility in Indonesia

Pima V. Tresna & Irwan A. Ekaputra
Department of Management, Faculty of Economics and Business, Universitas Indonesia, Depok, Indonesia

ABSTRACT: This research examines the impact of foreign institutional and individual ownership on the one step ahead monthly stock return volatility. This research uses monthly data of foreign and domestic ownerships from the Kustodian Sentral Efek Indonesia (KSEI) database. This study employs Fama-MacBeth cross-sectional regression based on Rhee and Wang (2009). The results of this research indicate that foreign institutional ownership lowers future stock return volatility. Meanwhile, foreign individual ownership increases future stock return volatility. Changes of foreign institutional and individual ownerships also display consistent results with the level of foreign institutional and individual ownerships. Further investigations that separate foreign institutional ownership from foreign financial institutional ownership and non-financial institutional ownership indicate that foreign non-financial institutional ownership consistently lowers future stock return volatility. Meanwhile, foreign financial institutional ownership exhibits mixed results.

1 INTRODUCTION

Many studies about capital market liberalization have been done in the past years in many points of views. One of the studies examines the impact of capital market liberalization on variability of stock returns. Variability of stock return, or known as stock return volatility, has an important role in emerging capital market. Many investors use capital market volatility as one of determinants in making investment decisions because high volatility in capital market is equal with a high cost of capital. Additionally, high volatility is also considered equal with high instability in capital market (Bekaert and Harvey, 1997).

At first, much of the research about the impact of capital market liberalization on stock return volatility was conducted by event studies. Event studies were done by examining changes of volatility on the period before, during and after capital market liberalization. However, the type of the event studies of research only examines one period of time that is not enough to describe the impact of capital market liberalization on stock return volatility. This is because the process of capital market liberalization not only occurs in one period of time but also occurs gradually (Bae *et al.*, 2004).

After the type of the event studies of research, the research is developed into panel studies. One of the panel studies includes research about the impact of foreign investor investability on stock return volatility (Bae *et al.* 2004). It is assumed that the higher degree the foreign investor investability, the higher degree the capital market liberalization in that country (Bae *et al.*, 2004).

After the type of the event studies of research, the research is developed into panel studies. One of the panel studies includes research about the impact of foreign investor investability on stock return volatility (Bae *et al.*, 2004). It is assumed that the higher degree the foreign investor investability, the higher degree the capital market liberalization in that country (Bae *et al.*, 2004).

Foreign investor investability is the maximal percentage of foreign ownership that is allowed in one country's capital market. However, foreign investor investability cannot fully represent capital market liberalization in one country. There are possibilities that one firm stock has a high degree of foreign investor investability, but it has a low percentage of foreign ownership. Increasing firm stock investability does not directly increase foreign ownership in that stock (Li *et al.*, 2011).

Because foreign ownership cannot fully represent capital market liberalization in one country, research is developed further by using another variable as the proxy for capital market liberalization. Another variable that is considered sufficient to represent capital market liberalization is foreign ownership. Different from foreign investor investability, the percentage of foreign ownership of firm stock represents the actual proportion of foreign ownership. The higher the foreign ownership percentage in one country capital market, the higher the foreign investor participations and activities at the capital market are (Li *et al.*, 2011).

The first research about the impact of foreign ownership on stock return volatility was done by Li *et al.* (2011). Li *et al.* (2011) did research by using Large Foreign Ownership (LFO) as a proxy for capital market liberalization. As further examination, Li *et al.* (2011) also categorized LFO into large foreign ownership in financial institution and non-financial institution. Li *et al.* conclude that examining heterogeneity and understanding each role of foreign investors are important.

Stiglitzt (2000) also emphasizes the importance of examining heterogeneity of foreign investors in capital markets. Stiglitz (2000) highlights the importance of examining the impacts of various types of capital flows coming into capital markets. One of the types of capital flows is various categories of foreign investors in capital markets. If the role of each category of foreign investors is already known, it is expected to provide different treatment tailored to the role of each category of foreign investors.

This research refers to the importance of knowing the impact of heterogeneous foreign investors on stock return volatility. Heterogeneous foreign investors are further categorized into foreign institutional and individual investors. Additionally, to examine the impact of foreign institutional investor heterogeneity, foreign institutional investors are also further categorized into foreign financial institution investors and foreign non-financial institution investors, and we examine its impact on stock return volatility. Thus, this research aims to examine the impact of foreign institutional and individual ownership on stock return volatility.

2 MAIN HYPOTHESES

Institutional and individual ownership is assumed to have different characteristics and impacts on capital markets. Stiglitz (2000) conveys arguments where long-term investments from Foreign Direct Investment (FDI) provide a stabilizing impact on capital market's volatility. This is because foreign investors who are doing long term investment in a country not only provide the flow of funds but also provide technology, access to global markets, and improved human resources. Meanwhile, short-term investments by foreign investors give impact on increasing the volatility of the capital markets.

Stiglitz (2000) argues that the short-term investment by foreign investors is an unstable form of investment due to information imbalance for foreign investors. In addition, due to lack of attachment between companies and short-term foreign investors, this can lead to foreign investors easily withdrawing funds from the capital market if the economic condition is declining.

The results of research conducted by Li *et al.* (2011) show that foreign institutional investors give negative impacts or stabilize stock return volatility. Li *et al.* (2011) argue that foreign institutional investors have similar characteristics as the characteristics of Foreign Direct Investment (FDI). Based on above arguments, the first hypothesis for this research is described as follows:

Hypothesis-1: Foreign institutional ownership gives an impact that stabilizes stock return volatility in Indonesian capital market.

Meanwhile, according to the research conducted by Foucault *et al.* (2011), I ndividual investors's trading activities give a positive impact or increase stock return volatility in

Euronext Paris capital market. Foucault *et al.* (2011) argue that individual or retail investors have a role as noise traders in the capital market. Although individual investors who use the kontrarian strategy are supposed to stabilize the capital market volatility, the capital market stabilizing effect is covered by volatility increasing the effect that occurs with momentum trading by individual investors.

This argument is also corroborated by the research conducted by Barber and Odean (2009) who found that the herding behavior of individual investors increases volatility in the United States capital market. Meanwhile, Chen *et al.* (2013) found that the trading activity by foreign investors leads to fluctuations on the capital markets of developing countries. Chen *et al.* (2013) found that individual investors are less rational and tend to give an excessive reaction to news and circulation of market information. Based on the above arguments, the second hypothesis for this research is described as follows:

Hypothesis-2: Foreign individual ownership gives an impact that increases stock return volatility in Indonesian capital market.

3 DATA

This research uses the following data: Foreign and domestic investor ownership of firm's stocks that are listed on Indonesian Stock Exchange (IDX) since the enactment of SID (Single Identity Investor) regulation i.e. February 2012-January 2015 are obtained from the Indonesian Central Securities Depository (KSEI) database. In addition, this research uses the stock price, market capitalization, transaction value, the number of shares outstanding, the interim financial statements of companies listed on the Indonesian Stock Exchange (IDX) since the enactment of the SID (Single Investor Identity) regulation i.e. February 2012-January 2015 obtained from Thomson Reuters database.

4 VARIABLES

4.1 *Dependent variables*

On this research, the firm's stock return volatility is represented by two proxies. Two of these proxies are monthly Standard Deviation (SD) of stock returns volatility and volatility of stock returns based on Chen *et al.* (2013) and Li *et al.* (2011). The formula of the monthly Standard Deviation (SD) calculation used in this research is as follows:

$$SD_i = \sqrt{\frac{\sum\left(x_{i,t} - \overline{x_{i,t}}\right)^2}{(T-1)}} \tag{1}$$

where $x_{i,t}$ is daily stock return in one month period, $\overline{x_{i,t}}$ is average daily stock return in one month period, and T is the sum of stock return in one month period.

The formula of monthly Volatility calculation used in this research is as follows:

$$VL_i = \frac{1}{T}\sum \ln\left(x_{i,t}^{\ 2}\right) \tag{2}$$

where $x_{i,t}$ is daily stock return in one month period, and T is the sum of stock trading day in one month period.

4.2 *Independent variables*

Because this research aims to examine the impact of foreign institutional and individual investor ownership, the independent variable used in this study is monthly foreign ownership which is divided into foreign institutional ownership (FORINST) and foreign individual ownership (FORIND). Given the importance of knowing the impact of foreign investors that are heterogeneous, Li *et al.* (2011) further divide foreign institutional investors into foreign financial institution investors and foreign non-financial institutions investors. Similar to Li *et al.* (2011), this research also divides foreign institutional ownership into foreign financial institution ownership (FOREIGN_FIN) and foreign non-financial institution ownership (FOREIGN_NONFIN).

4.3 *Control variables*

In accordance with the research conducted by Chen *et al.* (2013) and Li *et al.* (2011), this research also uses several control variables including the following: Lagged stock return volatility (Lag (SD) and Lag (VL); Domestic ownership; Firm size (SIZE), which is measured as natural logarithm of market capitalization of stock market value in one month period; Turnover (TURNOVER), measured as monthly transaction value divided by its market capitalization; Leverage ratio (LEVERAGE), measured as total liabilities divided by a total asset taken from firm's interim financial statements.

4.4 *Interaction variables*

Similar to the research conducted by Rhee and Wang (2009), in addition to examine the impact of each independent and control variables on stock return volatility separately, this research also uses an interaction variable between an independent variable and a control variable. The interaction variables used are interaction between foreign ownership (FORINST and FORIND) and firm's size (SIZE) or stock turnover (TURNOVER). By examining the impact of interaction between foreign ownership and firm's size or stock turnover, it can be known the combined impact which is not visible if we examine the impact of foreign ownership and firm's size or stock turnover separately (Rhee and Wang 2009).

4.5 *Non-Linear variables*

Similar to the research conducted by Rhee and Wang (2009), this research also adds the non-linear variable as a regressor. The non-linear variables used are the squared form of foreign ownership and domestic ownership. The aim for this addition is to examine the non-linear impact of foreign ownership on stock return volatility.

5 METHODOLOGY

Generally, the method implemented in this research refers to the method used by Rhee and Wang (2009) and Wang (2013). To examine the causality impact of foreign ownership on stock return volatility, this research performed regression using foreign institutional and individual ownership during one month period before stock return volatility as an independent variable (Rhee and Wang, 2009). Cross-sectional regression was done for each month during the 36-month period using lagged foreign ownership (FORINSTi, t–1 and FORINDi, t–1) as independent variables. Meanwhile, present stock return volatility (SDi, t and VLi, t) is used as a dependent variables.

After each monthly regression has been performed, the average coefficient is calculated from the coefficient deriving from the 36-month period regressions. In the meantime, to find out the significance of the coefficient, according to Rhee and Wang (2009), the standard error (SE) is calculated using the modified Fama-Macbeth formula by Cochrane (2001). The following is the formula of the modified Fama-Macbeth standard error:

$$SE = \frac{SD(\beta)}{\sqrt{T}}\left(\frac{1+\rho(1)}{1-\rho(1)}\right) \quad (3)$$

Description: SE is modified Fama-Macbeth standard error by Cochrane (2001); SD(β) is the standard deviation of coefficients (β) resulting from each month estimation; T is the number of the months; ρ(1) is the first-order auto-correlation of coefficients (β) resulting from each month estimation.

Meanwhile, empirical models used in this research are as follows:

5.1 *Impact of foreign institutional and individual ownership on stock return volatility empirical model*

The following is the regression model to examine the impact of foreign institutional and individual ownership on stock return volatility:

$$\begin{aligned} VOLAT_{i,t} = {} & \beta_0 + \beta_1 \, Lag\left(VOLAT_{i,t}\right) + \beta_2 FORINST_{i,t-1} + \beta_3 FORIND_{i,t-1} \\ & + \beta_4 CONTROL_{i,t-1} + \beta_5 INTERACTION_{i,t-1} + \beta_6 NON \\ & - LINEAR_{i,t-1} + \varepsilon_{i,t} \end{aligned} \quad (4)$$

Description: CONTROLi, t−1 is the control variable on period t−1; FORINSTi, t−1 is the sum of foreign institutional ownership divided by shares outstanding on period t−1; FORINDi, t−1 is the sum of foreign individual ownership divided by shares outstanding on period t−1; INTERACTIONi, t−1 is the interaction variable on period t−1; NON-LINEARi, t−1 is a non-linear variable on period t−1; VOLATi, t is the standard deviation/volatility of daily stock return volatility in one month period of t.

5.2 *Impact of foreign financial institution and non-financial institution ownership on the stock return volatility empirical model*

The following is the regression model to examine the impact of the foreign financial institution and non-financial institution ownership on stock return volatility:

$$\begin{aligned} VOLAT_{i,t} = {} & \beta_0 + \beta_1 \, Lag\left(VOLAT_{i,t}\right) + \beta_2 FOREIGN\_FIN_{i,t-1} \\ & + \beta_3 FORIENG\_NONFIN_{i,t-1} + \beta_4 CONTROL_{i,t-1} + \beta_5 INTERACTION_{i,t-1} \\ & + \beta_6 NON - LINEAR_{i,t-1} + \varepsilon_{i,t} \end{aligned} \quad (5)$$

Description: CONTROLi, t−1 is the control variable on period t−1; FOREIGN_FINi, t−1 is the sum of foreign financial institution ownership divided by shares outstanding on period t-1; FOREIGN_NONFINi,t−1 is the sum of foreign non-financial institution ownership divided by shares outstanding on period t−1; INTERACTIONi, t−1 is the interaction variable on period t−1; NON-LINEARi,t-1 is the non-linear variable on period t−1; VOLATi, t is the standard deviation/volatility of daily stock returns volatility in one month period of t.

6 RESULTS AND ANALYSIS

6.1 *The impact of foreign institutional and individual ownership on stock return volatility*

Table 1 is the results of regression that examines the impact of foreign institutional and individual ownership on stock return volatility.

According to model IC on Table 1, foreign institutional ownership gives a negative impact on future stock return volatility. This result corroborates the initial hypothesis where foreign institutional ownership is assumed to give a negative impact on stock return volatility. The result of this study is consistent with the research conducted by Li *et al.* (2011) and Wang (2013), but is not consistent with the research conducted by Chen *et al.* (2013) and Ekaputra (2015).

Table 1. Regression results of the impact of foreign institutional and individual ownership on stock returns volatility (coefficients and standard errors).

	Model 1A		Model 1B	
	SD	VL	SD	VL
	(Std. Error)	(Std. Error)	(Std. Error)	(Std. Error)
FORINSTi,t–1	–0.0011	–3.6E–05	–0.002	0.00042
	(0.00084)	(0.0001)	(0.00646)	(0.00082)
FORINDi,t–1	0.00102	9.5E–05	0.01672**	0.00162**
	(0.00111)	(0.0001)	(0.00748)	(0.00079)
FORINSTi,t–1*SIZEi,t–1			8.8E–05	–3E–05
			(0.00042)	(5.7E–05)
FORINDi,t-1*SIZEi,t–1			–0.001**	-9E–05*
			(0.00052)	(5.2E–05)
R^2	0.26296	0.20598	0.28865	0.24133
	Model 1C		**Model 1D**	
	SD	VL	SD	VL
	(Std. Error)	(Std. Error)	(Std. Error)	(Std. Error)
FORINSTi,t–1	–0.0075***	–0.00062**	–0.0073	–0.0001
	(0.00274)	(0.00027)	(0.0072)	(0.00089)
FORINDi,t–1	0.00475**	0.00045**	0.02319***	0.00222***
	(0.00219)	(0.00022)	(0.00793)	(0.00084)
FORINSTi,t–1*SIZEi,t–1			–7E–05	–4E–05
			(0.00036)	(5.4E–05)
FORINDi,t–1*SIZEi,t–1			–0.0013**	–0.0001**
			(0.00051)	(5E–05)
$(FORINSTi,t–1)^2$	0.00751***	0.00068***	0.00875***	0.00082***
	(0.00227)	(0.00025)	(0.00229)	(0.00027)
$(FORINDi,t-1)^2$	–0.0049*	–0.00052**	–0.0039	–0.0003
	(0.00253)	(0.0002)	(0.00304)	(0.00023)
R^2	0.27229	0.21595	0.29832	0.25112

Description: * two-sided significance at 10%; ** two-sided significance at 5%; *** two-sided significance at 1%. SE (Standard Error) is re-calculated using the Cochrane (2001) method.

Li *et al.* (2011) give two arguments on why foreign institutional investors can give a stabilizing impact on stock return volatility of emerging countries' capital markets. The first argument is that foreign institutional investors have a commitment to their investments in the capital market and they aim to conduct long-term investment than short-term investment. Foreign institutional investors as shareholders give strong incentive in the form of financial assistance to the companies which have a reducing impact on stock return volatility.

The second argument is that foreign institutional investors have a supervision role towards the company. With its role of supervision, foreign institutional investors can pursue company's management transparency, more accountable management, and also a good risk management. In general, the increase of company performance can contribute to the decline of firm's stock return volatility.

In addition, Wang (2013) argues that foreign ownership could expand the investor basis of the company, so broader risk-sharing can give a decreasing impact on stock return volatility. Wang (2013) also argues that foreign investors do better supervision on the company than local investors, foreign analysts do a better analysis that local analysts, which causes the depletion of stock's information asymmetry. With reduced information asymmetry, it can give an impact on stabilizing stock return volatility.

Meanwhile, according to models IB, IC, and ID of Table 1, foreign individual ownership gives a positive impact on future stock return volatility. This result also corroborates the initial hypothesis where foreign individual ownership is assumed to give a positive impact on stock return volatility. The result of this study is consistent with the research conducted by Foucault *et al.* (2011), but it is not consistent with the research conducted by Chen *et al.* (2013).

Foucault *et al.* (2011) argue that individual or retail investors have a role as noise traders in the capital market. Although individual investors who use the kontrarian strategy are supposed to stabilize the capital market volatility, the capital market stabilizing effect is covered by the volatility increasing effect that occurs with momentum trading by foreign investors. This argument is also corroborated by the research conducted by Barber and Odean (2009) who found that the herding behavior of individual investors increases volatility in the United States capital market.

Meanwhile, the regression results indicate that the interaction between foreign individual ownership and firms' size gives a negative impact on stock return volatility. Although lagged foreign individual ownership gives an increasing impact on stock return volatility, if the company has a large market capitalization, then it can reduce the increasing impact on the stock return volatility. Thus, it can be concluded that firm's size can help reduce the increasing impact of foreign ownership on stock return volatility.

Additionally, according to models IC and ID of Table 1, the non-linear variable of lagged foreign institutional ownership gives an increasing impact on stock return volatility. Similarly, according to models IC and ID of Table 1, the non-linear variable of lagged foreign individual ownership gives a decreasing impact on stock return volatility.

The result of this study corroborates the result of the research conducted by Wang (2013) where foreign institutional and individual ownership gives a non-linear impact on stock return volatility. The result of this research is also consistent with the result of the research conducted by Ekaputra (2015). It can be concluded that foreign institutional and individual ownership gives a non-linear impact on stock return volatility.

6.2 *The impact of the foreign financial institution and Non-Financial institution ownership on stock return volatility*

Estimation results of regression that examine the impact of foreign financial and non-financial institution investor ownership on future stock return volatility are presented in table 2.

According to models 3C and 3D of Table 2, foreign financial institution ownership gives a stabilizing impact on future stock return volatility. However, model 3A gives a different result where the result of regression shows that foreign financial institution ownership gives a destabilizing impact on future stock return volatility. Meanwhile, according to models 3B and 3D of Table 2, foreign non-financial institution ownership gives a stabilizing impact on future stock return volatility. Thus, it can be concluded that foreign non-financial ownership gives a more consistent impact rather than foreign financial institution ownership.

These results are in line with the results of the research conducted by Li *et al.* (2011) where foreign non-financial institution investor ownership gives a more consistent impact in decreasing stock return volatility. Li *et al.* (2011) argue that foreign non-financial institution investors have a greater commitment in decreasing stock return volatility than foreign financial institution investors.

7 ROBUSTNESS TEST

In addition to regression using the level form of foreign institutional and individual ownership as the independent variable, this research also uses the change of the level form of foreign institutional and individual ownership. According to the research conducted by Rhee and Wang (2009), the aim of using the change of the level form of foreign ownership is to find out the impact of the change of foreign ownership in the past on stock return volatility.

Table 2. Regression results of foreign financial and non-financial institution ownership on stock return volatility (coefficient and standard error).

	Model 3A		Model 3B	
	SD	VL	SD	VL
	(Std. Error)	(Std. Error)	(Std. Error)	(std. error)
FOREIGN_FINi,t−1	0.00124*	0.00018**	0.00035	0.0007
	(0.00071)	(8.3E-05)	(0.00382)	(0.00055)
FOREIGN_NONFINi,t−1	3E-06	1.6E-05	−0.0133***	−0.0015***
	(0.00046)	(6.8E-05)	(0.00328)	(0.00044)
R^2	0.26098	0.20466	0.28549	0.23803
	Model 3C		Model 3D	
	Coefficient	Coefficient	Coefficient	Coefficient
	(Std. Error)	(Std. Error)	(Std. Error)	(Std. Error)
FOREIGN_FINi,t−1	−0.0065**	−0.0006**	−0.0092**	−0.0002
	(0.00311)	(0.00029)	(0.00446)	(0.00045)
FOREIGN_NONFINi,t−1	−0.0004	−0.0001	−0.0148***	−0.0017**
	(0.00311)	(0.00051)	(0.00497)	(0.00084)
R^2	0.27106	0.21697	0.29561	0.2502

Description: * two-sided significance at 10%; ** two-sided significance at 5%; *** two-sided significance at 1%. SE (Standard Error) is re-calculated using the Cochrane (2001) method.

In addition, using the change of the level form of foreign ownership in the regression is also aimed as a comparison to regression which uses the level form of foreign ownership (Rhee and Wang, 2009).

Generally, the result of the regression model which examines the impact to the change of the level form of foreign institutional and individual ownership on future stock return volatility supports the result of the previous regression model which used the level of foreign institutional and individual ownership as the independent variable (Table is available upon request). The regression results show that the change of the level form of foreign institutional ownership supports the previous results which show that the level form of foreign institutional ownership gives a stabilizing impact on future stock return volatility.

In addition, the change of the level form of foreign individual ownership also supports the previous regression model results which show that the level form of foreign individual ownership gives a destabilizing impact on future stock return volatility. Meanwhile, the regression model with the change of the level form of foreign financial institution and non-financial institution ownership also supports the regression model with the level form of foreign financial institution ownership as the independent variable.

8 CONCLUSION

The results of this research indicate that foreign institutional ownership lowers future stock return volatility. Changes of foreign institutional ownership display consistent results with the level of foreign institutional ownership. Meanwhile, foreign individual ownership increases future stock return volatility. Changes of foreign individual ownership also display consistent results with the level of foreign individual ownerships.

Further investigation that separates foreign institutional investors from foreign financial and non-financial institution investors displays that foreign financial institution investors give both stabilizing and destabilizing impacts on future stock return volatility. This result is supported by regression models which use the change of foreign financial institution investors as the independent variable.

Meanwhile, foreign non-financial institution ownership gives a stabilizing impact on future stock return volatility. This result is supported by the regression model using the change of foreign non-financial institution investors as the independent variable. Thus, it can be concluded that foreign non-financial institution ownership gives a more consistent impact than foreign financial institution ownership.

REFERENCES

Bae, Kee Hong, Kalok Chan, and Angela Ng. 2004. "Investibility and Return Volatility." *Journal of Financial Economics* 71 (2): 239–63. doi:10.1016/S0304–405X(03)00166–1.

Barber, Brad M., Terrance Odean, and Ning Zhu. 2009. "Do Retail Trades Move Markets?" *Review of Financial Studies*. 22(1): 151–186. doi:10.1093/rfs/hhn035.

Bekaert, Geert, and Campbell R. Harvey. 1997. "Emerging Equity Market Volatility." *Journal of Financial Economics* 43 (1): 29–77. doi:10.1016/S0304–405X(96)00889–6.

Bekaert, Geert, Campbell R. Harvey, and Christian Lundblad. 2007. "Liquidity and Expected Returns: Lessons from Emerging Markets." *Review of Financial Studies* 20 (6): 1783–1831. doi:10.1093/rfs/hhm030.

Brooks, Chris. 2008. *Introductory Econometrics for Finance. Finance*. doi:10.1111/1468–0297.13911.

Chari, Anusha, and P.B. Henry. 2001. "Stock Market Liberalizations and the Repricing of Systematic Risk." *Working Paper*. doi:10.3386/w8265.

Chen, Zhian, Jinmin Du, Donghui Li, and Rui Ouyang. 2013. "Does Foreign Institutional Ownership Increase Return Volatility? Evidence from China." *Journal of Banking and Finance* 37 (2): 660–69. doi:10.1016/j.jbankfin.2012.10.006.

Cochrane, John H. 2001. *Asset Pricing*. Princeton University Press.

Ekaputra, Irwan Adi. 2015. "Foreign Institutional Ownership and Stock Return Volatility in Indonesia." *Jurnal Keuangan Dan Perbankan, 19*(3): 357–367.

Foucault, Thierry, David Sraer, and David J. Thesmar. 2011. "Individual Investors and Volatility." *Journal of Finance* 66 (4): 1369–1406. doi:10.1111/j.1540–6261.2011.01668.x.

Li, Donghui, Quang N. Nguyen, Peter K. Pham, and Steven X. Wei. 2011. "Large Foreign Ownership and Firm-Level Stock Return Volatility in Emerging Markets." *Journal of Financial and Quantitative Analysis* 46 (4): 1127–55. doi:10.2469/dig.v42.n2.44.

Rhee, S. Ghon, and Jianxin Wang. 2009. "Foreign Institutional Ownership and Stock Market Liquidity: Evidence from Indonesia." *Journal of Banking & Finance* 33 (7): 1312–24. doi:10.1016/j.jbankfin.2009.01.008.

Stiglitz, Joseph E. 2000. "Capital Market Liberalization, Economic Growth, and Instability." *World Development* 28 (6): 1075–86. doi:10.1016/S0305–750X(00)00006–1.

Wang, Jianxin. 2007. "Foreign Equity Trading and Emerging Market Volatility: Evidence from Indonesia and Thailand." *Journal of Development Economics* 84 (2): 798–811. doi:10.1016/j.jdeveco.2006.05.001.

Wang, Jianxin. 2013. "The Impact of Foreign Ownership on Stock Volatility in Indonesia." *Asia-Pacific Journal of Financial Studies* 42 (3): 493–509. doi:10.1111/ajfs.12022.

Competition and Cooperation in Economics and Business – Gani et al. (Eds)
© 2018 Taylor & Francis Group, London, ISBN 978-1-138-62666-9

Empirical analysis of the government spending and disparities of education outcomes at the district level in Indonesia

T. Jasmina & H. Oda
Graduate School of Policy Science, Ritsumeikan University, Osaka, Japan

ABSTRACT: In 2003, the government of Indonesia under the Law of Education System implemented a nine year basic education and a requirement of central and local governments to allocate a minimum of 20% of their budget for education. Both central and local governments have managed to allocate 20% of their budget for education in 2009. At national level, education outcomes such as net enrolment ratios have improved in the past ten years. However, several studies show that education outcomes at the district level remain an issue. On the other hand, Indonesia faces challenges of globalisation, especially in the implementation of the ASEAN Economic Community in 2016. Enhancing human resources through education is essential for Indonesia to compete in the globalised world. This paper aims to analyse the impact of government spending on education and other socio-economic factors which cause disparities of education among districts in Indonesia. Government spending on education will indeed increase the education sector. However, there is a limit on that. Other prominent factors must also be considered, so that the spending can be translated into education outcomes. The paper applies an econometric approach by employing a cross section analysis using a set of updated secondary data of districts in Indonesia during the period of 2010 to 2014.

1 INTRODUCTION

As part of an effort to enhance education, the government of Indonesia issued the 2003 Law of National Education System, which among others applied a nine year compulsory basic education (primary and junior secondary education) and an allocation of 20% of government spending on education. Due to local government decentralisation in 2001, the implementation of this basic education is under the authority of local government at the district level. Furthermore, the central government sets a minimum service standards on education services that has to be followed by the local government.

The central government budget for education has increased by more than double from IDR142.4 trillion in 2007 to IDR375.5 trillion in 2014. A significant increase took place in 2008, when the budget for education increased by 35.1% from IDR154.2 trillion to IDR208.3 trillion in 2009. The increase was partly caused by the increase of spending for education managed by the central government (for early childhood education, higher education, and Islamic education), and partly due to the increase of spending that was transferred to local governments at the district level (for elementary and secondary education). Since 2009, on average about 60% of central government education spending has been transferred to local, district level governments in the form of a school operational assistance program (*Bantuan Operasional Sekolah*) for primary and junior secondary education, special allocation fund for education (*Dana Alokasi Khusus*), and additional allowances for teachers (*Tunjangan Profesi Guru and Tambahan Penghasilan*). In addition, local government spending on education at the district level has also increased. In 2007, local governments spent on average 26% of their budget on education, and this increased to 31.9% in 2014. From 2010 to 2014, local governments at district level allocated on average 32.8% of their spending on education.

The education outcomes in Indonesia have improved significantly in the past ten years. The net enrolment ratio has improved during 2003 to 2014, from 92.6 in 2003 to 95.6 for primary education, from 63.5 to 75.2 for junior secondary education, and from 40.6 to 60.4 for senior secondary education. However, if we look closely into the district level, there are disparities of education among districts in Indonesia. In 2014, approximately 43% of the districts had a net enrolment ratio of primary education below the national level. The condition was even worse for the net enrolment ratio in junior and senior secondary school, with around 50% and 42% of the districts below the national level, respectively.

Based on the reports by the World Bank (2009, 2012, 2013a, 2013b), the Ministry of Education and Culture of Indonesia (2013), OECD and ADB (2015), education at district level in Indonesia remains an issue. There are challenges in enhancing education at district level in Indonesia, such as: (i) disparities in student access, educational quality, and teachers; (ii) low access to secondary and tertiary education, especially in remote areas; (iii) low association between the number of teachers and the quality of education and learning outcomes; and (iv) poor performance and lack of transparency of local governments in managing financial resources for education.

On the other hand, Indonesia faces challenges of globalisation, especially in the implementation of the ASEAN Economic Community in 2016. Enhancing human resources through education is vital for Indonesia to compete in a globalised world. According to Tullao et al. (2015), Indonesia is one of the ASEAN countries which spends the most on education, with 18.1% of the government budget allotted to education in 2012. Another ASEAN country that allocates high government spending on education is Malaysia, with 20.9% in 2012. However, compared to country GDP, the Indonesian government's funding on education was only 3.6% of GDP in 2012; while in the same year Malaysia allocated approximately 5.6% of GDP.

There have been many empirical studies showing the positive impact of government spending on education outcomes, such as prominent studies conducted by Gupta et al. (2002), and Rajkumar and Swaroop (2008). However, there are studies that show little or no impact of government spending on education outcomes. Mingat and Tan (1998), for example, found that higher budget allocation in education makes a relatively small contribution to the differences in resources of education. Hanushek (2002) shows that there is limited evidence of a consistent relation between education resources and student performance.

In the case of Indonesia, there are several studies on the relationship between government spending and education at district level, especially after the decentralisation policy in 2001. For example, a study of Simatupang (2009) found a positive impact of decentralisation on education services at district level. A study by Kristiansen and Pratikno (2006) suggests that in order to enhance education outcomes, the government has to allocate more funds to primary and secondary education. Their study is supported by findings of Arze del Granado et al. (2007) which show a positive and significant impact of government spending on enrolment rate. Furthermore, Zufri and Oey-Gardiner (2012) show that central government spending has a significant impact on education outcomes. Another study by Suryadarma (2012) shows that public spending on education is more effective in improving education outcomes in the less corrupt districts. Al-Samarrai and Cerdan-Infantes (2013) point out that despite the increase of access to education in Indonesia, the quality of education and the capacity of local governments in allocating resources remains an issue.

With a significant amount of government spending on education in Indonesia, it is interesting to further elaborate on how this spending has affected education outcomes at the district level in Indonesia. This paper, as part of an ongoing comprehensive study on the impact of government spending on education, aims to empirically analyse how government spending affects education outcomes among districts in Indonesia. This paper is distinct from the previous studies in at least two ways. Firstly, it analyses the impact of government spending on education after 2009, when the government fully allocated 20% of its budget for education. Secondly, it uses a comprehensive data of government spending on education in Indonesia, both central and local, from 2010 to 2014. The structure of the paper is as follows. Section 2 lays out an empirical analysis that includes data, model specification, empirical results, and discussions; then section 3 provides concluding remarks.

2 EMPIRICAL STUDIES

This section examines the impact of central and local government education spending, and other socio-economic factors on education outcomes. The analysis uses available data from 491 districts in Indonesia from 2010 and 2014.

2.1 *Data and model specification*

The empirical analysis performed here is a cross section regression at the district level in Indonesia on the impact of government spending in 2010 on the change of education outcomes during the period of 2010 to 2014. As an indicator of education outcome, this paper uses net enrolment ratio at the district level as done by some earlier studies (see for example studies by Barro (1991) and studies by Arze del Granado et al. (2007), Zufri and Oey-Gardiner (2012), and Suryadarma (2012) for the district level in Indonesia). In order to better capture the impact of government spending on education at a certain period, this paper uses the change of net enrolment ratio between 2010 and 2014 as the dependent variable.

The paper analyses the impact of government spending on the nine year basic education in Indonesia, which consists of six years of primary education and three years of junior secondary education. Hence, there are two dependent variables applied in the regressions representing two education levels, which are: (i) the change of net enrolment ratio of primary education of 2010 and 2014; and (ii) the change of net enrolment ratio of junior secondary education of 2010 and 2014. The data is collected from the National Socio Economic Survey 2010 and 2014 of BPS-Statistics of Indonesia.

The independent variables consist of government spending on education and other relevant socio-economic factors. In order to understand the impact of government spending on education, the government spending on education is categorised into: (i) total government spending on education, which is the sum of central and local government spending; (ii) central government spending on education; and (iii) local government spending on education (all in terms of ratio to Gross Regional Domestic Product (GRDP). Data on the central government spending on education, unlike the previous studies, is an aggregation of central government spending that is transferred to local governments at the district level in the form of: (i) the school operational assistance program for primary and junior secondary education; (ii) special allocation funds for education; and (iii) additional allowances for teachers. Whereas, data on the local government spending on education is data on allocated local budget for education. The data is collected from the Ministry of Finance and the Ministry of Education and Culture of Indonesia.

For the socio-economic factors as control variables of the regression, the study applies several indicators as in the previous researches such as Gupta et al. (2002), Arze del Granado et al. (2007), Rajkumar and Swaroop (2008). The socio-economic variables include: initial net enrolment ratio, per capita GRDP in logarithmic form, poverty headcount ratio, share of population below 15 years old, share of households living in urban areas, life expectancy ratio, and a dummy variable of less developed and remote districts as defined by the government (*Daerah Tertinggal, Terdepan, dan Terluar*). Sources of the socio-economic data at the district level are the National Socioeconomic Survey 2010 and 2014, and BPS-Statistics Indonesia 2010–2014. Summary statistics of the variables are presented in Table 1.

In general, the estimated regressions with the subscript i that represents districts are as follows:

$$NER\_DIFF_i = \beta_0 + \beta_1 GOV\_ALL_i + \beta' X_i + \varepsilon_i \quad (1)$$

$$NER\_DIFF_i = \beta_0 + \beta_1 GOV\_LOCAL_i + \beta_2 GOV\_CTRL_i + \beta' X_i + \varepsilon_i \quad (2)$$

where the dependent variables for each district i (NER_DIFF_i) is the change of net enrolment ratio in 2010 and 2014 of primary and junior secondary education. The independent variables are government spending on education as follows: (i) total government spending on

Table 1. Summary statistics of the variables.

Variables	Mean	Standard deviation	Minimum	Maximum
Net enrolment ratio for primary 2010	0.941	0.075	0.118	1.000
Net enrolment ratio for primary 2014	0.956	0.061	0.309	1.000
Net enrolment ratio for junior secondary 2010	0.657	0.117	0.090	0.883
Net enrolment ratio for junior secondary 2014	0.751	0.104	0.160	0.949
Total government spending to GRDP 2010	0.051	0.036	0.001	0.354
Local government spending to GRDP 2010	0.037	0.026	0.001	0.262
Central government spending to GRDP 2010	0.014	0.011	0.000	0.102
GRDP per capita 2010 (in ln)	16.807	0.716	14.352	20.362
Poverty headcount ratio 2010	0.156	0.094	0.017	0.496
Share of population below 15 years old 2010	0.313	0.048	0.202	0.457
Share of households living in urban area 2010	0.362	0.311	0.000	1.000
Life expectancy ratio 2010	68.213	3.878	52.650	77.370

Source: Authors' calculation.

education to GRDP (GOV_ALL_i) as in Equation 1; and (ii) disaggregation of local government spending on education to GRDP (GOV_LOCAL_i), and central government on education to GRDP (GOV_CTRL_i) as in Equation 2. A set of control variables in the regression is captured in X_i, and ε_i is an error term. All regressions employ the same socio-economic factors as the control variables. In performing the regressions, heteroscedasticity is detected and it is overcome by using robust standard errors in the estimations.

2.2 *Results and discussion*

Table 2 presents results of the regressions for the change of net enrolment ratio of primary and junior secondary education. The first column shows regression of the change net enrolment ratio for primary education with total government spending (Eq. 1.1), whereas the second column shows regressions with local government spending and central government spending (Eq. 1.2). The third and fourth columns show regression results of the change of net enrolment ratio for junior secondary education with total government spending (Eq. 2.1), and with local government spending and central government spending (Eq. 2.2), respectively. All the regressions employ the same control factors as the independent variables. For discussion in this paper, we will focus on the impact of government spending on education.

As can be seen from Eq. 1.1 and Eq. 1.2, for primary education, central and local government spending on education, combined together or separately, do not have a significant impact on the change of the net enrolment ratio. During the course of four years from 2010 to 2014, the net enrolment ratio increased from 94.1 to 95.6. At the primary level, net enrolment ratio is relatively high and disparity of the ratio among districts is relatively low (see Table 1). Hence, it is possible that spending more on this level of education will not significantly change the net enrolment ratio.

The results are more interesting for junior secondary education. As can be seen from Eq. 2.1 in Table 2, the total government spending on education does not have a significant impact on net enrolment ratio of junior secondary education at the district level. However, as we disaggregate the government spending in Eq. 2.2, the local government spending has a significant negative impact on the change of net enrolment ratio, whereas the central government spending has a significant positive impact on the change of net enrolment ratio of junior secondary education. If the local government spending increases by 1%, the change of net enrolment ratio will decrease by 0.69%. On the other hand, if the central government spending increases by 1%, the change of net enrolment ratio will increase by 2.24%. The results confirm previous studies, such as by Zufri and Oey-Gardiner (2012), which show that the central government spending in the form of the school operation assistance program has a

Table 2. Regression results.

Dependent variable	Change of NER primary education 2010 & 2014		Change of NER junior secondary education 2010 & 2014	
Independent variables	Eq. 1.1	Eq. 1.2	Eq. 2.1	Eq. 2.2
Total government spending to GRDP	–0.0808		–0.1647	
	(0.0644)		*(0.1633)*	
Local government spending to GRDP		–0.2201		–0.6862**
		(0.1451)		*(0.3074)*
Central government spending to GRDP		0.5575		2.2362**
		(0.4647)		*(0.9501)*
Initial NER primary education	–0.2978***	–0.3061***		
	(0.6350)	*(0.0653)*		
Initial NER junior secondary education			–0.4650***	–0.4823***
			(0.0416)	*(0.0405)*
GRDP per capita (ln)	–0.0031	–0.0002	–0.0062	0.0044
	(0.0025)	*(0.0029)*	*(0.0057)*	*(0.0064)*
Poverty headcount ratio	–0.0799***	–0.0834***	–0.1101**	–0.1181**
	(0.0201)	*(0.0207)*	*(0.0526)*	*(0.0521)*
Share of population below 15 years old	–0.0392	–0.0517	–0.2356***	–0.2861***
	(0.0347)	*(0.0371)*	*(0.0881)*	*(0.0904)*
Share of households living in urban areas	–0.0135**	–0.0135**	0.0365***	0.0367***
	(0.0058)	*(0.0057)*	*(0.0122)*	*(0.0119)*
Life expectancy ratio	–0.0003	–0.0004	0.0003	0.0001
	(0.0005)	*(0.0005)*	*(0.0013)*	*(0.0013)*
Dummy – less developed & remote districts	–0.0046	–0.0052	–0.0059	–0.0089
	(0.0035)	*(0.0035)*	*(0.0078)*	*(0.0078)*
Constant	0.4012***	0.3667***	0.5750***	0.4225***
	(0.0774)	*(0.0796)*	*(0.1352)*	*(0.1428)*
Adj. R-squared	0.2559	0.2614	0.3697	0.3827
Observation	473	473	473	473

Source: Authors' estimates.

Notes: *** and ** denote statistical significance at the 1% and 5% level respectively. Standard errors in parentheses-standard errors are heteroscedasticity-robust standard errors. DKI Jakarta as a special region is excluded from the analysis, since the decentralisation only applied in the provincial level. Some outliers are excluded from the analysis, so that total number of observations is 473 out of 491 available district data.

significant impact on net enrolment ratio of junior secondary education, whereas the local government spending does not have significant impact.

If we further elaborate spending on education by the local governments, according to Al-Samarrai and Cerdan-Infantes (2013), about three quarters of the local government education spending are for teachers' salaries. Furthermore, previous reports on education in Indonesia such as by the Indonesian Ministry of Education and Culture (2013) and by the World Bank (2013a) show that there is no relation between spending more on teachers' salaries and education outcomes. These previous empirical findings might explain the negative relation of local government spending and the change of net enrolment ratio. Increasing local government spending on education that cannot be fully transformed into education resources might have no or even negative impact on education outcomes.

Another alternative discussion in explaining the finding of this paper is by following studies by Rajkumar and Swaroop (2008), and Suryadarma (2012). Their studies explore the effectiveness of government spending on education by linking the government spending, governance, and education outcomes. Suryadarma (2012) found that when corruption is not taken into account, the effect of local government spending on net enrolment ratio of junior secondary is statistically insignificant. However, after taking corruption into account, the study shows a negative impact of local government spending on net enrolment ratio in the districts with high corruption. In addition, a report by the World Bank (2013b) concludes that, among others, poor performance and lack of transparency of local governments in managing financial resources for education might affect the impact of spending on education at the district level. Although this paper does not capture the issues of governance in the local government at district level, the negative relation between local government spending and net enrolment ratio of junior secondary education might imply that there is an issue of governance in the local government spending on education.

On the other hand, the results show that the central government spending has a significant positive impact on the change of net enrolment ratio. The central government spending on education applied in this paper consists of specific spending that is transferred to the district level, which are: the school operational assistance program, special allocation funds for education, and additional allowances for teachers. The funds are transferred from the central government to the district level for specific purposes with guidelines and supervisions from the central government. The result indicates that allocation of financial resources from the central government with specific purpose and oversight seems to have a positive impact on the education outcomes at the district level.

One interesting finding worth mentioning from the control variables is the consistency of significant and negative impact of poverty ratio to both the change of net enrolment ratio of primary and junior secondary education. This implies that different levels of poverty at the district level might cause disparities of education outcomes among districts. Hence, government spending specifically for the poor might enhance access of primary and junior secondary education at the district level.

3 CONCLUDING REMARKS

This paper finds that there is no significant relation between government spending on education, both local and central governments, on the change of net enrolment of primary education during the period of 2010–2014. However, a mixed result is found in junior secondary education, where local government spending shows a significant negative impact on the change of net enrolment ratio, whereas the central government spending shows a significant positive impact. The central government spending on education transferred to the districts with specific purposes and measurements seems to have a positive impact on education outcomes.

The findings seem to be in line with previous studies showing that most local government spending on education at the district level is allocated for teachers' salaries, and there is still a question on the relation between the increase of spending for teachers and education outcomes. The issue of governance might also affect the relation between local government spending and education outcomes. Hence, a question of capacity and governance of local governments in allocating spending for education might be intriguing to be explored in the next study.

This paper also finds a significant negative impact of poverty ratio on both net enrolment ratio of primary and secondary education. Specific government spending on education for the poor should be included in further analysing the impact of government spending on education at the district level in Indonesia.

REFERENCES

Al-Samarrai, S. & Cerdan-Infantes, P. (2013). Where did all the money go? Financing basic education in Indonesia. In D. Suryadarma & G.W. Jones (Eds.), *Education in Indonesia* (pp. 109–138). Singapore: ISEAS Publishing.

Arze del Granado, F.J., Fengler, W., Ragatz, A. & Yavuz, E. (2007). Investing in Indonesia's education: Allocation, equity, and efficiency of public expenditures. *Munich Personal RePEc Archive*, no. 4372: 1–43. doi:10.5897/JAERD12.088.

Barro, R.J. & Lee, J.W. (2001). International data on educational attainment: Updates and implications. *Oxford Economic Papers*, *53*(3), 541–563. doi:10.1093/oep/53.3.541.

Barro, R.J. (1991). Economic growth in a cross-section of countries. *The Quarterly Journal of Economics*, *106*(2), 407–443. doi:10.2307/2937943.

Basu, P., Bhattaraif K. & Ravikumar, B. (2012). Government bias in education, schooling attainment, and long-run growth. *Southern Economic Journal*, *79*(1), 127–143. doi:10.4284/0038-4038-79.1.127.

Blankenau, W.F. & Simpson, N.B. (2004). Public education expenditures and growth. *Journal of Development Economics*, *73*(2), 583–605. doi:10.1016/j.jdeveco.2003.05.004.

Blankenau, W.F., Simpson, N.B. & Tomljanovich, M. (2007). Public education expenditures, taxation, and growth: Linking data to theory. In *American Economic Review*, *97*, 393–397. doi:10.1257/aer.97.2.393.

BPS Statistics Indonesia. (2010 & 2014). National Socioeconomic Survey 2010 and 2014. Jakarta: BPS Statistics Indonesia.

BPS Statistics Indonesia. (2010–2014). Retrieved from https://www.bps.go.id.

Devarajan, S., Swaroop, V. & Zou, H. (1996). The composition of public expenditure and economic growth. *Journal of Monetary Economics*, *37*(2), 313–344. doi:http://dx.doi.org/10.1016/S0304-3932(96)90039-2.

Gupta, S., Verhoeven, M. Tiongson, E.R. (2002). The effectiveness of government spending on education and health care in developing and transition economies. *European Journal of Political Economy*, *18*(4), 717–737. doi:10.1016/S0176-2680(02)00116-7.

Hanushek, E.A. (2002). *Publicly provided education*. NBER Working Paper Series (8799).

Kristiansen, S. & Pratikno. (2006). Decentralising education in Indonesia. *International Journal of Educational Development*, *26*(5), 513–531. doi:10.1016/j.ijedudev.2005.12.003.

Lin, S.. (1998). Government education spending and human capital formation. *Economics Letters*, *61*(3), 391–393. doi:10.1016/S0165-1765(98)00193-1.

Mingat, A. & Tan, J.P. (1998). The mechanics of progress in education: Evidence from cross-country data. Policy Research Working Paper WPS2015.

Ministry of Education and Culture. (2013). *Overview of the education sector in Indonesia 2012: Achievements and challenges*. Jakarta: Ministry of Education and Culture Republic of Indonesia.

OECD & ADB. (2015). *Education in Indonesia: Rising to the challenge*. *Far Eastern survey*. Vol. 20. doi:10.1525/as.1951.20.15.01p0699q.

Pritchett, L. (2001). Where has all the education gone? *World Bank Economic Review*, *15*(3), 367–391. doi:10.1016/S0140-6736(94)92201-2.

Rajkumar, A.S. & Swaroop, V. (2008). Public spending and outcomes: Does governance matter? *Journal of Development Economics*, *86*(1), 96–111. doi:10.1016/j.jdeveco.2007.08.003.

Schulze, G.G., Sjahrir, B. S & Hill, H. (2014). Decentralization, Governance and Public Service Delivery. In H. Hill (Ed.), *Regional Dynamics in a Decentralized Indonesia* (pp. 186–207). Singapore: ISEAS.

Simatupang, R.R. (2009). Evaluation of decentralization outcomes in Indonesia: Analysis of health and education sectors. *ProQuest Dissertations and Theses*. 128. doi:10.1080/00074918.2011.585955.

Suryadarma, D. (2012). How corruption diminishes the effectiveness of public spending on education in Indonesia. *Bulletin of Indonesian Economic Studies*, *48*(1), 85–100. doi:10.1080/00074918.2012.654485.

Tullao Jr, T.S., Borromeo, M.R. & Cabuay, C.J. (2015). *Framing the ASEAN socio-cultural community (ASCC) Post 2015: Quality and equity issues in investing in basic education in ASEAN*. ERIA Discussion Paper Series ERIA-DP-2015-65.

World Bank. (2009). *Investing in Indonesia education at the district level: An analysis of regional public expenditure and financial management*. Jakarta: World Bank Office.

World Bank. (2012). *Indonesia economic quarterly: Redirect spending*. Jakarta: World Bank Office.

World Bank. (2013a). *Spending more or spending better: Improving education spending in Indonesia*. Jakarta: World Bank Office.

World Bank. (2013b). *Local governance and education performance: A survey of the quality of local education governance in 50 Indonesian districts*. World Bank Office.

Zufri, D. & Oey-Gardiner, M.O. (2012). *Analisis dampak dan determinan belanja pendidikan pemerintah daerah terhadap pendidikan tingkat sekolah menengah pertama (Analysis of the impacts and determinants of local government spending on junior secondary education)*. Jurnal Kebijakan Ekonomi FEB UI 8 (1).

Competition and Cooperation in Economics and Business – Gani et al. (Eds)
© 2018 Taylor & Francis Group, London, ISBN 978-1-138-62666-9

Intra-household decision-making and educational attainment in Indonesia

H. Haryani & T. Dartanto
Department of Economics, Faculty of Economics and Business, Universitas Indonesia, Depok, Indonesia

ABSTRACT: Many studies on the determinants of children's education attainment have found that parental background and family income are the most important factors. However, the current research shows the importance of intra-household decision-making on children's educational attainment. An intense discussion between parents and a child on the best path of education that the child should take, will probably have a future consequence on the child's educational attainment. This study aims at analysing the impact of children's involvement in decision-making of their educational attainment. We separate the decision-making of children's schooling choices into three types: decided by parents, decided by both parents and children, and decided by children. Which type of decision-making, regarding children's education, has the best outcome for future educational attainment? This study uses two waves of the 2000 and 2007 Indonesian Family Life Survey (IFLS) to examine this issue. Applying econometric estimations, this study confirms that the type of decision-making on children's schooling choices in 2000 has a significant effect on the future children's educational choice, and that the schooling choices decided by children has the highest impact on children's educational attainment compared to the other types of decision-making. This result suggests that parents should listen to their children in deciding their education choice.

1 INTRODUCTION

'Education is the most powerful weapon which you can use to change the world'
(Nelson Mandela, 1918–2003)

Many studies on the determinant of children's educational attainment have found that parent's background and family income are the most important factors (Becker, 1964; Leibowitz, 1974; Becker & Tomes, 1986; Behrman & Rozsenzweig, 2005. However, current research shows the importance of intra-household decision-making on children's educational attainment (Fleisher, 1977; Rangel, 2006). Conventional wisdom perceives that parents are always trying to provide the best for their children (Becker, 1981; Becker & Tomes, 1986), so that they sometimes have a dominant role in the decision of the children's schooling. The differences in parents' resources allocation for their children will cause differences in the level of achievement in children (Bloome, 2015).

Changes in environment, culture, and information have changed intra-household decision-making from a traditional type of household decision-making (parent centre) to a more democratic type of decision-making that allows all family members to actively discuss household issues (Mikkelsen, 2006; Lundberg et al., 2007). Moreover, the United Nation Convention on the Rights of the Child (1989), ratified by the Indonesian Government through Presidential Decree No. 36 (1990), stated in articles 12 & 13 that the children shall have the right to freedom of expression. In the case of education, an intense discussion, between parents and children, about the best educational path that the child should take, would probably have optimal impact on the quantity and quality of their children's education (d'Addio, 2007), while the children would be fully responsible for the decision made by both parties.

The study aims at analysing the impact of children's involvement in decision-making for their educational attainment. We separate the decision-making of children's schooling choices into three types: decided by parents, decided by both parents and children, and decided by children. Which type of decision-making regarding children's education, has the best outcome for future educational attainment?

2 THEORETICAL FRAMEWORK

This study uses a model of collective household (Chiappori, 1992; Browning & Chiappori, 1998), seeking to maximise the utility of family consisting of parents and children utility. Simplified, this study uses the assumption that a household consists of two parents and one child (Lundberg et al., 2007). Parents' decisions are unitary models with the assumption that they share the same preference and/or decision or only one parent makes the decision.

The utility functions of a family are formed from the variable of other commodity consumption, the quantity, and quality of children in the family (Becker, 1981). Other commodity consumption includes all family consumption like food, clothing, and housing in general. The quantity of children is defined as the number of children in a family. Furthermore, the quality of children in this study refers to the educational attainment of children.

The utility function of a family is a bundle utility of parents and their children. $U = g(U_c, U_p)$ where $U_c = U_c(x,n,q)$, $U_p = U_p(x,n,q)$ and $U_c \neq U_p$. The cooperative outcome is the solution to:

$$\max U = \alpha U_c(x,n,q) + (1-\alpha)U_p(x,n,q) \tag{1}$$

subject to the budget constraint:

$$p_x x + p_n n + p_q q = y \tag{2}$$

where, U is a family utility consisting of child utility (U_c) and parents' utility (U_p),x is the general commodity consumption for all members of the family, n describes the number of children in the family, and q is usually the quality of child in the family (whereas in this study the quality of the child is described by the child's educational attainment). The income and expenditure constraint consists of the total income of the family, personal consumption, expenditures for each child, and the investment in a child's education costs. α is the contribution of children utility in the family utility besides parents.

Specifically, a child's quality function is:

$$q^* = f\left(p_x^*, p_n^*, p_q^*, y^*, \alpha^*\right) \tag{3}$$

It can be shown that the model produces standard demand functions for consumption goods, leisure, and education (a formal proof is contained in Browning and Chiappori (1998)). These functions depend on the price of education, wages, household resources, the distribution of power, and household characteristics (observable and unobservable) (Mazzocco, 2007).

To analyse the impact of children's involvement in decision-making on educational attainment, we separated the decision-making choice of school children into three types: liberal, authoritarian and democratic. Bargaining power of children within household can be observed from from the value of α, namely:

1. If α value is 1 this means that the decision in education is entirely in the authority of children (liberal/freedom);
2. If α value is 0 this means that the decision is entirely in the authority of parents (authoritarian);

3. If α value is between 1 and 0, then the decision is made by both parents and children together (democratic).

Involving children in decision-making in education is a strategy that may be done by parents to maximise the satisfaction of the family in improving the children's educational attainment in the final education results. This happens because of communication between parents and children. Involving children in decision-making in the household regarding their education impacts the motivation of children positively because they psychologically feel the full support from their parents, so their potential will be maximal (Smart & Pascarella, 1986; Papalia, 2004).

Based on Equation 1, an important variable that needs to be considered is the quality of the children. Thus, parents should pay attention to the quality of children in the family aside from other variables. Using Equation 3, it can be determined that decision-making affects the children (α) in a family and the quality of children reflected in the educational attainment of children (q). In the theory described above, decision-making in education involving the children positively affects the educational attainment of children. In other words, if the predominance of children in decision-making in education increases, the educational attainment of the children also increases. However, based on the literature, educational attainment of children is also affected by family income, family expenditure for each child (for health care, clothing, and so on, which is accumulated according to the number of children), the consumption of other commodities, and household characteristics (children and parents/family) that are observable and unobservable (Becker, 1981; Mazzocco, 2007).

The contribution of children's utility (α) describes the bargaining power of children in a family and their involvement in decision-making in the family. From the conclusion above, it can be seen that if children are included in decision-making regarding the choice of their school/education, it will have a positive impact and produce better educational attainment.

3 RESEARCH METHODOLOGY

In our empirical analysis, we write the model to be estimated in a log-linear form:

$$q_{it} = \beta_0 + \beta_1 dm_{it-1} + \beta_2 h_{cit-1} + \beta_3 h_{pit-1} + \epsilon_i \tag{4}$$

where, q_{it} is the educational attainment (years of schooling) in 2007, which measures child quality, dm_{it-1} is the decision-making in the children's education in 2000, h_{cit-1} is a vector of children background variables (age, number of siblings, and cognitive abilities) in 2000, h_{pit-1} is a vector of parents/families background variables (the age of the head of household, families living, parents education, and families income), and ϵ_i is a random error term that is decided by children distributed across families but may be correlated across siblings. Table 1 shows the operational definitions of the variables used in this study:

This study uses two waves of the 2000 and 2007 Indonesian Family Life Survey (IFLS) to examine this issue. The IFLS is a comprehensive longitudinal socioeconomic survey that represents an area including 13 of Indonesia's 26 provinces and 83 percent of its population (Frankenberg & Thomas, 2000). The IFLS contains, among other measures, detailed information on family structure and composition, marriage, school enrolment and completion, parents' employment, income of each family member, total real expenditure, and within household decision-making. IFLS has numerous strengths: nearly everyone in the household was interviewed directly so that the data is both comprehensive and largely self-reported.

To see the effect of the role of the child's education decision makers, we used the data of 2000, and the educational attainment of children who are dependent variables used 2007 data. The observations in this study were children between 11–18 years old and with unmarried status, in the household and interviewed in 2000. Table 2 shows the summary statistics, including the mean and standard deviation of each variable. These are 3,908 children between 11–18 years old in the sample.

Table 1. Variable operational definition.

Dependent variable (2007)	The meaning of the variable
Child's education (q)	Child's years of schooling
Explanatory variable (2000)	**The meaning of the variable**
Children involvement in decision-making on children's education (decided by both parents and children or only by children) (democratic)	Dummy (1 if children involvement in decision-making on children's education, 0 if not)
Decision-making on children's education decided by both parents and children (dm1)	Dummy (1 if decision-making on children's education decided by both parents and children, 0 if not)
Decision-making on children's education decided by children (dm2)	Dummy (1 if decision-making on children's education decided by both parents and children, 0 if not)
Child vector variable	
Child's cognitive ability (c_kog00)	Child's cognitive in analysis and maths.
Age of child (c_age00) and child's age square	Age of child and child's age square
Sex of child (c_sex00)	Dummy (1 if female and 0 if male)
Child's status (c_stat00)	Dummy (1 if the status of child is adopted or stepchild, 0 if real child)
Child's schooling status (c_sch00)	Dummy (1 if the child is still schooling, 0 if not)
Family vector variable	
Residence of family (c_res00)	Dummy (1 if living in rural, 0 if living in urban)
Number of siblings (num_sib00)	Number of siblings, both real siblings and adopted/stepsiblings
Age of household's head (krt_age00) and age of household's head square (krt_agesq00)	The age of household's head and household's head age square
Father's education (pf_yos00)	Father's years of schooling
Mother's education (pm_yos00)	Mother's years of schooling
Log total real family expenditure (lnf_rtotal00)	Log total real family expenditure
Log total expenditure for education (lnxeducall00)	Log total expenditure for education

Source: Authors.

4 ANALYSIS OF RESULTS

In Figure 1 we can see the average educational attainment based on education decision makers. Educational decision-making by children has resulted in educational attainment that is the higher than when decided by others. The average educational attainment based on the type of decision by parent decision-making becomes the lowest kind of decision-making, compared with the two other types of decision-making. The results are consistent with the theoretical framework above.

From Table 3, it can be seen that if children are involved in decision-making regarding their educational choice (democracy) in 2000, it statistically has a positive impact on educational attainment of the children in 2007. This means that if the children are involved in decision-making regarding their education, then the educational attainment is better than if the children are excluded. The magnitude of all decision-making coefficients in education involving children (democracy) has the same direction and shows consistent results. These results are consistent with the theoretical framework that has been presented in the previous section.

Table 2. Statistics summary.

Variable	Obs	Mean	Std. Dev.	Min	Max
Child's education in 2007 (c_yos07)	3,908	10.385	3.298	0	18
Decision-making on children's education decided by parents and children (dm1)	3,908	0.099	0.299	0	1
Decision-making on children's education decided by children (dm2)	3,908	0.018	0.134	0	1
Decision-making on children's education decided by parents (dm3)	3,908	0.883	0.322	0	1
Decision-making on children's education (c_dm)	3,908	2.783	0.607	1	3
Children's involvement in decision-making on children's education (decided by parents and children or decided by only children) (demokrasi)	3,908	0.117	0.322	0	1
Child's education in 2000 (c_yos00)	3,908	7.031	2.515	0	16
Child's cognitive ability (c_kog00)	3,908	63.619	26.799	0	100
Age of child (c_age00)	3,908	14.353	2.253	11	18
Child's age square (c_agesq00)	3,908	211.093	65.171	121	324
Sex of child (c_sex00)	3,908	0.478	0.499	0	1
Child's status (c_stat00)	3,908	0.024	0.154	0	1
Child's schooling status (c_school07)	3,908	0.131	0.337	0	1
Residence of family (c_res00)	3,908	0.549	0.498	0	1
Number of siblings (num_sib00)	3,908	3.990	1.758	1	11
Age of head household (krt_age00)	3,908	46.190	8.256	25	105
Age of head household square (krt_age00)	3,908	2,191.053	817.667	625	11,025
Father's education (pf_yos00)	3,908	6.343	4.329	0	18
Mother's education (pm_yos00)	3,908	5.094	3.971	0	17
Log total real family expenditure (lnf_rtotal00)	3,908	13.929	0.663	11.082	16.607
Log total expenditure for education (lnxeducall00)	3,717	11.056	1.154	5.809	15.093

Source: IFLS, authors.

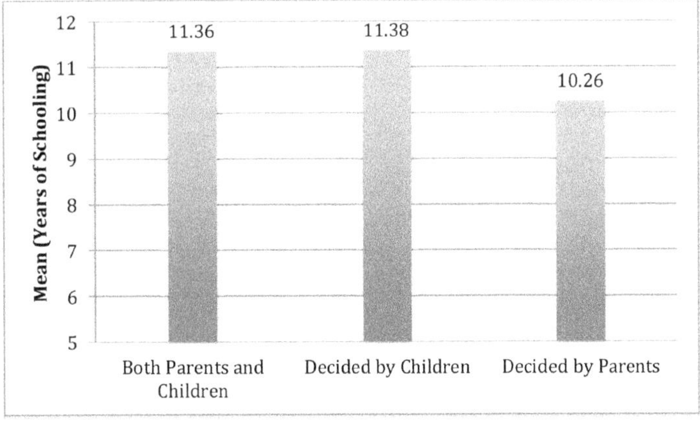

Figure 1. The type of decision-making and years of schooling.

Furthermore, to obtain in-depth and detailed information, the separation of decision makers in education is divided into two dummy variables, namely decision-making in education decided jointly by the children and parents (dm1) and decision-making in education decided by children alone (dm2). Decision-making in education by parents is used as a basis group.

Table 3. Regression result with 1 dummy variable of decision-making.

	Dependent variable: Child's educational attainment on 2007 (years of schooling) (c_yos07)					
	Child's age 11–18 years		Child's age 11–14 years		Child's age 15–18 years	
Explanatory variables	Model 1	Model 2	Model 1	Model 2	Model 1	Model 2
Children's involvement in decision-making on their education attainment	0.4942*** (0.130)	0.4396*** (0.114	0.4271*** (0.166)	0.3093*** (0.160)	0.5120*** (0.168)	0.5349*** (0.156)
Decision-making in children's education decided by parents and children (dm1)	-----	-----	-----	-----	-----	-----
Decision-making in children's education decided by parents and children (dm2)	-----	-----	-----	-----	-----	-----
Children's cognitive ability (c_kog00)	0.0293*** (0.002)	0.0251*** (0.002)	0.0320*** (0.003)	0.0280*** (0.003)	0.0282*** (0.002)	0.0237*** (0.002)
Age of children (c_age00)	1.1554*** (0.261)	0.8676*** (0.256)	3.6298*** (1.3233)	2.8973** (1.287)	1.3361 (1.989)	−0.0461 (1.960)
Age of child square (c_agesq00)	−0.0323*** (0.009)	−0.0228** (0.009)	−0. 1349** (0. 053)	−0.1060** (0.051)	−0.0404 (0.060)	0.0022 (0.060)
Sex of children (c_sex00)	0.3393*** (0.080)	0.3377*** (0.078)	0. 1967* (0. 107)	0.2072** (0.103)	0.4890*** (0.121)	0.4895*** (0.119)
Children's status (c_stat00)	−0.4016 (0.257)	−0.2876 (0.242)	−0.6915** (0. 344)	−0.3561 (0.337)	−0.0757 (0.381)	−0.1298 (0.356)
	Child's age 11–18 years		Child's age 11–14 years		Child's age 15–18 years	
	Model 3	Model 4	Model 3	Model 4	Model 3	Model 4
Children's schooling status (c_school07)	1.3149*** (0.117)	1.2830*** (0.114)	1.4233*** (0.126)	1.3571*** (0.125)	1.4873*** (0.288)	1.3696*** (0.270)
Residence of family (c_res00)	−0.3165*** (0.086)	−0.0626 (0.086)	−0.2151* (0.117)	0.0253 (0.116)	−0.4189*** (0.128)	−0.1849 (0.128)
Number of siblings (num_sib00)	−0.1604*** (0.025)	−0.2299*** (0.024)	−0.1102*** (0. 035)	−0.1664*** (0.034)	−0.2154*** (0.036)	−0.2895*** (0.035)
Age of household head (krt_age00)	0.0374 (0.042)	0.0354 (0.045)	0.0071 (0.055)	−0.0266 (0.053)	0.0719 (0.070)	0.1868** (0.081)
Age of household head square (krt_agesq00)	−0.0003 (0.0004)	−0.0002 (0.0004)	−0.0004 (0.0005)	−0.0003 (0.0005)	−0.0005 (0.0007)	−0.0015* (0.0008)
Father's education (pf_yos00)	0.1856*** (0.013)	0.1512*** (0.013)	0.1804*** (0.018)	0.1547*** (0.017)	0.1852*** (0.019)	0.1396*** (0.019)
Mother's education (pm_yos00)	0.1272*** (0.014)	0.1031*** (0.014)	0.1033*** (0.019)	0.0818*** (0.018)	0.1508*** (0.021)	0.1204*** (0.020)
Log total real family expenditure (lnf_rtotal00)	0.6413*** (0.075)	-----	0.3561*** (0.102)	-----	0.9250*** (0.109)	-----
Log total expenditure for education (lnxeducall00)	-----	0.7538*** (0.043)	-----	0.6215*** (0.057)	-----	0.8999*** (0.065)
Constant	−12.686*** (2.251)	−9.239*** (2.052)	−22.901 (8.299)	−18.670** (7.971)	−18.292 (16.414)	−6.463 (16.223)
F-Statistic	227.63	239.33	113.47	123.16	122.71	123.59
R-Squared	0.4294	0.4671	0.4112	0.4412	0.4543	0.4917
# of observations	3,908	3,717	2,071	2,030	1,837	1,687

Note: t-statistics are in parentheses: ***p<0.01, **p<0.05, *p<0.1.
Source: authors.

Table 4. Regression result with 2 dummy variables of decision-making.

	Dependent variable: Child's educational attainment on 2007 (years of schooling) (c_yos07)					
	Child's age 11–18 years		Child's age 11–14 years		Child's age 15–18 years	
Explanatory variables	Model 3	Model 4	Model 3	Model 4	Model 3	Model 4
Children's involvement in decision-making on their education attainment	-----	-----	-----	-----	-----	-----
Decision-making in children's education decided by parents and children (dm1)	0.4898*** (0.130)	0.3917*** (0.123)	0. 4045** (0.182)	0.2601 (0.174)	0.5396*** (0.181)	0.4904*** (0.169)
Decision-making in children's education decided by parents and children (dm2)	0.5187* (0.272)	0.7088*** (0.258)	0.5546 (0.358)	0.5943* (0.353)	0.3619 (0.389)	0.7790** (0.356)
Children cognitive ability (c_kog00)	0.0293*** (0.002)	0.0251*** (0.002)	0. 0320*** (0.003)	0.0280*** (0.003)	0. 0282*** (0.002)	0.0237*** (0.002)
Age of children (c_age00)	1.1548*** (0.261)	0.8609*** (0.256)	3.6456*** (1.3255)	2.9295** (1.288)	1.3570 (1.991)	–0.0733 (1.962)
Age of child square (c_agesq00)	–0.0323*** (0.009)	–0.0225** (0.009)	–0. 1356** (0. 053)	–0.1073** (0.052)	–0. 0410 (0. 060)	0.0031 (0.060)
Sex of children (c_sex00)	0.3393*** (0.080)	0.3382*** (0.078)	0. 1967* (0. 107)	0.2067** (0.103)	0. 4879*** (0. 121)	0.4913*** (0.119)
Children's status (c_stat00)	–0.4014 (0.257)	–0.2850 (0.242)	–0. 6896** (0. 344)	–0.3514 (0.338)	–0. 0763 (0. 381)	–0.1280 (0.356)
Children's schooling status (c_school07)	1.3147*** (0.116)	1.2813*** (0.114)	1.4337*** (0.126)	1.3566*** (0.125)	1.4893*** (0. 288)	1.3651*** (0.270)
Residence of family (c_res00)	–0.3165*** (0.086)	–0.0613 (0.086)	–0.2158* (0.117)	0.0243 (0.116)	–0. 4300*** (0. 128)	–0.1814 (0.128)
Number of siblings (num_sib00)	–0.1604*** (0.025)	–0.2297*** (0.024)	–0. 1098*** (0. 035)	–0.1657*** (0.034)	–0. 2154*** (0. 036)	–0.2896*** (0.035)
Age of household head (krt_age00)	0.0373 (0.042)	0.0342 (0.045)	0.0066 (0.055)	–0.0279 (0.053)	0. 0725 (0.070)	0.1855** (0.081)
Age of household head square (krt_agesq00)	–0.0003 (0.0004)	–0.0002 (0.0004)	–0.0004 (0.0005)	–0.0003 (0.0005)	–0.0005 (0.0007)	–0.0015* (0.0008)
Father's education (pf_yos00)	0.1856*** (0.013)	0.1514*** (0.013)	0.1805*** (0.018)	0.1548*** (0.017)	0.1848*** (0.019)	0.1398*** (0.019)
Mother's education (pm_yos00)	0.1272*** (0.014)	0.1027*** (0.014)	0.1033*** (0.019)	0.0817*** (0.018)	0.1511*** (0.021)	0.1197*** (0.020)
Log total real family expenditure (lnf_rtotal00)	0.6412*** (0.075)	-----	0.3566*** (0.101)	-----	0.9264*** (0.109)	-----
Log total expenditure for Education (lnxeducall00)	-----	0.7555*** (0.043)	-----	0.6214*** (0.057)	-----	0.9022*** (0.065)
Constant	–12.6791*** (2.254)	–9.1822*** (2.054)	–22.9907*** (8.308)	–18.8482** (7.978)	–18.4956 (16.444)	–6.2390 (16.240)
F-Statistic	212.43	223.47	105.92	114.96	1148	115.52
R-Squared	0.4294	0.4673	0.4112	0.4413	0.4544	0.4919
# of observations	3,908	3,717	2,071	2,030	1,837	1,687

Note: t-statistics are in parentheses: ***p<0.01, **p<0.05, *p<0.1.
Source: authors.

The regression results in Table 4 show that decision-making in education is decided jointly by parents and children (dm1) and if it is decided by the children alone (dm2), it statistically has a significant positive impact on the educational attainment of the children when compared with the basis (decision-making decided by parents).

The division of age groups is required to view the role of decision maker for both age groups more specifically. Parents are still in control of the decision-making in education, where the children may be considered not able to determine the best option for their education. Whereas with children's age, parents increasingly give trust and listen to the children's voice by involving children in decision-making in education. In fact, parents fully entrust the decision-making of children's education to the children.

4.1 *Decision-making on children's education decided by parents and children*

Based on Table 4, it can be seen that the decision-making in education decided solely by children (dm2) in the children's schooling choice in 2000 produces better educational attainment than the other types of decision-making in education in 2007. It is suggested that this type of decision-making produces the highest educational attainment compared to other types of decision-making. Overall the regression results still show consistent results with the theory presented in the previous section.

Decision-making in children's education that is handed over to the children in question may have an impact on the release of their fullest potential. Children who are able to decide and choose their own school/education are identified as children who have a more mature and positive self-concept. According to Smart and Pascarella (1986), the children who have a mature and positive self-concept know their capabilities, purposes and the direction of their life, including the capability to decide the best school/education for themselves. Thus, these children will be responsible for the decisions they take (Papalia, 2004).

The results are consistent across all specifications. The coefficients for the impact of children's involvement on decision-making on their educational attainment are positive. This implies that more bargaining power to the children is associated with the highest educational attainment. In the case of the decision-making on children's education choice decided only by children, this has the highest impact on children's educational attainment. This implies that the distribution of power within the household has to affect the children's educational attainment in the household. It is mainly affected by variables other than the distribution of power within the household, especially the age of children (Lundberg et al., 2007).

4.2 *The results of control variable*

Additional information on the result of this study is the characteristics of children and the characteristics of parents or family that have different impacts for each category to educational attainment. The variables of children's cognitive abilities, children's age, children with female sex, head household's age, paternal education (both father and mother), the total real expenditure of family, and the total education expenditure, have a positive impact on educational attainment of children. This means that as more the cognitive abilities of children increase, so more will increase the educational attainment of children. Although, there are other capabilities that could affect educational attainment in addition to the cognitive abilities, such as emotional ability, and motivation of the children to learn (Pajares, 1996).

By increasing the age of children, educational attainment of children should also increase. However, along with the increasing age of children, educational attainment of children tends to stagnate. This could be because the children are no longer continuing their education, or they are already at the peak of the highest formal educational attainment. This fact also implies that at the age 17.88 years (calculate: $\frac{\textit{coefficient of children's age}}{2x\,\textit{coefficient of children's age squared}}$ taken from the proxy of log total real expenditure of family) the average children in Indonesia quit or do not continue their education. It is most likely that at that age, the children have graduated to the level of high school (SMA-Sekolah MenengahAtas or in English Senior High School).

Another implication to Indonesia is that 17.88 years is an age at which it is effective to ask the children to discuss their education, or even hand to over the decision in education to the children independently. These results also reflect that the decision-making in children's education is still dominated by the parents, so that the parents begin to listen to the children's voice at the age of 17.88 years; where at this age, children may have graduated from high school.

Furthermore, the higher education of the father and mother has an impact on the higher attention to children's education, so that children's education is also increased (Gang & Zimmerman, 2000; Maralani, 2008). Likewise, family income has a positive impact on the educational attainment of children. It means that with the increasing income of a family, the greater the allocation of finance to their children's education, which affects the increasing education attainment of the children (Becker & Tomes, 1986; Haveman & Wolfe, 1995). Another interesting thing is that female children get a larger portion than male children, so that female children get better opportunities in educational attainment.

The number of siblings of children in a family, the children's residence, and children with the status of adopted/step children negatively affect the educational attainment of children. This study confirms a previous study which states that the number of children in a family negatively affects the children's educational attainment (Maralani, 2004). This means that by the increasing number of children living in a family, the chances of a child to study will decrease (Maralani, 2004). Likewise, children who live in a rural area have a limitation to increase their formal education, encountering a variety of obstacles (Maralani, 2008). Furthermore, the adopted/step children are included in a marginalised group, for which their education is given less attention by their adopted/step parents.

5 CONCLUDING REMARKS

Applying econometric estimations, this study confirms that the type of decision-making on children's schooling choices in 2000 has a significant and positive effect on the future children's educational attainment in 2007. The type of decision-making on children's education choice, decided only by children, has the highest impact on children's educational attainment compared to the other types of decision-making. The decision-making decided by children or decided by both parents and children will be made based on the age of children. These results also prove that in the second generation, the children are more concerned about the achievements of their education. This result suggests that parents should hear their children's voice when deciding their educational choice.

ACKNOWLEDGEMENT

This research is a part of Haryani's master thesis under the supervision of Teguh Dartanto, PhD The author would like to thank the generous financial support of PITTA Grant-Universitas Indonesia.

REFERENCES

Baumrind, D. (1966). Effects of authoritative parental control on child behavior. *Child Development*, *37*(4), 887–907.

Becker, G.S. & Tomes, N. (1986). Human capital and the rise and fall of families. *Journal of Labor Economic*, *4*, S1–S39.

Becker, G.S. (1964). *Human capital.* New York: Columbia University Press.

Becker, G.S. (1981). *A Treatise on the family.* Cambridge: Harvard University Press.

Becker, G.S., Murphy, K & Tamura, R. (1990). Human capital, fertility, and economic growth. *Journal of Political Economy*, *98*, S12–S37.

Behrman, J.R. & Rosenzweig, M.R. (2005). Does increasing women's schooling raise the schooling of the next generation? *American Economic Review*, *95*, 1745–1751.

Behrman, J.R. (1997). Intrahousehold distribution and the family. In M.R. Rosenzweig & O. Stark (Eds.), *Handbook of Population and Family Economics* (pp. 125–187). Amsterdam: Elsevier.

Behrman, J.R., Foster, A.D., Rosenweig, M.R. & Vashishtha, P. (1999). Women's schooling, home teaching and economic growth. *Journal of Political Economy, 107*, 682–714.

Bloome, D. (2015). Income inequality and intergenerational income mobility in the United States. *Social Forces March, 93*(3), 1047–1080.

Browning, M. & Chiappori, P.A. (1998). Efficient intra-household allocations: A general characterization and empirical tests. *Econometrica, 66*,1241–1278.

Chiappori, P.A. (1992). Collective labour supply and welfare. *Journal of Political Economy, 100*(3), 437–467.

Currie, J. & Moretti, E. (2003). Mother's education and the intergenerational transmission of human capital. Evidence from college openings. *Quarterly Journal of Economics, 118*(4), 1495–1532.

d'Addio, A.C. (2007). *Intergenerational transmission of disadvantage: mobility or immobility across generations?* OECD Working Paper, 52.

Ermisch, J. & Francesconi, M. (2001). Impact of family background on educational attainments. *Economica, 68*(270), 137–156.

Fleisher, B.M. (1977). Mother's home time and the production of child quality. *Demography, 12*(2),197–212.

Frankenberg, E. & Thomas, D. (2000). *The Indonesia family life survey (IFLS): Study design and results from waves 1 and 2.* Santa Monica: CA: RAND. CA. DRU-2238/1-NIA/NICHD.

Gang, I.N. & Zimmermann, K. (2000). Is child like parent? Educational attainment and ethnic origin. *The Journal of Human Resources, 35*(5) 550–569.

Hamilton, L.C. (2013). *Statistics with STATA: Updated for Version 12.* (8thed.). Boston, USA: Brookc/ Cole, Cengage Learning.

Hanushek, E.A. & Kimko, D.D. (2000). Schooling, labor-force quality, and the growth of nations. *The American Economic Review, 90*(5), 1184–1208.

Haveman, R. & Wolfe, B. (1995). The determinants of children's attainments: A review of methods and findings. *Journal of Economy Literature, 33*(4) 1829–1878.

Holmlund, H., Lindahl, M. & Plug, E. (2008). The causal effect of parent's schooling on children's schooling: A comparison of estimation methods. *IZA Discussion Paper Series, 3630.*

Leibowitz, A. (1974). Home investment in children. In *Economics of the family: Marriage, children, and human capital*, by T.W. Schultz (pp. 432–456). UMI.

Lundberg, S., Romich, J. & Tsang, K.P. (2007). Decision-Making by children. *IZA DP No. 2952. Discussion Paper Series.*

Maralani, V. (2004). Family size and educational attainment in Indonesia: A cohort perspective. *California Center for Population Research,* CCRP 017-04.

Maralani, V. (2008). The changing relationship between family size and educational attainment over course of socioeconomic development: Evidence from Indonesia. *Demography*, 45(2), 693–717.

Mazzocco, M. (2007). Household intertemporal behavior: A collective characterization and a test of commitment. *Review of Economic Studies, 74*(3), 857–895.

Mikkelsen, M.R. (2006). Children's influence on family decision-making in food buying and consumption. *Denmark: The National Institute of Public Health Denmark and MAPP-Centre for Research on Customer Relations in the Food Sector. Aarhus School of Business Denmark.*

Pajares, F. (1996). Self-Efficacy beliefs in academic settings. *Review of Educational Research, 66*(4), 543–578.

Papalia, D.E. (2004). *Human Development* (9th Edition). New York: McGraw Hill.

Rangel, M.A. (2006). Alimony rights and intrahousehold allocation of resources: Evidence from Brazil. *The Economic Journal, 116*(513), 627–658.

Republik Indonesia. (2014). Undang-Undang No.35 Tahun 2014 tentang Perubahan atas Undang-Undang Nomor 23 Tahun 2002 tentang Perlindungan Anak (Law No.35/2014 revision of UU No.23/2002 about Child Protection).

Smart, J.C. & Pascarella, E.T. (1986). Self-Concept development and educational degree attainment. *Higher Education, 15*(1/2), 3–15.

Strauss, J., Witoelar, F., Sikoki, B. & Wattie, A.M. (2009). *The fourth wave of the Indonesia family life survey: Overview and field report volume 1.* Santa Monica, CA: RAND Labor and Population.

Teachman, J.D. (1987). Family background, educational resources, and educational attainment. *American Sociological Review, 52*(4), 548–557.

The United Nation (1989). *Child Convention*. UN: New York.

Competition and Cooperation in Economics and Business – Gani et al. (Eds)
© 2018 Taylor & Francis Group, London, ISBN 978-1-138-62666-9

Empowering business incubator in creating technology based entrepreneurs

Hardiana & Hera Susanti
Department of Economics, Faculty of Economics and Business, Universitas Indonesia, Depok, Indonesia

ABSTRACT: We attempted to assess the performance of business incubators in Indonesia, focusing on their weaknesses. Questionnaire data were gathered from 8 business incubators, 39 incubates and 5 experts. Paired t-sample test was conducted and revealed that there was a gap between the tenants' expectations of business incubator services and the actual services provided by the business incubators. This study exposes the weaknesses of business incubators regarding space, personnel, operational funding, and networks. Strategic Assumption Surfacing and Testing (SAST) was used to develop strategic policies to overcome the weaknesses of business incubators in Indonesia. The strategic policies are proposing space expansion, hiring full-time managers, urging the institution head's commitment to providing funding, and promoting business incubators.

1 INTRODUCTION

Micro, Small, and Medium Enterprises (MSMEs) have an important role in the economic sector of many countries, including in Indonesia. A report by the Ministry of Cooperatives and SMEs (2015) notes that 57.9 million MSMEs absorbed 97% manpower, 89% of which were working in the micro enterprises. When compared to large enterprises, MSMEs contributed about 60% to GDP. Furthermore, at constant 2000 price, the growth of MSME sector's contributions to GDP was 5.89%, which was 0.33% higher than the growth of contributions made by big companies.

However, MSME sector is vulnerable to the numerous challenges. Tambunan (2007) concludes that the smaller the company, the bigger the challenges. Data from the Central Bureau of Statistics (2003) in Tambunan (2007) show the main obstacles for micro and small enterprises in Indonesia are the limited access to business capital (34.78%), difficulties in marketing the products (30.63%), and difficulties in obtaining supplies of raw materials (20.50%).

Another hurdle to MSMEs is related to utilization of technology to boost productivity. Data from the Central Bureau of Statistics in 2013 present a comparison of the output productivity per worker in the micro and small segment (20.1%) to that in the middle and large-scaled industries (79.9%). This condition has affected the quality of products, in which MSMEs' products have not been able to match big industry's quality standards. Furthermore, the export capacity of MSME segment was only 15.68% of the total output, far lower than the exports recorded by big industry at 84.32% of the total output (Ministry of Cooperatives and SMEs, 2015).

Business incubators (BIs) have the capacity to overcome these obstacles. The International Business Incubation Association/INBIA (2016) defines BI as an institution that provides supports in the business process to accelerate the growth of start-up companies that have limited access to business supports and resources.

However, in line with the development of BIs, people started to see weaknesses in this system, as reported by Al-Mubaraki and Busler in their study (2013), which found BIs in five developing economies (China, Bahrain, Jordan, Morocco and Syrian Arabic Republic) have limitation in terms of financing for the company, expertise in entrepreneurship, personal economic resources, and technology literacy. To that end, this study aims to identify the capacity of BIs to assist tenants in carrying out effective business and at the same time to assess the weaknesses of BIs in Indonesia.

2 METHODOLOGY

This study was conducted using a combination of qualitative and quantitative approaches through surveys. There were three stages in this study. First, questionnaires were distributed to BI managers to gather information about BI's resources. Second, questionnaires were sent to tenants who received 18 incubation items from the BIs and measured them based on the Likert scale. The questions in the questionnaire were divided into two groups. In the first group, the tenants were asked to express their expectations regarding the scope and extent of services provided by the BIs, and in the second group, they were asked to assess the services that have been received from the BIs during the incubation period. Third, questionnaires were also distributed to experts (represented by the BI management, government, and BI associations), in an effort to collect alternative policies to improve BI services.

Data sampling was conducted using purposive method targeting 15 BIs in the development and maturity phases (where in these phases BIs already provide adequate services) and 10 tenants from each BI located in Java and Bali, Indonesia between April and July 2016. Responses from the tenants were processed for preliminary mapping of the gap of the needs of these tenants, done by comparing means of each services. Meru and Struwig (2011) applied a paired t-test method to compare means between expected and perceived BIs' services in Kenya.

After that, a Strategic Assumption Surfacing and Testing (SAST) was applied to gather alternative policies from the perspectives of experts, in an effort to overcome gaps of tenants' needs. Mitroff and Mason (1981) explains that SAST is a very helpful tool to explore critical assumptions that become basis of a model or strategy. SAST is built through four stages, which are group formation, specifying assumptions and assessments, dialectic debate, and final synthesis.

3 RESULT

Responses were received from 8 BIs (50% university BIs, 25% public BIs, and 25% private BIs), 50% of which were in the development phase, while the rest were already in the maturity phase. Tenant questionnaire responses were collected from 39 tenants (97.44% males and 2.56% females). 35.9% of them held bachelor degrees, 30.8% held below bachelor degrees and the remainder had other backgrounds. The majority of the tenants (74.36%) received in-wall incubation, which means tenants enjoyed mutual work space and office facilities. Expert questionnaire responses were gathered from 5 experts (20% BI associations, 40% government, and 40% BI management personnel).

3.1 *Space*

The minimum standard of space area for a BI institution that applies in Indonesia, as mentioned by Syamas *et al.* (2015) is 500 m^2. Table 1 shows that only public BIs have 100% fulfilled or even exceeded the minimum requirement for space area. It is no surprise that financial sup-

Table 1. Total space.

Space	University	Government	Private	Total
<500m^2	75.0%	0%	50.0%	50.0%
>500m^2	25.0%	100.0%	50.0%	50.0%

Source: Survey, 2016.

port from the state budget has helped these institutions own adequate facilities. Mean while in the category of university BIs, most of them could not meet the minimum standard. This should be a concern because the majority of BIs in Indonesia are university BIs.

3.2 *Management personnel*

The majority of management personnel in university BIs had post-graduate degree. On the other hand, a lot of management personnel in government and private BIs held undergraduate degree. Overall, around 62.5% of the BI institutions already had full-time managers, while the remainder were led by part-time managers. Based on the institution type, 100% of private BIs and also government BIs assigned full-time managers, while only 25% of university BIs were managed by full-time managers. The low percentage of full-time manager assignment in university BIs can be accounted to the lack of available time of the BI personnel who also have the responsibility as lecturers. (Table 2)

Only a small number (37.5%) of BI institutions already engaged in internship or a benchmarking program with more advanced BIs. Both private and government BIs surveyed in the study already conducted internship at other BIs, but only 1 of 4 university BIs engaged in such a program involving other incubators with better performance. Regarding reward, 50% of BI personnel received monthly salary of between Rp5–Rp10 million. Unfortunately 50% of university BIs offered very small remuneration for their staff members, which was <Rp 1 million/month. However, personnel of university BIs had their main income source from working as lecturers.

3.3 *Operational funding*

BIs need financial support to carry out their activities. According to the survey, several sources of funding for BIs included funding from within the organization, grant, space leasing income, and profit sharing. Until the data were collected, BIs did not charge tenants for the services they provided.

The entire BIs surveyed in this study received a grant from the government under certain conditions. Governments BIs were 100% committed to providing budget on location for their activities, but some universities or private BIs did not provide budget on location for incubation activities (Table 3).

3.4 *Network*

All BIs in this study have been registered as members of an association, either nationally or internationally such as AIBI, in BIA, ANDE and IASP. Each BI has attended training,

Table 2. Management personnel.

Deskripsi	University	Government	Private
Educational Background			
<Bachelor	17.6%	12.8%	11.2%
Bachelor	8.8%	66.0%	55.6%
Master	58.8%	14.9%	27.8%
Doctoral	14.7%	6.4%	5.6%
With full-time manager	25.0%	100.0%	100.0%
With full-time staff	50.0%	100.0%	100.0%
Internship experience	25%	0%	100.0%
Salary/month			
<1 Jt	50.0%	0%	0%
Rp 1 Jt - Rp 3 Jt	25.0%	0%	0%
Rp 3 Jt - Rp 5 Jt	0%	50.0%	0%
Rp 5 Jt - Rp 10 Jt	25.0%	50.0%	100.0%

Source: Survey, 2016.

Table 3. Sources of operational funding.

Funding source	University	Government	Private	Total
Internal of organization	50.0%	100.0%	50.0%	62.5%
Government	100.0%	100.0%	100.0%	100%
Space leasing	25.0%	100.0%	50.0%	50.0%
Service providing	0%	0%	0%	0%
Profit sharing	50.0%	100.0%	50.0%	62.5%

Source: Survey, 2016.

Tabel 4. Networking.

Description	University	Government	Private	Total
Joining association	100.0%	100.0%	100.0%	100.0%
Attending training, workshops, exhibitions	100.0%	100.0%	100.0%	100.0%

Source: Survey, 2016.

Table 5. Paired t-sample statistics showing a gap in the Tenants' needs.

Supports and services	Received (Mean)	Expected (Mean)	Mean difference	t-value	Significance (2-tailed) (p-value)
Working space	4.46	4.56	–0.10	.681	.500
Laboratorium	3.56	4.41	–0.85	3.451	.001*
Office shared	4.36	4.49	–0.13	.777	.442
Internet	3.62	4.69	–1.08	4.352	.000*
Access to finance	3.36	4.59	–1.23	5.792	.000*
Access to investor	3.15	4.54	–1.38	5.246	.000*
Business plan services	3.77	4.56	–0.79	4.312	.000*
Training	3.90	4.46	–0.56	3.367	.002*
Mentorship program	3.46	4.33	–0.87	3.611	.001*
Product design and development	3.33	4.21	–0.87	5.157	.000*
Marketing and sales services	3.54	4.36	–0.82	4.391	.000*
Shows and exhibitions	3.51	4.41	–0.90	4.202	.000*
Bookkeeping services	3.23	4.15	–0.92	4.349	.000*
University collaboration	3.64	4.21	–0.56	2.641	.012*
R&D department collaboration	3.59	4.44	–0.85	4.748	.000*
Technology transfer	3.36	4.46	–1.10	5.593	.000*
Patent and copyright protection	3.28	4.41	–1.13	6.379	.000*
Legal services	3.05	4.26	–1.21	6.073	.000*

* Significant at $p<0.05$.

workshops and exhibitions organized by the government at the least (Table 4). These aim at improving BIs' skills as well as broadening their network.

3.5 *The need gap*

The calculation results of tenants' needs gap as stated in Table 5 show that only 2 out of 18 service items that did not have significant disparities between the expectation and the reality in terms of the provided service from the BIs. It indicates tenants' satisfaction with these two BI services, namely working space (p-value >0.05) and shared office equipment (p-value

>0.05). The remaining calculation results, on the other hand, show significant disparities between expectation and reality related to the provided services. The results suggest a gap between what tenants expected from the provided services and what the BIs offered to them.

3.6 *Empowerment of BIs*

While the tenants were satisfied with the working space provided by the BIs, the interview with the AIBI chairman showed that the majority of building dimension of BIs in Indonesia has not met the standards yet. Furthermore, based on the survey we conducted, there were merely 50% of the BIs that complied with the minimum standards.

Based on the gap of tenants' needs and the staggering amount of below-minimum standard BI space dimension, we classified four weaknesses of BIs as follows: space, personnel, operational funding, and networking. Space is tenants' place to work. Personnel are closely related to how the qualification and capacity of BI personnel in assisting tenants during the incubation. Operational funding reflects BI's capability of funding its activities. Meanwhile, network is about how BI can successfully bridge or connect tenants to various stakeholders. Seventeen experts' assumptions that have been identified using the SAST method are shown in Table 6.

3.6.1 *Working space barrier*

There are many BIs constrained by below-minimum standard building dimension. Considering these two identified assumptions (A1 and A2), Figure 1 shows that A1 is deemed the most essential, with high assurance and is possibly effective to get to the bottom of space barrier.

3.6.2 *Personnel barrier*

The commitment borne by BI manager which is represented by the institution's vision, mission, and goal is appropriate. The entire BIs are committed to creating technopreneurs as well as to improve the cooperation between academic, business and government. Each personnel member is responsible for his or her roles according to the job descriptions.

Table 6. Strategy assumption on BIs' role improvement.

Role improvement	Strategy assumption
Space	Proposing building expansion to the institution head
	Cooperating with state-owned enterprises/private sectors on CSR and government funding utilization for building expansion
Personel	Appointing dedicated and committed managers
	Assigning full-time managers
	Implementing junior–senior team-up
	Implementing junior–mentor team-up
	Delegating incubator manager interns in established incubators (national/ international)
	Signing up incubator managers for training on capacity building improvement
	Offering decent monthly income
Operational Funding	Financial support from institution head
	Seeking for government funding
	Cooperation on CSR program with state-owned enterprises and private sectors
	Generating income from tenants through space rent and profit sharing
	Formulating measured performance achievement target in order to attract investors
Networking	Identifying and understanding the stakeholders
	Proactively promoting BIs to stakeholders through business gatherings, workshops, exhibitions and training
	Signing up for incubator association as members

Source: Survey, 2016.

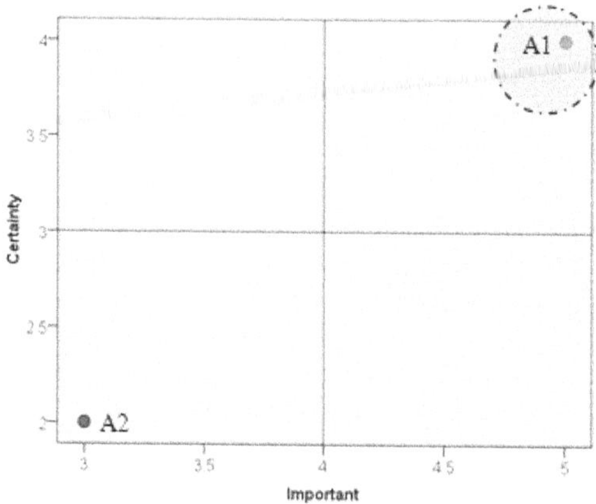

Figure 1. Priority working space enhacement.

Tenants' success is vastly determined by the performance of the BI manager who is in charge of the whole incubation activities. University BIs employing full-time managers are lower in proportion compared to the other two types. Full time is interpreted as someone who works 40 hours/week or 37.5 hours/week (for civil servants in particular). It has become a concern among the experts because the total number of BIs in Indonesia is mainly university BIs (72.8%). This outlook is in line with Scaramuzzi (2002) who views that BIs have to place a proportional number of full-time managers to match the size of the BIs, tenants and their activities.

BIs also have assistants. One of their tasks is to monitor tenants' growth as well as obstacles to be discussed with BI managers or mentors. These assistants have to be capable of understanding the general aspects of incubation management as well as tenant industry aspect in particular. In several BIs that we observed, there were many younger assistants with different background qualifications as required by the tenants. This phenomenon is accentuated by an opinion of one of the tenants who ran Oyster Mushroom cultivation saying that the assistant barely had any idea about mushroom. Another tenant in essential oil derivative sector also expressed a similar concern. According to him, the BI still had no specialist assistants in sales area, while sales function is undeniably vital to his business. Therefore, experts disclosed an assumption on how important it is to do junior assistant–senior assistant and junior assistant–mentor team-up.

It is possible to improve the ability of BI managers through capacity building programs. In 2015 a working group on BI enhancement was set up in accordance with the Coordinating Ministry of Economic Affairs decree No. 184/2015, where one of the functions is to conduct capacity building training for BI managers and tenants. Today there have been plenty of programs launched by both the government and private sectors to boost the managers' and tenants' capacity.

Another attempt, benchmarking against established BIs for an instance, has the potential to improve the quality of BI's personnel. Another advantage of benchmarking is to inspire managers to set off groundbreaking programs that improve service quality to tenants.

Providing decent income is deemed capable of boosting managers' motivation in doing their part. Unfortunately we explored no further than how much they earned on a monthly basis, and we did not seek further information if there was bonus sharing scheme, which is a common practice in private sector companies and state-owned enterprises.

Among seven assumptions which were identified as attempts to enhance management personnel's abilities, Figure 2 shows that A4 is deemed the most essential alternative with the highest assurance intensity to effectively improve the quality of BIs.

3.6.3 *Operational funding barrier*

In order to meet the financial demand for their operational activities, it is possible for BIs to seek funding from various sources. All government BIs are ensured to receive funding

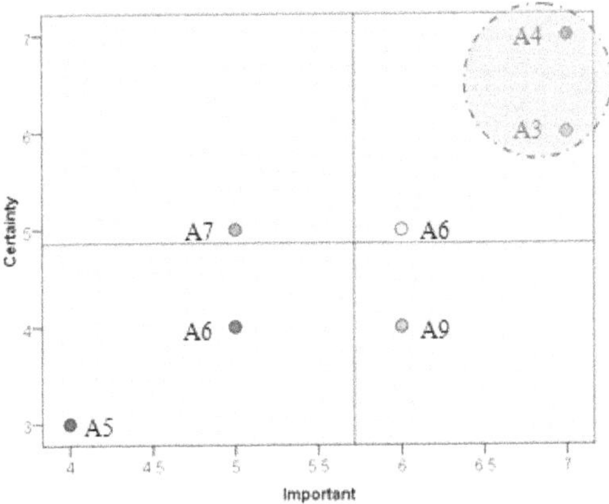

Figure 2. Priority personnel enhancement.

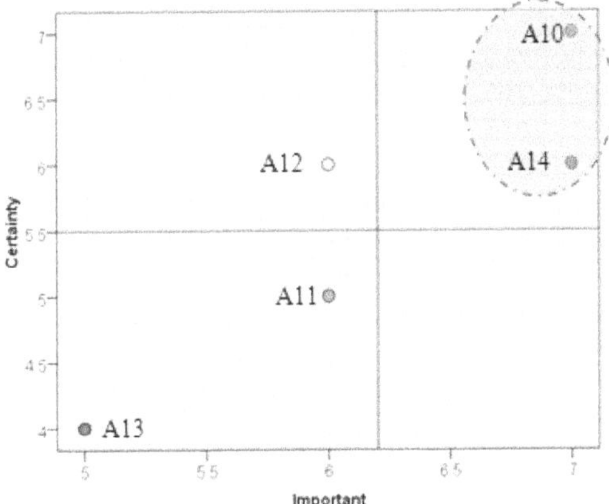

Figure 3. Priority operational funding enhacement.

from the state budget through their core institution. However, not all BIs receive operational funding from the core institution such as university or private BIs. For example, a non-profit organization private BI received funding from grants such as the Lemelson Foundation, which provides aid to investors and invention-based enterprises in developing countries.

Another instance is a university BI in the maturity phase which already had the ability to generate its own income through government incentives, tenants, or state-owned enterprises as well as private sector companies' CSR. This is why despite the absence of internal funding, the BI was able to incubate more than 5 tenants using diverse funding sources within one year. On the other hand, university BIs in the development phase still relied their operations on grants and space rents, all of which raised only limited amount of fund. As a result, this incubator only incubated fewer than 5 tenants this year.

All BIs observed received funding from the government. However, it is important to retain that the grants offered by the government are competitive, which means that grants are only given to the BIs that meet the requirements in a certain period. The grant should be an additional source for operational funding: however, it is not meant to be the main source.

Therefore, from A10-A14 identified assumptions, it is visible in Figure 3 that A10 with the highest assurance intensity fortifies BIs' operational funding. This review is based on the fact that the majority of BIs in Indonesia are still on the startup and development phases, where their capacity to gather external funding is still limited, implying that internal funding

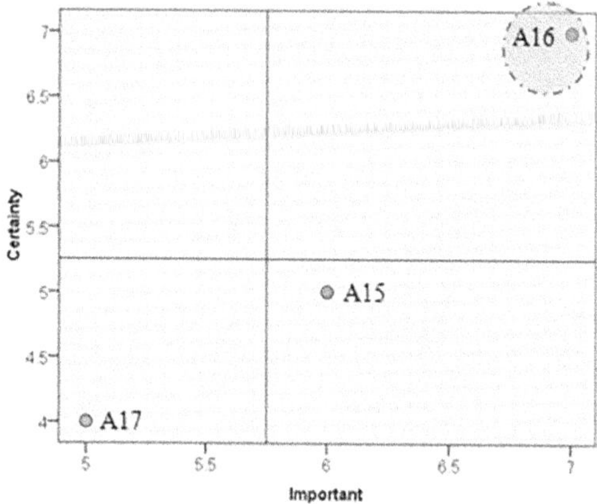

Figure 4. Priority network enhancement.

is highly prioritized. Back then Bank Indonesia (2006) also suggested the same concern, that commitment on continuous internal funding support from the chairman, mostly in universities, plays a critical role in the BI's sustainability.

3.6.4 *Networking barrier*

It is impossible for a BI to operate on its own in furnishing a range of programs for the tenants. Todorovic and Moenter (2010) emphasizes the significance of building network as well as increasing the number of stakeholder contacts, which the tenants might need in the future. Beate Pettersen *et al.* (2016) explains that potential networks to build are, among others, R&D divisions, public bodies, law firms, regional network organizations, investor groups, technology transfer office and diverse industry contacts and networks.

The next follow-up step is how to build a network with the above stakeholders. BIs have to be proactive in maintaining the relationships with the stakeholders. 100% of the surveyed BIs are members of BI associations in Indonesia, and some are even members of international associations. Registering to an association enables them to easily communicate with other BIs as well as nurture communication channels with the stakeholders.

Actively endorsing BIs is deemed potential to widen the networks. BIs' activity and achievement promotions through website page is a low-cost promotion tool. About 75% of the BIs studied already had their own website. Some of them even shared information on tenants' activities as well as their products. The remaining 25% of the BIs which did not have their own website were university BIs. Other promotional efforts that were already implemented by all the BIs are attending and enlisting to exhibitions, workshops or training. Even some BIs conducted business gatherings to bring tenants together with stakeholders from industry, government and investors.

The above efforts are in accordance with the experts' judgment. Out of the three identified alternatives, it is visible in Figure 4 that the experts believe that the policy to promote BIs to stakeholders is an utmost alternative with highest intensity of significance and assurance.

4 CONCLUSION

The findings of this research have shown that the services provided by BIs are still below tenants' expectation. It has been identified that there are four BI weaknesses in providing tenant assistance, which are space, personnel, operational funding, and networking. To overcome the space issue, the experts suggested that BIs submit a proposal on space expansion to the institution head. Furthermore, as for improving BI managers' capacity, each BI is suggested to hire a full-time manager. This becomes an essential step to do, considering the lack of full-time managers in university BIs, while the majority of BIs in Indonesia stand under the shelter of universities.

BI's ability to gather funding sources accordingly relates to the number of the managed tenants. The priority highlights the fact that the majority of BIs in Indonesia are still on the startup and development phases, while operational funding issue can be resolved through funding support commitment from the institution head. To improve networking, BIs have to proactively promote their services by advertising their activities on their websites and organizing business gatherings, workshops, exhibitions and training.

Finally, to improve BIs' role in Indonesia, the government needs to focus on endorsing their existence by conducting development programs such as granting programs to BIs on the startup and development phases. University BIs also need to be prioritized, considering the fact that their proportion is the biggest out of all the BIs in Indonesia.

REFERENCES

Aernoudt, Rudy. 2004. "Incubators: Tool for Entrepreneurship?" *Small Business Economics* 23 (2): 127–35. doi:10.1023/B:SBEJ.0000027665.54173.23.

Al-Mubaraki, Hanadi Mubarak, Ali Husain Muhammad, and Michael Busler. 2015. "Categories of Incubator Success: A Case Study of Three New York Incubator Programmes." *World Journal of Science, Technology and Sustainable Development* 12 (1): 2–12. doi:10.1108/WJSTSD-06-2014-0006.

Al-Mubaraki, Hanadi Mubarak, and Michael. Busler. 2010. "Business Incubators: Findings from a Worldwide Survey, and Guidance for the GCC States." *Global Business Review* 11: 1–20. doi:10.1177/097215090901100101.

Bank Indonesia. 2006. *Kajian Inkubator Bisnis dalam Rangka Pengembangan UMKM [Business Incubator Research on MSME Development]*. Jakarta: Author. http://www.bi.go.id/id/umkm/penelitian/nasional/kajian/Pages/riil6.aspx.

Beate Pettersen, Inger, Jarle Aarstad, Øystein Stavø Høvig, and Anita Ellen Tobiassen. 2016. "Business Incubation and the Network Resources of Start-Ups." *Journal of Innovation and Entrepreneurship* 5:7 (1): 1–17. doi:10.1186/s13731-016-0038-8.

Bungin, H.M. Burhan. 2007. *Penelitian kualitatif: Komunikasi, ekonomi, kebijakan publik dan ilmu sosial lainnya [Qualitative research: Communication studies, economics, public policy and other social sciences]*. Jakarta: Kencana.

Calvo, José L. 2006. "Testing Gibrat's Law for Small, Young and Innovating Firms." *Small Business Economics* 26 (2): 117–23. doi:10.1007/s11187-004-2135-5.

Coordinating Ministry for Economic Affairs. 2015. *Laporan Kelompok Kerja Pengembangan Inkubator Wirausaha [A Working Group Report on Business Incubator Development]*. Jakarta: Author.

Hasbullah, Rokhani, Memen Surahman, Ahmad Yani, Deva P. Almanda, and Elisa N. Faizaty. 2015. "Role of University Business Incubators on the Improvement of Food SMEs Business Performances." *Jurnal Ilmu Pertanian Indonesia 20*(1): 59-65. http://ilkom.journal.ipb.ac.id/index.php/JIPI/article/download/9290/7288.

INBIA [International Business Inovation Association]. 2016. *Business incubation FAQ*. www.inbia.org/resources/business-incubation-faq.

Lalkaka, Rustam. 2001. "Best Practices in Business Incubation: Lessons (yet to be) Learned." Paper presented at European-Union- Belgion Presidency International Conference on Business Centers: Actors for Economic and Social Development, Brussels, November 14-15, 2001. doi:10.1111/j.1537-2995.2009.02262.x.

Meru, Abel Kinoti, and Miemie Struwig. 2011. "An Evaluation of the Entrepreneurs' Perception of Business-Incubation Services in Kenya." *International Journal of Business Administration 2*(4): 112. http://dx.doi.org/10.5430/ijba.v2n4p112.

Ministry of Cooperatives and SMEs. 2015. *Laporan Tim Hasil Pelaksanaan Kerja: Tim Kelompok Kerja Pengembangan Inkubator Wirausaha*. Report from business incubator development working group. Jakarta, December 31, 2015. https://www.ekon.go.id/ekliping/download/1965/1401/pengembangan-inkubator-usaha.pdf.

Mitroff, Ian I., and James. R. Emshoff. 1979. "On Strategic Assumption-Making: A Dialectical Approach to Policy and Planning." *Academy of Management Review* 4 (1): 1–12. doi:10.5465/AMR.1979.4289165.

Mitroff, Ian I., and Richard O. Mason. 1981. "Dialectical Pragmatism: A Progress Report on an Interdisciplinary Program of Research on Dialectical Inquiring Systems." *Synthese* 47(1): 29-42. http://www.jstor.org/stable/20115616.

Mitroff, Ian I., James R. Emshoff, and Ralph H. Kilmann. 1979. "Assumptional Analysis: A Methodology for Strategic Problem Solving." *Management Science* 25 (6): 583–93. doi:10.1287/mnsc.25.6.583.

M'Chirgui, Zouchaier. 2012. "Assessing the Performance of Business Incubator: Recent France Evidence." *Business and Management Research 1*(1): 62. http://dx.doi.org/10.5430/bmr.v1n1p62.

Scaramuzzi, Elena. 2002. *Incubators in Developing Countries: Status and Development Perspectives*. Washington DC: World Bank. http://documents.worldbank.org/curated/en/186751468770425799/pdf/266370WP0Scode090incubators0Infodev.pdf.

Schaper, Michael, and Thierry Volery. 2007. *Entrepreneurship and Small Business*. Australia: John Wiley & Sons.

Syamas, A., F., *et al.* (2015). Model Inkubasi Bisnis Teknologi di Lingkungan Industri Kecil Menengah pada Perusahaan Pemula Berbasis Teknolgi [Technology business incubation model among technology-based startup SMEs]. the Ministry of Research, Technology and Higher Education: Jakarta.

Tambunan, Tulus. 2007. "Entrepreneurship Development: SMEs in Indonesia." *Journal of Developmental Entrepreneurship* 12(01): 95-118. http://dx.doi.org/10.1142/S1084946707000575.

The World Bank. 2010. *Innovation Policy: A Guide for Developing Countries*. Author: Washington, DC. https://openknowledge.worldbank.org/bitstream/handle/10986/2460/548930PUB0EPI11C10Dislosed061312010.pdf?sequence = 1.

Todorovic, Zelimir William, and Katherine Moenter. 2010. "Tenant Firm Progression within an Incubator: Progression toward an Optimal Point of Resource Utilization." *Academy of Entrepreneurship Journal* 16 (1): 23–40.

Tola, Alessio, and Maria Vittoria Contini. 2015. "From the Diffusion of Innovation to Tech Parks, Business Incubators as a Model of Economic Development: The Case of 'Sardegna Ricerche.'" *Procedia - Social and Behavioral Sciences* 176 (February): 494–503. doi:10.1016/j.sbspro.2015.01.502.

Xu, Lilai. 2010. "Business Incubation in China: Effectiveness and Perceived Contributions to Tenant Enterprises." *Management Research Review* 33 (1): 90–99. doi:10.1108/01409171011011599.

——. 2013. "The Effect of Business Incubation in Developing Countries." *European Journal of Business and Innovation Research* 1 (1): 19–25.

——. 2014. "Beyond Incubators Mechanisms: Innovation, Economic Development and Entrepreneurship." Paper presented at the *Proceedings of the 9th European Conference on Innovation and Entrepreneurship (*ECIE), September 18–19, 2014. Belfast, United Kingdom.

Competition and Cooperation in Economics and Business – Gani et al. (Eds)
© 2018 Taylor & Francis Group, London, ISBN 978-1-138-62666-9

Formal sector is not so formal anymore: Informality of work in the formal sector among ethnics

O. Herwantoko
Ministry of Manpower, Republic of Indonesia

D. Handayani
Department of Economics, Faculty of Economics and Business, Universitas Indonesia, Depok, Indonesia

R. Indrayanti
Institute of Demography, Faculty of Economics and Business, Universitas Indonesia, Depok, Indonesia

ABSTRACT: This paper analyses the informality of work in the formal sector in the era of globalisation. The concept of informal worker in this research is different from a number of previous studies using a sectoral-based economy definition. Informal workers or workers who experience informality of work in the formal sector are defined as workers with employee status, have working hours below 35 hours a week, are paid less than the minimum provincial wages, do not have employment contracts/have employment contracts but without a clear time duration, and do not have social security. By using Indonesian Family Life Survey (IFLS) 2014 data and binary logistic regression analysis, this study aims to examine the influence of ethnic identity and socio-demographic factors (such as gender, education level, marital status, age group, economic sector, rural-urban and Java–Outer regional) towards the tendency to experience informality of work in the Indonesian formal sector. The results indicate that ethnic identity and socio-demographic factors are simultaneously significant in determining the tendency of experiencing informality of work in the formal sector. The policy implication that can be suggested are in refocusing labour policy related to working time, minimum wages, employment contracts and social security, to workers with the following characteristics: non-local ethnic, young, less educated and married women, and engaged in the secondary and tertiary sectors located in the urban-Java region.

1 INTRODUCTION

The discourse about informal workers is very dynamic, particularly in the context of socio-economic development in the Third World countries (developing countries). Entering the 2000s, there was a fairly fundamental change in terms of informal workers. According to Hussmanns (2001), informal workers are not only understood to be workers engaged in the informal sector, but also workers who are working in informal forms of employment/jobs. These forms can also be found in the economic sectors considered formal. Workers doing informal work are defined as follows:

> '*Employees are considered to have informal jobs if their employment relationship is not subject to standard labor legislation, taxation, social protection or entitlement to certain employment benefits. Reason may include the following: the employee or the job is undeclared; the job is a causal or of a short duration; the hours of work or wages are below a certain threshold (e.g. for social security contribution); or the employer is an unregistered enterprise or a person in a household; or the employee's place of work is outside of premises of the employer's or the customer 'enterprise (e.g. outworkers without employment contract).*' (Hussmanns, 2001).

In a sense like this, the notion about workers in the informal sector should be separated from the notion of workers doing informal jobs. This is because the forms, characteristics, and informal labour relations are not only in the informal sector, but also contained in the formal sector. Thus, it is less appropriate if informal workers are only identified as workers in the informal sector.

A more dynamic and analytical notion about the informality of work in the formal sector is presented by Chang (2009). Based on Hussmanns's (2001) definition, Chang saw symptoms of informalisation of work in line with the globalisation process. Informalisation of work is a phenomenon which refers to the expanding informality of work (forms of informal employment/jobs), even in a number of sectors that are considered formal. Informalisation of work is the consequence of capital movement in the context of spatial mobility between regions and between countries in the contemporary era. Capital movement stimulates the need for labour market flexibility on the demand side, where the form of informal employment/jobs can be more appropriate and more necessary to the development of flexibility compared to formal employment/jobs.

Based on Hussmanns's (2001) and Chang's (2009) perspective, the concept of informality of work in the formal sector in this study at least can be approximated operationally by indicators/dimensions as follows:

- status of workers as employees,
- work under normal working time (less than 35 hours a week),
- paid less than the Provincial Minimum Wages (UMP),
- without an employment contract, or with an employment contract but without a clear time duration, and
- lack of social security.

Regarding the conditions of the informality of work, besides considering determinants from the demand side, we also must consider determinants from the supply side. On the supply side, according to Williams and Lansky (2013), the recent development of theoretical perspectives on informal workers or informality of work is not only concerned about economic issues per se, but also cross-disciplinary factors including socio-cultural ones, as expressed by the postmodern perspective. Related to that condition, a number of socio-demographic factors can be identified in the supply side, such as: age group, gender, rural-urban, marital status, and education level (Shehu & Nilsson, 2014; Warouw, 2008; Chen, 2012; Anker et al., 2003). In addition to these factors and in the context of Indonesia, informality of work in the formal sector is also related to socio-cultural factors, one of which is ethnicity. For example, according to Soedjatmoko (1984), Javanese people tend to want to work as employees, particularly in the public sector (government), because it is associated with their views or attitudes of life and the structure of social stratification of their society. Moreover, social exclusion based on ethnicity is also related to informality of work in the context of contemporary regional autonomy (Minza, 2016).

2 RESEARCH GOALS, DATA AND METHOD

Based on that background, by using the Indonesian Family Life Survey (IFLS) 2014 data and binary logistic regression analysis, this paper aims to examine the influence of ethnic identity and socio-demographic factors (such as gender, education level, marital status, age group, economic sector, rural-urban and Java—Outer regional) towards the tendency to experience informality of work in the Indonesian formal sector. In this study, categorisation of ethnic identity is related to the dimensions of 'homeland' (Green, 2016). This is because of the condition of ethnicity in Indonesia, where most ethnics have a 'homeland', such as the Javanese in Central and East Java, the Sundanese in West Java, the Batak people in North Sumatra, the Dayak people in Kalimantan. It is also because of the social exclusion of non-local ethnics that are related to informality of work (Minza, 2016). However, because of limited data, the categories of ethnicity-based homeland cannot be done at the district/city level, but at the regionality-island level (Lan et al., 2010; Schiller & Caglar, 2009). Under these conditions, ethnicity variables in this study are divided into categories:

1. Local Ethnics: people from an ethnic group living in a region which is an island of their ethnic homeland, such as: the Javanese or Sundanese people living on the regional island of Java;
2. Non-Local Ethnics: individuals from an ethnic group living in a region which is not an island of their ethnic homeland, such as: the Javanese or Sundanese people living on the island of Sulawesi.

The unit of analysis in this study is workers with employee status and aged between 15 and 64. According to IFLS 2014 data, the total number of employees with that status and having complete information related to this study is 9,369 people.

3 MAIN FINDINGS

The main findings in this study are that at a significance level of 5 per cent, ethnic identity and socio-demographic factors are simultaneously significant in determining the tendency to experience informality of work in the formal sector. Workers who tend to experience informality of work have the following characteristics: non-local ethnics, local ethnics in the secondary sector, young less educated and married women, workers who are engaged in the secondary and tertiary sector, and located in the urban-Java region.

Table 1 shows that ethnicity becomes significant in determining the tendency of experiencing informality of work after interaction between the variables of ethnicity and economic sectors are included in the analysis. This condition implies that the ethnic variable cannot stand alone as a determinant of labour informality, but must interact first with economic sectors as a determinant. In general, non-local ethnic workers have 0.7 times lower opportunities to experience the informality of work than the local ethnic workers. However, local ethnic workers in the secondary sector have 2.2 times higher opportunities to experience the informality of work than others.

Table 1. Binary logistic regression output.

Determinant/Categorical scale	B	S.E.	Wald	df	Sig.	Exp(B)
Ethnicity (1 = Local Ethnic, 0 = Non-Local Ethnic)	−0.337	0.117	8.361	1	0.004	0.714
Ethnicity and Economic Sector (1 = Local Ethnic in secondary sector, 0 = Other)	0.774	0.206	14.188	1	0.000	2.169
Gender Identity (1 = Woman, 0 = Man)	−1.344	0.077	302.105	1	0.000	0.261
Preschool Level (1 = Preschool, 0 = Other)	0.116	0.092	1.572	1	0.210	1.122
Basic Education Level (1 = Basic, 0 = Other)	0.911	0.107	72.436	1	0.000	2.488
Intermediate Education Level (1 = Intermediate, 0 = Other)	0.563	0.080	49.626	1	0.000	1.756
Age group (1 = 15–24/youth, 0 = Other)	−0.336	0.136	6.090	1	0.014	0.714
Age group and gender identity (1 = Woman in youth group, 0 = Other)	0.431	0.181	5.679	1	0.017	1.539
Marital Status (1 = Married, 0 = Other)	0.433	0.126	11.830	1	0.001	1.542
Marital Status and gender identity (1 = Married woman, 0 = Other)	0.699	0.167	17.576	1	0.000	2.013
Rural-urban (1 = Urban, 0 = Rural)	0.592	0.066	80.321	1	0.000	1.808
Regional (1 = Java, 0 = Outer Java)	0.514	0.068	57.220	1	0.000	1.672
Secondary sector (1 = Secondary, 0 = Other)	0.465	0.074	39.506	1	0.000	1.593
Tertiary sector (1 = Tertiary, 0 = Other)	0.472	0.191	6.131	1	0.013	1.603
Constant	−3.529	0.266	176.339	1	0.000	0.029

Source: IFLS 2014, analysed with Statistical Package for Social Science (SPSS).

There are a number of explanations for this. The first explanation is the existence of social exclusion in the socio-economic structure of the sectors which are segregated by ethnic identity lines. Such conditions generally make non-local ethnics experience labour informality and in particular local ethnics in the secondary sector. That condition is portrayed in Minza's study (2016) in Pontianak. This study describes the logic of ethnicity in the labour market in the city of Pontianak, where the young people are not so free in choosing a preferred job. Those conditions are not merely a matter of individual capacity, but also related to structural factors that shape the patterns and aspirations of their employment opportunities.

Furthermore, according to Minza (2016), in the era of regional autonomy, young people from the Malay ethnicity begin to feel concerned about the ethnic Dayak control of the government bureaucracy and access networks to obtain employment in that sector. In fact, before the era of regional autonomy, the Malay ethnicity was the dominant ethnic group in the government bureaucracy. In this present era, though able to enter the public sector, the people of Malay ethnicity worry that they will only get jobs that are 'not important', while the 'important work' is controlled by the ethnic Dayak.

Meanwhile, the Chinese ethnics in the city tend to be reluctant to try to work in the government bureaucracy. They think that it is only natives who can become civil servants (PNS), and that non-natives, such as Chinese, are obviously going to be rejected. Therefore, the majority of Chinese people in the city are found to be in private companies, particularly in the trade sector. They also have to rely on a network of friends of their ethnicity to enter as workers into a private company (Table 2).

Social networks in the process of recruiting workers determine the structure of employment opportunities and play an important role in the presence of social exclusion based on ethnicity in the formal sector. This does not only happen in the city of Pontianak. In Indonesia generally, social networking (especially in ethnic networks formed through friendship or relatives) has become an important instrument to get a job in the Indonesian formal sector (Table 3).

The mechanism of ethnic social networks in the process of worker recruitment, will make the individuals of a particular ethnicity tend to cluster in certain economic sectors as well. This creates segregation of the labour market along the lines of ethnicity. Consequently, if an individual works in the economic sector in which her/his ethnicity is not dominant, then she/he would get a 'less important job', so that the chance to experience informality of work is greater.

Such situations not only happen in Pontianak, but also in Indonesia in general. In Table 4, it can be seen that local ethnics in Indonesia tend to be dominant in the tertiary sector (52.66 per cent), while non-local ethnics tend to be dominant in the primary (27.94 per cent) and secondary sector (27.85 per cent). It can be said that the tertiary sector 'belongs' to the local ethnic, while the primary and secondary sectors 'belong' to non-local ethnics. This explains the greater chances for local ethnics to experience informality of work in the secondary sector in Indonesia. This happens because those individuals are not working in the sector in which their ethnicity is dominant, considering the strong ethnic social-based networking during recruitment or during work placement.

Table 2. Job aspirations of youth in Pontianak by ethnicity.

	Civil servant		Entrepreneur		Private worker		Total	
Ethnicity	N	%	N	%	N	%	N	%
Malay	140	91.5	10	6.5	3	2.0	153	100.0
Dayak	34	77.3	8	18.2	2	4.5	44	100.0
Madurese	19	95.0	1	5.0	0	0.0	20	100.0
Javanese	39	97.5	1	2.5	0	0.0	40	100.0
Chinese	2	4.4	30	66.7	13	28.9	45	100.0
Other	21	77.8	3	11.1	3	11.1	27	100.0
Total	255	77.5	53	16.1	21	6.4	329	100.0

Source: Minza (2016).

Table 3. Ethnicity characteristics according to instruments to get a job.

	Non-local ethnic		Local ethnic	
Instrument to get a job	N	%	N	%
Government job fair	45	4.29	700	9.05
Private job fair	20	1.91	174	2.25
Academic job fair	15	1.43	79	1.02
Advertise	28	2.67	222	2.87
Contact the company	212	20.23	1,634	21.13
Friendship or relatives	527	50.29	3,564	46.08
Contacted by a company	194	18.51	1,288	16.65
Outsourcing	7	0.67	69	0.89
Manpower bureau	0	0.00	4	0.05
Total	1,048	100.00	7,734	100.00

Source: IFLS 2014.

Table 4. Ethnicity characteristics according to the economic sector.

	Non-local ethnic		Local ethnic	
Sector	N	%	N	%
Primary	311	27.94	1,820	22.04
Secondary	310	27.85	2,088	25.29
Tertiary	492	44.20	4,348	52.66
Total	1,113	100	8,256	100

Source: IFLS 2014.

The second explanation is related to skills. According to Warouw (2016), in Cilegon City local residents face several problems to get the position of permanent workers in the secondary sector (especially manufacturing) because they do not have the skills. The various permanent worker positions tend to be filled by migrants from Sumatra, while local workers only work as temporary workers. These migrant workers live in groups and stay separated from the surrounding community. Consequently, ethnic networks are rebuilt and this recreates socio-economic segregation in Cilegon.

A third explanation relates to the cultural aspects of life that become the frame of reference for members of ethnic groups. One example of this condition is in the case of the Javanese. In the Javanese social stratification, the *priyayi* status is higher than the *santri* and *abangan* statuses (Endraswara, 2015; Geertz, 2013). The *priyayi* are identified, mainly, to work as public servants (bureaucrats), employees, scholars or teachers. Being a *priyayi* is a general orientation for the Javanese people. This explains the high desire of the Javanese to become civil servants in the contemporary era (Soedjatmoko, 1984). Empirically it is quite evident. From a total of 1,559 government employees in the unit of analysis in this study, the majority of government employees are Javanese (33.10 per cent). The orientation to become *priyayi* could prevent the Javanese from experiencing informality of work. This is because the percentage of government employees who experience informality of work is relatively lower than that of private employees. Based on IFLS 2014 data, from the total of 1,559 government employees, workers who experience informality of work comprise only about 11.67 per cent. Meanwhile, from the total of 7,810 employees in the private sector, workers who experience informality of work comprise around 14.29 per cent.

Regarding the influence of gender identity on the informality of work, the opportunity to experience the informality of work is 0.3 times lower for female workers than it is for male.

Based on IFLS 2014 data, this is likely to be due to the higher tendency for men to have additional work outside their main job than women. Such conditions are likely to reduce the allocation of time spent in the main job (under 35 hours/week). As a result, companies are more selective in moving male workers who meet the formality aspects of work to permanent positions. This condition has been strengthened by the conditions in which the majority of male workers who were respondents in this study were indicated as being manual workers. According to Nugroho and Tjandraningsih (2012), less skilled workers such as blue-collar workers tend to have a relatively limited choice compared to skilled workers. The bargaining power they have is relatively weaker. Furthermore, at the corporate level, unskilled workers usually are non-permanent workers with low wages who do not have adequate working facilities or employment benefits.

However, for female workers aged 15–24 years, the chances of experiencing informality of work is actually 1.5 times higher than for others, and for married women, the opportunity to experience the informality of work is 2 times higher than others. One of the things that could explain this condition is the existence of a patriarchal culture in Indonesian society. In the patriarchal society, there tends to be a stereotype of the women's role. In the context of the labour market, Anker et al. (2003) says that the stereotype of the central role of women in the family influences their participation and their conditions in the job market. In addition, the division of traditional roles in which women are positioned as a wife and a mother while men work, also affects the differences in levels of education between men and women, in which men tend to have higher education levels than women. In fact, as portrayed in Table 1, higher education tends to reduce the risk of experiencing informality of work.

The Anker et al. (2003) explanation of women's education has indeed been proven empirically. Based on IFLS 2014 data, it can be seen that the higher the education level, the larger the percentage of male workers relative to female workers. At preschool level, the percentage of male workers is about 48.11 per cent, while female workers reach 51.89 per cent. But at the level of high education this percentage is reversed, with male workers reaching 67.79 per cent, whereas female only 32.21 per cent.

Furthermore, according to Suryakusuma (2012) the gender-based occupational segregation is very important in the Indonesian manufacturing sector. It is necessary to ensure that the scale of the wages of men is higher than that of women. This is considered reasonable because the man is the head of the household. In addition, the condition of women as temporary workers is also vulnerable. This non-permanent worker status can make companies shy away from the responsibilities of wage, social security and the most basic rights related to welfare, such as maternity leave and health benefits.

Related to rural-urban, regionality and economic sector, workers who live in the urban area and Java region, particularly those working in the secondary and tertiary sectors, have a higher chance of experiencing informality of work. Explanations for this matter can also be inferred from the results of Warouw's study (2016) in Cilegon City. According to him, the open market in the global era has made the practice of hiring temporary workers through third parties more important for industries in Cilegon. Such third parties may be formal, such as outsourcing, and informal, such as through networks of friends, the pattern of patron-client through figures in the surrounding communities, and youth organisations.

These non-permanent workers do routine work, ranging from the non-productive duties (office boy, janitor, gardener) to work in factories. They are generally not protected, in the sense that their employment relationship is outside the labour laws. They do not have access to a normal promotional mechanism and decent work conditions, such as: work safety facilities, uniforms, and food, as well as bonuses, allowances, pension funds or even the minimum wages earned by permanent workers. This situation is more common in the urban-Java region, because this region, historically, is the main motor of economic growth in Indonesia (Rahardjo, 1986; van Zanden & Marks, 2012; Kaur, 2004; Dick et al., 2002).

Regarding such conditions, Chang (2009) explains that informality of work is attached to the mobility of capital in the global era. Meanwhile, sectors that are relatively more associated with capital mobility are the secondary and tertiary sectors. In the *'global factory'*, a number of different manufacturing industries are connected in the global production chain. Fluidity of capital movement in the global era are facilitated by advances in the tertiary sector, particularly technological advances which then push the financial services sector growth. Changes in the global level then requires the presence of a more flexible labour market. This gives rise to the phenomenon of informality of work in the global era and it is not surprising that workers in the secondary and tertiary sectors experience that condition. Such situations happen because the secondary sector and the tertiary sector are relatively more globalised than the agricultural sector. And in the case of Indonesia, this sector is usually found in the urban-Java region.

Considering these conditions, it is interesting to look at the arguments of Bradley and Healy (2008) about the types of labour market flexibility that are associated with nonstandard/non-formal employment relationships. The types of flexibility are:

- Functional flexibility: the ability to move workers among a number of types of work and functions,
- Numerical flexibility: the ability to add and reduce workers in response to fluctuations on the demand side, and
- Financial flexibility: the ability to raise and lower the income of workers in response to price fluctuations on the demand side.

According to Bradley and Healy (2008), functional and numerical flexibility is usually found in the tertiary sector and this condition is likely to be experienced by ethnic minorities, women (especially married or with children) and young workers. Bradley and Healy's (2008) argument conforms with the conditions of Indonesia, where informality of work is experienced more by non-local ethnic, women and younger age groups, especially in the tertiary sector. That means that the type of labour market flexibility in Indonesia is functional and numerically flexible.

4 CONCLUSION

Based on the analysis, a number of important points can be generated as a conclusion. The first is related to socio-cultural factors. In the case of ethnicity, this factor does not stand alone in influencing the informality of work, but interacts with the economic sector variables. In addition, the systems of regional autonomy and culture in Indonesia also become important contextual factors to be observed. Second, the age group variable does not affect the informality of work directly, but must interact with gender identity variables as socio-cultural factors. Third, marital status variables do not affect informality of work directly, but must interact with gender identity variables. From such conditions, it can be seen that their gender identity, especially for women, is very important in influencing the informality of work. This is consistent with Chen's (2012) argument that informality is closely related to women workers. Fourth, the importance of the contextuality factor in Indonesia. There are three elements of contextuality aspects: globalisation (labour market flexibility in terms of functional and numerical flexibility), decentralisation, and the local-culture system. Therefore, it can be said that the informality of work is a phenomenon that reflects the dynamics between macro-level (globalisation), meso-autonomy regional and the local socio-cultural system. Based on that conclusion, the policy implications that can be established are as follows:

- Refocusing labour policy related to working time, minimum wages, employment contracts and social security, to workers with the following characteristics: non-local ethnics, young less educated and married women, engaged in the secondary and tertiary sectors located in the urban-Java region.
- Building an integrative incentive scheme for the secondary and tertiary sectors in the Java region to increase their competitiveness in order to increase the working conditions for the workers.
- Disseminating gender mainstreaming, equal employment opportunities and decent work in local governments related to manpower development, workers and employers.
- Multicultural mainstreaming in the labour market.

REFERENCES

Anker, R., Melkas, H. & Korten, A. (2003). *Gender-Based occupational segregation in the 1990's*, Geneva: International Labour Organization (ILO).

Bradley, H. & Healy, G. (2008). *Ethnicity and gender at work: Inequalities, careers and employment relations*. New York: Palgrave MacMillan.

Chang, D.O. (2009). Informalising labor in Asia's global factory. *Journal of Contemporary Asia*, *39*(2) 161–179.

Chen, M.A. (2012). *The informal economy: Definitions, theories and policies*. USA: WIEGO.

Dick, H., Vincent J.H. Houben, J. Thomas Lindblad, Thee Kian Wie. (2002). *The emergence of a national economy: An economic history of Indonesia, 1800-2000*. Australia: Allen & Unwin.

Endraswara, S. (2015). *Etnologi Jawa (Java Ethnology)*. Jakarta: Center for Academic Publishing Service.

Geertz, C. (2013). *Agama Jawa: Abangan, Santri, Priyayi dalam kebudayaan Jawa (The Religion of Java: Abangan, Santri, Priyayi in the Javanese Culture)* (Aswab Mahasin & Bur Rasuanto, Penerjemah). Jakarta: Komunitas Bambu.

Green, E.D. (2016). *Redefining ethnicity*. UK: Development Studies Institute, London School of Economics.

Hussmanns, R. (2001). *Informal sector and informal employment: Elements of conceptual framework*. Geneva: ILO.

Kaur, A. (2004). *Wage labour in Southeast Asia since 1840: Globalization, the international division of labour and labour transformation*. New York: Palgrave MacMillan.

Lan, T.J., Dedi S. Adhuri, Achmad Fedyani Saifuddin, Zulyani Hidayah. (2010). *Klaim, kontestasi dan konflik identitas: Lokalitas vis a vis nasionalitas (Claim, contestation, and conflict of identity: Locality vis a vis nationality*). Jakarta: Institut Antropologi Indonesia.

Minza, W.M. (2016). *Etnisitas dan cita-cita kerja orang muda di Pontianak (Ethnicity and working ambition of young people in Pontianak)*. In G. van Klinken & W. Berenschot (Eds.), *In search of Middle Indonesia: Kelas Menengah di Kota-Kota Menengah* (Edisius Riyadi Terre, Penerjemah.) (pp. 141–167). Jakarta: KITLV-Jakarta dan Yayasan Pustaka Obor Indonesia.

Nugroho, H. & Tjandraningsih, I. (2012). *Rezim fleksibilitas pasar kerja dan tanggung jawab negara (Labour market flexibility regime and state responsibility). Memetakan Gerakan Buruh*, 100–133.

Rahardjo, M.D. (1986). *Transformasi pertanian, industrialisasi dan kesempatan kerja (Agriculture transformation, industrialisation, and job opportunity)*. Jakarta: UI-Press.

RAND. Indonesian Family Life Survey (IFLS). Available: http://www.rand.org/labor/FLS/IFLS.html.

Schiller, N.G. & Caglar, A. (2009). Towards a comparative theory of locality in migration studies: Migrant incorporation and city scale. *Journal of Ethnic and Migration Studies*, *35*(2), 177–202.

Shehu, E. & Nilsson, B. (2014). *Informal employment among youth: Evidence from 20 school-to-work transition surveys*. Geneva: ILO.

Soedjatmoko. (1984). *Etika pembebasan: Pilihan karangan tentang agama, kebudayaan, sejarah dan ilmu pengetahuan (Liberation ethic: Selected essays on religion, culture, history and science)*. Jakarta: LP3ES.

Suryakusuma, J. (2012). *Agama, seks, & kekuasaan* (*Religion, sex, & power*). Jakarta: Komunitas Bambu.

van Zanden, J.L. & Marks, D. (2012). *Ekonomi Indonesia 1800–2010: Antara drama dan keajaiban pertumbuhan (The Indonesian Economy 1800-2010: Between drama and miracle of growth)*. Jakarta: Penerbit Buku Kompas.

Warouw, N. (2008). Industrial workers in transition: Women's experience of factory work in Tangerang. In M. Ford & L. Parker (Eds.), *Women and Work in Indonesia* (pp. 104–119). New York: Routledge.

Warouw, N. (2016). Meninjau kembali kelas pekerja: Relasi kelas di kota-kota menengah Indonesia. In G. van Klinken & W. Berenschot (Eds.), *In Search of Middle Indonesia: Kelas Menengah di Kota-Kota Menengah* (Edisius Riyadi Terre, Penerjemah) (pp. 61–83). Jakarta: KITLV-Jakarta dan Yayasan Pustaka Obor Indonesia.

Williams, C.C. & Lansky, M. (2013). Informal employment in developed and developing economies: Perspectives and policy responses. *International Labour Review*, *152*,(3–4), 355–380.

Competition and Cooperation in Economics and Business – Gani et al. (Eds)
© 2018 Taylor & Francis Group, London, ISBN 978-1-138-62666-9

Determinant factors for a successful collaboration in the leading research program between university and industry

M. Mutiara & H. Susanti
Department of Economics, Faculty of Economics and Business, Universitas Indonesia, Depok, Indonesia

ABSTRACT: Various empirical studies have shown that University-Industry Collaboration (UIC) is one of the driving forces in accelerating economic growth as indicated by the shifting role of university as a producer of knowledge. Therefore, this study aims to analyze the framework and pattern of collaboration between university and industry under the RAPID program (Riset Andalan Perguruan Tinggi dan Industri/ Leading Research between University and Industry) as well as the factors that determine the success of such collaboration. Through a qualitative and quantitative analysis, the study reveals: (1) that the collaboration under the RAPID program is predominantly distributed among universities in Java in comparison to universities located outside Java, (2) it also underlines the importance of having a database that matches the needs of industries and the research offered by universities to strengthen the networking between universities and industries and to further encourage collaboration between them, and (3) the importance of government support by providing research funds as an incentive to foster collaboration between university and industry.

1 INTRODUCTION

The impending ASEAN Economic Community (AEC) or Masyarakat Ekonomi ASEAN (MEA) sends a clear signal to Indonesia that it needs to shift its economic base from being heavily dependent on natural resources and inexpensive labor to knowledge based economy. Track records of a number of countries that experienced a rapid economic growth show that these countries have banked their future on science and technology. In dealing with the challenges posed by AEC, Indonesia needs to reflect on the fact that those countries have successfully boosted their economic growth. One of the key factors of such a success is the role played by technology as indicated by the growing collaboration between university and industry.

In 2015, Indonesia gained the opportunity to increase the collaboration between university and industry. Under President Joko Widodo, the function of the Ministry of Research and Technology was merged with Higher Education (Kemenristekdikti). Such merger implicitly indicates that a university is no longer seen as a provider of knowledge; instead, it is also a contributor to the country's economic growth. The newly stated demand that universities should play a role in advancing economic growth is in line with the objectives of Riset Andalan Perguruan Tinggi and Industri, RAPID/Leading Research between University and Industry, a program under the auspices of the Directorate of Research and Community Dedication of the Ministry of Research, Technology and Higher Education.

Unfortunately, even though the program has been going on for 10 years, its impact is not significant enough. The expected results in the form of products or technologies that are ready for commercialization are only a few in numbers. Admittedly, a number of products have made it; however, most of them are still in the basic stages. Only a handful of the products are ready to be applied commercially by the relevant industries. Therefore, the impact of the RAPID program has not yet been materialized until now.

With that in mind, this study will attempt to answer the following questions: (1) What is the framework and pattern of collaboration between university and industry under the

RAPID program? (2) What are the factors that will determine the collaboration between university and industry under the RAPID program?

2 LITERATURE REVIEW

2.1 *Previous study*

The most common format of collaboration between university and industry is in the form of cooperative association, in which the university contributes the knowledge, and the industry assists the university during the innovation process by using its resources (Fiaz and Naiding 2012). Therefore, collaboration may be defined based on two fundamentals, the theory on mutual dependence and the theory on interaction. There are a number of perspectives on the theory of mutual dependence in literatures; however, Barringer dan Harrison (2000) have developed six perspectives of the theory that are widely in use today, among others. Nonetheless, this study will only cover three theories that are relevant to the aspect of collaboration under the RAPID program, and they are as follows:

1. Transaction Costs Economics (TCE)
   TCE starts with the assumption that a transaction is the basic unit in the analysis of economic relationship between organizations. The objective of such a relationship is to reduce production cost and improve efficiency.
2. Resources Dependency (RD)
   RD Theory explains the underlying motives of the university and industry to form a collaboration based on mutual dependency as a result of limited resources, wherein the resources needed by one party are owned by the other.
3. Organizational Learning (OL)
   Theory puts the emphasis on the role of knowledge in establishing and maintaining a competitive advantage. An organization that wishes to acquire a certain skill will seek a better opportunity to achieve its objective by forming a relationship with another organization worthy to be emulated.

2.2 *Empirical model*

The determining factors for RAPID's collaboration program refers to resource dependency theory (Barringer and Harrison 2000). Organizations will collaborate to reduce production costs (TCE), to overcome the limitation of resources (RD), and to learn specific skills from other organizations (OL). These theories only explain why university and industry interact with each other in terms of the relationship between organizations. Therefore to build the model, this study refers to previous studies that also analyze the university-industry collaboration and then adjusted it to university-industry collaboration under RAPID program.

Different approaches for different studies are required to measure the level of collaboration between university and industry. In contrast to Fiaz (2013) which uses a variable number of university R&D collaboration with industry as a proxy for the level of collaboration, this research uses other proxies to describe the level of university and industry collaboration in the RAPID program. This study uses the activity approach (Perkmann *et al.*, 2011) as a proxy for the following activities: financial assistance, collaborative research, transfer of knowledge, and transfer of technology.

Each collaboration shall be assigned its own weighted value that will be adjusted according to the hierarchal priority based on the inputs provided by AHP respondents as shown in Table 1.

After receiving its weighted value, the depth of collaboration will be measured using a modified Subramanyam (1983) method. The level of collaboration is the ratio between the number of collaborations between university and industry participating in the RAPID program and the total number of collaborations between university and industry for a given period, wherein the total number of collaborations will have a value of 100; therefore, the formula will be as follows:

Table 1. Hierarchal priority of collaboration.

Format of collaboration	Weighted value based on the hierarchal priority provided by AHP respondents
Transfer of Technology (TG): Developing a Commercial Product System Development	40
Collaborative Research (KP): Research Agreement between Institutions Infrastructure and Facility Support	30
Financial Assistance (BK): Financial Assistance	20
Transfer of Knowledge (TP): Training Consultation Human Resources Exchange Internship	10

$$C=\frac{1}{3}\sum_{t=1}^{3}\left[0.4\sum_{i=1}^{n}TGit+0.3\sum_{i=1}^{n}KPit+0.2\sum_{i=1}^{n}BKit+0.1\sum_{i=1}^{n}TPit\right] \quad (1)$$

where:

C = depth of collaboration between university and industry participating in the RAPID program, with a value between 0 to 1

TG = collaboration between university and industry in the form of transfer of technology in the RAPID program

KP = collaboration between university and industry in the form of research collaboration in the RAPID program

BK = collaboration between university and industry in the form of financial assistance in the RAPID program

TP = collaboration between university and industry in the form of transfer of knowledge in the RAPID program

t = year of collaboration in the RAPID program; t = 1–3

i = the number of collaborations for each type of collaboration (TG, KP, BK, TP); i = 1 – n

There are three levels of collaboration; low (C = 0.1–0.4), moderate (C = 0.5–0.7), and high (0.8–1). Measuring the level of collaboration is flexible because the weight value of each form of collaboration refers to the justification of respondent regarding to the importance of collaboration. Therefore when the weights of respondents changed, then the level of collaboration will change.

Referring to the study by Fiaz (2013), this reserach uses perceptions held by the university and industry with regard to the attribute of such collaboration, for instance, the perception held by the university and industry related to the objectives and orientations of each institution, the perception with regard to a promotional cost that must be paid to market a certain product, and the perception that there will be risks involved in research. A case specific to Indonesia is the study conducted by Moeliodihardjo, *et al.* (2012), which argues that one of the obstacles in the collaborative efforts between university and industry is the institutional lack of trust caused by the different objectives and orientations. Furthermore, Darwadi, *et al.* (2012) argue that one of the determinant factors before the two parties will collaborate is the cost for promoting a product followed by the possibility that the necessary research may fail because of the lengthy time and process.

A study by Fiaz (2013) states that one of the factors for a successful collaboration between university and industry is state support, either in the form of R&D funds or through other supporting policies. For that reason, this study will limit its focus on the impact of incentive funds as a form of government financial support for collaboration between university and industry under the RAPID program.

Soh and Subramanian (2014) in their study argue that one of the factors that will determine the success of collaboration between university and industry is the characteristics of that industry, namely the size of that industry and its age. The size of the industry in this study is based on the number of its workers/employees and will refer to the figures published by the National Statistical Bureau; meanwhile, the industry age is measured from the year of its incorporation until 2015. Besides the characteristics of industry, Perkmann *et al.* (2011) also states that the higher the frequency of interaction between university and industry, the stronger the collaboration is between them. Therefore, this research uses a variable interaction using a proxy number of meetings or coordination in universities and industry for one year.

3 RESEARCH METHODOLOGY

The study uses a combination of primary and secondary data. The secondary data were obtained by conducting literature research and reviews to obtain the general picture with regard to the collaboration between university and industry. Meanwhile, the primary data were obtained by conducting a two-stage questionnaires survey. At the first stage, the respondents were 47 universities and 48 companies as the recipients of RAPID funds for the 2013–2015 budget years of Kemenristekdikti, followed by the second stage survey (*Analytical Hierarchy Process*/AHP Survey) and indepth interviews to a number of universities and industries. Thirty-four of the total first questionnaires were returned, and they were from 26 universities and 8 companies. The subjects for the interviews and respondents for the second stage survey were selected from the respondents that returned the first stage survey. Based on those criteria, the chosen respondent was one manager from the industry partner and six head researchers from the universities. The selection was intended to ensure that the samples would represent the actual condition.

After collecting data, an analysis was conducted using a descriptive analysis and a multiple regression analysis (Gujarati and Porter, 2012a, 2012b) to determine the factors of university—industry collaboration. The model that will be built in this study is a modification of the study conducted by Fiaz (2013), Soh and Subramanian (2014), and Perkmann *et al.* (2011), and adjusted according to the actual condition of university and industry collaboration under the RAPID program, as follows:

$$\begin{aligned}\text{KOLUI} = \alpha_1 + \text{GOAL}\beta_1 + \text{PROM}\beta_2 + \text{RISK}\beta_3 + \text{FUNDS}\beta_4 + \text{SIZE}\beta_5 \\ + \text{AGE}\beta_6 + \text{INTERACTION}\beta_7 + \varepsilon\end{aligned} \tag{2}$$

where:

KOLUI (UI Collaboration) = the depth of collaboration between university and industry in the RAPID program with a value of 0–1.

GOAL = the perception held by the university/industry with regard to the objectives/orientations of the collaboration between university and industry with a scale of 1–4.

PROM (Promotion) = the perception held by the university/industry with regard to the cost to market the product or technology produced by the collaboration between university and industry with a scale of 1–4.

RISK (Risk) = the perception held by the university/industry with regard to the risk failure of the research/product produced by the collaboration of university—industry with a scale of 1–4.

DANA (R&D Fund) = financial assistance provided by the government in the form of R&D funds under the (Riset Andalan Perguruan Tinggi dan Industri (RAPID)/Leading Research between University and Industry) (in billion Rupiah).

SIZE (Size of the Industry) = classification of the industry according to the number of workers/employees based on the figures provided by BPS.

AGE (Age of the Industry) = the age of the industry calculated from its establishment until the year 2015 (year).

INTERACTION = the frequency of interaction between university and industry per year.

4 RESULT AND DISCUSSION

4.1 *Descriptive analysis of RAPID program*

The RAPID program began in 2004 as a forum, which provides an opportunity for realizing a synergic working relationship between one institution that produces concepts and technology or the university and the manufacturing institution or the industry (Direktorat Penelitian and Pengabdian kepada Masyarakat, 2013). Participation in the RAPID program is expected to bring and nurture the growth of a culture of research in the industrial world, wherein the ultimate objective is to create a continuous product invention. Furthermore, the program is also expected to nurture the growth of industrial culture that is responsive toward time within the university circle. Based on the surveys, the framework of collaboration between university and industry under the RAPID program can be discerned from Figure 1.

Under the RAPID program, the collaborating institutions will receive a grant ranging from Rp. 300,000,000 – Rp. 400,000,000. On the other hand, the industry is required to provide an in-cash fund at least for 25% of the total amount of the contract in addition to the in-kind payment, which will be formalized in a duly stamped document. The university may also contribute an in-cash fund of at least 15% of the total amount of the contract's value funded by the government in addition to the in-kind payment, which will be formalized in a duly stamped document. The formats of collaboration between university and industry under the RAPID program were divided into four classification (Table 2) where its activity such as meetings or coordination via e-mail or telephone. The frequency of the interaction ranges from once a month to once every 2–3 months in a year.

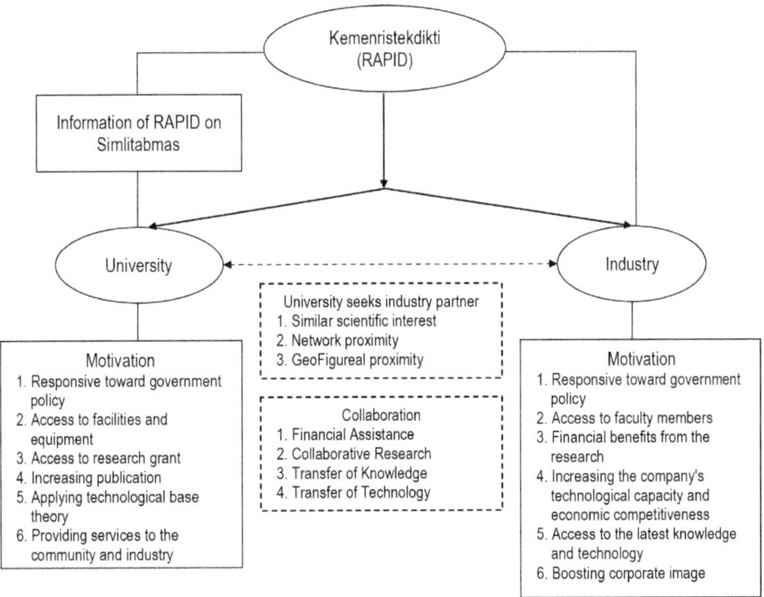

Figure 1. The framework of collaboration between university and industry under the RAPID program. Source: Survey RAPID, 2016.

Table 2. Formats of Collaboration between University and Industry under the RAPID program.

Format of Collaboration under the RAPID program	Remarks
Financial Assistance: Financial Assistance	What is referred to as financial assistance here is support in the form of funds, whether such fund is provided by industry or other party, such as the university or government.
Collaborative Research: Research agreement between institutions Infrastructure and facility support	What is referred to as collaborative here is an agreement between the university and industry to collaborate in research that encompasses the task and role of each institution including the expected benefits.
Transfer of knowledge: Training Consultation Human resources exchange Internship	What is referred to as transfer of knowledge is the sharing of knowledge by each institution in the form of training, consultation, exchanging human resources, and an internship program.
Transfer of technology: Developing commercial product Developing system	What is referred to as the transfer of technology is the common efforts to develop a certain product or system development up to the commercialization stage.

Source: Survey RAPID, 2016.

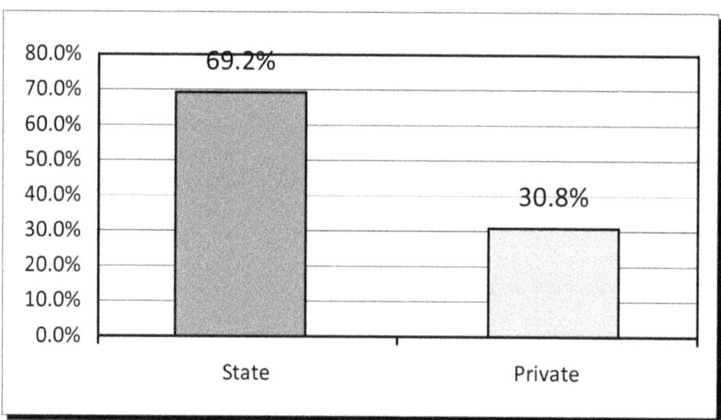

Figure 2. Profile of universities.
Source: RAPID questionnaires, 2016.

Based on data analysis, the distribution of RAPID recipients is dominated by state universities (Figure 2). This is because the major state universities, such as ITB, IPB, ITS, UI, UGM, and Unpad (Figure 3), already have a strong tradition of research and development.

From the perspective of the region where the recipients are located, the majority are from Java, specifically from the region of West Java at 41.2 percent (RAPID Survey 2016). This is mainly because of different capacity with regard to education infrastructure (Analytical and Capacity Development Partnership [Program Kemitraan untuk Pengembangan Kapasitas dan Analisis Pendidikan/ACDP, 2013]). Hence, universities in Java are stronger than their counterparts from outside Java. In addition to that, the lack of information also led to a smaller number of participants from outside of Java. Even though the information on the RAPID program was sufficiently socialized on the web through Sistem Informasi Manajemen Penelitian dan Pengabdian Masyarakat (Simlitabmas), nonetheless, it was not clear enough; hence, only a handful of participants from outside of Java took part in the program.

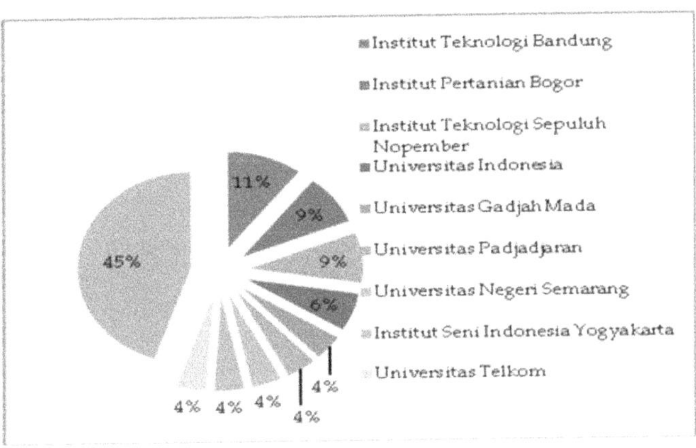

Figure 3. The big nine universities receivers of RAPID grants.
Source: RAPID questionnaires, 2016.

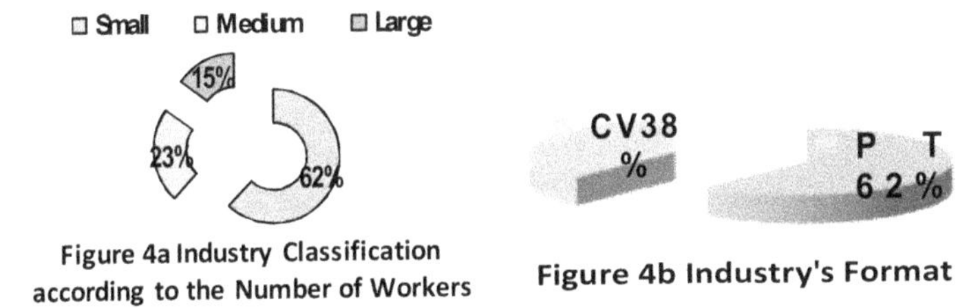

Figure 4. Characteristics of the industries participating in the RAPID program.
Source: Survey RAPID, 2016 (processed).

Figure 4 shows that the participating industries in the RAPID program were predominantly small industries (62 percent), while large-scale industries that took part in the RAPID program was around 15 percent. There were two main reasons for large-scale industries to take part in the program; besides demonstrating their understanding and responsibility for their active role in the world of education, the research offered by participating universities has a potential to be developed further and may provide the answers to the industry's needs.

Table 3 shows the output realization of each collaboration performed under the RAPID program for the 2013–2015 budget year. Even though the distribution of RAPID funds for collaboration between university and industry continues to rise each year, the expected benefit does not necessarily follow the trend, and this is because the RAPID program is actually a long-term endeavor (ACDP, 2013). Until now, the output realized by the program is still limited to basic research, such as an industrial product design, national and international publication, or being published as a textbook. However, many of them have to struggle when it comes to applied research or when preparing a prototype because for many of them it is a new venture.

Since the realized output under the RAPID program is still in a basic stage, therefore, the expected or desired output and benefit of the program have yet to materialize because the final products or technology that is ready for commercialization (applied research) is still in incubation. Even though some products have been produced commercially, there is no study that specifically analyzes the impact of the RAPID program. This is the result of the current evaluation and monitoring (monev) system used by the Kemenristekdikti that only covers the period of three years, i.e. while the collaboration is still ongoing. Therefore, to measure the impact of the RAPID program, it is deemed necessary to extend the monitoring and evaluation period for at least one year after the end of the program.

Table 3. Realization of RAPID output.

Collaboration between University and Industry under the RAPID program	Achieved	Under target	Over target	Remarks
1.	√			Patent Process
2.	√			Product Development Process
3.	√			International publication and prototype achieved
4.	√			Patent and Prototype Process
5.			√	Patent Process fulfilled and publication over target
6.			√	Some product has been manufactured
7.	√			Production achieved and other output still undergoing process
8.	√			Patent: Registration number has not been informed (Winner of HKI KemenristekDikti 2015)
9.			√	Patent fulfilled, over target.
10.			√	Prototype process
11.			√	Patent fulfilled, over target.
12.	√			Publication target achieved
13.	√			Patent process
14.			√	Publication over target
15.	√			Patent process
16.	√			Patent process
17.	√			Prototype process
18.	√			Patent process
19.	√			Prototype process
20.	√			Patent process
21.	√			Patent process
22.			√	Preparation is underway to file a product patent
23.	√			Prototype and publication achieved
24.	√			Prototype process
25.	√			Patent and trade mark achieved
26.	√			Prototype process

Source: Survey RAPID, 2016.

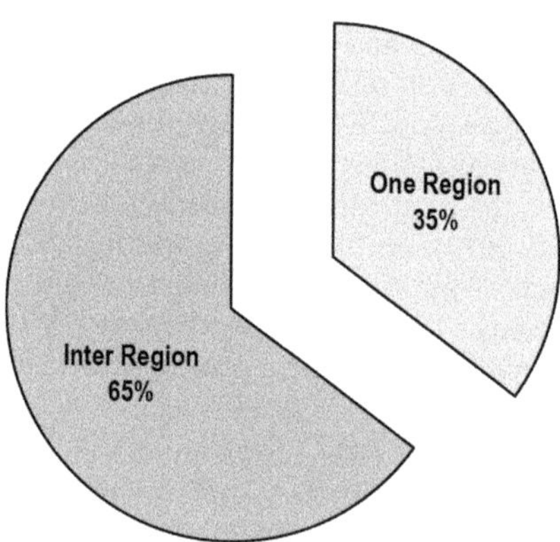

Figure 5. Collaboration between university and industry under the RAPID program (inter region).

Generally speaking, the pattern of collaboration between university and industry under the RAPID program is somewhat random, wherein the university has to seek an industry willing to establish a collaboration with them. Some of the universities selected their industry

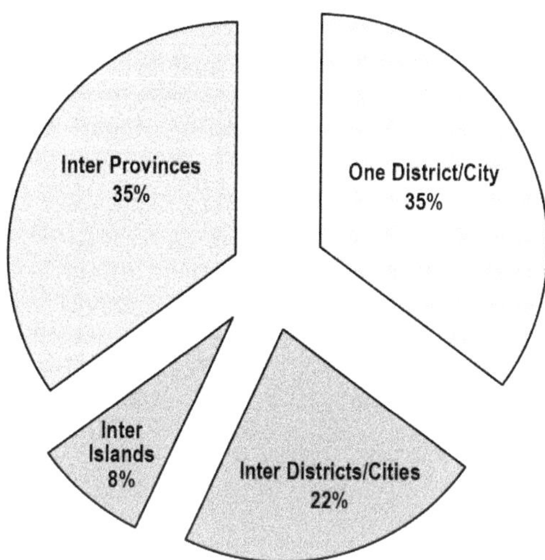

Figure 6. Collaboration between university and industry under the RAPID program (across islands).
Source: *Direktorat Riset dan Pengabdian Masyarakat, 2015 (processed).*

partners simply based on geographical proximity, while others selected their collaboration partners from other regions. Figure 5 shows that around 65 percent of the collaboration between universities and industries were from different regions (inter region). The remaining thirty five percent were formed by universities and industries located in the same region. As for inter region collaborations, 35 percent of them were across provinces, 22 percent across districts/cities, and 8 percent of them were across islands (Figure 6).

4.2 *Factors that determine the success of collaboration between university and industry under the RAPID program*

Using ordinary least squares (OLS), several models obtained are as follows (Table 4):

The results in model 6 (Table 4) shows that the perception related to the objectives and orientations of each institution does not have a significant effect on the degree of collaboration between university and industry under the RAPID program. This result is unlike the findings of the previous study that states the main issue concerning the collaboration between university and industry in Indonesia is the lack of trust on both sides, in which the university or industry has their own orientations and objectives. However, in the case of RAPID, both parties, the university and the industry, share the same interest and trust to participate in the RAPID program, where each party knows its task and role. Therefore, there is no issue regarding different orientations or objectives since both parties share the same vision and missions.

Perceptions regarding promotion costs (cost of product marketing) have positive and significant impacts on the level of collaboration. It is in line with the previous study that states that one of the considerations for a university and industry to collaborate is the potentially high cost for promoting a product. The theory on Transaction Costs Economics (TCE) explains why university and industry choose to form a relationship for the purpose of minimizing the development cost of technologies. The survey even shows that all parties involved (100%) agree to do it again if they are given another opportunity to collaborate under the RAPID program. The other form of collaboration that both the university and industry (16.7 percent) would like to pursue further is marketing promotion.

Perceptions regarding the risk of failure of the research/product have negative and significant impacts on the level of collaboration. This is because investing in a commercialization venture of a new product has its own uncertainty and is highly risky from the financial perspective. Hence, an industry as a profit oriented institution needs to calculate their every move carefully, including the cost of investing in research. Furthermore, during the testing stage it is not uncommon

Table 4. Estimation results using Multiple Regression Models (OLS).

Dependent variable: University—Industry Collaboration (KOLUI)						
Independent variable	Model 1	Model 2	Model 3	Model 4	Model 5	Model 6
Goal & Orientation Perception (GOAL)	–0.4080 (0.0581)	–0.3439* (0.1028)	–0.3086* (0.1255)	–0.3115* (0.1217)	–0.2999 (0.1532)	–0.1633 (0.4212)
Promotion Cost Perception (PROM)	0.4049*** (0.0438)	0.4094*** (0.0363)	0.4288*** (0.0227)	0.3842*** (0.0443)	0.3925*** (0.0474)	0.4368*** (0.0290)
Risk Perception (RISK)	–0.0393	–0.2124	–0.1440	–0.1472	–0.1477	–0.4071***
Goal & Orientation Perception (GOAL)	–0.4080 (0.0581)	–0.3439* (0.1028)	–0.3086* (0.1255)	–0.3115* (0.1217)	–0.2999 (0.1532)	–0.1633 (0.4212)
Risk Perception (RISK)	(0.8626)	(0.3873)	(0.5037)	(0.4934)	(0.5005)	(0.0424)
R&D Fund (DANA)			0.0286*** (0.0251)	0.0275*** (0.030)	0.0277*** (0.0332)	0.0339*** (0.0066)
Size of Industry (SIZE)				0.0525 (0.2941)	0.0545 (0.2918)	0.0441 (0.4079)
Age of Industry (AGE)					–0.0008 (0.8029)	–0.0030 (0.2830)
Interaction (INTERAKSI)		0.0085* (0.1080)				0.0118*** (0.0027)
Constanta	0.5361*** (0.0047)	0.4438*** (0.0189)	0.3632*** (0.0495)	0.2704 (0.1799)	0.2627*** (0.2058)	0.1105 (0.6276)
Number of Observation	31	31	31	31	31	31
R^2	0.2701	0.3404	0.4004	0.4268	0.4283	0.5463
Adj-R^2	0.1890	0.2389	0.3082	0.3121	0.2854	0.4083
F	3.3317	3.3553	4.3418	3.7230	2.9969	3.9578

Note: significant at * $p < 0.15$, ** $p < 0.1$, *** $p < 0.05$.

for a miss between the university and industry to occur. Based on the interviews, many universities found it challenging to explain the testing process, due to its theoretical nature, to their industry partner. On the other hand, from the industry's perspective, they want something practical that will take less time. Consequently, even though the collaborating university and industry understand their respective tasks and roles, including the consequences, sometimes the industry changes its commitment because of the risk factor associated with the research.

R & D funds have positive and significant effects on the level of collaboration. The incentive in the form of research funds has provided a positive impact in boosting the collaboration between university and industry under the RAPID program. This is because of the realization that it may take a large amount of money to bring forth the desired output in the form of technology/commercial products from the research. Therefore, in the last few years, there have been numerous efforts by the government to build the synergy between the university and industry through incentive program, such as the RAPID program, Hi-link, etc.

The result of the regression analysis shows that the size or age of the industry has no significant impact on the collaboration between university and industry under the RAPID program. These contradictory findings are caused by the fact that in the RAPID program, the university seeks its own industry partner. The survey shows that one of the factors for selecting a specific industry partner is that the partner belongs to the same network—which means that university and industry are familiar with each other; thus, the size or age of the industry is not considered as crucial in the selection of an industry partner.

More interactions between university and industry have positive and significant impacts on collaboration. The higher the intensity of interaction between the parties, the better the social familiarity and the more the opportunities to expand collaboration. As the RAPID survey shows, one of the reasons for establishing a collaboration is the factor of familiarity or networking.

5 CONCLUSION

Based on this research it is clear that the distribution of RAPID funds is concentrated in universities located in Java (especially West Java). This is because universities in Java, such as ITB, IPB, ITS, UI, UGM, and Unpad have strong R&D traditions. Moreover, the collaboration between university and industry under the RAPID program is mostly founded on networking between the participants. This is the result of a lack of database that identifies the industry requirements and the research offered by the universities. Most of the outputs of each collaboration under the RAPID program have been realized. However, the majority of them are still in the form of basic research. Since the majority of the outputs are still in a form of basic research, then the full impact of the RAPID program has not been felt in full because RAPID by nature is a long-term program.

The result of the regression analysis reveals that the perception about the promotional cost that needs to be spent by university and industry to commercialize a product, incentive in the form of RAPID funds, and the frequency of interactions between university and industry have a positive impact towards increasing the collaboration between university and industry under the RAPID program. It shows the importance of the government's support through research funding to encourage more collaboration. Meanwhile, the perception about research risks has a negative impact towards the collaboration between university and industry under the RAPID program. The risk of failure with regard to the research is responsible for the lack of commitment on the part of the industry since it will take quite a long time and a tremendous amount of money to commercialize a research product.

REFERENCES

Barringera Bruce R, and Jeffrey S Harrisona. 2000. "Walking a Tightrope: Creating Value Through Interorganizational Relationships" *Journal of Management* 26 (3): 367–403.

Darwadi, A, Soenarso, Jusron, dan Sutanto. 2012. "Peningkatan Peran Puspiptek dalam Proses Alih Teknologi." *Teknovasi Indonesia* 1. Asdep Budaya dan Etika Iptek, Deputi Kelembagaan Iptek, Kementerian Riset dan Teknologi.

Direktorat Penelitian dan Pengabdian Kepada Masyarakat. 2013. *Panduan Pelaksanaan Penelitiandan Pengabdian Kepada Masyarakat di Perguruan Tinggi Edisi IX*. Direkotat Jendral Pendidikan Tinggi.

Fiaz, Muhammad dan Naiding, Yang. 2012. "Exploring the Barriers to RnD Collaborations: A Challange for Industry and Faculty for Sustainable U-I Collaboration Growth." *International Journal Science and Technology* 5 (2).

Fiaz, Muhammad. 2013. "An Empirical Study of University–Industry R&D Collaboration in China: Implications for Technology in Society." *Journal of Technology in Society, Elsevier* 35(3): 191–202. http://dx.doi.org/10.1016/j.techsoc.2013.03.005.

Gujarati, Damodar N. dan Dawn C Porter. 2012a. *Dasar—dasar Ekonometrika*. Buku 1 Edisi 5 (Translated by Eugenia Mardanugraha, SitaWardhani, Carlos Mangunsong, Penerjemah). Jakarta: Salemba Empat.

Gujarati, Damodar N. dan Dawn C Porter. 2012b. *Dasar—dasar Ekonometrika*. Buku 2 Edisi 5 (Translated by Raden Carlos Mangunsong, Penerjemah). Jakarta: Salemba Empat.

Moeliodihardjoa, Bagyo Y, Biemo W Soemardib, Satryo S Brodjonegoroc, Sachi Hatakenakad. 2012. "University, Industry, and Government Partnership: Its Present and Future Challanges in Indonesia." *Journal of Social and Behavioral Science, Elsevier* 52: 307–316.

Perkmann, Markus, Neely, Andy, and Walsh Kathryn. 2011. "How Should Firms Evaluate Success in University–Industry Alliances? A Performance Measurement System." *R&D Management* 41 (2): 202–216.

Program Kemitraan untuk Pengembangan Kapasitas dan Analisis Pendidikan/ACDP. 2013. *Pengembangan Strategi Kemitraan Perguruan Tinggi, Industri, dan Pemerintah di Indonesia*. Jakarta: ACDP.

Subramanyam, K. 1983. "Bibliometrics Studies of Research Collaboration: A Review." *Journal of Information Science* 6 (1): 34.

Soh, Pek-Hooi, and Annapoornima M. Subramanian. 2014. "When Do Firms Benefit from University—Industry R&D Collaborations? The Implication of Firm R&D Focus on Scientific Research and Technological Recombination." *Journal of Business Venturing, Elsevier* 29 (6): 807–821.

Competition and Cooperation in Economics and Business – Gani et al. (Eds)
© 2018 Taylor & Francis Group, London, ISBN 978-1-138-62666-9

Indonesia's academic readiness in facing a freer flow of skilled labour in 2015 ASEAN Economic Community (AEC) implementation

B.Y. Gitaharie & L. Soelistianingsih
Department of Economics, Faculty of Economics and Business, Universitas Indonesia, Depok, Indonesia

ABSTRACT: The ASEAN Economic Community (AEC) has been established since 2015. One of its visions is to develop a single market and production base among the ASEAN member countries through its five pillars, one of which is freer flow of skilled labour, that is often referred to as university graduates. Because of its large population size, Indonesia is considered as a large market, yet with potential. However, the unemployment rate in Indonesia is 5.7% and the university graduate unemployment is 4.3% in 2014. In addition to that, even before the AEC effectuation, foreign workers in Indonesia have increased in number—from 59,577 in 2009 to 68,957 in 2013 – and mostly work as professionals, advisors, consultants, managers, directors, supervisors, technicians and commissioners in agriculture, industry, trades and service sectors. Are Indonesia's university graduates ready to face the tighter competition in the labour market? Performing primary data collection, this study analyses the graduates' quality in the five major state universities in Indonesia and their readiness to compete in the regional labour market. The study finds that there is a gap between what the universities produce and what the labour market expects. The findings could contribute to the provision of a policy recommendation to the government on how to meet the future challenges in the labour market.

1 INTRODUCTION

The ASEAN Economic Community (AEC) has been established since 2015. One of the four visions of the regional economic integration is to develop a single market and production base among the ASEAN member countries through the five pillars of freer flow of goods, services, investment, capital, and skilled labour (Arifin et al., 2009, 15–17). Because of its large population size, Indonesia is often considered as a country with a large potential market. Of its large population (254.5 million in 2014), 125.3 million are in the labour force with 94.3% working and 5.7% unemployed. From the proportion of the unemployed, 9.1% are high school, 5.9% are diploma, and 4.3% are university graduates.

Even before the AEC effectuation, foreign workers in Indonesia have increased in number—from 59,577 in 2009 to 68,957 in 2013 – and mostly work as professionals, advisors, consultants, managers, directors, supervisors, technicians and commissioners in agriculture, industry and trade, and service sectors. They come from China (22%), Japan (15%), South Korea (7%), India (7%), Malaysia (5%), and the US (5%), and are expected to increase as the AEC is implemented and the Indonesian economy grows. The freer labour inflows from foreign and neighbouring countries are mostly skilled labour and represent formal job types. This condition implies that Indonesia will have tighter competition in the labour market, particularly in the formal economic and business sectors. Is Indonesia's skilled labour, hence university graduates, ready to face the tighter competition that comes not only from local but also foreign graduates? Collecting primary data, this study describes the existing condition of the five major state universities in Indonesia and analyses their preparedness to compete in the regional labour market.

2 LITERATURE REVIEW

The AEC's implementation implies a more challenging competition in all markets, including the labour market. To produce a regionally/internationally competitive labour force, it cannot be separated from their education process, which in this case is higher education. Higher education is provided by universities whose graduates nowadays are facing tighter competition.

There is a difference between national and global competition, but both are complements (Marginson, 2006). National competition focuses on positional competition in higher education, whose focus is in granting academic titles to graduates, that in turn may provide access to social and income prestige. Providing universities can maintain their prestige, potential students will compete to be accepted and admitted into the institutions despite the tuition fees. On the contrary, higher education institutions with global competition are not only concerned with how to maintain the prestige, but also in considering their roles in the global market. The institutions must consider the possibilities of short-term student movement/exchange, global student flows, and segmentation of global competition.

Van Damme (2001) concludes that internationalisation of higher education is characterised by: student and lecturer mobility; internationalised curriculums that focus on quality assurance, meeting minimum international standards, such as criteria of professionalism and accreditation; campus branches; agreements on institutional cooperation and network development; Mutual Recognition Arrangements (MRAs); transnational university networks; and transnational virtual delivery of higher education to eliminate geographical barriers. The most important factor in the processes of higher education internationalisation is the quality that requires higher education institutions to have a code of practice or code of conduct. Marginson (2004) adds that in a globalised environment a higher education institution should consider global, national, and local dimensions together, better known as 'glonacal'.

Levin et al. (2006) adds that to be a World-Class University (WCU), a university must provide high quality education for students and conduct productive research and community services. To meet the requirements, a higher education institution must be distinctive, in terms of providing academic freedom and facilities, funding, diversity in students and lecturers, and programmes with internationalised curriculum; and keep the instructional quality, research quality, and student quality. A WCU faces challenges in financial resources, governance and strategic planning, quality assurance, research atmosphere/environment, and political condition.

ASEAN has already prepared what is called the ASEAN University Network (AUN). The agreement among the AUN member universities is: to enhance cooperation and solidarity among professionals, academics, scientists, and students in the ASEAN region; to develop academic and professional resources in ASEAN; and to enhance information dissemination, including electronic/digital library networks, exchanging and sharing information among the academic community, policy makers, students, and other relevant users (Agreement on the Establishment of the ASEAN University Network). Invited university members of the AUN are those: known by their status as centres of excellence in priority disciplines; that have strength in academic staff quality; that have experience in regional or international programmes; and that have laboratories, libraries, and other facilities in sufficient numbers (Agreement on the Establishment of the ASEAN University Network). Recently there are 30 universities from 10 ASEAN country members (6 plus Cambodia, Laos, Myanmar, and Vietnam) that have joined the AUN membership. In the context of Indonesia, Universitas Indonesia (UI), Institut Teknologi Bandung (ITB), Universitas Gadjah Mada (UGM), and Universitas Airlangga (UA) are already AUN members (AUN Member Universities, http://www.aunsec.org/aunmemberuniversities.php).

Students, according to Donald and Denison (2001), are the central focus in evaluating the quality of education. Several criteria are used to perceive student quality including:

firstly, generic skills and abilities—which cover openness and flexibility in learning, independence, responsibility, being able to make analysis and synthesis, and to think critically; secondly, academic performance—which covers academic achievement during study time at the university, completing the programme promptly, and intelligence; thirdly, employment competence—which covers employability performance, ability to secure a job after graduating, and expertise; fourthly, specific skills—which covers competence in a second language and basic mathematics; lastly, academic readiness—which covers readiness from high school graduation to enter university, and readiness for general academic requirement and certain programmes.

The qualities of university graduates influence their employability and competitiveness in the labour market. Raza and Naqvi (2011) state that graduate qualities in terms of generic skills, quality, and capability can be perceived in employers, Employers are more practical as they are normally more concerned with generic skills compared to academic grades. Therefore, it is important to accommodate generic skills into the curriculum. It is also necessary to direct and challenge the lecturer's competence to meet the skills.

3 METHODS

The study employs primary data collected from five selected State Universities (SUs) in Indonesia—UI in Depok, Institut Pertanian Bogor (IPB) in Bogor, ITB and Universitas Padjadjaran (UNPAD) in Bandung, and UGM in Jogjakarta. All of the university samples are located in Java.

Respondents include undergraduate active students, management of the universities/departments/undergraduate programmes, and undergraduate lecturers. The study also interviewed a few multinational companies as graduate users, and alumni who have at least five years' work experience in order to capture external perspectives. The limited samples of users and alumni should be regarded with caution as they only represent their perspectives on the questions asked. The study uses structured questionnaires and in-depth interviews, and employs SWOT analysis. The number of respondents is shown in Table 1.

As a preliminary and case study on graduate readiness to compete in the ASEAN market, samples of universities/faculties/departments/study-programmes within the SUs are selected based on the most competitive university entrance criteria, and are summarised in Table 2 below. Further research may expand the samples to include more faculties/departments/study-programmes and employ random sampling method with, of course, larger sample size. To capture the international aspects of users, authors also interviewed multinational company users in the fields of informatics and technology, finance, and manufactures.

Table 1. Types and distribution of respondents.

Respondents	Number of respondents (people)	Percentage (%)
Active students	364	85.25
Lecturers	30	7.03
University/ Department/ Programme Management	20	4.68
Users	6	1.41
Alumni	7	1.64
Total	427	100.00%

Source: Survey data.

Table 2. Selected faculties/departments/programmes for interviews.

The Five State Universities				
UI	IPB	UNPAD	ITB	UGM
(1) School of Medicine—Undergraduate Programme	(1) Faculty of Human Ecology—Department of Community Nutrition	(1) Faculty of Medicine—Undergraduate Programme	(1) Faculty of Mining and Petroleum Engineering—Petroleum Engineering Undergraduate Programme	(1) Faculty of Social and Political Sciences—Communication Study Programme
(2) Faculty of Engineering—Department of Electrical Engineering	(2) Faculty of Economics and Management—Department of Management	(2) Faculty of Dentistry—Undergraduate Programme	(2) School of Electrical Engineering and Informatics	(2) Faculty of Social and Political Sciences—International Relations Study Programme
(3) Faculty of Social and Political Sciences—Department of International Relations	(3) Faculty of Mathematics and Natural Sciences—Department of Computer Sciences	(3) Faculty of Pharmacy—Undergraduate Programme	(3) Faculty of Industrial Technology	(3) Faculty of Engineering—Informatics Engineering Study Programme

4 RESULTS AND ANALYSIS

This subsection discusses three issues—SU education profiles, graduate quality, and readiness to compete. Graduate competitiveness cannot be separated from the existence of the curriculum, teaching and learning process, student and lecturer quality, and international activities.

4.1 *Higher education profiles of the five SUs*

4.1.1 *Curriculum*

The selected SUs have the vision for internationalisation and have benchmarked their curriculum to some universities in the US, the Netherlands, England, Germany, France, Australia, and Japan. The management agrees that their curriculum is at par with the curriculum of other universities in ASEAN. Using the Likert scale of 5, the strengths of the SUs curriculum are shown in Table 3. To compete among ASEAN graduates, the management agrees that the curriculum should: (1) not only put a stress on hard skills but also on soft skills (57.9%); (2) consider global and national dimensions/aspects (21.1%); and (3) meet employment competence (15.8%).

4.1.2 *Teaching and learning process*

Teaching and learning process is one of the most essential factors in delivering education. All subjects offered in the surveyed programmes have a syllabus that can be accessed easily as they are distributed in class (73.3%) or made available in the school/department websites (43.3%). Most of the lecturers (86.7%) have included and disclosed their Expected Learning Outcomes (ELOs) on the syllabus. The syllabuses are reviewed regularly every academic year.

Around 43% of lecturers practice mixed methods of lecturing (problem-based learning, collaborative learning, experiential learning). Only some (16.7%), who teach classes with large numbers of students, employ conventional-lecturing methods. In delivering the subjects, most of the lecturers use textbooks (89.9%) in English. The students add that they also read articles

Table 3. Average scores of curriculum strength.

Curriculum strength	Management
Curriculum is at par with those of universities abroad.	4.37
Curriculum focuses on case studies.	4.37
Curriculum emphasizes on *Student Centred Learning* (SCL).	4.37
Curriculum is competence based.	4.32
Curriculum based on international standards.	4.32
Curriculum applies inductive methods.	3.68

Notes: 1: Strongly disagree; 2: Disagree; 3: Undecided; 4: Agree; 5: Strongly agree.
Source: Data processed.

published in international journals. This indirectly implies that students are forced to have at least passive English skills. However, in facing a tighter global competition in the labour market, course delivery in English for regular programmes has not yet become a common policy in all SUs surveyed. IPB does not have such a policy yet, but ITB does. ITB has a common policy that applies to all faculties/schools: if there is a foreign student in a class, the course should be delivered in English. All SUs, except IPB, have policies for regular undergraduate programmes to deliver courses of 6–15 credits in English. In addition to that, the SUs offer cross-faculty courses and the policy varies across SUs. Cross-faculty courses in ITB are mandatory and students are required to take cross-faculty courses for their minors of 12–18 credits.

To support the learning process, the SUs have prepared references that are easily accessed by students, for example in the libraries and by using online systems. Some subjects use local references—modules prepared by the lecturers, textbooks in Bahasa Indonesia, and national journals. According to students, international journals are easier to access compared to national journals, because only international journals are available online.

Mid-term and final term exams are already scheduled before the academic year begins. Almost all lecturers (93.3%) publish their final marks, but there are still 40% who do not publish their mid-term marks. In addition to that, the SUs have not had official appeal procedures. They have complaint mechanisms that vary across schools.

4.1.3 *Students*

The five SUs surveyed are among the best in the country. High school graduates vie to be admitted to one of the five SUs. Due to the limited capacity, less than 1% of high school graduates are admitted. There are several schemes for student admissions: 1) government invitation for top ranked high school students; 2) a national entrance test to SUs; and 3) an individual university entrance test. Admitted students to these five SUs are definitely among the best high school graduates in Indonesia.

Referring to Donald and Denison (2001), student quality is not only determined by Grade Point Average (GPA) but also by other skills and abilities, one of which is student participation in research. Most lecturers (93%) involve students in their research. However, they do not take GPA as the main consideration in recruiting students, but are more concerned in students' integrity, work ethics, teamworking ability, and English skills. Some students have their English test (TOEFL) scores ranging from 451–500. However, this score is relatively lower than that is expected by users who are facing the AEC, which is 550.

4.1.4 *Lecturers*

Most of the faculty members or lecturers have master or doctoral degrees. Most of them are permanent lecturers and obtained their degrees abroad. Around 80–90% of them have had teaching certificates from the Ministry of Research Technology and Higher Education and other relevant certificates—some issued by international institutions. They also participate in trainings/workshops organised by the university to improve their capacity in certain areas.

Lecturers have also developed and maintained networks and collaboration in research, teaching, and community engagement. More than half of the lecturers (60%) have organised joint research with European, Asian, and Australian universities/research institutes, and participate (83%) in seminars/workshops/trainings in those countries and also the US.

4.1.5 *Internationalisation*

Lecturer/student mobility is one indicator for internationalisation and is assessed by international accreditations and AUN. The advantages of such mobility are to create, develop, and maintain networks and joint academic activities, to broaden students'/lecturers' insights of knowledge development and dissemination, and to improve students/lecturers in their academic performance in order to be acknowledged internationally/regionally. In the last five years, lecturers in the SUs have participated in lecturer mobility, mostly to ASEAN countries, Europe, Australia, and the US. Student mobility is even more in numbers. Students go for exchange to Japan, South Korea, Hungary, the Netherlands, and Austria. This exchange activity gives significant added values as students learn about other countries' cultures, academic atmosphere, study/work ethics, and they become more confident. Students spend around one month to one year for the exchanges.

Some programmes in the SUs have international students, mostly from Asian countries. As an example, UNPAD Faculty of Medicine has one special class for students from Malaysia where they use English to communicate in the class. However, they use mixed language—Bahasa Indonesia and English with their local peers. Their presence in UNPAD adds somewhat to the international exposure for local students, at least in terms of language and interactions.

From the interviews with students, not more than 12% of the students participate in international seminars/conference, workshop/training, and competition. Most of the activities take place in Asia (Thailand, Hong Kong, Japan, South Korea, Malaysia, Singapore, and Taiwan), Europe (Austria, the Netherlands, Denmark, Hungary, England, German, Norway, and France), the US, and Australia. Some also participate in summer schools abroad and cultural missions.

4.2 *Graduate quality*

The survey also asked users about employment competence. According to users, graduates must have soft skills and capabilities (66.7%), work performance (14.3%) and specific skill (14.3%). Employers consider more on accountability (50%), openness and flexibility in learning something new (33.3%), and professionalism (16.7%), and not just academic performance. Employers also consider leadership skills (33.3%), ability to work in groups (33.3%), and willingness to accept challenges (33.3%). Mastering good communication (83.3%), foreign language (16.7%), and basic maths (16.7%) are also valuable. In a globalised and borderless labour market, good oral communication and foreign language skill, other than English, become very important. Basic maths skills support graduates in logical and systematic thinking that will assist them in making decisions. Users agree that SU graduates have sufficient knowledge and capabilities for problem-solving, but are lacking in good communication skills in English.

Another indicator of student quality is how fast a graduate gets a job after he/she finishes school. Based on the interview with the school/department/programme management of each SU, it only takes three to six months to get a job after he/she graduates. Some take less than three months. This implies that SU graduates are in high demand in the labour market, particularly those from UI School of Medicine and ITB Faculty of Mining and Petroleum Engineering. According to the school/department/programme management and the lecturers, their graduates have broad knowledge, are able to adapt quickly to new circumstances, including new cultures, and know how to apply their knowledge; but they lack ability in foreign languages other than English.

In order to compete in the ASEAN labour market, graduates should have the skills of communication, presentation, and basic computer applications. Graduates, according to respondents, must be able to work together in groups and have loyalty, persistence and commitment, and good professional work ethics. International certificates on professions are additional values for graduates. Graduates should also have some basic knowledge on: Informatics, Computers, and Technology (ICT); communication; maths; community and cultures; macroeconomics; marketing; and international law in order to compete internationally.

4.3 *Readiness to compete*

Lecturers are confident that their graduates are ready to compete both in domestic and regional labour markets. However, graduates need to improve their English skills and broaden their horizons to compete in the international market. From the students' perspectives, they (91%) are ready to compete in the domestic labour market, but more than half of them (62%) are not confident to compete in the international labour market, realising their shortcomings in English language skills. Some of the students (32.1%) have had certificates in certain fields that give additional value to compete in the market. They also agree that in order to compete graduates should have good communication skills (98.6%), knowledge on information and communication technology (96.7%), and knowledge on community and cultures (85.4%).

According to alumni and users, graduates are ready to compete in the domestic market, but not in regional or international markets. The main constraint is insufficient English and other soft skills. They are not prepared to face regional/international labour market competition yet. Based on alumni experience, the infrastructures in the SUs are not adequate to support the requirements for competition. Alumni also add that graduates are very good with textbook-knowledge but lack practice and creativity. They also admit the superior quality of lecturers, but their delivery method is quite monotonous. Practical experiences and solving cases should be enhanced and improved. According to users, graduates need to be equipped with the ability to communicate effectively and to make active and positive arguments. Graduates should know how to put themselves in their position.

Table 4. SWOT summaries.

Strength	Weaknesses	Opportunity	Threats
SUs have:	Graduates:	AEC provides larger job opportunity in the regional market. Local SU graduates have higher comparative advantage in knowing Indonesia than foreign graduates.	Tighter competition in labour market, particularly with graduates from Singapore, Malaysia, and the Philippines.
Strong curriculum Excellent intake student quality Wide networks with universities abroad	Lack English and interpersonal skills Lack self-esteem Lack argument skills Low survival skills when stationed in outer java Lack necessary soft skills for international exposures		
Graduates have:	SUs have insufficient infrastructures to support learning process		
Superior theoretical knowledge and academic performance Strong quantitative and analytic capabilities Good work ethics, work discipline, and are independent Ability to quickly adapt to new circumstances and are fast learners			

Source: Processed from in-depth interviews.

Table 4 summarises the results of in-depth-interviews in the form of a SWOT matrix, from which we can derive a strategic policy recommendation.

5 CONCLUSION AND POLICY RECOMMENDATION

This study concludes that there is a gap between the perspectives of graduates from the five SU graduates and those of users. SU graduates have excellent academic performance, strong knowledge in their relevant fields, and are ready to compete in regional labour market. Their main constraint is in the ability to communicate in English. In addition to that, users also endorse that graduates should have adequate generic and specific skills and meet employment competence. The study draws up some recommendations to meet the AEC challenges:

1. Graduate competence should include both hard and soft skills. Foreign language, particularly English, and basic computer skills must be considered as hard skills and improved significantly. Delivering hard skills should be conducted innovatively and attractively rather than in conventional ways. Improving soft skills should include oral communication and presentation skills, making objective argumentation and expressing ideas, negotiating, and networking. This will increase graduates' self-esteem in facing tighter competition in the labour market.
2. To improve human resource, lecturers are endorsed to pursue their doctoral degree and to participate more in international exposures and activities, for example to participate in international conferences as speakers/presenters and to take part in lecture mobility.
3. Universities should improve their infrastructures. Laboratories should be furnished with modern equipment. For that purpose, universities should expand networks and collaboration with relevant industries.

ACKNOWLEDGEMENTS

The authors would like to express their gratitude to the Directorate of Higher Education, Ministry of National Education, for granting the 2011 National Strategy Grants, high appreciation to Ms. Metri Sriwati for her full assistance in every stage of the research and Mr. Sulistiadi Dono Iskandar for his assistance in the survey to the five SUs.

REFERENCES

Agreement on the Establishment of the ASEAN University Network. https://web.archive.org/web/20080704150447/http://www.aseansec.org/9621.htm Retrieved on 21 October 2016 at 15:12. AUN Member Universities. Retrieved on 21 October 2016 at 15:25 http://www.aunsec.org/aunmemberuniversities.php

Arifin, S., Djaafara, R.A. & Budiman, A.S. (2009). 2015 ASEAN Economic Community (AEC): Strengthening ASEAN synergy in the midst of global competition (*Masyarakat ekonomi ASEAN 2015: Memperkuat sinergi asean di tengah kompetisi global*). Jakarta: PT. Elex Media Komputindo.

Donald, J.G. & Denison, D.B. (2001). Quality assessment of university students. *Journal of Higher Education*, *72*(4), 478–502.

Levin, H.M., Jeong, D.W. & Ou, D. (2006). What is a world class university? Paper presented on *The 2006 Conference of the Comparative and International Education Society*, Honolulu, Hawaii. www.tc.columbia.edu/centers/coce/pdf_files/c12.pdf

Marginson, S. (2004). Competition and markets in higher education: A glonacal analysis. *Policy Futures in Education*, *2*(2).

Marginson, S. (2006). Dynamics of national and global competition in higher education. *Higher Education*, *52*(1), 1–39.

Raza, S.A. & Naqvi, S.A. (2011). Quality of Pakistani university graduates as perceived by employers: Implications for faculty development. *Journal of Quality and Technology Management*, *VII*(1), 57–72.

van Damme, D. (2001). Quality Issues in the Internationalization of Higher Education. *Higher Education*, *41*(4), 415–441.

Competition and Cooperation in Economics and Business – Gani et al. (Eds)

Financial inclusion: The impacts of bank loans on household welfare

B.Y. Gitaharie & L. Soelistianingsih
Department of Economics, Faculty of Economics and Business, Universitas Indonesia, Depok, Indonesia

T. Djutaharta
Institute of Demography, Faculty of Economics and Business, Universitas Indonesia, Depok, Indonesia

ABSTRACT: The relationship between financial inclusion and poverty alleviation has been much discussed, but only few studies further explore the impacts of financial inclusion on household welfare. As a follow-up to the study of Gitaharie et al. (2013), this study is to analyse the differences in business development between business-owning households with bank loans and those without, and to analyse how business development affects the household welfare. The study also constructs a household welfare index from wealth, education, housing condition, and asset ownership indicators. The study collected primary data from business-owning households. The study employs two-stage regressions and finds that business-owning households with bank loans have a higher welfare index. The findings of this study could contribute to government policies on how to improve business-owning households and empower the owners to further improve their welfare.

1 INTRODUCTION

When the Asian crisis hit Indonesia in 1997–1998, Small and Medium Enterprises (SMEs) were discovered to have better resistance towards crisis pressures when many large companies were forced to close their business and be declared bankrupt. From time to time, SMEs have also proven to contribute positively to Indonesia Gross Domestic Product and have helped expand the labour market.

Like other developing countries, Indonesia has a higher number of SMEs compared to developed countries in general. The density of SMEs in Indonesia is more than 51 per 1,000 people, whereas in the United States the densities range between 21–30 per 1,000 people (Kushnir et al., 2010). In Indonesia, almost all firms (99%) are micro and SMEs in all economic sectors, and employ over 95% of the population (Bellefleur et al., 2012). However, not all micro firms and SMEs in Indonesia have access to finance, particularly to banks. In practice, their income is remarkably low so that they can hardly save, let alone apply for loans. They are not commercially attractive to banks. Only a small number of banks provide services in microfinance. Moreover, there is a large disparity between the banks' collateral requirement and the actual condition of these micro enterprises, as they are more likely to only own small-value assets which are considered insufficient as collateral.

With the predicament in the banking sector, the microfinancing service is filled by informal lenders, which typically provide loans with a very high interest rate but more straightforward and simpler terms and conditions, and importantly without collaterals. There are only 3.5% of households in Indonesia obtaining loans for their business from banks, 4.3% from non-bank institutions (e.g. co-operatives), and 1.9% from individuals, including relatives, neighbours, and informal money lenders; whilst 90% do not obtain loans (BPS, 2012b). Loans to micro and small enterprises are in fact crucial as they assist them in increasing their business capacities so that they can eventually improve their welfare. However, very few studies have linked loan access and household welfare (Appleton, 1996; Pitt et al., 2003; Shetty, 2008).

This study focuses on business-owning households—those that run informal businesses with fewer than 20 employees. The objective of this study is to analyse the differences in business development between business-owning households with bank loans and those without; and to analyse how sales turnover (revenue) affects the household welfare. The study takes a municipality in DKI Jakarta and two regencies in West Java as the case.

2 LITERATURE REVIEW

A business-owning household is defined as the smallest social unit where human and economic resources are administered (Carter et al., 2015). Such business depends very much on family labour, management, and ownership. The purpose of households owning a business is to improve their economic and social well-being. The terms well-being and welfare are often used interchangeably.

Welfare was previously only measured by economic aspects, such as preference, expenditure, and income. Preference is determined by individual behaviour in making choices between consumption and leisure. Thus, household welfare depends on goods/services consumed, leisure time allocation, household member composition, and access to public services (Kokoski, 1987). Meanwhile, the income approach considers all monetary aspects, including ownership of housing, land, and others in terms of monetary value. Further studies explain that because welfare is a multi-dimensional and dynamic concept, its measurement can be expanded and developed further. The measurement can include human development indicators, such as health, education, nutrition, fertility, child mortality, and access to public services (Grootaert, 1981).

Measuring welfare through the household expenditure and income is very unstable because both measures are liable to fluctuations. To offset the fluctuations, it can be proxied by indicators, such as housing conditions, access to amenities, age composition, and education attainment of household members, and productive asset ownership. Sahn and Stifel (2000) add that asset values in poor countries are relatively small and tend to be more easily measured. Physical assets, such as land, housing, and human capital, are typically used in a lot of research. In its development, asset accumulation becomes an important aspect in the discussion of welfare. Asset accumulation is often identified by the ownership of household savings, the use of savings, and changes of assets stock.

2.1 *Roles of financial inclusion*

Hannig and Jansen (2010) conclude that several factors indicate how more affordable access to finance for depositors and borrowers with low incomes can reduce the likelihood of systemic effects of financial system risk. The success of the financial inclusion programme can be measured through four perspectives, namely: 1). Access to use financial products and services from formal institutions, measured by calculating the proportion of the account numbers in financial institutions in the total population; 2). Quality of financial products, measured by whether the products of financial institutions suit the needs of the people; 3). Usage of services, measured by the frequency of use of financial products and the number of active accounts to total existing bank accounts; and 4). The impact of using financial products and services on the lives of consumers. Supporting Hannig and Jansen (2010), Hameedu (2014) states that the provision of accessible banking services is one of the main objectives of public policy. Policies to improve financial inclusion are certainly expected to improve the financial condition and living standards of the poor, the vulnerable, and the disadvantaged.

Dhawan (2014) concludes that the financial inclusion policy in India does not show any significant progress in attracting banking products. The Central Bank of India's financial inclusion policy introduces a limited range of financial products and services with emphasis on rural areas, and forces commercial banks to meet the Central Bank's directions. Apparently, top-down directive policies limit the success of India's financial inclusion policy. The failure is likely due to two main factors: the provision of products is not designed based on the needs of the poor, and banks' lack of enthusiasm for the programme.

Ruiz (2013) explains that the opening of a bank branch in Mexico, Banco Azteca, has had a significant impact on the regional economy in the short time compared to opening more than a thousand retail stores in areas without opening any bank branch. Households in areas with bank branches have better ability to smooth their consumption and to accumulate durable goods. When formal bank credit is available, households have a better capacity not to rely on savings as a buffer. Khandker (2000) concludes that the presence of microfinance facilities in the rural Bangladesh has increased the number of households that save their money from 17% to 60% – both voluntary and compulsory savings. Similarly, the presence of microfinance loans induces an increasing number of borrowing households. This reduces the incidence of households borrowing from informal lenders. The increasing number of these micro loans ultimately increases the amount of household savings. Microcredit (McKernan, 2002) for poor households contributes positively to poverty alleviation. Microcredit for the purpose to purchase capital inputs, generates a positive impact on programme participation and self-employment profits.

3 RESEARCH METHODOLOGY

3.1 *Data and sampling*

This research takes place in the provinces of DKI Jakarta and West Java, which are selected based on a stratified purposive sampling method. Municipality of East Jakarta, in DKI Jakarta province, represents the urbanised area and is considered because it had the highest number of self-employed workers, as many as 208,235 people or 27.8% of the total self-employed workers in 2012 (BPS, 2012a). Bogor Regency and Karawang Regency represent rural areas in West Java province. Two sub-districts are selected for each district/city in West Java and one in Jakarta. There are 2–3 villages selected in each district.

Each region has the following number of respondents: 93 respondents from East Jakarta, 108 respondents from Bogor, and 107 from Karawang Regency. Respondents are business-owning households with fewer than 20 employees (BPS, 2013). The total number of respondents is 308. Information collected from the field survey includes: 1) demographic characteristics and health condition of each household member; 2) demographic characteristics of households; 3) business identity and characteristics; 4) the capital structure and business costs; 5) the value of production and sales turnover; and 6) access to banks.

3.2 *The model and variable description*

There are two endogenous variables in the equations of this study: first, the amount of business sales turnover (B_i) as a proxy of business development, and second, household welfare index (IW_i) (see Appendix 1 for household welfare index construction). The amount of sales turnover (*Bi*) is affected by the access to bank loans, as a proxy for financial inclusion, with control variables of business characteristics, business experience, business management, human resources development, and demographic characteristics and asset ownership of the business-owning households.

Household welfare index (IW_i) is affected by the predicted value of sales turnover ($\hat{B}_i$) and the control variables, which include: the total income, proxied by total household expenditure; the demographic characteristics of business-owning households; and the number of other dependents outside the household.

Below are the equations to be estimated using the two-step regression method:

$$B_i = \beta_0 + \beta_1 X_{1i} + \beta_2 X_{2i} + \beta_3 X_{3i} + \beta_4 X_{4i} + \beta_5 X_{5i} + \beta_6 X_{6i} + \beta_7 X_{7i} + \beta_8 X_{8i} + \beta_9 X_{9i} + \varepsilon_i \quad (1)$$

$$IW_i = \gamma_0 + \gamma_1 \hat{B}_{1i} + \gamma_2 Z_{2i} + \gamma_3 Z_{8i} + \gamma_4 Z_{4i} + v_i \quad (2)$$

where ε, υ are the error terms and β, γ are the estimation parameters.

The exogenous variables are:

X_{1i} = access to bank loans in the last five years (1 = business-owning households have bank loans, 0 = no bank loans)
X_{2i} = business sector (1 = non-agricultural sector, 0 = other sector)
X_{3i} = business experience (years in business)
X_{4i} = financial separation (1 = separate business and household finances, 0 = others)
X_{5i} = participation in courses (1 = business owners or household members who help in the business attend courses/trainings, 0 = never attend)
X_{6i} = business location (1 = urban, 0 = others)
X_{7i} = sex (1 = male, 0 = others)
X_{8i} = education of business owners is grouped into: X_{8i}_SMA (1 = high school graduate, 0 = others); and X_{8i}_PT (1 = at least one-year diploma program graduate, 0 = others)
X_{9i} = vehicle ownership (1 = at least own one vehicle, 0 = others)
Z_{2i} = household income (proxied by household expenditure)
Z_{3i} = location of residence (1 = urban, 0 = rural)
Z_{4i} = dependents outside main household members (1 = at least one dependent, 0 = others).

4 RESULTS AND ANALYSIS

The survey results show that 63% of 253 business owners are females. The average age of business owners is 43 with approximately 13 years of experience. The education data summary shows that 55% of the business-owner respondents do not finish elementary school, 17% junior high school graduates, 21% are high school graduates, and only 4% are educated up to one-year diploma programme. Most of the business owners (85%) are engaged in construction/trade/restaurant/service/other as their main line of business. Most respondents (64%) run only one business and employ 1–2 workers, most of whom (80%) are not paid. More than half (60%) of business owners' residence is merged with their business place. These characteristics suit micro enterprises.

The study indicates that access to banks affects the value of the sales turnover of micro enterprises as shown by higher monthly sales turnover for business owners with bank loans (IDR29,687,594) than those without (IDR16,717,701). Those with access to bank loans also have a higher household welfare index (5.59 compared to 5.14). Table 1 below shows that business owners with access to bank loans have a higher score for health and the asset ownership; however, there is a slight difference for the education and the housing conditions indicators between the two groups.

Table 1. Sales turnover and household welfare index comparison.

	Without access to bank loans		With access to bank loans		Overall	
	Mean	SD	Mean	SD	Mean	SD
Sales turnover (IDR/month)	16,717,701	50,818,663	29,687,594	61,022,842	21,851,617	55,346,636
Household welfare index	5.14	1.72	5.59	1.60	5.31	1.68
Health indicator	0.95	0.72	1.16	0.65	1.03	0.70
Asset indicator	1.36	0.62	1.54	0.58	1.43	0.61
Education indicator	0.19	0.40	0.16	0.37	0.18	0.38
Housing indicator	2.75	1.06	2.77	1.09	2.76	1.07

Note: SD: Standard Deviation.
Source: Primary data processed.

4.1 *Effect of bank credits to business sales turnover*

Table 2 summarises Equation 1 estimation results. Bank loans significantly and positively affect sales turnover. Business owners who have bank loans have 65% higher sales turnover than those who do not. This is supported by the descriptive statistics showing that 83% of those with bank loans utilise the fund to increase their working capital; hence, it indicates that working capital for SMEs is a necessary condition in order to improve and develop their business. The field survey result shows that the amount of bank loans that most business owners apply is IDR10 million, but only 20% of them are approved with house or land (30%) and motor vehicles (14%) as the collaterals.

Male business owners have 35% higher sales turnover than female ones. This finding is similar to the one of Appleton (1996), and DeGraff and Bilsborrow (1993). Business owners with high school diploma and those with minimum one-year diploma programme have higher turnover than those who only graduate from junior high school and below, about 46% and 95% higher respectively. Those who possess vehicles receive turnover as much as 63% higher than those who do not. The descriptive statistics show that the percentage of household expenditures for transportation ranks second (11.2%) after food (59.6%). Owning their own vehicles, business owners can run their business more efficiently and therefore generate greater sales turnover from their business.

Business sectors, business experience, financial record separation, co-operation with other micro enterprises, participation in courses/trainings, and business location do not have any

Table 2. Business turnover estimation—Equation 1.

Dependent variable: Logturnover	Model 1	Model 2
Constant	14.470***	14.495***
	(0.436)	*(0.184)*
Access to bank loans in the last five years (1 = have bank loans)	0.652***	0.629***
	(0.186)	*(0.177)*
Business sector (1 = non-agricultural sector)	–0.165	
	(0.355)	
Business experience	0.098	
	(0.088)	
Financial separation (1 = separate business and household finances)	–0.098	
Financial separation (1 = separate business and household finances)	*(0.239)*	
Course participation (1 = participate in courses)	0.049	
	(0.468)	
Business location (1 = urban)	0.092	
	(0.232)	
Sex (1 = male)	0.346*	0.376**
	(0.182)	*(0.175)*
Education1 (1 = high school graduate)	0.948***	0.973***
	(0.244)	*(0.236)*
Education2 (1 = at least one-year diploma programme graduate)	0.0462**	0.464***
	(0.232)	(0.213)
Vehicle (1 = own vehicle)	0.628***	0.680***
	(0.192)	*(0.185)*
R Square	0.181	0.175
Adjusted R Square	0.150	0.159
F-stat	5.788***	11.302***
N (Missing data in Model 1)	273	288

Note:
The numbers in parentheses show the standard errors.
*** significance level of $\alpha = 0.01$;
** significance level of $\alpha = 0.05$;
* significance level of $\alpha = 0.1$.

significant impact on the sales turnover. The insignificant effect of business sector on sales turnover is supposedly due to micro enterprises' activities that are still considerably simple with very slight difference in productivity. Similarly, business experience does not provide a significant impact on the sales turnover, presumably because there is less learning process needed in the daily business operation to further expand their business. This is also supported by the fact that business-owner education levels are generally low. Occasional courses and trainings, that supposedly aim to improve their capabilities, have little impact on developing their notion of how to improve their business.

To monitor business development, one should have proper financial bookkeeping and separate household and business finances. This study finds that financial record separation does not significantly affect the sales turnover. Field survey data indicates that most respondents (86%) do not record their business financial flows. There are quite a number of respondents who could not answer the questions related to the structures of cost and capital in the field. Co-operation among SMEs does not significantly affect the sales turnover either. It is likely due to the existing partnerships which are not effective yet, so that potential market expansion, co-operation in utilisation of labour and other inputs, as well as in financing business capital cannot be realised.

4.2 *Business sales turnover effects on household welfare index*

Higher sales turnover signifies increasing household welfare index. Table 3 shows that 1% increase in sales turnover increases household welfare index by 0.86 points. Increasing sales turnover implies the rise of non-labour income which is normally utilised to purchase vehicles and to improve housing conditions. This finding is supported by the increasing score of vehicle ownership and improved sanitation quality as business sales turnover increases (see Appendix 2). The increasing vehicle ownership reflects the response about high transportation expenditure which ranks second after food.

Table 3. Household welfare index estimation—Equation 2.

Dependent variable: Welfare index	Model 1	Model 2
Constant	–12.688***	–13.748***
	(4.066)	*(3.997)*
Household income	0.310	0.374
	(0.242)	*(0.230)*
Location of residence (1 = urban)	0.275	0.289
	(0.340)	*(0.324)*
Education1 (1 = high school graduate)	–0.368	–0.347
	(0.411)	*(0.407)*
Education2 (1 = at least one-year diploma programme graduate)	0.832	0.867
	(0.360)	*(0.346)*
Dependents outside main household members	–0.489	–0.390
(1 = at least one dependent)	*(0.336)*	*(0.320)*
Predicted business turnover	0.860***	0.863***
	(0.227)	*(0.227)*
R Square	0.219	0.224
Adjusted R Square	0.191	0.199
F-stat	7.892***	8.773***
N (Missing data in Model 1)	273	288

Note:
The numbers in parentheses show the standard errors.
*** significance level of $\alpha = 0.01$;
** significance level of $\alpha = 0.05$;
* significance level of $\alpha = 0.1$.

Table 4. Welfare index components based on sales turnover per month.

Indicators		Sales turnover (million IDR per month)						
		< 1	1 – < 5	5 – < 100	10 – < 20	20 – < 50	> = 50	Total
Household welfare index	Mean	4.96	4.94	5.37	5.24	6.09	6.00	5.27
	SD	1.66	1.59	1.53	1.73	2.04	1.36	1.68
Health indicator	Mean	1.15	0.95	1.05	1.04	1.03	1.08	1.02
	SD	0.60	0.70	0.74	0.66	0.78	0.70	0.70
Asset indicator	Mean	1.44	1.28	1.33	1.53	1.69	1.60	1.42
	SD	0.58	0.63	0.63	0.58	0.54	0.50	0.61
Education indicator	Mean	0.13	0.09	0.22	0.18	0.32	0.19	0.17
	SD	0.34	0.29	0.42	0.39	0.48	0.40	0.38

Source: Processed data.

For all levels of household income, there is no significant difference in the household welfare index. Business owners who live in urban areas or rural areas do not have any significant difference on the welfare index either. Education level of household heads does not significantly influence the welfare of households because the educational level cannot be used to entirely project the differences in skills and human resource quality. The existence of dependents outside the household does not make any difference in the household welfare index either.

Table 4 further demonstrates that household welfare index increases as sales turnover increases. It also shows that health, asset, and education indicators improve as sales turnover rises. This implies that micro enterprises should be supported not only for the purpose of business expansion but also to encourage them to have better life-attainment.

5 CONCLUSION AND RECOMMENDATION

The study finds that the average monthly sales turnover is higher for business owners with access to bank loans than for those without. Male business owners with the education level of high school and above and owning vehicles have significantly higher sales turnover. Business sector, business experience, bookkeeping activity, co-operation with other SMEs, participation in courses/trainings, and location do not significantly affect sales turnover.

Sales turnover has significant positive effects on household welfare. The increasing sales turnover will also push the asset ownership and housing conditions indices upward. Meanwhile, all controlling variables in the model do not have any significant impact on the household welfare.

This study shows that access to bank loans increases sales turnover. However, the main obstacle of SMEs is to expand working capital. For that purpose, it is necessary to encourage banks to give clearer information on banking products and services; to provide larger opportunities for micro entrepreneurs to access bank loans, preferably without collateral; and to simplify the borrowing procedures for micro-entrepreneurs.

Noting that most micro-entrepreneurs do not record their financial transactions in an orderly and correct way; the approval of bank loans for micro-business has to be accompanied with continuous assistance on financial management. Furthermore, as many micro-entrepreneurs do not have co-operation/networking with other SMEs, facilitation/assistance from banks should also include training for marketing their products. Improving networking among micro-businesses and SMEs can strengthen connections among them and will also create valuable social capital in order to produce strong and solid groups of SMEs, and encourage them to access business credits collectively.

However, this study still has limitations in developing the household welfare index. It assigns equal weights for all indicators forming the household welfare index; and it only considers house, vehicle, and saving ownership as the indicators of asset index, and does not include assets in the forms of farm land, fish ponds and livestock. Further studies in this area may consider the possible endogeneity between household welfare index, particularly asset ownership indicators, and access to banks.

ACKNOWLEDGEMENTS

The authors would like to express their gratitude to Universitas Indonesia for providing the UI Research Grants 2014, high appreciation to Ms. Metri Sriwati and Ms. Esti Riyani for their administrative assistance, and to the great survey team. The authors would also like to express their deep condolences to the late Umar Dhani, Bogor field coordinator for the survey, who passed away in July 2015.

REFERENCES

Appleton, S. (1996). Women-headed households and household welfare: An empirical deconstruction for Uganda. *World Development*, *24*(12), 1811–1827.

Bellefleur, D., Murad, Z. & Tangkau, P. (2012). A snapshot of Indonesian entrepreneurship and micro, small, medium sized enterprise development. Retrieved from https://crawford.anu.edu.au/acde/.../20120507-SMERU-Dan-Thomson-Bellefleur.pdf on November 20, 2016 at 14:59.

Badan Pusat Statistik. (2012a). National labour force survey (*Survey angkatan kerja nasional*). Jakarta: Badan Pusat Statistik.

Badan Pusat Statistik. (2012b). National socioeconomic survey (*Survey sosial ekonomi nasional*). Jakarta: Badan Pusat Statistik.

Badan Pusat Statistik. (2013). Indonesian Statistics (*Statistik Indonesia*). Jakarta: Badan Pusat Statistik.

Carter, S., Alsos, G. & Ljunggren, E. (2015). The irrational benefits of small business ownership: Constructing economic well-being in business-owning households. In S.L. Newbert (Ed.), *Small Business in a Global Economy.* California: Praeger ABC-CLIO, LLC.

DeGraff, D.S. & Bilsborrow, R.E. (1993). Female-Headed households and family welfare in rural Ecuador. *Journal of Population Economics*, *6*(4), 317–336.

Dhawan, R. (2014). Why is financial inclusion in India not improving? New numbers, new approaches. http://blog.microsave.net/why-is-financial-inclusion-in-india-not-improving-new-numbers-new-approaches/at June 28, 2014 on 09:55.

Gitaharie, B.Y., Soelistianingsih, L. & Djutaharta, T. (2013). Financial inclusion: Household access to credits in Indonesia using national labour force survey data (*Inklusi finansial: Akses rumah tangga terhadap layanan kredit di Indonesia dengan menggunakan data SUSENAS.*) Unpublished manuscript, Universitas Indonesia.

Grootaert, C. (1981). The conceptual bases of measures of household welfare and their implied survey data requirements. Retrieved from http://www.roiw.org/1983/1.pdf on September 10, 2014 on 18:28.

Hannig, A. & Jansen, S. (2010). Financial inclusion and financial stability: Current policy issues. *Asian Development Bank Institute. Working Paper Series*, 259.

Hameedu, M.S. (2014). The emerging development model in India differently-abled entrepreneurs. *International Journal of Scientific and Research Publication*, *4*(1).

Khandker, S.R. (2000). Savings, informal borrowing, and microfinance. *The Bangladesh Development Studies*, *26*(2/3), 49–78.

Kokoski, M.F. (1987). Indices of household welfare and the value of leisure time. *The Review of Economics and Statistics*, *69*(1).

Kushnir, K., Mirmulstein, M.L. & Ramalho, R. (2010). Micro, small, and medium enterprises around the world: How many are there, and what affects the counts? *MSME Country Indicator*, World Bank: IFC.

McKernan, S.M. (2002). The impact of microcredit program on self-employment profits: Do noncredit program aspects matter? *The Review of Economics and Statistics*, *84*(1), 93–115.

Pitt, M.M., Khandker, S.R., Chowdhury, O.H. & Millimet, D.L. (2003). Credit programs for the poor and the health status of children in rural Bangladesh. *International Economic Review*, *44*(1), 87–118.

Ruiz, C. (2013). From pawn shops to banks. The impact of formal credit on informal households. Policy research working paper; no. WPS 6634. Washington, DC: World Bank. http://documents.worldbank.org/curated/en/465221468286796985/From-pawn-shops-to-banks-the-impact-of-formal-credit-on-informal-households.

Sahn, D.E. & Stifel, D. (2000). Assets as a measure of household welfare in developing countries. Working Paper (00–11), A paper presented at the *Inclusion in Asset Building: Research Policy Symposium*, organized by The Center for Social Development, Washington University.

Shetty, N.K. (2008). The microfinance promise in financial inclusion and welfare of the poor: Evidence from Karnataka, India. *Working Paper* (205). Institute for Social and Economic Change. Retrieved on December 15, 2013 at 15:15 http://ideas.repec.org/p/sch/wpaper/205.html.

APPENDIX 1

Household Welfare Index Construction.

The household welfare index (*IW*) is constructed from a composite of four indicators, namely health and housing conditions, education of household members, and asset ownership. All indicators are assumed to have equal weight. The IW range is 0–10.

The health indicator comprises of health insurance ownership and existing health complaints for each member of the household. The total score for health indicators is scaled from 0–2. The housing conditions indicator includes: 1) floor area per person of the residential building; 2) sanitary and water conditions; 3) the source of drinking water; and 4) installed electric power. The total score for the housing conditions indicator ranges from 0–4. The education indicator is a combination of the age and education attainment of each household member. The total score for the education indicator is from 0–1. The asset ownership indicator includes the ownership of: 1) residential buildings; 2) vehicles; and 3) financial assets. The total score for asset ownership indicators is 0–3.

APPENDIX 2

Housing Index and Asset Index components based on sales turnover per month.

		Sales turnover (million IDR per month)						
Indicators		< 1	1 – < 5	5 – < 100	10 – < 20	20 – < 50	> = 50	Total
Floor area per person (m²)	Mean	0.81	0.74	0.78	0.72	0.73	0.79	0.75
	SD	0.40	0.44	0.42	0.46	0.45	0.41	0.43
Sanitation quality	Mean	0.11	0.17	0.21	0.21	0.13	0.32	0.19
	SD	0.32	0.37	0.41	0.41	0.34	0.48	0.39
Ownership of residence	Mean	0.85	0.77	0.66	0.79	0.72	0.72	0.75
	SD	0.36	0.42	0.48	0.41	0.46	0.46	0.44
Vehicle ownership	Mean	0.56	0.51	0.66	0.72	0.94	0.88	0.66
	SD	0.51	0.50	0.48	0.45	0.25	0.33	0.47
Ownership of deposit/giro account	Mean	0.04	0.00	0.02	0.02	0.03	0.00	0.01
	SD	0.19	0.00	0.13	0.15	0.18	0.00	0.12

Source: Processed data.

Competition and Cooperation in Economics and Business – Gani et al. (Eds)
© 2018 Taylor & Francis Group, London, ISBN 978-1-138-62666-9

Trouble in paradise: How women's intra-household bargaining power affects marital stability

M.V. Dayanti & T. Dartanto
Department of Economics, Faculty of Economics and Business, Universitas Indonesia, Depok, Indonesia

ABSTRACT: The substantial improvement in socio-economic conditions in Indonesia has been followed by an increase in women's independence and empowerment as well as the divorce rate. This study aims at examining the relationship between women's bargaining power and marital stability. According to Nash Equilibrium derived from mixed strategy, the relationship between women's bargaining power and marital stability could be both positive and negative depending on the level of women's bargaining power. We conducted an online survey among 752 married women to find out who acted as decision maker on their household expenditure, which indicated women's bargaining power, and to identify occurrence of households conflict which indicated stability. By controlling such variables as education, income, ethnicity, religion and divorce experience, our econometric estimations confirm the U-shaped relationship between women's bargaining power and stability. As the women's bargaining power increases, household conflict decreases until it reaches some level beyond which the conflict starts to increase. Marital instability tends to be higher when the wives have experienced divorce, are in mixed unions, and were married because of other possible reasons (marriage by accident). This study suggests that a smaller socio-economic gap between couples will guarantee less household conflict and more stability in marriages.

1 INTRODUCTION

Socio-economic conditions in Indonesia have been improving rapidly in the past three decades. The World Bank reported that the per capita GDP (constant, 2010 US$) of Indonesia had jumped from $1,095 (1980) to $3,834 (2015), while in terms of women's empowerment, gender parity index for gross enrolment ratio in tertiary education had substantially increased from 0.465 (1982) to 1.11 (2013). Latest figures indicate that women in Indonesia have become more empowered and independent. However, some improvements in women's empowerment have resulted in numerous adverse impacts on marital stability; hence the rise of divorce rate. BBC (2009) reported that the divorce rate had spiked from an average of 20,000 cases a year to more than 200,000 a year during 1999–2009. Women were the main driver behind this phenomenon. Their economic independence and greater awareness of their rights would be some of the reasons for divorce. Arijaya (2011) elaborated on the reason behind a sharp increase in divorce rate. Of all the factors, nearly 40% of couples were divorced due to disharmonious marriage factors; one of them was the prolonged spousal dispute. Women have started to understand that they have a right to terminate the marriage too. The large number of divorce cases is initiated by women – 57% of divorce cases compared to only 28% that were filed by men. Women's growing independence is the culprit of this unfortunate issue. This research then aims at elaborating the relationship between women's bargaining power and marital stability.

In the realm of Indonesia patriarchal society, Indonesian women have long struggled in obtaining greater involvement in the nation's development and in getting their voices heard. The increasing exposure to equal opportunities of men and women could be indicated, to a lesser extent, by the rise of women's employment and educational attainment. Schaner and Das (2016) have reported that there is evidence of increasing participation of younger women living in the urban areas in wage employment. The study also found a narrowing trend in gender gap in wage employment—in 2011, the median women earned 84% as much as their male counterparts compared to 57% in 1990. In 2011, almost 30% of women obtained senior secondary and post-secondary degrees compared to merely 11% of women in 1990 and around 8% of women had tertiary degree (Schaner & Das, 2016). Consequently, greater participation in wage employment has enabled them to be more economically independent, while better access to education has enriched them with better understanding of their worth.

The surging divorce rate and the increasing women's independence—hence, their bargaining power—have awakened interests in the causes of marital dissolution. Past empirical studies have tried to explain that the impact of women's increasing bargaining power may increase marital instability. Yet the results are mixed. Booth et al. (1984), Hiedemann et al. (1998), and Winslow (2011) found a positive correlation between wives' economic independence and the likelihood of marital dissolution. Meanwhile, Weagley et al. (2007) suggest that the decrease in wives' housework time increased the odds of divorce. Furthermore, Ono (1998) proved that the relationship between the wife's bargaining power and the risk of dissolution followed the U-shaped curve. Risman and Johnson-Summerford (1998) confirm that the risk of divorce reached the lowest point in egalitarian unions. On the other hand, Heckert et al. (1998) and Rogers (2004) found the exact opposite result, arguing that the association between wives' bargaining power as depicted by their economic resources and the likelihood of marital dissolution formed an inverted U-shaped curve.

Due to inconclusive previous evidence, as well as very little evidence/empirical studies examining Indonesia's case, we attempt to investigate the relationship between women's bargaining power within family and marital stability. However, instead of using women's economic resources as the proxy of women's bargaining power, we define bargaining power as the person who has the authority to control expenditure in family decision-making. This definition follows Agarwal (1997) that relative intra-household bargaining power could be defined as the person who participates in household decision-making processes. Furthermore, instead of using divorce rate as the proxy of instability, this study uses ten-item questions of Marital Instability Index. We apply both game-theoretic and econometric models to theoretically and empirically provide clear evidence of the relationship between women's bargaining power and marital stability. This study will then contribute to the growing literature on household economics as well as providing valuable information on how to prevent marital dissolution, especially in Indonesia.

2 THEORETICAL MODEL AND EMPIRICAL SPECIFICATION

2.1 *The spousal bargaining game*

This study establishes a simple game-theoretic model to analyse the relationship between women's relative intra-household bargaining power and marital stability. The spousal bargaining game consists of two rational players, the wife and the husband. The strategy of the wife is to decide whether to actively participate in the household decision-making process or to obey her husband. The strategy of the husband is also to decide whether to actively participate in the household decision-making process or to obey his wife. The idea of the spousal bargaining game follows Dartanto's (2010) work on corruption game.

		Husband	
		Active	Obedient
Wife	Active	$w + rM - D(I), Z - rM + H - E(I)$	$w + rM, Z - rM$
	Obedient	$w + R(I), Z - E(I)$	$w + R(I), Z$

Source: Authors.

To start, let w be the wife's benefit for being married, r is the fraction/percentage of the wife's participation in household decision-making, M is a value of marital conflict which can be measured using the value of wife's hypothetical situation outside marriage, such as the wife's expected income, career opportunity, and inheritance benefits. The higher value of marital conflict borne by the wife will increase the attractiveness of dissolution. $D(I)$ is the cost borne by the wife when both spouses actively participate in household decision-making. $R(I)$ is the reward received by the wife when she obeys her husband's decisions, such as the husband's extra love, or the husband being extra nice. Hence, the benefits from relatively higher women's bargaining power if her husband is obedient are the sum of the wife's benefits for being married and the gain from being active; hence, $w + rM$. On the contrary, the benefits if both players jointly participate in the household decision-making process are the sum of the wife's benefits for being married and the gain from her increase in bargaining power minus the costs inflicted, $(w + rM - D(I))$. Furthermore, the benefits to the wife for being obedient to her husband are the sum of her benefits of being married and the reward she receives for being obedient $(w + R(I))$. The condition that is considered acceptable for the wife's participation in household decision-making is when the gain from her higher bargaining power is greater than the reward she receives for being obedient, $rM > R(I)$. Under those conditions, the wife's active participation in household decision-making is economically rational.

Let us assume that D is a continuous increasing function of the women's bargaining power index (I), $\frac{\partial D(I)}{\partial I} = D'(I) > 0$. This suggests that the costs paid by the wife, in a household with higher women's relative intra-household bargaining power compared to their husbands, are larger than that of households with relatively lower women's intra-household bargaining power. Hereafter, we assume that the second derivative of $D(I)$ is negative, $\frac{\partial^2 D(I)}{\partial I^2} = D''(I) < 0$. On the other hand, $R(I)$ is a continuous decreasing function in I, $\frac{\partial R(I)}{\partial I} = R'(I) < 0$. These assumptions imply that in a household with relatively lower women's intra-household bargaining power will receive better reward from their husbands.

On the husband's side, Z is the utility from his domination, H represents the husband's pride, and $E(I)$ is the cost of enforcement paid by the husband as a function of the women's bargaining power. If the husband performs an active participation in household decision-making as does his wife, he will get benefits of $Z - rM + H - E(I)$. We assume that the benefit received from the husband's domination is larger than the cost of enforcement, $Z > E(I)$. Accordingly, the husband's active participation in household decision-making is economically rational. Let E be a continuous increasing function in I, $\frac{\partial E(I)}{\partial I} = E'(I) > 0$ and the negative second derivative, $\frac{\partial^2 E(I)}{\partial I^2} = E''(I) < 0$. The high value of I represents more cost of enforcement needed for negotiating with his partner.

The Nash Equilibrium derives from assumptions $rM - D(I) < R(I)$ and $rM < Z$. The first assumption means that the net benefits of actively participating in household decision-making when their husbands are also active decision makers are smaller than the net benefits for being obedient. It suggests that the wife will choose to actively participate in household decision-making if the net benefits of being active are larger than that of being obedient. The second assumption means that the value of the husband's domination is greater or equal to the gain for a wife's higher bargaining power. If these assumptions are violated, there is no rational reason for the husband's active participation in the household decision-making process.

In this bargaining game, a mixed strategy for the husband is the distribution function $(p, 1 - p)$, where p is the probability for being actively participative in the household decision-making process and $1 - p$ is the probability for being obedient to his wife, for $0 \leq p \leq 1$. A mixed strategy for the wife is the distribution function $(q, 1 - q)$, where q is the probability for being actively participative in decision-making and $1 - q$ is the probability for being obedient, for $0 \leq q \leq 1$. From these explanations, the solution of p can be derived by following the expected benefits of the wife as shown below:

$$\begin{aligned} E(\pi_{active}) &= p(w + rM - D(I)) + (1-p)(w + rM) \\ &= -pD(I) + (w + rM) \end{aligned} \tag{1}$$

$$E(\pi_{obedient}) = p(w + R(I)) + (1 - p)(w + R(I)) \\ = w + R(I) \quad (2)$$

By substituting Equation 2 into Equation 1, we obtain:

$$p = \frac{rM - R(I)}{D(I)} \quad (3)$$

or,

$$M = \frac{pD(I) + R(I)}{r} \quad (4)$$

From Equation 3, it can be inferred that an increase in the gain of a wife's higher bargaining power (rM) is also an increase in the probability of a husband's participation in the household decision-making processes. The husband's active participation in household decision-making is needed to lessen the wife's dominance or active participation. Thus, the higher the costs endured by the wife and the higher the reward she receives, the lower the probability of the husband's active participation. Equation 4 implies that the husband's active participation will increase the rate of marital conflict because it deters the wife's preferences in household decision-making. The higher cost of the wife's bargaining power and higher reward received by the wife for being obedient also increases the value of marital conflict (M).

Based on Equation 4, we can derive the impact of women's bargaining power from conflict. The first order condition of Equation 4 is expressed below:

$$M' = \frac{pD'(I) + R'(I)}{r}. \quad (5)$$

According to the first order condition depicted in Equation 5, the value of M' will depend on the level of women's bargaining power, that is I. If we assume that $D'(I) > 0$ and $R'(I) < 0$, the value of M' can be both negative and positive depending on I. If $pD'(I) > R'(I)$ then M' will be positive and if $pD'(I) < R'(I)$ then M' will be negative. M' will be zero if $pD'(I) = R'(I)$. Such condition is considered as the turning point. Indeed, the relationship between marital conflict and women's bargaining power will be a positive one, which means higher bargaining power will increase the rate of marital conflict when the reward she receives from her husband cannot outweigh the increase in costs she endures. Yet, the relationship will be negative which means that higher women's bargaining power will reduce the rate of marital conflict when the reward she receives exceeds the cost of higher bargaining power.

2.2 *Model specification*

The data used in this research is obtained through online survey by spreading a four-part questionnaire. The subject is limited to only married women. This study adapts the list of questions from IFLS Book 3A Section *Pengambilan Keputusan* (Decision-Making) as the proxy for women's bargaining power and questions from Marital Instability Index as the proxy for instability. The section *Pengambilan Keputusan* consists of a list of questions of couples' participation in the household decision-making process, and the Marital Instability Index consists of a list of questions that measure the tendency to terminate the marriage. This research tested four models to investigate the relationship between women's participation in intra-household decision-making and the occurrence of intra-household conflict. Each of these models was demonstrated in the following form:

$$\boldsymbol{Conflict = \beta_0 + \beta_1 BPower + \beta_2 BPowerSQ + \beta_3 ControlVar} + \epsilon$$

3 RESULTS AND ANALYSIS

Based on the regression results, the coefficient of BPower is negative in value, but the coefficient of BPowerSQ is positive. This indicates the existence of a non-linear relationship between those two, that is that women's bargaining power negatively impacts the marital instability while its squared shifts in the opposite direction. Women's higher bargaining power does not solely encourage or discourage the household's marital stability. Yet, both effects may prevail at the same time depending on the degree of women's bargaining power. Both the coefficient of BPower and BPowerSQ prove the theoretical prediction made in the spousal bargaining game that higher women's bargaining power can only increase stability up to a certain point. Surpassing this point implies that higher women's bargaining power imposes a threat to marriage stability. Table 1 below shows the regression results.

3.1 *Regression results*

Following the U-shaped curve (Figure 1), in households where the wives have a relatively lower or higher bargaining power than their husbands, the instability is higher compared to those where both spouses have nearly equal bargaining power. The graph presents a simple illustration that the increasing women's bargaining power could alleviate the instability within the family up to a certain point. At this state, the intra-family conflict arises due to the wife's true

Table 1. The regression results.

	Model 1 conflict index	Model 2 conflict index	Model 3 conflict index	Model 4 conflict index
Bargaining Power	−1.539***	−1.444***	−1.263***	−1.140***
	(0.311)	(0.312)	(0.308)	(0.307)
Squared Bargaining Power	0.288***	0.271***	0.238***	0.217***
	(0.051)	(0.051)	(0.05)	(0.05)
Wife's Education (1 = higher than husband)		0.040*	0.031	0.033
		(0.023)	(0.022)	(0.022)
Wife's Income (1 = higher than husband)		0.055*	0.047*	0.036
		(0.029)	(0.028)	(0.028)
Ethnicity (1 = different)			0.026	0.028
			(0.019)	(0.019)
Religion (1 = different)			0.105*	0.101*
			(0.0608)	(0.0604)
Parental divorce (1 = divorced)			0.0679*	0.0657*
			(0.0373)	(0.037)
Divorce experience (1 = divorced)			0.266***	0.254***
			(0.046)	−0.046
Taaruf (1 = taaruf)				−0.055
				(0.035)
Arranged Marriage (1 = arranged marriage)				0.046
				(0.068)
Others (1 = others)				0.211***
				(0.062)
Observations	752	752	752	752
R-squared	0.112	0.121	0.173	0.189
F-stat	47.3	25.69	19.43	15.71
Standard errors in parentheses				

*significant at 10%; **significant at 5%; ***significant at 1%.

interest, which in part suggests her well-being is not fully perceived in the household decision-making processes. Hence, the husbands' interests mostly dominate the household's utility. According to Agarwal's (1997) definition of bargaining power, which is the person who participates in the decision-making processes and determines what decisions to make, the relatively low women's bargaining power implies low participation in the household decision-making processes. Women are not keen to urge hard bargains with their husbands and therefore cannot get more than they would (England & Killbourne (1994) in Pollak). This less-perceived well-being in household decision-making processes can incite conflict among spouses.

However, as a woman's bargaining power starts to increase, the household moves to reach stability, that is, the higher the woman's intra-household bargaining power, the less the occurrence of intra-family conflict. The increase of women's bargaining power means that they have more involvement in household decision-making processes. They are getting more involved in determining the allocation of household expenditure together with their spouses. As a consequence, their active participation in household decision-making processes drives down the intra-family conflict because their true interests and preferences are embedded in the household's joint decisions. Another reason as to why marital instability and women's bargaining power move in the opposite direction before surpassing the turning point is that the husbands are willing to give more reward to their wives in order to keep them obedient. As long as the reward that the wife receives outweighs both the costs inflicted from higher bargaining power and the husband's active participation in household decision-making processes, the marital instability will go down.

Yet, beyond the conflict-minimising point, the increasing women's bargaining power, ironically, hampers the household's stability. It occurs when their bargaining power is relatively higher than their spouses, resulting in increasing spousal conflict. The relatively higher women's bargaining power means that they are prone to dictate the decisions made within the family or, at an extreme level, dominate the decisions made in the household. Indeed, this will hurt the husbands' domination as well as their pride.

The U-shaped curve form indicates that the degree of marital conflict will attain the lowest point when the women's intra-household bargaining power index ranges from 2.62–2.67. This point is considered the turning point. This turning point implies that the wife participates in at least five areas of decision-making in the household. This compelling evidence emphasises an important conclusion: when women's bargaining power is relatively too low and too high compared to their spouses it imposes a threat to marriage stability. It is only a matter of bargaining power level that becomes the determining factor to attain the stability. The wives can alleviate marital conflict by increasing participation in household decision-making process in allocating resources to achieve higher family welfare. However, when this optimum level is surpassed, the increasing women's bargaining power becomes hazardous as it raises the marital instability.

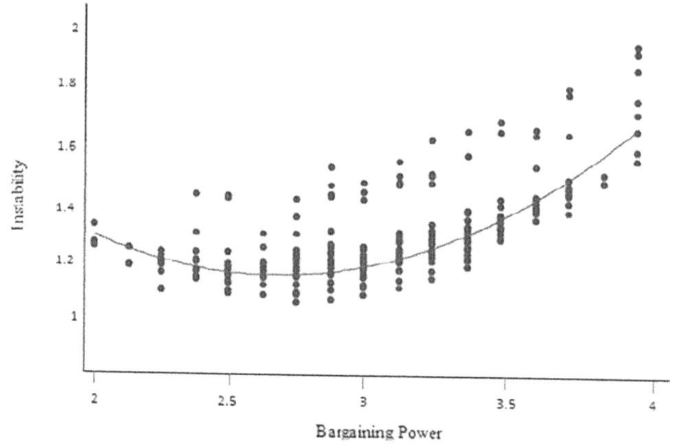

Figure 1. The Relationship between instability and bargaining power.
Source: Author's calculation.

The regression result suggests that wives who are 'marrying down' suffer relatively higher marital instability compared to those whose educational attainment is lower or equal to their husbands, ceteris paribus. This suggests the importance of assortative mating as spouses with equal level of income and education result in lower instability. The mixed unions—spouses with different religious beliefs or ethnicity—suffer relatively higher marital instability, ceteris paribus. Past studies suggested that spousal dissimilarity in religious beliefs and ethnicity elevated the risk of dissolution (Lehrer & Chiswick, 1993; Vaaler et al., 2009; Clarkwest, 2007).

Another substantive finding is that the wives' experience of parental divorce can elevate the occurrence of spousal conflict, ceteris paribus, though not statistically significant. As compelling as parental divorce suggests, the wives' divorce experiences also suffer from higher proneness to conflict in remarriages, ceteris paribus. According to several studies, children of divorce are more likely to end up in divorce compared to those from intact families. They are more likely to experience lower marital quality, lower marital commitment and confidence, and have a higher tendency to end up in divorce (Glenn & Kramer, 1987; Whitton et al., 2008). An individual's divorce experience also contributes to the high probability of instability within remarriage. Becker (1981) gave a plausible explanation concerning this issue. He argued that remarried individuals were inclined to have characteristics that diminished the gain from marriage compared to those in first marriages.

Last, the regression result poses an interesting finding on the importance of choosing our potential partner. Marriages that were united by the so-called *taaruf*, which is the Islamic-based marriage arrangement, result in lower marital conflict compared to those whose marriages are based on personal choice, ceteris paribus. Some plausible reasons that explain why religion-based marriage arrangement can lower the risk of dissolution are because it enhances the spouse's moral values that could lead to higher fidelity within marriages, encourages the bond between spouses and God which discourages any actions which would displease God, and increases marital quality as they believe that marriages are sacred in the eye of God (Dollahite & Lambert, 2007). However, family-based arranged marriages (e.g. marriage arranged by the parents) and other possible reasons (e.g. marriage by accident) endure higher instability compared to those whose marriages are based on personal choice, ceteris paribus. Traditionally, parents took a major role in marrying their children off. Yet, from the early 1900s, there has been a substantial change regarding the nature of marriage as an institution due to government regulation and some social movements, such as the legislation on minimum age of marriage, the abolishment of polygamy, and the rise in the importance of female education (Hull, 2006). As a matter of fact, the growing independence generated from those changes has offered more freedom for children to seek their partners based on mutual interests such as love, compatibility, and other romantic attractions. These brief explanations can be used as a rational argument on how arranged marriages have resulted in higher intra-family conflict because, nowadays, the freedom of choice has been an important factor in determining one's well-being.

4 CONCLUDING REMARKS

The promising trend of greater participation of women in wage employment and the narrowing trend of gender discrepancy in education between men and women are considered to have precarious effects on marriage stability. However, several past studies focusing on this issue found contradicting results—a positive correlation, a U-shaped curve, and an inverted U-shaped curve. Departing from these mixed results, this study employs a game-theoretic model which later will be proven by econometric estimation to analyse the relationship between women's bargaining power and marital stability.

Based on the spousal bargaining game, marital stability can be positive or negative, depending on the level of women's bargaining power. This non-linear relationship suggests that there exists an optimum level of bargaining power which will minimise instability. Consequently, there should be an incentive to make a joint agreement between interacting parties to achieve an optimum level of women's bargaining power in order to maintain long run stability. The econometric estimation also reveals that the difference in spouses' socio-economic status—

differences in religion and ethnicity, as well as where the wives earn more and have a higher education than their counterparts—leads to higher instability. Thus, it can be inferred that when the spouses have equal socio-economic status, the marriages are likely to be more stable.

ACKNOWLEDGEMENTS

The author wishes to thank Dr Teguh Dartanto for his expert advice and also to the anonymous reviewer(s) for their useful suggestions. This research is supported by Publikasi Internasional Terindeks untuk Tugas Akhir Mahasiswa (PITTA) grant from Universitas Indonesia.

REFERENCES

Agarwal, B. (1997). 'Bargaining' and gender relations: Within and beyond the household. *Feminist Economics*, *3*(1), 1–51.

Arijaya, R. (2011). Why is divorce in Indonesia increasing? *The Jakarta Post*. http://www.thejakartapost.com/news/2011/09/12/why-divorce-indonesia-increasing.html#sthash.lIm45McH.dpuf.

BBC. (2009). Indonesian divorce rate surges. BBC. http://news.bbc.co.uk/2/hi/asia-pacific/7869813.stm.

Becker, S.G. (1981). *A treatise on the family*. Cambridge, MA: Harvard University Press.

Booth, A., Johnson, D.R., White L. & Edwards, J.N. (1984). Women, outside employment, and marital instability. *American Journal of Sociology, 90*(3), 567–583.

Clarkwest, A. (2007). Spousal dissimilarity, race, and marital dissolution. *Journal of Marriage and Family*, *69*(3), 639–653. doi:10.1111/j.1741-3737.2007.00397.x.

Dartanto, T. (2010). The relationship between corruption and public investment at the municipalities' level in Indonesia. *Munich Personal RePEC Archive.*

Dollahite, D.C. & Lambert, N.M. (2007). Forsaking all others: How religious involvement promotes marital fidelity in Christian, Jewish, and Muslim couples. *Review of Religious Research*, 290–307. http://www.jstor.org/stable/20447445.

Glenn, N.D. & Kramer, K.B. (1987). The marriages and divorces of the children of divorce. *Journal of Marriage and Family*, *49*(4), 811–825. doi:10.2307/351974.

Heckert, D.A., Nowak, T.C. & Snyder, K.A. (1998). The impact of husbands' and wives' relative earnings on marital disruption. *Journal of Marriage and the Family*, *60*(33815), 690–703. doi:10.2307/353538.

Hiedemann, B., Suhomlinova, O. & O'Rand, A.M. (1998). Economic independence, economic status, and empty nest in midlife marital disruption. *Journal of Marriage and Family*, 219–231.

Hull, T.H. (2006). The marriage revolution in Asia. *Population Association of America.*

Lehrer, E.L. & Chiswick, C.U. (1993). Religion as a determinant of marital stability. *Demography*, *30*(3), 385–404.

Ono, H. (1998). Husbands' and wives' resources and marital dissolution. *Journal of Marriage and the Family*, *60*(3), 674–689. doi:10.2307/353537.

Pollak, R.S. (1994). For better or worse: The roles of power in models of distribution within marriage. *The American Economic Review*, *84*(2) 148–152.

RAND. Indonesian Family Life Survey (IFLS). Available: http://www.rand.org/labor/FLS/IFLS.html

Risman, B.J. & Johnson-Sumerford, D. (1998). Doing it fairly: A study of postgender marriages. *Journal of Marriage and the Family*, *60*(1), 23–40. doi:10.2307/353439.

Rogers, S.J. (2004). Dollars, dependency, and divorce: Four perspectives on the role of wives' income. *Journal of Marriage and Family*, *66*(1), 59. doi:10.1111/j.1741-3737.2004.00005.x.

Schaner, S. & Das, S. (2016). Female labor force participation in Asia: Indonesia country study. *ADB Economics Working Paper Series.*

Vaaler, M.L., Ellison, C.G. & Powers, D.A. (2009). Religious influences on the risk of marital dissolution. *Journal of Marriage and Family, 71*(4), 917–934. doi:10.1111/j.1741-3737.2009.00644.x.

Weagley, R.O., Chan, M.L. & Yan, J. (2007). Married couples' time allocation decisions and marital stability. *Journal of Family and Economic Issues*, *28*(3), 507–525. doi:10.1007/s10834-007-9070-y.

Whitton, S.W., Rhoades, G.K., Stanley, S.M. & Markman, H.J. (2008). Effects of parental divorce on marital commitment and confidence. *Journal of Family Psychology: JFP: Journal of the Division of Family Psychology of the American Psychological Association (Division 43)*, *22*(5), 789–793. doi:10.1037/a0012800.

Winslow, S. (2011). Marital conflict and the duration of wives' income advantage. *International Journal of Sociology of the Family*, *37*(2), 203–225.

Competition and Cooperation in Economics and Business – Gani et al. (Eds)
© 2018 Taylor & Francis Group, London, ISBN 978-1-138-62666-9

Analysis on the effects of social norms on local property tax compliance

S. Sukono & S. Djamaluddin
Department of Economics, Faculty of Economics and Business, Universitas Indonesia, Depok, Indonesia

ABSTRACT: The goal of this study is to determine the effects of social norms toward property tax compliance, either by direct or indirect influence. The data used are primary and secondary data. The primary data are from 156 respondents of Jakarta property taxpayers obtained by questionnaire and the secondary data are from the Tax Office of Jakarta Province. Data were analyzed using Structural Equation Modeling (SEM) with LISREL 8.7 software. Social norms refer to the identification of Cialdini and Tross (1998), which consists of descriptive norms (what people actually do), injunctive norms (the moral rule of the group), subjective norms (the perception of those closest) and personal norms (the standard behavior of individuals). The results of this study indicate that the property taxpayers' injunctive norms affect subjective norms, property taxpayers' subjective norms affect personal norms and personal norms have a significant effect on tax compliance. The relationship between social norms in tax compliance is indirect through personal norms, while injunctive and subjective norms have no direct effect on tax compliance.

1 INTRODUCTION

Property tax (PBB-P2) is expected to give significant contribution to the DKI Jakarta government revenue since it became a local tax. Data shows that the contribution of PBB-P2 is the highest among other local taxes for the last three years. In 2015 revenue from PBB-P2 was about Rp 3.6 billion, Rp 5.8 billion in 2014, and Rp 6.7 bilion in 2015. Data from the Jakarta Tax Office (Dinas Pelayanan Pajak Provinsi DKI Jakarta) shows that the realization of the PBB-P2 provided a substantial contribution to the local tax revenue, with an increase of 15% in 2013 to 20% in 2015.

However, revenue of PBB-P2 fluctuated between 2013 and 2015. Revenue from PBB-P2 decreased by 4.27% from 2013 to 2014 and increased by 1.77% from 2014 to 2015. The Jakarta PBB-P2 also did not achieve the revenue target of PBB-P2 in 2013, 2014 and 2015. Many factors

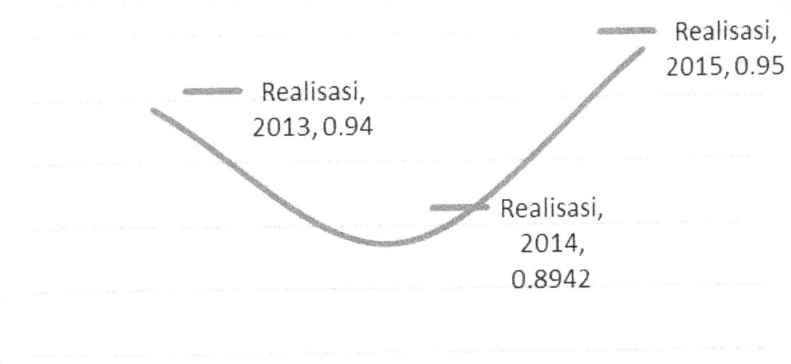

Figure 1. Realization of DKI Jakarta PBB-P2.

can affect tax revenue. Tax system performance and tax compliance play important roles in increasing tax revenue. This study will analyze tax compliance with regard to social norm factors.

In the early theory of tax compliance, researchers frame compliance as a rational decision based on the expected utility (Allingham and Sandmo, 1972). Compliance is modeled as a game between the taxpayer and the tax authority where compliance is influenced by the amount of potential tax evasion, the audit level, and the amount of fines to be paid if caught. This model is known as an economic-of-crime model, where the taxpayer is viewed and treated as a potential criminal. The emphasis is on the prevention of illegal behavior through audit levels and severe punishment.

The traditional economic model of compliance, which includes detection variables and enforcement variables, states that the rate of compliance is much higher than expected based on the audit and penalties based on the current level (Andreoni *et al.*, 1998). Then Andreoni *et al.* (1998) propose a study to explore the effect of psychological, social and moral norms on tax compliance, in order to explain and understand the level of this compliance-is-more-than-estimate issue. Research on morals by Bobek and Hatfield (2003), ethics research by Wenzel (2004), Torgler and Schaltegger (2005) and the study of social norms by Alm *et al.* (1999), Wenzel (2004) and Bobek et al. (2007) respond to the proposal. The results of these studies show that social norms, ethics and morals have strong influence on the behavior of tax compliance.

Accounting researchers who study the effect of non-economy on tax compliance focus on individual psychological variables such as morals and attitudes (Bobek and Hatfield, 2003). Meanwhile, economists are trying to ascertain the influence of social norms on tax compliance models, though not identified with appropriate social norms (Alm *et al.*, 1999).

This study adopts the study by Bobek *et al.* (2013) that analyzes the role of social norms in tax compliance behavior. Bobek *et al.* (2013) utilizes social norms identified by Cialdini and Trost (1998). Research on the influence of social norms in tax compliance has been conducted in various countries using income tax as the tax object, but comprehensive research on social norms has yet to be conducted on the taxable PBB-P2. This study will examine empirically and comprehensively how social norms are considered to affect the behavior of Jakarta PBB-P2 taxpayers' tax compliance.

1.1 *Research purposes*

The purpose of this study is to investigate the influence of social norms which consists of personal norms, subjective norms, injunctive norms and descriptive norms on tax compliance of Jakarta PBB-P2 taxpayers in 2015. The research objective is obtained by discovering the direct and indirect influences of social norms on tax compliance decisions (there are interactions of social norms before they impact tax compliance decisions).

1.2 *Research hypotheses*

The hypotheses of this study are derived from the model in Figure 2 for the indirect effects and Figure 3 for the direct effects as follows:

H12: PBB-P2 taxpayers' descriptive norms will affect their injunctive norms towards tax compliance

H13: PBB-P2 taxpayers' descriptive norms will affect their subjective norms towards tax compliance

H14: PBB-P2 taxpayers' descriptive norms will affect their personal norms on tax compliance

H23: PBB-P2 taxpayers' injunctive norms will affect their subjective norms towards tax compliance

H24: PBB-P2 taxpayers' injunctive norms will affect their personal norms on tax compliance

H34: PBB-P2 taxpayers' subjective norms on tax compliance will affect their personal norms towards tax compliance.

H45(a): PBB-P2 taxpayers personal norms towards tax compliance will directly affect their tax compliance decisions, with the relationship between subjective norms, injunctive

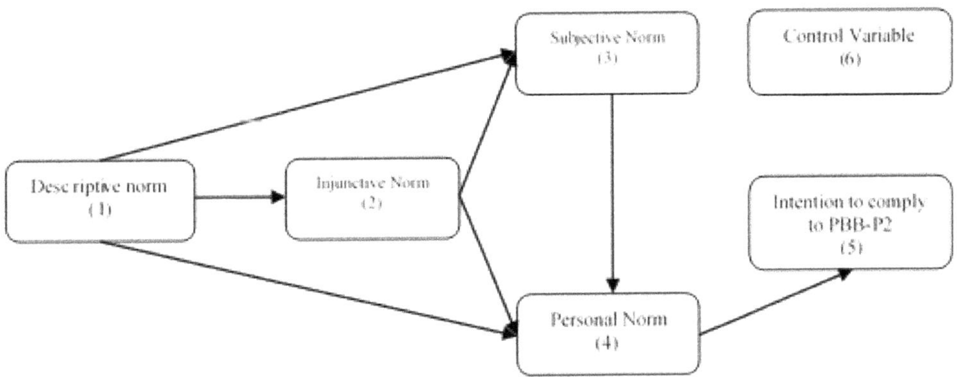

Figure 2. The research hypotheses to model indirect effects.

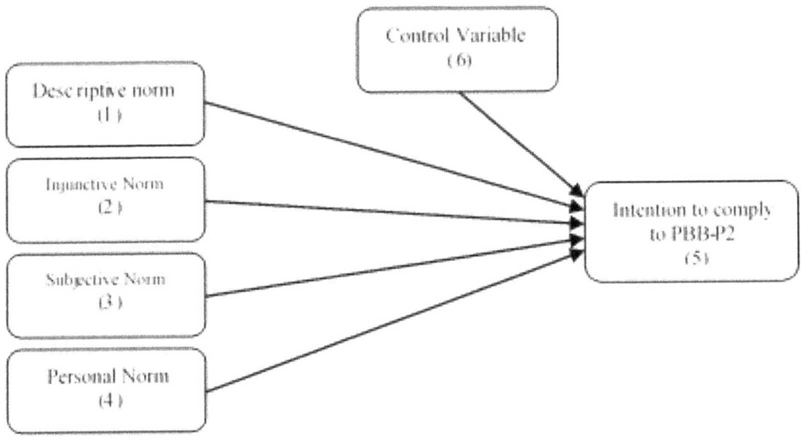

Figure 3. The research hypotheses to model direct effects.

norms and descriptive norms towards tax compliance only occurring indirectly (through personal norms).

H15: PBB-P2 taxpayers' descriptive norms towards tax compliance will directly influence the decision of their tax compliance.

H25: PBB-P2 taxpayers' injunctive norms towards tax compliance will directly influence the decision of their tax compliance.

H35: PBB-P2 taxpayers' subjective norms towards tax compliance will directly influence the decision of their tax compliance.

H45(b): PBB-P2 taxpayers' personal norms towards tax compliance will directly influence the decision of their tax compliance.

2 LITERATURE REVIEW

2.1 *Tax compliance*

According to Nowak in Zain (2007), taxpayer compliance means: "A climate of awareness of compliance and fulfillment of tax obligations, reflected in situations where taxpayers:

1. Understand or try to understand all the provisions of tax legislation
2. Fill out tax forms completely and clearly

3. Calculate the amount of tax payable correctly
4. Pay taxes owed on time."

2.2 *Social norms theory*

Social norms are defined as "the rules and standards that are understood by members of the group, and guide and/or limit social behavior without the force of law" (Cialdini and Trost, 1998, p. 152). Four constructions of social norms identified by Cialdini and Trost (1998) are descriptive norms, injunctive norms, subjective norms and personal norms.

Descriptive norms emerge or evolve from standards of observation of how others really behave in certain situations. These norms influence the behavior of individuals in a social group. Descriptive norms are based on the actual actions of the other members of the group and are sometimes contrary to the behavior of a group that has been approved (Cialdini and Trost, 1998). Injunctive norms, on the other hand, determine what should be done and are therefore moral rules of the group. These norms are mostly a reference when they discuss social norms in general (Allison, 1992). In contrast to the descriptive norms that describe how people actually act, injunctive norms represent how people should act (Kallgren *et al.*, 2000). Subjective norms specifically relate to expectations of significant others (effect) on us (e.g. family, friends, colleagues, and so on). Subjective norm is a particular type of injunctive norm. Personal norms are the standard behavior of someone that may arise as part of the internalization of injunctive norms. Personal norms evolve through the internalization of social norms of the group—a group which identified individuals (Wenzel, 2004).

2.3 *Pajak bumi dan bangunan perdesaan dan perkotaan (PBB-P2)*

According to the Law No. 28 of 2009 on Local Taxes and Levies and Jakarta Local Regulation No. 68 Year 2011 on Rural and Urban Land and Building Tax, PBB-P2 is defined as a tax on land and/or buildings owned, controlled, and/or utilized by an individual or agency, except the area used for plantation, forestry, and mining. Land is the earth's surface that includes soil and inland waters and territorial sea of the district/city. Building is a construction that is planted or permanently attached to the land and/or inland waterway and/or sea.

3 RESEARCH METHODS

3.1 *Structural Equation Modeling (SEM)*

This study uses SEM to analyze the hypotheses. SEM is an appropriate model used for multivariate analysis in social research in addition to financial or variables used to a nominal scale/ratio. This study uses latent variables (variables that cannot be measured directly), which are social norms. Social norms variable cannot be measured directly, so it must use an indicator or questionnaire. This is in contrast to direct measurable variables such as Gross Domestic Product, the value of export, import values, etc. If using regression analysis, then each of these variables are assumed to be measured directly, so we use the average score or the total of these items. However, this method ignores the presence of measurement error. If we do not take into account the measurement error of the path coefficients, the results can be biased (Smith and Langfield-Smith, 2004). In addition, SEM is capable of testing complex research and many variables simultaneously. SEM analysis can be completed with one estimate that others solved with some regression equation. SEM can perform factor analysis, regression and pathways at once.

3.2 *Model framework research*

SEM models for indirect and direct effects are shown in Figure 4 and Figure 5.

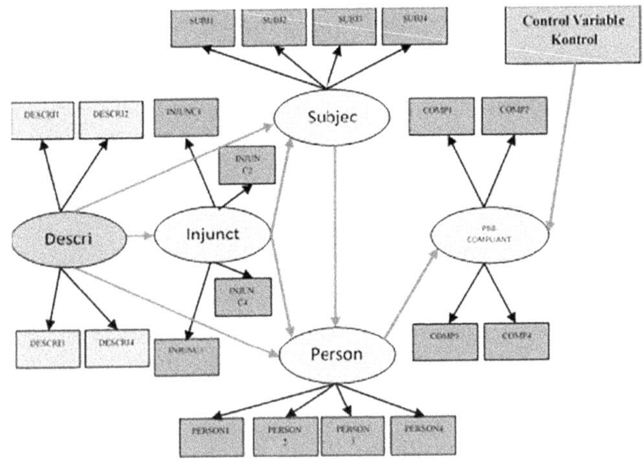

Figure 4. SEM Model for indirect influence.

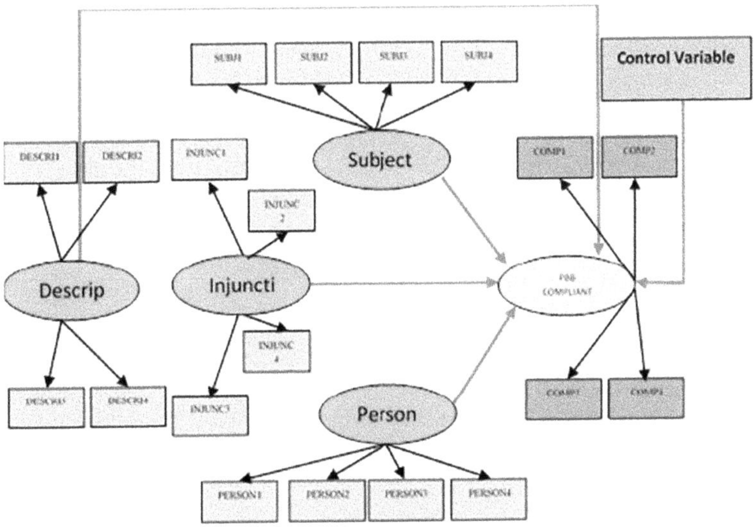

Figure 5. SEM Model for direct influence.

3.3 *Types and sources of data*

Data used in this study are primary and secondary data. Primary data were obtained from the questionnaire answered by Jakarta PBB-P2 taxpayers. The questionnaire was adopted from Bobek et al. (2013) and Jimenez (2013) with modifications adapted to the conditions of the PBB-P2 taxation system in Jakarta. The primary data to be processed are the perception of PBB-P2 taxpayers to the questions in the questionnaire. Secondary data obtained from the Jakarta Tax Office are the PBB-P2 target and realization data from 2013 to 2015 and other data required in this study.

3.4 *Sample selection methods*

In this study, the population is PBB-P2 taxpayers in the area of Central Jakarta City Administration. The reasons for selecting the area of Central Jakarta City Administration to take a sample of respondents are its relatively smaller area and considerations of time and cost. The sampling technique used is incidental sampling based on chance, meaning that anyone who by chance/incidentally met with investigators can be used as a sample if they meet the sample requirement, i.e. individual PBB-P2 taxpayers (not a legal entity). The sampling location is in the village office in the area of Central Jakarta City Administration.

3.5 *Data collection and processing techniques*

This study is a cross-sectional study that only examines a number of respondents at a certain time. Data collection techniques used in this study are:

1. PBB-P2 taxpayers who happened to be in the village office and were willing to become respondents were given data collector instrument in the form of a questionnaire to be filled. The data collection was conducted over 1.5 (one and a half) months commencing from February 8 to March 19, 2016.
2. Study of literature, which is studying and citing theories or concepts from a number of literatures, such as text books, journals, magazines, and other scientific works which are relevant to the research problem.

3.6 *Data analysis method*

Based on the model of the influence of social norms on tax compliance behavior, the analysis method is Structural Equation Modeling (SEM). SEM is processed in two stages (two-stage approach). The first is an analysis of the measurement model to test compatibility (GOFI), validity and reliability. After measurement analysis indicates its fit the process continues to the second phase, namely structural analysis, which includes the overall suitability test models, analysis of causal relationships and the results of hypothesis testing.

4 RESULTS AND DISCUSSION

4.1 *Demographic characteristics*

The number of samples collected and used in this study were 156 samples from the territory of Central Jakarta City Administration. The demographic data of respondents are shown in Table 1.

4.2 *Data processing results SEM with LISREL 8.7*

The results of the analysis of the measurement model to the model of this study are as follows:

1. Descriptive latent variables are not included in the analysis of the structural model because they do not meet the reliability test.
2. 6 (six) observed variables are removed from the model and not included in the analysis of structural models. They are the variables of INJUNC3, INJUNC4, DECRI1, DESCRI2, DECRI3 and DESCRI4, leaving 14 (fourteen) other observable variables.
3. 6 (six) control variables are included in the structural model.

If descriptive latent variables are still included in the analysis of the structural model, they would potentially create a problematic model and could change coefficient directions of the hypotheses.

The path diagrams for structural models with direct and indirect influence are shown in Figure 6a and Figure 6b.

Testing the hypotheses with direct and indirect influence was conducted and resulted in four (4) significant paths. Table 2 summarizes the results of testing the hypotheses.

Control variables consisting of age, gender, education, employment, income and revenue sources were not included in the summary Table 2 above because none of these variables were

Table 1. Characteristics of respondents.

	Compliant	Not compliant
PBB-P2 Compliance	108	48
Gender		
Male	57	29
Female	51	19
Age		
<30	5	1
30–39	30	8
40–49	39	23
>50	34	16
Education		
Below Elementary School	5	2
Junior High School	12	7
Senior High School	50	24
University Graduate	41	15
Income level		
≤ Rp. 2,700,000,-	46	18
Rp. 2,700,001,- to Rp. 8,000,000,-	48	29
≥ Rp. 8,000,001,-	14	1
Occupation		
Civil servant	4	0
Privately owned company employee	31	15
Entrepreneur	24	11
Not working	6	4
Retired	6	6
State owned company employee	2	2
Neighborhood unit chief	25	9
Others	10	1
Income source		
Interest	1	0
Wages	46	18
Retirement benefits	7	7
Profit	26	14
Other	28	9
Inheritance	0	0
Location		
Menteng	10	15
Gambir	21	5
Senen	18	8
Kemayoran	21	6
Sawah Besar	5	2
Johar Baru	10	5
Cempaka Putih	5	1
Tanah Abang	18	6

significant. This result is different from the literature or previous studies on the control variables' effect on tax compliance as described in the literature in the previous section.

4.3 *Discussion*

1. Hypothesis H23 predicts that PBB-P2 taxpayers' injunctive norms will affect their subjective norms towards tax compliance. This hypothesis is proven with an estimated value of 0.72 and t-value of 8.57. This can be interpreted to mean that the moral rules of tax

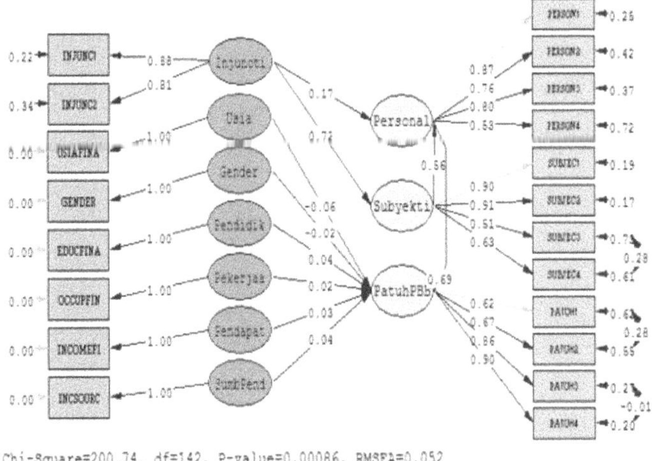

Figure 6a. The indirect effects of social norms on tax compliance.

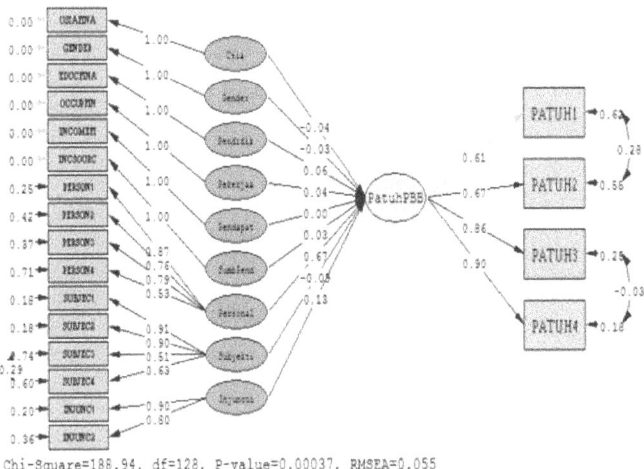

Figure 6b. The direct effects of social norms on tax compliance.

Table 2. Summary of results of testing the hypotheses of direct and indirect influence.

Hypothesis	Path	Estimation	t-values	Conclusion	Type of effect
H23	Injunctive -> Subjective	0.72	8.57	Significant	Indirect
H24	Injunctive -> Personal	0.17	1.49	Not Significant	Indirect
H34	Subjective -> Personal	0.56	4.79	Significant	Indirect
H45 (a)	Personal -> PBB Compliant	0.69	5.71	Significant	Indirect
H25	Injunctive --> PBB Compliant	0.13	1.09	Not Significant	Direct
H35	Subjective --> PBB Compliant	–0.09	–0.65	Not Significant	Direct
H45 (b)	Personal --> PBB Compliant	0.67	4.49	Significant	Direct

compliance in a group of PBB-P2 taxpayers have significant effect on what other PBB-P2 taxpayers should do in relation to tax compliance.

Social values of tax compliance (which is stated in the questionnaire) have been the moral rules in a group of PBB-P2 taxpayers. Therefore, to build a social relationship with the group of PBB-P2 taxpayers, others should adapt these values to avoid social stigma. Another driving factor is their common background with the group of PBB-P2 taxpayers in terms of tax compliance. This interaction causes the moral rules to affect the decision of the other PBB-P2 taxpayers' compliance.

On the statements related to injunctive norms, 80.66% of respondents (the average number of answers comprising somewhat agree, agree and strongly agree) give positive responses to the questions provided. The average score on the statements related to injunctive norms is 5.54. This shows that according to the perception of respondents, the Jakarta PBB-P2 taxpayers agree to comply with the PBB-P2. These data reinforce the influence of injunctive norms on subjective norms.

2. Hypothesis H24 predicts that the injunctive norms of the Jakarta PBB-P2 taxpayers will affect the taxpayers' personal norms on tax compliance. This hypothesis is disproven with an estimated value of 0.17 and t-value of 1.49. This can be interpreted to mean that the moral rules of tax compliance in a group of Jakarta PBB-P2 taxpayers have no significant effect on the standard of behavior of individuals/PBB-P2 taxpayers related to tax compliance.

   The social values of tax compliance (which is stated in the questionnaire) that have become moral rules in a group of Jakarta PBB-P2 taxpayers, cannot affect the behavior of the individual/PBB-P2 taxpayer. This could be due to the absence of interaction or communication of the individual/PBB-P2 taxpayer with Jakarta PBB-P2 taxpayer groups so that these values are not internalized in the individual/PBB-P2 taxpayer to become standard behavior.

   The result of this study cannot prove a theory by Schwartz (1977) described in the literature, that individuals develop behavioral standards themselves (personal norms) of behaviors expected by people who learned through social interaction (injunctive norms). Individuals may initially follow the injunctive norms to avoid social stigma, but over time, individuals may continue to follow these norms for internal reasons, such as improving their self-image. In this study the individual/Jakarta PBB-P2 taxpayer is not affected by the moral rule of a group of overall Jakarta PBB-P2 taxpayers. Individuals may interact to build social relationships with those closest alone.

   Another fact that emerged from this study is the Jakarta PBB-P2 taxpayers are individualists. They do not interact with the surrounding community and are busy with their individual activities. They only interact and communicate with the people closest to them either at home or in the workplace, which gives the highest impact.

3. Hypothesis H34 predicts that the subjective norms of Jakarta PBB-P2 taxpayers on tax compliance would affect their personal norms towards tax compliance. This hypothesis is proven with an estimated value of 0.56 and t-value of 4.79. The significance of this hypothesis is the perception of the people closest to taxpayers in relation to tax compliance and the significant positive effect on taxpayers' standard behavior or what PBB-P2 taxpayers believe he/she should do in relation to tax compliance.

   It can also be seen from the respondents' answers related to subjective norms that more than 80.26% (the average number of answers comprising somewhat agree, agree and strongly agree) of PBB-P2 taxpayers agree to those closest to them on PBB-P2 tax compliance. The average score for statements related to subjective norms is 5.64. This shows that those closest to PBB-P2 taxpayers influence their standard of behavior.

   The result of this study is the same as the opinion of Blanthorne and Kaplan (2008) described in the literature. Individuals/Jakarta PBB-P2 taxpayers internalize the values of tax compliance from those closest to them. They may be surrounded by family members and friends who support tax compliance, making it possible to show the ethics that support this behavior. Jakarta PBB-P2 taxpayers more often refer to the norms of those closest to them regarding tax compliance rather than the overall PBB-P2 taxpayer group norms. It is also proven that the relationship between injunctive norms and personal norms is not significant. In addition to the interaction with people, they make tax compliance become central/important.

4. Hypothesis 45 (a) predicts that the PBB-P2 taxpayers' personal norms on tax compliance will directly influence the decision of their tax compliance, while the relationship between subjective norms, injunctive norms, and descriptive norms on tax compliance will only

occur indirectly (through personal norms). This hypothesis is proven by the path from the personal norms toward the intention to comply with these taxes which has a significant value of 0.69 and t-value of 5.71. This means the standard compliant behavior of PBB-P2 taxpayers is affected by the perception of those closest to them about tax compliance, and those closest to the taxpayers are affected by moral rules in a group of PBB-P2 taxpayers. This result asserts that the decision of tax compliance is only a personal decision; personal norm is the only direct social norm affecting tax compliance.

On the answers related to personal norms, 85.27% of respondents (the average number of answers comprising somewhat agree, disagree and strongly agree) agree with the given questions. The average score on respondents' answers to the statements of personal norms is 5.80. This indicates that Jakarta PBB-P2 taxpayers adhere to compliance standards that reinforce the norms on tax compliance and will directly influence the decision of their tax compliance. The results show that the influence of injunctive norms to the willingness to pay taxes is not significant and that the hypothesis is disproven with an estimated value of 0.13 and their behavior follows the above explanation.

The result of this study is the same as the result of research by Bobek *et al.* (2013) which states that the path of personal norms' influence on the intention to comply with the PBB-P2 is significant. The result of this study is also the same as Schwartz (1977), which presupposes that personal norm is the only type of norm that directly affects the behavior in this study, namely the behavior of PBB-P2 tax compliance.

5. Hypothesis H25 predicts that Jakarta PBB-P2 Jakarta taxpayers' injunctive norms are not significant with an estimated value of 0.13 and t-value of 1.09. These results can be interpreted to mean that the moral rules of tax compliance among Jakarta PBB-P2 taxpayers do not directly influence the willingness to comply with PBB-P2.

   The result of this study is the same as the result of Bobek *et al.* (2013) and Jimenez (2013) where the injunctive norm has no direct effect on tax compliance. This hypothesis is to test the opinion of Cialdini and Trost (1998) claiming that various types of norms can be activated together with each other in a certain situation. Thus, other types of social norms can influence personal behavior outside the norm. Although the behavior of tax compliance is relatively private, there is a possibility that other norms are also influential.
6. Hypothesis H35 predicts that Jakarta PBB-P2 taxpayers' subjective norms towards tax compliance will directly affect their tax compliance decisions. The result shows that this path is not significant with an estimated value of –0.09 and t-value of –0.65. The meaning of this hypothesis is that the perception of compliance of the people closest to the Jakarta PBB-P2 taxpayers has no direct effect on the willingness to comply with PBB-P2.

   The result of this study is different from the result of research by Bobek et al. (2013) in which subjective norms directly affect tax compliance and is the same as the result of research by Jimenez (2013) in which subjective norms have no direct effect on tax compliance.
7. Hypothesis H45 (b) predicts that the PBB-P2 taxpayers' personal norms on tax compliance will directly influence the decision of their tax compliance. The result shows that this path is not significant with an estimated value of 0.67 and t-value of 4.49. This means the standard behavior of PBB-P2 taxpayers' compliance has a positive and significant direct effect on the willingness to comply with PBB-P2. Personal norms in this model are not affected by other social norms like the model with indirect influence. Another factor that might influence the decision of the taxpayers is their trust on the tax authorities or the government.

   The result of this study is the same as the result of research by Bobek *et al.* (2013) and Jimenez (2013) where personal norms directly and significantly impact tax compliance. The insignificant path between injunctive norms and subjective norms on tax compliance indicates that the relationship between the norms and full tax compliance behavior is mediated by the relationship between social norms. The significant relationships between personal norms and tax compliance show that tax compliance is a purely personal decision.

8. These results also reaffirm the opinion of Cialdini and Trost (1998) described in the hypothesis of the direct influence of injunctive norms on the above tax compliance of Jakarta PBB-P2 taxpayers. Personal norm is the only norm that directly affects tax compliance behavior based on this research.

5 CONCLUSION

Based on the testing of the hypotheses of the study, we can conclude the following:

1. The standard of tax compliance behavior (personal norms) of Jakarta PBB-P2 taxpayers is influenced by the perception of tax compliance of those closest to them (subjective norms), while the people closest to the taxpayers are strongly influenced by the moral rules of tax compliance in a group of Jakarta PBB-P2 taxpayers (injunctive norms).
2. The standard of Jakarta PBB-P2 taxpayer's behavior (personal norms) is the only norm that directly or indirectly impacts the tax compliance behavior of Jakarta PBB-P2 taxpayers.
3. Moral rules of tax compliance in the group of Jakarta PBB-P2 taxpayers (injunctive norms) and the perception of the people closest to the taxpayers (subjective norms) have no direct effect on tax compliance behavior, but must interact with other norms to be able to influence the behavior of tax compliance.

6 RECOMMENDATIONS

Based on these results, we recommend the following:

1. The DKI Jakarta Government needs more optimized internet-based media socialization which has been used by the Jakarta Tax Office (Facebook, Twitter, Instagram, Pinterest, Google+ and website) to raise awareness of the importance of paying PBB-P2 and the impact if taxpayers do not dutifully pay PBB-P2. Moreover, Jakarta PBB-P2 taxpayers are more likely to be individuals, so there is less possibility for communication between them. This method aims to target the personal norms of Jakarta PBB-P2 taxpayers.
2. The DKI Jakarta Government must communicate the goals and the realization of PBB-P2 and always keep them updated in language that is easily understood by all Jakarta PBB-P2 taxpayers, so that they are encouraged to pay their obligations on time. The government should also publish the names of Jakarta PBB-P2 taxpayers who commit violations. This method aims to target PBB-P2 taxpayers' injunctive norms.
3. The Jakarta Tax Office must conduct a persuasive tax campaign by identifying Jakarta PBB-P2 taxpayers as dutiful taxpayers, creating a feeling of being part of a group of upstanding citizens. Thus, Jakarta PBB-P2 taxpayers will adjust to these conditions to establish and maintain social relationships.

REFERENCES

Ananda, Sely, Desi Nurmalasasi, and Alita Destiyamda Pertiwi. 2013. "Pengaruh Sosial Konformitas, Compliance, dan Obidien." Universitas Mercubuana. http://www.slideshare.net/aiananda/psikologi-sosial-pengaruh-sosialkelompok-10.

Allingham, Michael G, and Agnar Sandmo. 1972. "Income tax evasion: A theoretical analysis." *Journal of Public Economics* 1: 323–338.

Alm, James. 1999. Tax Compliance and Administration. https://www.researchgate.net/profile/James_Alm/publication/247699437_Tax_Compliance_and_Administration/links/54ac05e80cf2bce6aa1de74a.pdf.

Alm, James, and Benno Torgler. 2011. "Do Ethics Matter? Tax Compliance and Morality." *Journal of Business Ethics* 101 (4): 635–51. doi:10.1007/s10551-011-0761-9.

Allison, Paul D. (1992) "The Cultural Evolution of Beneficent Norms". Social Forces, Vo. 71, No.2. Pg. 279–301. http://statisticalhorizons.com/wp-content/uploads/AllisonSF92.pdf

Andreoni, James, Brian Erard, and Jonathan Feinstein. 1998. "Tax Compliance." *Journal of Economic Literature* 36 (2): 818–60. doi:10.1016/j.jebo.2003.02.003.

Ariany, Nany. 2010. "*Analisis Progresivitas Pajak BumidDan Bangunan (PBB) Bagi Wajib Pajak Orang Pribadi di Jakarta SelatansSerta Hubungannya dengan Ketidakmampuan Membayar PBB.*" Master's thesis, Universitas Indonesia.

Badan Pajak dan Retribusi Daerah, Provinsi DKI Jakarta. 2015. "Tata Cara Permohonan Keberatan Pbb (Pergub 203 tahun 2012)." http://dpp.jakarta.go.id/59235/tata-cara-permohonan-keberatan-pbb-pergub-203-tahun-2012/.

Badan Pengawasan Keuangan dan Pembangunan. 2014. "PBB Hilang Rp11 Miliar." http://app.bpkp.go.id:11000/kliping/c_index/view_berita_pusat/168.

Baron, Robert A, and Donn Byrne. 2004. *Psikologi Sosial.* 10th ed. Jakarta: Penerbit Erlangga.

Blanthorne, Cindy, and Steven Kaplan. 2008. "An Egocentric Model of the Relations among the Opportunity to Underreport, Social Norms, Ethical Beliefs, and Underreporting Behavior." *Accounting, Organizations and Society* 33 (7–8): 684–703. doi:10.1016/j.aos.2008.02.001.

Bobek, Donna D, Robin W Roberts, and John T Sweeney. 2007. "The social norms of tax compliance: Evidence from Australia, Singapore, and the United States." *Journal of Business Ethics* 74(1): 49–64.

Bobek, Donna D, and Richard C Hatfield. 2003. "An Investigation of The Theory of Planned Behavior and The Role of Moral Obligation in Tax Compliance." *Behavioral Research in Accounting* 15 (1): 13–38. doi: http://dx.doi.org/10.2308/bria.2003.15.1.13.

Bobek, Donna D, Amy M. Hageman, and Charles F. Kelliher. 2013. "Analyzing the Role of Social Norms in Tax Compliance Behavior." *Journal of Business Ethics* 115 (3): 451–68. doi:10.1007/s10551-012-1390-7.

Brown, Abraham, and Crawford Moodie. 2009. "The Influence of Tobacco Marketing on Adolescent Smoking Intentions via Normative Beliefs." *Health Education Research* 24 (4): 721–33. doi:10.1093/her/cyp007.

Budiningrum, and Endah Wening. 2014. *"Pengaruh Norma-Norma Sosial terhadap Perilaku Kepatuhan Pajak Usaha Mikro, Kecil, dan Menengah (UMKM)."* Unpublished master's thesis, Universitas Gajah Mada.

Candra Fajri Ananda E S. 2012. *Analisa Dampak Pengalihan Pemungutan BPHTB ke Daerah Terhadap Kondisi Fiskal Daerah.* Jakarta: Dirjen Perimbangan Keuangan, Kementerian Keuangan Republik Indonesia.

Cialdini, Robert B, and Melanie R Trost. 1998. "Social Influence: Social Norms, Conformity and Compliance." *The Handbook of Social Psychology* 2. doi:10.2307/2654253.

Clapp, John D, and James E Lange, Cristel Russell, Audrey Shillington, and Robert B. Voas. 2003. "A Failed Norms Social Marketing Campaign." *Journal of Studies on Alcohol* 64 (3): 409–14. doi:10.15288/jsa.2003.64.409.

Devano, Sony, and Siti Kurnia Rahayu. 2006. *Perpajakan: Konsep, Teori dan Isu*, Jakarta: Kencana.

Data PBB-P2 Dinas Pelayanan Pajak Provinsi DKI Jakarta.

Chau, Gerald, and Patrick Leung. 2009. "A Critical Review of Fischer Tax Compliance Model: A Research Synthesis." *Journal of Accounting and Taxation* 1 (2): 034–040. doi:10.5897/JAT09.021.

Goldstein, Noah J, Robert B Cialdini, and Vladas Griskevicius. 2008. "A Room with a View point: Using Social Norms to Motivate Environmental Conservation in Hotels." *Journal of Consumer Research* 35 (3): 472–482. doi: 10.1086/586910.

Gujarati, Damodar N. 2003. *Basic Econometrics.* 4th ed. New York: McGraw Hill.

Hair, Joseph F, Bill Black, Barry Babin, Rolph E Anderson, and Ronald L Tatham. 2007. *Multivariate Data Analysis.* 6th ed. New Jersey: Pearson International Edition.

Hechter, Michael, and Karl-Dieter Opp. 2001. *Social Norms (Eds.).* New York: Russell Sage Foundation.

Hyun, Jin Kwon. 2005. *Tax Compliance in Korea and Japan: Why are they different?* Department of Economics Ajou, University Suwon, South Korea.

Iriawan, Her Ovita Trianggono. 2010. *"Pengaruh Pelayanan Prima Terhadap Kepatuhan Wajib Pajak Di Kantor Pelayanan Pajak Pratama Jakarta Gambir Empat."* Unpublished master's thesis, Universitas Indonesia.

Ismhi, Nurul. (n.d.). "Pengertian Pajak." http://isma-ismi.com/pengertian-pajak.html.

Jimenez, Peggy D. 2013. "Tax Compliance in a Social Setting: The Influence of Norms, Perceptions of Fairness, and Trust in Government on Taxpayer Compliance." Dissertation, University of North Texas.

Kallgren, Carl A, Raymond R Reno, and Robert B Cialdini. 2000. "A Focus Theory of Normative Conduct: When Norms Do and Do Not Affect Behavior." *Personality and Social Psychology Bulletin* 26 (8): 1002–12. doi:10.1177/01461672002610009.

Kementerian Keuangan Republik Indonesia. 2014. *Budget in brief APBN 2015.* Jakarta: Direktorat Penyusunan APBN, Direktorat Jenderal Anggaran.

Kementerian Keuangan Republik Indonesia. 2014. *Pedoman Umum Pengelolaan Pajak Bumi dan Bangunan Perdesaan dan Perkotaan.* Jakarta: Kementerian Keuangan Republik Indonesia, Direktorat Jenderal Perimbangan Keuangan.
Kementrian Keuangan. 2005. *Bab IV Bagaimana Fungsi Pajak dalam Pembangunan?* Direktoral Jenderal Pajak. http://www.pajak.go.id/sites/default/files/BAB%20IV%20Bagaimana%20 Fungsi%20 Pajak%20Dalam%20Pembangunan.pdf.
Kementrian Keuangan Replubik Indonesia. 2013. *Undang-Undang Republik Indonesia Nomor 28 Tahun 2007 Tentang Perubahan Ketiga atas Undang-Undang Nomor 6 Tahun 1983.* http://www.kemenkeu.go.id/sites/default/files/UU-KUP%20Mobile.pdf
Kementrian Keuangan Replubik Indonesia. *Undang-Undang Nomor 28 Tahun 2009 sebagai pengganti Undang-Undang Nomor 18 Tahun 1997 dan Undang-Undang Nomor 34 Tahun 2000 tentang Pajak Daerah dan Retribusi Daerah.* http://www.djpk.depkeu.go.id/attach/post-no-28-tahun-2009-tentang-pajak-daerah-dan-retribusi-daerah/UU-427–973-UU_28_Tahun_2009_Ttg_PDRD.pdf
Keputusan Kepala Dinas Pelayanan Pajak Provinsi DKI Jakarta Nomor 2885 Tahun 2015 Tentang Penghapusan Sanksi Administrasi Pajak Bumi dan Bangunan Perdesaan dan Perkotaan (PBB-P2) Terutang Tahun 2013, Tahun 2014 dan Tahun 2015.
Klein, Katherine, and Franklin Boster. 2005. *Subjective, Descriptive and Injunctive Norms Three Separate Constructs.* Michigan: Michigan State University, Department of Communication.
Manning, Mark. 2009. "Relative Influence of Descriptive and Injunctive Norms on Behavioral Intentions." Doctoral's dissertation, University of Massachusetts Amherst.
Mengapa Menggunakan Sem (Structural Equation Modeling)? 2015. http://budiwidyatama.blogspot.co.id/2015/05/mengapa-menggunakan-sem-structural.html.
Mengapa Menggunakan Sem (Structural Equation Modeling)? 2014. http://yrasemsi.blogspot.co.id/2014/01/mengapa-menggunakan-sem-structural.html.
Menteri Keuangan Republik Indonesia. 2007. *Peraturan Menteri Keuangan Nomor: 192/KMK.03/2007 tentang Tata Cara Penetapan Wajib Pajak dengan Kriteria Tertentu dalam Rangka Pengembalian Pendahuluan Kelebihan Pembayaran Pajak.*
News replubika.co.id. 2014. "Sisi Positif dan Negatif Kenaikan NJOP DKI Jakarta." http://www.republika.co.id/berita/nasional/jabodetabek-nasional/14/03/11/n29p57-sisi-positif-dan-negatif-kenaikan-njop-dki-jakarta
Nurmantu, Safri. 2003. *Pengantar Perpajakan.* Jakarta: Granit.
Onu, Diana, and Lynne Oats. 2014. "Social Norms and Tax Compliance." *TARC Tax Administration Research Center*, 6–14.
Peraturan Daerah. 2011. *Peraturan Daerah Nomor 16 Tahun 2011 tentang Pajak Bumi dan Bangunan Perdesaan dan Perkotaan.*
Peraturan Gubernur. 2015. *Peraturan Gubernur DKI Jakarta Nomor 134 Tahun 2015 tentang Pengurangan Pokok dan Penghapusan Sanksi Administrasi Piutang Pajak Bumi dan Bangunan Perdesaan dan Perkotaan Tahun Pajak Sebelum Dikelola oleh Pemerintah Daerah.*
Pommerehne, Werner W, Albert Hart, and Bruno S Frey. 1994. "Tax Morale, Tax Evasion and the Choice of Policy Instruments in Different Political Systems." *Public Finance.*
Prasetyo, Kristian Agung. 2014. "Quo Vadis Tax Ratio Indonesia?" http://www.bppk.kemenkeu.go.id/publikasi/artikel/167-artikel-pajak/12643-quo-vadis-tax-ratio-indonesia.
Rahardjo, Soemarso S. 2007. *Perpajakan: Pendekatan Komprehensif.* Jakarta: Salemba Empat.
Rosdiana, Haula, and Rasin Tarigan. 2005. *Perpajakan teori dan aplikasi.* Jakarta: PT. Raja Grafindo Persada.
Sarwono, Sarlito W, and Eko A Meinarno. 2012. *Psikologi Sosial.* Jakarta: Salemba Humanika.
Schwartz, Shalom H. 1977. "Normative Influences on Altruism." *Advances in Experimental Social Psychology* 10 (C): 221–79. doi:10.1016/S0065-2601(08)60358-5.
Sekundina, Kartika. 2009. "Faktor-Faktor yang Berpengaruh terhadap Pembayaran Pajak Bumi dan Bangunan di Kota Depok." Unpublished master's thesis, Universitas Indonesia.
Smith, David Alan; Langfield-Smith, Kim M. "Structural equation modeling in management accounting research: critical analysis and opportunities". Journal of Accounting Literature, Vol. 23, 2004, p. 49–86.
Slemrod, Joel, and Jon Bakija. 2008. "Taxing Ourselves: A Citizen's Guide to the Debate over Taxes." *MIT Press Books* 1. https://ideas.repec.org/b/mtp/titles/0262693631.html.
Soemitro, H Rochmat 1998. *Asas dan Dasar Perpajakan I.* Bandung: Refika Aditama.
Sugiyono. 2007. *Metode Penelitian Administrasi.* Bandung: Alfabeta.
Sriharati, Anggraeni. 2014. "Analisis Hubungan Persepsi Wajib Pajak Atas Faktor Struktur Pajak Terhadap Kepatuhan Wajib Pajak Dengan Pendekatan Logistik Model (Studi Kasus: Wajib Pajak Bumi dan Bangunan Perdesaan dan Perkotaan di Provinsi DKI Jakarta Tahun 2014." Unpublished master's thesis, Universitas Indonesia.

TATA CARA PERMOHONAN KEBERATAN PBB (Pergub 203 tahun 2012). 2015. http://dpp jakarta.go.id/59235/tata-cara-permohonan-keberatan-pbb-pergub-203-tahun-2012

Taylor, Shelley E, Letitia Ane Peplau, and David O Sears. 2009. *Psikologi Sosial*. Jakarta: Kencana.

Torgler, Benno and Christoph A. Schaltegger. 2005 Tax Moral and Fiscal Policy. http://leitner.yale.edu/sites/default/files/files/resources/docs/taxmorale.pdf

Torgler, Benno. 2007. *Tax Compliance and Tax Morale*. Northampton, MA: *Edward Elgar Publishing*.

Torgler, Benno. 2004. "Moral Suasion: An Alternative Tax Policy Strategy? Evidence from a Controlled Field Experiment in Switzerland." *Economics of Governance* 5 (3): 235–53. doi:10.1007/s10101-004-0077-7.

Torgler, Benno. 2002. "Speaking to Theorists and Searching for Facts: Tax Morale." *Journal of Economic Surveys* 16 (5): 657–83. doi:10.1111/1467-6419.00185.

Torgler, Benno. 2001. "Is tax evasion never justifiable?" *Journal of Public Finance and Public Choice* 19 (2/3): 143–168.

Wangsa, Sudira. 2009. "Tinjauan Moral dan Etika dan Dampaknya terhadap Kepatuhan Pajak" Unpublished master's thesis, Universitas Indonesia.

Wenzel, Michael. 2004. "An Analysis of Norm Processes in Tax Compliance." *Journal of Economic Psychology* 25 (2): 213–28. doi:10.1016/S0167-4870(02)00168-X.

Wenzel, Michael. 2005. "Motivation or Rationalisation? Causal Relations between Ethics, Norms and Tax Compliance." *Journal of Economic Psychology* 26 (4): 491–508. doi:10.1016/j.joep.2004.03.003.

Wijanto, Setyo Hari. 2015. *Metode Penelitian menggunakan Structural Equation Modelling dengan Lisrel 9*. Jakarta: Lembaga Penerbit FEBUI.

Winarno, Sigit, and Sujana Ismaya. 2003. *Kamus Besar Ekonomi,* Bandung: Pustaka Grafika.

Wu, Shih-Ying, and Mei-Jane Teng. 2005. "Determinants of Tax Compliance—A Cross-Country Analysis." *FinanzArchiv/Public Finance Analysis* 61 (3): 393–417. doi:10.1628/001522105774979001.

Yamin, Sofyan. 2014. *Rahasia Olah Data Lisrel*. Jakarta: Mitra Wacana Media.

Zain, Mohammad. 2007. *Manajemen Perpajakan*. Jakarta: Salemba Empat.

http://www.perpustakaan.kemenkeu.go.id/FOLDERJURNAL/2014_kajian_pprf_Pajak%20Potensi%20dan%20Pengumpulannya.pdf.

Competition and Cooperation in Economics and Business – Gani et al. (Eds)
© 2018 Taylor & Francis Group, London, ISBN 978-1-138-62666-9

Financial inclusion: Household access to credit in Indonesia

B.Y. Gitaharie & L. Soelistianingsih
Department of Economics, Faculty of Economics and Business, Universitas Indonesia, Depok, Indonesia

T. Djutaharta
Institute of Demography, Faculty of Economics and Business, Universitas Indonesia, Depok, Indonesia

ABSTRACT: Financial inclusion has been an important issue recently, and has become one of the programmes of the Millennium Development Goals (MDG) of the United Nations (UN) to alleviate property all over the world. Indonesia has taken part in the programme. Financial inclusion aims to open access to formal financial services, especially bank services, to the poor. This access could leverage the poor's financial ability to provide capital for their business activities and improve their welfare with more affordable interest charged compared to non-financial institutions or informal moneylenders. The World Bank (2010) reports that only 21% of the Indonesian population have access to banks, which is considered a very low number. Using the National Economic Social Survey (SUSENAS) data of 2008 and 2012, this study identifies household profiles and analyses factors that determined households' access to loans from banks, non-bank institutions, and individual (informal) sources. The possibility for households to obtain loans is influenced by several factors such as demographics, including sex, socio-economic condition, education, and the effectiveness of government programmes on financial inclusion. The biggest constraint to obtaining bank loans is mainly the collateral. Most of the poor have no quality assets to pledge for bank collateral. The findings in this study could help the Indonesian government formulate more effective policies for poverty alleviation.

1 INTRODUCTION

Financial inclusion has become one important programme of the UN Millennium Development Goals (MDG) to open access to financial services. Financial inclusion aims to connect people with banks (Swamy, 2010) and provide them with more access to bank accounts, savings, credits, and other financial services. Band et al. (2012) stress that access to financial services by the poor and vulnerable groups is a prerequisite for poverty reduction. With broader access to financial services, these groups of people will have a larger opportunity to obtain more funds or capital to start up and run their businesses.

The World Bank (2010) reports that only 21% of Indonesia's population have access to banks and another 2% engage in formal non-bank financial services. In order to broaden access to financial services, the central bank of Indonesia, Bank Indonesia (BI), has promoted a programme called 'Tabunganku'. This programme introduces an inexpensive and safe banking product. 'Tabunganku' has the mandatory features of a very modest initial deposit of only IDR 20,000 (less than USD2), bearing no administrative fees, and paying returns based on customers' daily balances. To attract more customers, the central bank allows commercial banks that participate in the programme to provide customised/optional features on top of the mandatory features. Despite all the facilities, the programme has only attracted 3.2 million customers with total savings of IDR3.2 billion, or an average of IDR 1,000,000 per customer as of March 2012. This figure is yet smaller compared to the total savings of IDR 2,800 billion and around 101 million customers.

The National Economic Social Survey (SUSENAS) data, issued by Statistics Indonesia (Badan Pusat Statistik (BPS)), indicates that households obtain loans from formal (banks and non-bank) and informal (personal, relatives, neighbours, friends) institutions. Households with access to banks increased by 1.2% whereas access to non-bank institutions went up by 12.5% per year from 2008–2012. What has restrained the access to banks?

The objectives of this research are to analyse households' determinant factors to obtain loans. To complement and enrich the results, the study also carries out in-depth interviews with several informants representing the banking regulator, banks, and business-owning households. The study also provides some recommendations to policymakers in improving access to banks.

2 LITERATURE REVIEW

2.1 *Financial sector and poverty alleviation nexus*

King and Levine (1993) and Levine (1997, 2005) have thoroughly demonstrated the positive relationship between the financial sector and economic growth, both in theoretical and empirical contexts. Through an efficient financial system, the financial sector channels funds to more productive uses and allocates risks to those who have the capabilities to bear risks (Demirguc-Kunt & Honohan, 2008). In its development, the discussion on the financial sector-economic growth nexus has extended to more multidimensional issues, covering social and economic welfare. Honohan (2004) relates the financial development-economic growth with poverty and shows that financial depth is negatively correlated to headcount poverty rates. The discussion then goes deeper into whether economic growth and the intensive use of financial products-and-services have been pro-poor. Claessens (2006) points out that in many developing countries financial access has not reached all people. Beck and Demirguc-Kunt (2008) further state that financial exclusion may retard economic development. The poor's limited access to finance may contribute to impeding physical and human capital accumulation, which in turn will slow down economic growth and raise income inequality.

2.2 *Financial inclusion experience*

Access to finance, according to Band et al. (2012), is a necessary condition to lower the poverty rate. India has had a financial inclusion programme since 2004. The programme has been considered successful by having good coordination between the government and the central bank, and by taking into account the role of Indian women and human resource development (Bagli & Dutta (2012); Band et al. (2012)). In order to reach out to more people and attract more bank customers, Indian commercial banks are allowed to work together with civil society groups, micro-institutions, and society organisations as business facilitators to act as intermediaries for rural people to access banks. The Indian government also provides campaign funds for the programme and recommends banks to open a branch for every 10,000 people. Previously, almost 60% of the population did not have bank accounts and almost 90% of the population did not have access to bank loans. Only in two years after the policy recommendation was put into effect, banks have opened their branches in 1,237 rural areas. The Reserve Bank of India has set a target of 600 million bank accounts in 2020 (Nalini & Mariappan, 2012).

The financial sector in Uganda is not well developed yet and tends to be dualistic. Rural areas, mostly dominated by agricultural activities, have almost no access to financial services. In fact, credit services in rural areas are important as the agricultural sector is the main contributor to the Ugandan economy. Loans in rural Uganda mainly come from co-operatives, government programmes, relatives, local communities, and credit associations. Rural people rarely obtain loans from banks due to the distance of banks from rural areas. The demand for loans in Uganda (Mpuga, 2004) is found to positively correlate with demographic factors (education level, marital status, location where people live, age, and occupation in the household), interest rates, and distance to financial institutions. Mpuga also finds that young individuals tend to save and invest more than older individuals, and men tend to have control over assets. Individuals with higher income save more, and hence they have more assets to be pledged as collateral when they borrow money for business.

Likewise, around 40 million people in Indonesia do not have any access to financial services. BI (2012) identifies the problems of lack of access to financial services, which are due to geographical conditions, designs and patterns of services that often do not match the people's needs, and the information gap. On the other hand, many banks tend to provide loans to larger-scale entrepreneurs. Other factors such as level of education, legal issues, and self-exclusion are a few reasons that hold back the poor from bank access. BI has also launched campaign activities since 2008 to disseminate bank intermediary roles, products and services, benefits, risks and fees They all aim to increase Indonesia's marginal propensity to save and its domestic funds, and to promote a saving culture for Indonesians, particularly the productive poor.

3 RESEARCH METHODS

Two approaches are employed in this research—the analysis of quantitative data, and in-depth interviews. The 2008 and 2012 SUSENAS data is employed. The use of the two-year data is to capture whether the implementation of the BI programme has a positive and significant effect on households' loans/credit obtainment.

Applying the multinomial logit regression (Cameron & Trivedi (2005); Gujarati (2003)), the study investigates what factors determine the households' loans/credit obtainment. To complement the secondary data analysis, in-depth interviews (Jonker et al., 2011) are also conducted with several informants representing the regulator, financial institutions—both bank and non-bank—and productive poor households (micro-entrepreneurs).

Loans or credit in this study are defined as funds borrowed by households to run their business from various sources—banks, non-bank institutions, and non-formal personal sources. Credit obtained is determined by households' characteristics and socio-economic factors. Households' characteristics are represented by sex, age, education, location, marital status, and household size. Socio-economic factors include poverty status, employment status, employment sector, home ownership, and access to technology—fixed-line ownership, cell phone ownership, and computer ownership. This study also controls the implementation of public education programmes for banking by the use of year dummy. The multinomial logit model is as follows:

$$\begin{aligned} log\left(\frac{p_j}{p_0}\right) = {} & \beta_0 + \beta_1 sex + \beta_2 age + \beta_3 age^2 + \beta_4 educ + \beta_5 location + \beta_6 marrital \\ & + \beta_7 HHsize + \beta_8 emplsect + \beta_9 emplstatus + \beta_{10} houseown \\ & + \beta_{11} telown + \beta_{12} celltelown + \beta_{13} compown + \beta_{14} povstatus \\ & + \beta_{15} dcampaign + \varepsilon \end{aligned} \quad (1)$$

Equation 1 is estimated by *maximum likelihood*. The variable definition is described in Appendix 1 of this paper.

4 RESULT AND DISCUSSION

4.1 *Household profiles*

Most Indonesian households, based on 2008 and 2012 SUSENAS data, live in rural areas with low education, and work in agricultural and service sectors as own-account workers or as employers assisted by labour. Most households have real income of below IDR 750,000. The data indicate that 25% of the households have insufficient income to meet their daily needs.

Figure 1 depicts how they make up for the insufficiency. Most of them borrow from relatives (69%) and from neighbours/friends (54%). Only a very small percentage of households come to formal institutions to close their deficit—co-operatives (4%) and banks (2%). This preliminary finding indicates that formal financial institutions are not the households' main option from whom to borrow money. Banks specifically seem detached from households.

Figures 2–4 display the distribution of households that get loans based on their poverty status. BPS classifies poverty status into poor, near poor, vulnerable poor, and non-poor. This study combines 'very poor' and 'poor' classifications into poor, and near poor and vulnerable poor into near poor (see Table A1 in Appendix 1). Figure 2 shows that non-poor households have the largest opportunity to obtain loans from all sources, particularly banks. Near poor households obtain loans mostly from non-bank institutions and individuals (Figure 3). This group of households is vulnerable to degrade to poor, so the government provides more financial assistance to this group. Figure 4 displays the decreasing proportion of households that obtain loans from individuals (relatives, neighbours/friends), from 2.02% in 2008 to 1.89% in 2012, except for poor households. Poor households prefer borrowing from individuals to other sources because it is much quicker, easier in terms of not requiring legal documents, and requires no collateral to bepledged.

4.2 *Determinant factors of households' loans*

The estimation results (Table 1) indicate that households in urban areas are less likely to get loans from all sources relative to those in rural areas. This finding is interesting, as the fact

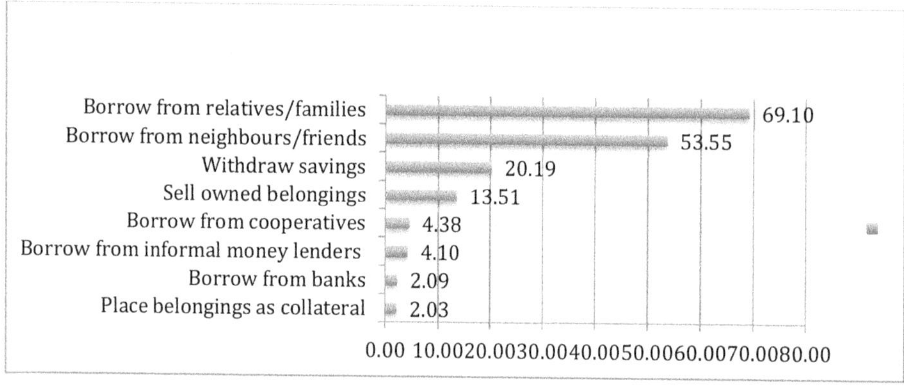

Figure 1. Source of funds of households in meeting their daily needs (%), 2012.
Notes: Sample number (n) = 286,113.
Source: Author's own estimation from 2012 SUSENAS data.

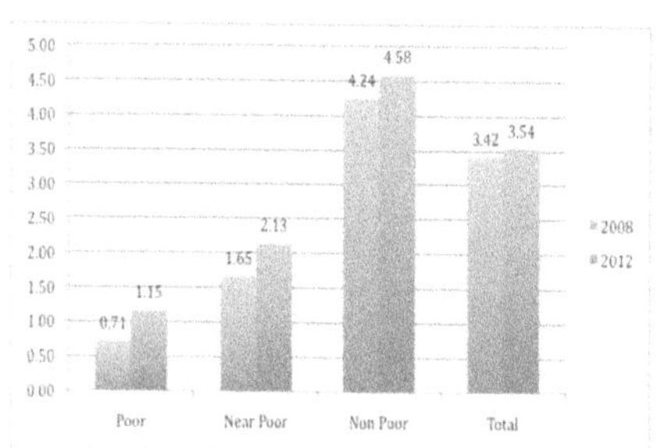

Figure 2. Distribution of households who obtain bank loans based on poverty status (%).
Source: Author's own estimation from 2008 and 2012 Susenas.

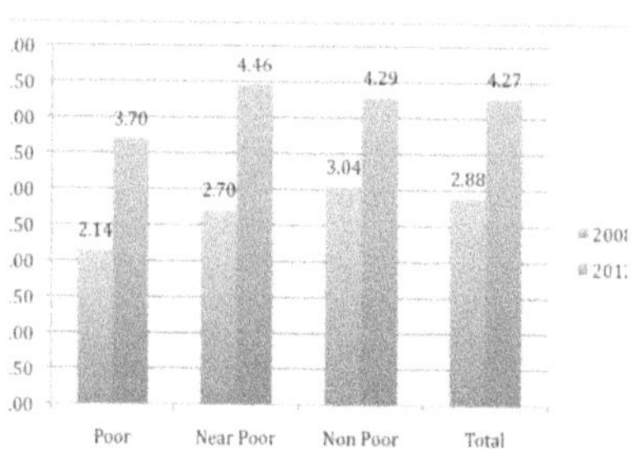

Figure 3. Distribution of households who obtain non-bank loans based on poverty status (%).
Source: Author's own estimation from 2008 and 2012 Susenas.

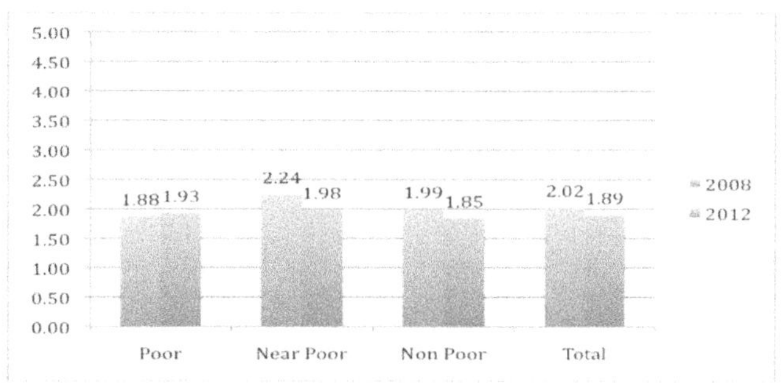

Figure 4. Distribution of households who obtain individual loans based on poverty status (%).
Source: Author's own estimation from 2008 and 2012 Susenas.

Table 1. Multinomial logistic regression results.

Number of observations	459,786		
LR chi2	63,970.99		
Prob> chi2	0.0000		
Pseudo R2	0.1938		
Log likelihood	−133,055.28		
	Regression results		
Dependent Variable	Model 1	Model 2	Model 3
$log(\frac{p_j}{p_0})$, base outcome 0 = do not obtain	*1 bank*	*2 non-bank*	*3 non-formal personal*
Independent Variables	RRR – Relative Risk Ratio *(P>|z|)*	RRR – Relative Risk Ratio *(P>|z|)*	RRR – Relative Risk Ratio *(P>|z|)*
Household characteristics			
Location (1 = urban)	0.528***	0.542***	0.819***
	(0.000)	*(0.000)*	*(0.000)*
Sex (1 = male)	1.034	0.608***	0.860*
	(0.639)	*(0.000)*	*(0.088)*
Marital status (1 = married)	1.530***	1.419***	1.085
	(0.000)	*(0.000)*	*(0.331)*
Age	1.144***	1.099***	1.935***
	(0.000)	*(0.000)*	*(0.000)*
Age squared	0.999***	0.999***	0.993***
	(0.000)	*(0.000)*	*(0.000)*
Household size	1.018**	1.054***	1.507***
	(0.011)	*(0.000)*	*(0.000)*
Education (reference category: did not graduate from elementary school)			
Elementary education	1.330***	1.122***	1.025
	(0.000)	*(0.000)*	*(0.510)*
Secondary education	1.637 ***	1.359***	0.479***
	(0.000)	*(0.000)*	*(0.000)*
Tertiary education	1.912***	0.976	2.093***
	(0.000)	*(0.525)*	*(0.000)*
College/university education	2.790***	0.926	0.961
	(0.000)	*(0.215)*	*(0.644)*

(*Continued*)

Table 1. (*Continued*).

Employment sector (reference category: other sector)			
Agricultural sector	0.391***	0.764**	0.882
	(0.000)	*(0.001)*	*(0.308)*
Mining sector	0.398***	0.895	0.949
	(0.000)	*(0.325)*	*(0.755)*
Manufacturing sector	0.812**	1.08672	1.287*
	(0.003)	*(0.362)*	*(0.055)*
Service sector	0.814**	0.93272	1.300**
	(0.000)	*(0.215)*	*(0.644)*
Employment status (reference category: casual employee)			
Own-account worker or assisted employer	1.776***	1.239***	3.641***
	(0.000)	(0.000)	(0.000)
Employee	1.029	0.986	0.790***
	(0.571)	(0.750)	(0.000)
House, communication, and technology			
Home ownership (1 = own home)	1.447***	1.394***	0.224***
	(0.000)	*(0.000)*	*(0.000)*
Fixed-line telephone ownership (1 = own fixed-line telephone)	1.330***	1.038	0.524***
	(0.000)	*(0.409)*	*(0.000)*
Cellular telephone ownership (1 = own cell phone)	3.023***	1.407***	0.583***
	(0.000)	*(0.000)*	*(0.000)*
Computer ownership (1 = own computer)	1.046	0.531***	0.0716***
	(0.154)	*(0.000)*	*(0.000)*
Poverty status (reference category: poor)			
Near poor	2.001***	1.598***	0.592***
	(0.000)	*(0.000)*	*(0.000)*
Non-poor	2.713***	1.446***	6.176***
	(0.000)	*(0.000)*	*(0.000)*
Campaign program (1 = year 2012, after program implementation)	0.126***	0.394***	1.512***
	(0.000)	*(0.000)*	*(0.000)*
Constant	0.0001***	0.002***	0.000***
	(0.000)	*(0.000)*	*(0.000)*

Source: Authors' own estimation.
Note: * significant at alpha 10%, ** significant at alpha 5%, and *** significant at alpha 1%.

that it is harder to get loans in urban areas, particularly from banks, seems counter-intuitive. Inequality and urban poverty may explain the finding. Indonesia's Gini coefficient, that measures income inequality, slightly increased from 0.35 in 2007–2008, to 0.41 in 2011–2013, indicating that the equality gap between the haves and the poor has widened. Over two decades, the urban Gini coefficient in Indonesia has been larger than that of the rural (Badan Pusat Statistik, 2012, p. 417). BPS indicates that 11.6% people lived below the urban poverty line in 2008, and the figure declined to 8.4% in 2012. The urban poor lack access to basic services and also to business credit (World Bank, 2013).

Sex does not significantly affect the probability of households obtaining loans from banks. Women are more likely to get loans from non-banks and individuals, as banks often hesitate to lend women money because they lack the required collateral, as their property are not registered under their own names (World Bank, 2013). They prefer to go to lending-saving co-operatives, pawnshops, or participate in *arisan* (a gathering). It is estimated that 43% of formal Small and Medium-Sized Enterprises (SMEs) in Indonesia are women-owned (World Bank, 2013).

Being married indicates that couples engage in formal households and have a larger number of family members, hence a larger demand for finance. Married couples, that tend to own joint bank accounts and assets in the family, have a larger opportunity to get loans from banks and non-banks. Age positively and significantly affects the probability of households getting loans from all sources.

The probability of households getting loans from banks is larger for households with college/university degrees, while those with lower education level—secondary level—are more likely to get loans from non-bank institutions. The education level is less important for informal money lenders to provide loans. This finding may inform policymakers in choosing the means to attract bank customers—that is, policymakers should use easily-understood ways to communicate with unbanked people as they are mostly of lower level education.

Employment sectors are less likely to affect the probability of households obtaining loans from banks and non-bank institutions. This may be because they are not too confident to borrow from formal institutions due to the procedures and legal requirements. They, particularly those working in manufacturing and service sectors, are more likely to borrow from individuals as probably they need the loan immediately and not in a large amount. Households with own-account workers or assisted employers are more likely to have to get loans from banks, non-banks, and individuals, compared to those with an employment status as employees. Being an employee may provide a more fixed and regular income than being an own-account worker.

Home ownership provides a higher probability for households to get loans both from banks and non-bank institutions. Homes (ownership documents of them) can be placed as collateral for credit to cover risks for the loan providers. SUSENAS data (Badan Pusat Statistik, 2008 and 2012) indicate that 80% of households have their own homes. However, from the in-depth interviews with micro-entrepreneurs, they are not raring to place their houses as collateral as a home may be their only and most valuable asset.

Information may be accessed more easily nowadays via the use of telephones, both fixed lines and cell phones, and on-line computers. Through these media, information is retrieved faster and cheaper and can be accessed by more people. Banks use these media to promote and market their products. Customers' accessibility to these media becomes important. SUSENAS data (Badan Pusat Statistik, 2008 and 2012) shows that less than 10% of households have fixed-line telephones, but nearly 80% of households have cell phones. From the multinomial logit estimation, it is found that telephone ownership—both fixed lines and cellular—provides a larger possibility for households to get loans from banks. Households with cell phone ownership in particular are three times more likely to get loans from banks. Households' familiarity with cell phones will facilitate the implementation of mobile payment services.

SUSENAS data (Badan Pusat Statistik, 2008 and 2012) also indicates that computer ownership, both laptops and desktops, doubled in 2012. Information nowadays is profoundly circulated in cyberspace. The increasing percentage of computer ownership indicates that households are getting more familiar with the use of computers. However, computer ownership is a less important consideration in obtaining loans.

Poverty status is an important consideration for formal institutions to provide loans. Both near poor and non-poor households have at least twice the probability of the poor in obtaining loans from banks and non-bank institutions. They have wider access to finance as they have more money to save and more assets to pledge as collateral if they want to borrow money from those formal institutions.

On the other hand, the poor are often excluded from formal financial institutions. Based on the in-depth interviews, the poor have insufficient income to save, have insufficient income to repay loans, have no valuable assets (except home ownership documents) to put as collateral if they want to borrow money, perceive that the documents to fill in are onerous, and often lack citizen identity cards (Kartu Tanda Penduduk (KTP)), whereas this identity card is a mandatory requirement to open savings accounts, and hence loans.

To support their financial needs, some households run micro-scale businesses. This group of businessmen is longing for additional capital. It implies that they actually need access to banks. From the field, we find that some are banked micro-entrepreneurs; however, they use their accounts to simply keep revenues from their customers. They only practice simple banking transactions—save, transfer, and withdraw. Only a few of them have experience in borrowing money from banks. Some of them, who promptly repay their loans, face no difficulties, but some have had bad experiences with banks—particularly debt collectors—when they fail to repay the loans.

Some of the micro-entrepreneurs borrow money from individuals/informal moneylenders (*rentenir* in Indonesian language). In their opinion borrowing from *rentenir* is more practical, easier, and faster. *Rentenir* can provide immediate cash without collateral requirement. *Rentenir* come to the business sites to promote and market their lending services. They impose much higher borrowing rates. Some of the micro-entrepreneurs are trapped in huge debt to *rentenir*. However, due to the urgent need for cash to keep running the business, some are less concerned with how much they have to repay on their loans as long as the cash needed is immediately met. *Rentenir* occasionally prepare a less burdensome method of repayment. The repayment is made on a daily or weekly basis, whichever is more convenient for the micro-borrowers.

The market for such informal funding is still wide open, as long as the supply and demand exists. This may explain why the national campaign/movement to go to banks and to save, introduced by BI, is less likely to affect the probability of obtaining loans from formal institutions.

5 CONCLUSION AND RECOMMENDATIONS

Most Indonesian households, based on 2008 and 2012 SUSENAS data, live in rural areas with low education, and work in agricultural and service sectors as own-account workers or as employers assisted by labour. Most households have real income per capita of below IDR 750,000 and 25% of them have insufficient income to meet their daily needs. Only a very small percentage of them come to formal institutions to make up for their deficit.

Based on the multinomial logit estimation, the probability of households obtaining loans from banks, non-bank institutions and individuals, relative to those who do not, is significantly affected by age, marital status, household size, employment status (own-account worker or assisted employer), home ownership, telephone ownership, and computer ownership. Households with college/university graduates are more likely to get loans from banks, whereas secondary and elementary school graduates are more likely to get loans from non-bank institutions and individuals respectively. In regard to poverty status, non-poor households have a higher probability to get loans from banks and individuals relative to the poor, whereas the near poor have a higher probability to get loans from non-bank institutions relative to the poor. The public campaign for banking is less likely to increase the probability of households getting loans from both banks and non-bank institutions. For comparison, further study may estimate the probability of households obtaining loans from banks and non-bank institutions, relative to those who obtain loans from individuals.

Based on the secondary and primary data findings, it is recommended for the central bank to urgently create lending programmes for the unbanked people with minimum collateral requirement as additional funds/capital are urgently required for them to run and keep their businesses. For that, the central bank should collaborate with commercial banks to make breakthroughs and create products that suit the poor's needs. Secondly, the issue of branchless banking, better known as mobile payment services, should be introduced and gradually implemented. A consideration found in the study is that nearly 80% of households in Indonesia own cell phones, so there is no doubt that they are technologically updated.

ACKNOWLEDGEMENT

The authors would like to express their gratitute to Universitas Indonesia for providing the UI Research Grants 2013, high appreciation to Ms. Metri Sriwati for her administrative assistance and to Ms. Dwinda Andaninggar Hangonowati for her assistance in initial data processing.

BIONOTE

Beta Yulianita Gitaharie is currently a lecturer at Faculty of Economics and Business, Universitas Indonesia. Her research interests are in financial inclusion, economic development, monetary and macroeconomic issues.

Lana Soelistianingsih is currently a lecturer at Faculty of Economics and Business, Universitas Indonesia. Her research interestsare in financial and capital market.

Triasih Djutaharta is a PhD student at Faculty of Economics and Business, Universitas Indonesia and also a researcher at Lembaga Demografi, Faculty of Economics and Business, Universitas Indonesia. Her research interests are in public health, tobacco, and nutrition economics.

REFERENCES

Badan Pusat Statistik. (2007). *Statistik Indonesia*. Jakarta: BPS.
Badan Pusat Statistik. (2008). *Statistik Indonesia*. Jakarta: BPS.
Badan Pusat Statistik. (2009). *Statistik Indonesia*. Jakarta: BPS.
Badan Pusat Statistik. (2010). *Statistik Indonesia*. Jakarta: BPS.
Badan Pusat Statistik. (2011). *Statistik Indonesia*. Jakarta: BPS.
Badan Pusat Statistik. (2012). *Statistik Indonesia*. Jakarta: BPS.
Badan Pusat Statistik. (2008). *Statistik Kesejahteraan Rakyat*. Jakarta: BPS.
Badan Pusat Statistik. (2008). *SUSENAS*. Raw Data. Unpublished.
Badan Pusat Statistik. (2012). *SUSENAS*. Raw Data. Unpublished.
Bagli, S. & Dutta, P. (2012). A study of financial inclusion in India. *Radix of International Journal of Economics and Business Management*, *1*(8).
Band, G., Naidu, K. & Mehadia, T. (2012). Opportunities and obstacles to financial inclusion. *Arth Praband Journal of Economics and Management*, *1*(1).
Bank Indonesia. (2012). *Booklet Perbankan 2012*.
Beck, T. & Demirguc-Kunt, A. (2008). Access to finance: An unfinished agenda. *World Bank Economic Review*, *22*(3), 383–396. doi:10.1093/wber/lhn021.
Cameron, A.C. & Trivedi, P.K. (2005). *Micro econometrics: Methods and applications*. London: Cambridge University Press.
Claessens, S. (2006). Access to financial services: A review of the issues and public policy objectives. *World Bank Research Observer*. doi:10.1093/wbro/lkl004.
Demirguc-Kunt, A., Beck, T. & Honohan, P. (2008). Access to finance and development: theory and measurement. In F. Bourguignon & M. Klein *Finance for all: Policies and pitfalls in expanding access*. World Bank, ISBN: 978-0-8213-7291-3.
Gujarati, D. (2003). *Basic econometrics*. International Edition. New York: McGraw Hill.
Honohan, P. (2004). Financial development, growth and poverty: How close are the links? *World Bank Policy Research Working Paper* 3203. http://econ.worldbank.org.
Jonker, J., Pennink, B.J.W. & Wahyunil, S. (2011). *Metodologipenelitian. Panduan Untuk Master Ph.D di bidang Manajemen*. Jakarta: Salemba Empat.
King, R.G. & Levine, R. (1993). Finance and growth: Schumpeter might be right. *The Quarterly Journal of Economics*, *108* (August), 717–738. doi:10.2307/2118406.
Levine, R. (1997). Financial development and economic growth: Views and agenda. *Journal of Economic Literature*, *35*(2), 688–726. doi:10.1126/science.151.3712.867-a.
Levine, R. (2005). Chapter 12 finance and growth: Theory and evidence. *Handbook of Economic Growth*. doi:10.1016/S1574-0684(05)01012-9.

Mpuga, P. (2004). Demand for credit in rural Uganda: Who cares for the peasant? Presented on *Conference on Growth, Poverty Reduction and Human Development in Africa,* Centre for the Study of African Economies.

Nalini, G.S. & Mariappan, K. (2012). Role of banks in financial inclusion. *The International Journal of Commerce and Behavioural Science*, *1*(4), 33–36.

Swamy, V. (2010). Bank-Based financial intermediation for financial inclusion and inclusive growth. *Banks and Bank Systems*, *5*(4), 1–12.

World Bank. (2010). *Improving access to financial services in Indonesia.* Available from: http://www.bi.go.id/id/perbankan/keuanganinklusif/berita/Documents/World%20Bank%20Report%20-%20Improving%20 Access%20to%20Financial%20Services%20 in%20Indonesia.pdf accessed on April01, 2013.

World Bank. (2013). *Indonesia: Urban poverty and program review*. Available from: http://ifcextapps.ifc.org/ifcext%5Cpressroom%5Cifcpressroom.nsf%5C0%5C0CEFD2A433E3DE2685257FBD0026B7FA accessed on October 01, 2016.

APPENDIX 1

Below is the description of variables used in the model (Equation 1).

j	=	source of credit with p_0 = the probability of households do not obtain credit (reference category), p_1 = the probability of households obtain credit from banks, p_2 = the probability of households obtain credit from non-bank institutions, p_3 = the probability of households obtain non-formal personal credit and i
sex	=	1 if male, 2 if female (reference category)
age	=	Respondent's age of 15 or above
educ	=	Education attainment of head of household measured from highest diploma attainment, 1 if do not graduate from elementary school (reference category), 2 if graduate from elementary school, 3 if graduate from junior high school (secondary), 4 if graduate from high school, 5 if graduate from college/university
location	=	1 if urban areas, 2 if rural areas (reference category)
marrital	=	Marital status, 1 if married, 2 if not married (never been married, life divorced, death divorced) (reference category)
HHsize	=	Household size
povstatus	=	Poverty status*, 1 if poor (reference category), 2 if near poor, 3 if non poor
emplsect	=	Employment sector in which respondents worked the last one week, 1 if agricultural sector, 2 if mining/quarrying sector, 3 if manufacture sector, 4 if service sector, 5 if other sectors (reference category)
emplstatus	=	Main employment status in the last one week, 1 if own account worker or employer assisted by temporary/unpaid/permanent workers, 2 if employee, 3 if casual employee (reference category)
houseown	=	House ownership, 1 if own, 2 if do not own house (reference category)
telown	=	Fixed line communication ownership, 1 if own fixed line telephone, 2 if do not own fixed line telephone (reference category)
celltelown	=	Cellular phone ownership, 1 if own cell phone, 2 if do not own cell phone (reference category)
compown	=	Computer ownership, 1 if own computers (laptop/desktop), 2 if do not own computers (laptop/desktop) (reference category)
dcampain	=	Year of bank campaigns by BI, 1 if year is 2008 prior to the program implementation (reference category), 2 if year is 2012 after the program implementation

Note:*is explained below.

BPS definition of poverty status in this study is simplified into three categories only—poor, near poor, and non-poor. BPS' very poor category is combined with poor and vulnerable poor with near poor. Table A1 summarizes the modified poverty status definition.

Table A1. Definition of poverty status.

	BPS definition		Authors' simplified definition	
No.	Poverty status	Ranges of poverty line (PL)**	Poverty status	Ranges of poverty line (PL)
1	Very Poor	expenditure per capita $< 0.8 \times PL$		
2	Poor	$0.8 \times PL \leq$ expenditure per capita $< 1.0 \times PL$	Poor	$0.8 \times PL <$ expenditure per capita $< 1.0 \times PL$
3	Near Poor	$1.0 \times PL \leq$ expenditure per capita $< 1.2 \times PL$	Near Poor	$1.0 \times PL \leq$ expenditure per capita $< 1.6 \times PL$
4	Vulnerable Poor	$1.2 \times PL \leq$ expenditure per capita $\leq 1.6 \times PL$		
5	Non Poor	expenditure per capita $> 1.6 \times PL$	Non Poor	expenditure per capita $> 1.6 \times PL$

**The poverty line in this study is based on the provincial rural-urban areas.

Table A2. In-depth interview summaries.

Micro business type	Sales per month (in million IDR)	Bankable	Source of loans	Willingness to borrow from banks in the near future	Constraints to get bank loans	Knowledge of *tabunganku* product
(1)	(2)	(3)	(4)	(5)	(6)	(7)
1. Traditional Snack	1.5	Yes	NA	No	Never knows	No
2. Catering Services	2.5–3	No	NA	Very much wanted to expand the business, particularly to buy/rent land.	Never borrows from banks, but knows banks require collateral	No
3. Lady's Taylor	3–3.5	Yes	Cooperatives, Bank	Wanted if necessary	No constraint	Yes
4. Convection K1	4	Yes	Money lender	Very much wanted	**No collateral**	No
5. Convection K2	2–3 (gross)	Yes	Bank	No information	No constraint	No
6. Convection K3	0.3 (gross)	Yes	Family/ Friends	No	**No collateral**	Yes
7. Daily Utility Shop 1	6	Yes	Bank	No	No constraint	Yes
8. Daily Utility Shop 2	3.6	No	NA	No	**Increasing charges**	No
9. Furniture Shop F1	50	Yes	Bank, Cooperatives, Money lender	No	**Increasing charges**	No
10. Furniture Shop F2	7	Yes	Bank	Willing to borrow in 2014	No constraint	Yes
11. Agricultural Shop A1	60–90 (gross)	Yes	Bank	No plan, still has bank borrowing	No constraint	No
12. Agricultural Shop A2	1 (gross)	Yes	NA	No information	**No collateral**	**No**

Source: IDI results with micro enterprises.

Competition and Cooperation in Economics and Business – Gani et al. (Eds)
© 2018 Taylor & Francis Group, London, ISBN 978-1-138-62666-9

Household characteristics and demand for private tutoring in Indonesia

W. Wahyuni & H. Susanti
Department of Economics, Faculty of Economics and Business, Universitas Indonesia, Depok, Indonesia

ABSTRACT: This research aims to identify the effects of household characteristics in household demand for private tutoring function by using the Tobit model. The household characteristics are represented by a number of variables such as household income rate, parents' educational level, and number of family members. The results show that the demand for private tutoring increases along with the higher level of household income, children's education level, household location in urban areas and the lower number of family members. The impact of the fathers' educational level on household demand for private tutoring is higher than that of the mothers' educational level.

1 INTRODUCTION

Indonesian parents nowadays are dealing with anxiety about their children's academic achievement as an impact of the educational system applied in Indonesia. In order to maintain the national educational system, the Indonesian government, through the Government Regulation Number 19 Year 2005, stipulates a minimum standard of academic competence as a requirement for promotion to a higher grade, graduation and application to a higher educational level. More examinations arranged both by the government and schools, in order to measure competence and categorise students according to their academic competence level, may lead to competition among students in terms of academic achievement. Moreover, they also need to compete in order to get into favourite universities, as one of the preparations needed to enhance their competence in the labour market competition. This condition motivates both parents and children to seek more lessons, in addition to the ones they get from their schools. It indicates an increase of the households' demand for education (Bray & Kwok, 2003).

The increase of such households' demand for education is used as an opportunity to develop private tutoring businesses, either by schools, individuals or private tutoring agencies. This is in line with Takayama *et al.* (2013) who explain that the high level of academic competition in South Korea boosts the growth of private tutoring industries. On the other hand, Finland, a country that runs an educational system without any examinations to determine promotion to a higher grade and higher educational level, shows that private tutoring is not necessarily needed. Private tutoring is a supplementary instruction given after or outside of school hours, using a method where the tutors deliver materials in more innovative and flexible ways, and the parents must pay for their services (Dang & Rogers, 2008). In general, private tutoring concentrates more on students' success in school examination or admission to favourite universities (Kim & Lee, 2010; Mori & Baker, 2010). This sector has become a rapidly growing industry and is called 'shadow education', due to its presence that grows along with the development of the educational system applied in a country as a supplement for the lack of existing formal education (Bray *et al.*, 2014).

The growth of private tutoring in Indonesia always increases year after year. Figure 1 shows the increasing number of agencies that provide private tutoring, commonly known as tutoring agencies. The large number of tutoring agencies indicates high household demand for

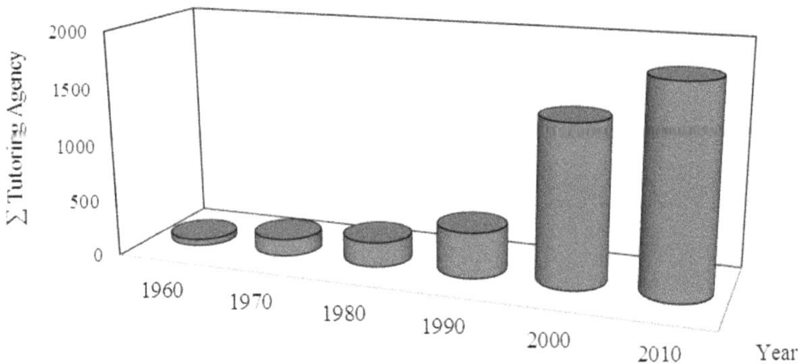

Figure 1. Growth of total tutoring agencies registered in Indonesia.
Source: Ministry of Education and Culture of Republic of Indonesia.

additional instruction given after or outside school hours and is also considered a response to the inability of the existing formal educational system in accommodating household needs for education (Dang, 2007; Kim & Lee, 2010). Data regarding the presence and activities of private tutoring, whether provided by agencies, schools or individuals in Indonesia, are generally hard to obtain. Therefore, according to Bray and Kwok (2003), in order to see the demand for private tutoring, it will be easier for us to use household data.

Several empirical studies that examined household total expenditure on education in Indonesia were conducted by Kristiansen and Praktikno (2006) and Novita (2008). So far, we have not found the specific reference for the private tutoring. Several previous empirical studies show that private tutoring is considered an essential commodity in a household (Psacharopoulos & Papakonstantinou, 2005; Dang, 2007). In accordance to the explanation above, this study aims to identify the effects of household characteristics on household demand for private tutoring.

2 DETERMINANT OF HOUSEHOLD DEMAND FOR PRIVATE TUTORING

Several studies find that household demand for private tutoring is affected by household characteristics, some of which are household income rate, parents' educational level, age of the head of the household and number of family members. The higher the household income rate, the more disposable income can be spent and the more spending allocation options can be chosen by the household members. A household that has a higher preference or higher expectation on children's education is likely to spend more money taken from the income, to invest on children's education. One of the ways is to spend it on private tutoring.

Parents with higher education level commonly have a higher preference and expectation on returns for children's education in the future. This leads them to allocate more expenditure on children's private tutoring (Dang & Rogers, 2008). Previous studies show that the effect of mothers' education on the expenditure on private tutoring in Turkey and Vietnam is more dominant than fathers' education (Tansel & Bircan, 2006; Dang, 2007). Meanwhile, fathers' education does not significantly affect the amount of household expenditure on private tutoring in Hong Kong (Bray *et al.*, 2014), and fathers' and mothers' education do not significantly affect the amount of household expenditure on private tutoring in Malaysia (Jelani *et al.*, 2014).

A larger number of family members causes a decrease in household expenditure rate for private tutoring in Turkey, Vietnam and Korea (Tansel & Bircan 2006; Kim & Lee 2010). This happens due to the fact that households face budget constraints that should be allocated for all family members. A different result is seen in which a larger number of family members does not reduce the household expenditure on private tutoring in Hong Kong

(Bray *et al.,* 2014) and Malaysia (Jelani *et al.,* 2014). As the head of a household grows older, it is assumed that there will be more experience. Moreover, at a certain age, usually a household's economic condition is becoming settled. In that case, the awareness level of the parents on the importance of children's education is becoming higher. It may lead to increasing the household expenditure on private tutoring at a decreasing rate (Dang, 2007; Tansel & Bircan, 2006; Davies, 2004).

Corresponding to children's gender, none of the previous studies have found any discriminations of gender on household expenditure on private tutoring (Bray *et al.*, 2014; Jelani *et al.,* 2014; Dang, 2007). The demand for private tutoring increases along with the increase in education level and types of school. Public schools in Hong Kong positively affect the increase in household demand for private tutoring compared to private schools (Bray *et al.*, 2014).

3 THEORETICAL AND EMPIRICAL MODEL

3.1 *Theoretical model*

This study utilises the basic theory of Becker and Lewis (1973), that households, in maximising their utility function, are affected by the number of children, n, the quality of children, q, and the consumption of a number of goods, x, with the determined price (*P*x) equal to 1, since it is assumed that the price determined for all households is the same. In maximising their utility function, households face a budget constraint, I.

$$Max\ U(n, q, x) \tag{1}$$

$$P_x X + n(p_q q) = I \tag{2}$$

Furthermore, in order to analyse and explain households' decision in deciding to invest on children's education through private tutoring, Dang and Rogers (2016) developed that theory by adding the quality of children, and formal education constraint, formulated as follows:

$$q = e_u + e_r \tag{3}$$

$$e_u \le \overline{e}_u \tag{4}$$

The quality of children (q) is the total of formal education (e_u) and private tutoring (e_r) which the household invests for the children. Meanwhile, p_q is the existing price in household investment for the quality of children, which is $_q = _u$ or $_r$. The existing formal education is assumed to have a limited capacity in improving the quality of children according to the expectations of the household ($\overline{e}_u$). The number and quality of children, and the amount of products consumed by the household, are assumed to be non-negative ($n \ge 0$, $q \ge 0$ and $y \ge 0$).

From the result of maximising households' utility function, which shall deal with the three constraints, and by assuming that the marginal of household utility with budget constraints (λ_1) is positive, we obtain an equation described as follows:

$$q^* = f(I, p_u, p_r) \tag{5}$$

$$e_r^* = f(I, p_u, p_r) \tag{6}$$

$$E_r = p_r e_r^* = p_r\ f(I, p_u, p_r) \tag{7}$$

$$\lambda_2(\overline{e}_u - e_u) = 0 \tag{8}$$

In this case, *Er* is household demand for private tutoring. According to the Equation 8, it is implied that the marginal of household utility on formal education constraint (λ_2) is:

1. If $\lambda_2 = 0$, it is assumed that the household does not maximally consume the quality of existing formal education ($e_u \leq \overline{e}_u$). This occurs in households with lower demand and preference for education.
2. If $\lambda_2 > 0$, it is assumed that the household has maximally consumed all of the quality of existing formal education ($e_u = \overline{e}_u$). It may lead to an implication that for households with higher demand and preference for education, in order to improve the quality of their children, they do not have any other choice other than investing their expenditure on private tutoring. Since $e_u = \overline{e}_u$, then private tutoring is the only one option to maximise household demand as a supplement for education.

3.2 *Empirical model*

According to the theoretical model and other previous empirical studies, a model was developed as follows:

$$E_r = P_r f(I, P_r, P_u, Z) \tag{9}$$

where

E_r = household demand for private tutoring;
I = household permanent income;
P_r = price of private tutoring;
P_u = price of formal education;
Z = variable of household characteristics vector.

Therefore, the empirical model utilised in this study can be described as follows:

$$\mathrm{Er}_{ij} = \beta_1+\beta_2 \mathrm{I}_i+\beta_3 \mathrm{Pr}_{ij}+\beta_4 \mathrm{Pu}_{ij}+\beta_5 \mathrm{Z}_j+\varepsilon_{ij} \tag{10}$$

The problem in this study is that many households choose zero spending on private tutoring. Several previous studies that have a similar type of data utilise the Tobit estimation method proposed by Amemiya (1974). To cope with this simultaneously, the household total expenditure and private tutoring expenditure are estimated using Two Stage Least Squares Models (Smith & Blundell, 1986) by using current income as the instrument variables (Liviatan, 1961).

4 DATA DESCRIPTION AND PRIVATE TUTORING SITUATION IN INDONESIA

This study uses secondary data from the National Socio-Economic Household Survey (SUSENAS) in 2012 (SUSENAS, 2012). The majority of households that still have children who attend school spend about 1% to 5% of their total income for private tutoring. The percentage number of richest households (quintile 5) whose spending for private tutoring is larger than the poorest household (quintile 1) is shown to be 32.51% and 15.66% respectively. The higher the level of household income, the more the households choose to spend on private tutoring (Table 1).

The average of fathers' and mothers' education level in households with positive expenditure on private tutoring is higher than households with zero expenditure on private tutoring. The average level of fathers' education for both households with positive or zero expenditure is higher than the average of mothers' education level (Figure 2). An increase in the number of family members initially increases household spending on private tutoring, up to the point when the number of family members reaches an average of four people. Then households start to reduce their spending on private tutoring (Figure 3).

Table 1. Percentage of households that consume private tutoring according to percentage of private tutoring expenditure of total household expenditure and total household expenditure groups (%).

Percentage of private tutoring expenditure (%)	Total household expenditure groups					
	Quint.1	Quint.2	Quint.3	Quint.4	Quint.5	Total
1–5%	14.66	14.69	17.07	21.07	32.51	100.00
	91.21	86.98	80.79	71.56	69.74	76.83
>5%	4.69	7.29	13.45	27.78	46.79	100.00
	8.79	13.02	19.21	28.44	30.26	23.17
Total	12.35	12.97	16.23	22.63	35.82	100.00
	100.00	100.00	100.00	100.00	100.00	100.00

Source: Calculation from SUSENAS 2012.

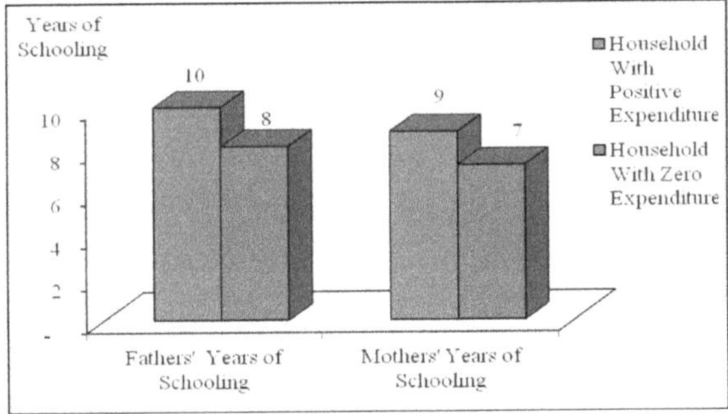

Figure 2. Average level of parents' education for households with positive expenditure and zero expenditure.

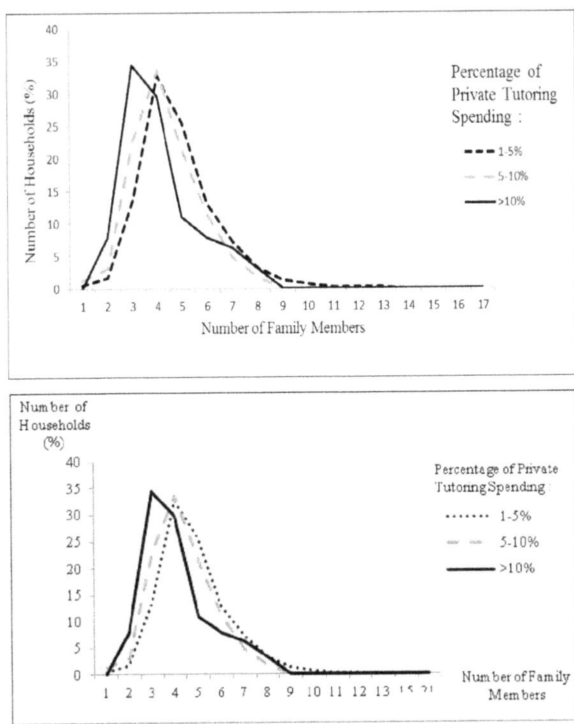

Figure 3. Number of family members and percentage of private tutoring spending by household.

Table 2 shows that our estimation results are fairly robust for alternative specifications and the main model for analysis in this study is Model 5. Table 3 shows that the total household expenditure positively and significantly affects household expenditure on private tutoring. This happens due to the increasing amount of budget-share allocated for educational spending, including private tutoring expenditure. The simultaneous increase of household demand for private tutoring and increase of household income indicates that private tutoring has functioned as normal goods. Similar cases are also found in several studies conducted by Bray and Kwok (2003), Jelani *et al.* (2014), Tansel and Bircan (2006), Dang (2007); Bray *et al.* (2014), Dang and Rogers (2008), and Kim and Lee (2010).

Table 2. Estimation result with Tobit model.

Dependent Variable: Ln. Households Expenditure on Private Tutoring (ln_Les).					
Independent variable	Model 1	Model 2	Model 3	Model 4	Model 5
Ln. Total Household Expenditure	4.022***	3.841***	3.595***	3.088***	2.835***
	(0.387)	(0.661)	(0.688)	(0.718)	(0.749)
Ln. Formal Education Cost	−0.951***	−0.594***	−0.588***	−0.619***	−0.624***
	(0.161)	(0.168)	(0.169)	(0.169)	(0.170)
Ln. Formal Education's Cost Square	0.186***	0.142***	0.141***	0.143***	0.144***
	(0.012)	(0.013)	(0.013)	(0.013)	(0.013)
Number of Tutoring Agency (Bimbel)	0.142***	0.133***	0.132***	0.0989***	0.0979***
	(0.011)	(0.011)	(0.011)	(0.011)	(0.011)
Middle School Education Level		4.367***	4.198***	4.339***	4.168***
		(0.329)	(0.333)	(0.328)	(0.332)
High School Education Level		5.200***	4.996***	5.197***	4.969***
		(0.354)	(0.359)	(0.354)	(0.359)
Father's Years of Schooling		0.101**	0.118**	0.0793*	0.0912*
		(0.038)	(0.039)	(0.037)	(0.038)
Mother's Years of Schooling		0.0854*	0.0883*	0.0693	0.0728*
		(0.036)	(0.038)	(0.036)	(0.037)
Number of Family Members		−1.199***	−1.246***	−1.104***	−1.154***
		(0.107)	(0.107)	(0.109)	(0.110)
Female			0.835**	0.783**	0.790**
			(0.272)	(0.271)	(0.270)
Household Head's Age			0.177*		0.203*
			(0.082)		(0.082)
Household Head's Age Square			−0.00132		−0.00161
			(0.001)		(0.001)
Urban				4.335***	4.377***
				(0.349)	(0.352)
Scholarship				2.275***	2.246***
				(0.431)	(0.431)
Public Schools				1.979***	1.983***
				(0.399)	(0.399)
_cons	−101.9***	−95.15***	−96.58***	−87.17***	−88.59***
	(6.258)	(10.073)	(10.302)	(10.927)	(11.173)
sigma_cons	18.65***	18.32***	18.31***	18.14***	18.12***
	(0.090)	(0.089)	(0.089)	(0.087)	(0.087)
N	66,234	66,234	66,234	66,234	66,234
pseudo R-sq	0.0345	0.0422	0.0427	0.0472	0.0475

Robust standard errors in parentheses, *significant at 10%, **significant at 5%, ***significant at 1%.

Table 3. Estimation result and marginal effect of Model 5.

Dependent Variable: Ln. Households Expenditure on Private Tutoring (ln_Les).

		Marginal effect	
Independent variable	Coefficient	Unconditional	Conditional
Ln. Total Household Expenditure	2.835***	0.1693***	0.4107***
	(0.749)	(0.0448)	(0.1085)
Ln. Formal Education Cost	–0.624***	–0.0373***	–0.0905***
	(0.170)	(0.0102)	(0.0247)
Ln. Formal Education's Cost Square	0.144***	0.0086***	0.0208***
	(0.013)	(0.0008)	(0.0019)
Number of Tutoring Agencies (Bimbel)	0.0979***	0.0058***	0.0142***
	(0.011)	(0.0007)	(0.0016)
Middle School Education Level	4.168***	0.2490***	0.6039***
	(0.332)	(0.0196)	(0.0477)
High School Education Level	4.969***	0.2969***	0.7201***
	(0.359)	(0.0213)	(0.0516)
Father's Years of Schooling	0.0912*	0.0054*	0.0132*
	(0.038)	(0.0023)	(0.0055)
Mother's Years of Schooling	0.0728*	0.0043*	0.0105*
	(0.037)	(0.0022)	(0.0054)
Number of Family Members	–1.154***	–0.0689***	–0.1672***
	(0.110)	(0.0065)	(0.0158)
Female	0.790**	0.0472**	0.1144**
	(0.270)	(0.0161)	(0.0392)
Household Head's Age	0.203*	0.0121*	0.0294*
	(0.082)	(0.0049)	(0.0119)
Household Head's Age Square	–0.00161	–0.0001	–0.0002
	(0.001)	(0.0000)	(0.0001)
Urban	4.377***	0.2615***	0.6342***
	(0.352)	(0.0208)	(0.0506)
Scholarship	2.246***	0.1342***	0.3254***
	(0.431)	(0.0257)	(0.0624)
Public Schools	1.983***	0.1185***	0.2874***
	(0.399)	(0.0239)	(0.0578)
_cons	–88.59***		
	(11.173)		
sigma _cons	18.12***		
	(0.087)		
N	66,234		
pseudo R-sq	0.0475		
Obs. summary:	61,419 left-censored observations at ln_Les <=0		
	4,815 uncensored observations		
	0 right-censored observations		

Robust standard errors in parentheses, *significant at 10%, **significant at 5%, ***significant at 1%.

Since the household expenditure for private tutoring and total household expenditure are presented in logarithmic form, the coefficient of estimation reflects the income elasticity rate of household expenditure for private tutoring. The result shows that household demand for private tutoring is inelastic on the change of household permanent income. Therefore, it can be concluded that private tutoring is normal goods needed by the household. Similar cases are also found in two studies conducted in Greece (Psacharopoulos & Papakonstantinou, 2005) and Vietnam (Dang, 2007). On the other hand, Tansel and Bircan (2006) find that private tutoring in Turkey is unitary elastic on the change of household permanent income.

The education level of fathers and mothers positively and significantly affects household expenditure for private tutoring. The effect of fathers' education level is stronger than mothers' education level. The results are different from the study conducted by Tansel and Bircan (2006), Dang (2007), Bray *et al.* (2014), and Jelani *et al.* (2014). This happens due to the fact that on average, fathers' education level is higher than mothers' education level. The higher one's education level, it is assumed that the preference and expectation on returns for children's education in the future will also be higher. A head of the household with higher education level usually gets a job with a higher salary. Therefore, this household group will tend to spend more for private tutoring (Dang & Rogers, 2008).

The number of family members negatively and significantly affects household expenditure for private tutoring. A larger number of family members means more household spending allocation that needs to be fulfilled under limited available household income. Thus, it will cause a decrease in household expenditure for private tutoring. Similar cases are found in some studies conducted by Tansel and Bircan (2006), and Kim and Lee (2010).

The higher number of tutoring agencies available in an area increases household demand for private tutoring, because it provides more price options for households. Meanwhile, tutoring agencies become more accessible with lower transportation cost so that the tutoring agencies are relatively cheaper than those in areas with a lower number of tutoring agencies. Higher cost of formal education causes a decrease in household expenditure for private tutoring with an increasing reduction rate, along with expensive formal education costs. This occurs due to the fact that there are gradual reductions of part of the income that can be allocated for private tutoring expenditure. The relationship between formal education costs and household expenditure for private tutoring reflects cross-price elasticity level. The coefficient of estimation result (Table 3) shows that formal education and private tutoring cannot replace each other. Instead, both factors are complementary to each other (Dang & Rogers, 2008).

The age of the head of a household positively and significantly affects household expenditure for private tutoring. This is because the advancement in the age of the head of a household usually goes along with the betterment of the household's economic condition. Moreover, the older the head of the household, the experience and knowledge of the importance of investment for children's education is also increasing. This affects the increase in spending for private tutoring with the reduction on the increasing level. Increasing effects of the household head's age to household spending on private tutoring follows the pattern of life cycle of the household head's expenditure (Tansel & Bircan, 2006). Table 3 shows that the increase in spending on private tutoring follows the age of the household head and reaches the optimum point at the age of 63, which is near the optimum level of the age of productive labour in Indonesia, which is 64 years.

Households living in urban areas spend a larger share of their income for private tutoring, as many as 63.4% larger than households living in rural areas. This happens because there are more private tutoring agencies available in urban areas, so that families living in urban areas may get easier access to that particular service compared to families living in rural areas (Tansel & Bircan, 2006).

Demand for private tutoring is increasing along with the higher education level of children, where at high school level the necessity for private tutoring is considered the highest compared to the levels below. This is due to the high competition for admission to the best universities. Children and parents assume that getting admitted to the best universities can improve children's competitiveness in the labour market. Children who get scholarships will have more opportunities to participate in private tutoring. This is due to the fact that by getting scholarships, they have the opportunity to pay for private tutoring services. Moreover, children who get scholarships must maintain their academic achievement so that they need to get private tutoring.

Households spend 11.44% more for female children's private tutoring. It is estimated that daughters tend to have more learning motivation and a higher level of anxiety about their academic achievement compared to sons, so that parents feel the need to send their daughters for private tutoring. This assumption is contradictory to the studies conducted by Bray *et al.* (2014), Jelani *et al.* (2014), and Dang (2007), who found that there is no difference in demand for private tutoring among females and males. Demand for private tutoring increases by about 28.7% for children who attend public schools. This indicates that the quality of private schools is better than public schools. Commonly, private schools provide a better and more complete education package than public schools. Therefore, some households tend to choose to send their children to private schools, instead of supplying their needs to get better education from private tutoring (Bray *et al.*, 2014).

6 CONCLUSION

A more settled condition of a household will motivate the household's spending on children's private tutoring. This is signified by the higher education level of the parents, total household expenditure, age of the head of the household and the number of family members. The high education level of parents correlates to the types of occupations in which the salary level is also higher. It leads to a high total of household income, which leads to more resources that can be allocated such as for private tutoring expenditure. The spending on private tutoring follows the life cycle expenditure pattern of the head of the household.

Besides the household characteristics explained previously, household expenditure level for private tutoring is also significantly affected by children's education level and household domicile, whether a household lives in an urban or rural area. The high level of children's education will motivate the parents in spending more for private tutoring. This indicates that a household's high necessity for education is in line with the high level of education being attended by the children. Meanwhile, households living in urban areas tend to spend more of their income for private tutoring than households living in rural areas. Households living in urban areas on average have a higher education level and higher income compared to households living in rural areas. These are the factors that support the high demand for private tutoring in urban areas.

REFERENCES

Amemiya, T. (1974). Multivariate regression and simultaneous equation models when the dependent variables are truncated normal. *Econometrica*, 42(6), 999–1012. doi:10.2307/1914214.

Becker, G.S. & Lewis, H.G. (1973). On the interaction between the quantity and quality of children. *The Journal of Political Economy*, 81(2), S297–A288.

Bray, M. & Kwok, P. (2003). Demand for private supplementary tutoring: Conceptual considerations, and socio-economic patterns in Hong Kong. *Economics of Education Review*, 22(6), 611–620. doi:10.1016/S0272-7757(03)00032-3.

Bray, M., Zhan, S., Lykins, C., Wang, D. & Kwo, O. (2014). Differentiated demand for private supplementary tutoring: Patterns and implications in Hong Kong secondary education. *Economics of Education Review*, 38, 24–37. doi:10.1016/j.econedurev.2013.10.002.

Dang, H.A. (2007). The determinants and impact of private tutoring classes in Vietnam. *Economics of Education Review*, *26*(6), 683–698. doi:10.1016/j.econedurev.2007.10.003.

Dang, H.A. & Rogers, F.H. (2008). The growing phenomenon of private tutoring: Does it deepen human capital, widen inequalities, or waste resources? *World Bank Research Observer*, 23(2), 161–200. doi: 10.1093/wbro/lkn004.

Dang, H.A.H. & Rogers, F.H. (2016). The decision to invest in child quality over quantity: Household size and household investment in education in Vietnam. *World Bank Economic Review*, 30(1), 104–142. doi:10.1093/wber/lhv048Davies, S. (2004). School choice by default? Understanding the demand for private tutoring in Canada. *American Journal of Education*, 110(3), 233–255.

Jelani, J., Tan, A.K.G. & Mohd-Zaharim, N. (2014). Demand for extracurricular activities amongst primary school students: Exploratory evidence from survey data in Penang (Malaysia). *The Asia-Pacific Education Researcher*, 24(1), 125–135. doi:10.1007/s40299-013-0165-y.

Kim, S. & Lee, J.H. (2010). Private tutoring and demand for education in South Korea. *Economic Development and Cultural Change*, 58(2), 259–296. doi:10.1086/648186.

Kristiansen, S. & Pratikno, P. (2006). Decentralising education in Indonesia. *International Journal of Educational Development*, 26(5), 513–531. doi:10.1016/j.ijedudev.2005.12.003.

Liviatan, N. (1961). Errors in variables and Engel curve analysis. *Econometrica*, 29(3), 336.
Mori, I. & Baker, D. (2010). The origin of universal shadow education: What the supplemental education phenomenon tells us about the postmodern institution of education. *Asia Pacific Education Review*, 11(1), 36–48. doi:10.1007/s12564-009-9057-5.
Novita, M. (2008). Pengaruh karakteristik rumah tangga dalam fungsi permintaan Pendidikan. Magister's thesis, Program Pasca Sarjana Ilmu Ekonomi Fakultas Ekonomi dan Bisnis, Universitas Indonesia.
Psacharopoulos, G. & Papakonstantinou, G. (2005). The real university cost in a 'free' higher education country. *Economics of Education Review*, 24(1), 103–108.
Smith, R.J. & Blundell, R.W. (1986). An exogeneity test for a simultaneous equation Tobit model with an application to labor supply. *Econometrica*, 54(3), 679–685.
Takayama, K., Waldow, F. & Sung, Y.K. (2013). Finland has it all? Examining the media accentuation of 'Finnish education' in Australia, Germany and South Korea. *Research in Comparative and International Education*, 8(3).
Tansel, A. & Bircan, F. (2006). Demand for education in Turkey: A Tobit analysis of private tutoring expenditures. *Economics of Education Review*, 25(3), 303–313.

Competition and Cooperation in Economics and Business – Gani et al. (Eds)
© 2018 Taylor & Francis Group, London, ISBN 978-1-138-62666-9

The income and consumption profiles of public-sector employees: Indications of unreported income

A.L. Tobing & A. Kuncoro
Department of Economics, Faculty of Economics and Business, Universitas Indonesia, Depok, Indonesia

ABSTRACT: This study uses data from the Indonesian Family Life Survey (IFLS) to see whether there are differences between the wages of observably comparable public and private sector employees. This study also displays indications of unreported income among public-sector employees. This is shown by using private-sector employees as a control group and comparing the consumption profiles of both types of employees; we find that public employees consume more of their reported income than a comparable private employee. Finally, this paper reports that households with at least one public employee are more prone to report consumption levels above their income levels compared to households without public employees, thereby strengthening the case that public employees receive more income than they report.

1 INTRODUCTION

As of 2015, a survey conducted by Transparency International showed that from all national institutions, Indonesia's civil servants and public officials were perceived to be some of the most corrupt. The respondents perceived the civil service to be cleaner only to the judiciary, elected officials, political parties, and the police but more corrupt relative to the media, NGOs (Non-Governmental Organizations), religious bodies, and the private sector. What this research attempts to see is whether this perception of illegal behaviour by public-sector employees is backed up by data that does not rely on public perception.

As it may be impossible to truly demonstrate the existence of corruption through econometric analysis, we will follow the method previously used by Pissarides and Weber (1989), Gorodnichenko and Peter (2007), Gorodnichenko et al. (2009), and Hurst et al. (2014), to explore whether public-sector employees show indications of having 'unreported income'. This research is based on the logic that if someone has received an income above what they report to data collectors, they will then use that 'extra' income for consumption, wherever that extra income may have come from. Thus, the main finding of this paper will demonstrate that public-sector employees (whom we suspect of having unreported income) do indeed consume more than an otherwise comparable control group given the same levels of reported income. The control group we have chosen are employees in the private sector.

For this method to work, we must first confirm that there are negligible levels of income shifting between family members or between income sources. Due to the fact that labour income is reported at the individual level whereas expenditure is reported at the household level, the logical framework described above will not effectively show the existence of unreported income if there is income shifting, as households who untruthfully ascribe which members get how much income (or from what source) will have reported actual total household income that may already include income from corruption. The way this paper will show that there does not seem to be an attempt at income shifting is by presenting public-private income differentials at both the individual and household levels; were the differentials to be different to each other (say, by having a different sign or of a different order of magnitude), this may indicate that income shifting has indeed systematically occurred.

We will show that the differentials at both levels of analysis are not very different from each other, along with the finding that public employees are associated with higher income levels than observably comparable private employees. The method which we use to highlight income differentials between the two types of employee stem from a rich literature concerning income differentials (see Anton and Bustillo (2013) in Spain, Azam and Prakash (2010) in India, Bargain and Melly (2008) in France, Borjas (2002) in the USA, Hyder and Rilley (2005) in Pakistan, Melly (2005) in Germany, Mueller (1998) in Canada, Nielsen and Rosholm (2001) in Zambia). This paper will also show that a majority of the households in the data we will use report expenditure levels above their income levels and that households with at least one public employee are more likely to report expenditures above their income than households with no public employees.

The main questions that this research has attempted to answer are: (1) Is there a difference of the income levels of observably comparable public and private employees at both the individual and household levels?; (2) Does substituting a public employee for a private employee, in an otherwise observably comparable household and given the same income, become associated with higher levels of consumption?; and (3) Does substituting a public employee with a private employee in an otherwise observably comparable household become associated with a higher probability to report expenditure levels higher than income levels? This paper will not dwell too much on public-private pay differentials, as that line of research will have to focus on controlling issues concerning endogeneity and a whole host of other problems regarding compensating pay differentials, which is beyond the scope of this paper.

2 DATA AND METHODOLOGY

The data that this research will use is the third, fourth, and fifth waves of the Indonesian Family Life Survey (IFLS) (from Strauss et al., 2004, 2009, 2016; henceforth called the IFLS), from the years 2000, 2007, and 2014, respectively. At the individual level, this author has omitted any public or private-sector employees without at least a high school or equivalent diploma, outside the prime working age (17–65), and who does not receive a regular salary. Unfortunately, the IFLS data does not allow us to distinguish between civil servants and employees of state-owned enterprises. Although our analysis at the household level will only use households that comprise at least one individual from our individual-level analysis, there are still income earners within the household who are not part of the individual-level analysis, income earners who consist of self-employed workers, casual workers, or employees not meeting the criteria for the individual-level analysis described above.

There are two key variables with which this study will be occupied: income and consumption. The IFLS asks respondents for their last month's total salary received from their primary and secondary jobs that includes the value of all benefits that respondents receive. The income of self-employed workers is taken from the approximate value of profits they receive. Other sources of income include income from pensions and assets. The individual level analysis will only use the salary from an employee's primary job, whereas the household analysis will include all forms of income. Note that the individual level analysis does not include self-employed workers and employees not meeting the main criteria stated above, whereas the household analysis does.

The other key variable in our analysis is consumption. The IFLS asks respondents for the approximate money value of all consumption expenditures that a household makes. These expenditures include the weekly consumption of various types of food, quarterly non-durable consumption other than food, yearly expenditures on durable goods, and any monthly ritual and tax expenditures a household may make. This research omits any consumption that was produced by the household or given from outside the household so as to capture the total of only market purchases a household may make. This research converts these values to their monthly values.

There is a total of four empirical models that we will use to answer the research problems:

$$\ln w = \delta_1 + X\delta_2 + S\delta_3 + \varepsilon_\delta \tag{1}$$

Model (1) above will be used to answer whether there are any differences between the pay of public and private-sector employees at the individual level. ln w is the log of last month's wages, X is a vector of covariates comprising of a gender indicator, a region indicator (rural vs. urban), highest level of education attained (high school or equivalent, tertiary vocational school, bachelor's degree, or master's and above), experience (defined as age minus years of schooling minus six), squared experience, tenure, squared tenure, a series of indicators denoting provincial location, and ε_δ is an error term with a mean of zero. Model (1) above is essentially an extension of a standard Mincer earnings function taken from Gorodnichenko and Peter (2007). The parameter of primary interest here is δ_3 which will show us whether or not there are wage differentials between the two employment sectors at the individual level.

To estimate the effects of substituting a private-sector employee for a public employee on household income, the empirical model is:

$$\ln Y = \alpha_1 + X\alpha_2 + N^{private}\,\alpha_3 + N^{other}\,\alpha_4 + N^{IE}\,\alpha_5 + \varepsilon_\alpha \qquad (2)$$

where ln Y is the natural log of total household income, $N^{private}$ is the number of private-sector wage earners (labelled as 'Private Sector' in regression result tables below), N^{other} is the number of other sector wage earners, and N^{IE} is the number of income earners. X is a set of covariates as described in Model (1) but averaged across all private- and public-sector employees in the household, and ε_α is the residual with a mean of zero.

The target parameter for Model (2) is α_3. Because we have controlled the number of wage earners and individuals working in places other than the private and public sector, α_3 will identify the effect of substituting one public-sector worker with a private-sector worker upon the income of the household, with other factors held constant.

The primary interest of this study is the next model, which is expected to capture the effects of substituting a private employee for a public employee within a household upon consumption, ceteris paribus, the specification of which is given by:

$$\ln C = \beta_1 + Z\,\beta_2 + N^{private}\,\beta_3 + N^{other}\,\beta_4 + N^{IE}\,\beta_5 + \ln Y\,\beta_6 + \varepsilon_\beta \qquad (3)$$

where ln C is the natural log of a household's total consumption (or non-durable consumption), ln Y is the natural log of total reported household income, and Z is a vector of covariates which we expect to influence consumption levels that include: the number of household members, the number of children in the household, the proportion of men between income earners, the proportion of married couples between income earners, the average years of schooling between income earners, the average age of all household members, the average age of all income earners, a region dummy (rural = 0, urban = 1), a series of dummies denoting provincial location, and ε_β is the residual with a mean of zero. The parameter of interest here is β_3, which will show the effect of substituting a private employee for a public employee, as in Model (2).

As will later be discussed in more detail, a substantial number of households in the IFLS data report consumption have expenditures larger than their total income. To test whether households have a stronger or weaker probability of reporting expenditures larger than their incomes when a private employee is substituted for a public employee, we will estimate the following model:

$$[\Pr(C > Y) = 1] = \gamma_1 + V\,\gamma_2 + N^{private}\,\gamma_3 + N^{other}\,\gamma_4 + N^{IE}\,\gamma_5 + \varepsilon_\gamma \qquad (4)$$

This empirical model will be estimated by a Linear Probability Model (LPM). The dependent variable equals one when reported expenditure is larger than reported income or equals zero when reported expenditure is smaller than reported income. The parameter of interest here is γ_3, the sign of which will show us whether substituting a public-sector employee with a private employee will result in a decreasing or increasing probability that a household has an expenditure larger than their income, ceteris paribus. V is a vector of control variable that includes the region, gender, the average age of income earners, and the log of total income; ε_γ is the error term with a mean of zero.

3 RESULTS AND DISCUSSION

3.1 *Descriptive statistics*

Before we discuss the main findings of our analysis, it is worth taking a look at the summary differences between employees in either sector.

As can be seen in Tables 1 and 2, public employees are generally older than private employees; they also tend to be more educated than the private sector. Although the public-sector employs less than 30% of our sample, yet roughly half of all university graduates are in the public sector, whereas the great majority of private employees in the sample are high school graduates.

Another significantly different characteristic between public and private workers is the various fringe benefits public employees enjoy. The majority of public employees receive pension benefits and health insurance from their employers, whereas only 37% percent of private employees receive health insurance and less than 16% receive pension benefits from their employer. Roughly the same proportion in either sector receive credit from their employer. Whether the greater fringe benefits received by public employees are a function of their higher labour market endowments or not is outside the scope of this study, nor whether the greater benefits serve as a substitute or a complement to their wages.

Lastly, public-sector employees have significantly longer tenure than private employees: the average public employee stays at their job more than twice as long as the average private employee. This fact is telling of the inherent differences not just between the two employers but also the kinds of people they employ. The longer tenure of government workers may be due to the fact that civil servants or employees of state-owned enterprises are relatively safer from any economic fluctuations which might otherwise adversely affect private employees; another explanation may be that the average public employee tends to be more risk-averse than the average private employee, so they tend to adhere to their jobs longer than their private counterparts; or it may also be due to the relatively greater observable and unobservable amenities that public employees receive, which makes staying at their current posts more desirable.

Table 1. Summary statistics of continuous variables, by sector.

	Public sector			Private sector		
Variables	N	Mean	Standard deviation	N	Mean	Standard deviation
Age	2,623	39.17	9.563	6,469	31.07	8.315
Years of Schooling	2,618	14.32	2.019	6,452	13.04	1.748
Tenure	2,623	12.08	9.111	6,469	5.303	5.829

Table 2. Summary statistics of categorical variables, by sector.

	Public sector		Private sector	
Variables	N	Proportion	N	Proportion
Employer Pension	1,832	0.651	5,050	0.151
Health Insurance	2,622	0.682	6,468	0.368
Employer Credit	2,623	0.417	6,469	0.418
Senior High School	2,623	0.405	6,469	0.728
Bachelor's Degree	2,623	0.504	6,469	0.23
Female	2,623	0.408	6,469	0.366

Note: Employer pension data is only available for 2007 and 2014.

3.2 *Individual and household income regression results*

Table 3 below shows the estimation results of Model (1). We find that public employees are associated with higher wages than their observably comparable private counterparts. The sectoral gap in the year 2000 is 21 log points (19 per cent), 12 log points (11 per cent) in 2007, and 22 log points (19 per cent) in 2014).

The adjusted wage gaps are consistently in favour of public employees, and this finding holds true for either gender and for every year. For both genders, it seems that the year 2007 was when the wage gap was lowest between the two sectors, falling to 0.8 log points for men. By 2014, the wage gap increased again to more than 15 log points for men and 31 log points for women.

As confirmed in Table 4, the pay gap estimations at the individual level analysis is apparently consistent with the income gap estimations at the household level. Holding all other factors constant, we find that substituting a private employee for a public employee is associated with statistically significant higher household income, and this holds true for all three periods under review. That being said, the gap has somewhat lessened. This may be due to the slightly different model specification, but may also indicate either that households with no public employee may have systematically higher levels of income from sources other than the primary jobs of public and private employees. At the very least, the results in Table 4 show the same sign as the results in Table 3.

3.3 *Household consumption regression results: First indications of unreported income*

Now that we have seen that public- and private-sector income differences are consistent at both the individual and household levels, we can now move on to the main part of this study: demonstrating that there are indications of unreported income among public-sector employees. A look at Figures 1(a), 1(b), and 1(c) below will be instructive for our discussion. In the year 2000 there did not seem to be a significant difference in the consumption

Table 3. Adjusted individual public-private wage gaps, by year.

Last month's wage (log)	2000	2007	2014
Sector	–0.213***	-0.142***	-0.232***
	(0.0386)	(0.0354)	(0.0394)
Observations	2,186	3,111	3,712
R-squared	0.342	0.319	0.268

Note: Robust standard errors in parentheses. ***$p < 0.01$, **$p < 0.05$, *$p < 0.1$. Not shown above are controls for education, experience, tenure, gender, provinces, and regional location.

Table 4. Household level income gap, yearly and panel estimates.

Total monthly income	2000	2007	2014	PLS
Private Sector	–0.136***	–0.112***	–0.159***	–0.138***
	(0.0386)	(0.0352)	(0.0431)	(0.0224)
Observations	1,115	1,886	1,899	4,900
R-squared	0.422	0.349	0.270	0.496

Note: Robust standard errors in parentheses. ***$p < 0.01$, **$p < 0.05$, *$p < 0.1$. The dependent variable uses total income constructed from the value of last month's wages or profits received from primary and secondary jobs. Not shown above are controls for number of income earners, number of other sector workers, provincial location, experience, squared experience, tenure, squared tenure, and regional location.

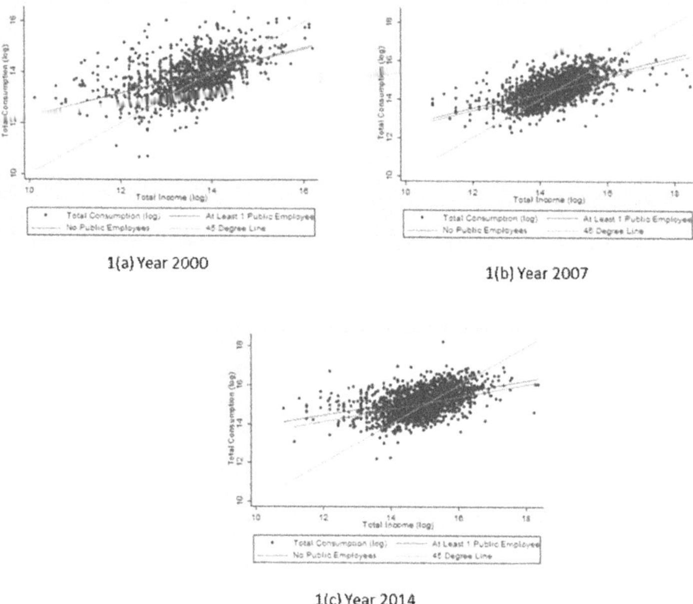

Figure 1. Consumption to income linear fits by household type. Years 2000, 2007, and 2014. Note: Consumption is here defined as total market-based expenditures, and income is constructed using last month's wages or profits from primary and secondary jobs and all non-labour income.

to income levels of households with no public employees and households with at least one public employee. This means that given the same income, there is no apparent difference in the consumption levels of both types of household. Yet, a gap clearly exists if we compare the consumption to income profile of the year 2007 in Figure 1(b) with that of year 2000 in Figure 1(a), and an even larger gap when we see the profile of year 2014 in Figure 1(c). This indicates that there has apparently been a behavioural evolution within at least one of the two types of households: generally speaking, households with at least one public employee seem to increasingly be consuming more of their reported income compared to households without public employees, even though the differences in their reported income has stayed roughly constant over the three periods.

The finding can also be summarised in Figure 2. During the year 2000, households with no public employees have higher mean and median levels of APC (Average Propensity to Consume; in logarithm) than households with at least one public employee, as is expected of a group with lower incomes. However, theoretically, what we would expect to find is for APC to grow smaller as incomes rise, because as people get richer (in income terms) they will spend a smaller proportion of their income on consumption. This theoretical proposition holds true for households with no public employees. Nevertheless, households with at least one public employee seem to have rising APC values over the course of three years. These figures are in themselves quite suspicious, as those show that households with at least one public employee have systematically different behaviours compared to our control group.

The pattern of increasingly larger consumption differences between sectors is confirmed in Table 5. In the year 2000, there did not seem to be a statistically significant difference between the consumption levels within a comparable household were we to substitute a public employee for a private employee. However, by the year 2007, a difference of 0.05 log points (4 per cent) was detected in consumption were we to substitute a private employee for a public employee, and by the year 2014, an even stronger difference in consumption was detected. This is our first indication that public-sector employees receive income above what they have actually reported. In the year 2014, even after controlling household and income

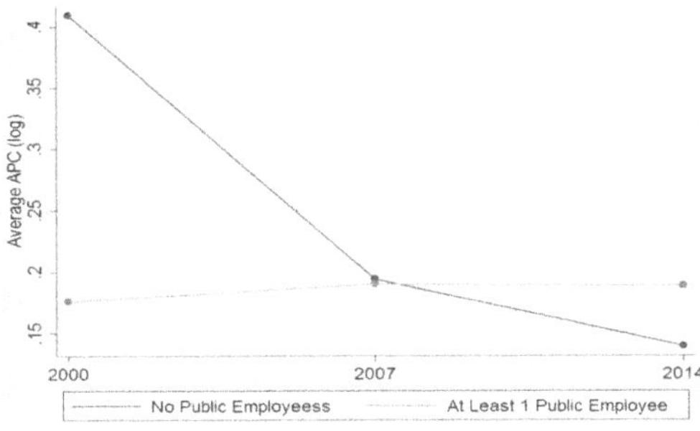

Figure 2. Mean APC by household type and year.

Table 5. Household total consumption difference, by year.

Total consumption	2000	2007	2014
Private Sector	0.00933	−0.0537**	−0.152***
	(0.0351)	(0.0251)	(0.0274)
Observations	1,173	1,941	2,006
R-squared	0.364	0.443	0.372

Note: Robust standard errors in parentheses. ***p < 0.01, **p < 0.05, *p < 0.1. Not shown above are controls for number of income earners, number of other sector workers, total income, provincial location, household size, regional location, average years of schooling of employees, the proportion of men between employees, average age of household members, and average age of employees.

earner characteristics, including the level of total household income, substituting a public-sector employee with a private-sector employee was associated with a 0.15 log points (14 per cent) difference in total consumption.

What this author has further decided to do is to divide the sample into five groups according to the income of the household's head (defined as the person with the largest income within a household) in each year, resulting in 15 different groups. Upon each group, we will use empirical Model (3) to estimate differences in consumption levels for each income group. The results as seen in Table 6 are more nuanced. A majority of the results show that substituting a private employee for a public employee is associated with smaller consumption levels, with all other factors including income held as constant. Nonetheless, some income groups show the opposite sign, and this means that rather than among public employees, there are actually indications of unreported income among private employees. That being said, most of the results show that having a public employee in a household instead of a private employee is associated with higher consumption levels, ceteris paribus, and this finding is strongest in the year 2014.

3.4 *The probability of reporting consumption higher than income: Results*

However, look back again at Figures 1 and 2 which show that there is still a lack of components in our analysis. Theoretically, we would expect to see a value of APC between zero and one, meaning that consumption does not surpass income. In logarithmic terms, this would mean that the normal values of the log of APC are negative. Nevertheless, Figure 2 shows that the average log of APC is above 1. Even if we were to allow for the fact that some households might be in debt, it seems unlikely that the number of indebted

Table 6. Total consumption differences, by year and quintile of income.

Year		1st Quintile	2nd Quintile	3rd Quintile	4th Quintile	5th Quintile
2000	Private Sector	−0.232*	−0.155*	0.165**	−0.0604	−0.0596
		(0.130)	(0.0833)	(0.0786)	(0.0754)	(0.0565)
	Observations	283	232	193	232	233
	R-squared	0.268	0.214	0.249	0.135	0.300
2007	Private Sector	−0.175**	−0.0397	−0.0673	−0.0783	−0.128***
		(0.0699)	(0.0635)	(0.0507)	(0.0490)	(0.0458)
	Observations	402	412	401	348	379
	R-squared	0.177	0.234	0.252	0.244	0.205
2014	Private Sector	−0.175**	−0.0397	−0.0673	−0.0783	−0.128***
		(0.0699)	(0.0635)	(0.0507)	(0.0490)	(0.0458)
	Observations	402	412	401	348	379
	R-squared	0.177	0.234	0.252	0.244	0.205

Note: Robust standard errors in parentheses. ***$p < 0.01$, **$p < 0.05$, *$p < 0.1$.
Not shown above are controls for total household income, number of income earners, number of other sector workers, household size, years of schooling, the proportion of men among sector employees, proportion of married people among sector employees, provinces, region, average household age, and average age of sector employees.

households and the magnitude of their debt is as large as Figure 2 suggests. There are only two explanations for this. The first one is that consumption has been over-reported. To a certain extent this may be due to imperfections in survey collection: respondents may have—for whatever reason—upwardly biased the quantity or price of the goods and services they have consumed. The other is that income has been under-reported, which is exactly what this study hopes to prove.

The next step in our study is to see whether or not it is more likely for households with at least one public employee to report expenditures larger than their income compared to households without a public employee. Figures 3(a), 3(b), and 3(c) are a good starting place to see this point. In those figures, the author has divided households into five groups according to the income of the household head, with Q1 having the lowest income and Q5 the highest, and further dividing them into households without a public employee (0 PE) and with at least one public employee (> 0 PE). With the few exceptions being some income groups in the year 2000, the group of households with at least one public employee always had higher proportions of households that report their consumption above their incomes than households with no public employees.

The LPM regression results presented in Table 7 further confirm the pattern of behavioural change discussed in the preceding sub-chapter. In the year 2000, there did not seem to be a statistically significant difference in the likelihood a household reports expenditures above income when we substitute a private employee for a public employee. Yet, by the year 2014, substituting a private for a public employee appeared to significantly decrease that probability across the income distribution.

3.5 *Analysis and discussion*

The results of this study suggest that public employees surveyed in the IFLS consume more than a private employee given the same income, providing a somewhat weak indication that there was unreported income among public employees, at least in the years 2007 and 2014. The results of study where we find that government employees consume more of their income can still be explained without invoking corruption or even unreported income. Government employees in Indonesia receive greater fringe benefits, higher employment security, and they may even have inherently different behavioural attitudes towards risks compared to their private-sector counterparts. All of this serves as a potential explanation why government employees are more consumptive than private employees.

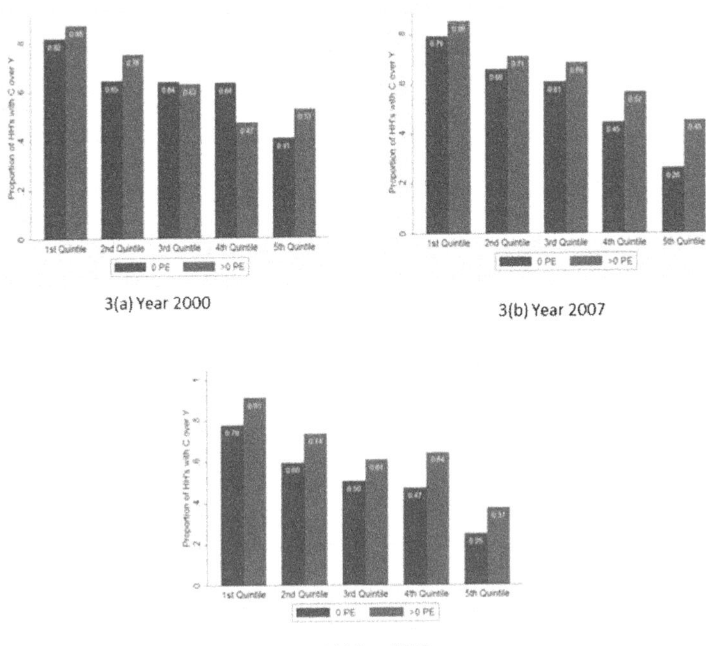

3(a) Year 2000

3(b) Year 2007

3(c) Year 2014

Figure 3. Proportion of households with expenditures above income, by income quintiles of household head and household type. Years 2000, 2007, and 2014.

Table 7. Probability of reporting consumption over income by year and quintile of income.

Year		1st Quintile	2nd Quintile	3rd Quintile	4th Quintile	5th Quintile
2000	Private Sector	−0.0757	−0.0197	0.154**	0.161***	−0.0431
		(0.0608)	(0.0615)	(0.0607)	(0.0577)	(0.0493)
	Observations	286	233	194	232	235
	R-squared	0.187	0.194	0.149	0.088	0.092
2007	Private Sector	−0.0320	−0.0244	−0.0182	−0.0562	−0.0996**
		(0.0441)	(0.0519)	(0.0469)	(0.0444)	(0.0399)
	Observations	404	413	405	350	380
	R-squared	0.165	0.183	0.115	0.066	0.091
2014	Private Sector	0.00600	−0.00699	−0.159***	−0.0904*	−0.0912**
		(0.0316)	(0.0696)	(0.0578)	(0.0474)	(0.0417)
	Observations	406	417	388	410	394
	R-squared	0.203	0.229	0.215	0.174	0.155

Note: Robust standard errors in parentheses. ***p < 0.01, **p < 0.05, *p < 0.1.
Not shown above are controls for total household income, number of income earners, number of other sector workers, household size, years of schooling, the proportion of men among sector employees, proportion of married people among sector employees, provinces, region, average household age, and average age of sector employees.

However, the fact that households with public employees are clearly more prone to reporting expenditures above their reported income compared to households without them does strengthen the case of public employees under-reporting their true income. The finding that substituting a public employee for a private employee is strongly associated with a higher probability that a household reporting expenditures above income cannot be taken trivially.

Particularly in the highest quintiles of income, the argument of public employees having higher employment security, greater fringe benefits, nor inherently different attitudes towards risks is unable to explain that finding.

All this being said, there are many weaknesses of this paper stemming either from the limitations of the available data or the author. Concerning the data, the author was not able to distinguish between employees of state-owned enterprises (national or regional), civil servants, or elected public officials. Although the IFLS data allows us to distinguish between different types of occupations (blue-collar, administrative, and professional/technical), yet the resulting size of the sub-samples (particularly for the white-collar jobs) the author judges to be insufficient as a basis for a meaningful statistical analysis. This is also the reason why the author decided not to do the same analysis for each provincial region, even though the differences in income levels and standards of living between provinces would undoubtedly have an effect on the consumption behaviour of public employees.

The diminutiveness of the data is also the reason why this paper did not distinguish the time when the subjects were surveyed. The timing of observations is of importance because it is very common for employees (whether public or private) to receive more than 12 times their salaries per year, as there are various bonuses that a person may receive during different times of the year. Most Indonesians usually receive an extra salary after the fasting season each year, while many organisations (whether state-owned enterprises or private firms) also disburse bonuses to their employees at the end of the year. These weaknesses thus mean that extreme caution must be taken in interpreting the results hitherto presented.

4 CONCLUSION

Although the individual wage or household income differences between public and private employees stayed roughly constant between the years 2000 to 2014, we have found that public employees increasingly consume more compared to private employees given the same levels of income. This serves as our first indication that Indonesian public employees may have received income other than what was reported by the IFLS. To a certain extent, this behavioural difference may be explained by the fact that public employees receive greater fringe benefits or higher employment security, which may affect their attitudes towards risks and thus their consumption behaviour.

A stronger indication of unreported income that we have found is that public employees are more prone to reporting expenditure levels above their income than private employees, a fact that has grown stronger over time and more pronounced as we get higher up the income distribution. The finding that public employees have a higher probability of reporting expenditures levels higher than income cannot be explained by the fact that public employees have higher employment security or greater fringe benefits, and thus this serves as this study's second and stronger indication that public employees receive higher incomes than what they have reported.

REFERENCES

Antón, J.I., and de Bustillo, R.M., 2015. Public-private sector wage differentials in Spain. An updated picture in the midst of the Great Recession. *Investigación económica, 74*(292), pp.115–157.

Azam, M., & Prakash, N. (2015). A Distributional Analysis of the Public-Private Wage Differential in India. *Labour, 29*(4), 394–414.

Bargain, O., & Melly, B. (2008). Public Sector Pay Gap in France: New Evidence Using Panel Data.

Borjas, G. (2002). The Wage Structure and the Sorting of Workers into the Public Sector. *National Bureau of Economic Research*.

Christofides, L.N. & Michael, M. (2013). Exploring the public-private sector wage gap in European countries. *IZA Journal of European Labor Studies, 2*(1), 1–53. doi:10.1186/2193–9012–2-15.

Dur, R. & Zoutenbier, R. (2015). Intrinsic motivations of public sector employees: Evidence for Germany. *German Economic Review, 16*(3), 343–366. doi:10.1111/geer.12056.

Fisman, R. & Gatti, R. (2002). Decentralization and corruption: Evidences across countries. *Journal of Public Economics*, *83*(3), 325–345. doi:10.1016/S0047–2727(00)00158–4.
Gorodnichenko, Y., Martinez-Vazquez, J. & Peter, K.S. (2009). Myth and reality of flat tax reform: Micro estimates of tax evasion response and welfare effects in Russia. *Journal of Political Economy*, *117*(3), 504–554. doi:10.1086/599760.
Gorodnichenko, Y. & Peter, K.S. (2007). Public sector pay and corruption: Measuring bribery from micro data. *Journal of Public Economics*, *91*(5–6), 963–991. doi:10.1016/j.jpubeco.2006.12.003.
Gourinchas, P.O. & Parker, J.A. (2002). Consumption over the life cycle. *Econometrica*, *70*(1), 47–89. doi:10.1111/1468–0262.00269.
Haider, S. & Solon, G. (2006). Life-Cycle variation in the association between current and lifetime earnings. *American Economic Review*, *96*(4), 1308–1320. doi:10.1257/aer.96.4.1308.
Hurst, E., Li, G. & Pugsley, B. (2014). Are household surveys like tax forms? Evidence from income underreporting of the self-employed. *The Review of Economics and Statistics*, *96*(1), 19–33. doi:10.3386/w16527.
Hyder, A., & Reilly, B. (2005). The Public and Private Sector Pay Gap in Pakistan: A Quantile Regression Analysis. *The Pakistan Development Review*, 271–306.
Mueller, R.E. (1998). Public–Private sector wage differentials in Canada: Evidence from quantile regressions. *Economics Letters*, *60*(2), 229–235. doi:10.1016/S0165–1765(98)00110–4.
Nielsen, H.S., & Rosholm, M. (2002). The Public-Private Sector Wage Gap in Zambia in the 1990's: A Quantile Regression Approach. *Economic Applications of Quantile Regressions*, 169–182.
Pissarides, C.A. & Weber, G. (1989). An expenditure-based estimate of Britain's black economy. *Journal of Public Economics*, *39*(1), 17–32. doi:10.1016/0047–2727(89)90052–2.
Saha, S., Roy, P. & Kar, S. (2014). Public and private sector jobs, unreported income and consumption gap in India: Evidence from micro-data. *North American Journal of Economics and Finance*, *29*, 285–300. doi:10.1016/j.najef.2014.07.002.
Strauss, J., Witoelar, F. & Sikoki, B. *The fifth wave of the Indonesia family life survey (IFLS5): Overview and field report.* March 2016. WR-1143/1-NIA/NICHD.
Strauss, J., Witoelar, F., Sikoki, B. & Wattie, A.M. *The fourth wave of the Indonesian family life survey (IFLS4): Overview and field report*. April 2009. WR-675/1-NIA/NICHD.
Strauss, J., Beegle, K., Sikoki, B., Dwiyanto, A., Herawati, Y. & Witoelar, F. *The third wave of the Indonesia family life survey (IFLS): Overview and field report*, March 2004. WR-144/1-NIA/NICHD.

Competition and Cooperation in Economics and Business – Gani et al. (Eds)
© 2018 Taylor & Francis Group, London, ISBN 978-1-138-62666-9

The role of industrial estate characteristics in the export decision of manufacturing firms in Indonesia

Meindra Sabri, Nachrowi D. Nachrowi, Widyono Soetjipto &
Maddaremmeng A. Panennungi
Department of Economics, Faculty of Economics and Business, Universitas Indonesia, Depok, Indonesia

ABSTRACT: A firm's location determines access to certain economic inputs and trade facilities. The presence of adequate infrastructure and agglomeration economies may boost a firm's export potential. We investigated whether export performance of firms in industrial estates was better than those firms outside industrial estates. We then assessed how industrial estate characteristics determined the export decision of firms in industrial estates. The treatment effect method was used to examine the export performance of firms, and the probit model was developed to explain how industrial estate characteristics (port distance, port capacity, electricity, water, number of tenants and fiscal incentive) affected a firm's export decision. To strengthen the quantitative analysis, we conducted a qualitative analysis through in-depth interviews with the industrial estate top-level managers and related institutions. The results showed that export performance of firms in industrial estates was better than those outside industrial estates. The role of many industrial estate characteristics was significant to the firms' decision to export especially in the Greater Jakarta area. Infrastructure such as electricity and water positively affected the firms' decision to export, but had an opposite direction effect in regions outside Greater Jakarta. Agglomeration economies, which were represented by the number of manufacturing firms, had encouraged firms to export. In addition, government fiscal policy was effective in improving export decision of firms in industrial estates.

1 INTRODUCTION

Locations have comparative advantages that play a non-negligible role in the potential of individual firms to export. Being located in the same location as other firms can facilitate the firm's capacity to deal with external markets and also to become more successful. Marshall (1890) introduced localization economies where firms get benefit from the availability of specialized input providers, the access to a large pool of similar and specialized labor and the production of new ideas. In addition, Jacobs (1969) argues that firms may benefit from externalities arising in regions with diverse industrial structures or from urbanization economies. While industrial districts were spontaneous localizations of industries through market forces in a delimited area, many governments in the world built some planned counterparts.

The Indonesian government has been trying to boost export of manufacturing industry through the development of industrial estates. An industrial estate is an industrial activity center which is equipped with facilities and infrastructures provided by the industrial estate management. The goal is to locate manufacturing firms in a proper, correct, and environmentally friendly area, so the industry becomes more effective and efficient, hence facilitating future development (the Ministerial Industry Decree No. 35 2010).

We investigated whether the policy was effective to encourage the export performance of firms in industrial estates. Previous studies indicate that special economic areas have positive influence on the export of firms. Schminke and Van Biesebroeck (2013) show that firms in special areas such as Economic and Technological Development Zones (ETDZs) and the Science and Technology Industrial Parks (STIPs) in China have a greater total value of exports than firms outside of the two regions. Yi and Wang (2012) state that firms in an economic zone are more likely to export compared to firms outside it.

The problem that arises when comparing the performance of export firms in an industrial estate and those outside of industrial estates is the endogenous nature of firms' location decisions. Being located in an industrial estate might be more advantageous for firms that would have performed better than average anywhere else. This self-selection into industrial estate might result in the upwardly biased estimate of the causal effects on firm's export decision. To solve this problem, we used statistical analysis of cause and effect based on the framework of potential outcomes that refer to Rubin Causal Model (RCM). We need to estimate the potential outcome of treated firms (in the industrial estate) if they had located elsewhere. This counterfactual is inherently unobservable, but we can find the average treatment on the treated (ATT) value base on similar observable covariates under the unconfoundedness and overlap assumptions.

We then assessed how the characteristics of industrial estates determined the export decision of the firms in the industrial estates. From a policy perspective, identifying such determinants is important since industrial estate characteristics influence the costs of exporting through the availability of skills, transport costs, and infrastructure. In line with the theory, firms will choose to export if the expected profit when exporting is higher than the expected profit when operating only on the domestic market. The location in the industrial estates can influence the expectations of firm profits through cost savings from trade and economies of scale. Trade costs are costs in addition to production costs incurred until the goods are received by consumers, and economies of scale are obtained savings through the firm's size, output or scale of operations that cause the cost per unit of output to decline.

This study contributes to the empirical literature in the comparison of the export performance of firms inside and outside of special economic areas in Indonesia. Moreover, using industrial estate characteristics such as distance to the seaport, capacity of the seaport, public infrastructures such as electricity and water, local agglomeration, and incentive fiscal is a new approach to studying the role of location as determinants of firms' decision to export. This approach and the results can be used for future expansion of industrial estates.

This paper is organized as follows. Introduction in Section 1, after that Section 2 reviewed previous studies that discussed determinants of exporting and the role of location characteristics. In Section 3 we outlined our methodology to control firms self-selecting into industrial estates and firm's decision to export. The data is described in Section.4, followed by the estimation results and discussion in Section 5. We closed with conclusions on firm behavior and policy in Section 6.

2 LITERATURE REVIEW

Economic zones, namely Industrial Park, Special Economic Zone (SEZ), Industrial Estate (IE), Eco Industrial Park (EIP), Technology Park (TP), and Innovation District (ID), are promoted by many governments around the world to spur economic growth and competitiveness. Firms in the economics zone have advantages through economies of scale and agglomeration, which can influence their decision to export (The World Bank, 2009). Yi and Wang (2012) found that firms in an economic zone are more likely to export compared to firms outside the economic zone. Schminke and Van Biesebroeck (2013) show that export performance of firms in Economic and Technological Development Zones (ETDZs) and the Science and Technology Industrial Parks (STIPs) was better than those of firms outside of both areas in China.

A firm's decision to export is determined by both internal and external factors. The internal factors, firm-specific determinants, are associated with internal economies of scale such as age, size, and productivity (Aitken, Hanson, and Harrison, 1997; Bernard and Jensen, 1999; Greenaway and Kneller, 2007; Roberts and Tybout 1997; Zhao and Zou 2002). The external factors involve external economies of scale in terms of agglomeration (Bernard and Jensen

2004; Greenaway and Kneller 2008; Koenig, 2009). Agglomeration is the benefit that firms would acquire by being located close to each other in a city or industrial estates as compared to separately located (Glaesere, 2010). In those studies, firm location was merely indicated by a regional dummy variable. The characteristics of the regions were not fully explored as determinants of firms' decision to export.

In the recent approach, Farole and Winkler (2011) and Rodriguez-Pose *et al.* (2013) found that regional characteristics influence a firms' decision to export, but they are unable to capture the differences among industrial estates that are located within the same regional area. In a region, there can be more than one industrial estate where each has different transportation costs, infrastructures, and numbers of tenants that can affect the costs of exporting. The industrial estate characteristics that influence a firm's decision to export need to be identified because they are related to cost savings from exporting in each industrial estate.

Transportation costs are part of the trading costs that can affect a firm's decision to export (Krugman, *et al.*, 2012). We used the distance from the firm location to the seaport and seaport capacity as the proxies for the transportation costs. The farther the distance, the higher the costs of transportation; therefore, we expected that the distance to the seaport would negatively affect a firm's decision to export. Yi and Wang (2012) found that firms that are located near coastal areas are more likely to export than those located on the mainland. The seaport capacity was thus expected to positively affect a firm's decision to export through a reduction in transportation costs. Abe and Wilson (2011) found that a 10% increase in port capacity reduces transport costs by up to 3% in East Asia.

Porter (1998) states that the development of a region or industrial clusters needs government involvement, particularly for providing adequate infrastructure. The state-owned electricity enterprise (PLN) and the state-owned water supply company (PAM) are part of the infrastructure provided by the government. According to the state regulations, the availability of sources of electricity to industrial areas is guaranteed to be uninterrupted. The fact that their prices are relatively cheaper than those of the privately-owned power supply enterprises can reduce the production costs. PAM plays a crucial role in the industrial estates that do not have natural water sources such as rivers. The industrial estates that are not served by PAM distribution have to build their own water treatment plants, and this can significantly increase production costs.

Various types of industries in an industrial estate can form economic urbanization as part of agglomeration. As the number of firms operated in the industrial estate increase, the competition will also increase, which make the firm more productive and tend to export, Long and Zang (2011). Agglomeration can influence firm's decision to export through reducing costs of exporting, Aitken *et al.* (1997) and raising productivity. Agglomeration can make production costs lower through sharing infrastructure and information and lower transportation and transactions costs through the better relation of suppliers and customers (Malmberg 2009). However, Krugman (1991) stated that in addition to the forces that drive agglomeration (centripetal forces), there is also the forces that against agglomeration (centrifugal forces) and characterized by congestion costs. Congestion costs can raise production costs by the higher price of inputs (land, capital, and labor) and transportation and transactions costs, through longer waiting times (e.g. mobility of intermediate inputs or licenses). These costs may counterbalance the gains from agglomerations, and the net effect can be either positive or negative.

Firms that located in an EPZ or bonded zone which is part of an industrial estate received the fiscal incentive through various taxes. Exemption of import tariff and tax for intermediate input and machinery are given to firms that produce for export market. This fiscal incentive allowed the exploitation of scale economies in production. More firms that have fiscal incentive is expected to influence other firm's decision to export in the industrial estate. This effect is known as export spillover. Previous studies showed positive effects such as Koenig (2009), Farole and Winkler (2011), Rodriguez-Pose *et al.* (2013), but Bernard and Jensen (2004) didn't find a significant export spillover effect.

Other determinants of export are, but no limited to, export status in the previous year, productivity, amount of labor, source of capital, firm's age, and information availability. Export status in the previous year as a proxy for sunk entry costs has a positive influence on a firm's decision to export as Robert and Tybout (1997), Rodriguez *et al.* (2013), Narjoko and Atje (2007), and Farole and Winkler (2011) have stated. Melitz (2003) shows that only firms with a high productivity can export. Labor is also an important factor of production; more labor implies more output. Sources of capital, especially from abroad bring technological transfer. The higher the foreign share, the higher the export opportunities. Sjöholm and Takii (2008) found that in Indonesia firms that use foreign capital are more likely to export than domestically funded firms. However, the influence of a firm's age on its export decision is still inconclusive. Naudé *et al.* (2013) state that the starting time to export is an important strategy for firms. Firms often delay this until information is available and requirements such as learning ability, innovation, productivity growth, and access to finance are met. Johanson and Vahlne (1977) argue that making the decision to export is a long process, but adequate transportation facility may shorten that process.

3 METHODOLOGY

Treatment effect was used to analyze the export performance of the firms in the industrial estates. Regression adjustment method and inverse probability weighted regression adjustment method, better known as double robust, were used to measure the firms' export intensity and decision to export. These methods could answer the question of whether the firms' performance in the industrial estates was better than those outside of the industrial estates.

To analyze the role of industrial estate characteristics in the firms' decision to export, a combination of quantitative and qualitative analyses was used. For the quantitative analysis, probit regression method was employed and for the qualitative analysis, in-depth interviews with the industrial estate managers and other stakeholders were conducted.

3.1 *Treatment effect*

y is the outcome variable of performance, and w is the binary variable of treatment. There are two potential outcomes even though only one can be observed for each firm. If firm i is in an industrial estate, its potential outcome is y_{i1}, and the potential outcome for the same firm when it is located outside of an industrial estate is y_{i0}. The average treatment effect (ATE) measures the effect of treatment on a random sample of a particular population, i.e., the effect of an average over the entire population. However, not all firms are willing to enter into an industrial estate, so the measurement of treatment effect needs to be focused on firms that are located in industrial estates. This measurement is called the average treatment effect on the treated (ATT).

ATT is defined as

$$\tau_{att} = \mathrm{E}((y_1 - y_o) \mid w = 1) \tag{1}$$

To get the value y_o which is a potential outcome when firms in an industrial estate are located outside, we need two assumptions:

1. Unconfoundedness or ignorability of treatment

$$(y_1 - y_0) \perp w \mid x \tag{2}$$

This first assumption of unconfoundedness or ignorability of treatment, conditioned on a set of covariates x, makes a firm's presence in the industrial estate become random. Covariate variables in x must be effective in separating the correlation between the potential performance of a firm when it is not located in the industrial estate and the actual value of the firm's performance in the industrial estate.

Weaker assumption of (2) known as independent mean is used, so

$$\mathrm{E}[y_0 \mid x, w] = \mathrm{E}[y_0 \mid x] \tag{3}$$

2. Overlap

When conditioned on a set of control variables **x**, each unit in the population has the potential to get a treatment

$$\forall x \in \chi, 0 < P(w = 1 \mid x) < 1 \tag{4}$$

where χ is a collection of the covariate. These assumptions guaranteed that for each firm in the sample, there was a counterpart firm outside of industrial estates to be used as a control with the same covariate.

This study used the firms' internal characteristics such as productivity, number of employees, percentage of foreign capital, age, and location as the covariates. The location of the firms was represented by a dummy variable that distinguished whether they were located in Greater Jakarta or outside Greater Jakarta. Export performance was measured by export intensity (total export per output) and the firms' decision to export as a dummy variable which has the value of 1 if they exported and 0 otherwise.

In estimating the average treatment effects, ATT was defined as

$$\hat{\tau}_{att} = \frac{1}{\Sigma_i w_i} \sum_{i=1}^{N} w_i [\hat{m}_1(x_i) - \hat{m}_0(x_i)] \tag{5}$$

where the functions $\hat{m}_1(.)$ and $\hat{m}_0(.)$ are the predicted value of the variable performance for firms in industrial estates and those outside of industrial estates using the same covariates but can have different coefficients.

3.2 *Firm's decision to export*

The model of firm's decision to export refers to Farole and Winkler's (2011). The decision to export for firm i at time t depends on the expected revenue R and production costs c and sunk entry costs S when it began to export.

$$\text{P}(Exp_{it} = 1) = \text{P}(R_{it} > c_{it} + S(1 - Exp_{it} - 1)) \tag{7}$$

Exp = export status, where the firm would export if expected profit $\pi_{it} > 0$.

S = sunk entry costs, 0 if the firm exported in period t – 1 and 1 otherwise. Sunk entry costs can be either gathering information about the conditions of demand or building the distribution system when entering the export market.

Expected profit π^*_{it} was assumed to be influenced by the characteristics of the industrial estates, firms' internal characteristics, and sunk entry costs that could increase or decrease the revenue R and/or costs c.

Furthermore, the equation (1) was translated into

$$\text{P}(Exp_{it} = 1) = \text{P}(\pi^*_{it} = \beta \text{I}_{kt} + \gamma F_{it} + S(1 - Exp_{it-1}) > 0 \tag{8}$$

π^*_{it} was approximated as a linear combination of I_{it} (factor characteristic industrial estate), F_{kt} (internal characteristic factor), and sunk entry costs, which could be observed in period t where subscript k indicates the industrial estate, i individual firm and regional r. Then, using the previous year's export status as a proxy for sunk entry costs, the empirical model of firm decision to export could be written as

$$\begin{aligned} Exp_{irk,t} = {} & \alpha_0 + \beta_1\, lnportdistance_{kt} + \beta_2\, lnportcap_{kt} + \beta_3\, electricity\_pln_{kt} \\ & + \beta_4\, water\_pam_{kt} + \beta_5\, tenant_{kt} + \beta_6\, fiscalincentiveb_{kt} + \gamma_1\, Exp_{irk,t-1} + \gamma_2\, Inprodtk_{irkt} \\ & + \gamma_3\, Inemployment_{irkt} + \gamma_4\, foreigncapital_{irkt} + \gamma_5\, age_{irkt} + \gamma_6\, age\,2_{irkt} + \varepsilon_{irkt} \end{aligned} \tag{9}$$

where

- Export: dummy export status, 1 if firm exports and 0 otherwise
- lnportdistance: distance to the sea port in logarithms
- lnportcap: seaport capacity, approached by export volume (thousand tons) in logarithms
- electricity pln: percentage of PLN electricity usage by the total kWh in an industrial estate
- water pam: dummy water source, 1 if PAM water and 0 otherwise
- tenant: number manufacturing firms in logarithms
- fiscalincentive: number of the firms that have bonded area facilities divided by the total number of firms in the industrial estate
- Export t−1: dummy last year export status, 1 if firm exports and 0 otherwise
- lnprodtk: productivity, calculated from the output value divided by the amount of labor in logarithms
- lnemployment: number of workers in logarithms
- foreign capital: percentage of foreign capital
- age: firm's age (years)
- age2: squares of firm's age

4 DATA

The database in this study was mainly built from three data sets, (i) industrial estate directory and tenants from the Ministry of Industry, (ii) Large and Medium Industries Survey (IBS) of Statistics Indonesia (BPS) and (iii) export statistics directory from BPS. Information on industrial estate characteristic variables such as distance to the seaport, source of water and number of tenants was sourced from industrial estate directory data, while information on port capacity was sourced from transportation statistics. The percentage of PLN electricity usage by the total kWh in an industrial estate was sourced from IBS, and the number of the firms that had bonded area facilities in the industrial estate was sourced from the Customs Department. Meanwhile, all of the firms' internal characteristics were sourced from IBS.

There were some data adjustments for data processing. We aggregated the capacity of seaports in Batam as a single value. The capacity of seaport that we used was export volume (thousand tons) from the seaport. We could not find the exact capacity of seaports due to the differences in measurement. The export status of the firm was matched with export statistics if it was empty.

The survey of industrial estate directory conducted by the Ministry of Industry in 2012 covered 59 industrial areas, but after checking the existence of large and medium industries, only 48 industrial estates had data of large and medium industrial firms. The reduction occurred due to (i) the unavailability of tenant data in an industrial estate because it was still under construction and (ii) no matching firms in the IBS data, which can occur when an industrial estate contains many warehouses or offices.

4.1 *Statistics summary*

There were 1583 firms located in the industrial estates at the national level. Based on the concentration of the industrial estates, we divided the national level into two regions the Greater Jakarta area with 24 industrial estates that were located in Jakarta, West Java, and Banten and outside the Greater Jakarta area with 24 industrial estates that were located in 9 provinces (North Sumatera, West Sumatera, Riau, Riau Islands, Lampung, Central Java, East Java, South Sulawesi, and East Kalimantan). Table 1 provides an overview of the variables at the regional level.

The percentage of exporting firms in the Greater Jakarta and outside the Greater Jakarta are about 51% and 49% respectively. These indicate that the firm's decision to export in the industrial estates in the Greater Jakarta is higher than outside the Greater Jakarta.

Table 1. Descriptive statistics of variables.

	National (1583 firms)				Greater jakarta (1117 firms)				Outside greater jakarta (466 firms)			
Variables	Mean	Std. Dev	Min	Max	Mean	Std. Dev	Min	Max	Mean	Std. Dev	Min	Max
Export	0.51	0.5	0	1	0.51	0.5	0	1	0.49	0.5	0	1
Port distance	3.32	0.87	0	5.2	3.57	0.73	2	5.2	2.73	0.89	0	4.09
Port capacity	8.99	0.86	5.48	10.42	9.3	0.45	7.06	9.39	8.24	1.1	5.48	10.42
Electricity PLN	0.67	0.34	0	1	0.63	0.36	0	1	0.78	0.29	0	1
Water PAM	0.34	0.47	0	1	0.19	0.39	0	1	0.7	0.46	0	1
Tenant	4.18	1.07	0	5.68	4.45	0.99	0	5.68	3.54	0.99	0	4.8
Fiscal incentive	0.17	0.18	0	1	0.2	0.2	0	1	0.2	0.2	0	1
Export last year	0.49	0.5	0	1	0.5	0.5	0	1	0.5	0.5	0	1
Productivity	12.91	1.32	7.85	18.15	13	1.3	7.8	18.1	12.7	1.3	9.2	17.8
Employment	4.84	1.35	3	10.17	4.9	1.4	3	9.1	4.8	1.3	3	10.2
Foreign capital	41.65	47.08	0	100	45.7	47.1	0	100	31.8	45.5	0	100
Age	15.83	9.17	1	100	15	7.5	1	67	17.8	12.1	1	100
Age2	334.51	536.84	1	10000	280.2	331.7	1	4489	464.6	832.1	1	10000

The port distance to the industrial estate in the Greater Jakarta area on average is farther away from the seaport than outside the Greater Jakarta area. It has higher average that caused by many industrial estates located in Bekasi, Karawang, and Purwakarta. The average of seaport capacity is higher in the Greater Jakarta than the outside the Greater Jakarta area. The Greater Jakarta region has fewer ports, but has a relatively large capacity greater than outside the Greater Jakarta.

For infrastructure, the average use of PLN power in the industrial estates in the Greater Jakarta area is 63%, lower than the outside of the Greater Jakarta that reached 78%. It shows the dependence of the industrial estates on PLN power outside the Greater Jakarta is higher. The PAM water is only used by 19% of firms in the industrial estates in the Greater Jakarta area, much lower than outside the Greater Jakarta where 75% of firms. The average number of tenants in the Greater Jakarta is higher than outside the Greater Jakarta. Fiscal incentive on average in the industrial estate is higher than outside the Greater Jakarta. It means there are more firms accept fiscal incentive.

The export status in the previous year shows that in both regions 50% of firms were exporting. The average productivity in the Greater Jakarta is higher than outside the Greater Jakarta. The number of labor is relatively indifferent between the two regions. The average of foreign capital in the Greater Jakarta is higher than outside the Greater Jakarta area, It shows more firms with higher foreign capital are located in the Greater Jakarta. Firm's age in the Greater Jakarta tends to be younger than outside the Greater Jakarta.

5 RESULTS

5.1 *Export performance of firm in the industrial estate*

The firms' export performance was measured using two variables, export intensity and export decision. By using regression adjustment and inverse-probability-weighted regression adjustment, we obtained the value of Average Treatment Effects on the Treated (ATT) as follows:

The firms located in an industrial estate had a higher level of export intensity and export decision than the control firms. Both regression methods gave similar results in that the difference was positive and highly significant at the 1% level as shown in Table 2. Using regression adjustment, the export intensity of firms in the industrial estates was 3.5% higher than

Table 2. Export performance.

ATT	Regression adjustment	Inverse-probability-weighted regression adjustment
Export intensity	0.0353**	0.0493***
Export decision	0.355***	0.355***

the average export intensity of firms outside the industrial estates and 4.9% higher using inverse-probability-weighted regression adjustment. Firms' decision to export in the industrial estates had 0.36 propensity score higher than the control firms outside the industrial estates for both methods. We can say that export performance of firms in the industrial estates was better than that of firms outside of the industrial estates.

5.2 *The role of industrial estate characteristics on firm's export decision*

At the national level, there were 1583 firms. Greater Jakarta had 1117 firms or 71% of the total observations, while the regions outside Greater Jakarta had 466 firms or 29% of total observations. By using probit models (see Table 3), the estimation of various sets of data was robust to heteroscedasticity. The standard error was clustered by the seaports so that errors could be correlated between the firms that accessed the same seaports. The regressions showed that the model used was good enough, where the Hosmer-Lemeshow test showed that the goodness-of-fit statistic $\chi 2$ value was not significant. Furthermore, the significance of the regression coefficients for each variable was based on 10%, 5%, and 1% level.

The distance to the seaport has a significant positive effect in the Greater Jakarta means that a greater distance of an industrial estate from sea port will increase firm's decision to export, but it was insignificant for outside the Greater Jakarta. These results contradicted to the hypothesis that stated the greater the distance will increase transportation costs and lower the firm's export decision. The possible reason is the economic benefit from production costs outweighed the effect of increasing transportation costs. From the interview, all industrial estates have highway access to the seaports. The toll road is relatively smooth so that the more distance is unnecessarily increase the transportation costs in the Greater Jakarta area.

Seaport capacity has no significant effect on the firm's export decision in the industrial estate in the Greater Jakarta area. There are only two seaports in the Greater Jakarta, and most of the firms access the same port (Tanjung Priok). For the outside the Greater Jakarta, seaport capacity has a significant negative effect on the firm's exports decision. This is caused by non-optimal exporting of manufacturing firms in the outside Greater Jakarta. Port with larger capacity is accessed by fewer exporting firms. A Large share of the port capacity used by economics sectors that were based on natural resources such as agriculture and mining. We didn't find the effect of costs saving as in Abe and Wilson (2011) who found that increasing the port's capacity by 10% will reduce transport costs of up to 3% in East Asia.

PLN power source has a positive effect for the greater Jakarta area. A higher percentage of the PLN electricity in the industrial estate will improve firm's decision to export. The existence of alternative power sources that has excellent reliability made PLN improve their quality of services by providing premium power. The quality is guaranteed from shutting down or flicker. Based on the experiences of respondents who used the premium PLN power, flicker might turn off the large and sensitive production machine. Production costs will increase because of the large machines need time to be able to work normally after a flash happens.

In the outside Greater Jakarta, PLN electricity has a negative effect which means that the greater portion of PLN power in an industrial estate will reduce firm decision to export. This result was influenced by the industrial estate that has a high percentage PLN power usage is mostly occupied by firms that do not export. Conversely, the industrial estate that has its power source is mostly occupied by firms that export. This result indicates that PLN has not been able to meet the capacity or quality of the power required by exporting firms in the industrial estate. Industrial estates have to build its power source according to the needs of its tenants.

Table 3. Results of regression at regional level.

Variable	National	Greater jakarta	Outside greater jakarta
Port distance	0.104	0.174***	0.144
Port capacity	0.0301	0.0493	–0.198**
PLN Electricity	–0.113	0.0348***	–0.879***
PAM Water	0.183*	0.265***	–0.365***
Tenants	–0.0224	0.0508***	0.204
Fiscal incentive	1.045**	1.644***	–0.217
Export last year	3.194***	3.223***	3.551***
Productivity	0.0727**	0.048	0.222*
Employment	0.0531	–0.00793	0.282***
Foreign capital	0.00329***	0.00201***	0.0121**
Age	0.0374***	0.0346***	0.00962
Age2	–0.000721***	–0.000240***	–0.000669*
Cons	–3.843***	–4.228***	–4.693**
N	1583	1117	466

* p<0.1, ** p<0.05, *** <0.01.

PAM water has a positive significant effect on firm's decision to export in the industrial estate in the Greater Jakarta area, which means that firms in the industrial estate that use PAM water are more likely to export than firms in the industrial estate that does not use PAM water. PAM limitations regarding water supply for the industrial estate causing many industrial estates treat their water needs by utilizing river water or ground water, especially in the area of Bekasi, Karawang, and Purwakarta. Water treatment independently by operators of industrial estates may increase production costs due to the price of water is more expensive. On the other hand, PAM water has a significant negative effect on outside the Greater Jakarta area where more exporting firms located in the industrial estate that has an independent water source. Conversely, in the industrial estate that used PAM water, fewer firms are exporting. Like the PLN power, there is a tendency that industrial estate whose many tenants are exporting build water installations independently.

The number of tenants has a positive significant effect on decision-export company in the industrial estate in the Greater Jakarta area. More tenants can improve interaction between firms so that the positive effects of agglomeration becomes more prominent as described Marshall (1890) and Jacob (1969). From the interviews, we obtained information that the industrial estate that has many tenants usually have a major firm or industry (anchor). For example, automotive firms get components from supplier firms that located in the same industrial estate. Likewise, fiscal incentives have a positive influence on firm's export decisions. More firms that received fiscal facilities (bonded zone) in the industrial estate made firm's export decision increase. This result shows the positive effect of the export spillover through the existence of the bonded zone in the industrial estate. Both of variables, the number of tenants in an industrial estate and the number of firms in the bonded zone showed that economic agglomeration in the industrial estate significantly influences the firm's decision to exports.

Exports last year is a proxy of sunk entry cost has a significant positive effect on firm's decisions to export in the Greater Jakarta and outside the Greater Jakarta. Firms that export in the previous year tends to re-export. The tendency to maintain exports is greater in firms that located outside the Greater Jakarta area. The first time export of firm that located outside the Greater Jakarta needs greater sunk entry cost than the Greater Jakarta area. The results previous export role, support the empirical study of firm's decision to export such as Rodriguez *et al.* (2013), and Farole and Winkler (2011).

Productivity has no significant impact on exporting firms in the Greater Jakarta area. This result showed that firms with low productivity also export. From the interview, we get information that in the Greater Jakarta area, many firms were not only supplying components to a larger company, but also export. For outside the Greater Jakarta area, productivity has a significant positive effect following Melitz (2003) which states that only high-productivity firms could export. The number of workers is insignificant to firm's decision to export in the greater Jakarta area. It showed that not only the big firms do export but also medium-sized firms. Different result occurred at the outside Greater Jakarta where big firms have more tendency to export than middle size firms.

Foreign capital has a significantly positive influence in both the greater Jakarta area and outside Greater Jakarta areas. A high portion of foreign capital encouraged the firm to export in the industrial estate. These results support Sjöholm and Takii (2008) who found that foreign capital firms are more likely to export than domestically funded firms in Indonesia. The main reason of why foreign capital might encourage export is global production network. Multinational companies established factories in several countries that have different comparative advantages as Athukorala (2010), Jones and Kierzkowski (2001). From interviews, we obtained information that the ownerships of almost all industrial estates in the greater Jakarta area are a joint venture company, in which most of the foreign ownership came from Japan and South Korea. The presence of this foreign ownership can foster the confidence of investors to invest in Indonesia and located in the industrial estate which they managed.

Firm's age has a significant positive effect in the Greater Jakarta and outside Greater Jakarta area. Older firms have a greater probability to export. This result is consistent with Jan and Vahlne (1977) who shows that export is a gradual development process. But when we use squared of age, the coefficient is negative. It means that firms age is positively increasing the probability to export, but the marginal increase is decreasing.

6 CONCLUSION

Exporting firms performed better when located inside rather than outside the industrial estates. At the regional level, industrial estate characteristics gave different results across the regions. Almost all of the industrial estate characteristics had a significant effect on export in the Greater Jakarta area, which is consistent with the hypothesis. Local agglomeration had a significantly positive effect, and the fiscal policy shows that the number of firms in bonded zone had a positive influence on firms in the industrial estates to export.

The policy of development of industrial estates is appropriate and should be continued to encourage the export of the manufacturing industry. Different roles of the industrial estate characteristics across regional areas add to the evidence that the Greater Jakarta area is still an excellent choice of location. The industrial estates in Greater Jakarta are typically managed by private owners in collaboration with foreign parties which have a significant role in the development of industrial estates in Indonesia.

REFERENCES

Abe, Kazutomo and John S. Wilson. 2011. "Investing in Port Infrastructure to Lower Trade Costs in East Asia." *Journal of East Asian Economic Integration* 15 (2): 3–32. https://ssrn.com/abstract = 2318301.

Aitken, Brian, Gordon Hanson, and Ann E. Harrison. 1997. "Spillovers, Foreign Investment, and Export Behavior." *Journal of International Economics* 43: 103–32. doi:10.1016/S0022–1996(96)01464-X.

Athukorala, Prema-chandra. (2010). Production Networks and Trade Patterns in East Asia: Regionalization or Globalization? The ADB Working Paper Series on Regional Economic Integration. No. 56 August 2010.

Bernard, Andrew B, and J Bradford Jensen. 2004. "Why Some Firms Export." *The Review of Economics and Statistics* 86 (2): 561–69. doi:10.1162/003465304323031111.

Farole, Thomas, and Deborah Winkler. 2011. "Firm Location and the Determinants of Exporting in Developing Countries." *The World Bank. Policy Research Working Paper* 5780: 67–102. http://dx.doi.org/10.1596/9780821398937_CH03.

Glaeser, Edward L. (2010). Agglomeration Economics. The University of Chicago Press. ISBN: 0-226-29789-6.
Greenaway, David, and Richard Kneller. 2007. "Firm Heterogeneity, Exporting and Foreign Direct Investment." *The Economic Journal* 117 (517): F134–61. doi:10.1111/j.1468–0297.2007.02018.x.
Greenaway, D. and R. Kneller. 2008 "Exporting, Productivity and Agglomeration." *European Economic Review* 52 (5): 919–39. doi:10.1016/j.euroecorev.2007.07.001.
Jacobs, Jane. 1969. The Economics of Cities. Vintage, New York.
Johanson, Jan, and Jan-Erik Vahlne. 1977. "The Internationalization Process of the Firm—A Model of Knowledge Development and Increasing Foreign Market Commitments." *Journal of International Business Studies* 8 (1): 23–32. doi:10.1057/palgrave.jibs.8490676.
Jones, R.W. Kierzkowski, H. (2001). Globalization and the Consequences of International Fragmentation. In R. Dornbusch, G. Calvo, and M. Obstfeld, eds. Money, Factor Mobility and Trade: The Festschrift in Honor of Robert A. Mundell. Cambridge, Mass: MIT Press. pp. 365–381.
Koenig, Pamina. 2009. "Agglomeration and the Export Decisions of French Firms." *Journal of Urban Economics* 66 (3): 186–95. doi:10.1016/j.jue.2009.07.002.
Krugman, Paul R., Maurice Obstfeld, and Marc J. Melitz. 2012. *International Economics : Theory and Policy*. 9th Ed. New York: Pearson Publishing Ltd.
Krugman, Paul. 1991. "Increasing Returns and Economic Geography." *Journal of Political Economy* 99 (31): 483–99. doi:10.1086/261763.
Long, Cheryl, and Xiaobo Zhang. 2011. "Cluster-Based Industrialization in China: Financing and Performance." *Journal of International Economics* 84 (1): 112–23. doi:10.1016/j.jinteco.2011.03.002.
Malmberg, Anders. 2009. "Agglomeration." In *International Encyclopedia Of Human Geography*, edited by Rob Kitchin and Nigel Thrift, 48–53. USA: Elsevier Ltd.
Marshall, Alfred. 1890. *Principles of Economics*. Available from http://eet.pixel-online.org/files/etranslation/original/Marshall,%20Principles%20of%20Economics.pdf Accessed on August 17, 2016.
Melitz, Marc J. 2003. "The Impact of Trade on Intra-Industry Reallocations and Aggregate-Industry Productivity." *Econometrica* 71 (6): 1695–1725. doi:10.1111/1468–0262.00467.
Narjoko, Dionisius A., and Raymond Atje. 2007. "Promoting Export: Some Lessons from Indonesian Manufacturing" *Asia-Pacific Research and Training Network on Trade Working Paper Series* 32: 1–47.
Naudé, Wim, Thomas Gries, and Bilkic Natasa. 2013. *Firm-Level Heterogeneity and the Decision to Export: A Real Option Approach. IZA Discussion Paper* 7346: 1–29.
Porter, Michael E. 1998. "Clusters and the New Economics of Competition Harvard Business Review." *Harvard Business Review* 76 (6): 77–90. doi:10.1042/BJ20111451.
Roberts, Mark J., and James R. Tybout. 1997. "The Decision to Export in Colombia: An Empirical Model of Entry with Sunk Costs." *American Economic Review* 87 (4): 545–64. doi:10.2307/2951363.
Rodrıguez-Pose, Andres, Vassilis Tselios, Deborah Winkler, and Thomas Farole. 2013. "Geography and the Determinants of Firm Exports in Indonesia." *World Development* 44: 225–240. http://dx.doi.org/10.1016/j.worlddev.2012.12.002.
Schminke, Annette, and Johannes Van Biesebroeck. 2013. "Using Export Market Performance to Evaluate Regional Preferential Policies in China." *Review of World Economics* 149 (2): 343–67. doi:10.1007/s10290–012–0145-y.
Sjöholm, Fredrik, and Sadayuki Takii. 2008. "Foreign Networks and Exports: Results from Indonesian Panel Data." *Developing Economies* 46 (4): 428–46. doi:10.1111/j.1746–1049.2008.00072.x.
The World Bank.(2009). World Development Report. Reshaping Economic Geography.
Yi, Jingtao, and Chengqi Wang. 2012. "The Decision to Export: Firm Heterogeneity, Sunk Costs, and Spatial Concentration." *International Business Review* 21 (5): 766–81. doi:10.1016/j.ibusrev.2011.09.001.
Zhao, Hongxin, and Shaoming Zou. 2002. "The Impact of Industry Concentration and Firm Location on Export Propensity and Intensity: An Empirical Analysis of Chinese Manufacturing Firms." *Journal of International Marketing* 10 (1): 52–71. doi:10.1509/jimk.10.1.52.19527.

Competition and Cooperation in Economics and Business – Gani et al. (Eds)
© 2018 Taylor & Francis Group, London, ISBN 978-1-138-62666-9

Interdependence and contagion in five ASEAN countries and five developed countries in the area of financial linkages

N. Trihadmini
Universitas Katolik Indonesia Atma JayaJakarta, DKI Jakarta, Indonesia

T.A. Falianty
Department of Economics, Faculty of Economics and Business, Universitas Indonesia, Depok, Indonesia

ABSTRACT: Financial globalisation brings great benefits but also results in the vulnerability of the economy. This study aims to analyse whether there were contagion effects of the 2008 global financial crisis or mere interdependence between five countries of the Association of Southeast Asian Nations (ASEAN) and five developed countries through financial channels. The analysis used mixed-frequency data and included foreign exchange markets, stock markets, policy interest rates, and money markets in the period 1990–2016. Cross-market correlation, Dynamic Conditional Correlation (DCC), and vector auto regression (VAR) were also used in the analysis.

The DCC method indicated that there was no extreme increase in the DCC coefficient in the period after the crisis at variables of (1) exchange rates, (2) stock indices, and (3) money market rates; thus, the phenomenon of the relationship was interdependence and not contagion. However, the policy interest rate variable showed a significant increase in the DCC coefficient in the three periods of analysis (1990–1997, 1998–2007, 2008–2015). This indicates the presence of a strong relationship between the monetary policies of each country and those of the other countries and the increasing coordination of monetary policies among the five ASEAN countries in this study. Although the US monetary policy was used as a reference by the five ASEAN countries and the five developed countries still had a dominant influence on the ASEAN financial markets, the linkages through financial channels among the five ASEAN countries were relatively weak.

1 INTRODUCTION

Financial globalisation has brought enormous benefits to the economy, but these are comparable to the economic vulnerability caused. Countries worldwide are increasingly connected to each other through financial channels, trade, or other economic activities. Interdependence through financial channels results in capital and financial flows across borders. Liquidity constraints and loss on the financial markets result in recomposition and global turnaround assets. This asset turnover creates opportunities and threats, as a result of the close interconnection between countries.

In line with the increasingly integrated economy in the era of globalisation, financial crisis in one country can easily spread to other countries and become a global financial disaster within a short period of time. An economic crisis caused by financial turmoil leading to instability, especially when it happens continuously for a long time, will interfere with other sectors in the economy. When the financial system is large, the risk of volatility or crisis is also bigger; so, it can lead to vulnerability in the stability of a financial system.

Economic crises and their spread have resulted in huge economic costs, as already experienced by most countries worldwide, including five countries of the Association of Southeast Asian Nations (ASEAN; Indonesia, Malaysia, Thailand, Singapore, and Philippines).

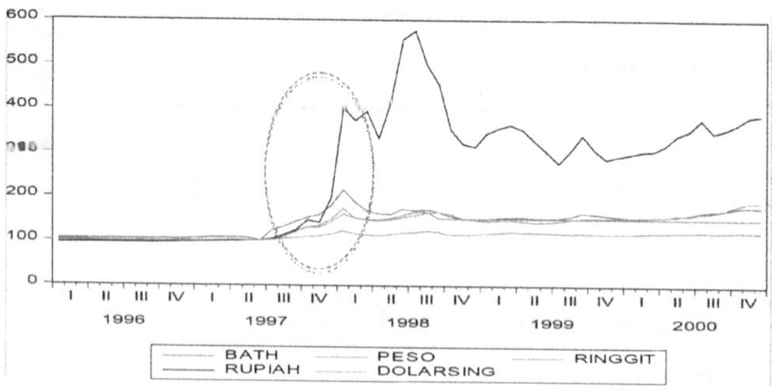

Figure 1. Exchange rate movement in five ASEAN countries during the 1997–98 financial crisis. Source: finance.yahoo.com (data processed).

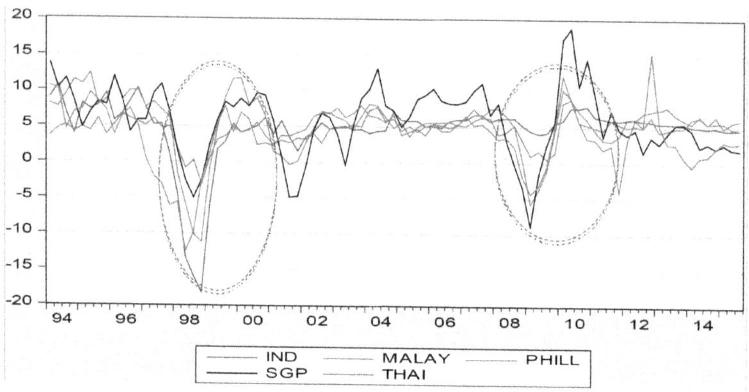

Figure 2. Economic growth in five ASEAN countries during the financial crisis in 1997 and 2008. Source: IFS, IMF (data processed).

Figure 1 shows that most of the five ASEAN countries experienced a sharp depreciation during the financial crisis of 1997–98. The depreciation had a widespread impact on the economy, which decreased purchasing power, aggregate demand, and ultimately output. In line with these conditions, Figure 2 shows the declining economic growth in five ASEAN countries as a result of the global financial crisis in 2008. The decrease in the national income lowered the quality of life in society as a whole. During these crises, interdependence and contagion of financial markets played a crucial role in the high level of integration between countries and financial markets. The interdependence became a major source of spillover effects, which led to extreme volatility and involved contagion effects.

Based on this background, this study aims to determine whether there were contagion effects of the 2008 global financial crisis or whether there was mere interdependence between five ASEAN countries and five developed countries through financial channels. This research is highly important because understanding the relationship or interconnection of financial channels allows analysis of the global mapping of risk, also known as concentrations of systemic risk, identification of transmission propagation of shocks, and improvement of the supervision of implementation in macro prudential policy.

2 LITERATURE REVIEW

2.1 *Interdependence and contagion*

There is a fundamental difference between interdependence and contagion. Referring to Forbes and Rigobon (2002), interdependence is co-movement during a stable period, driven by strong linkages between markets. It emphasises real linkages and fundamental integration as channels for transmission shocks between countries and markets in both crisis and non-crisis periods (Shen et al., 2015). Interconnection or financial linkage is a network of credit exposure, trade relations, or other linkages between economic activities and the interdependence of the financial agent (Gray, 2015). Finance interconnection is very important because it can serve as a channel of transmission (crisis). The impact of the failure of large interconnected entities can spread widely and rapidly throughout the financial system, thereby causing instability in the financial system worldwide. Financial interdependence can occur through the relationship between financial institutions, both banks and non-banks.

Meanwhile, Dornbusch et al. (2000) define contagion as a significant increase in cross-market linkages after a shock to an individual country or a group of countries, measured by the degree to which asset prices and financial flows move together across markets relative to this co-movement in tranquil times. An increase in co-movement does not reflect irrational behaviour of investors. Forbes and Rigobon (2001) explain a more specific definition of contagion, which is a significant increase in cross-market correlation after a shock in one country or a group of countries. Cross-market linkages can be measured by the correlation of asset returns, the probability of a speculative attack to transmit shocks or volatility. Cleassens and Forbes (2004) state that when two countries are located in separate geographical areas, have different economic structures, and almost do not have a direct connection through the channels of trade for example, the propagation from one country to another is called contagion. Forbes and Rigobon (2002) propose the use of the definition of 'shift contagion', which is not merely a contagion to emphasise the occurrence of a significant change in cross-market linkages after a shock. Some economists argue that if a shock is transmitted from one country to another, even though there is no significant change in the cross-market linkage, this transmission is also called contagion, but it is not the shift-contagion.

Kaminsky et al. (2003) state that contagion is an episode where there is an instant or immediate significant effect in the aftershocks in several countries. This results in excess co-movement in the economic and financial variables across countries as a response to the common shock in one country or a group of countries.

2.2 *Financial channel*

Several studies have been conducted to investigate contagion through financial channels. Gómez and Rivero (2014) found that the integration of bilateral debt and bond yields is an important channel that transmits a shock, although it is not proven that the integration of foreign direct investment will drive asset prices together. Stock market volatility is the variable most commonly found to spread out the crisis, as stated by Luchtenberg, K.F. & Vu, Q.V. (2015), Morales, L. & O'Callaghan, B.,A., (2014), Kenourgios (2014), and Podlich, N., & Wedow, M. (2014). Ambiguity between interdependence and contagion was also investigated by Morales, L. & O'Callaghan, B.,A., (2014), and Ahlgren, N. & Antell, J. (2010) who state that there is no contagion effect of the US stock market on the global financial markets, but volatility is derived from the transmission of economic linkages between countries. They also state that there are short-term relationships between developed countries (United States, Germany, Japan, United Kingdom) and emerging markets (Hong Kong, Korea, Mexico), but that is not contagion. The short-term relationship has important implications on portfolio diversification.

Unsal, D.F. & Jain-Chandra, S. (2012) states that global financial shocks are transmitted to domestic economies through three channels separately, namely (1) financial channels, (2) channels of trade, and (3) investors' perception channels through substitution effects. Financial channels can occur through financial spill over and also connectedness asset markets such as herding behaviour, which only exists or occurs in a crisis condition, which is generally called 'pure contagion' (Kaminsky and Reinhart, 2000).

Dungey and Tambakis (2003) also state that financial link transmission can occur through the currency, equity, and bond markets, where the patterns of interaction and transmission in one country and between countries can be described as follows.

1. Interaction between currency and equity markets in the country.

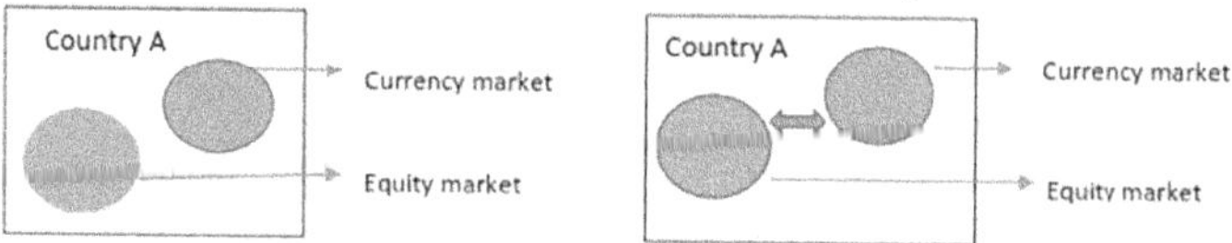

2. Several countries with similar patterns of interaction between currency andequity markets.

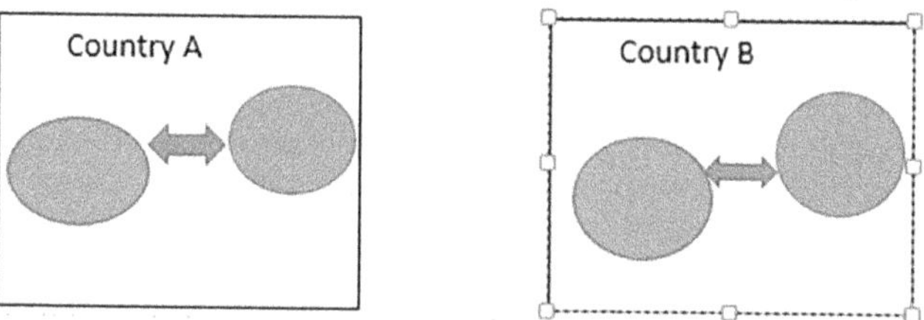

3. Interaction between country A and country B and their currency markets

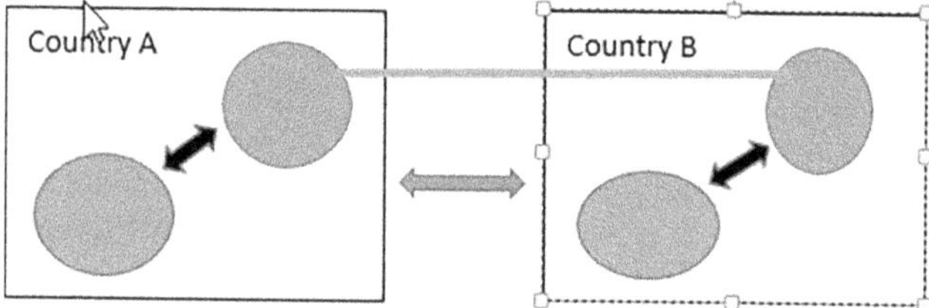

4. Interaction between the currency and equity markets of country A and country B.

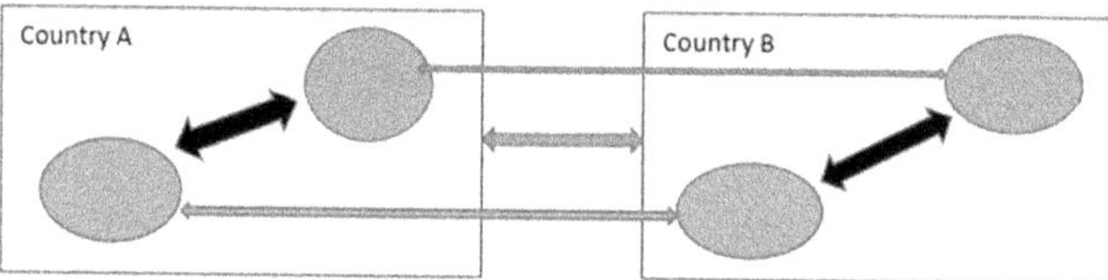

5. Interaction involving more than two countries.

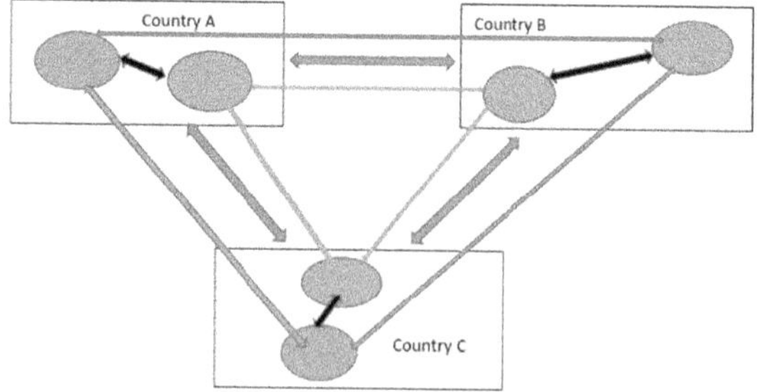

3 METHODOLOGY

The aim of this research is to analyse whether contagion effect occurs in times of crisis or whether there is only interdependence among five ASEAN countries (Indonesia, Malaysia, Thailand, Singapore, and Philippines) and five developed countries (United States, Europe Union, Japan, China, and Korea). The selection of the five ASEAN countries to represent the ASEAN profile is based on the consideration that the five countries have good economic growth compared with other ASEAN member states. In addition, they have high economic openness, so exposure to global shock could be propagated through them. The selection of the five developed countries is based on the consideration that these countries have great economic power and relative dominance, making them a potential source of shock in the global economy.

In analysing the dynamic relationships through financial linkages, financial variables such as exchange rates, stock indices, central bank policy rates, and money market rates were used. The selection of these research variables is based on the economic theory and the review of literature relating to contagion and propagation. For the purposes of the analysis, several analytical tools were used, namely (1) Dynamic Conditional Correlation (DCC), (2) Granger causality test, (3) Engle–Granger cointegration test, and (4) variance decomposition analysis. This study took different periods for the variables, between the years 1990 and 2016, and used monthly data published by the International Monetary Fund, World Bank, Asian Development Bank, and other relevant resources.

4 EMPIRICAL RESULTS

4.1 *Currency market*

Figure 3 illustrates the rolling standard deviation of the exchange rate movement in the five ASEAN countries and five developed countries. As seen on the Figure 3 graphs, there was similar movement, especially the one occurring in 2008, as a result of the global financial crisis. The South Korean Won, Indonesian rupiah, Philippine peso, Singaporean dollar, and Thai baht experienced the same fluctuation as the euro, Japanese yen, and Chinese yuan. All these currencies were already in US dollar units, so the movement of the US dollar is already included in the movement of the currency of each of these countries.

Exchange rate fluctuation in the same direction indicates the presence of global currency interconnection. There is still debate between contagion and interconnection. Forbes and Rigobon (2002) state that if the correlation coefficient between two or more countries is relatively high in the period of crisis and non-crisis, it indicates the occurrence of interdependence and not contagion.

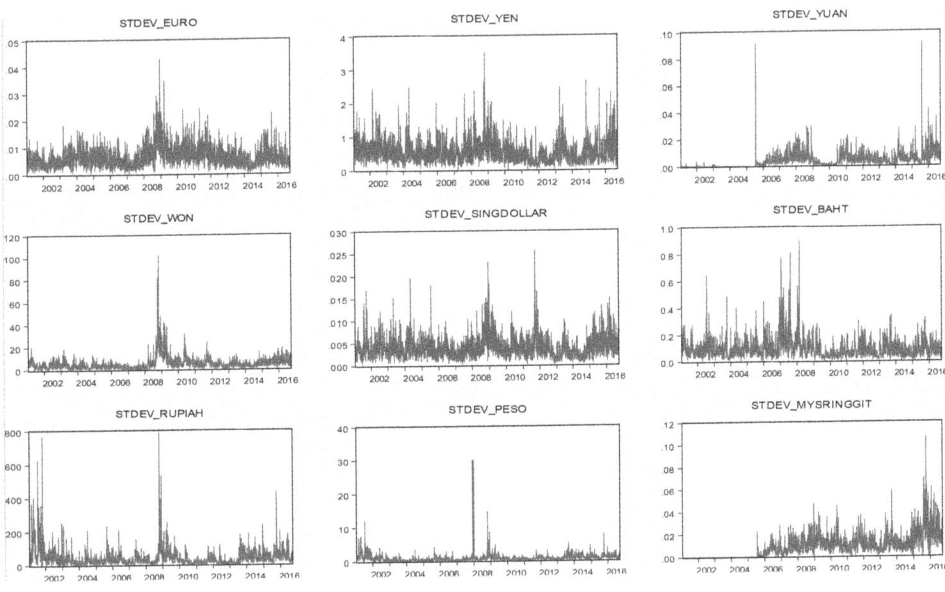

Figure 3. Rolling standard deviation of exchange rate movement (2000–16). Sources: IFS, World Bank (data processed).

The statistical tests using DCC in Tables 1 and 2 show that for the ASEAN and developed countries, the relationship was interdependence and not contagion. The average coefficient of DCC was 12.86% prior to 2008 and 26.09% after the 2008 financial crisis. Although there was an increase in the coefficient of DCC in the period after the crisis, the increase did not occur significantly; thus, it cannot be stated as the occurrence of contagion.

Table 1. Dynamic Conditional Correlation (DCC) of exchange rates (January 2000–December 2007).

	STDEV_ EURO	STDEV_ YEN	STDEV_ YUAN	STDEV_ WON	STDEV_ RUPIAH	STDEV_ PESO	STDEV_ SINGDO LLAR	STDEV_ BAHT	AVG_ GLOBAL
STDEV_ EURO	1.0000	0.2410	–0.0537	0.1951	–0.0100	0.0342	0.3373	0.0583	
STDEV_ YEN	0.2410	1.0000	0.0714	0.3593	0.0618	0.0682	0.4124	0.2056	
STDEV_ YUAN	–0.0537	0.0714	1.0000	–0.0834	–0.0963	0.1492	0.0843	0.2632	
STDEV_ WON	0.1951	0.3593	–0.0834	1.0000	0.1069	–0.0087	0.3934	0.1673	
STDEV_ RUPIAH	–0.0100	0.0618	–0.0963	0.1069	1.0000	0.2320	0.1372	–0.0212	
STDEV_ PESO	0.0342	0.0682	0.1492	–0.0087	0.2320	1.0000	0.0547	0.0336	
STDEV_ SINGDO LLAR	0.3373	0.4124	0.0843	0.3934	0.1372	0.0547	1.0000	0.2092	
STDEV_ BAHT	0.0583	0.2056	0.2632	0.1673	–0.0212	0.0336	0.2092	1.0000	
Sum	0.8022	1.4198	0.3347	1.1300	0.4105	0.5633	1.6285	0.9159	
Average	0.1146	0.2028	0.0478	0.1614	0.0586	0.0805	0.2326	0.1308	0.1286

Note: STDEV, standard deviation; AVG, average.
Source: IFS, World Bank (data processed).

Table 2. DCC of exchange rates (January 2008–August 2016).

	STDEV_ EURO	STDEV_ YEN	STDEV_ YUAN	STDEV_ WON	STDEV_ RUPIAH	STDEV_ PESO	STDEV_ SING DOLLAR	STDEV_ BAHT	AVG_ GLOBAL
STDEV_ EURO	1.000	0.312	0.073	0.418	0.217	0.194	0.553	0.205	
STDEV_ YEN	0.312	1.000	0.063	0.386	0.244	0.258	0.279	0.154	
STDEV_ YUAN	0.073	0.063	1.000	0.056	0.033	0.034	0.176	0.141	
STDEV_ WON	0.418	0.386	0.056	1.000	0.398	0.355	0.527	0.145	
STDEV_ RUPIAH	0.217	0.244	0.033	0.398	1.000	0.871	0.328	0.147	
STDEV_ PESO	0.194	0.258	0.034	0.355	0.871	1.000	0.307	0.156	
STDEV_ SING DOLLAR	0.553	0.279	0.176	0.527	0.328	0.307	1.000	0.274	
STDEV_ BAHT	0.205	0.154	0.141	0.145	0.147	0.156	0.274	1.000	
Sum	1.972	1.696	0.576	2.286	2.238	2.174	2.445	1.223	
Average	0.2818	0.2423	0.0823	0.3265	0.3197	0.3106	0.3492	0.1747	0.2609

Source: IFS, World Bank (data processed).

Table 3 indicates that bidirectional relationship between two currencies does not always occur. There was causality between the currencies of the five developed countries and those of the five ASEAN countries, except for the Malaysian ringgit. The Malaysian ringgit was relatively independent from the currencies of the developed countries, but there was causality with the regional currencies. Currency movements of the five ASEAN countries (except Malaysia) were relatively more influenced by the strong currencies of the developed countries compared with that of neighbouring countries in the region. Based on the Granger causality test, the exchange rate relationship can be mapped as shown in Table 3.

4.2 *Equity market*

Figure 4 shows that all the stock indices declined sharply during the global financial crisis, indicating a strong interconnection in the stock market. The Engle–Granger cointegration test showed that there was cointegration between all stock indices in the five ASEAN countries (except Thailand) and the five developed countries, both before and after the global financial crisis in 2008, as shown in Table 4.

Thus, Table 4 shows that cointegration or long-term relationship on the stock market occurred not only in the period after the crisis but also in the period before the crisis. This suggests that there is a naturally strong relationship to the stock market.

Table 3. Causality relationship of exchange rates (9 January 2001–25 August 2016).

	YUAN	YEN	WON	EURO	SING DOLLAR	RUPIAH	RINGGIT	BAHT	PESO
YUAN			V	—	V	V	—	V	V
YEN	V		V	—	V	V	—	V	V
WON	V	V		V	V	V	—	V	V
EURO	—	V	V		V	V	V	—	V
SING DOLLAR	—	V	V	—		V	V	—	—
RUPIAH	V	V	V	V	V		V	V	V
RINGGIT	V	—	V	—	—	V		V	—
BAHT	—	—	—	—	—	—	V		V
PESO	—	V	V	V	V	V	—	—	

Source: IFS, World Bank (data processed).

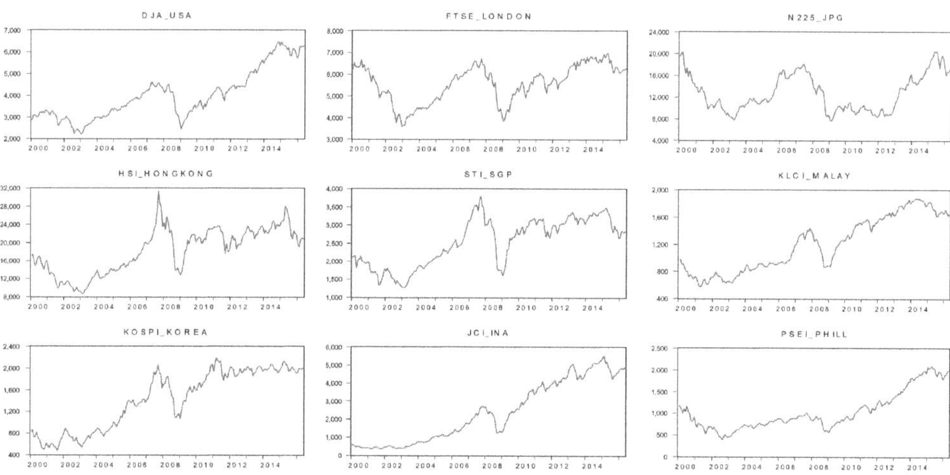

Figure 4. Movement of stock indices (2000M01–2016M05). Source: IFS, World Bank (data processed).

Table 4. Engle–Granger cointegration test (2000M01–2007M12 and 2008M01–2016M05).

Dependent	τ-Statistic	Probability*	z-Statistic	Probability*
Sample (adjusted): 2000M02–2007M12[a]				
VOLDJA	−11.04717	0.0000	107.1666	0.0000
VOLFTSE	−11.77804	0.0000	−112.2529	0.0000
VOLN225	−8.721248	0.0000	−84.34860	0.0000
VOLHSI	−8.118686	0.0000	−79.19224	0.0000
VOLSTI	−12.06771	0.0000	−117.4155	0.0000
VOLKLCI	−10.38735	0.0000	−101.0770	0.0000
VOLKOSPI	−8.258850	0.0000	−80.96877	0.0000
VOLJCI	−8.707544	0.0000	−85.31091	0.0000
VOLPSIE	−9.848600	0.0000	−96.96465	0.0000
Sample: 2008M01–2016M05[b]				
VOLDJA	−8.980313	0.0000	−88.82707	0.0000
VOLFTSE	−14.68942	0.0000	−136.9682	0.0000
VOLN225	−9.486608	0.0000	−92.76510	0.0000
VOLHSI	−14.13411	0.0000	−133.3122	0.0000
VOLSTI	−11.86782	0.0000	−117.3961	0.0000
VOLKLCI	−11.49213	0.0000	−113.4567	0.0000
VOLKOSPI	−11.11427	0.0000	−110.4453	0.0000
VOLJCI	−9.721341	0.0000	−97.43779	0.0000
VOLPSIE	−9.264620	0.0000	−93.04608	0.0000

Notes: [a]Observations included: 95 after adjustments. Null hypothesis: Series are not cointegrated. Cointegrating equation deterministics: C. Automatic lags specification based on Schwarz criterion (maxlag = 11). *MacKinnon (1996) *p*-values. [b]Observations included: 101. Null hypothesis: Series are not cointegrated. Cointegrating equation deterministics: C. Automatic lags specification based on Schwarz criterion (maxlag = 12) *MacKinnon (1996) *p*-values. VOL: Volatility, Source: finance.yahoo.com (data processed).

Table 5. Stock indices variance decomposition of four ASEAN countries (2000–16).

Period	SE	VOLDJA	VOLFTSE	VOLN225	VOLHSI	VOLSTI	VOLKLCI	VOLJCI	VOLKOSPI	VOLPSIE
Variance decomposition of VOLSTI										
1	114.9600	47.69286	11.43424	0.399156	2.222508	38.25124	0.000000	0.000000	0.000000	0.000000
5	138.1525	34.60195	9.310296	2.406261	3.683320	28.93443	5.732481	9.580951	4.344760	1.405559
10	160.5433	27.27902	10.52134	4.094994	3.695616	25.66420	9.201067	7.750331	9.359941	2.433493
Variance decomposition of VOLKLCI										
1	40.66464	28.25837	0.202599	0.283383	0.346999	6.537785	64.37087	0.000000	0.000000	0.000000
5	48.23097	24.31232	2.171990	2.543803	1.832226	8.162921	48.98572	2.087606	7.394239	2.509173
10	54.08151	20.68463	2.228770	2.873557	3.480435	7.418176	45.72194	4.899124	9.604965	3.088402
Variance decomposition of VOLJCI										
1	139.4159	31.92667	0.750433	0.396513	10.90902	10.11281	1.363412	44.54114	0.000000	0.000000
5	164.9735	28.56517	3.637239	4.997737	8.122945	10.12948	4.221935	34.22995	3.952738	2.142801
10	189.9228	24.52058	4.619489	5.524466	9.937179	12.86817	6.201101	26.98155	4.049529	5.297934
Variance decomposition of VOLPSIE										
1	47.85577	75.47052	2.585713	2.855917	0.002269	0.992603	0.210547	0.031210	0.164168	17.68705
5	56.39542	56.99567	3.507121	4.533257	1.312654	4.407288	9.884506	2.821910	2.180939	14.35666
10	64.47664	46.38718	4.802660	9.713338	2.484115	6.124848	9.963242	2.880605	5.412743	12.23127

Notes: SE, standard error. Cholesky ordering: VOLDJA VOLFTSE VOLN225 VOLHSI VOLSTI VOLKLCI VOLJCI VOLKOSPI VOLPSIE.VOL: Volatility.
Source: IFS, World Bank (data processed).

As shown in Table 5, in the period 2000–16, the stock indices of four ASEAN countries were predominantly influenced by the stock of the developed countries, especially the United States, with a contribution of between 20% and 75% with a forecast horizon of 10 periods (monthly). Contributions from the regional stock indices to the movement of stock of ASEAN countrieswere still relatively small, with a maximum of about 12.86%. Long-term relationship (cointegration) develops between the stock indices of developed countries and that offour ASEAN countries, in the periods before and after the 2008 financial crisis.

To determine whether there was a contagion effect, DCC testing in the period before and after the 2008 financial crisis was conducted, and the empirical results show that there was no significant increase in the DCC coefficient, indicating that contagion effect did not occur. What actually occurred was interdependence, as shown in Tables 6 and 7.

Table 6. DCC (2000–07).

	SDDJA_ USA	SDFTSE_ LDN	SDN225_ JPG	SDKLCI_ MALAY	SDHSI_ HKG	SDJCI_ INA	SDSTI_ SGP	SDPSEI_ PHIL	SDKOSPI_ KOREA	
SDDJA_ USA	1.000	0.726	0.217	0.242	0.122	0.097	0.422	0.348	0.175	
SDFTSE_ LDN	0.726	1.000	0.353	0.018	0.245	−0.017	0.334	0.502	0.155	
SDN225_ JPG	0.217	0.353	1.000	0.089	0.265	−0.044	0.256	0.388	0.145	
SDKLCI_ MALAY	0.242	0.018	0.089	1.000	0.370	0.396	0.652	0.111	0.205	
SDHSI_ HKG	0.122	0.245	0.265	0.370	1.000	0.636	0.607	0.137	0.479	
SDJCI_ INA	0.097	−0.017	−0.044	0.396	0.636	1.000	0.521	−0.330	0.481	
SDSTI_ SGP	0.422	0.334	0.256	0.652	0.607	0.521	1.000	0.147	0.461	
SDPSEI_ PHIL	0.348	0.502	0.388	0.111	0.137	−0.330	0.147	1.000	0.029	
SDKOSPI_ KOREA	0.175	0.155	0.145	0.205	0.479	0.481	0.461	0.029	1.000	
Sum	3.349	3.315	2.670	3.082	3.860	2.740	4.401	2.332	3.131	
Average	0.2936	0.2894	0.2088	0.2603	0.3575	0.2175	0.4251	0.1665	0.2663	0.2761

Source: IFS, World Bank (data processed). SD: Standard Deviation.

Table 7. DCC (2008–16).

	SDDJA_ USA	SDFTSE_ LDN	SDN225_ JPG	SDKLCI_ MALAY	SDHSI_ HKG	SDJCI_ INA	SDSTI_ SGP	SDPSEI_ PHIL	SDKOSPI_ KOREA	
SDDJA_ USA	1.000	0.469	0.424	0.445	0.324	0.297	0.425	0.542	0.280	
SDFTSE_ LDN	0.469	1.000	0.033	0.704	0.534	0.523	0.678	0.257	0.621	
SDN225_ JPG	0.424	0.033	1.000	0.150	0.274	0.261	0.155	0.328	−0.235	
SDKLCI_ MALAY	0.445	0.704	0.150	1.000	0.619	0.542	0.768	0.253	0.636	
SDHSI_ HKG	0.324	0.534	0.274	0.619	1.000	0.418	0.787	0.408	0.450	
SDJCI_ INA	0.297	0.523	0.261	0.542	0.418	1.000	0.588	0.272	0.300	
SDSTI_ SGP	0.425	0.678	0.155	0.768	0.787	0.588	1.000	0.330	0.617	
SDPSEI_ PHIL	0.542	0.257	0.328	0.253	0.408	0.272	0.330	1.000	0.229	
SDKOSPI_ KOREA	0.280	0.621	−0.235	0.636	0.450	0.300	0.617	0.229	1.000	
Sum	4.206	4.819	2.389	5.118	4.815	4.203	5.349	3.620	3.898	
Average	0.401	0.477	0.174	0.515	0.477	0.400	0.544	0.328	0.362	0.3893

Source: IFS, World Bank (data processed).

Based on the DCC after the 2008 financial crisis, there was an increase of interdependence between the stock indices of the four ASEAN countries and the developed countries, except United Kingdom. It shows that after the 2008 financial crisis, the openness of capital flows across countries increased, resulting in the greater spread of global risk through portfolio diversification across countries. On the one hand, this increase provides convenience for global funding; on the other hand, the economic vulnerability also increases.

4.3 *Policy interest rates*

Based on Figure 5, the Central Bank policy interest rates in the five ASEAN countries and the United States tend to decrease, especially after the 1997 financial crisis. According to the Engle–Granger cointegration test policy rates in Table 8, a long-term equilibrium relationship did not occur in all the five ASEAN countries, but only cointegration occurred in Singapore and Malaysia.

As seen in Table 9, there was an increase in the DCC coefficient in the three periods of the analysis. This indicates the presence of a strong relationship between the monetary policies in each country and those in other countries. It also indicates the increasing coordination of monetary policies among the five ASEAN countries. The correlation coefficient of the US policy rates increased sharply, especially after the 2008 financial crisis. These indicate that the US monetary policy was still used as a reference by the five ASEAN countries.

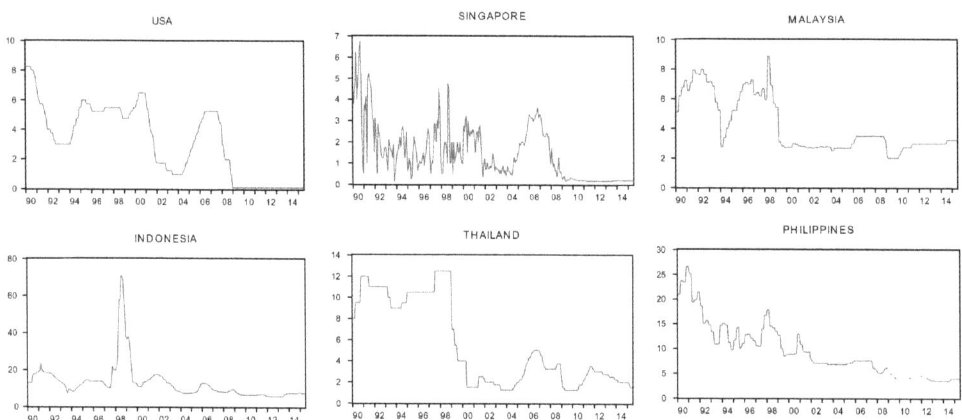

Figure 5. Movement policy interest rates. Source: IFS, World Bank (data processed).

Table 8. Engel–Granger cointegration test policy rates in five ASEAN countries and the United States.

Dependent	τ-Statistic	Probability*	z-Statistic	Probability*
United States	–3.215680	0.6367	–22.01180	0.5429
Singapore	–8.755051	0.0000	–119.7866	0.0000
Malaysia	–4.499010	0.0946	–37.95018	0.0833
Indonesia	–3.995860	0.2492	–34.02247	0.1440
Thailand	–4.195686	0.1753	–32.93174	0.1678
Philippines	–4.016446	0.2404	–28.86497	0.2766

Notes: Series: United States, Singapore, Malaysia, Indonesia, Thailand, Philippines.
Sample: 1990M01–2015M06. Included observations: 279. Null hypothesis: Series are not cointegrated. Cointegrating equation deterministics: C. Automatic lags specification based on Schwarz criterion (maxlag = 15). *MacKinnon (1996) *p*-values.
Source: IFS, World Bank (data processed).

Table 9. DCC of policy interest rates (1990–2015) on average from each country.

	DCC_ CHINA	DCC_ IND	DCC_ JPG	DCC_ KOR	DCC_ MALAY	DCC_ PHIL	DCC_ SGP	DCC_ THAI	DCC_ UK	DCC_ USA
1990M01 1997M07	0.0322	0.0897	0.1278	0.3410	0.0239	0.3488	0.3420	0.1867	0.2686	0.0100
1998M01 2007M12	0.4289	0.2845	0.3753	0.3836	0.4956	0.3758	0.4470	0.3885	0.3262	0.0998
2008M01 2015M05	0.7104	0.5539	0.7877	0.8442	0.7682	0.7691	0.8477	0.6764	0.8222	0.8084

Source: IFS, World Bank (data processed).

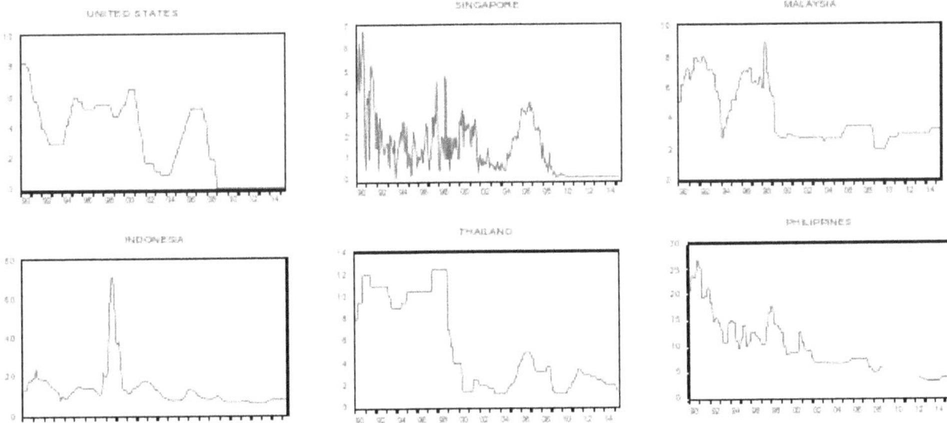

Figure 6. Money market rates in the five ASEAN countries and the United States. Source: IFS, World Bank (data processed).

Table 10. Engel–Granger cointegration test of money market rates.

Dependent	τ-Statistic	Probability*	z-Statistic	Probability*
United States	−4.052550	0.7464	−24.27781	0.8566
European Union	−3.994672	0.7695	−26.37325	0.7898
Japan	−4.122106	0.7174	−27.51796	0.7483
China	−5.543007	0.1374	−44.72646	0.1292
Korea	−2.134344	0.9993	−11.78542	0.9983
Singapore	−5.097402	0.2750	−38.26990	0.3087
Malaysia	−3.476455	0.9224	−176.8311	0.0000
Indonesia	−3.022637	0.9790	−60.78869	0.0035
Philippines	−5.434343	0.1651	−42.44309	0.1810
Thailand	−5.712595	0.1014	−46.43230	0.0984

Notes: Null hypothesis: Series are not cointegrated. Cointegrating equation deterministics: C. Automatic lags specification based on Schwarz criterion (maxlag = 11).*MacKinnon (1996) *p*-values.
Source: IFS, World Bank (data processed).

4.4 *Money market*

Figure 6 shows that the movement of money market interest rates does not have the same pattern. This is reinforced by the cointegration test results, which in the long run shows there is no relationship of money market interest rates in the five ASEAN countries and those in the five developed countries (see Table 10). The absence of cointegration relationships shows

Table 11. Correlation coefficient of money market rates in three periods.

Period	Correlation coefficientin average
1994Q1–1997Q3	0.35
1998Q1–2007Q4	0.41
2008Q1–2014 Q2	0.56

Source: IFS, World Bank (data processed).

Table 12. Variance decomposition of money market rates (1994Q1–2014Q2).

Period	SE	D(USA)	D(EU)	D (JAPAN)	D (CHINA)	D (KOREA)	D (SING)	D (MALAY)	D(INA)	D(PHIL)	D(THAI)
Variance decomposition of D(SING)											
1	0.248879	21.18238	8.783368	11.02646	7.128446	3.709676	48.16967	0.000000	0.000000	0.000000	0.000000
5	0.479858	20.81597	10.47808	29.14958	6.182966	2.814854	18.69895	2.459190	2.208520	1.363422	5.828464
10	0.615699	23.85886	12.08395	20.44473	10.29808	2.603831	18.38128	2.074972	3.185010	1.953092	5.116194
Variance decomposition of D(MALAY)											
1	0.151211	0.423378	13.50496	14.49259	2.196494	40.20865	5.827138	23.34679	0.000000	0.000000	0.000000
5	0.416577	9.198588	5.907749	16.46720	10.28322	11.78560	29.76712	11.52986	1.847372	0.435574	2.777717
10	0.663802	26.96817	4.464784	9.784050	8.775853	8.542639	25.92369	5.921986	5.288041	1.521555	2.809228
Variance decomposition of D(INA)											
1	2.054912	29.30270	0.500169	8.72E–05	1.241160	1.354811	54.48772	0.837615	12.27574	0.000000	0.000000
5	3.966181	21.67978	3.101293	4.983198	10.42426	11.82745	25.60559	4.625691	7.847991	1.856753	8.047991
10	5.780639	24.47387	2.838444	7.335125	10.26036	7.872738	27.65734	4.336255	7.742558	1.733036	5.750272
Variance decomposition of D(PHIL)											
1	0.705387	2.341395	0.008007	30.27046	6.625379	4.807545	38.29211	7.076287	3.880918	6.697903	0.000000
5	1.352914	5.077809	2.196161	29.00651	9.307887	7.453509	33.26384	5.567594	2.602860	4.088540	1.435291
10	1.603841	7.300863	3.331738	26.37032	10.49192	9.103535	28.35473	4.725600	3.190950	3.930991	3.199350
Variance decomposition of D(THAI)											
1	0.508058	5.844559	24.72042	0.233492	1.379816	5.036296	25.17341	1.329327	2.492153	2.921947	30.86858
5	1.579542	10.63436	5.700098	17.65382	15.10879	7.911774	25.31946	7.401335	2.243197	2.503745	5.523430
10	1.906556	16.39848	5.410812	13.87307	14.85211	7.812366	24.97486	6.046350	3.609355	3.129570	3.893024

that in the long term movement of money market rates does not converge to a certain value, but it is more likely to be determined by the supply and demand for money in each country.

Based on the correlation coefficients in the three different periods as shown in Table 11, there was no increase in the correlation coefficient on a large scale after the 2008 financial crisis. Thus, contagion effect did not occur and only interdependence was noted.

Regionally, money market rates in Singapore had the most impact on movements in the five ASEAN countries; globally, the US money market rates were still dominant, as shown in Table 12.

5 CONCLUSIONS

1. Exchange rate fluctuations had the same direction in the five ASEAN and five developed countries, indicating a strong global relationship among currencies. However, empirical evidence indicated the occurrence of interdependence but not contagion.
2. Stock indices in the four ASEAN countries were predominantly influenced by the stock indices from the developed countries, especially the United States. Contributions from regional indices to the movement of stocks of ASEAN countries were relatively small. There was a long-term relationship (cointegration) between the stock of the developed countries and that of the ASEAN countries, in the periods before and after the 2008 financial crisis. The DCC coefficient showed no contagion effect and only interdependence among the stock markets.

3. Correlation of policy interest rates between the five countries of ASEAN and the United States increased after the crisis in 1997 and still continued after the 2008 financial crisis. The increase indicates that the US monetary policy was still used as a reference by the five ASEAN countries.
4. The movement of money market interest rates did not have the same pattern for the five ASEAN and the five developed countries. It shows that in the long term, the movement is more likely to be determined by the supply and demand of money in each country.

REFERENCES

Ahlgren, N. & Antell, J. (2010). Stock market linkages and financial contagion: A cobreaking analysis. *Quarterly Review of Economics and Finance*, 50(2), 157–166. doi:10.1016/j.qref.2009.12.004.

Claessens, S., & Forbes, K., (2004). International Financial Contagion: The Theory, Evidence and Policy Implications. The IMF's Conference: Role in Emerging Market Economies: Reassessing the Adequacy of its Resources' organized by RBWC, DNB and WEF.

Dornbusch, R., Park Y.C. & Claessens, S. (2000). Contagion: Understanding how it spreads. *The World Bank Research Observer*, 15(2), 177–197. doi:10.1093/wbro/15.2.177.

Dungey, M. & Tambakis, D. (2003). Financial contagion: What do we mean? What do we know? Available at: http://hpi.uw.hu/pdf/Dungey-Tambakis%20Presentation%20Dubai%2003.pdf (accessed May 2017).

Forbes, K. & Rigobon,R.(2001). Contagion in Latin America: Definition, measurement, and policy implications.*Economia Journal of The Latin America and Caribbean Economic Association*, 1(2),1–46.

Forbes, K. & Rigobon, R. (2002). No contagion, only interdependence: Measuring stock market comovements.*Journal of Finance*, 57(5), 2223–2261.

Gómez, M., & Rivero, S. (2014). Causality and contagion in EMU sovereign debt markets. *International Review of Economics and Finance*, 33(2014), 12–27.

Gray, A.(2015).Understanding interconnectedness risk, to build amore resilient financial system. A white paper to the industry.

Kaminsky, et.all., (2003). The unholy trinity of financialcontagion. *Journal of Economic Perspectives—Volume 17, Number 4— Fall 2003— Pages 51–74.*

Kaminsky, G.L., & Reinhart, C.M. (2000). On crises, contagion, and confusion. *Journal of International Economics*, 51, 145–168. doi:10.1016/S0022-1996(99)00040-9.

Kenourgios, D. (2014). On financial contagion and implied market volatility. *International Review of Financial Analysis*, 34(C), 21–30.

Luchtenberg, K.F. & Vu, Q.V. (2015). The 2008 financial crisis: Stock market contagion and its determinants.*Journal of Research in International Business and Finance*, 33(2015), 178–203.

Morales, L. & O'Callaghan, B.,A.,. (2014). The global financial crisis: World market or regional contagion effects? *International Review of Economics and Finance*, 29, 108–131. doi: http://dx.doi.org/10.1016/j.iref.2013.05.010.

Podlich, N. & Wedow, M. (2014). Crossborder financial contagion to Germany: How important are OTC dealers?*International Review of Financial Analysis*, 33, 1–9. doi:10.1016/j.irfa.2013.07.008.

Shen, P.L., Li, W., Wang, X.-T., and Su, C.-W. (2015). Contagion effect of the European financial crisis on China's stock market: Interdependence and pure contagion. *Economic Modelling*, 50(2015), 193–199.

Unsal, D.F. & Jain-Chandra, S. (2012). The effectiveness of monetary policy transmission under capital inflows: Evidence from Asia. IMF Working Paper WP/12/265.

Competition and Cooperation in Economics and Business – Gani et al. (Eds)
© 2018 Taylor & Francis Group, London, ISBN 978-1-138-62666-9

Fiscal and monetary dynamics: A policy duo for the Indonesian economy

E.Z.W. Yuan & C. Nuryakin
Department of Economics, Faculty of Economics and Business, Universitas Indonesia, Depok, Indonesia

ABSTRACT: Research on how monetary and fiscal authorities can and should interact has been abundant internationally; however, the amount of research is still small in Indonesia. One consensus that converges from the research is the importance of coordination between monetary and fiscal authorities for an optimal inflation rate and economic growth as well as to minimise welfare losses. What has not yet been observed is the optimal level of monetary and fiscal policy pertaining to monetary and fiscal policy interaction, which is the focus of this research. In a non-cooperative game theory model, we used a loss function of monetary policy, which uses the Surat Berharga Bank Indonesia (SBI) rate as its instrument, and fiscal policies with government spending as their tool, as the payoff for each authority. In general, the result shows that the actual SBI rate and government expenditure have yielded in non-Nash equilibrium and non-Pareto efficiency equilibrium. Thus, there is much room to improve the policies, especially the smoothing of government expenditure throughout the year; that is, improving government expenditure absorption in the second quarter and moderating it in the third and fourth quarters, as well as lowering SBI rates.

1 INTRODUCTION

A good policy mix between monetary and fiscal policies is required for an economy to experience stable economic growth along with a low and stable inflation rate. A previous study by Taylor (1994) found that high economic stability generates high economic growth and a low inflation rate in the long run. Nevertheless, there is no single market economy in which monetary and fiscal policies are governed by a single institution. Reigning over different institutions, each monetary and fiscal authority has different objectives, with the monetary authority focusing on price stability and the fiscal authority focusing on output level stability. On the other hand, as empirically witnessed and theoretically proven, there is a trade-off relationship between price and output levels; high output is often followed by high price level. Thus, a good policy mix between the two authorities heavily relates to whether or not the institutions coordinate.

Conflicting results were found in previous research on the coordination between monetary and fiscal policies. Some of these studies are from Rogoff (1985), Kydland and Prescott (1977), and Bartolomeo and Giocchino (2004). Specifically, Rogoff (1985) argues that society will be better off when the central banker agent does not share the social objective function; instead, the only concern is inflation rate stabilisation. However, Bartolomeo and Giocchino (2004) found that economic stability could not be achieved under either authority. Under monetary leadership, the central bank is not forced to bail out fiscal deficits that cannot guarantee fiscal stability, although fiscal dominance has forced the central bank to monetise public debts, which deprioritises monetary stability.

On the other hand, more recent research has discovered the opposite. Nordhaus (1994), Petit (1989), Beetsma and Bovenberg (1997), Faure (2003), and Javed and Sahinoz (2005) are some researchers who are proponents of the beneficial effect of monetary and fiscal policy coordination. Nordhaus (1994) theoretically explains that the separation of monetary and fiscal authority will provoke high fiscal deficit and interest rates that are too high to promote a healthy level of private investment and adequate long-term growth of potential output. Petit (1989) empirically found that the coordination between monetary and fiscal policies has a positive impact on the macroeconomic condition in Italy. Beetsma and Bovenberg (1997) also proposed the need for coordination between monetary and fiscal authorities.

Faure (2003) found that in the European Union, the coordination of monetary and fiscal policies yields a higher welfare by disciplining the government and stabilising the employment level, even if the inflation rate increases ex post. Javed and Sahinoz (2005) also found that coordination between the authorities generated a positive impact in Turkey.

Notable research on the interaction between monetary and fiscal policies in Indonesia has been implemented by Mochtar (2004), Simorangkir and Goeltom and (2012), and Santoso (2012). Santoso (2012) and Simorangkir and Goeltom (2012) found that, in the event of negative output shock, the coordination between monetary and fiscal policies in Indonesia results in fewer welfare losses than if no coordination exists. Simorangkir and Goeltom (2012) used the game theory model and empirical studies using data from 1970 to 2002, whereas Santoso (2012) used the Dynamic Stochastic General Equilibrium (DSGE). While suggesting that a coordinated policy can be strengthened through a coordinated institution, such as Dewan Moneter, Santoso (2012) also found that the policy coordination in response to inflationary shock has not been as efficient as the response to output shock in Indonesia.

Mochtar (2004) measured the quasi-fiscal activities of the central bank, which indicates the dominance of monetary authority over fiscal authority. Using a vector autoregressive estimation, Mochtar (2004) ran the data from 1998 to 2003 and found that, after the financial crisis in 1997–98, fiscal authority played a dominant role, although at a small scale. All in all, research on interactions of policies in Indonesia proves that there is coordination between monetary and fiscal policies.

Regarding optimal policy literature in Indonesia, recent research on optimal monetary policy has been conducted by Widjaja and Mardanugraha (2009), Brouver et al. (2006), Ramayandi (2007), and Kuncoro and Sebayang (2013). Ramayandi (2007) used a simple Policy Reaction Function (PRF) applied to an econometric model to predict the optimal monetary policy rates using data from 1989 to 2004 and found that, during that period, the optimal monetary policy rates were lower than the actual policy rates. Brouver et al. (2006) noted that for Indonesia the interest rates in 2005 seemed to still be higher than what the rule suggested. Kuncoro and Sebayang (2013) also argue for a generally lower optimal interest rate in the 1999–2010 period; they also used the PRF framework to predict optimal policy rates. In general, these studies argue that the monetary policy interest rates in 1989–2010 were too high.

On the other hand, Widjaja and Mardanugraha (2009) used the Guender model to predict the optimal nominal interest rate for Bank Indonesia's (BI) target of operation and found that from 2005 to 2008 the actual interest rates implemented by BI were lower than the predicted optimal policy rates. The optimal policy rates are predicted using the gap between actual inflation and the targeted inflation rate. In contrast to the findings of Brouver et al. (2006) and Kuncoro and Sebayang (2003), this research argues that the monetary policy interest rates from 2005 to 2008 were too low.

Kendrick and Amman (2011) prove theoretically that small quarterly changes in a fiscal policy provide a less volatile path for the economy than large annual changes. A quarterly instead of annual approach to evaluate a fiscal policy calls for time distribution of government expenditure. Kuncoro and Sebayang (2003) found that in Indonesia the deviation of the implemented fiscal policy from their optimal values in 1999–2010 fluctuates within a one-year period.

Despite the importance of coordination between monetary and fiscal policies, and the abundant research on an optimal policy, there is still a lack of research on the optimal policy with a comprehensive model regarding monetary and fiscal interactions. The independence of BI that was granted in 1999 and the implementation of a price targeting the framework in 2005 have changed the nature of the interaction between monetary and fiscal policies in Indonesia,which affects the optimal policy rates between the two. Thus, this research aims to find the optimum policy mix between fiscal and monetary policies by simulating certain levels of policy instruments using game theory.

2 THE BASIC FRAMEWORK

The literature on coordination between monetary and fiscal policies has used the DSGE method, the mathematical approach with an additional game theory method, and empirical studies to prove the mathematical model, to determine whether coordination between monetary and fiscal authorities generates higher output. Here, output is not limited to economic output, but it also includes social welfare, inflation, and the utility level of monetary and fiscal authorities.

Our framework is based on game theory prediction for local optimal monetary and fiscal policies. The local term for optimal policy rates in this study means that we only analyse optimal strategies of each agent that are very close to the existing policy rates. This also means that we look for optimal policy rates that, if implemented, require the least drastic changes from existing policy rates.

First, we assume that BI and the Government of Indonesia (GOI) have the objective to minimise welfare losses in the following function:

$$L_{BI}\left(r_i, g_j\right) = \alpha_{BI}\left(\hat{y}-\overline{y}\right)^2 + \beta_{BI}\left(\hat{p}-\overline{p}\right)^2 + \gamma_{BI}\left(r-\overline{r}\right)^2 \quad (1)$$

$$L_{GOI}\left(r_i, g_j\right) = \alpha_{GOI}\left(\hat{y}-\overline{y}\right)^2 + \beta_{GOI}\left(\hat{p}-\overline{p}\right)^2 + \delta_{GOI}\left(g_i-\overline{g}\right)^2, \quad (2)$$

where BI indicates monetary authority, GOI is the fiscal authority, and r_i and g_j denote any level of policy instruments for both monetary and fiscal policies. The payoff function shows that even though both BI and GOI have an inflation and output element in their payoffs, the weight that the fiscal and monetary authorities attribute to final targets in terms of output and inflation differ; that is, $\alpha_{BI} < \alpha_{GOI}$; $\beta_{BI} > \beta_{GOI}$. This assumption is based on the theoretical framework proposed by Dixit and Lambertini (2001) and has been used by Simorangkir and Goeltom (2012). The payoffs show the loss function of the economy, which represents the stabilisation objective of both policies. For BI, its loss function consists of deviation of the predicted price level from the optimum price level, deviation of the predicted output from the optimum output, and deviation of the real interest rate from its optimum, which is set to 2%. For GOI, the loss function consists of deviation of the predicted price level from the optimum price level, deviation of the predicted output from the optimum output, and deviation of the fiscal policy, which is government expenditure g_j, from its optimum level.

Thus, the framework of the game is as follows:

Player: $N = \{BI, GOI\}$

Strategy: $S = \{r_i, g_j\}$, where $i, j > 0$.

Payoff:

$$L_{BI}\left(r_i, g_j\right) = 1\left(\hat{y}-\overline{y}\right)^2 + 2\left(\hat{p}-\overline{p}\right)^2 + 1\left(r-\overline{r}\right)^2 \quad (3)$$

$$L_{GOI}\left(r_i, g_j\right) = 2\left(\hat{y}-\overline{y}\right)^2 + 1\left(\hat{p}-\overline{p}\right)^2 + 1\left(g_i-\overline{g}\right)^2 \quad (4)$$

The weight on output and price levels in BI and GOI ($\alpha_{BI} = 1$, $\alpha_{GOI} = 2$, $\beta_{BI} = 2$, and $\beta_{GOI} = 1$) are according to Simorangkir and Goeltom (2012).

The game model used in this research is a static game of complete information. A static game has two distinct features. First, each player simultaneously and independently chooses an action. Both players, which in this research are BI and GOI, choose their action at the exact same moment. Second, the condition on the players' choices of actions and payoffs are distributed to each player. Once BI and GOI make their choices of the government expenditure level and SBI rate, their choices will result in a particular loss value from the predicted price and output level or a probabilistic distribution over the loss value. In addition to this static game, a game of complete information requires that all the players in the game have common knowledge among them in four components: (1) the possible actions of all players, (2) all the possible outcomes, (3) how each combination of actions of all players affects which outcome will materialise, and (4) the preferences of each and every player over outcomes.

3 EMPIRICAL MODEL

To quantify the welfare loss in Equations 3 and 4, we estimate the output level (y') and the price level (p'). To predict the effect of the fiscal and monetary policies (i.e. government expenditure and SBI rate) on the stability of macro economy indicators, which include the price and output levels or Gross Domestic Product (GDP), one needs to find the model best fit to predict the price and output levels. There are numerous models for predicting output and price levels. However, we used the model from Mohanty and John (2015) to predict the price level and modified models from Grossman (1988) to predict the output level. The predicted price level is the prediction when the monetary authority sets r_i and the fiscal authority sets g_j. The predicted output level is the prediction when the fiscal authority sets g_j and the monetary authority sets r_i. The predictions of the price and the output levels in certain policy rates used models that are estimated first using econometrics.

The fiscal theory of the price level shows how fiscal policy affects the price level, in contrast to the conventional view that inflation is always a monetary phenomenon. An aggregate supply equation and the quantity theory of money can explain how the output gap and quantity of money in the economy affects the price level, as mentioned in Mohanty and John (2015), in addition to the monetary policy measured by interest rates. Thus, the regression model for the price level is estimated in Equation 5.

$$p_t = \gamma_0 + \gamma_1 m_t + \gamma_2 o_t + \gamma_3 r_t + \gamma_4 g_t, \tag{5}$$

where p_t is the consumer price index level in period t (using a base year of 2002); m_t is the quantity of money in the economy in period t (in hundreds of trillions); and o_t is the output gap of the economy in period t.

Equation 6 is estimated by the Ordinary Least Squares (OLS) method. The output gap is the gap between the actual output level y_t and its potential output $\bar{y}$, which requires the calculation of $\bar{y}$. Several methods estimate the potential output as well as the output gap. Nasution and Hendranata (2014) used Indonesia's data from 1983 to 2013 and found that the best output gap measurement for Indonesia is the band pass filter, outperforming the Hodrick–Prescott filter, autoregressive integrated moving average, production function, and structural vector autoregressive estimation. For the output level, the modification models in Grossman (1988) were used, which incorporated the Solow growth model with Keynes and the neoclassical model of investment. The regression model for the output level is estimated by the two-stage leastsquares model.

$$\begin{aligned} y_t &= \beta_0 + \beta_1 k_t + \beta_2 l_t + \beta_3 g_t + \varepsilon_t \\ k_t &= \beta_4 + \beta_5 r_t , \end{aligned} \tag{6}$$

where y_t is the natural logarithm of GDP in period t; k_t is the natural logarithm of capital in period t; l_t is the natural logarithm of labour in period t; g_t is the natural logarithm of government expenditure in period t; and r_t is the monetary policy rates in period t.

The monetary policy instrument r is chosen based on which policy instrument best fits the model. Monetary policy instruments considered in this model are BI rates, Fasilitas BI rates, repurchase order rates, SBI rates, and Jakarta interbank overnight rate.

As mentioned in Equations 3 and 4, the payoffs are in the form of a loss function, which is deviation of the price level, output level, interest rate, and government expenditure from their optimum values. Optimum values of the output level, price level, and government expenditure are obtained through finding their trends, where the regression models are as follows:

$$y_t^d = \theta_0 + \theta_1 t \tag{7}$$

$$p_t^d = \delta_0 + \delta_1 t \tag{8}$$

$$g_t^d = \varphi_0 + \varphi_1 t, \tag{9}$$

where y_t^d is the deseasonalised data for y in period t; p_t^d is the deseasonalised data for p in period t; g_t^d is the deseasonalised data for g in period t; and t is the period, $t = 1$ for the first quarter of 2005.

Equations 7, 8, and 9 estimate the optimum value of the output level, price level, and government expenditure for each quarter in 2014 and 2015. These equations are estimated using the OLS method.

4 RESULTS AND ANALYSIS

In the years 2014 and 2015, GOI was struggling to boost economic growth in order for the Indonesian economy to graduate from the 'middle income trap'. Since President Joko Widodo began his term, various economic stimulant packages were implemented to boost consumption and investment. The fiscal policies implemented were considered ambitious, owing to a 5.7% economic growth target in 2015 (Ministry of Finance, n.d.).

On the other hand, the global crisis as a result of the global financial crisis and the aftermath of the Greek crisis forced BI to keep the benchmark interest rate (i.e. BI rate) as high as it could to prevent possible capital outflow from Indonesia. Many economists debated the stance that BI took at that time, arguing that the domestic interest rate was not a significant pulling factor of capital flow in Indonesia. Thus, lowering the domestic interest rate should help boost economic growth and achieve economic stability, without having a significant capital outflow to disrupt the balance of the economy. The policy mix that happened during 2014 and 2015 was generally represented by a loose fiscal policy but a tight monetary policy.

Figures 1 and 2 show that during 2014–15, the actual price was generally lower than the optimal price, except for the fourth quarter of 2014 when the actual price jumped against its optimal level. A sudden price increase in the fourth quarter of 2014 was related to the cut on electric subsidy by the government that was fully implemented in November 2014. There was also a global oil price increase at the end of 2014; at the same time, there was a cut on government subsidies for oil and an elimination of the subsidy on gases by the newly elected President Joko Widodo in October 2014. The cost of gasoline (premium) was raised from Rp 6,500 to Rp 8,500 per litre, whereas the cost of diesel was raised from Rp 5,500 to Rp 7,500 per litre.

The output level has more volatility across quarters within a one-year period, which also shows the business cycle in a one-year period. The output level in the first and fourth quarters of the years 2014 and 2015 was below the optimum level, but it jumped above the optimum level in the third quarter of both years. The economic growth in 2014 and 2015 slumped relatively to 2012 levels due to weaker foreign demand, slower investment growth, and the global oil price dropping in 2015. The investment slump, especially in 2014, was due to the common investor behaviour to wait and see the performance of President Joko Widodo after the election before investing in Indonesia.

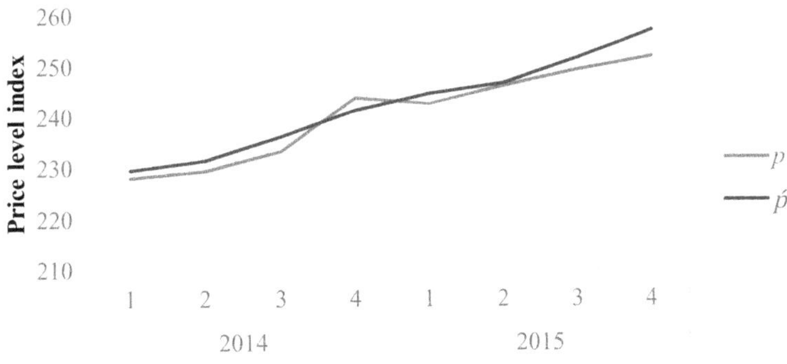

Figure 1. Actual and optimum price levels in 2014 and 2015. Source: Authors' calculation and Biro Pusat Statistik (2016).

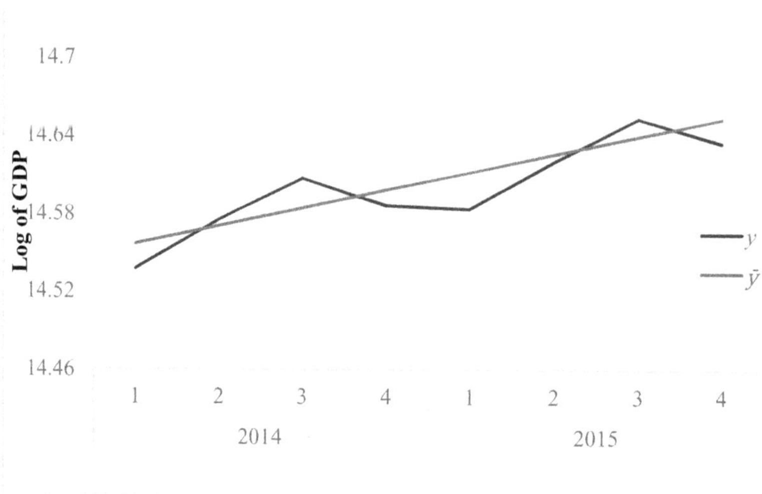

Figure 2. Actual and optimum output levels in 2014 and 2015. Source: Authors' calculation and *Biro Pusat Statistik* (2016).

Table 1. Regression estimation for output and price model.

	Dependent variable: Y		Dependent variable: p	
Constant	(1)	(2)	(3)	(4)
	12.3752***	10.3605***	−331.217**	−132.9626***
G	0.1794***	0.0598***	44.0694***	20.5252***
R	−0.0322***	−0.1965***	−4.3734***	−1.3494***
K	.	0.2496***	.	.
M	.	.	.	28.4722***
O	.	.	.	−4.6842***
d1	.	0.0032	.	9.423***
d4	.	−0.0251*	.	−6.7367**
Adjusted R2	0.7081	0.9735	0.8077	0.9667
Probability > F	0.0000	0.0000	0.0000	0.0000
BP test		0.22		0.29
BG test		0.17		0.12
VIF		2.68		4.64

Significance level *10%, **5%, ***1%.
"." (dots) means that the corresponding variables are not included in the regression.
BP test = Breusch-Pagan Test.
BG test = Breusch-Godfrey test.
VIF = Variance inflation factor.
Source: Authors' calculation.

4.1 *Game simulation*

Using the model predicted by econometrics, the payoff function for monetary and fiscal authorities with each policy rate is as follows (see the regression results in Table 1):

$$L_{BI}\left(r_i, g_j\right) = 1\left(\hat{y}\left(r_i, g_j\right) - \overline{y}(t)\right)^2 + 2\left(\hat{p}\left(r_i, g_j\right) - \overline{p}(t)\right)^2 + 1\left(r - \overline{r}\right)^2 \tag{10}$$

$$L_{GOI}\left(r_i, g_j\right) = 2\left(\hat{y}\left(r_i, g_j\right) - \overline{y}(t)\right)^2 + 1\left(\hat{p}\left(r_i, g_j\right) - \overline{p}(t)\right)^2 + 1\left(g_i - \overline{g}(t)\right)^2, \tag{11}$$

where output and price levels are predicted for each value of r_i and g_j, and $t = [37, 44]$.

$$\hat{y}_t(r_i, g_j) = 10.361 + 0.249k_t + 0.059g_j - 0.025d_4, \\ k_t = 14.383 + 0.249r_i \quad (12)$$

$$\hat{p}_t(r_i, g_j) = -132.99 + 28.472m_t - 4.684o_t - 1.349r_i + 20.525g_j + 9.423d_1 - 6.736d_4 \quad (13)$$

Moreover, regression results for the optimum level of output, price, and government expenditure at t are as follows:

$$\overline{y_t} = 14.044 + 0.014t \quad (14)$$

$$\overline{p_t} = 121.493 + 2.865t \quad (15)$$

$$\overline{g_t} = 11.762 + 0.032t \quad (16)$$

The strategy for each authority includes the policy rates (SBI rates and government expenditure level) that are possible to be implemented by either authority, based on the actual implementation for the last five years. The games are formed in a 50 × 14 matrix. There are 50 rows of SBI rates ranging from 4% to 8.9%, with a 0.1 interval. As for the government expenditure in a form of natural logarithm, there are 14 different strategies ranging from 12.1 to 13.4, with a 0.1 interval.

The simulation of the game shows that there are multiple Nash equilibria for each quarter. The equilibria shown in Table 2 are the local Pareto efficiency equilibrium (in italics) and

Table 2. Results of game simulation for 2014 and 2015.

Quarter		GOI	BI	LGOI	LBI	Total L
2014						
1	Actual	12.3	7.2	1.895	13.784	15.679
	Nash 1	12.5	7	0.025	10.085	10.11
	Nash 2	12.4	5.5	0.03	10.088	10.118
2	Actual	12.5	7.1	4.428	20.389	24.817
	Nash 1	12.8	6.9	0.008	11.586	11.594
	Nash 2	12.7	5.3	0.01	11.589	11.599
3	Actual	13.3	7	13.301	36.694	49.995
	Nash 1	12.7	6.4	0.013	9.828	9.841
	Nash 2	12.8	7.4	0.016	9.833	9.849
4	Actual	13.3	6.9	4.515	21.829	26.344
	Nash 1	12.9	6.5	0.007	12.838	12.845
	Nash 2	13	7.3	0.012	12.841	12.853
2015						
1	Actual	12.5	6.7	5.972	22.342	28.314
	Nash 1	12.5	6.3	0.074	10.451	10.525
2	Actual	12.6	6.7	0.098	10.805	10.903
	Nash 1	12.6	5.9	0.026	10.664	10.69
	Nash 2	12.7	7.1	0.028	10.671	10.699
3	Actual	13.4	6.8	27.514	66.082	93.596
	Nash 1	12.8	5.4	0.035	11.105	11.14
	Nash 2	12.9	6.9	0.038	11.108	11.146
4	Actual	13.4	7.1	5.892	25.182	31.074
	Nash 1	*12.9*	*5.2*	*0.01*	*13.449*	*13.459*
	Nash 2	*13*	*6.7*	*0.015*	*13.455*	*13.47*

Note: Local Pareto efficiency equilibrium represented in italics.
Note: L_{BI} = Loss of Bank Indonesia (monetary sides).
Note: L_{GOI} = Loss of Government of Indonesia (fiscal sides).
Source: Authors' calculation.

the local Nash equilibrium point that is the closest to the actual policy rates. The local optimal point, in opposition to the global optimal point, is the solution that is optimal within a neighbouring set of candidate solutions. On the other hand, the global optimum point is the optimal solution among all possible solutions (Floudas et al., 2013). Theoretically, the actual SBI rates and the government expenditure level should be one of the Nash equilibria in the same period if each player acts independently and rationally, without coordination at all. Nevertheless, the simulation in 2014 and 2015 does not find any period that has a Nash equilibrium point equal to both monetary and fiscal policies implemented. The Pareto efficiency between monetary and fiscal policies can be attained when both authorities coordinate optimally (Dixit, 2001).

Graphic representation of Table 2 is shown in Figures 3 and 4. Figure 3 shows the actual SBI rate, its closest Nash equilibrium, and the Pareto efficiency equilibrium in 2014–15. The Nash equilibria move close to their actual levels, with a maximum deviation of 0.4%; these have a 0.3% average difference with the actual rates.

Figure 4 shows the actual government expenditure level, its Nash equilibrium level, and the Pareto efficiency level in 2014–15. In contrast to the monetary sides, there are relatively large

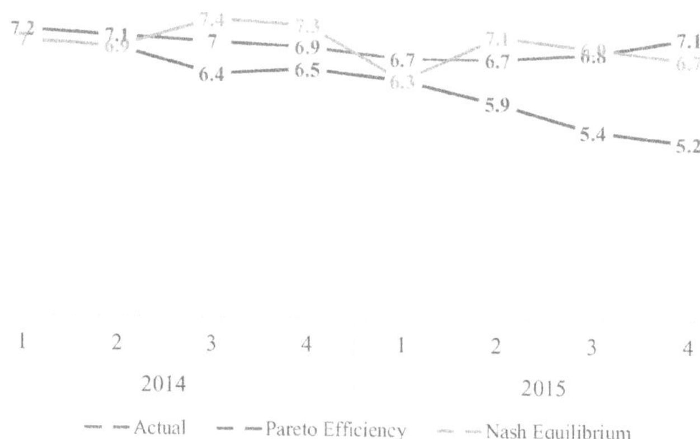

Figure 3. Actual equilibrium, Nash equilibrium, and Pareto efficiency of SBI in 2014 and 2015. Source: Authors' calculation.

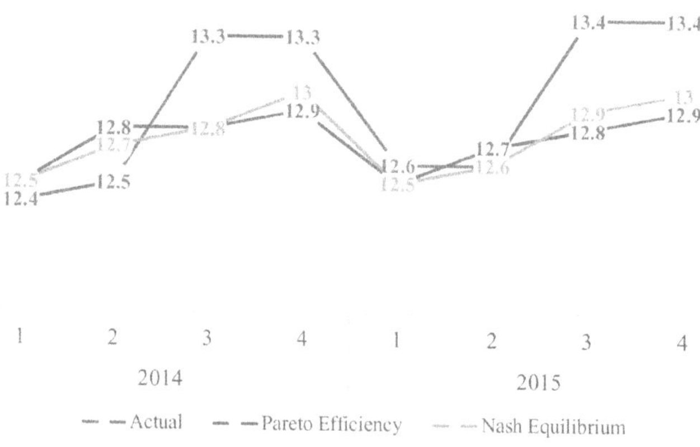

Figure 4. Actual equilibrium, Nash equilibrium, and Pareto efficiency of the government expenditure level in 2014 and 2015. Source: Authors' calculation.

differences between the actual government expenditure levels and even the closest Nash equilibrium level, with the exception of the first and second quarters of 2015. On average, deviation of the actual government expenditure from the closest Nash equilibrium is Rp 900 trillion.

Deviation of the actual level of policy rates from its Nash equilibrium policy rates may be caused by several possible factors. First, BI and/or GOI may implement policies not based on stabilisation, which is represented by a loss function, but they only aim for low inflation or high economic growth. Second, there may be a modelling error, which means the weight of each price and output level stabilisation might not be captured by a coefficient as suggested in this study; that is, 2 for output growth and 1 for price level in the GOI loss function, and 1 for output growth and 2 for price level in the BI loss function. Third, the game theory model used in this study (i.e. the non-cooperative single game) assumes that there is common knowledge and symmetric information. However, asymmetric information may have occurred between GOI and BI regarding the economic condition that may affect the loss function between the two, leading to the result of the policies deviating from the true Nash equilibrium. Finally, a possible cause is that in real life there are other factors, such as socio-political factors, that are involved in fiscal and monetary policymaking. For example, the fiscal authority does not act rationally in response to the monetary authority, which may be caused by the rigidity of the national budget planned almost one year ahead. The business cycle within one year forces government expenditure to follow the economic cycle. The government is also often forced to accelerate the absorption of the government budget in the third and fourth quarters to spend the budget.

The Pareto efficiency of SBI rates generally shows that the optimum point of SBI rates falls below the actual level in 2014 and 2015. This implies that SBI rates implemented in 2014 and 2015 were too high for the economy to stabilise. Even though the Pareto efficiency of SBI rates in 2014 and 2015 shows a decreasing trend, in the fourth quarter of 2014, optimum SBI rates increased from 6.3% to 6.4%. This may have happened because of the sudden jump in the price level in the fourth quarter of 2014 that was caused by the increase in global oil prices and the withdrawal of electric and oil government subsidies.

Lower optimum SBI rates may be because the actual price levels in 2014 and 2015, except for the fourth quarter of 2014, were still lower than their optimum levels, leaving some room for lower SBI rates. This may also be caused by the optimum level of government expenditure that is also generally lower than its actual level; thus, high SBI rates are not needed to stabilise the economy.

The Pareto efficiency of government expenditure levels has made visible patterns. The first and second quarters of 2014 and 2015 have actual government expenditures below the optimum level, whereas the third and fourth quarters of both years have actual government expenditure levels above the optimum level. This means that there should be redistribution of government expenditure throughout quarters within one year, which proves the Taylor theory of fiscal policy in the study by Kendrick and Amman (2011).

Figure 5 shows that the predicted inflation is proved to be more stable than the actual inflation, if GOI and BI apply the optimal policy rates. Even though the actual inflation was lower than the predicted inflation when the policies were optimal in the third quarter of 2014 and fourth quarter of 2015, the actual inflation in the first, second, and fourth quarters of 2014 and the first, second, and third quarters of 2015 was higher than the inflation predicted in the optimal policy rates. For example, the actual inflation in the fourth quarter of 2014 became as high as 8.96%, but if GOI and BI had coordinated optimally and implemented the Pareto efficiency, the inflation would have only gone to 4.82%.

On the other hand, the Pareto efficiency is predicted to have higher output growth than the actual output growth. Figure 6 shows that in the first quarter of 2014 the actual output growth was 4.04% and in the third quarter it hit 5.53%, whereas the stabilisation policies yielded as high as 5.95% in the first quarter of 2014 and 5.76% in the third quarter of 2014. The optimal policy rates yield higher output growth.

In general, stabilisation of the economy in 2014 and 2015 required time redistribution of government expenditure, lower level of government expenditure in the annual total, and lower SBI rates. With the implementation of optimal policy rates, the output growth is predicted to

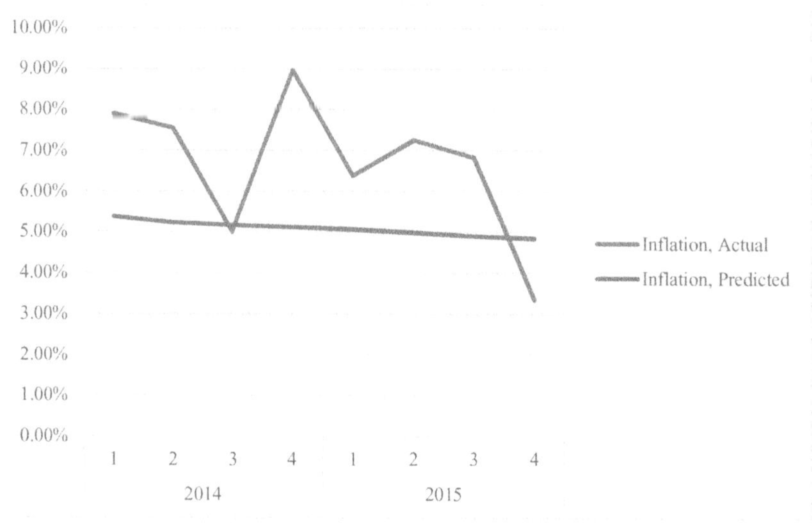

Figure 5. Actual and predicted inflation for optimal policy rates. Source: Authors' calculation.

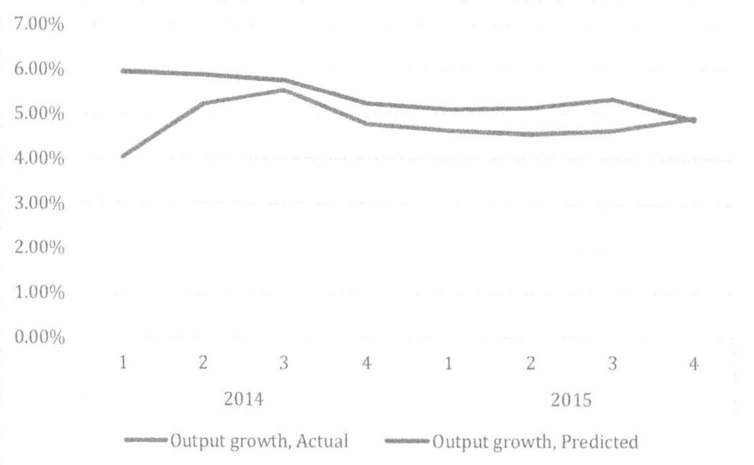

Figure 6. Actual and predicted output growth for optimal policy rates. Source: Authors' calculation.

be higher than the actual growth, but inflation rates are predicted to be more stable and lower than the actual inflation rates.

5 CONCLUSIONS

Policy coordination between monetary and fiscal authorities has been the object of many studies in Indonesia. Whereas a monetary policy aims for low and stable inflation, a fiscal policy aims for output growth and low unemployment. The difference in their objectives has become increasingly apparent since the independence of BI in 1999.

Although earlier research argued that coordination between monetary and fiscal policies has a negative impact on social welfare, more recent research, especially in Indonesia, has proved that coordination between policies lessens the social welfare loss. Furthermore, in search for optimal policy rates in Indonesia, previous research used a modified econometric model to estimate the optimality of either fiscal or monetary policy. The present research finds the optimal policy rates for both fiscal and monetary authorities using the game theory model approach to capture the interaction between monetary (BI) and fiscal (GOI) authorities.

Using a loss function to depict economic stabilisation, the game simulates the economic condition if monetary and fiscal authorities take different policy actions in the period 2014 Q1–2015 Q4 and analyses the actual policies taken (i.e. the Nash equilibrium and the Pareto efficiency).

There are generally slight differences between the actual SBI rates, the monetary policy, and their Nash equilibrium, whereas the differences in the fiscal policy are wider. Deviation of the actual monetary and fiscal policy rates from their Nash equilibrium might be caused by factors such as the objective of BI and GOI not being stable, but rather the output and price levels; the modelling error, especially the weighting of the objectives; asymmetric information between BI and GOI; and socio-political factors that exogenously affect the decision of the authorities.

While the simulation shows the importance of coordination between the two authorities, the Pareto efficiency shows that, in order to stabilise the economy in 2014 and 2015, SBI rates had to be lower and government expenditure had to be redistributed throughout the year. GOI must implement time distribution of government expenditure to smooth the output and price levels in the economy. The simulation predicts higher output growth and lower inflation rate in the implementation of optimal policy rates.

Some caveats in this research include the assumptions constructed in the model of loss function, in that exogenous factors, such as the socio-political and business cycle factors, cannot be captured in the model, which also causes deviation between the actual and Nash equilibrium of the policies. There are also differences in the time frame between the monetary and fiscal authorities taking action; while BI decides on interest rates every month, GOI plans the national budget almost one year ahead; thus, its implementation is greatly affected by the plan.

The game theory model in this research used a simultaneous single game. Yet, there is still the need to use a repeated game model or a sequential (or dynamic) game model between monetary and fiscal authorities. Further research is suggested to use the dynamic game model, based on the Saulo et al. (2013) theoretical framework.

REFERENCES

Bartolomeo, G.D. & Giocchino, D.D. (2004). Fiscal-monetary policy coordination and debt management: A two stage dynamic analysis. Universita Degli Studi di Roma 'La Sapienza' Working Paper 74.

Beetsma, R.M.W.J. & Bovenberg, A.L. (1997). Central bank independence and public debt policy. *Journal of Economic Dynamics and Control*, 21(4–5), 873–894.

Biro Pusat Statistik (2016). Available at www.bps.go.id

Brouver, G.D., Ramayandi, A. & Turvey, D. (2006). Macroeconomic linkages and regional monetary cooperation: Steps ahead.*Asian Economic Policy Review*, I(2), 284–301.

Dixit, A. & Lambertini, L. (2001). Monetary-fiscal policy interactions and commitment versus discretion in a monetary union. *European Economic Review*, 45(4–6), 977–987. doi:10.1016/S0014-2921(01)00134-9.

Faure, P. (2003). Monetary and fiscal policy fames and effects of institutional differences between the European Union and the rest of the world. *Revue economique*, 937–959.

Floudas, C.A., Pardalos, P.M., Adjiman, C., Esposito, W.R., Gümüs, Z.H., Harding, S.T., Klepeis, J.L., Meyer, C.A. & Schweiger, C.A. (2013). *Handbook of test problems in local and global optimization*, vol. 33. New York: Springer Science & Business Media.

Goeltom, M.S. (2007). Monetary and fiscal policy synergy in Indonesia. In M.S. Goeltom (Ed.), *Essays in macroeconomic policy: The Indonesian experience* (pp. 106–123). Jakarta: PT Gramedia Pustaka Utama.

Grossman, P. (1988). "Growth in Government and Economic Growth: the Australian Experience." *Australian Economics Papers 27: 33–45.*

Javed, Z.H. & Sahinoz, A. (2005). Interaction of monetary and fiscal policy in case of Turkey. *Journal of Applied Sciences*, 5(2), 220–226.

Kendrick, D.A. & Amman, H.M. (2011). A Taylor rule for fiscal policy. Tjalling C. Koopmans Research Institute Discussion Paper 11–17.

Kuncoro, H. & Sebayang, K.A.D. (2013). The dynamic interaction between monetary and fiscal policies in Indonesia. *Romanian Journal of Fiscal Policy IV*, 4(1), 47–66.

Kydland, F.E. & Prescott, E.C. (1977). Rules rather than discretion: The inconsistency of optimal plans. *The Journal of Political Economy*, 85(3), 473–492.

Ministry of Finance (n.d.). Available at: www.kemenkeu.go.id. Accessed 29 June 2016, 05.10 WIB.

Mochtar, F. (2004). Fiscal and monetary policy interaction: Evidences and implication for inflation targeting in Indonesia. Working Paper Bank Indonesia (Penerbit Kanisius), 111–140.

Mohanty, D. & John, J. (2015). Determinants of inflation in India. *Journal of Asian Economics*, 36, 86–96.

Nasution, D. & Hendranata, A. (2014). *Estimasi output gap Indonesia (Estimation of Output Gap in Indonesia)*. Jakarta: Badan kebijakan Fiskal, Kementerian Keuangan Republik Indonesia.

Nordhaus, W. (1994). 'Policy games: Coordination and independence in monetary and fiscal policies.' *Brookings Papers on Economic Activity*, 5(2), 139–216.

Petit, M.L. (1989). Fiscal and monetary policy coordination: A differential game approach. *Journal of Applied Econometrics*, 4(2).

Ramayandi, A. (2007). Approximating monetary policy: Case study for the ASEAN-5. Working Papers in Economics and Development Studies.

Rogoff, K. (1985). The Optimal Degree of Commitment to an Intermediate Monetary Target. *Quarterly Journal of Economics*, 100, 1169–1190.

Santoso, W. (2012). Interaksi Kebijakan Moneter dan Fiskal di Indonesia. (Interaction of Monetary and Fiscal Policy in Indonesia) In Sri Adiningsih (Ed.), *Koordinasi dan Interaksi Kebijakan Fiskal-Moneter: Tantangan ke Depan (Coordination and Interaction of Fiscal—Monetary Policy: Future Challenges)* (pp. 225–262). Yogyakarta: Penerbit Kanisius.

Saulo, H., Dos, S.B., Rêgo, L.C. & Divin, J.A. (2013). Fiscal and monetary policy interactions: A game theory approach. *Annals of Operations Research*, 206(1), 341–366.

Simorangkir, I. & Goeltom, M.S. (2012). Peranan Koordinasi Kebijakan Moneter dan Fiskal terhadap Perekonomian Indonesia. (Roles of Monetary and Fiscal Policy Coordination in Indonesian Economy) In Sri Adiningsih (Ed.), *Koordinasi dan Interaksi Kebijakan Fiskal-Moneter: Tantangan ke Depan (Coordination and Interaction of Fiscal—Monetary Policy: Future Challenges)* (pp. 83–110). Yogyakarta: Penerbit Kanisius.

Taylor, J.B. (1994). Stabilization policy and long-term economic growth. Paper presented at the Center for Economic Policy Research Conference.

Widjaja, M. & Mardanugraha, E. (2009). The optimal instrument rule of Indonesia monetary policy. *International Journal of Economic Policy Studies*, 55–75.

Competition and Cooperation in Economics and Business – Gani et al. (Eds)
© 2018 Taylor & Francis Group, London, ISBN 978-1-138-62666-9

Contract stability and farmer characteristics of contract chilli farming

P.A. Muchtar & C. Nuryakin
Department of Economics, Faculty of Economics and Business, Universitas Indonesia, Depok, Indonesia

ABSTRACT: Contract farming has long been applied in Indonesia but has just revealed its ascending trend in recent times through private sector initiatives. The purchase guarantee and price certainty have induced farmers to join; however, deviation still occurs among them, shadowing the stability in its development process. In this study, an effort has been made to shed some light on the stability of a contract practised by farmers and a private agro-industry in the Jember chilli farming areas. Specifically, this study attempts to find the determinants of farmers' deviation from the contract on the basis of a survey involving a multiple-price list method to elicit individual characteristics drawn from the perspective of behavioural economics. Using probit regression controlled by demographic characteristics, this study found that present-bias preference and individual discount factor significantly affected the likelihood of deviation. The results also suggest that farmers' decision was significantly affected by several farming characteristics, including coordinator's decision, land size, expected price, and chilli farming experience.

1 INTRODUCTION

Price uncertainty has been a major issue in the agriculture sector and may affect both consumers and farmers (Security & OECD, 2011). As the price unpredictably fluctuates over time, contract farming (CF), which offers a fixed-price system, can be applied in order to address such issues (Eaton & Shepherd, 2001). In developing countries, including Indonesia, the implementation of CF has revealed its ascending trend in recent times through private sector initiatives (Prowse, 2012). The purchase guarantee and price certainty have induced farmers to join.

Numerous studies have been conducted to investigate the working of contractual partnership in the agriculture market, which mostly discuss two main issues: impact on farmers and participation motives (Wang et al., 2014). The former focuses exclusively on deliberating how farmers' welfare is affected by CF (Abdulai, 2016; Bellemare, 2012; Igweoscar, 2014; Miyata et al., 2009), whereas the latter focuses primarily on farmers' decision to participate in a contract (Masakure & Henson, 2005; Opoku-Mensah, 2012).

The studies emphasised the various aspects behind contract establishment, yet lacked explanations of any act that might have happened during a contract, such as a decision to deviate by one or both parties. In fact, a lot of deviations exist among CF practices, shadowing the stability in its developing process. If the market price is considerably low compared with the agreed price, farmers honour the agreement and comply. However, when the market price is rocketing, there is a big incentive for farmers to gain significant profit by deviating from the contract and selling their products in the open market. This act may or may not be rational behaviour, as they do not necessarily know the future price.

Variation in market price and future expected price affects farmers' decision to join a contract, whereas deviation may be induced by price shock. In the discipline of behavioural economics, a deviation of choice involving periodical time may be classified as a case of dynamic inconsistency. Farmers may have a self-control problem in committing to a contract because they naively perceive the future with bias. When they tend to underweight the future by ignoring price certainty from the contract, present-bias preference is probably involved and leads them to impatiently attempt short-term payoff. Such a decision might not be rational because it may deviate from the maximum expected utility.

This study aims to analyse the stability of contract practised by farmers and an agriculture firm. Specifically, it attempts to find the significant determinants of farmers' decision to deviate towards a contract on the basis of a survey of farmers contracting with a private agro-industry in the Jember chilli farming areas. The theoretical underpinning for this study was drawn from the perspective of behavioural economics; therefore, to investigate the deviation in CF, special attention was given to its association with individual characteristics, including time preference and individual discount factor (IDF). Farmer characteristics were also involved to support and enrich the analysis, controlled by the demographic characteristics.

In Indonesia, behavioural economics has not yet developed and lacks empirical local research. However, in many developed countries, this discipline has long been applied by academicians, practitioners, or even governments (Rabin, 2004). This research may not only explain the farmers' deviation phenomenon but also become a stepping stone for further behavioural economics studies.

The subsequent section deliberates the underpinning in the specific field of behavioural economics and reviews several prior studies. Section 3 presents the methodology used to answer the research questions, including the description of the elicitation design as well as the suggested econometric framework. Section 4 presents and discusses the results. Conclusions and suggestions are presented in the last section.

2 ANALYTICAL FRAMEWORK

2.1 *CF*

CF is a special way to connect the gap between farmers and the agriculture industry whose primary production is not centred on big players but rather handled by small-to-medium scale farmers through agreement with downstream firms (White, 1999). It takes place in the supply chain of agricultural production as part of marketing, providing purchase guarantee and price certainty to farmers. By contracting with an agribusiness firm, farmers can generate higher profits and reach wider end markets (Barrett et al., 2012).

The establishment of a contract between farmers and firms provides not only opportunities but also challenges. On the one hand, as the contract helps farmers manage price and production risks, they may benefit by focusing on improving the production quality instead of dealing with uncertainty. These lead to more efficient use of land and processing capacities. Some studies state that the major determinants explaining the reason why farmers join a contract are reduction on risk (Allen & Lueck, 1995) and transaction cost (Costales & Catelo, 2008; Murrell, 1983).

On the other hand, there is obviously a risk for farmers to suffer potential loss (or be unable to benefit significantly) when the market price dramatically increases. There is also another risk that firms fail to purchase as promised. Sambuo (2014) found that the farming experience, along with farm group and age, has significant influence on farming participation. This may explain how the understanding about the risks of CF, growing with experience, leads to their willingness to join a contract.

The system of CF should be perceived as a mutual trust between farmers and agribusiness. To achieve its stability, CF requires long-term obedience to the agreement from both parties. Manipulative arrangements and exploitative agreements by firms are likely to step down the contract duration to only a limited period and may cause failure in investments. Similarly, farmers need to consider that committing to the contract is likely to be to their long-term benefit. Any dishonest act can lead to a failure of contract, and contract reconstruction after deviation is not likely to happen (Dhillon et al., 2006).

CF in developed countries is considerably different from that in developing countries. Farmers achieve more success in improving welfare in developing countries than in developed countries, especially when much attention is paid by third parties such as the government and non-governmental organisations (Wang et al., 2014). In developing countries, farmers are able to obtain more advanced technology, reach scale economies, and improve their welfare by gaining more return (Miyata et al., 2009; Tripathi et al., 2005).

2.2 *Dynamic inconsistency*

A standard model in economics assumes that people maximise their own utility rationally, considering the payoff level of their choice, in a particular probability and consistency over a time period (Bailey, 2010). This maximisation is formulised by Rabin (2004) and modified by DellaVigna (2007) in the form of an inter-temporal utility function:

$$\max_{x_i^t \in X_i} \sum_{t=0}^{\infty} \delta^t \sum_{s_t \in s_t} p(s_t) U(x_i^t \mid S_t) \tag{1}$$

Individual i at time $t = 0$ maximises expected utility subject to probability distribution $p(s)$ of the states of the world $s \in S$. The utility function $U(x|s)$ is determined from the payoff x_i^t that individual i receives. The discount factor δ^t underweights the future utility and is always consistent over time.

Deviation from rationality refers to inconsistency in preference, belief, or choice, and each of these has its own dimension (DellaVigna, 2007). Despite these many kinds of inconsistencies, this study only focuses on the issue of time preference, so-called dynamic inconsistency, because CF involves inter-temporal decisions. Time preference can be disturbed by a self-control problem, in which the value of the discount factor δ^t varies among time periods.

The accentuation on inter-temporal decision is also pointed out in the game theory through the concept of infinitely repeated game, a multistage game model where the same set of agents repeatedly plays the same game, so-called stage game, over a long time horizon (Osborne, 2000). This concept resembles time preference as it also emphasises the salience of the discount rate to check whether the player will stick to the same decision in the repeated-stagegame, such as harvesting in CF. Players (farmers and firms) find subgame-perfect equilibrium in the form of sustainable contract as long as there is no profitable incentive to deviate. Such cooperative strategy 'to stay on the contract' can sustainably continue if players are sufficiently patient (do not discount future payoffs too extreme) (Huang, 2010; Osborne, 2000; Tadelis, 2013). Thus, the following equation should be satisfied:

$$v_i^c + \frac{\delta v_i^c}{1-\delta} \geq v_i^d + \frac{\delta v_i^*}{1-\delta} \tag{2}$$

where the four components of Equation 2 represent current payoff from cooperate, future payoff if cooperate, current payoff from best defect, and future payoff if defect, respectively.

After a short algebra simplification, the inequity above may be presented as:

$$\delta \geq \frac{v_i^d - v_i^c}{v_i^d - v_i^*} \tag{3}$$

2.3 *Present-bias preference*

Present-bias preference is a tendency towards inconsistency in a time preference that overweights immediate payoff and underweights long-term payoff. In daily life, it is usually called procrastination or naivety. Present-bias preference is modelled with the value of the discount following a quasi-hyperbolic pattern (Laibson, 1997; O'Donoghue & Rabin, 1999).

$$U_t = u_t + \beta\delta u_{t+1} + \beta\delta^2 u_{t+2} + \beta\delta^3 u_{t+3} + \cdots \tag{4}$$

This model is slightly different from the inter-temporal utility function since present bias involves $\beta\delta^t$ as a discount factor at time $t > 0$. Coefficient β is a value between zero and one ($0 < \beta < 1$) which indicates that the discount factor between the present and the next period is higher than the discount between two periods in the future.

Unless $\beta = 1$, individuals act irrationally; however, amidst the people who suffer present bias, some are not fully naive. Thus, O'Donoghue & Rabin (2001) revised their model and allowed a new parameter $\hat{\beta} \geq \beta$, whereas the full naive person possesses $\hat{\beta} \geq \beta$.

To explain the application of the present bias (β, δ) model, let us assume that an action has two consequences: immediate payoff b_1 at time $t = 1$ and postponed payoff b_2 at time $t = 2$. There are some actions that demand particular effort at present and deliver enjoyment later—so-called immediate cost (Rabin, 1999), with payoff features $b_1 < 0$ and $b_2 > 0$. It applies to exercising, searching for a job, and doing homework. Conversely, an immediate reward, such as watching a movie or hanging out, has the features $b_1 > 0$ and $b_2 < 0$.

Then, the model can be written as $b_1 + \beta\delta b_2 \geq 0$. It means that a naive person will tend to avoid 'immediate cost' action ($b_2 > 0$) and be more likely to choose 'immediate reward' action ($b_2 < 0$). Based on the issue explained in Section 1, it indicates that farmers may have present bias as they choose to gain profit by deviating from the contract at the present and ignore price consistency in the future.

In one of the earliest studies about dynamic inconsistency, Strotz (1956) proved that if the optimal plan of future behaviour chosen as a given time is inconsistency then the conflict might not be recognised wisely by the individual, as present bias exists. If, however, the conflict is recognised, the inconsistency can be solved by the strategy of pre-commitment or consistent planning. In a study of credit card holders, Meier and Sprenger (2007, 2010) found that individuals with present-bias preferences have significantly higher amounts of credit card debt, even after controlling for disposable income, credit constraints and other socio-demographic characteristics. Moreover, Laibson (2015) implied that present-biased agents often do not make a commitment because they perceive benefits of commitment as sufficiently low. Their benefit of committing does not surpass the (modest) direct price of commitment and the indirect losses as their flexibility in farming reduces.

3 METHODOLOGY

In order to achieve the aims of the study, a quantitative field research was conducted on the basis of a primary survey of farmers contracting with a private agro-industry. The study population was chilli farmers in the Jember agricultural area who participated in CF with PT Heinz ABC; the unit of analysis was individual farmers. Jember was chosen because it had been reported by the Ministry of Agriculture (2015) as one of the most productive areas in producing chilli. Meanwhile, PT Heinz ABC was chosen as it has the largest share in the chilli sauce market, which is approximately 52%. CF between PT Heinz ABC and the farmers in Jember is a pioneer of vertical integration in the agricultural industry and can be a suitable case to be studied.

3.1 *Data*

Since we needed to assume that the data were normally distributed, we took 60 chilli farmers from the total of approximately 450 members of Koperasi Hortikultura Lestari's (KHL) network.[1] Although the sample size seems to be limited, it is actually more than 10% of the population. We did not take other samples from other practices of contract chilli farming because it had not been widely implemented in every region in Indonesia. Even if there were other CF practices, they may use different rules and systems that could not be compared.

1. Koperasi Hortikultura Lestari (KHL) was appointed by PT ABC Heinz as the representative to organise CF in Jember. KHL assigned several coordinators to recruit farmers who were willing to join CF. The coordinator's task was to lead farmers in producing and supplying the agreed amount of chilli to PT Heinz ABC. In return, the firm paid the purchase with a determined price. From 2010 to 2015, the proposed contract set the price at IDR 8,000 (approximately USD 0.75). The contract price would not change unless the market price exceeded IDR 15,000 (approximately USD 1.1). PT Heinz ABC provided particular incentives of additional price to compensate the potential loss the farmers had to suffer (Appendix 1).

Personal data of farmers were not available in the administration system; therefore, we could not conduct random sampling or stratified sampling. To alter this issue, we decided to use a snowball sampling technique in which a KHL officer pointed out several coordinators who led us to their subordinate farmers. These samples included farmers who had been expelled from the contract because of dishonest behaviour (deviation from contract). Using direct interviews, we collected data from samples based on the designed questionnaire.

All 60 farmers participated in the survey and were provided a gift package worth IDR 20,000 (approximately USD 1.5) for the time they devoted to this study. The surveyors visited the farmers' place and asked their willingness to participate. They were seated and interviewed to fill the data in the questionnaire. They were not given any chance to look at another person's answers and were also instructed not to do so. The survey took about 30 minutes for each participant.

3.2 *Elicitation design*

We identified the farmers' decision to deviate from CF based on their confessions to indirect questions in the middle of the interview. We also checked their status with their coordinator or KHL officer to determine whether they were still on the contract. Other personal characteristics data were taken normally from the interview, except for individual time preference data that were elicited using a hypothetical experiment, called a multiple-price list method (Coller & Williams, 1999; Frederick et al., 2002; Harrison et al., 2000). Since it was hypothetical, the farmers would not benefit or suffer from the real payoff over their choice.[2]

In the elicitation experiment, we analysed subjects' decision on a questionnaire containing two multiple-price lists in two distinguished periods (see Appendix 1).[3] The subjects were asked to make a series of choice between a smaller payoff (IDRX) in period t and a larger payoff (IDR $Y >$ IDRX) in period τ. The value of Y was constant while the value of X varied in two time frames. Y amounted to IDR 800,000 (approximately USD 70), which represented farmers' revenue on each chilli picking time (one harvesting period has several picking times) for a common-sized farming field (approximately 0.25 Ha).

In time frame 1, t indicates the present time ($t = 0$) and τ is a planting period[4] ($\tau = 1$). In time frame 2, t is the next six planting periods ($t = 6$) and τ is the next seven planting periods ($\tau = 7$). The length difference, d, between t and τ is one planting period for both time frames. IDF was measured depending on respondents' choice at point X^*, where they decided to move from a lower incentive in period t to a higher incentive in period τ. IDF was calculated using the formula $X^* \approx IDF^d \times Y$, where d represents length difference. Since the difference is one planting period, $IDF \approx (X^*/Y)^{1/1}$. Using this formula to calculate two time-frame conditions results in two discount measures $IDF_{\tau,\tau}$: $IDF_{0,1}$ and $IDF_{6,7}$ In the main analysis, we used the average of both calculations as the IDF variable.

In order to identify the dynamic inconsistency, we compared discount measures yielded from two time frames. Dynamic inconsistency was distinguished into two kinds, present bias and future bias. An individual is classified as present bias if impatience drives him/her to take

2. We readily acknowledge the weaknesses of using the hypothetical elicitation method that may cause bias, although another study proved insignificant difference between the hypothetical method and the actual method in experiment choice (see Carlsson & Martinsson, 2001). In order to reveal the truthful answer, the surveyors used familiar terms related to farming activities in eliciting time preference; for instance, the time sign used in the questionnaire was planting period instead of month. The use of familiar language was inspired by List and Gallet (2001) who found that true preference is more likely to be stated when performing a familiar hypothetical task rather than an unfamiliar one.

3. Since this experiment was more or less representing the contract framework, the enumerators avoided any conversation that may lead farmers to associate this experiment with the contract.

4. One planting period equals approximately four months. In the survey, we mentioned the time indicator in months to make the instructions easier to understand for the farmers.

Table 1. Variable description.

Notation	Variable	Description
dec	Decision towards contract	Binary variable of farmer experience on obeying the contract in the past three years. 1 is to comply and 0 is to defect.
idf	Individual discount factor	The averages of $IDF_{0,1}$ and $IDF_{6,7}$ that were taken from the experiment, ranging continuously between 0 and 1.
Pbias	Present bias	Based on IDF calculation, this is a dummy variable of time inconsistency. 1 for someone exhibiting presentbias preference, otherwise 0.
coor_d	Coordinator decision towards contract	Dummy variable of coordinator's obedience on the contract. 1 is to comply and 0 is to defect.
dem_ch	Demographic characteristics	Demographic characteristics include age (*age*, in year), education level (*edu*, in year), monthly expenditure (*m_exp*, in Indonesia rupiah), and dummy of expenditure shock (*exp_s*, 1 = exist).
far_ch	Farming characteristics	Farming characteristics include land size (*lsize*, in hectare), farming experience (*famexp*, in year), and subjective expected price (*se_price*, in Indonesia rupiah).

Source: Authors' compilation.

a smaller reward sooner in the present time ($t = 0$). Formally, present bias is $IDF_{0,1} < IDF_{6,7}$ and future bias is $IDF_{0,1} < IDF_{6,7}$ (see Appendix 2).

3.3 *Regression model*

To answer the main research question, we conducted a logit regression analysis using the data from the survey. The following model was constructed to find the likelihood of farmers to defect from the contract. The estimated determinants of behaviour were IDF and time preference, yet we still used demographic characteristics as control variables.

$$P(dec) = \frac{1}{1+e^{-Z'}} \tag{5}$$

where

$$Z = \beta_0 + \beta_1 idf + \beta_2 pbias + \beta_3 coor\_d + \beta_4 \mathrm{dem\_ch} + \beta_5 \mathrm{fam\_ch} + \mathrm{e} \tag{6}$$

The specification of the variables is explained in Table 1.

4 RESULTS AND DISCUSSION

We excluded five farmers from the analysis since their participation could not be processed due to some missing data and improper answers to the questionnaire. Some of them also lacked communication with the surveyors because of language barrier. From the remaining 55 observations, 14 farmers were identified as deviating from the contract, so-called deviators, and punished by being prohibited from joining any further contract. The remaining 31 farmers decided to comply throughout the contract period, so-called compliers, and stayed on the agreement no matter how extreme the market price fluctuation. They were not given

Table 2. Descriptive statistics.

Variable	Mean (standard deviation)		Minimum		Maximum	
	Compiler	Deviator	Compiler	Deviator	Compiler	Deviator
Time preference (present bias = 1)	0.365(0.487)	0.5714(0.513)	0	0	1	1
IDF	0.630(0.184)	0.618(0.173)	0.375	0.375	0.9375	0.9375
Short-term IDF	0.591(0.212)	0.566(0.215)	0.375	0.375	0.9375	0.9375
Long-term IDF	0.669(0.222)	0.669(0.225)	0.375	0.375	0.9375	0.9375
Coordinator's decision towards contract (deviate = 1)	0.146(0.357)	0.857(0.363)	0	0	1	1
Monthly expenditure (IDR)	1,362,927 (768,736.8)	1,725,714 (804,044.7)	400,000	900,000	3,000,000	3,000,000
Expenditure shock (exist = 1)	0.365(0.487)	0.285(0.468)	0	0	1	1
Years spent on education	8.634(3.533)	10.428(2.208)	0	6	16	12
Age (years)	42.75(3.53)	40.71(7.48)	20	29	65	52
Subjective expected chilli price (IDR)	7,560.976 (3,581.105)	10,642.86 (4,651.007)	3,000	2,500	15,000	17,500
Years experienced in planting chilli	8.87(10.55)	17.78(10.32)	1	1	40	33
Total land size	0.970(0.909)	0.875(0.859)	0.25	0.25	4	3

Note: Complier and deviator indicate behaviour towards the contract.
Source: Authors' compilation.

special rewards but were supervised in order to maintain product quality. Further descriptive statistics are shown in Table 2.

4.1 *Regression results*

The variables were checked using a multicollinearity test, resulting in a low correlation between all the independent variables (Appendix 3). Any misspecification was treated by obtaining an estimate of robust variance, although some parameters reported larger standard error than in the previous specification. The robust treatment excluded the agenda to test the possibility of the heteroscedasticity problem. The fitness of the model was tested by predicting the outcome of the dependent variable based on regression results (Appendix 4). The model correctly predicted 92.73% of the values and the rest were misclassified.

As depicted in Table 3, the general test for all the models [ordinary least square (OLS), logit, and probit] shows that the dependent variable was significantly affected by the independent variables, as their *p*-value scores were less than 0.01. However, we decided to focus on the probit regression results since the two others have some weaknesses. OLS was excluded since we could not interpret the magnitude of the coefficient nor were we able to find the partial effect, and logit had sufficiently higher standard error than probit.

In the probit column, some variables (i.e. time preference, IDF, and experience) significantly affected the farmers' decision towards the contract at a significance level of 0.01. Another association was also seen at significance levels of 0.05 and 0.10 on the coordinator's behaviour, monthly expenditure, unexpected expenditure, expected chilli price, and land owned. Education and age were not significant in all the models, yet they were important for controlling the variation of the dependent variable. Almost all the independent variables had a positive correlation with the dependent variable except unexpected expenditure, age, and land use.

Since the coefficient value in the regression results cannot be interpreted in binary outcome regression, marginal effect estimation is needed. Table 4 informs the magnitude of

Table 3. Regression results.

Behaviour towards contract (deviate = 1)	OLS coefficient	Logit coefficient	Probit coefficient
Time preference (present bias = 1)	0.117(0.0964)	2.273(1.430)	1.3008*(0.7297)
IDF	0.295(0.3065)	16.137***(5.821)	9.252***(3.094)
Coordinator's decision towards contract (deviate = 1)	0.502***(0.1498)	5.939***(1.692)	3.441***(0.9024)
Monthly expenditure	8.95e–08(8.33e–08)	2.10e–06(1.79e–06)	1.18e–06*(6.87e–07)
Expenditure shock (exist = 1)	–0.126(0.1032)	–5.050***(1.715)	–2.943***(0.913)
Years spent on education	–0.00108(0.0186)	0.195(0.224)	0.1175(0.103)
Age	–0.0494(0.0064)	–0.1543(0.1193)	–0.0856(.0598)
Subjective expected chilli price	0.0000157(0.0000137)	0.000278(0.0002114)	0.000158*0.0000861
Years experienced in planting chilli	0.0044(0.00558)	0.1206(0.0838)	0.0688*(0.0385)
Total land used	–0.0841(0.0757)	–2.466**(1.338)	–1.404**(0.5964)
Constant	–0.1127(0.3554)	–14.817(4.996)	–8.654***(2.729)
Probability > F / Probability > χ^2	0.0000***	0.0040***	0.0009***
R^2 / Pseudo R^2	0.5247	0.6366	0.6404

Note: Standard errors are in parentheses. *$p < 0.05$, **$p < 0.01$, ***$p < 0.001$.
Source: Authors' compilation.

Table 4. Marginal effect results.

Decision towards contract (deviate = 1)	Probitmarginal effect at the means	Probit average marginal effect
Time inconsistency (present bias = 1)	0.0399(0.0579)	0.1461**(0.072)
IDF	0.2838(0.3360)	1.039***(0.2872)
Coordinator's behaviour towards contract (deviate = 1)	0.1055(0.130)	0.3865***(0.0659848)
Monthly expenditure	3.63e–08(4.21e–08)	1.33e–07*(7.55e–08)
Expenditure shock (exist = 1)	0.0036(0.0057)	–0.3306***(0.0912)
Years spent on education	–0.0903(0.1153)	0.0132(0.0127)
Age	–0.0026(0.0027)	–0.0096(0.0060)
Subjective expected chilli price	4.85e–06(6.18e–06)	0.0000178*(9.33e–06)
Years experienced on planting chilli	0.0021(0.0028)	0.0077**(0.0036)
Total land owned	–0.043(0.049)	–0.1577***(0.0607)

Note: Standard errors are in parentheses. *$p < 0.05$, **$p < 0.01$, ***$p < 0.001$.
Source: Authors' compilation.

marginal effect from the regression model, expressing additional probability of the dependent variable to occur successfully (farmers deviating from the contract). The table only lists the probit model results; however, both logit and probit actually show the same magnitude of marginal effect.

4.2 *Individual characteristics and farmers' decision towards contract*

The coefficient of time inconsistency shows a positive correlation indicating that the farmers with present-bias preference are more likely to deviate from the contract. On average, those exhibiting present-bias preference had 14.61% higher probability to disobey. Such individuals possess an ascending trend of IDF, or in other words, they discount short-term payoff sufficiently lower than long-term payoff. The impatience, therefore, leads them to ignore future price stability and to deviate from the contract in the present time instead.

Since the concept of time preference has never been studied to deliberate the working of agriculture contracts, a study of the financial industry can be a good comparison. Meier and Sprenger (2007) found many individuals exhibiting present bias among credit card holders who have significantly higher default rate than people with consistent time preference. The extension of the study also supports previous results, as Meier and Sprenger (2010) found that persons with inconsistent time preference are more likely to borrow and their debt level is significantly larger than dynamically consistent individuals. This finding is controlled by demographic characteristics, credit constraints, and interest rate.

IDF was found to have a positive association with farmers' decision to deviate from the contract. This seems to be unusual since we predicted that the larger the IDF, the more patient the individual to stay on the contract. Moreover, such a finding is actually quite surprising as previous research in the financial service area failed to prove the significance of IDF (Harrison et al., 2000; Meier and Sprenger, 2007). However, this could be rationalised by some arguments. First, farmers may perceive that the future value of deviating from the contract is sufficiently higher rather than accepting the contract agreement; so, if such future is discounted slightly, it may induce farmers to deviate. This argument is imperfect because it indirectly claims that farmers exhibiting small IDF, indicating impatience, tend to comply with the contract.

Second, IDF elicitation in this study might not exclusively reflect the behaviour of farmers in a contractual partnership. Some studies have used discount rate as predictive variables of particular behaviour related to time spans, including smoking (e.g. Harrison et al., 2010), exercise (e.g. Malmendier & DellaVigna, 2006), alcoholism (e.g. Vuchinich & Simpson, 1999), and credit card debt (Meier & Sprenger, 2010); however, Chabris et al. (2008) found that discount rates are weakly correlated with such field behaviour or even have no correlation at all.

Third, famers may not behave rationally as they do not always take decisions based on their preference. This study emphasises solely on the dimension of time preference, excluding the other types of preference, namely, risk preference and social preference. In addition, no attention was paid to another aspect of irrationality involving inconsistency of belief and choice (DellaVigna, 2007). Their decision of whether to comply with the contract might be determined by other important factors.

4.3 *Farming characteristics and farmers' decision towards contract*

Despite the ambiguous result of discount factor, more attention should be given to other important findings as farmers' decision to deviate from the contract is not based only on their consideration alone, but somehow considerably depends on their coordinator. On average, the farmers were 38.65% more likely to deviate if their coordinator also deviated from the contract. Such a percentage has the largest effect compared with other variables. As the coordinator represents the social capital of the contract establishment, this finding seems to be reciprocal with the explanation of the role of trust and social capital as studied by Polman and Slangen (2008). They found that besides farm and farmer characteristics, motivational issues, including the perception of institutional design, the use of extension services, and trust in the government, are important for agriculture contract development, although not all types of contracts are significantly influenced. Here, the coordinator has a salient role to maintain such motivational issues.

Land size coefficient, showing a negative sign, indicates that the wider the land used to crop chilli, the less likely the farmers are to deviate from the contract. Farmers with sufficiently large size of land are able to mitigate the risk of price volatility by land use diversification; that is, either by planting another kind of commodity in the same land or by not using the whole land for CF and sparing the rest for the open market instead. This finding resembles that of some contract participation studies. Farmers with larger farms are more likely to join CF (Barrett et al., 2012; Wang et al., 2011). It is commonly believed that they are more likely to be offered CF as firms want to reduce transaction costs.

The estimate result also shows a significant positive association between expected price and the decision towards the contract. On average, when the farmers perceived future expected price a thousand rupiahs higher, their probability to deviate from the contract increased by 1.78%. Nevertheless, we cannot interpret this for every level of expected price since the partial effect might be different.

Another significant variable is the number of years of farming. Farmers who have experience in chilli farming for longer periods are more likely to deviate. The probability of such a decision to occur increased by 0.77% as the experience increased by one year. Experience trains farmers to better understand the risk of price volatility and production; therefore, such an advantage leads them to leave the contract and to face the risk in the open market. In participation studies, the association of experience is found ambiguously in positive, negative, and quadratic relationships (Wang et al., 2014).

4.4 *Demographic characteristics and farmers' decision towards contract*

Monthly expenditure appeared to positively affect farmers' decision to deviate. Those who have larger expenditure are more likely to deviate. The more the expenditure, the more the income needed to overcome it. They may perceive that selling to the open market is more lucrative than achieving a constant return from the contract.

Expenditure shock had a negative sign on its coefficient indicating that farmers reporting to have spent their income on unexpected expenditure are less likely to deviate. On average, such individuals had 33.06% lower probability to deviate compared with those who had spent their income normally. Farmers experiencing shock probably do not want to bear more risk by dealing with the volatile price in the open market. Staying in the contract gives them a safe place if another shock happens unpredictably. This finding is not consistent with those of some studies that found a positive association between shock and change in inter-temporal decisions (e.g. Bíró, 2013; Smith & Love, 2007).

The results show that age did not significantly affect farmers' decision towards behaviour and neither did education level, measured by years spent in educational institutions. Such results were also found in several contract-participation studies, such as Bao et al. (2013), Bellemare (2012), Simmons et al. (2005), and Wang et al. (2011).

5 CONCLUSIONS

Despite the purchase guarantee and price stability provided by the firm in CF, dishonest behaviour among farmers who violate the agreement still occurs. Such decision does not only disturb the supply chain in the agricultural industry but also undermines potential welfare gained for farmers. This issue shades the stability of CF, which has only recently shown its emerging trend. Understanding farmer characteristics is very important for the development and mass implementation of CF; therefore, this study attempts to find significant determinants of farmers' decision towards the contract.

This study found that individual characteristics played a significant role in affecting farmers' deviation from the contract. Farmers with present-bias preference are more likely to defect from the contract and so are those who exhibit higher IDF. The former indicates that short-term impatience leads farmers to decide irrationally, whereas the latter has several explanations, including the perception of future payoff, lack of preference understanding, and distorted elicitation method.

Moreover, the results suggest that farming characteristics are also important and significant predictors in analysing farmers' deviation. Coordinators play a salient role in building social capital to achieve contract stability as their choice could skew farmers towards one decision. Other significant variables of farming characteristics were land size, expected price, and experience.

A broader conclusion from the results is that, in actual condition, showing that farmers have more narrow-sized land and the government induces more new farmers may reflect positively on the further development of CF. Concerns over quality when involving new farmers should not become an issue as the process of quality control is managed by KLH. Thus, cooperatives play an important role as an intermediary agent in this contract.

REFERENCES

Abdulai, Y. (2016). Effects of contract farming on small-holder soybean farmers' income in the eastern corridor of the northern region, Ghana. *Journal of Economics and Sustainable Development*, 7(2), 103–113.

Allen, D.W. & Lueck, D. (1995). Risk preferences and the economics of contracts. *American Economic Review*, 85(2), 447–451. doi:10.2307/2117964.

Bailey, E.M. (2010). Behavioral economics: Implications for antitrust practitioners. *The Antitrust Source*, 74, 1–7. Retrieved from: http://goo.gl/aWBNk.

Bao, Q., Tang, L., Zhang, Z.X. & Wang, S. (2013). Impacts of border carbon adjustments on China's sectoral emissions: Simulations with a dynamic computable general equilibrium model. *China Economic Review*, 24(1), 77–94. doi:10.1016/j.chieco.2012.11.002.

Barrett, C.B., Bachke, M.E., Bellemare, M.F., Michelson, H.C., Narayanan, S. & Walker, T.F. (2012). Smallholder participation in contract farming: Comparative evidence from five countries. *World Development*, 40(4), 715–730. doi:10.1016/j.worlddev.2011.09.006.

Bellemare, M.F. (2012). As you sow, so shall you reap: The welfare impacts of contract farming. *World Development*, 40(7), 1418–1434. doi:10.1016/j.worlddev.2011.12.008.

Bíró, A. (2013). Subjective mortality hazard shocks and the adjustment of consumption expenditures. *Journal of Population Economics*, 26(4), 1379–1408. doi:10.1007/s00148-012-0461-5.

Carlsson, F. & Martinsson, P. (2001). Do hypothetical and actual marginal willingness to pay differ in choice experi ments? *Journal of Environmental Economics and Management*, 41(2), 179–192. doi:10.1006/jeem.2000.1138.

Chabris, C.F., Laibson, D., Morris, C.L., Schuldt, J.P. & Taubinsky, D. (2008). Individual laboratory-measured discount rates predict field behavior. *Journal of Risk and Uncertainty*, 37(2–3), 237–269. doi:10.1007/s11166-008-9053-x.

Coller, M. & Williams, M.B. (1999). Eliciting individual discount rates. *Experimental Economics*, 2(2), 107–127. doi:10.1007/BF01673482.

Costales, A. & Catelo, M.A.O. (2008). Contract farming as an institution for integrating rural small-holders in markets for livestock products in developing countries: (I) Framework and applications. Pro-Poor Livestock Policy Initiative.

DellaVigna, S. (2007). Psychology and economics: Evidence from the field. NBER Working Paper 13420.

Dhillon, S.S., Singh, N. & Dhillon, S.S. (2006). Contract farming in Punjab: An analysis of problems, challenges and opportunities. *Pakistan Economic and Social Review*, 44(1), 19–38.

Eaton, C., & Shepherd, A.W. (2001), Contract Farming: Partnerships for Growth. A Guide. FAO Agricultural Services Bulletin 145.

Frederick, S., Loewenstein, G. & O'Donoghue, T. (2002). Time discounting and preference: A critical review. *Journal of Economic Literature*, 40, 351–401. doi:10.1126/science.151.3712.867-a.

Harrison, G.W., Lau, M.I. & Rutström, E.E. (2010). Individual discount rates and smoking: Evidence from a field experiment in Denmark. *Journal of Health Economics*, 29(5), 708–717. doi:10.1016/j.jhealeco.2010.06.006.

Harrison, G.W., Lau, M.I. & Williams, M.B. (2000). Estimating individual discount rates in Denmark: A field experiment. *American Economic Review*, 92(5), 1606–1617. doi:10.1257/000282802762024674.

Huang, Q. (Ed.)(2010). *Game theory*. Rijeka: Sciyo. http://doi.org/10.4135/9781412984317.

Igweoscar, O. (2014). Effect of contract farming on productivity and welfare of cassava-based farmers in South Eastern Nigeria. *European Journal of Business and Management* (Online), 6(7), 2222–2839.

Laibson, D. (1997). Golden eggs and hyperbolic discounting. *Quarterly Journal of Economics*, 112(2), 443–477. doi:10.1162/003355397555253.

List, J.A. & Gallet, C.A. (2001). What experimental protocol influence disparities between actual and hypothetical stated values? *Environmental and Resource Economics*, 20(3), 241–254. doi:10.1023/A:1012791822804.

Malmendier, U. & DellaVigna, S. (2006). Not to go to the gym paying. *The American Economic Review*, 96(3), 694–719. doi:10.1257/aer.96.3.694.

Masakure, O. & Henson, S. (2005). Why do small-scale producers choose to produce under contract? Lessons from nontraditional vegetable exports from Zimbabwe. *World Development*, 33(10), 1721–1733. doi:10.1016/j.worlddev.2005.04.016,
Meier, S. & Sprenger, C. (2007). Impatience and credit behavior: Evidence from a field experiment. Federal Reserve Bank of Boston Working Papers 07–3 (pp. 1–42). Retrieved from: http://hdl.handle.net/10419/55628.
Meier, S. & Sprenger, C. (2010). Present-biased preferences and credit card borrowing. *American Economic Journal: Applied Economics*, 2(1), 193–210. doi:10.1257/app.2.1.193.
Ministry of Agriculture (2015). *Agricultural Statistics of Food Crops*. Jakarta. Available at: http://epublikasi.setjen.pertanian.go.id/arsip-perstatistikan/160-statistik/statistik-pertanian/383-statistik-pertanian-2015.
Miyata, S., Minot, N. & Hu, D. (2009). Impact of contract farming on income: Linking small farmers, packers, and supermarkets in China. *World Development*, 37(11), 1781–1790. doi:10.1016/j.worlddev.2008.08.025.
Murrell, P. (1983). The economics of sharing: A transaction cost analysis of contractual choice in farming. *The Bell Journal of Economics*, 14(1), 283–293.
O'Donoghue, T. & Rabin, M. (1999). Doing it now or later. *American Economic Review*, 89(1), 103–124. doi:10.2307/116981.
O'Donoghue, T. & Rabin, M. (2001). Choice and procrastination. *The Quarterly Journal of Economics*, 116(1): 121–160. doi:10.1162/003355301556365.
Opoku-Mensah, S. (2012). Logistic analysis of factors motivating smallholder farmers to engage in contract farming arrangements with processing firms in Ghana. *Journal of Biology, Agriculture and Healthcare*, 2(11), 58–74. http://iiste.org/Journals/index.php/JBAH/article/view/3663.
Osborne, M.J. (2000). *An introduction to game theory*. Oxford: Oxford University Press.
Polman, N.B.P. & Slangen, L.H.G. (2008). Institutional design of agri-environmental contracts in the European Union: The role of trust and social capital. *Wageningen Journal of Life Sciences*, 55(4), 413–430. doi:10.1016/S1573-5214(08)80029-2.
Prowse, M. (2012). Contract farming in developing countries—A review. Available at: http://value-chains.org/dyn/bds/docs/830/12-VA-A-Savoir_ContractFarmingReview.pdf.
Rabin, M. (2004). Advance in behavioral economics. Princeton: Princeton University Press.
Sambuo, D. (2014). Tobacco contract farming participation and income in Urambo; Heckma's selection model. *Journal of Economics and Sustainable Development*, 5(28), 230–238.
Security, G.F. & OECD. (2011). Price volatility in food and agricultural markets: Policy responses. Retrieved from: http://www.oecd.org/agriculture/pricevolatilityinfoodandagriculturalmarketspolicyresponses.htm.
Simmons, P., Winters, P. & Patrick, I. (2005). An analysis of contract farming in East Java, Bali, and Lombok, Indonesia. *Agricultural Economics*, 33(Suppl. 3), 513–525. doi:10.1111/j.1574-0864.2005.00096.x.
Smith, P. and Love, D. (2007), *Does Health Affect Portfolio Choice?* Finance and Economics Discussion Series 45.
Strotz, R.H. (1956). Myopia and inconsistency in dynamic utility maximization. *The Review of Economic Studies*, 23(3), 165–180.
Tadelis, S. (2013). *Game theory: An introduction*. Princeton: Princeton University Press.
Tripathi, R.S., Singh, R. & Singh, S. (2005). Contract farming in potato production : An alternative for managing risk and uncertainty. *Agricultural Economics Research*, 18, 47–60.
Vuchinich, R.E. & Simpson, C.A. (1999). Delayed reward discounting in alcohol abuse. In *The economic analysis of substance use and abuse: An integration of econometrics and behavioral economic research* (pp. 103–132). Chicago: University of Chicago Press.
Wang, H.H., Wang, Y. & Delgado, M.S. (2014). The transition to modern agriculture: Contract farming in developing economies. *American Journal of Agricultural Economics*, 96(5), 1257–1271. doi:10.1093/ajae/aau036.
Wang, H.H., Zhang, Y. & Laping, W. (2011). Is contract farming a risk management instrument for Chinese farmers? Evidence from a survey of vegetable farmers in Shandong. *China Agricultural Economic Review*, 3(4), 489–505. http://doi.org/10.1108/17561371111192347.
White, B. (1999). Core and plasma: Contract farming and implementation of power in West Java uplands. In *Transforming Indonesian uplands: Marginality, power, and production* (pp. 293–306). Amsterdam: Harwood Academic Publisher.

Appendix 1: Pricing mechanism

If the market price exceeds IDR 15,000, PT Heinz ABC provides a special incentive.

$$Incentive = 0.7 \times (market\ price - grower\ price)$$

Since the incentive was equally divided between Koperasi Hortikultura Lestari (KHL) and the farmers, the contract price change becomes:

$$contractprice = (incentive : 2) + default\ contract\ price$$

For instance, the price of chilli per kilogramme in the market is IDR 19,500, whereas the grower price is IDR 13,500. PT Heinz ABC, therefore, provides a bonus of IDR 4,200. This changes the contract price agreed by farmers.

$$contract\ price = 0.7 * (19{,}500 - 13{,}500) + 8{,}000$$
$$contract\ price = 10{,}100$$

Appendix 2: Time preference elicitation method (multiple price list)

Appendix Tables A1 and A2 list two types of choices. For each decision, you are expected to choose between a smaller bonus in a shorter period (choice A) or a larger bonus in a longer period (choice B). Table A1 provides the decision choices in the time frames of 'today' and 'in 4 months', whereas Table A2 uses time frames of 'in 2 years' and 'in 2 years, 4 months'.

Table A1.

Time frame 1	Choice A (today)	Choice B (in 4 months)
Decision 1	Granted IDR 750,000	Granted IDR 800,000
Decision 2	Granted IDR 700,000	Granted IDR 800,000
Decision 3	Granted IDR 650,000	Granted IDR 800,000
Decision 4	Granted IDR 600,000	Granted IDR 800,000
Decision 5	Granted IDR 500,000	Granted IDR 800,000
Decision 6	Granted IDR 400,000	Granted IDR 800,000

The farmer is presented with the following questions: Based on your expectation, in the next two years, will you still work in agriculture sector? What could change at that time? How much will a kilogramme of chilli cost?

The enumerator should make casual conversation that leads the subject perception to the next two-year period.

Table A2.

Time frame 2	Choice A (in 2 years)	Choice B (in 2 years, 4 months)
Decision 1	Granted IDR 750,000	Granted IDR 800,000
Decision 2	Granted IDR 700,000	Granted IDR 800,000
Decision 3	Granted IDR 650,000	Granted IDR 800,000
Decision 4	Granted IDR 600,000	Granted IDR 800,000
Decision 5	Granted IDR 500,000	Granted IDR 800,000
Decision 6	Granted IDR 400,000	Granted IDR 800,000

Appendix 3. Matrix correlation

```
. *multikol
. corr $ylist $xlist
(obs=55)
```

	dev	tp	idf	kin	edu	uexp	exp	age	eprice	exprc	tland
dev	1.0000										
tp	0.1815	1.0000									
idf	-0.0298	0.0273	1.0000								
kin	0.6599	0.0371	-0.1163	1.0000							
edu	0.2373	0.1221	-0.0803	0.2984	1.0000						
uexp	-0.0734	0.1593	0.2862	-0.0178	0.0844	1.0000					
exp	0.2027	0.0160	-0.1777	0.2076	0.3889	0.1403	1.0000				
age	-0.1005	-0.1185	-0.0421	-0.0624	-0.4169	0.0929	-0.0565	1.0000			
eprice	0.3331	-0.0338	-0.1845	0.3777	0.0675	0.0800	0.0650	0.1265	1.0000		
exprc	0.3522	0.0290	-0.1017	0.3883	0.2041	-0.0200	0.4686	0.2703	0.1989	1.0000	
tland	-0.0473	-0.2097	-0.0577	0.0622	0.2803	0.0160	0.5609	-0.1426	-0.0122	0.2345	1.0000

Appendix 4. Goodness of fit test

```
. estat classification

Probit model for dev
```

Classified	True D	True ~D	Total
+	11	1	12
-	3	40	43
Total	14	41	55

```
Classified + if predicted Pr(D) >= .5
True D defined as dev != 0
```

Sensitivity	Pr(+| D)	78.57%
Specificity	Pr(-|~D)	97.56%
Positive predictive value	Pr(D| +)	91.67%
Negative predictive value	Pr(~D| -)	93.02%
False + rate for true ~D	Pr(+|~D)	2.44%
False - rate for true D	Pr(-| D)	21.43%
False + rate for classified +	Pr(~D| +)	8.33%
False - rate for classified -	Pr(D| -)	6.98%
Correctly classified		92.73%

Competition and Cooperation in Economics and Business – Gani et al. (Eds)
© 2018 Taylor & Francis Group, London, ISBN 978-1-138-62666-9

The impact of government expenditure on the real exchange rate: An empirical study in Indonesia

I. Prakoso
Department of Economics, Faculty of Economics and Business, University of Indonesia, Depok, Indonesia

ABSTRACT: In recent years, the government of Indonesia has intensively increased their spending for development, especially infrastructure. As a developing country with low technology, Indonesia needs to import the development component/material from foreign countries, which may result in the depreciation of the Indonesian rupiah (domestic currency). In contrast, the aggregate demand for local goods is higher because private spending would prefer to consume local goods. This will increase economic growth and appreciate domestic currency. In order to find an explanation for these contradictory outcomes, this study uses a structural vector autoregressive method with macroeconomic factors such as taxation, economic output, and trade balance as the transmission. This is because the increased spending does not directly affect foreign currency, but it is transmitted through government revenue, economic output, and trade balance. By using quarterly data, this study provides another view on related topics because previous research concludes that government expenditure shock will appreciate the foreign exchange rate highly.

1 INTRODUCTION

In recent years, the government of Indonesia is aiming to accelerate economic growth by increasing its expenditure, especially in the production sector. To stimulate economic growth, the government needs to fund its expenditure by increasing the tax revenue or by improving the tax/fiscal administration system, for example. In the long term, the cumulative balance of government expenditure should be equal to the tax revenue. In other words, an increase in government expenditure is a sign of future tax increase; hence, the response of the private sector may be varied in the long term. In the short term, most of the private sector will react positively, responding to government expenditure regardless of whether it experiences a direct impact from government spending. If the private sector does not get benefits from government expenditure, it will reduce demand for goods and vice versa.

Koethenbuerger (2011) argues that the Indonesian government may become revenue oriented or expenditure oriented. If the government is revenue oriented, the budget for expenditure will depend on how much revenue the government should be able to realise in future years. In most cases, both expenditure and revenue are adjusting intertemporally. For countries with a negative budget, it is expected that the tax revenue will increase in the future because of higher income per capita, which has been boosted by infrastructure development. However, if the government targets unrealistically high revenue, it can be categorised as expenditure orientation; in contrast, if the government has a realistic revenue target (whether achieved or not), it is revenue oriented. However, determining the orientation is difficult because in reality government revenue and expenditure adjust to each other (automatic stabilisers effect). Fernández and de Cos (2006) and Cebi and Culha (2013) assume in their research that there is no direct relationship between government expenditure and government revenue in the same period. They argue that any deficit budget will cause a rise in the revenue target in the next period.

Both government expenditure and tax revenue have a direct and indirect impact on the exchange rate. Expenditure will depreciate the foreign exchange rate if it is used to acquire foreign goods/service, payment of government debts, etc. Appreciation of the foreign exchange rate generally happens when expenditure is not used to acquire publicly traded goods. However, a rise in government expenditure may increase the aggregate demand as demand for the Indonesian rupiah is increasing because of higher prices of domestic goods.

An increase in tax revenue or an effective tax rate will lead to depreciation of the currency because the private sector's asset is low. However, Beck and Coskuner (2003) found that the impact of tax on the exchange rate depends on the type of tax when government expenditure is used to acquire non-tradable goods or service. Capital interest tax will depreciate the currency, but consumption and labour tax will appreciate the currency. Consumption tax will cause the demand for leisure to increase, and labour tax will cause the labour supply to decrease; hence, the wage rate will increase.

As discussed previously, the effect of an increase in government expenditure depends on the net-off of the effects of government expenditure and potential tax increase. In this study, it is expected that government expenditure will appreciate the currency because the Indonesian government is reluctant to increase the tax rate to attract investors (for more details, see KPMG, n.d.). In last five years, corporate tax rate in Indonesia (25%) has been higher than the average global tax rate (23.85%). Hence, the government is under pressure to maintain the tax rate to attract investors. Furthermore, the increase in government expenditure is mostly to acquire non-publicly traded goods for infrastructure.

The objective of this research is to explain the impact of government expenditure shock to the foreign exchange rate and other economic indicators by using impulse response function analysis. This study uses structural vector auto regression (SVAR) because the shock impact should be transmitted through other variables. The second reason is the relationship between government expenditure and tax revenue that adjust to each other but not in the same period. Furthermore, it is also interesting to perform an analysis on other macroeconomic indicators that are used as the transmission.

2 CONCEPTUAL FRAMEWORK

2.1 *Economic model*

An open economy country has three main economic agents: the household/private sector, the producer, and the regulator/government. It is assumed that if a country is able to produce goods/services of similar quality/specification as a foreign country, then the net export is the difference between its production and consumption. Hence, government expenditure consists of imported goods (g_{Mt}) and non-export domestic goods (g_{Dt}):

$$g_t = g_{Mt} + g_{Dt} \tag{1}$$

The demand for goods/services within the country (d_t) consists of private consumption (c_{Dt}) and government expenditure (g_D). When consumption is higher than supply, both the private sector and the government import to meet consumption levels ((c_{Mt}) and (g_{Mt})); thus, the demand equation is:

$$d_t = c_{Dt} + g_{Dt} + c_{Mt} + g_{Mt} \tag{2}$$

where $c_{Mt} = f\left(c_{Mt}, e_{Mt}, a\right)$ and $g_{Mt} = f\left(g_{Mt}, e_{Mt}, t_t\right)$. e_M is the relative price of export and import. α is a lifetime asset/wealth owned by the private sector, which is sensitive to the level of interest rate and tax. Different from the private sector, government expenditure depends on the tax revenue (T_t).[1] Government expenditure and private expenditure are distinguished because the government may acquire non-publicly traded goods.

1. Lifetime government budgeting shows that government expenditure and tax revenue are equal in the long run. Hence, it can be notated as follows: $\int_0^\infty e^{-R(t)} G_{(t)} dt \approx \int_0^\infty e^{-R(t)} T_{(t)} dt$, where $R(t)$ is the real interest rate on period t.

Government expenditure also has a direct impact on the trade balance (C_{At}) directly because the import balance (I_{Mt}) consists of private import and government import. Hence, increasing government expenditure may lower the trade balance because the export balance (EX_t) may reduce until the balance is lower than the import balance. Hence, the equations are as follows:

$$CA_t = EX_t - IM_t \tag{3}$$

$$IM_t = C_{Mt} + g_{Mt} \tag{4}$$

Equation 4 shows that the increase in government expenditure should be able to lower the foreign reserve if the expenditure is used to acquire foreign goods. However, when the government avoids acquiring foreign goods, the impact of government expenditure may depend on the portion of publicly traded goods the government acquires. If the government acquires less publicly traded goods, the aggregate demand will increase; hence, the foreign exchange rate will appreciate. An increase in other non-internationally traded goods, such as domestic goods, international aids, and international transfers of capital, will appreciate the foreign exchange rate.

2.2 *Government expenditure and tax revenue*

According to Dumrongrittikul and Anderson (2015), the differences in conclusions of previous research on the fiscal policy and exchange rate are caused by the estimation method. They accommodate more dynamic methods, such as SVAR with multiple countries and observation years. Their study found that government expenditure might appreciate the exchange rate because the government mainly acquires non-publicly tradable goods and services; hence, it does not have an impact on the overall aggregate demand.

Other research by Ricci et al. (2015) and Galstyan and Lane (2009) found similar conclusions to those of Dumrongrittikul and Anderson (2015). In more specific samples, Bouakez and Eyquem (2015) found that the impact of government expenditure shock is different across countries. In a small open economy with limited access to international capital flows and no active monetary policy, government expenditure shock will depreciate the foreign exchange rate. This is because government expenditure is unable to stimulate the economy (gross domestic product, GDP) appropriately.

In the context of two periods (i.e. intertemporal), Frenkel and Razin (1996), as explained in Balvers and Bergstrand (2002), divide the impact of government expenditure into two channels because of its ability to influence the private sector and the real exchange rate: resources withdrawal and consumption tilting. The resources withdrawal channel explains that increasing government expenditure on acquiring non-internationally traded goods (e.g. domestic goods, international aids, international transfers of capital) will appreciate the foreign exchange rate and vice versa. The effect is similar with negative supply shock. The consumption tilting channel explains that government expenditure on foreign goods may change intertemporally; hence, it depends on the characteristics of the utility function. For example, if the government prefers to import in the second period but the private sector prefers to import in the first period, the exchange rate is expected to hold in both the short and long term.

2.3 *Trade balance and foreign exchange*

A study by Kayhan et al. (2013) on deficit trade balance countries (Turkey) found the relationship (bi-directional causality) between government expenditure and trade balance by using the Toda–Yamamoto causality analysis. In other analysis, by using the frequency domain causality analysis, they found the negative impact of government expenditure on the trade balance in the short and medium term. They explain their findings by using the expenditure approach and the income approach as shown by the following equations:

$$c_t + I_t + g_t + (EX_t - IM_t) = C_t + S_t + t_t \tag{5}$$

$$(EX_t - IM_t) = (S_t - I_t) + (t_t - g_t) \tag{6}$$

Equation 5 shows GDP by using the expenditure approach (left side) and the income approach (right side). By modifying that equation, Equation 6 shows the trade balance equivalent to public saving and government saving. Hence, government expenditure and tax revenue have an impact on trade balance through GDP.

3 EMPIRICAL METHOD

This study uses time-series data that include quarterly time series from Q1 2010 to Q1 2016 obtained from Statistics Indonesia. This study defines the logarithm of the real exchange rate as the difference between the nominal exchange rate and the consumer price index for period t. For further analysis and transmission, other macroeconomics indicators, such as degree of openness (the ratio of total export and import to GDP), GDP, tax revenue to GDP ratio, and government expenditure to GDP ratio, are also included. All variables are defined in the logarithm form.

Similar to Fernández and de Cos (2006) and Cebi and Culha (2013), this research uses the SVAR method to explain the shock of macroeconomic indicators (e.g. government expenditure) to exchange rate. The relationship between residual and structural shocks is summarised in the following equations (see Table 1 for variable definitions):

$$u_t^q = \alpha_{qg} u_t^g + \alpha_{qt} u_t^t + \varepsilon_t^q \tag{7}$$

$$u_t^{nx} = \alpha_{nxg} u_t^g + \alpha_{nxq} u_t^q + \varepsilon_t^{nx} \tag{8}$$

$$u_t^x = \alpha_{xq} u_t^q + \alpha_{xnx} u_t^{nx} + \varepsilon_t^x \tag{9}$$

Both government expenditure and tax revenue adjust to each other (an automatic stabilisers effect) but do not directly impact each other in the same period. In the general government budgeting process, the balance is determined on how much revenue the government is expected to realise in future years. Theoretically, there should be some countries that start the budgeting process by overseeing government expenditure (expenditure oriented) and vice versa (revenue oriented).

Hence, the relationship between government expenditure and tax revenue is reflected in the residuals:

$$u_t^g \equiv \beta_{gt} \varepsilon_t^t + \varepsilon_t^g$$

$$u_t^t \equiv \beta_{tg} \varepsilon_t^g + \varepsilon_t^t$$

where if $\beta_{gt} = 0$ then the expenditure decision is prior to a tax and vice versa.

As seen in Equation 7, government expenditure and tax revenue are part of national income identity (GDP); hence, both of them affect GDP contemporaneously. Equation 8 shows that government expenditure affects the openness of foreign trading by importing goods and services for economic development (see Section 2.3 for more details). GDP also has an impact on

Table 1. Variable definitions.

Variable	Definition
G	Government expenditure
T	Tax revenue
Q	GDP growth
Nx	Net export
X	Real exchange rate (Indonesian rupiah to US dollar)

the openness of foreign trading because the fluctuation of economy causes domestic demand changes (Romer, 2012). The impact of shocks on government expenditure and tax revenue on the foreign exchange rate is indirect, through the channels of economic fluctuation and openness of economy (Equation 9).

The reduced-form vector auto regression (VAR) method can be written as:

$$\Gamma U_t = BV_t \tag{10}$$

where V_t is the vector containing the orthogonal shock.

Based on the previously explained theory, the restriction of this research can be summarised in the matrix below:

$$\Gamma = \begin{bmatrix} 1 & 0 & 0 & 0 & 0 \\ 0 & 1 & 0 & 0 & 0 \\ -\alpha_{qg} & -\alpha_{q\tau} & 1 & 0 & 0 \\ -\alpha_{nxg} & 0 & -\alpha_{nxq} & 1 & 0 \\ 0 & 0 & -\alpha_{xq} & -\alpha_{xnx} & 1 \end{bmatrix} \tag{11}$$

and

$$B = \begin{bmatrix} 1 & \beta_{gt} & 0 & 0 & 0 \\ \beta_{tg} & 1 & 0 & 0 & 0 \\ 0 & 0 & 1 & 0 & 0 \\ 0 & 0 & 0 & 1 & 0 \\ 0 & 0 & 0 & 0 & 1 \end{bmatrix} \tag{12}$$

4 EMPIRICAL RESULTS

4.1 *Tests of long-term relationship*

The unit root test in the VAR model has been used to determine whether the mean of the series and its variance is constant or whether the shock has a permanent or transitory effect. Different methods are used for this test, and this study uses the Phillip–Perron (PP) and augmented Dickey–Fuller (ADF) tests.

Table 2 presents the results of the PP and ADF tests, including their significance level with a constant value at the stationary level. According to the results, all the variables are stationary at 1% level of significance. Therefore, the variables in this study can be used to estimate the impact of government expenditure and other macroeconomic indicators on the exchange rate by using the VAR method.

Table 2. PP and ADF tests.

	PP test		ADF test	
Variable	*t*-value	Significance level (%)	*t*-value	Significance level (%)
(i) G	–13.6860	1	–9.6640	1
(ii) T	–3.8420	1	–3.9550	1
(iii) Q	–5.1920	1	–4.6800	1
(iv) Nx	–4.6540	1	–4.6540	1
(v) X	–4.4970	1	–4.5170	1

4.2 *Impulse response analysis*

As discussed previously, this research uses an impulse response function to determine how the shocks of government expenditure and other macroeconomic data affect real exchange rates and other variables. This analysis also shows how large the impacts are. Figure 1 shows the median impulse response of the real exchange rate over the years after the shock of government expenditure.

Consistent with the Ricardian equivalence, the findings show that the increase in government expenditure will raise the aggregate demand and may increase the tax rate in the future; hence, the private sector anticipates this by lowering consumption or causing no changes to the country's aggregate demand. However, as the decrease of the aggregate demand on the private sector is higher, the exchange rate depreciates on the first several quarters. Before converging to its natural level, the exchange rate appreciates because the government of Indonesia does not intend to increase the tax rate in order to maintain the investment level in Indonesia. The private sector learns that the government will increase tax revenues to offset expenditures preferably by improving the tax administration because the government also focuses on policies that stimulate the investment, especially from small-to-medium enterprises. Hence, the revenue effect is relatively small, although this study has included constraint on government revenue from taxes.

The last panel of Figure 2 shows that tax revenue shocks cause the exchange rate to depreciate. Consistent with the previous explanation, the private sector expects an increase in the tax rate initially. The shock does not hold because the private sector sees that the tax rate increase is not happening. Figure 2 also provides evidence that positive fluctuation of both economies (GDP) and trade openness cause the exchange rate to appreciate. This is because Indonesia has a net export position during the observation period. Hence, the openness is a good sign for the private sector in Indonesia. Economic growth is in line with the exchange rate. When the economy grows larger, the exchange rate will appreciate.

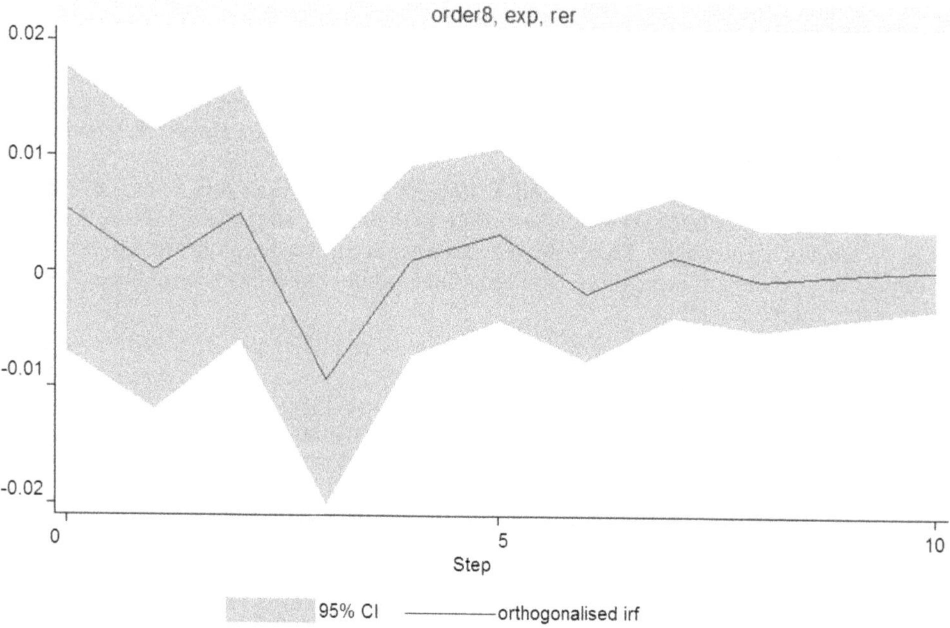

Figure 1. Government expenditure shock on the exchange rate.exp, government expenditure; rer, real exchange rate; CI, confidence interval; irf, impulse response function.

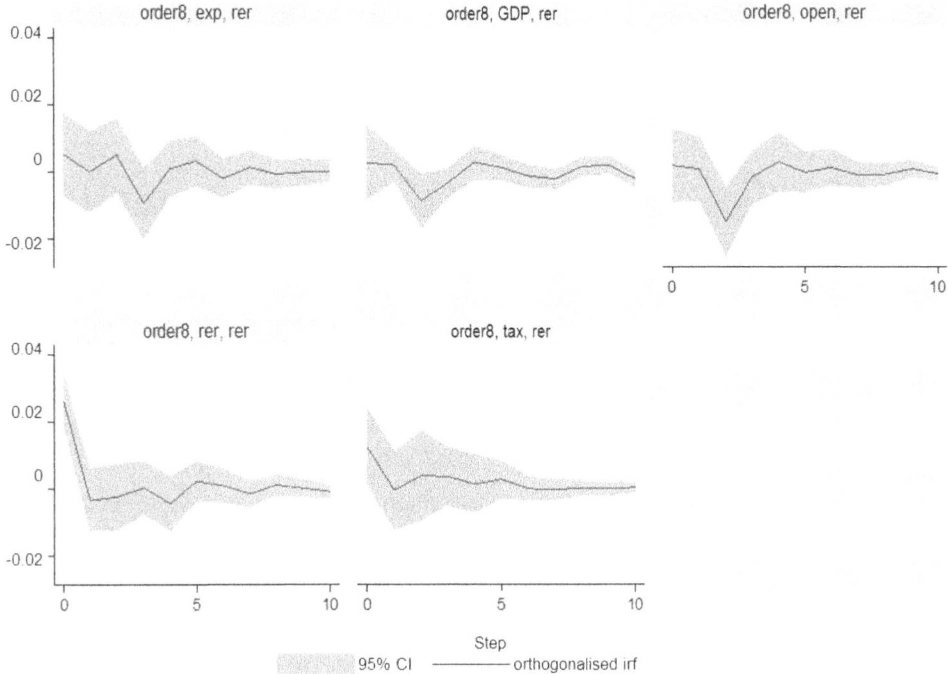

Figure 2. Other macroeconomic shocks on the exchange rate. exp, government expenditure; rer, real exchange rate; CI, confidence interval; irf, impulse response function.

4.3 *Robustness analysis*

Robustness analysis is used to assess whether the empirical results are sensitive to other factors, such as the research method and variable proxy. For robustness analysis, this study uses Cholesky decomposition on the restriction. It finds that there is no significant difference between using the SVAR method and VAR with Cholesky decomposition.

5 CONCLUSIONS

Sustainable economic growth in Indonesia attract global investor to invest hence Indonesia economy is growing more important in the global economy. Thus, exchange rate policies are getting more critical for the government. This study provides an analysis of the exchange rate movement, especially as an impact of the fluctuation of government expenditure.

The impulse response analysis from this study confirms the results expected by economic theories and found in other empirical literature. Based on the analysis, it is found that the exchange rate slightly depreciates when the government decides to increase its expenditure. This is because the private sector anticipates a potential tax rate increase (in any form) by smoothing their consumption and investment. In the next several quarters, the exchange rate tends to appreciate because the economic agent learns that the government will not raise the tax rate but optimise tax administration or even arrange an investment package to increase its tax revenue. Consistent with the expenditure shock, the tax revenue shock also appreciates the exchange rate because the private sector has learned the government policy is not to increase the tax rate.

This study has several limitations. First, it does not consider the qualitative aspect of government expenditure because the effectiveness of government expenditure may depend on factors including the level of socio-political instability and the credibility or accountability of the government. Second, this study has limited data series with only 23 observation periods; this should be overcome in the next research by using a group of similar countries (i.e. developing countries or Southeast Asian countries).

REFERENCES

Balvers, R.J. & Bergstrand, J.H. (2002). Government expenditure and equilibrium real exchange rates. *Journal of International Money and Finance*, 21(5), 667–692. doi:10.1016/S0261-5606(02)00015-3.

Beck, S. & Coskuner, C. (2003). Tax effects on the real exchange rate. University of Delaware Working Paper 2003–11.

Bouakez, H. & Eyquem, A. (2015). Government spending, monetary policy, and the real exchange rate. *Journal of International Money and Finance*, 56, 178–201. doi:10.1016/j.jimonfin.2014.09.010.

Cebi, C. & Culha, A.A. (2013). The effects of government spending shocks on the real exchange rate and trade balance in Turkey. Turkey Central Bank Working Paper 13/37.

Dumrongrittikul, T. & Anderson, H.M. (2015). How do shocks to domestic factors affect real exchange rates of Asian developing countries? *Journal of Development Economics,* 119, 67–85. doi: 10.1016/j.jdeveco.2015.10.004.

Fernández, F.d.C. & de Cos, P.H. (2006). The economic effects of exogenous fiscal shocks in Spain: A SVAR approach. European Central Bank Working Paper 647. Available at: http://www.pedz.uni-mannheim.de/daten/edz-ki/ezb/06/w-paper/ecbwp 647.pdf (accessed May 2017).

Frenkel, J. & Razin, A. (1996). *Fiscal policies and growth in the world economy* (3rd edn). Cambridge: MIT Press.

Galstyan, V. & Lane, P.R. (2009). The composition of government spending and the real exchange rate. *Journal of Money, Credit and Banking*, 41(6), 1233–1249. doi:10.1111/j.1538-4616.2009.00254.x.

Kayhan, S., Bayat, T. & Yüzbaşi, B. (2013). Government expenditures and trade deficits in Turkey: Time domain and frequency domain analyses. *Economic Modelling*, 35, 153–158. doi:10.1016/j.econmod.2013.06.022.

Koethenbuerger, M. (2011). How do local governments decide on public policy in fiscal federalism? Tax vs. expenditure optimization. *Journal of Public Economics*, 95(11–12), 1516–1522. doi:10.1016/j.jpubeco.2011.06.006.

KPMG (n.d.) Corporate tax rates table. Available at: https://home.kpmg.com/xx/en/home/services/tax/tax-tools-and-resources/tax-rates-online/corporate-tax-rates-table.html (accessed 30 September 2016).

Ricci, L.A., Milesi-Ferretti, G.M. & Lee, J. (2015). Real Exchange Rates and Fundamentals: A Cross-Country Perspective. *Journal of Money, Credit and Banking*, 45(3), 845–865. 10.1111/jmcb.12027.

Romer, D. (2012). *Advanced macroeconomics* (4th edn). New York: The McGraw-Hill.

Statistics Indonesia. Economic and Trade dataset. https://www.bps.go.id/index.php.

Competition and Cooperation in Economics and Business – Gani et al. (Eds)

Impact analysis of excise tariff on market share of cigarette companies, during the period 2009–2015

D.N. Sari & H. Susanti
Department of Economics, Faculty of Economics and Business, Universitas Indonesia, Depok, Indonesia

ABSTRACT: This research analyses the correlation between excise rates and cigarette companies in Indonesia recorded in the Directorate General of Customs and Excise. In order to find out how the market shapes the cigarette industry, the analysis was done using structure—conduct—performance frame work, that looked at the relationship of structure, behaviour, and performance in the cigarette industry. The cigarette industry in Indonesia has an oligopoly structure with a moderate level of competition. Of the two models in all companies, it has been found that the application of tax rates is significantly and negatively associated with the cigarette companies' market share and the sale price of cigarettes. The population growth of individuals aged 15–24 years has a positive and sinifcant impact on the cigarette market share. Meanwhile, for the five major companies, the other model (those a fed 15 years and above) has found that excise per selling price has a positive effect on the market share.

1 INTRODUCTION

A number of studies are correlating cancer with cigarette consumption making the global community try to reduce the rate of cigarette consumption through various health and tax policies. Cigarettes have an addictive substance and consumption of these products should be controlled. In Indonesia, under Article 2 of Law No. 39/2007 on excise, cigarettes qualify as goods subject to excise.

The government continues to increase tobacco excise rates significantly, reaching its upper limit of 57% of the retail price. In addition, the excise rates in Indonesia are still low compared with those of other developing or poor countries. In 2014, the proportion of taxes to cigarettes price in Indonesia was only 53.4%, compared with 73.13% in Thailand (Djoko, 2013). Since 2009, excise imposition of cigarettes in Indonesia has been set at a specific tariff, based on the type of cigarettes, class production, production restrictions, and limitations of the retail price. Tobacco products are divided into three types (1) clove cigarettes machine or sigaret kretek mesin (SKM) (2) white cigarettes machine or sigaret putih mesin (SPM), and (3) hand-rolled clove cigarettes or sigaret kretek tangan (SKT). Among these, the largest excise tariff is imposed on SKM, about 10% higher than the average tax rates on SPM and nearly three times the average SKT customs tariff.

In terms of the manufacturer, excise is viewed as a threat to business as it can reduce a company's profitability following the shrinking of people's buying power. To prevent the hike of excise tariff, the industry has also stated that tax has made the number of unemployed in this sector increase. However, on the other hand, the demand of cigarette products continues to increase each year, although a decrease was recorded in 2013–14 (see Figure 1).

Currently, the industry is producing SKM more than the two other cigarette types. Sumarno & Kuncoro (2002) state that the market domination among big players in the cigarette industry, such as PT Gudang Garam Tbk, PT HM Sampoerna Tbk, and PT Djarum, started to shift their production from SKT to SKM since the end of the 1990s. This can be seen by the increased SKM production in those companies.

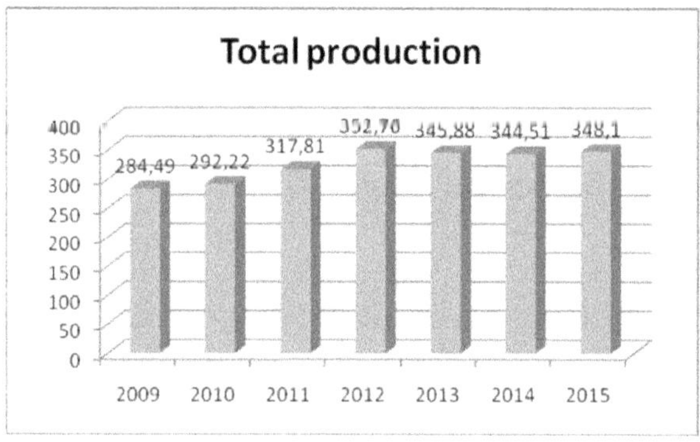

Figure 1. Cigarette production in Indonesia (in billion pieces), Source: Ministry of Finance, Indonesia.

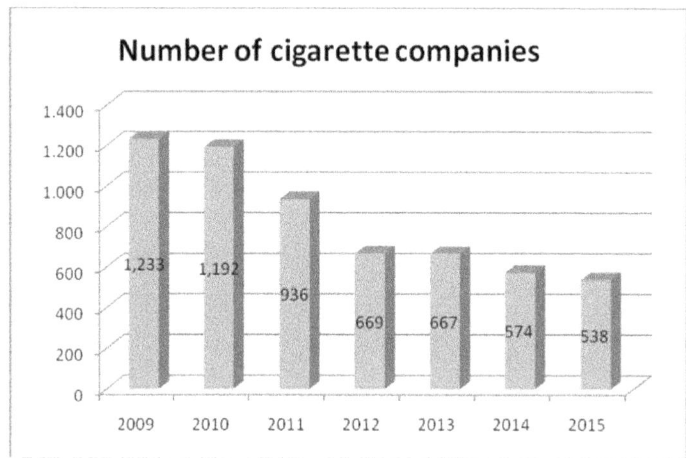

Figure 2. Number of cigarette companies in Indonesia, 2009–15. Source: Ministry of Finance, Indonesia.

On the other hand, a continual decline in the number of manufacturers was considered as the result of tax rates hike (see Figure 2). From 2009 to 2015, the decline in the number of companies was significant. Based on the data from the Directorate General of Customs and Excise, there were 88 companies in 2015, down from 1,233 companies in 2009. In order to survive, not infrequently, small cigarette companies make false excise bands so that the cigarettes are not imposed by excise and become illegal.

2 THEORETICAL FRAMEWORK

To explore the problems presented in this study, a structure—conduct—performance framework was used for analysis along with factors that affect the market share of a company, the price of goods themselves, the amount of production, and many types of products/services offered (Tauraz & Chaolupka, 2006; Babic & Kalic, 2014).

The increase in cigarette prices does not reduce the consumption of tobacco, but changes the pattern of cigarette consumption. Based on the consumption patterns of smokers, non—SKM cigarettes are a substitution of non—SPM goods. Meanwhile, the price of SPM cigarettes has a positive influence on demand for SKT cigarettes, where the SPM price increase causes an increase in demand of non—SKT products (Rudi, 2004: Herlambang, 2005).

In addition to the price, advertising is also believed to have an important role in the cigarette market share. As many as 60% of smokers start smoking at the age of 13 years. Children generally recognise cigarette brands from sports events on television that are sponsored by tobacco companies. Cigarette brands that often sponsor sports events are more likely to be chosen by teenagers to consume the cigarettes (Arnett & Terhanian, 1998; Bates & Rowell, 2012).

3 METHODOLOGY AND DATA

Data processing was done by using the structure—conduct—performance framework to determine the impact of policies that affect the tax structure, conduct, and performance of companies in the tobacco industry. We also conducted econometric measurements to determine the relationship between tax rates and corporate concentration.

The equation econometric model was estimated using the panel data balance. For data collection, first, nine companies were identified in the cigarette market of 2009–15 based on the classification in the application of specific tax rates. Five companies with the largest production, one company with more than 2 billion cigarettes, one with more than 350 million cigarettes and fewer than 2 billion tobaccos, one with more than 50 million cigarettes but not 350 million rods, and one with fewer than 50 million cigarette.

Afterwards, the annual market share was calculated per company. The excise rate or the price per cigarette was determined based on the best selling type of cigarette from each company. Diversified products were also included in order to see the effect on the market share of the existing companies. Using the data for 2009–15, 63 observations were made.

Excise rate data were obtained from the tobacco excise tariff from the Ministry of Finance in Indonesia. The selling price of cigarettes, taken from the top-selling cigarette brands, and the price obtained by interviewing informants were used. Data on the number of companies per year; they were obtained from the Directorate General of Customs and Excise. Data were also obtained from product diversification gathered from interviews with informants who have observed and understood the kinds of cigarettes.

The cigarette market is known to have big dominant players. In order to see the effect of tax on the large enterprise market share, this study could also be included in the regression of the same model, but only five big companies were included. Econometric measurement was done by developing a study to analyse the determinants of the market share, for this Babic and Kalic (2014) used an econometric model formulated as follows:

$$
\begin{aligned}
MS_{i...t} = {} & \alpha + \beta_1 \, Log\,(Ckphj_{it}) + \beta_2 \, Log\,(Hj_{it}) + \beta_3 Dvrs_{it} + \beta_4 \, Log\,(Pddk_t) + \beta_7 D\_A_{it} + \beta_8 D\_B_{it} \\
& + \beta_9 D\_C_{it} + \beta_{10} D\_D_{it} + \beta_{11} D\_E_{it} + \beta_{12} D\_F_{it} + \beta_{13} D\_G_{it} + \beta_{14} H\_E_{it} + \beta_{15} D\_I_{it} + e_{it} \qquad (1)
\end{aligned}
$$

In this study, the extent of sensitivity of the number of young people could affect the market share of a company. So, Pddk became Pddk2, which is the number of people aged 15–24 years from 2009 to 2015, considering the large number of people in this age group in the population of Indonesia and also the age of the potential market share for these companies (see Table 1).

3.1 *Structure analysis*

The cigarette industry in Indonesia is categorised in the relative competitive market with a moderate concentration level because the Herfindahl-Hirschman index (HHI) is in the range of 1,112 to 1,376 in 2009–15. The extent of market concentration of each manufacturer in 2009–15, can be described by concentration ratio of the eight largest companies (CR-8) in the industry, with the movement of the concentration ratio as shown in Table 2.

Of the eight major companies, GudangGaram, Sampoerna, and Djarum continue to be ranked the three highest orders of first, second, and third, respectively. Philip Morris Indonesia and Nojorono enlivened the eight largest companies by production of cigarettes in the tobacco

Table 1. Descriptive variables and research hypotheses.

Variable	Explanation	Hypothesis
(i) Ms	Market share (percentage)	
(ii) Ckphj	Proportion of excise tax to selling price	–
(iii) Hj	Selling price per cigarette	–
(iv) Dvrs	Number of variants products per company	+
Pddk	Population aged 15 years and above	+
(v) Pddk2	Population aged 15–24 years	+
(vi) D_A	Dummy variable A	
(vii) D_B	Dummy variable B	
(viii) D_C	Dummy variable C	
(ix) D_D	Dummy variable D	
(x) D_E	Dummy variable E	
D_F	Dummy variable F	
D_G	Dummy variable G	
(xi) D_H	Dummy variable H	
D_I	Dummy variable I	
I	(1,...,..., N) observant	
t	(1,...,..., t) 2009–15	

Source: Authors' compilation

Table 2. CR8 of cigarette companies in Indonesia, 2009–15.

Year	CR (%)
2009	70.98
2010	68.54
2011	68.46
2012	73.59
2013	77.21
2014	79.02
2015	81.24

Source: Ministry of Finance, Indonesia.

industry from 2009 to 2015. Meanwhile, in a period of seven years, several other companies, such as Asia Tobacco, PDI Tresno, Karyadibya Mahardika, Subur Safe, Bentoel Prima, and new player Indonesia Sampoerna Sembilan, had not previously been included in the eight companies.

3.2 *Behaviour analysis*

The imposition of excise tax on cigarettes is one of the biggest obstacles for these companies because the product's selling price must follow the rules set by the government. Although cigarettes are included in the inelastic goods category, frequent increase in prices does not arise from the cost of production, thus making the companies fear that the demand for cigarettes will decline. The tobacco companies have a variety of ways to overcome this. In terms of price, the companies raise the price of cigarettes before the excise increase, which usually occurs at the beginning of the year, in anticipation of losses that might occur due to rise in cigarette prices the next year, in line with the excise duty increase. In addition, tobacco companies also reduce the cost of production. The use of technology is adopted by tobacco companies, especially companies with large capitalisation, to suppress the average cost of production to achieve economies of scale efficiently. To maintain the affordability of cigarettes, tobacco companies also hire sales agents to sell cigarettes per piece or unit to lure the consumer.

The behaviour of tobacco companies in the responding market can also be determined by product differentiation. In relation to excise tax and the tobacco industry, a strong oligopoly structure would make the companies imitate patterns, and observe, and modify various products released by competitors. Cigarette companies generally have several product types, such as SKM, SKT, and SPM, with features that generally include flavour variations to distinguish products between one company and another.

Based on the data from the Directorate General of Customs and Excise, the SKM market share dominated more than 60% of total cigarettes market share each year (2009–15), and it could reflect that cigarette companies now focus more on premium products. Therefore, SKM has been categorised as a flagship product in line with profits generated. The companies keep producing new brands to give new experiences to their customers by providing new variants of flavours and ensuring a low level of nicotine and tar in response to the growing campaign about health and dangers of smoking. This has been the response amid impact of excise tariff hike to companies' performance and competition in the industry.

In order to become a major player in the tobacco industry, a new company should be able to conduct the same advertisements as competitors. Based on that fact, the government restricted cigarette advertising via Regulation No.109/2012 on the Control of Tobacco Products. Cigarette companies are not allowed to support restricted activities by placing their logo and product names in these. Due to the limited space in the tobacco industry as a result of regulations made by the government, cigarette companies have developed different strategies to keep improving brand awareness by supporting different activities. Currently, the companies implement other strategies, such as promoting musical events or sport activities that have a lot of influence, for promotion by disguising their logo.

3.3 *Performance analysis*

One performance measurement could be seen from the profit; according to Carden (2009), as company's goal, the volume of production is closely linked to income, where both are of paramount importance in making costs of a company's strategic plan. The volume of production is equal to the profit targets to be achieved. Based on the data released by the Directorate General of Customs and Excise, the trend in cigarette production increased from 2009 to 2015, although there was a decline in 2014. This is due to the application of a 10% cigarette tax in customs tariffs by the government, so the number of the supply manufacturers declined. Examining the dynamics in the tobacco industry through a number of existing productions, it can be said that the cigarette market continues to grow, although there has been a decline.

4 REGRESSION ANALYSIS RESULTS

4.1 *Regression results of the entire industry*

From the two models of age groups 15–24 years and 15 years above, we obtained similar results, except for residents. Big differences can be seen where the population variable in the age group 15–24 years has positive and significant results on the market share, whereas in the age group 15 years and obove the results are not significant. These results support further research, where the young population is a potential market for tobacco companies. From the viewpoint of the industry, the introduction of cigarettes to youth as early as possible will make them loyal consumers of the brand.

The results also show that the impact of tax on sales price and the selling price of cigarettes on the market share is elastic, different from existing research results where cigarette consumption is inelastic. Several factors are considered to affect the elasticity:

1. Types of cigarettes: Based on the consumption patterns of smokers, SKM is a substitution of SPM products. The price of SPM cigarettes has a positive influence on demand for SKT cigarettes, where the SPM price increase will cause demand of SKT to increase. Thus, if there is a change in consumption patterns, this will affect the market share of tobacco companies, as the demand of this type of cigarettes will change.

Table 3. Regression results of the industry.

	Fixed-Effect model: Aged 15–24 years			Fixed-effect model: Aged 15 years and above		
Variable	C	Probability	Result	C	Probability	Result
C	−685.8813	0.0859		31.13330	0.0222	
Log-(Ckphj)	−5.157254***	0.0750	Significant (−)	−0.720393*	0.0024	Significant (−)
Log-(Hj)	−8.847851**	0.0189	Significant (−)	−1.378618*	0.000	Significant (−)
Dvrs	0.434621***	0.0652	Significant (+)	0.349836*	0.0001	Significant (+)
Log-(Pddk)	42.60457***	0.0742	Significant (+)	−0.702992	0.2859	Not significant (−)
Dummy A	11.13600			9.609224		
Dummy B	10.41797			10.23074		
Dummy C	2.951478			2.080297		
Dummy D	−5.766594			−6.860440		
Dummy E	−5.408578			−6.454502		
Dummy F	1.685191			1.892945		
Dummy G	3.949806			5.220653		
Dummy H	−8.188779			−6.584814		
Dummy I	−10.77649			−9.134103		
	Adjusted R2 = 0.955913; F statistics = 90.34336; n = 63			Adjusted R2 = 0.988944; F statistics = 463.1451; n = 63		

Note: *α = 1%, **α = 5%, ***α = 10%.
Source: Authors' compilation.

2. Tastes of young smokers and adult: Young smokers generally prefer cigarettes produced by machines, with a low amount of nicotine and tar. Adult smokers prefer cigarettes with the possibility of a more pronounced flavour.
3. Government policy: One of the implications of excise duty is reduced tobacco consumption. In addition, to support the reduction of tobacco consumption, the government has also made a variety of policies ranging from limiting advertising in media, banning smoking in certain public places, to the latest ban on cigarette sponsorship for education, sports, and music events. Many of the policies have begun to give effects to the increasing awareness of the dangers of smoking.

From those two models (Table 3), different results have been obtained. In the group aged 15–24 years, excise tariff per selling price, selling price, and population do not affect cigarettes companies' market share. In the group aged of 15 years and above, excise tariff per selling price positively affects the market share. From the previous analysis of the market structure, the number of companies in the tobacco industry is known to continue to decrease each year. The shrinkage occurs in small companies that generally produce SK cigarettes. Individuals aged 15 years and above generally consume SKM cigarettes, so when excise tariff hike burdens production costs, the companies can suffer from bankrupty, and their market share will be taken up by a larger company. The selling price for the group aged 15 years and above is elastic. Elasticity at that age is possible because it increases awareness of the dangers of smoking (as described in the previous regression results). As for the impact of the group aged 15 years and above, population growth will reduce the market share of the five major companies because that age is not the primary target market of these companies, as they produce more machines-made cigarettes.

4.2 *Regression sesults of the five major companies*

From those two models, different results have been obtained. In the group aged 15–24 years, excise tariff per selling price, selling price, and population do not affect cigarettes companies' market share. In the group aged 15 years and above, excise tariff per selling price positively affects the market share. From the previous analysis of the market structure, the number of companies in the tobacco industry is known to continue to decrease each year (see Table 4). The shrinkage occurs in small companies that generally produce SKM cigarettes. Individuals aged 15 years and above generally consume SKM cigarettes, so when excise tariff hike

Table 4. Regression results of five major companies.

	Fixed-Effect Model: Aged 15–24 years			Fixed-Effect model: Aged 15 years and above		
Variable	C	Probability	Result	C	Probability	Result
C	–322.7474	0.2255		72.81254	0.0027	
Log-(Ckphj)	–0.581728	0.6412	Not significant (–)	1.849392***	0.0581	Significant (+)
Log-(Hj)	–4.048809	0.1291	Not significant (–)	–1.371495*	0.0016	Significant (–)
Dvrs	0.311782**	0.0256	Significant (+)	0.369493*	0.0003	Significant (+)
Log-(Pddk)	20.52535	0.2034	Not significant (+)	–2.712215*	0.0321	Significant (–)
Dummy A		8.212542			7.974007	
Dummy B		8.469831			8.576527	
Dummy C		0.495929			0.324335	
Dummy D		–8.713303			–8.735417	
Dummy E		–8.465000			–8.139453	
	Adjusted R2 = 0.991348; Fstatistics = 487.9812; n = 35			Adjusted R2 = 0.993787; Fstatistics = 680.8014; n = 35		

Note: *α = 1%, **α = 5%, ***α = 10%.
Source: Authors' compilation.

burdens production costs, the companies can suffer from bankruptcy, and their market share will be taken up by a larger company. The selling price for the group aged 15 years and above is elastic. Elasticity at that age is possible because it increases awareness of the dangers of smoking (as described in the previous regression results). As for the impact of the group aged 15 years and above, population growth will reduce the market share of the five major companies because that age is not the primary target market of these companies, as they produce more machines-made cigarettes.

5 CONCLUSIONS

Based on the amount of CR8 in the tobacco industry from 2009 to 2015, the industry is categorised in an oligopolistic market, where there are many players, but only the big players control the market. Moreover, the large companies continue to widen their market share by creating subsidiary or affiliated companies to maintain their position. Throughout 2009–15, the CR8 trend continued to rise, although it saw a decline in 2010 and 2011. However, from the value of HHI, which ranged from 1,254.73 to 1,376.41 between 2009 and 2015, the tobacco industry is categorised in the relatively competitive market with a moderate concentration level.

The regression results, show that the increase in excise duty does not significantly affect cigarette market share gains. In all companies in the industry, for both the age group 15 years and above and 15–24 years, the proportion of taxes per sale price decreases the market share, as well as the selling price. Elasticity is affected by the types of cigarettes, tastes of young smokers and adults, and the effectiveness of government policies. In the five major companies, the tax per selling price actually increases a company's market share due to higher excise duty rates, so it will reduce the cost of production, but in the long term it will make the company bankrupt. The market share left by the company who cannot survive in the market, taken up by a larger company. The cigarette demand that continues to increase from year to year, in line with the growth of the consumers, makes the company who still survive in the market grow up. Thus, it can be seen that the market share of large companies (in this case five main companies) continues to increase each year.

In addition to taxes, the selling price, product diversification, and the number of people who are potential consumers of cigarettes are other factors affecting a company's market share of cigarettes. Youth are believed to be the largest market because of the increasing population aged 1524 years the cigarette market share could also increase, considering that, at that age, they are smoking for the first time.

6 RECOMMENDATIONS

The government should not hesitate to raise excise tax because it does not affect the companies' performance. This has been proved by the increasing production, the profit generated by companies, as well as the growing market share, especially of large enterprises. Improved performance is also affected by the use of machines, so companies can work more efficiently. On the other hand, the use of machinery in the industry continues to decrease employment. Under these conditions, concern about the declining number of labourers because tax rates continue to rise should not be the government's focus.

Revenue from excise tariffs can be used for the health sector with a more precise amount, not only in revenue-sharing, as it exists today. For example, Thailand applies 2% of tax rates on the health sector. Health funds have been allocated to the locals. These funds are not only used as compensation for externalities caused by smoking but are also necessary for prevention activities in a more massive movement, given the increasingly young age people become active smokers. In addition, regulations are also important to limit cigarette consumption by adolescents. In terms of employment, excise funds can also be used to gradually shift workers who are no longer absorbed in the cigarette industry as they are being replaced by machines.

7 RESEARCH LIMITATIONS

The implementation of excise tariffs on cigarettes has made cigarette manufacturers behave fraudulently, such as by falsifying or even not buying a banderol according to regulations. These circumstances make this research underestimate and not fully cover the demand of cigarettes in the country. In order to obtain more accurate results, this research is expected to estimate production through a comparison between the sales and excise bands. It also needs to conduct a qualitative study on the smoking preferences of young people.

REFERENCES

Arnett, J. J. & Terhanian, G. (1998). *Adolescents' responses to cigarette advertisements: Links between exposure, liking, and the appeal of smoking.* Tobacco Control.

Babic, D. K. & Kalic, M. (2014). Market share modeling in airline industry: An emerging market economies application. University of Belgrade.

Bates & Rowell (2008). Tobacco advertising & youth: Marketing tactics. Available at: *.tobaccofreekids.org.*

Carden, A. (2009). "Profit and production." *Quarterly Journal of Austrian Economics*, 12-(2).

Herlambang, A. S. (2005). "Pengaruh Kebijakan Tarif Cukai Terhadap Konsumsi Rokok: Sebuah Kajian Kompeherensif" (The Impact of Excise Tax to Cigarette Consumption: a Comprehensive Review), Unpublished master's thesis. University of Indonesia, Indonesia.

Irvan, P. (2012). Sejarah Perkembangan Rokok Kretek di Indonesia (The History of Clove Cigarettes in Indonesia). Available at: http://id.netlog.com/irvandpoetra/blog/blogid = 135510.

Kuncoro, M. (2007). *Ekonomika Industri Indonesia.* (Indonesian Industrial Economics) Yogyakarta: Penerbit ANDI.

Learn, N. C. (2006). "Inventions in the tobacco industry" Tar Heel Junior Historia. 46,-(1). Available at: http://www.learnnc.org/lp/editions/nchist-newsouth/4402.

Lian, D. (2015). *Southeast Asian Tobacco Control Atlas Second Edition.* Southeast Asia Tobacco Control Alliance. Penang, Malaysia: Crown Print Associates.

Rudi, H. (2004). "Policy Analysis of the Impact of Increase in Excise Tariffs on Cigarette Consumption" (Unpublished master's thesis). University of Indonesia. Indonesia.

Sumarno, S. B. & Kuncoro, M. (2002). *Structure, Performance, and Cluster of Clove Cigarettes Industry: Indonesia, 1996–1999.* Jurnal Ekonomi dan Bisnis Indonesia.

Surjono, N. D. (2013). "Impact of Advalorem, Specific, and Hybrid Excise System on Cigarette Price and Consumption" (Unpublished dissertation). University of Indonesia. Indonesia.

Tauraz, P. & Chaolupka, F. J. (2006). The Role of retail prices and promotions in determining cigarette brand market share. Review of Industrial Organization. University of Illinois.

Competition and Cooperation in Economics and Business – Gani et al. (Eds)
© 2018 Taylor & Francis Group, London, ISBN 978-1-138-62666-9

Impact prediction of British Exit (Brexit) on the Indonesian economy

T.A. Falianty
Department of Economics, Faculty of Economics and Business, Universitas Indonesia, Depok, Indonesia

ABSTRACT: On 23 June 2016, an important moment in British history was the referendum on Brexit. In this referendum the majority of the people of the United Kingdom (51.9%) voted for Britain to leave the European Union (EU), after more than 40 years of being part of this giant political and economic block. Several analyses have been made regarding the impact on the global economy. Dhingra et al. (2016) for example, analysed the consequences of Brexit for the global economy especially for EU members and concluded that all EU members would be worse off. As an open economy Indonesia would also be impacted directly and indirectly. This paper quantitatively measures the impact of Brexit on the Indonesian economy. The methods used to measure the impact are Vector Auto Regression (VAR) method and Global Vector Autoregression (GVAR). VAR and GVAR method assess the impact of Brexit through the capital, financial, and foreign exchange markets (interest rate, exchange rate, and equity market) resulting from increased global uncertainty. Indonesia's economic relations with United Kingdom in the monetary and real sectors were assessed, and the result shows that the direct impact would be modest. The impact of Brexit is stronger in the monetary/financial channel than in the real sector channel (GDP, trade and investment). The global VAR model conducted a simple correlation analysis and found that the impact of Brexit on China is stronger than on the eurozone. Several policy anticipations could be derived from this research to minimise the loss of Brexit to Indonesian economy as well as Indonesian society, such as anticipating indirect effect from China and increasing credibility of government to manage market expectation because we found stronger financial response from Brexit (stock market, exchange rate, and interest rate).

1 INTRODUCTION

In 2015 David Cameron was elected as the Prime Minister of the United Kingdom (UK). At the time of his campaign, he promised to hold a referendum to decide if the UK should stay or leave the European Union (EU). He kept his promise and scheduled a referendum on British membership in the European Union for 23 June 2016, in the UK and Gibraltar. The results of the referendum showed that the UK public decided to leave the EU by a majority of 51.9% on a national turnout.

Booth (2016) points out that the primary reason for Brexit is that many people in different regions and occupations in the UK were worried about large-scale immigration from Eastern Europe. The global financial crisis in 2008 which impacted UK as the world's financial centre was also cited as a reason.

As a result of this issue 2016 many investors became worried and moved their assets to other countries for investment when they found out that UK would be holding a referendum. That is the reason why the UK's currency value (pound sterling) decreased after the Brexit vote. This decision make UK's economy go down and many people predicted that it would be the beginning of a recession in the UK. Besides this, it was predicted that UK would get different treatment from the EU. Further consequences would be that unemployment in EU would increase, the economy would slow down, stocks around the world would decrease and sterling would depreciate.

An analysis of the impact of Brexit was conducted by Dhingra et al. (2016) which included the effect on European countries, especially the issue of funding. Because of Brexit, EU loss of some part of their source of funding. At least the other European countries have to fill half of the amount from the shortfall caused by the loss of the British fund contributions. The amount of the UK's fund contributions to the EU in 2016 was €19.4 billion. This includes tariff cuts and imported goods tax. In return, the UK received around €7 billion for regional and agricultural subsidies. Germany, as the biggest country in EU, should provide the extra money to close this shortage of fund contributions.

According to the International Monetary Fund (IMF) (2016), Brexit could worsen the global economic outlook, which is still experiencing a slowdown. In addition, the Organization for Economic Cooperation and Development (OECD) which consists of high-income countries in 2016 (after Brexit happened) reminded their members that Brexit would threaten UK's economy and global economy. The Institute of Global Counsel (2016) stated that Brexit will have an impact on ten areas, namely, intra-European trade, direct investment, liberalization and regulation, industrial policy, immigration, financial services, trade policy, global influence, the budget and uncertainty as a result of the situation. They also assessed that the impact of Brexit would depend on new relationships between the EU and the UK. In the most likely scenario, Brexit would increase the cost of trade and affect world trade.

Indonesia, as a country with an open economy would be impacted directly and indirectly. According to the statement of the Governor of the Bank of Indonesia (BI) Agus Martowardo join a report in the mass media in Indonesia in 2016, Brexit's impact for Indonesia is not very significant because Indonesian's economic and trade relations to UK is not too extensive. In his opinion, the influence of Brexit for Indonesia is not very significant because Indonesian's economic and trade relations with the UK were not extensive.

This paper aims to quantitatively measure (using the Vector Auto Regression method and Global Vector Auto regression) the impact of Brexit on the Indonesian economy. One possible effect is the effect to the capital, finance and foreign exchange markets resulting from increased global uncertainty. The transmission of Brexit through financial and capital markets in Indonesia will be identified. The impact to the aggregate macro economy in the real sector, such as GDP and inflation, will be assessed too.

1.1 *Literature review*

The Brexit phenomenon is also related to the European crisis has occurred since 2011. Brexit is the side effect of the European countries economic slowdown following European crisis. Brexit is a form of opposition to combat the crisis. The power of Brexit voices and opinions become stronger after the European crisis. Galbraith (2014) concluded that the European crisis was more serious and more unstable than the American crisis. The interconnectedness of European countries and their institutional problems are part of the problem. Thus, the direct impact of Brexit should first of all acknowledge the EU-UK relationship.

According to Rassid Hussein Bank (RHB) analysis (2016), as Indonesia's economy is mostly driven by domestic consumption, they believe that Brexit will not have a considerable impact on the country's long-term economic growth. While Indonesia's direct trade to the eurozone is not significant in their view, the potential risk would mainly be the greater volatility of Indonesian Rupiah (IDR).

RHB believes the recent approval of the tax amnesty bill would strengthen Indonesia's budget stability and secure funding for infrastructure spending. Brexit's impact, according to RHB, is via two main channels. The first will be through the financial and currency markets. It is likely that a decrease in the value of global financial markets and weakness in sterling, the euro and emerging countries' currencies, including the Indonesian rupiah will be seen. The second is via trade and investment not just from the UK, but Europe as a whole. It is believed that the direct impact will be relatively mild and should not derail Indonesia's long-term economic growth trajectory, mainly due to Indonesia's greater dependence on domestic consumption, overall net export accounts of less than 2% of GDP, and the fact that Indonesia's direct trade with the eurozone is not significant.

In 2015 the proportion of Indonesia's direct exports to the UK was mall at around 1%, with a higher proportion to Europe at 11.4%. Based on RHB in Kristanto (2016) estimated, for every 1% drop in Europe's economic growth, Indonesia's exports will likely decline by 1% and its economic growth will be negatively impacted by 0.1–0.2%. They also assessed the impact on economic sectors. Indonesian rupiah depreciation could have a negative impact on consumer and healthcare sectors which have large exposure to raw material imports, and also sectors with high USD debt exposure, such as telecommunications and media.

Dhingra et al. (2016) estimated the effect of Brexit on the UK itself as well as the EU using modern quantitative trade models. Quantitative trade models incorporate the channels through which trade could affect consumers, firms and workers. The model used by Dhingra et al. (2016) is the most recent data in the World Input Output Database (WIOD) which divides the world into 35 sectors and 31 regions and allows for trade in both intermediate inputs and final output in both goods and services. According to Timmer (2012), The WIOD was developed to analyze the effect of globalization on trade patterns, environmental pressures, and socioeconomic development across a wide set of countries. WIOD combines national and international flows of products that require a high level of statistic harmonization across countries. The model takes into account the effects of Brexit on UK's trade with both the EU and the rest of the world. To make a projection, they had to make an important assumption about UK-EU trade relations. In the optimistic scenario, they assumed that the UK and the EU would continue to enjoy a free trade agreement and Brexit would not lead to any change in tariff barriers, where as in the pessimistic scenario where trade is governed by WTO rules, they assumed Most Favoured Nation (MFN) tariffs would be imposed on UK-EU trade.

In the optimistic scenario, there is an overall fall in income of 1.28% (see Table 1) that is largely driven by current and future changes in non-tariff barriers. Non-tariff barriers play a particularly important role in restricting trade in services, an area where the UK is a major exporter. In the pessimistic scenario, the overall loss increases to 2.61% (see Table 1). In addition to the above analysis, Dhingra et al. (2016) also analyzes the influence of Brexit on EU countries and other countries in the world. According to this analysis all EU members are worse off, and Ireland suffers the largest proportional losses from Brexit, alongside the Netherlands and Belgium. Countries that lose the most are those currently trading the most with the UK. Some countries outside the EU, such as Russia and Turkey, gain as trade is diverted towards them and away from the EU.

Dhingra et al. (2016) concluded that the economic consequences of leaving the EU would depend on what policies the UK adopts following Brexit. However, lower trade due to reduced integration with EU countries is likely to cost the UK economy far more than it gains from lower contributions to the EU budget.

Baig (2016) predicts that Asia's UK exposure is more financial than trade related. The direct impact of Brexit is not too great, but there is a secondary impact through currencies and a risk from slow down in investment. UK, however, is a small trading partner of Asia. Less than 4% of EM-Asia trade is with the UK and that share has been steadily falling in recent decades. Financial linkages are more significant according to Baig's research. As of mid-2015, UK investments to EM-Asia ranged between 5–13% of the respective countries' total portfolio investment liabilities, although shares have also fallen over the past ten years.

Table 1. The effects of Brexit on UK living standards according to Dhingra et al. (2016).

	Optimistic	Pessimistic
Trade Effects	−1.37%	−2.92%
Fiscal Benefit	0.09%	0.31%
Total Change in Income per Capita	−1.28%	−2.61%
Income Change per Household	−£850	−£1700

Notes on Model Assumptions: Optimistic scenario: Increase in EU/UK Non-Tariff Barriers (NTBs) (+2%) + exclusion from future fall in Non-Tariff Barrier NTBs within EU (−5.7%), saving of 17% of 0.53% lower fiscal transfer. Pessimistic scenario: MFN tariff + increase in EU/UK Non-Tariff Barriers (+6%) + exclusion from future fall in NTB within EU (−12.8%), saving of 0.31% net fiscal transfer. (MFN: Most Favored Nation)
Source: Dhingra et al. (2016).

Table 2. Monetary policy and fiscal policy space.

	Value added exports% of GDP	Monetary policy space	Fiscal space
Most exposed			
Malaysia	52.34	Medium	Low
Hong Kong	40.2	Low	Medium
Vietnam	49.3	High	Low
Singapore	58	High	High
Middle of the pack			
Thailand	41	High	High
Taiwan	38.5	High	High
Korea	29.8	High	High
Philippines	24.4	High	High
Least exposed			
Indonesia	21.6	High	Low
India	19	Low/Medium	Low
China	17.2	Medium	High

Source: Credit Suisse estimates (2016).

According to an analysis by Nasser (2016), in Credit Suisse Report, in order to face the impact of Brexit, monetary policy is likely to support growth, and the policy used is monetary easing. Bank Indonesia (BI) has done monetary easing through cutting BI policy rate. Although BI has cut 100 basis points (bps) year to date, falling inflation trend means that Indonesia still has further room to cut rates. They believe that BI monetary policy easing was not enough; BI can further cut rates by 50–75 basis points and by 50 bps to support growth. According to their analysis in Table 2, Indonesia has a high monetary policy space and is thus in a position to be the least exposed country from Brexit.

2 METHOD

When there is no confidence that a variable is actually exogenous, each variable can be treated symmetrically, using the basic method of the Vector Auto regression Model (Enders, 2015). The VAR model simultaneously processes all involved variables to link to each other directly and non-directly. However, this may not be the real case for the global economy.

VAR may be able to describe the interactions of variables in an economy, but might not be able to model the global economy (i.e. the interactions of several economies). Beside standard VAR for one (single) country model, it could be available in the Global VAR model (GVAR). Global VAR could elaborate several countries in the world.

A simple VAR model can be used for the case of a two-country model. Country 1 is Indonesia for example, and Country 2 is the UK. A two-country model (c1 & c2) consists of two macroeconomic variables (y & z), and one global variable (x). For simplicity of equation representation the optimum lag of '1' is used as an example.

Country 1:

$$y\_c_1 = C(1)*y\_c_1(-1)+C(2)*z\_c_1(-1)+C(3)*x(-1)+C(4) + e_1$$
$$z\_c_1 = C(5)*y\_c_1(-1)+C(6)*z\_c_1(-1)+C(7)*x(-1)+C(8) + e_2$$

Country 2:

$$y\_c_2 = C(9)*y\_c_2(-1)+C(10)*z\_c_1(-1)+C(11)*x(-1)+C(12) + e_3$$
$$z\_c_2 = C(13)*y\_c_2(-1)+C(14)*z\_c_1(-1)+C(15)*x(-1)+C(16) + e_4 \quad (1)$$

Notes: c_1: Indonesia
c_2: UK

The macroeconomy variables used would be GDP, inflation, nominal exchange rate and interest rate. Because the impact of Brexit on Indonesia is being focused up, Indonesia is Country 1. For Country 1 (Indonesia), x or variables that are typically used in looking at the macroeconomic conditions of the UK (one is an interest rate and the other is GDP). Shock will occur in GDP and interest rates from the UK.

On the other hand, the Global Vector Autoregressive (GVAR) approach has proven to be a very useful approach for analyzing interactions in the global macroeconomy and other data networks where both the cross-section and the time dimensions are large (Pesaran, 2014). GVAR was first developed by HashemPesaran in 2004. Each economy is interlinked to the global economy through many different channels in a complex way, for example: labour capital movements, cross-border trade, technological developments, even political relations, and many more. Similar to regular/standard VAR, the results of GVAR could be interpreted through Impulse-Response Function (IRF) or variance decomposition. The representations of the GVAR Model are as follows:

$$x_{it} = \delta_{i0} + \delta_{i1}t + \Gamma_{i1}x_{it-1} + \Gamma_{i2}x_{it-2} + \Lambda_{i0}x^*_{it} + \Lambda_{i1}x^*_{it-1} + \Lambda_{i2}x^*_{it-2} + \varepsilon_{it}$$

$$x^*_{it} = \sum_{j=1}^{N} w_{ij}x_{jt}$$

$$\sum_{j=1}^{N} w_{ij} = 1, \text{ and } w_{ii} = 0 \tag{2}$$

where x is a domestic macroeconomy variable and x* is a foreign macroeconomy variable and w is representing trade weight.

The VAR model will be employed for the first step analysis and make an impulse response analysis from Equation 1. The second step, we will use GVAR model (Equation 2). The period of the data used for both model are different. The data used for VAR analysis is from 1995Q1–2013Q1 and the data for GVAR analysis is from 1979Q1–2013Q1. The different period reason is for GVAR we used the database available in Pesaran (2014). Besides the VAR analysis, simple correlation and descriptive statistics from the current data (2010Q1–2015Q4) are used.

2 RESULTS AND DISCUSSION

2.1 *Indonesia-UK economy relationship in general*

To assess the impact of Brexit to the Indonesian economy the Indonesia-UK relationship in the context of external debt, remittances, and exports and imports will be plotted.

Based on Figure 1, the share of all variables used is below 1.2%, which means that Indonesia's direct exposure to Brexit is not very large. The share of Indonesian external debts to UK was on the decline during the period of observation. Other variables have an increasing trend, but the value is still low and the trend is volatile.

Based on the data from the Central Statistics Agency (BPS), the average portion of Indonesian value in non-oil exports to UK is only 1.2 percent of Indonesian total value in non-oil exports to the world. For example in 2015, Indonesia's value in non-oil exports to the UK amounted to USD1.53 billion, or 1.16 per cent of Indonesian's total value in non-oil exports which amounted to USD131.73 billion. In that year, UK was twenty-first among the non-oil exports destination countries for Indonesia. Indonesian exports to UK are less than Indonesian exports to the USA, China, Japan, India and Singapore. Compared to other member countries of the EU, Indonesian exports to UK are also less than those to the Netherlands, Germany and Italy.

Based on the data from the Investment Coordinating Board (BKPM), in 2015 Foreign Direct Investment (FDI) of the UK in Indonesia amounted to USD503 million, or 1.71% of FDI's total value of USD29.27 billion. The UK is included in the top ten countries with the

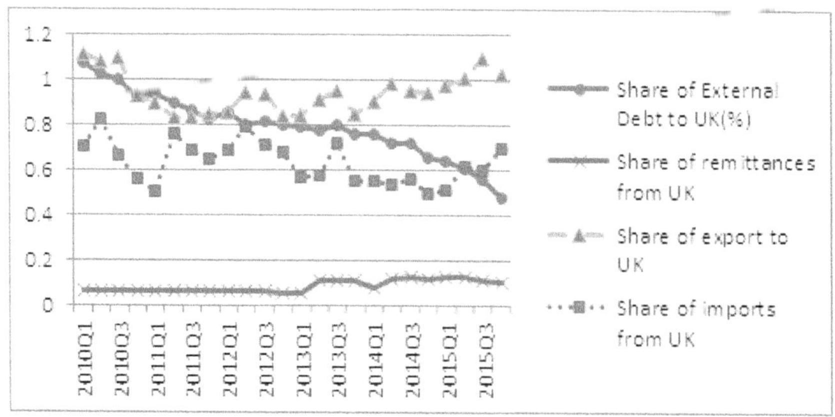

Figure 1. Indonesia-UK economy relationship 2010Q1–2015Q4.
Source: Processed data from the Central Bank of Indonesia's Statistics, external sector data.

Table 3. Indonesian FDI from UK, US, Japan and Singapore in USD millions.

Year	UK FDI	EU FDI	Japan FDI	US FDI	Singapore FDI	Total FDI
2010	276.2	1160.614	712.599	930.883	5565.0172	16,214.8
2011	419.0	2158.1456	1516.0631	1487.7873	5123.0449	19,474.5
2012	934.4	2303.3497	2456.9409	1238.2747	4856.3511	24,564.7
2013	1,075.8	2414.02119	4712.8927	2435.75025	4670.79902	28,617.5
2014	1,588.0	3764.2156	2705.1313	1299.5437	5832.1293	28,529.7
2015	503.2	2258.0658	2876.9901	893.1565	5901.1812	29,275.9

Source: Indonesia Investment Coordinating Agency (BKPM).

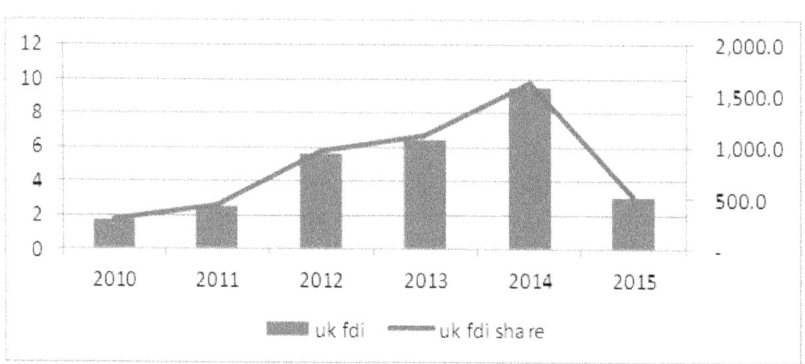

Figure 2. FDI level (USD million) and FDI share (%) from UK.
Source: Indonesia Investment Coordinating Agency (BKPM).
Note: LHS: Share (in percentage).
RHS: USD millions.

largest investment in Indonesia. The UK's investment value is still below Singapore, United States, Japan and the EU (Table 3). However, compared to EU countries, UK has the second largest investment after the Netherlands. Thus, in terms of investment, the UK's influence is relatively larger than the trade influence. FDI from the UK had an increasing trend from 2010–2014, but following global uncertainty it declined in 2015 (Figure 2).

2.2 *The impact on the financial market: Exchange rate volatility and Volatility Index (VIX)*

To assess the impact on the financial market the role of Brexit and the role of UK as a large financial centre in the world should be considered. All the UK must do to run this role is access to European countries. Many investors invest their business in UK because they want to access Europe's single market to bring jobs and investment. Thus, Brexit could create uncertainty for firms in London, less trade with Europe and fewer jobs. Brexit also endanger London's status as a global financial centre. A status of London as global financial centre was supported by passport simplification within Eurozone countries. The loss of London's passporting will require the issue of new passports for foreign firms operating in London and vice versa. This is an expensive process for financial transaction. According to Capital Economics (2016), British financial service trade with the EU increases almost every year. This is shown in Figure 3. Without simple passport procedure, it is predicted by Capital Economics (2016) that exports of financial services to the EU could fall by half.

Regarding the impact on the foreign exchange market, the exchange rate of three major international currencies—namely USD, euro, and pound—to the Swiss franc was compiled (note: Swiss franc as benchmark currencies). Indonesian rupiah (IDR) exchange rate to Swiss Franc was also summarized to assess the impact of Brexit to IDR volatility. Data from January 2010 to August 2016 was used (see Table 4) and divided into three phases. The first phase is the period before Brexit, but before Tapering off the Fed. The second phase is the period before Brexit but after Tapering off the Fed. Tapering off the fed is the period of Quantitative Easing exit from Federal Reserve US following the US economy recovery. Federal Reserve of US has started to contract their monetary policy or exit from Quantitative Easing policy since August 2013. The third phase is after Brexit happens. All of the currencies have declined in volatility after Brexit except for the pound sterling. Pound sterling to the Swiss franc exchange rate has increased in terms of volatility if the second phase (3.29) is compared with the third phase (3.99). The impact of Brexit on the exchange rate volatility of sterling can be found.

The impact of Brexit was also assessed through world financial market by plotting the Volatility Index Data. From Figure 4 it can be seen that even the VIX has increased after Brexit but only moderately. The VIX in June 2016 was still lower compared to VIX in 2011 when the European crisis hit its climax.

2.3 *Trade relations*

According to Table 5, Indonesian Trade with UK is weighted at 0.34%. This weight is lower than Indonesia's total trade to Singapore (10.30%), Malaysia (5.11%), Thailand (4.59%), Korea (3%) and Japan (3%).

From the UK side, Indonesia has a weight of 1.02%. This is lower than the UK's trade with Europe (18.69%), Norway (22.48%), and Australia, India, and Canada (around 3%).

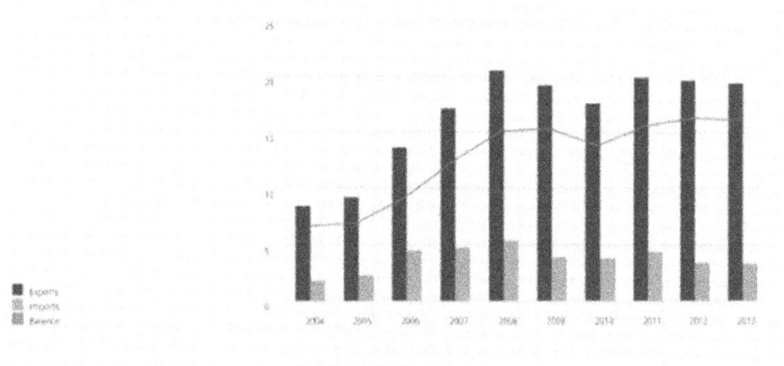

Figure 3. UK Financial services trade with the EU (£ billions).
Source: Thomson Datastream, Capital Economics.

Table 4. Descriptive statistics of the exchange rate to the Swiss franc, January 2010–August 2016.

Period		Variable	Obs	Mean	Std Dev	Min	Max	Variation coeff.
Before Brexit	January 2010–September 2013	IDR	940	9776.574	835.1505	7872.1	12788.8	8.5424
		USD	940	1.0566	0.0792	0.8609	1.3701	7.4955
		EURO	940	0.7937	0.0497	0.6726	0.9654	6.2601
		POUND	940	0.6718	0.0407	0.5874	0.8475	6.0569
	October 2013–May 2016	IDR	666	13427.53	608.5322	12311.6	15175	4.5320
		USD	666	1.0628	0.0466	0.9703	1.1778	4.3828
		EURO	666	0.8790	0.0578	0.8082	1.0227	6.5759
		POUND	666	0.6789	0.0224	0.6436	0.7784	**3.2977**
After Brexit	June 2016–August 2016	IDR	62	13547.03	207.1654	13170	13933	1.5292
		USD	62	1.0267	0.0107	1.0062	1.0452	1.0398
		EURO	62	0.9193	0.0052	0.9024	0.9274	0.5621
		POUND	62	0.7619	0.0304	0.7004	0.7996	**3.9860**

Source: Processed from Pacific Exchange Rate Database.

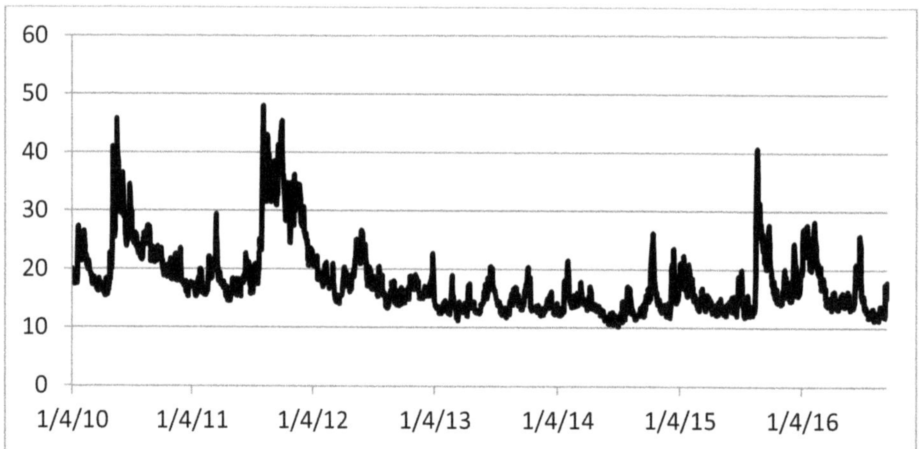

Figure 4. Global Volatility Index (VIX).
Source: Bloomberg

Table 5. Trade weight matrix Indonesia-UK.

Country	United Kingdom	Indonesia
Indonesia	0.0034	–
United Kingdom	–	0.0102

Source: GVAR database (Smith & Gelasi, 2014).

According to Rana (2016), the UK's major trading partners are the EU, the USA, China and India. China is an important trading partner for the UK.

We could also see Trade in Value Added (TiVA) of UK to get the picture of trade relations with UK partners. TiVA of UK considering the value added by each country partner in the production of goods and services in UK. TiVA indicator below presented share of

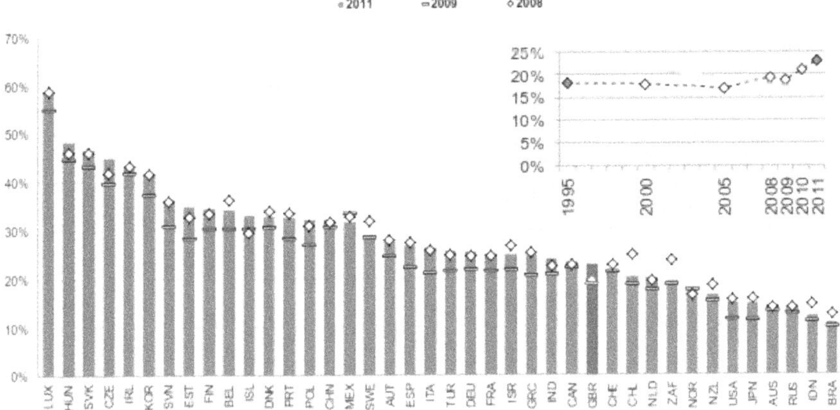

Figure 5. Foreign value added content of gross exports (UK TIVA).
Source: OECD-WTO trade in value added, October 2015.

Table 6. Indonesia-UK macro economy correlation 2010Q1–2015Q4.

Correlation	INA_ GROWTH	INA_ INFLATION	INA_ RATE	UK_ GROWTH	UK_ INFLATION	LIBOR
INA_GROWTH	1.000	−0.394	−0.789	0.021	0.903	0.427
INA_INFLATION	−0.394	1.000	0.579	0.255	−0.285	−0.630
INA_RATE	−0.789	0.579	1.000	0.302	−0.632	−0.477
UK_GROWTH	**0.021**	0.255	0.302	1.000	0.229	−0.293
UK_INFLATION	0.903	**−0.285**	−0.632	0.229	1.000	0.213
LIBOR	0.427	−0.630	**−0.477**	−0.293	0.213	1.000

UK partner contribution in UK exports. Indonesia has only around 15% share in the UK's gross exports (see Figure 5) which also supports the prediction of the low impact of Brexit on direct channels.

2.4 *Macroeconomy correlation*

A simple correlation analysis for Indonesia-UK macroeconomy relations was implemented (see Table 6). The data from 2010Q1–2015Q4 was used and it shows that Indonesian growth and UK growth only have a small correlation (0.021). UK inflation and Indonesian inflation have a small correlation as well. The Indonesian interest rate benchmark (called BI rate) has a negative and modest correlation with the London Interbank Offered Rate (LIBOR).

2.4.1 *Economic growth correlation*

According to Figure 6 and Table 7, Indonesia has a weak correlation in terms of economic growth with the UK at only 0.021. The biggest correlation Indonesia has is with China (0.63). UK growth has a strong correlation with both China and Europe growth. Surprisingly the correlation of the UK's growth with China's is bigger than that of Europe. Thus, even though the correlation between UK growth and Indonesian growth is weak (direct effect), but indirect effects from China can also occur. A second indirect effect could also arise from UK-Europe relationship. The correlation of Indonesian growth with Euro growth is bigger than that of the UK. However, it should be noted that this conclusion needs to be explored further and interpreted cautiously as it is a preliminary assessment. Future research could investigate the channel of direct and indirect effects more precisely.

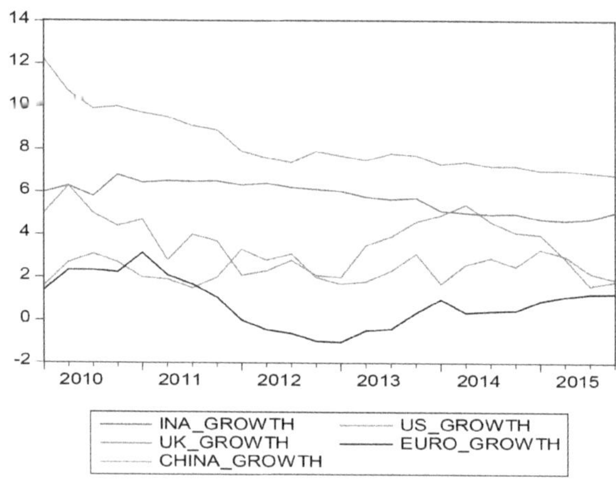

Figure 6. Economic Growth in Indonesia (INA), UK, China, US, and Euro.

Table 7. Correlation of Economic Growth between Indonesia (INA), US, UK, Europe, and China.

Correlation	INA_ GROWTH	US_ GROWTH	UK_ GROWTH	EUROPE_ GROWTH	CHINA_ GROWTH
INA_GROWTH	1.000	–0.192	0.021	0.157	0.630
US_GROWTH	–0.192	1.000	0.099	–0.073	–0.238
UK_GROWTH	**0.021**	**0.099**	1.000	**0.474**	**0.502**
EURO_GROWTH	0.157	–0.073	0.474	1.000	0.597
CHINA_GROWTH	0.630	–0.238	0.502	0.597	1.000

Source: Author's calculation.

To support the figures on economic growth correlation, trade relations between UK and other countries could be explored. The top five UK import-trading partners in 2015 were:

The United States: USD66.5 billion (14.5% of total UK exports)
Germany: 46.4 USD billion (10.1%)
Switzerland: 32.2 USD billion (7%)
China: 27.4 USD billion (5.9%)
France: 27 USD billion (5.9%)
(Data source, World Top Exports, 2016.)

USA is the first import-trading partner of UK, while Germany is in second place. China is in fourth place with a share of 5.9%. From the Chinese side, in 2015 the UK was the seventh import-trading partner, with a 2.6% share. In first place was USA, with Germany in fifth place. It can be concluded that one of the reasons for a high correlation between China and the UK's growth is related to trade relations.

2.4.2 *Interest rate correlation*

The Bank of Indonesia's rate (BI rate) has a negative correlation with LIBORs, but conclusions should be cautiously drawn. The coefficient correlation is relatively modest, higher than with the US prime rate, but the correlation is lower than with the Singapore Interbank Offered Rate. Many market analysts predict that the Bank of England will decrease the interest rate to zero, in anticipation of the lower economic growth of UK after Brexit (see Figure 7 and Table 8).

2.4.3 *Equity market correlation*

According to Table 8, the equity market volatility correlation shows that the Indonesian stock market (IHSG) has a relatively moderate correlation (0.342). The Indonesian stock

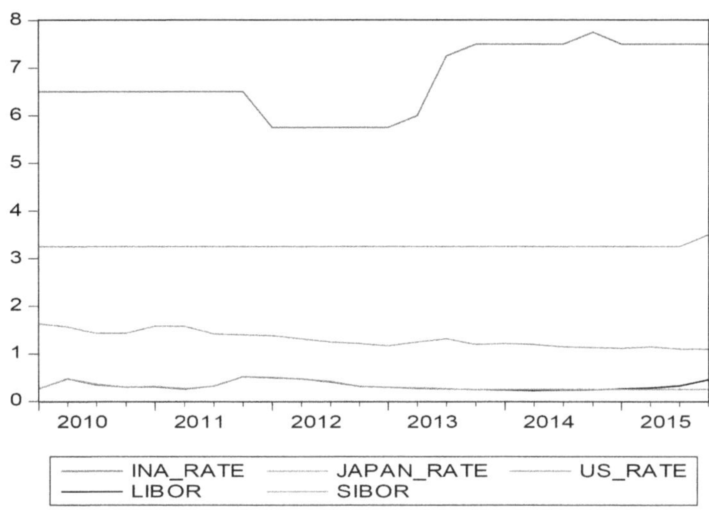

Figure 7. Overnight Interest Rate in Indonesia (INA), UK, China, US, and Euro.

Table 8. Correlation of overnight interest rate between Indonesia (INA), UK, China, US, and Euro.

Correlation	INA_RATE	JAPAN_RATE	US_RATE	LIBOR	SIBOR
INA_RATE	1.000	−0.447	0.226	−0.477	−0.636
JAPAN_RATE	−0.447	1.000	−0.259	0.217	0.397
US_RATE	0.226	−0.259	1.000	0.300	−0.155
LIBOR	**−0.477**	0.217	0.300	1.000	0.875
SIBOR	−0.636	0.397	−0.155	0.875	1.000

Table 9. Correlation of equity market volatility between Indonesia (INA), UK, China, US, and Euro.

Correlation	VOLFTSE	VOLIHSG	VOLDJ	VOLHANGSENG	VOLNIKKEI	VOLSTI
VOLFTSE	1.000	0.342	0.844	0.493	0.195	0.596
VOLIHSG	0.342	1.000	0.321	0.560	0.188	0.685
VOLDJ	0.844	0.321	1.000	0.498	0.236	0.577
VOLHANGSENG	0.493	0.560	0.498	1.000	0.307	0.719
VOLNIKKEI	0.195	0.188	0.236	0.307	1.000	0.195
VOLSTI	0.596	0.685	0.577	0.719	0.195	1.000

market has a higher correlation with the Hang Seng Index and Straits Times Index (STI) Singapore. However, the short-term impact of Brexit has been proven in the stock market reaction including Indonesia.

2.5 *Impulse response analysis from VAR*

A VAR analysis according to Equation 1 was run with the estimated model attached in the appendix.

2.5.1 *Impulse response of Indonesian macroeconomy variables from UK GDP shocks*

Figure 8 will be referred to so as to analyse the responses of the Indonesian exchange rate, interest rate, inflation and GDP to one standard deviation of UK output.

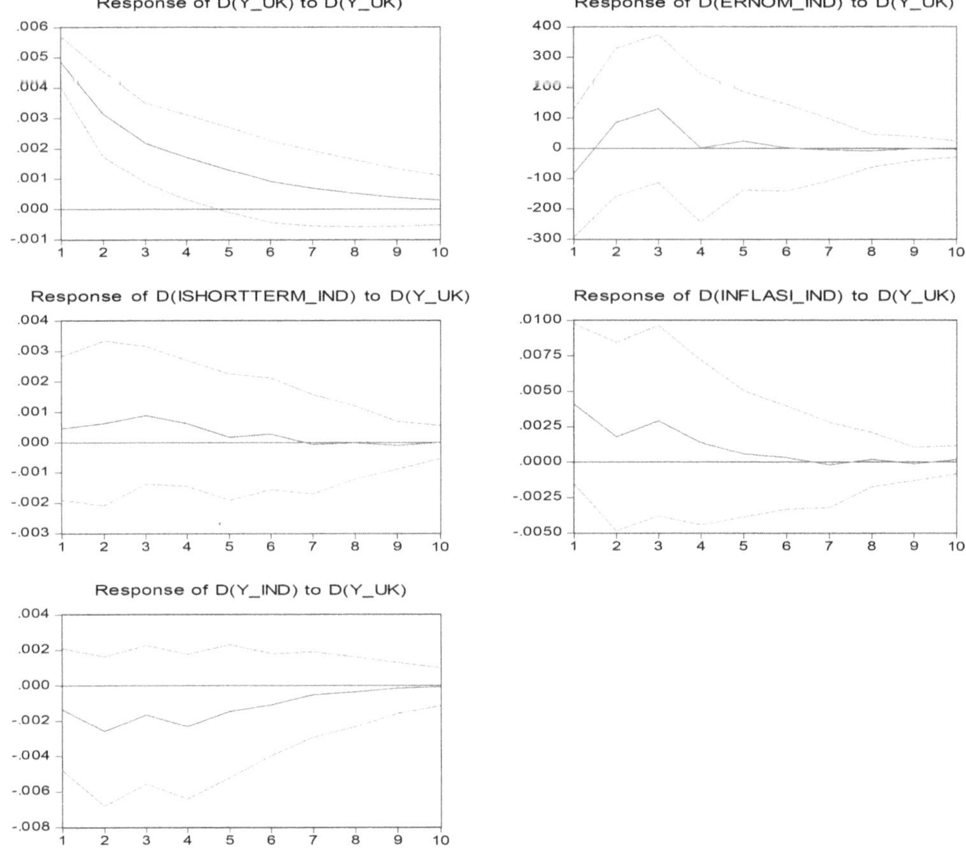

Figure 8. Impulse response of Indonesian exchange rate, interest rate, inflation and GDP to UK GDP shocks.
Notes: Y_IND = Indonesian GDP.
Y_UK = UK GDP.
Short-term_Ind = Indonesian short-term interest rate.
ERNOM_Ind = Indonesian nominal exchange rate.
INFLASI_Ind = Indonesian inflation.

2.5.1.1 *Exchange rate response*

One standard deviation of UK output would depreciate the Indonesian nominal exchange rate in the first period. But in the second and third periods, the exchange rate would appreciate. In the longer term, this change would approach equilibrium (back to zero). In the case of Brexit, a negative UK GDP shock would happen, so the effect on the Indonesian nominal exchange rate would be appreciation in the first period, and depreciation in the second and third periods.

2.5.1.2 *Short term interest rate response*

One standard deviation of UK output would increase the Indonesian short-term interest rate until the seventh period. In the case of Brexit, a negative UK GDP shock would happen, so the effect to the Indonesian interest rate would decrease the Indonesian short-term interest rate. A lower growth prediction of the UK has caused analysts to predict that the Bank of England would lower their interest rate and this could cause Indonesia to respond by lowering its interest rates too. A possible explanation is also from the prediction that Fed would delay the increase of the Fed Fund Rate in response to Brexit, and this delay of Fed fund rate increase could give the room for Indonesia to decrease its interest rate. Notes: usually Indonesian interest rate would increased in response to Fed fund rate increase according to theory of interest rate differential and capital flows in international finance.

2.5.1.3 *Inflation response*

Inflation is increasing in the short run as a result of positive UK GDP shocks and will move back to the equilibrium in the long run. Brexit gives a negative effect that is GDP shock in the UK, Indonesia inflation will decrease. Lower growth in the UK, Europe and China would decrease world demand and also world prices, so it would be possible for Indonesia to lower its inflation.

2.5.1.4 *GDP growth response*

One standard deviation of UK output would decrease Indonesian output (GDP). Surprisingly, the result is not as predicted. Positive output shocks in UK should increase Indonesian GDP, and negative output shocks in UK (Brexit) should decrease Indonesian GDP.

2.5.2 *Impulse response of Indonesian macroeconomy variables from UK long-term interest rate shocks*

To analyze the responses of the Indonesian exchange rate, interest rate, inflation, and GDP to one standard deviation of the UK interest rate, Figure 9 will be referred to.

2.5.2.1 *Exchange rate response*

One standard deviation of the UK interest rate would depreciate the Indonesian nominal exchange rate until the sixth period, but in the seventh period the exchange rate would appreciate. In the longer term, this change would approach equilibrium (back to zero). In the case of Brexit, a negative UK interest shock would happen, so the effect to the Indonesian nominal exchange rate would be appreciated until the seventh period, and would be depreciated in the seventh period to the ninth period. After the ninth period, the shock effect would disappear. Increased risk perception in the UK in the short term could become a positive sentiment for emerging markets like Indonesia in terms of capital inflows. Short-term capital inflow could appreciate the Indonesian rupiah.

2.5.2.2 *Short-term interest rate response*

One standard deviation of the UK interest rate would decrease the Indonesian short-term interest rate until the fifth period and would thereafter increase. In the case of Brexit, a negative UK interest rate shock would happen (interest rate decreases), so the effect to the Indonesian interest rate would be an increase to the Indonesian short-term interest rate until the fifth period which would decrease after the fifth period. There are several predictions that the Bank of England would lower interest rates. Indonesia may respond to a lower UK interest rate by lowering its interest rate too. The possible explanation would also be from the prediction that Fed would delay the increase of the Fed Fund Rate in response to Brexit (the clear explanation similar with sub-chapter 2.5.1.2). The decrease in the Indonesian interest rate will need time for adjustment. The first increase occurs as a result of increased perceptions of risk.

2.5.2.3 *Inflation response*

Indonesia's Inflation is increasing in the short term as a result of positive UK interest rate shocks and will be back to equilibrium in the long term. In the case of Brexit we should again make inverse interpretation because Brexit will decrease UK interest rate. The decrease in UK interest rate will responses by Indonesia's interest rate to decrease too. The decreased in interest rate will give the room for Indonesia's inflation to decrease. Which has a negative effect of UK interest rate shocks, Indonesian inflation will decrease. (The explanation also refers to sub-chapter 2.5.1.2).

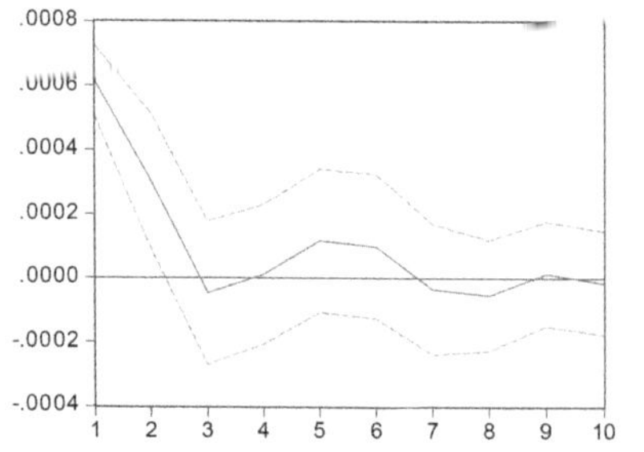

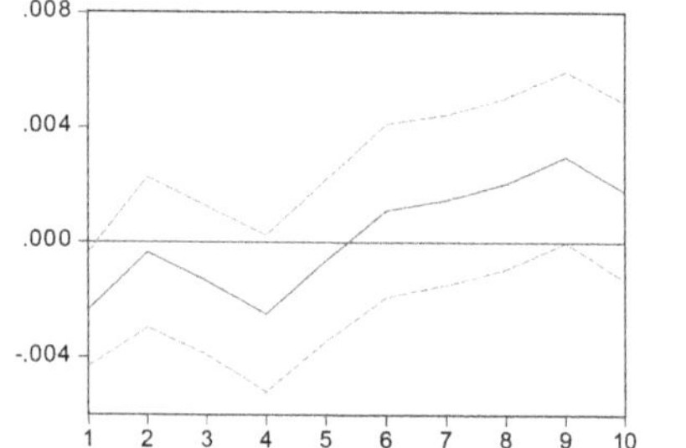

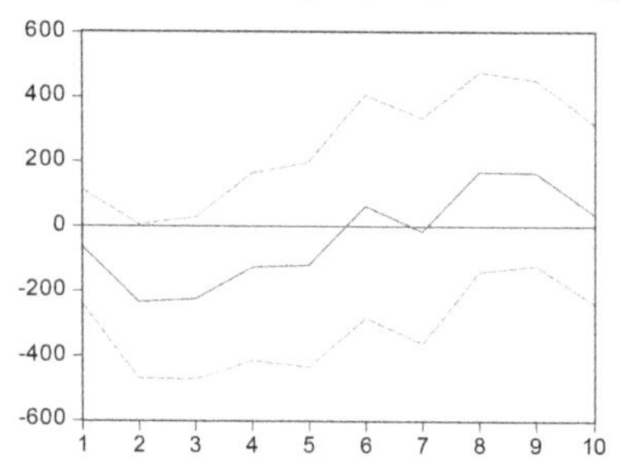

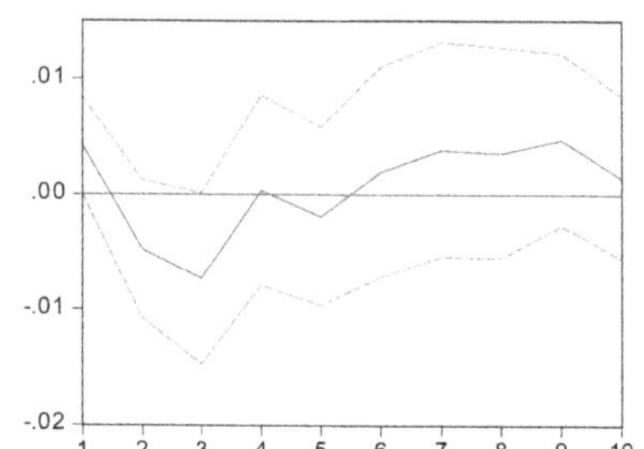

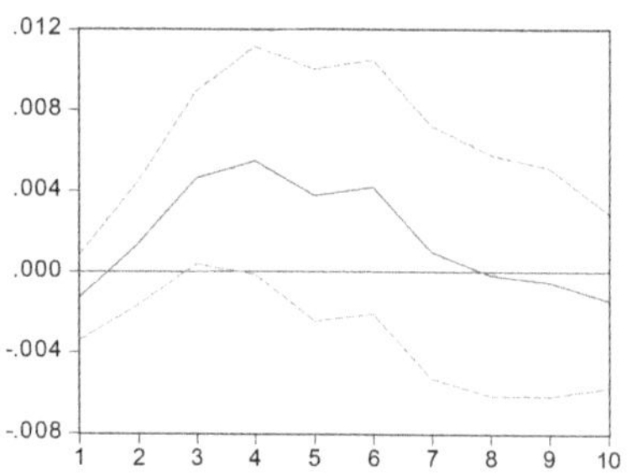

Figure 9. Impulse response of Indonesian exchange rate, interest rate, inflation and GDP to UK interest rateshocks.
Notes: Y_IND = Indonesian GDP.
Y_UK = UK GDP.
Short-term_Ind = Indonesian short-term interest rate.
ERNOM_Ind = Indonesian nominal exchange rate.
INFLASI_Ind = Indonesian inflation.

2.5.2.4 *GDP growth response*

One standard deviation of the UK interest rate would increase Indonesian output (GDP). In the case of Brexit, which has a negative effect of UK interest rate, this would decrease Indonesian output (GDP). This only gives a positive effect in the first period. Indonesian GDP should respond positively to a decrease in the UK interest rate.

2.6 *Global VAR results*

The results of GVAR from Equation 2 with the variables used in GVAR model are real exchange rate; GDP and interest rate are presented and discussed in the following part. The result would be in the form of impulse response of graph for one standard deviation of shocks from UK real exchange rate, UK GDP and UK interest rate.

2.6.1 *Impact of real exchange rate depreciation of pound sterling to several countries' GDP*

2.6.1.1 *Indonesia*

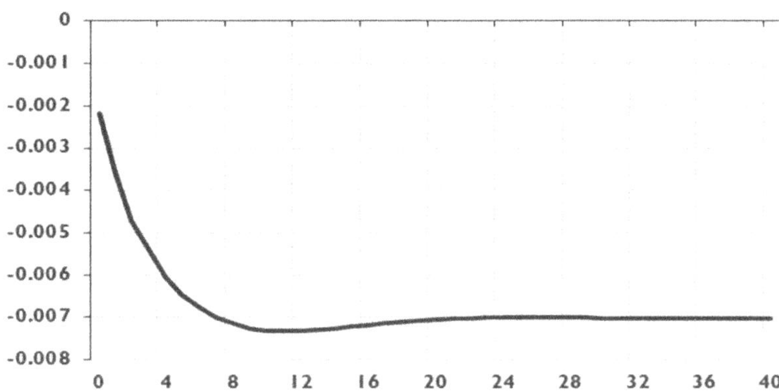

2.6.1.2 *China*

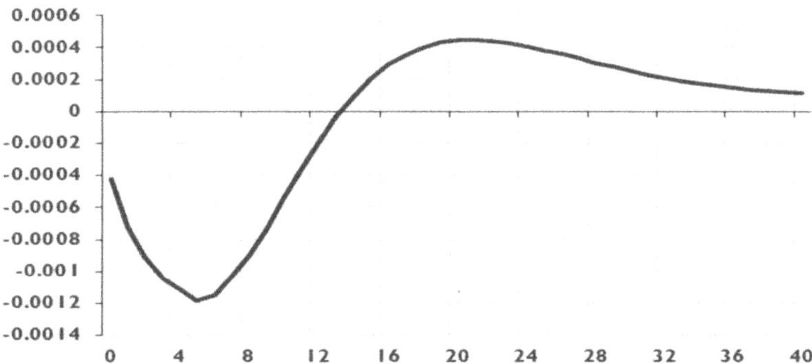

2.6.1.3 *Europe*

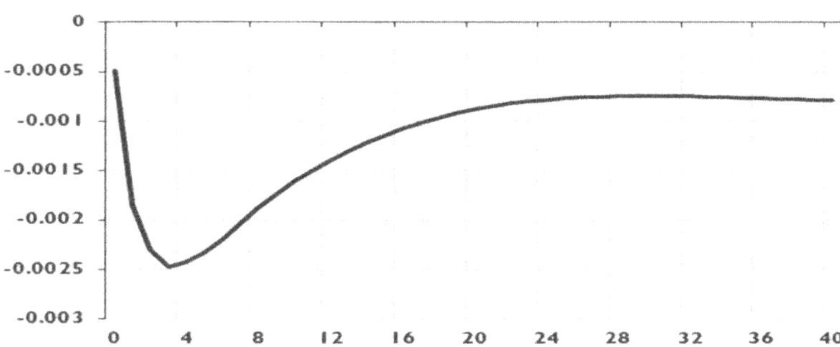

From the GVAR analysis using Mauro and Pesaran (2013) toolbox and data, it was found that the impact to Europe is larger compared to that of China and Indonesia. The impact to Indonesia is low but persistent.

2.6.2 *Impact of real exchange rate depreciation of pound sterling to several countries' real exchange rate*

2.6.2.1 *Indonesia*

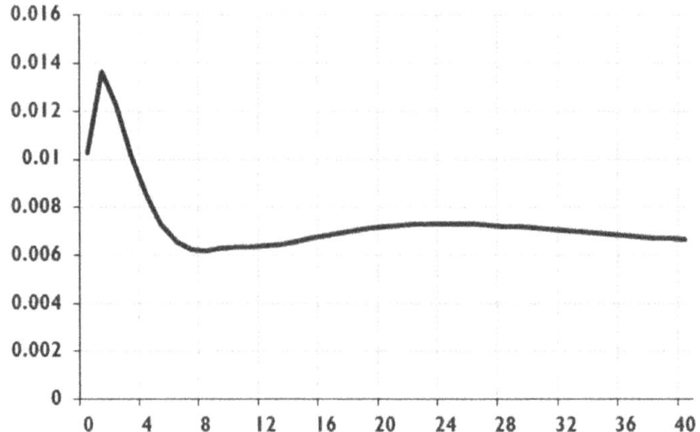

2.6.2.2 *China*

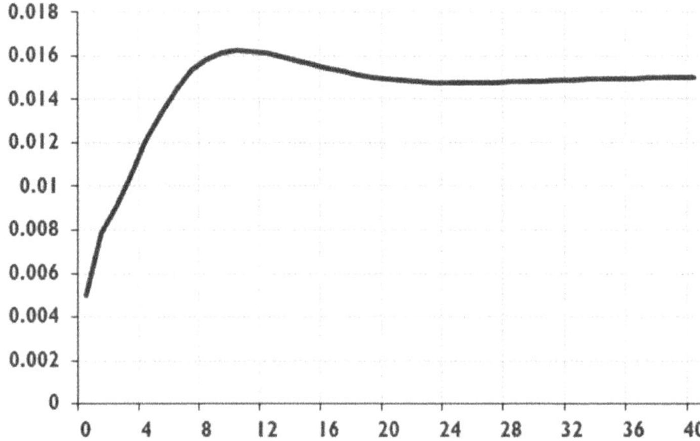

2.6.2.3 *Europe*

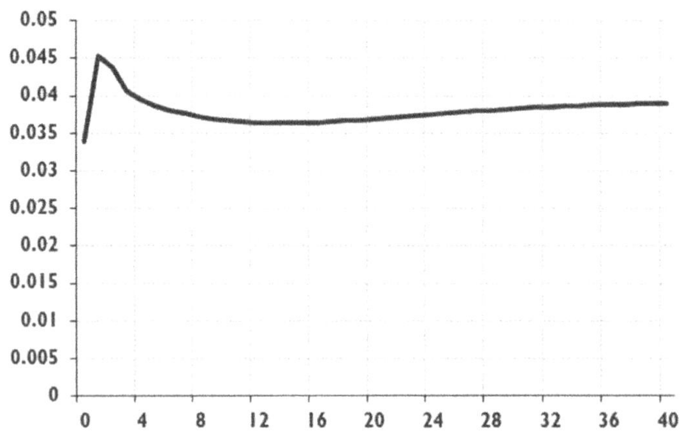

From the GVAR result, it can be seen that Indonesia and Europe both experience depreciation of the real exchange rate. China has a different path which experiences appreciation. Pound sterling is one of the Singapore Dollar (SDR) and international currencies. Decreasing sterling's credibility could increase the demand of Chinese currency in international transactions, so China's real exchange rate could appreciate, and this could then be related to potential support for Renminbi internationalisation.

2.6.3 *Impact of UK Positive GDP shocks to several countries' GDP*

2.6.3.1 *Indonesia*

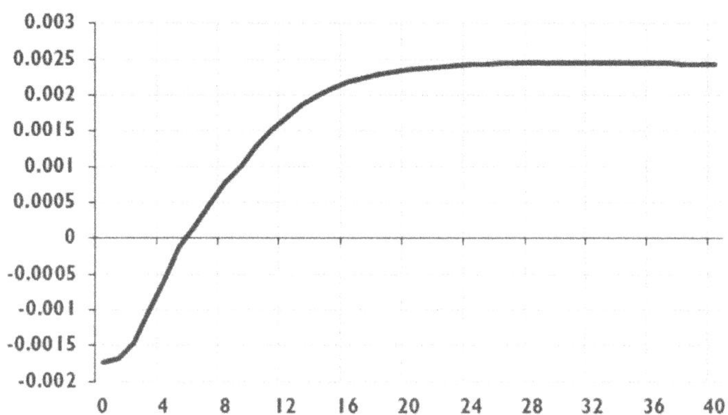

2.6.3.2 *China*

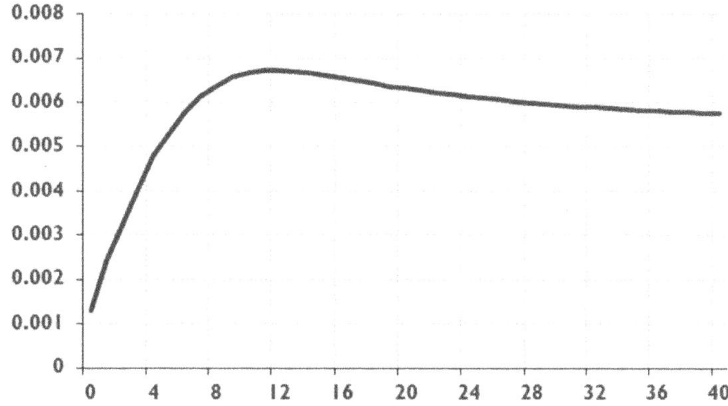

2.6.3.3 *Europe*

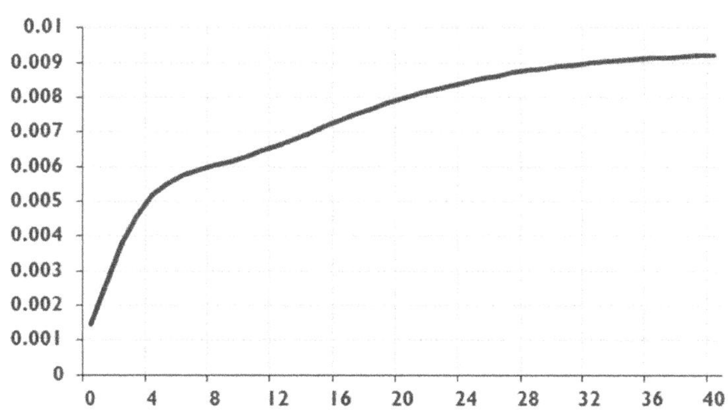

From the graph above, it can be seen that China's response to one positive shock to UK GDP is larger than the Europe, while the Indonesian GDP response is small. In interpreting the direction for Brexit (the negative shock of UK GDP), inverse interpretation should be applied because the default option in the available GVAR toolbox is positive shocks. All of the countries presented have a positive response to positive output (GDP) shocks of UK (Indonesia, China, and Europe). Related with inverse interpretation from GVAR result, in the case of Brexit, there are negative shocks to UK GDP. Hence, the response would be negative for the three sample countries. China has the biggest response. This result is consistent with the result of the correlation coefficient, which found that the China-UK correlation is bigger than the Euro-UK correlation. However, Europe correlation is still high compared to that of other countries.

2.6.4 *Impact of UK positive LIBOR shocks to several countries' Real Exchange Rate (RER)*

2.6.4.1 *Indonesia*

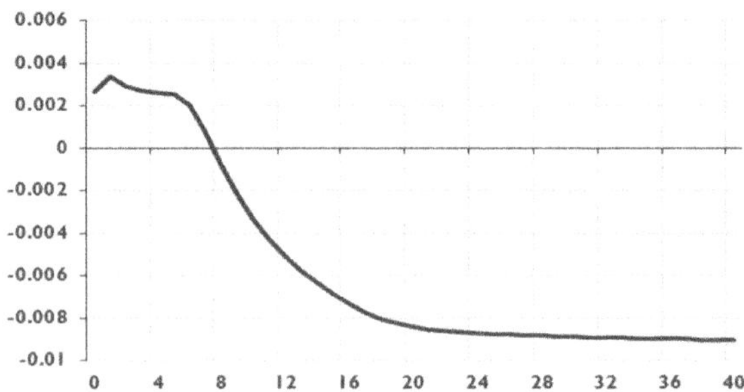

2.6.4.2 *China*

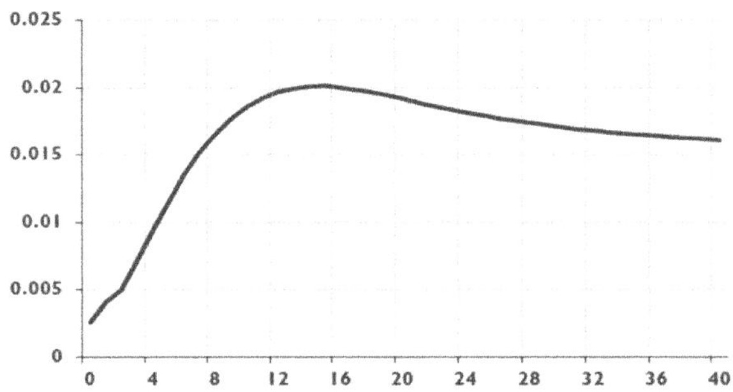

2.6.4.3 *Europe*

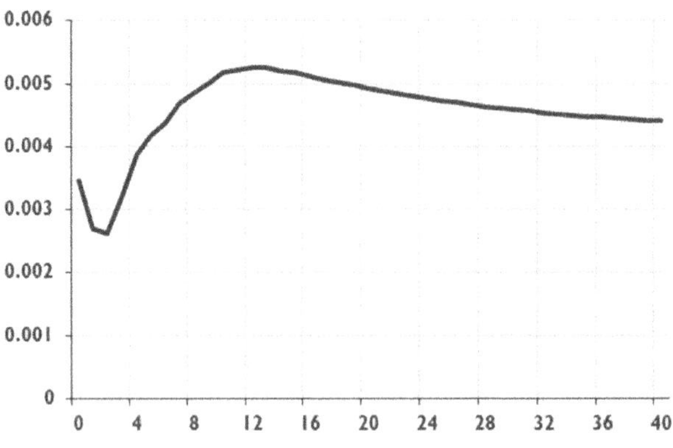

Indonesian real exchange rate (RER) has a different direction in response to positive shocks in LIBOR. Europe and China RER respond positively (in the same direction). Because it is predicted that Brexit will decrease the LIBOR rate, the results should be inversely interpreted. The decrease in the LIBOR rate will increase the Indonesian RER.

3 CONCLUSION

It can be concluded that the direct impact of Brexit on Indonesian economy would be modest especially in real sector channel (GDP, trade, and investment), but one should be careful predicting the indirect impact from the UK-EU and UK-China relationships. Indonesia has a moderate/relatively weak correlation in terms of economic growth and trade channel with the UK, but indirect effect from China could (possibly) work. The close economic relationship between UK-China and Indonesia-China could become the potential source of Brexit impact channeling to Indonesia. Beside UK-China relationship, the second indirect effect of Brexit to Indonesia could also come from the UK-EU relationship. The second important finding from this research is the evidence that Brexit impact is stronger in the monetary/financial channels (stock market, exchange rate, interest rate), than real sector channels (GDP, trade, and investment). The VAR results show some consistencies with analysts' predictions on the impact of Brexit on the Indonesian economy. However the VAR result related with impact to GDP still not consistent with common opinion. On the other hand, the GVAR result for the impact to GDP is consistent with market opinion and theoretical prediction. This preliminary assessment could be investigated further.

Several policy implications could be derived from these results. First, the possible indirect effect from China and eurozone countries should be anticipated. The anticipation is that policies will include increasing economic partnership with countries that currently have a low relationship with UK (non-traditional countries). Second, the credibility of government and the central bank to manage market expectation should be increased because a stronger financial response to Brexit was found (stock market, exchange rate and interest rate). Increased credibility could increase the benefits of the condition depreciated pounds sterling. Third, the potency of the domestic economy should be increased to compensate for the negative impact of Brexit. Indonesia's economic size and population size (represent demographic bonus and potential market) could be used beneficially to face the impact of Brexit.

REFERENCES

Baig, T. (2016). "Brexit loss could be emerging market gain." *Deutsche Bank Emerging Market Monthly Report*, July 1.

Booth, A. (2016). Brexit and ASEAN: What are the lessons? *Public Lecture Presentations at Faculty of Economics and Business*, November, 15, 2016.

Dhingra, S., G. Ottaviano, T. Sampson, and J. Van Reenen (2016). *The consequences of Brexit for UK trade and living standards*. Centre for Economic Performance Working Paper, London, March.

Enders, W. (2015). *Applied econometric time series*. New York: John Wiley & Sons Ltd.

Galbraith, J.K. (2014). *The end of normal: The great crisis and future of growth*. New York: Simon & Schuster.

Global Counsel. (2015). *Brexit: The impact to UK and EU*. https://www.global-counsel.co.uk/sites/default/files/special-reports/downloads/Global%20Counsel_Impact_of_Brexit.pdf, Retrieved at November 11, 2016.

International Monetary Fund. (2016). *World economic outlook update: Uncertainty in the aftermath of UK referendum*. July https://www.imf.org/external/pubs/ft/weo/2016/update/02/. Retrieved at August, 11, 2016.

Kristanto, H. (2016). *Regional Outlook: All About Brexit*. RHB Securities Market Report, June.

Mauro, F., and M.H. Pesaran (2013). *The GVAR handbook: Structure and applications of macro model of the global economy for policy analysis*. Oxford: Oxford University Press.

Nasser, J. and Kurniawan, B. (2016). "Brexit wounds: Not so serious" Credit Suisse Report in *Asian Daily*, June 27.

Pesaran, M.H., Schuermann, T. and Smith, V.L., (2008). *Forecasting economic and financial variables with global VAR*. Staff Reports. https://www.newyorkfed.org/research/staff_reports/sr317.html., Retrieved at August, 11, 2016.
Rana, P.B. (2016). *Brexit impact on Asia*. RSIS Commentary.
(Timmer, M., Erumban, A., Bart, R.G., Temurshoev, U. and De Vries, G.Z. (2012). *The world input output database (WIOD): Contents, sources, and method*. European Commision Working Paper, April. http://www.wiod.org/publications/source_docs/WIOD_sources.pdf, Retrieved at December, 5, 2016.

APPENDIX

1. VAR Result Estimation: UK GDP Shocks.

Vector Autoregression Estimates
Sample (adjusted): 1996Q3 2013Q1
Included observations: 67 after adjustments
Standard errors in ( ) & t-statistics in [ ]

	D(Y_UK)	D(ERNOM_ IND)	D(ISHORTTERM_ IND)	D(INFLASI_ IND)	D(Y_IND)
D(Y_UK(−1))	0.719692	3470.391	0.120835	0.933158	−0.442082
	(0.13610)	(24223.7)	(0.27226)	(0.65310)	(0.39750)
	[5.28797]	[0.14326]	[0.44382]	[1.42882]	[−1.11217]
D(Y_UK(−2))	0.058259	6947.003	−0.116715	0.010817	0.147697
	(0.13626)	(24251.8)	(0.27258)	(0.65385)	(0.39796)
	[0.42756]	[0.28645]	[−0.42819]	[0.01654]	[0.37114]
D(ERNOM_IND(−1))	1.08E-06	−0.047804	3.02E-06	1.38E-05	−5.60E-06
	(7.2E-07)	(0.12770)	(1.4E-06)	(3.4E-06)	(2.1E-06)
	[1.50000]	[−0.37435]	[2.10264]	[3.99736]	[−2.67250]
D(ERNOM_IND(−2))	1.60E-06	−0.385401	−2.36E-06	−1.25E-06	−1.08E-06
	(8.2E-07)	(0.14583)	(1.6E-06)	(3.9E-06)	(2.4E-06)
	[1.95811]	[−2.64290]	[−1.44071]	[−0.31879]	[−0.45076]
D(ISHORTTERM_ IND(−1))	−0.040854	20437.42	−0.051912	−0.216405	0.127111
	(0.06610)	(11764.3)	(0.13222)	(0.31718)	(0.19304)
	[−0.61809]	[1.73724]	[−0.39261]	[−0.68229]	[0.65845]
D(ISHORTTERM_ IND(−2))	−0.041485	32768.42	0.230465	1.111683	−0.676609
	(0.06050)	(10767.9)	(0.12102)	(0.29031)	(0.17669)
	[−0.68572]	[3.04316]	[1.90429]	[3.82926]	[−3.82927]
D(INFLASI_IND(−1))	−0.051350	5797.657	0.118219	−0.334443	−0.194388
	(0.02921)	(5199.55)	(0.05844)	(0.14019)	(0.08532)
	[−1.75775]	[1.11503]	[2.02292]	[−2.38573]	[−2.27830]
D(INFLASI_IND(−2))	−0.027358	906.6785	0.161706	0.009991	0.085124
	(0.02812)	(5005.80)	(0.05626)	(0.13496)	(0.08214)
	[−0.97275]	[0.18113]	[2.87415]	[0.07403]	[1.03630]
D(Y_IND(−1))	0.023817	−23162.84	0.127192	0.088600	0.120483
	(0.04267)	(7594.87)	(0.08536)	(0.20477)	(0.12463)
	[0.55814]	[−3.04980]	[1.49004]	[0.43269]	[0.96675]
D(Y_IND(−2))	0.005140	32591.46	−0.003890	0.683752	0.026929
	(0.03525)	(6274.37)	(0.07052)	(0.16916)	(0.10296)
	[0.14580]	[5.19438]	[−0.05516]	[4.04197]	[0.26156]
C	0.000344	39.17408	−0.001506	−0.012882	0.009917
	(0.00103)	(183.668)	(0.00206)	(0.00495)	(0.00301)
	[0.33364]	[0.21329]	[−0.72965]	[−2.60147]	[3.29060]
R-squared	0.542041	0.511802	0.375271	0.515292	0.548569
Adj. R-squared	0.460263	0.424623	0.263713	0.428736	0.467956
Sum squared residual	0.001317	41721009	0.005270	0.030327	0.011234

(*Continued*)

	D(Y_UK)	D(ERNOM_ IND	D(ISHORTTERM_ IND	D(INFLASI_ IND)	D(Y_IND)
S.E. equation	0.004850	863.1443	0.009701	0.023271	0.014164
F-statistic	6.628167	5.870747	3.363892	5.953336	6.804984
Log likelihood	267.9734	−542.0199	221.5179	162.8951	196.1632
Akaike AIC	−7.670849	16.50806	−6.284116	−4.534182	−5.527258
Schwarz SC	−7.308885	16.87002	−5.922152	−4.172218	−5.165294
Mean dependent	0.004531	110.1463	−0.000351	0.000207	0.009599
S.D. dependent	0.006601	1137.908	0.011306	0.030789	0.019418
Determinant residual covariance (dof adj.)		8.44E-11			
Determinant residual covariance		3.44E-11			
Log likelihood		331.7411			
Akaike information criterion		−8.260928			
Schwarz criterion		−6.451106			

2. VAR Result Estimation: UK Interest Rate Shocks.

Vector Autoregression Estimates
Sample (adjusted): 1997Q2 2013Q1
Included observations: 64 after adjustments
Standard errors in ( ) & t-statistics in [ ]

	D(ILONGTERM_ UK)	D(ISHORTTERM_ IND)	D(ERNOM_ IND)	D(INFLASI_ IND)	D(Y_ IND)
D(ILONGTERM_ UK(−1))	0.628060	−0.605708	−322511.1	−1.653192	2.830582
	(0.17197)	(2.25758)	(195770.)	(4.73427)	(2.38784)
	[3.65223]	[−0.26830]	[−1.64739]	[−0.34920]	[1.18542]
D(ILONGTERM_ UK(−2))	−0.579373	−1.917969	59761.87	−1.966487	−2.280800
	(0.18592)	(2.44081)	(211659.)	(5.11851)	(2.58163)
	[−3.11619]	[−0.78579]	[0.28235]	[−0.38419]	[−0.88347]
D(ILONGTERM_ UK(−3))	0.425561	−0.817133	−10714.78	1.740063	2.783842
	(0.19234)	(2.52504)	(218964.)	(5.29515)	(2.67073)
	[2.21255]	[−0.32361]	[−0.04893]	[0.32861]	[1.04235]
D(ILONGTERM_ UK(−4))	−0.019769	−0.279924	−290429.0	−2.380438	1.653449
	(0.17984)	(2.36101)	(204739.)	(4.95117)	(2.49723)
	[−0.10992]	[−0.11856]	[−1.41853]	[−0.48078]	[0.66211]
D(ILONGTERM_ UK(−5))	−0.165817	1.678360	146634.3	1.200267	−0.794150
	(0.16257)	(2.13420)	(185072.)	(4.47555)	(2.25734)
	[−1.01998]	[0.78641]	[0.79231]	[0.26818]	[−0.35181]
D(ISHORTTERM_ IND(−1))	0.019552	0.057823	28812.59	0.378003	0.012739
	(0.01406)	(0.18456)	(16004.9)	(0.38704)	(0.19521)
	[1.39076]	[0.31329]	[1.80024]	[0.97665]	[0.06526]
D(ISHORTTERM_ IND(−2))	−0.018292	0.357390	49250.00	2.449168	−1.282452
	(0.01384)	(0.18166)	(15752.9)	(0.38095)	(0.19214)
	[−1.32189]	[1.96737]	[3.12641]	[6.42912]	[−6.67457]
D(ISHORTTERM_ IND(−3))	0.018710	0.203519	42157.07	1.116395	−0.604879
	(0.01732)	(0.22741)	(19720.1)	(0.47689)	(0.24053)
	[1.08010]	[0.89495]	[2.13778]	[2.34101]	[−2.51479]
D(ISHORTTERM_ IND(−4))	0.005503	−0.190015	−14756.26	0.224327	0.408637
	(0.01571)	(0.20618)	(17879.0)	(0.43236)	(0.21807)
	[0.35042]	[−0.92162]	[−0.82534]	[0.51884]	[1.87387]

(*Continued*)

	D(ILONGTERM_ UK)	D(ISHORTTERM_ IND)	D(ERNOM_ IND)	D(INFLASI_ IND)	D(Y_ IND)
D(ISHORTTERM_ IND(–5))	–0.002941 (0.01202) [–0.24468]	–0.076143 (0.15782) [–0.48246]	–44505.32 (13685.9) [–3.25192]	–1.126690 (0.33096) [–3.40429]	–0.164779 (0.16693) [–0.98712]
D(ERNOM_ IND(–1))	–1.01E-07 (1.5E-07) [–0.67280]	–1.17E-06 (2.0E-06) [–0.59704]	–0.095277 (0.17051) [–0.55877]	4.81E-06 (4.1E-06) [1.16568]	–5.93E-06 (2.1E-06) [–2.84996]
D(ERNOM_ IND(-2))	2.01E-07 (1.6E-07) [1.29043]	–2.71E-06 (2.0E-06) [–1.32654]	–0.208176 (0.17720) [–1.17484]	–3.46E-06 (4.3E-06) [–0.80764]	9.02E-07 (2.2E-06) [0.41716]
D(ERNOM_ IND(–3))	4.21E-08 (1.5E-07) [0.27478]	–2.19E-06 (2.0E-06) [–1.08684]	–0.143621 (0.17439) [–0.82358]	–5.17E-07 (4.2E-06) [–0.12251]	–5.91E-07 (2.1E-06) [–0.27779]
D(ERNOM_ IND(–4))	2.68E-07 (1.3E-07) [2.10421]	1.97E-06 (1.7E-06) [1.18158]	–0.181736 (0.14482) [–1.25494]	1.06E-06 (3.5E-06) [0.30331]	1.81E-06 (1.8E-06) [1.02536]
D(ERNOM_ IND(–5))	–1.37E-07 (1.4E-07) [–1.00532]	6.69E-07 (1.8E-06) [0.37393]	0.133010 (0.15522) [0.85690]	–2.65E-06 (3.8E-06) [–0.70637]	–2.36E-06 (1.9E-06) [–1.24411]
D(INFLASI_ IND(–1))	–0.006333 (0.00518) [–1.22277]	–0.003233 (0.06800) [–0.04755]	3550.125 (5896.34) [0.60209]	–0.577225 (0.14259) [–4.04815]	–0.168747 (0.07192) [–2.34637]
D(INFLASI_ IND(–2))	0.007358 (0.00570) [1.29092]	0.036232 (0.07483) [0.48419]	–5874.667 (6488.97) [–0.90533]	–0.520993 (0.15692) [–3.32010]	0.046995 (0.07915) [0.59378]
D(INFLASI_ IND(–3))	–0.007672 (0.00560) [–1.37008]	–0.091881 (0.07351) [–1.24990]	–16400.11 (6374.66) [–2.57270]	–0.715948 (0.15416) [–4.64427]	0.230137 (0.07775) [2.95986]
D(INFLASI_ IND(–4))	–0.011196 (0.00664) [–1.68722]	–0.033390 (0.08711) [–0.38330]	–9576.115 (7554.06) [–1.26768]	–0.638774 (0.18268) [–3.49672]	0.239828 (0.09214) [2.60293]
D(INFLASI_ IND(–5))	0.000414 (0.00528) [0.07847]	–0.039698 (0.06932) [–0.57266]	–4214.686 (6011.44) [–0.70111]	–0.346874 (0.14537) [–2.38609]	0.093186 (0.07332) [1.27091]
D(Y_IND(–1))	0.014160 (0.00972) [1.45744]	–0.070591 (0.12755) [–0.55345]	–10071.46 (11060.5) [–0.91058]	0.142752 (0.26747) [0.53371]	–0.022409 (0.13491) [–0.16611]
D(Y_IND(–2))	0.005618 (0.00835) [0.67248]	–0.286533 (0.10967) [–2.61273]	12738.15 (9510.08) [1.33944]	–0.265011 (0.22998) [–1.15232]	0.086033 (0.11600) [0.74169]
D(Y_IND(–3))	0.007176 (0.00991) [0.72424]	0.409922 (0.13008) [3.15124]	–14331.82 (11280.4) [–1.27051]	0.059002 (0.27279) [0.21629]	0.314958 (0.13759) [2.28913]
D(Y_IND(–4))	–0.015507 (0.01083) [–1.43124]	0.068745 (0.14224) [0.48331]	–13618.83 (12334.4) [–1.10413]	0.149643 (0.29828) [0.50169]	–0.035972 (0.15044) [–0.23911]
D(Y_IND(–5))	–0.000494 (0.00966) [–0.05107]	0.122755 (0.12688) [0.96749]	4288.399 (11002.7) [0.38976]	–0.153892 (0.26608) [–0.57838]	0.288104 (0.13420) [2.14680]
C	–0.000258 (0.00020) [–1.32315]	–0.002629 (0.00256) [–1.02574]	296.5565 (222.228) [1.33447]	0.000918 (0.00537) [0.17088]	0.004592 (0.00271) [1.69428]

(*Continued*)

	D(ILONGTERM_ UK)	D(ISHORTTERM_ IND)	D(ERNOM_ IND)	D(INFLASI_ IND)	D(Y_ IND)
R-squared	0.528378	0.703824	0.780480	0.823879	0.879071
Adj. R-squared	0.218100	0.508972	0.636059	0.708010	0.799512
Sum squared residual	1.45E-05	0.002494	18754536	0.010968	0.002790
S.E. equation	0.000617	0.008101	702.5243	0.016989	0.008569
F-statistic	1.702918	3.612091	5.404199	7.110444	11.04931
Akaike AIC	−11.65187	−6.502371	16.23844	−5.021300	−6.390181
Schwarz SC	−10.77482	−5.625324	17.11549	−4.144254	−5.513135
Mean dependent	−0.000201	−0.000288	114.2078	−0.000110	0.009457
S.D. dependent	0.000698	0.011561	1164.517	0.031440	0.019137
Determinant residual covariance (dof adj.)		1.11E-13			
Determinant residual covariance		8.23E-15			
Log likelihood		583.7460			
Akaike information criterion		−14.17956			
Schwarz criterion		−9.794332			

Competition and Cooperation in Economics and Business – Gani et al. (Eds)
ISBN 978-1-138-62666-9

Dynamics of return on investment for highly educated workers

S. Nurteta
Statistics Indonesia of Jambi Province, Jambi, Indonesia

D. Handayani
Department of Economics, Faculty of Economics and Business, Universitas Indonesia, Depok, Indonesia

R. Indrayanti
Institute of Demography, Faculty of Economics and Business, Universitas Indonesia, Depok, Indonesia

ABSTRACT: This study aims to analyse the dynamics of the return on investment of higher education in Indonesia, by using cross section data of Indonesian Family Life Survey (IFLS) in 2000, 2007 and 2014 on the labour force aged 15–64 years. This study uses the Two Step Heckman method which calculates the probability of participation work with the Probit model to derive an inverse Mills ratio and then estimates earnings using the Mincerian Earning Function model with the inverse Mills ratio as one of the independent variables. The analysis shows that the rate of return in the investment of diploma is higher than S1/S2/S3 (undergraduate/graduate programmes). After being compared with the level of interest rates in 2000, 2007 and 2014, the investment in diploma is more profitable than the investment in S1/S2/S3. Thus, compared with the diploma level, in its development the S1/S2/S3 level shows that it has a higher rate of return to education than the diploma level as work experience increases.

1 INTRODUCTION

1.1 *Background*

Skill is needed to have a decent job. The more workers have skills, the more productive they are, and ultimately wellbeing will be achieved (ILO 2010). Indonesia's involvement in MEA in early 2016 forced Indonesia to prepare its workers in order to be able to compete with other countries. Besides this, the international migrant flow has increased, so Indonesian labour must compete with foreign workers (ILO, 2010). Workers with higher education are associated as workers with better skills (Ministry of Planning/Bappenas, 2015). Moreover, a further issue has been raised by Arkes (1999) that higher education is also a selection of tools in manpower recruitment.

High income is a pull factor affecting work participation. When the income being offered is high, so too is work participation. Increasing age creates a 'U' reversed curve on the income curve, meaning marginal revenue initially rises, but after a certain point it will decrease. The decrease describes the depreciation of human capital. And then as the wage decreases, the work participation also decreases (Handayani 2006).

Furthermore, some people decide to go to school as an investment to upgrade their skills, while other individuals decide to stop schooling at a relatively young age. Individuals who decide to go to school are ready to for go their present earnings in order to attain higher earnings in the future. This exchange position between present and future earnings also considers education fees. Education investment can be reflected in the return on investment of education. The measurement method for calculating the rate of return investment, namely the earning function to the level of education, is one of the explanatory variables. The education system is a system that interact with each other. Hence, the success of education investment is influenced by many factors that interact with each other.

1.2 *Research issues*

Based on the Labour Force Data, the Indonesian labour force is still dominated by people with basic and lower education, while the number of highly educated workers in the labour force is only about 10 per cent despite an increase every year (BPS 2016, 2011, 2010, 2001).

The wages growth of highly educated workers grew varied. Wages of individuals graduating from the diploma I/II/III programme continued to increase during the period of 2009–2015, with the largest growth in 2013 showing an increase of 13.56 per cent, and the lowest growth in 2014 with only 0.69 percent. As for the wages of workers with S1/S2/S3 level, continued to increase during the period of 2009–2015, except for 2010 their wages decreased by 0.71 percent. In comparison, wage growth for highly educated people is not as high as other levels of education, except in 2011 for educated workers graduating from D I/II/III, and 2015 for workers graduating from S1/S2/S3. Having a higher education does not directly increase wages annually, and the wages can even decrease. This is the next question: if higher education is assumed to be able to maximise earnings, perhaps this assumption needs to be examined again. If getting a higher income is usually the reason for workers to decide to continue their education, then why would some people decide to go to higher education, while others do not? The potential higher income does not necessarily mean that the education investment rate will also increase.

During a period of about 20 years, the investment rate of return to education in Indonesia, especially for higher education, shows varying trends. According to Purnastuti et al. (2013), the rate of return to education investment at diploma/bachelor degree level in 2007 shows a decline in both men (5.63 percent) and women (7.63 per cent) compared to 1993 with 9.78 per cent for men and 9.02 per cent for women. However, at graduate level it actually shows an increase from 6.04 per cent for men and 5.09 per cent for women in 1993, and it became 6.81 per cent for men and 8.67 per cent for women in 2007. However, Behrman and Deolalikar (1993) in Dumauli (2015) show that the rate of return to education in 1986 for higher education is still higher than in 2007 both in men (9.2–10.7 per cent) and in women (13.3–16.6 per cent). Thus, this research is trying to analyse the rate of return and development of higher education in Indonesia during 2000, 2007, and 2014.

2 LITERATURE REVIEW

2.1 *Literature review*

Human capital theory states that the skills learned at school can directly increase productivity and provide higher income. In addition, employers use educational attainment as a screening device for assessing workers' productivity quickly and cheaply. At the same time, workers also use education as a signal to employers for their own productivity. The higher the education, the higher productivity the workers are assumed to have. Therefore, individuals with higher education tend to earn higher incomes, not because the school makes them more productive, but because it has provided them with more productive credentials (Hungerford & Solon, 1987).

The purpose of investment in education is basically to improve the quality of human capital. Forms of this educational investment are divided into two, which are investments made by the individuals themselves (private investment), and investments made by the government/community, usually in the allocation of funds, counselling, training and construction of schools (Setyonaluri, 2002). Continuing education involves two different types of expenses. Each one year of schooling is equal to one year out of the labour force (or still working but receiving lower income), so university education forces workers to accept the lack of income; this is called the opportunity cost of schooling.

A person's decision to work is influenced by different factors. Becker (1976) states that individuals who participate in the labour market are faced with the choice between working and not working. In the model of labour supply, the decision to work or not work is based on rationality in order to obtain maximum satisfaction. Individual satisfaction depends on taste, the number of market goods (C), and leisure (L) which are consumed. To maximise satisfaction, individuals are faced with budget constraints (the price of market goods and leisure) and the total time (T) allocated to maximise the satisfaction of individuals (Borjas, 2007).

If analysing labour supply only uses the data from the existing individual earnings, then the estimation using Ordinary Least Squares (OLS) will be biased because it does not account for the samples of those who are not working. Therefore, the estimates of work participation will be calculated with the Probit model, where decisions to work or not work are taken into account.

2.2 *Theoretical framework*

Figure 1 above illustrates the theoretical framework used in this study. From the theory presented earlier, which begins with human capital theory (Becker 1976), three main variables in investment in human resources are presented: education, health and migration. One of the main variables is education, characterised by a diploma, as a signal for workers' productivity and also as a screening device for a company to establish workers' productivity quickly and easily. However, it also has a sheepskin effect that eventually provides a real assessment of workers' productivity for the company by their performance on the job. The real performance will affect earnings, which is used by Mincer as a proxy for assessing the return on investment in education.

The theory of Heckman (1979) argues that the sample selection to calculate the estimated earnings presented earlier by Mincer has shortcomings, namely, the existence of selectivity bias, because not all units have earnings. Heckman then offers a new method by first calculating the probability of participation works, which will then be obtained by the inverse Mills ratio as a correction of selectivity bias. The inverse Mills ratio is then included as one of the independent variables in the Mincer earnings function.

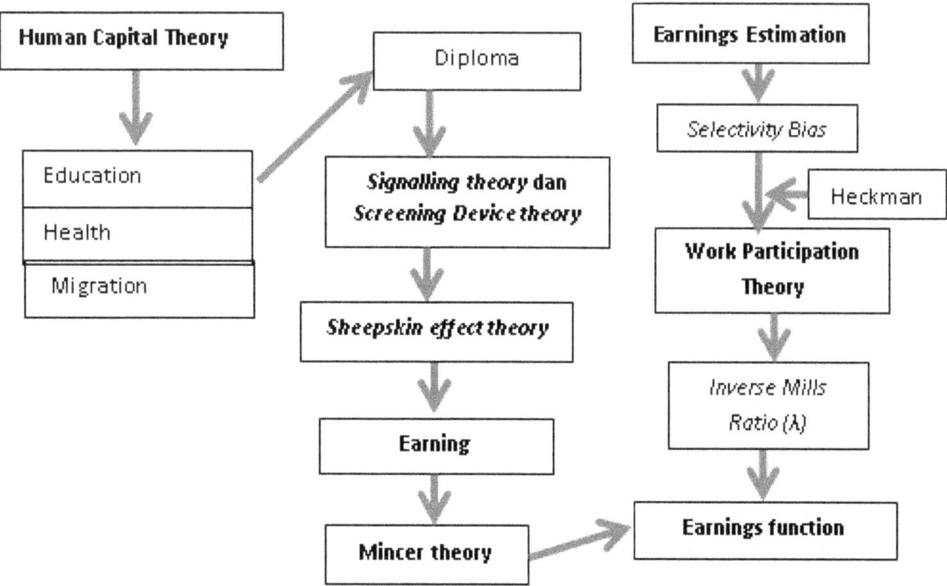

Figure 1. Theoretical framework scheme.

2.3 *Research hypotheses*

The hypotheses to be tested in this study are as follows:

1. The rate of return on investments in education for diploma and S1/S2/S3 graduate workers is higher than for other levels of education graduate workers.
2. The rate of return on investment for higher education has increased over the period 2000–2014.

3 RESEARCH METHODS

3.1 *Data sources and observation units*

The data used in this study are longitudinal data of Indonesian Family Life Survey (IFLS) in 2000, 2007 and 2014. All variables used in this study were taken from the book IIIA, book II, control books and proxy books on the questionnaire of IFLS year 2000, 2007 and 2014 by RAND Labor and Population, except for 2014 which does not use a proxy book because the book is not available.

The unit of the analysis in this research is the main activities of the past week of the labour force aged 15–64 years who are working, while some are not working but have a job, are unemployed, or looking for work. For the analysis of the probability of working, the number of observation units in 2000 was 17,444 people, 19,807 in 2007 and 23,324 in 2014. As for the analysis of income estimates, the number of working individuals in 2000 was 16,952, with 19,752 in 2007, and 22,569 in 2014.

3.2 *Framework analysis*

3.3 *Analysis method*

This study used two types of analysis to give a general overview of the labour force in Indonesia as shown in Figure 2; the descriptive analysis and inferential analysis. A picture is obtained by cross-tabulation between the dependent variables and the independent variables between years 2000, 2007, and 2014. The inferential analysis used the Two Step Heckman method to calculate the probability of work participation with the Probit model and used the Mincerian Earning Function model to calculate the estimated earnings. Both of these calculations are calculated for each of the years 2000, 2007 and 2014 and then calculated for the return on educational investment for each of the years 2000, 2007 and 2014 and its development during the same period.

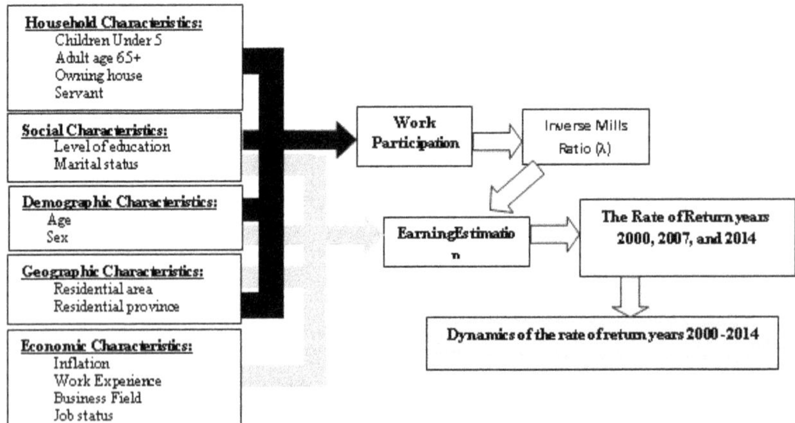

Figure 2. Framework analysis scheme.

3.4 *Research model*

In this study, the work participation functions are analysed using the Probit model with sample data that consists of people aged 15–64 years who are in the labour force, code 1 = is for participation code, while the code 0 = not participating.

$$
\begin{aligned}
Z_i = {} & \beta_{0t} + \beta_{1t} DIDIK_1 + \beta_{2t} DIDIK_2 + \beta_{3t} DIDIK_3 + \beta_{4t} DIDIK_4 + \beta_{4t} UMUR \\
& + \beta_{5t} UMUR^2 + \beta_{6t} JK + \beta_{7t} KAWIN + \beta_{8t} BALITA + \beta_{9t}\, LANSIA \\
& + \beta_{10t} RUMAH + \beta_{11t} PRT + \beta_{12t} DTT + \beta_{13t} JAWA + \varepsilon_t
\end{aligned} \tag{1}
$$

Furthermore, for earnings estimation, the Mincerian Earning Function models were used, namely:

$$
\begin{aligned}
\ln(W_t) = {} & \alpha_{0t} + \alpha_{1t} DIDIK_1 + \alpha_{2t} DIDIK_2 + \alpha_{3t} DIDIK_3 + \alpha_{4t} DIDIK_4 + \alpha_{4t} JK \\
& + \alpha_{5t} KERJA + \alpha_{6t} KERJA^2 + \alpha_{7t} LAPUS_1 + \alpha_{8t} LAPUS_2 + \alpha_{9t} STATUS \\
& + \alpha_{10t} DTT + \alpha_{11t} JAWA + \alpha_{12t} INFLASI + \lambda_t + \varepsilon_t
\end{aligned} \tag{2}
$$

where ln (Wt) is the natural logarithm of earning, while the definitions of the independent variables for both models above are as follows:

Table 1. Variable definition.

Variable	Operational definition
The highest education attainment	DIDIK1 = 1, if not in school; DIDIK1 = 0, if SMA (Senior High School). DIDIK2 = 1, if SD (Elementary School)/SMP (Junior High School) or equal level school; DIDIK2 = 0, if SMA. DIDIK3 = 1, if diploma I/II/III; DIDIK3 = 0, if SMA. DIDIK4 = 1, if DIV/SI/S2/S3; DIDIK4 = 0, if SMA.
Sex	JK = 1, if males; JK = 0, if females.
Marital status	KAWIN = 1, if married/have been married. KAWIN = 0, if not married.
Under five-year old child	BALITA = 1, if there is a child under 5 years old in a household. BALITA = 0, if there is not.
Adult aged 65 and above	LANSIA = 1, if there is an adult aged 65 and above. LANSIA = 0, if there is not.
Owning house	RUMAH = 1, if having a house. RUMAH = 0, if not having a house.
Servant	PRT = 1, if there is a servant. PRT = 0, if there is not.
Age	UMUR; Numeric variable
Work experience	KERJA, numeric variable
Business field	LAPUS1 = 1, if in the secondary business field. LAPUS1 = 0, if in the primary business field. LAPUS2 = 1, if in the primary tertiary field. LAPUS2 = 0, if in the primary business field.
Job status	STATUS = 1, if it is a formal status. STATUS = 0, if it is an informal status.
Residential area	DTT = 1, if living in an urban area. DTT = 0, if living in a rural area.
Residential province	JAWA = 1, if living in Java. JAWA = 0, if living in outer Java.
Inflation	INFLASI, Numeric variable

Sources: IFLS 2000, 2007, and 2014.

4 DESCRIPTIVE ANALYSIS

4.1 *General description*

Based on IFLS, there were 17,444 people in the labour force aged 15–64 years in 2000, 19,807 individuals in 2007, and 23,324 individuals in 2014. Next for workers (labour force who works and gets earning) fluctuated inter-annually. In 2000 there were 13,452 workers, the number increased to 16,819 workers in 2007, and declined to 14,076 workers in 2014. During the years 2000 to 2014, the labour force was dominated by individuals who had elementary education and lower. The average age was 35 years in 2000, 36 years in 2007 and 37 years in 2014, where the number of males in the labour force was slightly more than the number of females. More than 75 per cent were married or had been married, and the number increased over the period 2000–2014. Most of the labour force did not have toddlers or elderly people in their households, they already owned a home and they did not have a housekeeper. If viewed geographically, most of the labour force lived in urban areas. Then, by province of residence, the labour force mostly lived in the province on the island of Java.

An overview of individual work based on the data of IFLS 2000, 2007 and 2014 shows an average work experience of 22 years, the average earnings in 2000 amounted to 338,642.85 rupiah, and then increased in 2007 to 964,059.48 rupiah, and subsequently increased again to 1,100,198.00 rupiah in 2014. Another economic variable also used in the research is the variable of inflation, which in 2000 was at an average rate of 8.52, which then declined in 2007 to 6.80, after which it rose again to 8.08 in 2014. Most worked in the tertiary business sector largely as informal workers. In 2000, most lived in rural areas, but in 2007 and 2014 they mostly resided in urban areas. During the period 2000–2014, most of the working individuals were on the island of Java, and the number decreased in 2014.

4.2 *The average earnings of working individuals*

The average earnings for working individuals increased during the period 2000–2014, where the higher the education, the greater the earning. In 2000, working individuals with a graduate degree earned the highest among graduates from lower education levels, and the same pattern was evident in both 2007 and 2014.

In terms of gender, males earned the most and this increased every year over the period 2000–2014. The greatest average earning was for those who worked in urban areas. Based on the province of residence, variables showed an increase for each year. In 2014, earnings in Java were greater than in the outer Java, whereas in 2000 and 2007 the annual average earning was almost the same.

Based on the business sector, the biggest average earnings was for working individuals in the field of tertiary businesses, and the number continued to rise during the period 2000–2014. Another economic variable is the employment status at the main job. According to the average earnings, formal workers received higher earnings than informal workers.

5 INFERENTIAL

5.1 *Income analysis for working individuals*

The model earning estimates can be written in the following equation:

$$\begin{aligned}\text{Ln}\,(W_{2000}) = {} & 11{,}99021 - 0{,}7568586\,\textit{didik1}_{2000} - 0{,}\,4909115\,\textit{didik2}_{2000} \\ & + 0{,}4197155\,\textit{didik3}_{2000} + 0{,}5983906\,\textit{didik4}_{2000} + 0{,}4562337\,\textit{jk}_{2000} \\ & + 0{,}0379705\,\textit{kerja}_{2000} - 0{,}000608\,\textit{kerja}^2_{2000} + 0{,}2397666\,\textit{lapus1}_{2000} \\ & + 0{,}2574857\,\textit{lapus2}_{2000} - 0{,}0396675\,\textit{status}_{2000} - 0{,}0240359\,\textit{dtt}_{2000} \\ & - 0{,}1459589\,\textit{jawa}_{2000} + 0{,}029076\,\textit{Inflasi}_{2000} - 0{,}868339\,\lambda_{2000} \qquad (3)\end{aligned}$$

$$\begin{aligned}\text{Ln}\,(W_{2007}) = {} & 12{,}9357 - 0{,}6204434\, didik1_{2007} - 0{,}4193747\, didik2_{2007} \\ & + 0{,}3549153\, didik3_{2007} + 0{,}5919831\, didik4_{2007} + 0{,}3972653\, jk_{2007} \\ & + 0{,}0378645\, kerja_{2007} - 0{,}0006038\, kerja^2_{2007} + 0{,}3078005\, lapus1_{2007} \\ & + 0{,}3112447\, lapus2_{2007} + 0{,}3577648\, status_{2007} + 0{,}0329\, dtt_{2007} \\ & - 0{,}2092409\, jawa_{2007} - 0{,}0472361\, Inflasi_{2007} - 0{,}6883931\, \lambda_{2007} \end{aligned} \tag{4}$$

$$\begin{aligned}\text{Ln}\,(W_{2014}) = {} & 12{,}1805 - 0{,}5008455\, didik1_{2014} - 0{,}3799224\, didik2_{2014} \\ & + 0{,}3820961\, didik3_{2014} + 0{,}4129322\, didik4_{2014} + 0{,}5141261\, jk_{2014} \\ & + 0{,}0382537\, kerja_{2014} - 0{,}0006503\, kerja^2_{2014} + 0{,}1320962\, lapus1_{2014} \\ & + 0{,}1051192\, lapus2_{2014} + 0{,}6096873\, status_{2014} + 0{,}171429\, dtt_{2014} \\ & + 0{,}0789801\, Inflasi_{2014} - 0{,}2458593\, \lambda_{2014} \end{aligned} \tag{5}$$

Equation 3 above is the model selected for the estimated revenue in 2000, while Equation 4 represents the model selected in 2007, and Equation 5 is the model in 2014. The positive coefficient indicates that workers receive a bigger earning than those who worked for the reference variables, while the value of a negative coefficient indicates that workers receive a smaller earning than those who worked for the reference variables.

Next, the factors that affect the earning received by individuals working for each independent variable will be explained.

5.1.1 *Work experience*

For the variable 'work experience', it is evident that this variables affect increasing revenue, following the quadratic curve. This means that each additional year of work experience increased revenue by 3.80 per cent in 2000, 3.78 per cent in 2007 and 3.82 per cent in 2014. Then, after reaching its zenith further additional income will decline. This is consistent with the research of Moenjak and Worswick (2003) who stated that work experience factors affect the amount of earning for working individuals with longer work experience who are thus considered to have expertise and higher productivity levels. The highpoint of work experience can be obtained from the first derivative model of education return to variable work experience, which is 31 years for individuals working in 2000 and 2007, and 29 years for individuals working in 2014.

After reaching a peak point, the marginal revenue improvements decrease (diminish) along with the increased years of work experience, which is caused by ageing when activities are reduced because of lesser physical capabilities (Setiawan 2010), and eventually workers go into retirement. When compared over the period of 2000–2014, there was a slight shift in peak age of work experience. This indicates that higher education received diminishing return more rapidly also received decline additional earning more rapidly as well. According to the ILO (2010), besides education there is also a need for continued training of educated individuals in order to be able to continue learning throughout their career so as to improve productivity and achieve greater economic benefits.

5.1.2 *Sex*

The estimated earnings received by men are always larger than women. This relates to the role of men being responsible for the needs of families. Therefore, men have more working hours than women, and the number of working hours also has an effect on earnings (Setiawan 2010). Nevertheless, the role of women in the working world now days, is getting bigger and recognised. But it is undeniable that gender difference in the working world still exists.

5.1.3 *Business field*

The variable of 'business field' shows positive results for both secondary and tertiary business fields for the years 2000, 2007 and 2014. This means working in the business field will earn a larger income than working in the field of primary effort. Estimated income for individuals working in the field of tertiary and secondary businesses is almost the same, while the primary business field always has the lowest for 2000, 2007 and 2014. This is because the scales of the company in the field of primary business are generally smaller than the fields of secondary or tertiary business (Handayani 2006). Furthermore, when compared between years, income disparities that existed tended to fluctuate, and were reduced in 2014. This can be attributed to technological developments including in primary filed business, so the business increasingly advanced and revenue could also increased. For the estimation of earned income, income for the undertakings of primary, secondary and tertiary businesses was the greatest in 2014.

5.1.4 *Status*

'Employment status' variables have been proven to be statistically significant in affecting earnings. In 2000, those with a formal employment status earned 3.97 per cent less than informal workers. This is because in that year there was still the impact of the economic crisis where many companies were closed and many workers were affected by layoffs. Then in 2007 and 2014, individuals employed as formal workers received bigger earnings compared to informal workers. When compared, it appears that income differences for formal workers were greater than informal workers during the period 2000–2014. This indicates that formal economy had a bigger quality and also more secure than informal economy. It is also evident from the pattern of estimated earning, where the gap in earnings received by formal and informal workers was widening, and the largest earning was in 2014.

5.1.5 *Residential areas*

In 2000 and 2007, earning among working individuals in urban and rural areas did not differ significantly. However in 2014, the income of working individuals in urban areas was 17.14 per cent higher than in the countryside. This is because the company scale was larger in urban areas and also the number of jobs was higher than in rural areas (Handayani 2006). It is also evident from the pattern of earnings that was estimated for 2000, 2007 and 2014 where the difference between rural and urban earning was widening. This is because of a higher diversity of jobs and activities in the economic sector in urban areas (ILO 2015).

5.1.6 *Province of residence*

For the 'province of residence' variable, the years 2000 and 2007 show that working individuals in Java earned lower income than outside Java. During the period 2000–2007 the gap in earning differences between individuals working in Java and outside Java was widening. In 2000, the earnings of working individuals in outer Java were greater than in Java. This condition applied in all business fields, and for both statuses except for those in the field of secondary business, as well as the informal status in Java which earned greater than outside Java. As for the year 2007, formal and informal workers in every field of business in outer Java received a larger earning than in Java, except for the tertiary business field for both formal and informal statuses. Furthermore, in 2014 it shows that there is no difference of earning between those who lived in or outside Java. The above indicates that the largest earning in Java starting with 2000, tended to be informal in the secondary business field, which then shifted to the tertiary business field in 2007 that began to develop rapidly. This condition is picture of economic changes in Indonesia, begin with dominated by agriculture economy in the the village then towards to the economy with a larger share in the industry and services economy in urban areas (ILO 2015).

Estimated earnings for individuals living in Java were lower than those living outside Java, both for 2000 and for 2007. Furthermore, based on calculated earning comparisons across years for each province of residence, it shows that in 2007 earnings were higher than in 2000, both for individuals working in and outside Java.

5.1.7 *Inflation*

In 2000 the 'inflation' variable showed that every additional one per cent inflation would increase revenue by 2.91 percent. However, in 2007, precisely every additional one percent of inflation would reduce earnings by 4.72 percent. Then in 2014, every additional one percent inflation would increase revenue by 7.90 percent.

5.1.8 *Level of education*

Individuals who have not been educated in school and only study at elementary/junior high school earn lower earnings than those graduating from high school. Further to this, individuals with the level of diploma and S1/S2/S3 programmes earn more than high school graduates.

Figure 3 shows that during the years 2000–2014 the level of education had a significant effect on the amount of earnings received by working individuals. The higher the educational level, the greater the earning and the greater the difference in earnings from the low education level graduates. However, between the years 2000–2014 it appears that the earning difference for each level of education tended to be small, except for the level of diploma. This indicates that the level of education significantly affects the increase in earnings, but earning disparities with the low education levels are smaller.

Then, if compared to earnings across years for higher education (diploma and S1/S2/S3), it appears that earnings in 2014 were greater than in 2000 and 2007 (for more details see Figure 3 below):

5.1.9 *Estimated income for higher education*

For 2000 and 2007, working individuals for characteristics are S1/S2/S3 graduates, men, formal workers, work in the tertiary business field, live in urban areas, in Java, receive the highest earning compared with other characteristics. Though in 2014, exactly for the same characteristics workers as the previous two years, except for working in the field of secondary business, are received the highest earning compared with other characteristics.

The rate of return to education can be calculated by estimating equations, especially higher education as follows:

From the table above, it appears that the rate of return to education for diploma and S1/S2/S3 was greater than for high school, for all the years 2000, 2007 and 2014. The rate of return for diploma was always greater than S1/S2/S3. The diploma return rate shows the number fluctuates every year, whereas the return rate for S1/S2/S3 results steadily declines.

Furthermore, when compared with the interest rate deposits of commercial banks in 2000 and 2014, the return to education of diploma was still higher than the deposit interest rate, while the rate of return to education for S1/S2/S3 was lower. This means that in 2000 and in 2014, it was more profitable to invest in diploma level than deposits, but invest in S1/S2/

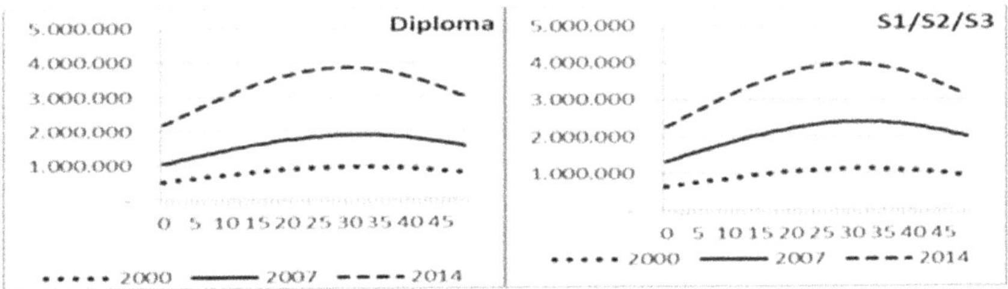

Figure 3. Earning estimation pattern by level of education, years and work experience. Sources: IFLS 2000, 2007, dan 2014, which have been processed.

Table 2. Rate of return to education for years 2000–2014.

	Rate of return (%)		
Level of education	2000	2007	2014
Not in school	–6.31	–5.17	–4.17
SD/SMP	–16.36	–13.98	–12.66
Diploma	20.99	17.75	19.10
S1/S2/S3	11.97	11.84	8.26

Sources: IFLS 2000, 2007 dan 2014, which have been processed.

S3 level is less profitable than deposits. Then in 2007, the return on investment for studying diploma and S1/S2/S3 programmes was higher than the deposit interest rate. This translates into meaning that it is more profitable to invest in higher education (either a diploma or S1/S2/S3) than to invest in deposits. In general, it can be illustrated that investing in education diploma is more profitable than investing in deposits. As for investing in education of S1/S2/S3, it would be more beneficial if investment was made in higher education than in deposits. This condition indicates that there is an occupational mismatch for higher education workers, that is, some workers cannot get a suitable job that can maximalize their ability and skills.

6 CONCLUSION

Direction of development in higher education is motivated by the enactment of the MEA, which implies the need for workers who are competent, creative and character-based to science. Based on this study, it was concluded that investment in higher education would be more advantageous for the diploma education level than S1/S2/S3 level. This condition imply the existence of occupational mismatch, or workers not getting jobs that can maximise their level of education and the capabilities they possess; it could also be interpreted as human resources not being utilised in the economy (Safuan & Nazara, 2005). In terms of employment, although educational attainment in society continues to increase, the proportion of jobs requiring skilled labour is not growing as fast as the increase in educational attainment (World Bank Indonesia 2010). Earnings are determined by human capital (education) and the characteristics of the job/position. Incompatibility of earnings as a result of occupational mismatch is more dependent on skill mismatch than job/position (Nordin et al., 2008). Hence, skills need to be boosted to fit the working world that is constantly evolving and dynamic in order to produce better graduates of higher education. Governments also need to expand their investment in education and skills training, so that highly educated workers can enjoy higher earnings and better employment opportunities (ILO 2015).

The return on investment of higher education during the period of 2000–2014 was higher for individuals with diploma than S1/S2/S3 education. When compared with the deposit interest rate of commercial banks, investing in diploma education is more profitable than investing in deposits, for 2000 and 2007. But for 2014, in the long term due to the decreasing trend rate of investment return of S1/S2/S3, it can be said that investment in education is still not profitable than investment in deposits.

REFERENCES

Arkes, J. (1999). What do educational credentials signal and why do employers value credentials? *Economics of Education Review, 18*(1), 133–41. doi:10.1016/S0272-7757(98)00024-7.

Badan Pusat Statistik. (2016). *Indikator Ekonomi* (Economic indicators): Februari/February 2016. Jakarta: BPS.

Badan Pusat Statistik. (2011). *Keadaan Angkatan Kerja di Indonesia* (Labor force situation in Indonesia): Agustus/August 2010. Jakarta: BPS.

Badan Pusat Statistik. (2010). *Indikator Ekonomi* (Economic indicators): Desember/December 2010. Jakarta: BPS.

Badan Pusat Statistik. (2001). *Keadaan Angkatan Kerja di Indonesia* (Labor force situation in Indonesia): Agustus/August 2000. Jakarta: BPS.

Becker, G.S. (1976). The University of Chicago Press. *The economic approach to human behaviour.*

Borjas, G.J. (2007). McGraw Hill, Inc. *Labor economics* (4th ed.).

Dumauli, M.T. (2015). Estimate of the private return on education in Indonesia: Evidence from sibling data. *International Journal of Educational Development, 42*: 14–24. doi:10.1016/j.ijedudev.2015.02.012.

Handayani, D. (2006). *Tingkat Pengembalian Investasi Pendidikan di Indonesia: Analisis Data Susenas 2004*. Master's thesis, Universitas Indonesia.

Heckman, J.J. (1979). Sample selection bias as a specification error. *Econometrica, 47*(1), 153–61. doi:10.2307/1912352.
Hungerford, T. & Gary Solon. (1987). Sheepskin effects in the returns to education. *The Review of Economics and Statistics, 69*(1), 175–177.
International Labour Organization (ILO). (2010). *Angkatan Kerja yang Terampil untuk Pertumbuhan yang Kuat, Berkelanjutan dan Seimbang/Kantor Perburuhan Internasional*. Jakarta: *ILO*, vi, 46.
International Labour Organization (ILO). (2015). *Tren Ketenagakerjaan dan Sosial di Indonesia 2014–2015: Memperluat Daya Saing dan Produktivitas Melalui Pekerjaan Layak*. Kantor Perburuhan Internasional. Jakarta: ILO.
Ministry of Planning/Bappenas. (2015). Seminar JEJAKMU: Jawaban Bagi Kaum Muda Pencari Kerja. *http://jejakmu.bappenas.go.id/berita/94-seminar-jejakmu-jawaban-bagi-kaum-muda-pencari-kerja*.
Moenjak, T. & Christopher Worswick. (2003). Vocational education in Thailand: A study of choice and returns. *Economics of Education Review, 22*(1), 99–107. doi:10.1016/S0272-7757(01)00059-0.
Nazara, S. & Sugiaharso Safuan. (2005). Identifikasi Fenomena Overeducation Di Pasar Tenaga Kerja Indonesia.pdf. *Jurnal Ekonomi Pembangunan Indonesia, VI*(1), 79–92.
Nordin, M., Inga Persson & Dan Olof Rooth. (2010). Education-occupation mismatch: Is there an income penalty? *Economics of Education Review, 29*(6), 1047–59. doi:10.1016/j.econedurev.2010.05.005
Purnastuti, L., Paul W. Miller, & Ruhul Salim. (2013). Declining rates of return to education: Evidence for Indonesia. *Bulletin of Indonesian Economic Studies, 49*(2), 213–36. doi:10.1080/00074918.2013.809842.
Setyonaluri, D. (2002). Tingkat Pengembalian Investasi Pendidikan Rumah Tangga Usaha Tani di Indonesia: Analisis Data Survey Aspek Kehidupan Rumah Tangga Indonesia (SAKERTI) 1997. Undergradute's thesis, Universitas Indonesia.
Setiawan, B. (2010). Efek Diploma dan Pengalaman Kerja pada Upah dan Jam Kerja di Indonesia. Master's thesis, Universitas Indonesia.
World Bank. (2010). *Laporan Ketenagakerjaan di Indonesia: Menuju Terciptanya Pekerjaan yang Lebih Baik dan Jaminan Perlindungan Bagi Para Pekerja*. Jakarta: Kantor Bank Dunia.

Competition and Cooperation in Economics and Business – Gani et al. (Eds)
© 2018 Taylor & Francis Group, London, ISBN 978-1-138-62666-9

Factors affecting selective sorting behaviour of household waste: The case of trash bank communities

R. Rahajeng & A. Halimatussadiah
Department of Economics, Faculty of Economics and Business, Universitas Indonesia, Depok, Indonesia

ABSTRACT: Through the issuance of Local Regulation No. 3 of 2013 on Waste Management a new regulation on was imposed that requires households to sort their trash. Trash banks have been established and provide incentives to the public to sort their household garbage and deposit the inorganic trash into trash banks. However, not all households have become members and participated in these trash bank activities. This study aims to identify the factors that influence the selective waste sorting behaviour of households in the trash bank communities. These factors include the respondents' knowledge and perceptions about Reduce-Reuse-Recycle (3R), household characteristics, community characteristics and characteristics of the trash banks. The survey was conducted on households in trash bank communities, for both trash bank members and non-members, and the heads of the trash banks. The results show that knowledge, environmental priorities, membership of the trash banks, level of household participation in the community, period of trash bank establishment, and education level of trash bank heads affect the behaviour of household waste sorting.

1 INTRODUCTION

As a major city Jakarta faces problems of population growth, economic growth and urbanisation, all of which in turn cause an increase in the quantity and types of waste generated. With a population of 10,012 million people and a growth rate of 1.43% (BPS, 2014), Jakarta produces on average 6,500 tonnes of rubbish/day. The pattern of waste management that still follows the paradigm of 'collect-haul-dispose' causes the volume of waste that must be disposed in landfills to become enormous, while the capacity of landfills is declining due to the limited land supply in urban areas and the inadequate and unsustainable/underdeveloped current waste processing technology owned by the government. In addition, the local government's capacity to collect and haul garbage still leaves 665 tonnes of waste unable to be transported to the landfill every day. The negative implications of the current condition are that garbage that cannot be managed correctly could be a transmission route for diseases (World Bank, 2012).

The waste management system in Indonesia is based on Law No. 18 of 2008, which is clarified further in the Jakarta Regional Regulation (Peraturan Daerah/Perda) No. 3 of 2013. This Perda puts forward the principle of 3Rs (Reduce, Reuse, Recycle) in waste management. The issuance of this regulation establishes the importance of community participation in order to realise the principles of the 3Rs of waste management, especially in sorting the household waste in accordance with its types. One form of public participation in waste management has been started with the formation of several trash banks in Jakarta. With the establishment of trash banks, it is expected that people of Jakarta will begin to apply selective waste sorting behaviour that has the potential to reduce the amount of waste that goes to landfill thereby reducing the severity of the waste problem as part of social problems that continue to rise. However, despite the availability of trash banks in Jakarta's communities, the fact remains that not all individuals in communities with trash banks participate in sorting the waste.

Many factors influence an individual's behaviour towards managing garbage, such as selective waste sorting. In developing countries, community participation has been proven to have an important role in waste management (Dokhikah & Trihadiningrum, 2012). Within a large population, people as a human resource have the power to change the paradigm in waste management. Public participation becomes an important factor in improving the waste management system because through public participation it is expected that the ability of each individual to engage in waste management programmes will continue to increase and sustain.

Knowledge on environment and the impact of environmental problem is the main factor that causes a person to have pro-environmental behaviour (Levine and Strube, 2012). Environmental knowledge will influence how people behave towards the environment, such as the decision to do selective waste sorting. The attitude of individuals or households is the main factor that influences recycling behaviour (Tonglet, 2004). Meanwhile, attitude itself is largely influenced by the perception of time and the availability of facilities (Nur Khaliesah, Sabrina Ho & Latifah, 2015; D'Ellia, 2008).

Previous studies reported that various socioeconomic and demographic factors influence pro-environmental behaviour. Education reportedly affects the behaviour of recycling (Grazhdani, 2015; Starr, 2015; Chanda, 1999). Home ownership status is also a factor that determines an individual's participation in recycling activities (Oskamp et al. 1991; Gifford and Nilson, 2014). Additionally, accessibility of a recycling facility is proven to significantly influence the behaviour of recycling (Berger, 1997). It is also reported that the distance between each house and recycling facilities affects the recycling behaviour of each household (Nur Khaliesah, Sabrina Ho & Latifah, 2015; Starr, 2015; Howenstine, 1993).

A 'trash bank' is a type of bank commonly founded by a community with the objective to accommodate household waste that can be recycled after being segregated by individuals/households. Every type of trash deposited in the trash bank such as plastic bottles, newspaper, aluminium bottles, cables, and others have different prices. The trash will be weighed and recorded in the trash bank passbook. With the recyclable waste collection centres in residential areas, it is expected that there will be an increase of community participation in household waste sorting (Omran et al. 2009). According to a previous study, the public will have a higher level of participation in selective waste sorting if there is a trash bank in their community (Dokhikah et al. 2015), therefore the availability of a trash bank is expected to change people's behaviour towards sorting their garbage. However, the fact remains that not all people who live in communities that have trash banks participate in selective waste sorting. Therefore, knowledge about factors that influence the selective sorting behaviour of society is required.

Based on that reason, the objective of this research is to determine the influencing factors of the selective sorting behaviour of household garbage based on the perceptions and knowledge of the respondents, characteristics of households, communities and trash banks. By understanding the significant factors that affect sorting behaviour of individuals/households, it is expected that policy formulation regarding waste sorting can be significantly improved to end the waste problem in Jakarta.

2 METHOD

2.1 *Research location*

Jakarta is the capital of the Republic of Indonesia with a population of over 10 million as of 2014. Jakarta is divided into five administrative cities/municipalities namely, Central Jakarta, North Jakarta, East Jakarta, West Jakarta, and South Jakarta, in addition to one administrative regency, named Thousand Islands. Overall, Jakarta has 44 districts and 267 *kelurahan* (administrative village) (BPS, 2014). This research was conducted in the area of East Jakarta Administrative City, Jakarta. East Jakarta is the largest administrative city in the Jakarta with an area of 188.03 km^2 and a population of 2,791,072 inhabitants in 2014. East Jakarta comprises ten districts, 65 *kelurahan*, 742 *RW* (*Rukun Warga*) and 7,755 *RT*

(*Rukun Tetangga*)(BPS, 2014). The main characteristic of East Jakarta city is that it consists mainly of residential areas; therefore, East Jakarta is also the most populous administrative area compared to other administrative cities and regions. The East Jakarta region produces an average of 1,920 tonnes of garbage/day with organic waste (63%) as the largest component, followed by inorganic waste and B3 (Hazardous and Toxic) waste.

In 2008, one private company began to initiate the establishment of a trash bank as one of their CSR programmes. The trash bank was first established in *RW* 3, *Kelurahan* Duren Sawit, East Jakarta, which since then has been replicated in other areas in Jakarta. The purpose of the trash bank establishment is to accommodate the collection of inorganic waste produced by community with the ultimate objective to facilitate waste recycling. With the availability of trash banks, people have an incentive to sort their household garbage. There are currently at least 38 trash banks in East Jakarta, but only about 21 banks actively conduct waste collection routine to accommodate the disaggregated trash.

In 2013, Jakarta Provincial Government issued a regulation, namely Regional Regulation (*Perda*) No. 3 of 2013, which requires Jakarta citizens to manage their household garbage at minimum by performing selective waste sorting according to its kinds. With the establishment of several trash banks, the trash bank programme can be considered as a pilot programme in communities that have implemented waste sorting. The existence of trash banks basically assists the provincial government of Jakarta to enforce the Regional Regulation No. 3 of 2013, which requires people to sort out their garbage. Nonetheless, up until now there has been no further research on the selective waste sorting behaviour people in Jakarta which could be used to recommend appropriate policies and measures in order to influence people's behaviour towards their garbage.

2.2 *Sampling method*

The study was conducted in Jakarta, as it is one of the provinces that has imposed the waste management regulation. The sample taken from East Jakarta district as it has quite a lot of waste bank communities. The sampling method used for this study is the cluster random sampling method in which the sample selection consists of a group of individuals. The cluster in this case is a community in which trash banks were available. There were a total of 21 communities with active trash banks in East Jakarta, and from those 21 communities, six communities were selected randomly as samples. The communities which were drawn as samples are trash bank communities in Cibubur, Bambu Apus, Malaka Sari, Susukan, Cipinang Melayu and Jatinegara. From the six clusters, several households were selected randomly to be the respondents.

The sample size determination was based on Malhotra (2010), which states that research using regression analysis or SEM (Structural Equation Modelling) should have a sample size ranging from 200 to 400 respondents. The samples in this study were 335 respondents from the six clusters that had been selected, with an average of 55 respondents per cluster.

2.3 *Questionnaires*

The first questionnaire was directed to those who lived in close vicinity to the trash banks, and the second questionnaire was directed to the heads of the trash banks. The questionnaire for the community members consists of six sections. The first section is regarding general information about the respondents' socio-demographic aspects, income, period of residence in the community, and other questions. The second part is about the respondents' connection to or relationship with the trash bank. The third section of the questionnaire includes questions related to selective waste sorting knowledge, including whether the respondents separate their trash, their knowledge of how to sort garbage, the extent and frequency/intensity of selecting garbage, their knowledge about the process of waste management and the impact of waste, as well as facilities required for waste sorting. The fourth part is about the respondents' perception of waste sorting activity and various environmental problems. The fifth part is about the level of participation of respondent in community activity.

The second questionnaire was directed to the heads of the trash banks and contains open-ended questions that probe deeper information on the trash bank's profile, development, activeness level, and also the board of managers.

2.4 *Model*

The logistic regression method was used to analyse the relationship between the response variable (Y) and one or more independent variables (X) where the response variable Y is a categorical variable, while binary logistic regression method is used to find the relationship between the response variable Y that is binary or dichotomous (e.g. y = 0 if it fails, y = 1 if successful) by the predictor variable X that is polychotomous (Hosmer & Lameshow, 2000). Logistic regression was used to determine factors that affect waste sorting activity in the previous study (Ekere et al., 2009) and community participation in waste reduction (Dokhikah et al., 2015). To analyse the relationship between the dependent variable and independent variables, the binary logistic regression method was used. The data was processed using SPSS version 22. The following model was used in this study:

$$\ln\frac{P_{SB}}{1-P_{SB}} = \beta_0 + \beta_1 Knowl_i + \beta_2 Tavail_i(1) + \beta_3 FacAvail_i(1) + \beta_4 EnvPrio_i(1) + \beta_5 Edu_i(1) + \beta_6 OwnH_i(1) + \beta_7 FamNum_i(1) + \beta_8 Stayt_i + \beta_9 Compart_i + \beta_{10} Soz_i(1) + \beta_{11} OwnH * Stayt_i(1) + \beta_{12} Dist_i(1) + \beta_{13} WBmem_i(1) + \beta_{14} WBlife_i(1) + \beta_{15} WBedu_i(1) + \mu_i$$

β_0 is an independent variable coefficient. In the binary logistic regression model, the interpretation of the model used Odd Ratio (OR), where OR = $\exp^B$.

Variable	Definition of variable	Explanation
Dependent variable		
SB	Waste sorting behaviour	Waste sorting behaviour (SB = 0 if respondent does not sort out the garbage; SB = 1 if respondent separates the garbage according to its types).
Perception and knowledge of respondents		
Knowl	Knowledge	The level of knowledge that includes the process of selective waste sorting and the impact of waste.
TAvail	Time availability to sort waste	Dummy variable for the time availability forsorting out the garbage (TAvail = 0 if the respondent does not have time to sort out the garbage, TAvail = 1 if the respondent has time to separate the garbage).
FacAvail	Availability of sorting facility	Dummy variable for public perception regarding the availability sorting facility (FacAvail = 0 if respondent perceives no facility available for waste sorting in the community, FacAvail = 1 if the respondent knows about the availability of waste sorting facilities in the community).
EnvPrio	Perception of environmental problem as a major issue	Dummy variable for perception of environmental problem as a major issue (EnvPrio = 0 if respondent does not recognise environmental problem as a major issue, EnvPrio = 1 if the respondent feels environmental problem is a major issue).
Characteristics of households		
Edu	Education level	Dummy variable for the highest level of education completed by a family member who lives in the same household (FamEdu = 0 graduated less than or equal to 9 years of education, FamEdu = 1 if the completed education is greater than 9 years).

(*Continued*)

Variable	Definition of variable	Explanation
OwnH	Home ownership status	Dummy variable for home ownership status (OwnH = 0 if the current occupied house is a rental property, OwnH = 1 if the occupied house is the respondent's private property).
FamNum	Number of family members	The number of family members living in the respondent's house.
Stayt	Length of stay	The respondent's period of residence in the current community area at the time of interview.
Soz	*Sosialisasi* (Advocacy and communication)	Dummy variable for whether the respondent participated in advocacy and communication programme regarding waste sorting (Soz = 0 if never participated, Soz = 1 if participated before).
OwnH * Stayt	Variable interaction	The variable interaction between home ownership and residence period.
Dist	Distance between home and the trash bank	Dummy variable for the distance between the location of trash bank and the house (Dist = 0 if the distance 100 m, Dist = 1 if the distance > 100 m).
Wbmem	Membership in the trash bank	Dummy variable for membership in the trash bank (Wbmem = 0 if the respondent is not a member of the trash bank, Wbmem = 1 if the respondent is a member of the trash bank).
Community characteristics		
Compart	Participation in community	The level of participation of the respondent's family members in existing activities in the neighbourhood.
Trash bank characteristics		
Wblife	Period trash bank	The length of period since trash bank establishment (in years).
WBedu	Education level of trash bank head	Dummy variable for the highest level of education completed by the head of the trash bank (Wbedu = 0 if the head of the trash bank only completed high school, Wbedu = 1 if the head of the trash bank completed S1/bachelor programme).

3 RESULTS AND DISCUSSION

3.1 *Profile of respondents*

The summary of the respondent profiles categorised by socio-demographic characteristics can be seen in Table 1. The respondents consisted of 75.8% women and 24.2% men. Those under 20 years old comprised 1.8% of the total respondents, while 34.6% were between 20 and 39 years old, 47.8% were 40–59 years old, and 60 years old and above made up 15.8%. Based on the highest level of completed education, the majority (51.6%) had only completed high school, followed by 20.9% who had completed elementary school and 17.9% who managed to finish junior high school. The remainder had completed higher education: 9.3% held a D3 (Diploma) or S1 (Bachelor) degree, while 0.3% held an S2 (Master), or S3 (PhD) degree. As for the occupation of the respondents, this was dominated by housewives (57.9%), followed by vendors (9.9%), self-employed (9.3%), private (8.7%), retired (7.2%), with the remaining balance consisting of students/university students, labourers and civil servants. From the category of home ownership status, approximately 73.1% of the total respondents owned the house that they occupied and the rest (26.9%) occupied a leased/rented property. Out of the total respondents 70.4% had between 3–5 family members in their households, while those with less than three and more than five family members were 10.2% and 19.4%, respectively. The average number of family members living in one household is 4.3.

Of all the respondents, 66.6% did household waste sorting in their home, while the other 33.4% did not. To further understand the type of sorting done by each household, each respondent was asked about the frequency of selective waste sorting in his or her household. The respondents who declared they always sorted their waste made up 27.5%, 19.4% said they often sorted their waste, 14.6% mentioned that they only sorted their garbage occasionally, and 5.1% rarely sorted

Table 1. Profile of respondents categorised by socio-demographic characteristics.

Variable	Category	Percentage
Gender	Female	75.8
	Male	24.2
Age	Less than 20	1.8
	20–39	34.6
	40–60	47.8
	More than 60	15.8
Education	SD (Elementary school)	20.9
	SMP (Junior High School)	17.9
	High School	51.6
	D3 (Diploma)/S1 (Bachelor)	9.3
	S2 (Master)/S3 (PhD)	0.3
Occupation	Housewife	57.9
	Vendor	9.9
	Self-employed	9.3
	Private company employee	8.7
	Retired	7.2
	Other	7.0
Home ownership status	Private ownership	73.1
	Rental/lease	26.9
Number of family members	Less than 3 people	10.2
	3–5 people	70.4
	More than 5 people	19.4

their trash. The fact that a large majority of the respondents followed a selective waste sorting process showed that people in the trash bank community have a good habit of sorting household garbage. Related to the main reason that incentivised them to sort their waste, more than half of the respondents (56.1%) reported wanting to contribute to environmental preservation as their main reason. Only 28.3% of them mainly selected their waste for monetary compensation, 11.7% admitted to doing it in order to adhere to their RT/RW regulation, and the remaining respondents stated that they sorted their trash to follow government regulations and to utilise their waste. The large percentage of the respondents that separated their garbage for environmental reasons revealed that most communities with trash banks were aware of the need to protect the environment through waste management. On the other hand, economic incentives can be used to strengthen the public's willingness to sort out their household garbage.

Nonetheless, the large number of respondents who sorted their household waste was not accompanied by the number of people who participated as members of trash banks. Respondents who were members of trash banks amounted to only 39.4%, while those who were not members of trash banks reached 60.4%. This means that more than half of the respondents were not members of trash banks in spite of them being available in their community. As for the most dominant reason for joining the trash banks as a member, most respondents stated that the trash banks helped them overcome waste problems in their environment (see Table 2).

3.2 *Waste sorting behaviour based on socio-demographic characteristics*

Based on the level of education of the household members, respondents with a D3/S1 degree or S2/S3 degree had a higher percentage in waste sorting behaviour compared with those with lower levels of education within the household. Household members with a higher level of education are expected to have greater environmental-related awareness as has previously been proved by other studies (Arcury & Christensen, 1993; Chanda, 1999; Hsu & Rothe, 1996; Starr, 2015). This is due to the level of knowledge about environmental conditions in general, specific issues related to the environment, and the impact of unmanaged and unprocessed

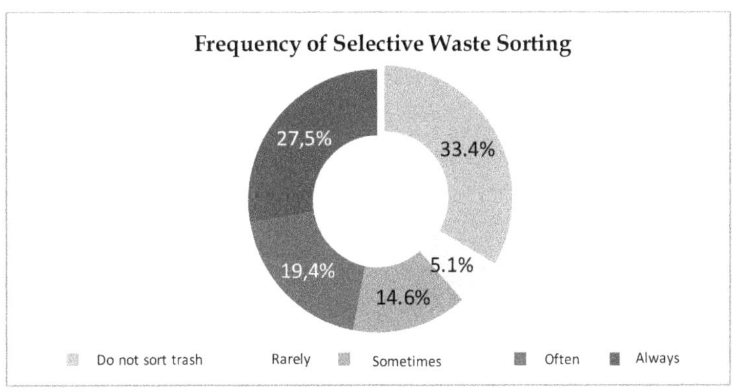

Figure 1. Waste sorting behaviour based on frequency.

Table 2. Profile of selective waste sorting behaviour.

Variable	Category	Percentage
Respondents' reason for sorting household waste	Monetary incentive	28.3
	Environmental preservation	56.1
	Rules of RT/RW	11.7
	Local government regulation	0.4
	Other	3.6
Is the respondent a member of a trash bank?	Yes	39.4
	No	60.6
Reason to be member of a trash bank	In order to make garbage available to be recycled	15.9
	Trash bank reduces waste problem	51.5
	To receive monetary incentive	20.5
	Because neighbour is also a member	3.8
	Do not know the reason	3.0
	Other	5.3

waste, which increases with a higher level of formal education. This knowledge leads to higher awareness by household members with a higher education level to prevent further adverse environmental impacts in their community compared to those with lower education.

Based on trash bank membership, respondents that were also members of trash banks had a higher percentage of always sorting their garbage (59%) than non-member respondents (Figure 3). People who want to join trash banks typically either have a higher environmental awareness or are quite attracted to the monetary incentives provided by the trash banks, making them compelled to sort their household garbage. However, from the results below, it was also found that several members of trash banks no longer sorted their trash (3%). Some reasons for this include the difficulties of bringing the sorted waste to the trash bank location because of the lack of transportation facilities to transport the sorted garbage, the perception of futility in sorting waste because their neighbours did not sort theirs, and waste handling and transport systems that had not accommodated the segregated trash.

A larger percentage of respondents who lived in their own house performed waste sorting (75.4%) compared to those who lived in rented houses (53.3%). Home ownership is known to be a factor influencing recycling activities according to previous studies (Oskamp et al., 1991; Gifford et al., 2014). This is because people who live in their own house tend to be more

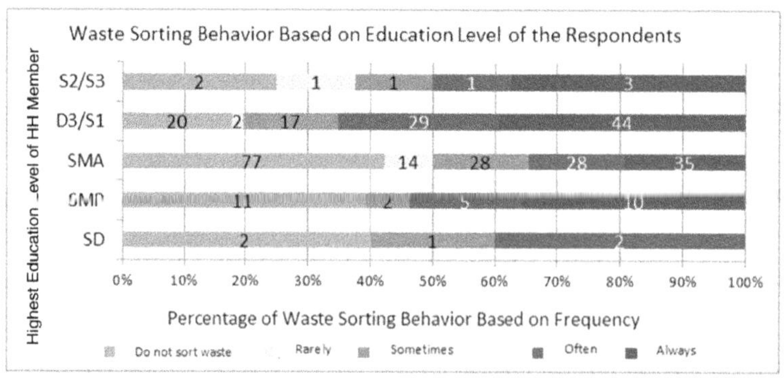

Figure 2. Waste sorting behaviour of respondents based on highest level of education completed.

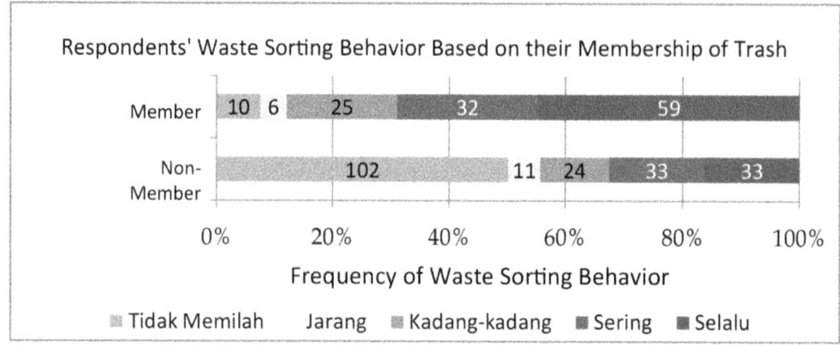

Figure 3. The comparison of trash sorting behaviour between trash bank members and non-members.

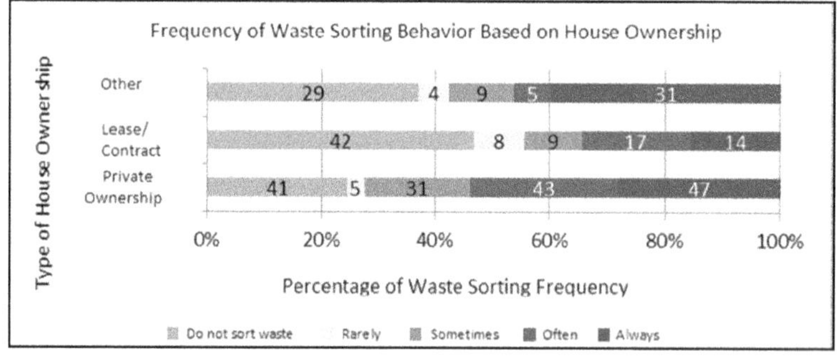

Figure 4. Respondents' sorting behaviour based on home ownership.

willing to take action to maintain the environmental sustainability since they would be more likely to settle in the region in comparison with people who live in rented/leased houses.

Based on the number of family members (Figure 5), the study found that respondents in households with more family members tended to sort their garbage. This could be due to the fact that a larger number of family members diminishes the burden of sorting garbage for each person as the responsibility for sorting trash would be divided among all the family members.

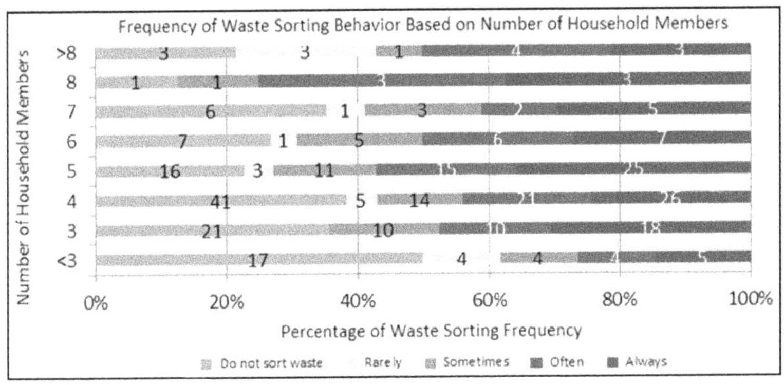

Figure 5. Profile of number of household members and sorting behaviour frequency.

3.3 *Factors affecting household waste sorting behaviour*

Waste sorting has a significant impact on economic efficiency, as garbage sorting is an important part in the process of reuse and recycle. The results of a binary logistic regression calculation of factors affecting waste sorting behaviour can be seen in Table 3. The variables that were expected to influence the behaviour of household waste sorting include knowledge, perception of time availability, availability of facilities, perception of the environment as a priority, education, household size, period of residence, interaction between home ownership and period of residence, membership of a trash bank, level of participation in the community, age of trash bank, and trash bank chief's education. The model can be considered to be quite accurate with a pseudo R-square 62.4%, and the significant chi-square at the 1% level.

Based on the results in Table 3, *knowledge* appeared to be a significant variable influencing waste sorting behaviour with a positive correlation. Knowledge of trash is highly correlated with the activity of managing waste; correct knowledge about the negative effects of waste can help change the attitude of the society to be more concerned about trash. Therefore, the more people who understand that waste is a problem for the environment, the bigger the likelihood that they will participate in waste management activities. Knowledge is a factor that positively affects selective waste sorting behaviour (Dokhikah et al., 2015; Levine & Strube, 2012).

Availability of time to sort out garbage significantly and positively influenced waste sorting behaviour. A respondent who was not constrained by time problems had a low opportunity cost to sorting garbage, making it easier for him or her to sort waste compared to those who felt they had no time to sort their garbage. Previous research has found limited time to be one of the factors that hinders public participation in sorting household garbage (Nur Khaliesah, Sabrina Ho & Latifah 2015; D'Ellia et al., 2008).

Availability of facilities was proven to have a positive and significant effect on waste sorting behaviour. This is due to the fact that the availability of facilities for waste sorting is often used as a reason for households to participate in recycling programmes (Nur Khaliesah et al., 2015; Kirakozian, 2014). Interviews with several respondents who did not perform household waste sorting, revealed that they would join waste sorting if they were provided with the supporting facilities.

Priority on environment appeared to be a significant factor influencing waste sorting behaviour with a positive correlation. The more a respondent perceived that environmental issues are major issues, the more likely the respondent would carry out actions in an attempt to maintain environmental sustainability. Someone with a greater level of attention to the environment is more motivated to preserve the environment including participating in recycling activities (Krajhanzl, 2010; D'Ellia, 2008).

Education was a significant factor positively influencing waste sorting behaviour. Education is known to correlate positively with recycling activities as seen in previous studies (Starr, 2015; Grazhdani, 2015), and in some countries where individuals have higher levels

Table 3. Results of binary logistic regression.

Variable	B	Odd ratio
Knowledge	.120*	1.128
Availability of time (1)	1.085***	2.960
Availability of sorting facilities (1)	1.879***	6.546
The environment as a priority (1)	1.015***	2.759
Education (1)	0.941**	2.563
Home ownership (1)	–0.185	0.831
Number of family members	0.260**	1.297
Period of residence	0.086***	1.090
Sosialisasi (Advocacy and communication programme) (1)	0.506	1.658
Interaction of home ownership with period of residence (1)	–0.065**	0.937
Participation in the community	.150***	1.162
Distance between home and the trash bank	–0.629	0.533
Membership of trash bank (1)	1.707***	5.513
Age of trash bank	0.282***	1.326
Education of the head of the trash bank (1)	–1.025**	0.359
Constant	–8.093	0.000
Pseudo R-square	0.624	
LR chi-square	200.006	
Prob (chi-square)	0.000	
Total Observation	335	

*Significant at the 10% level;
**Significant at the 5% level;
***Significant at the 1% level.

of education, general concern for the environment is proven to be higher (Arcury & Christensen, 1993; Chanda, 1999; Hsu & Rothe, 1996). Education within the family is important because household behaviour is not only determined by the education level of one person but the education level of the whole family. In this case, the behaviour of one family member may affect the behaviour of all family members, forming an emergent behaviour of the household.

Number of family members was proven to be a factor that positively and significantly influenced the behaviour of waste sorting. Past research has also suggested that the number of the family members is a factor that positively affects the behaviour of recycling (D'Ellia, 2008). This is because the burden of house chores will be smaller in households with a larger number of family members as the responsibility for waste sorting is shared among family members.

The period of residence appeared to have a positive and significant effect on waste sorting behaviour. This could be caused by a sense of belonging to the neighbourhood, which tends to be higher the longer a person lives in a particular environment. This sense of belonging encourages a greater level of participation in community activities. This result is in accordance with Ross (1967) who found a positive correlation between lengths of residence period with participation level in community activities. In another study, Nguyen (2015) found a negative correlation between lengths of residence period with one's intention to sort out his or her waste. The reason for this correlation is that the longer a person has lived in a community, the deeper the understanding a person has about the community such as the shared reluctance among the community members to participate in community activities.

The variable interaction between *home ownership and the length of* stay was found to significantly influence waste sorting behaviour in a negative correlation. This means that the respondents who owned their houses with a longer residence period were less likely to sort their waste compared to those who lived in rented houses with a shorter length of stay. This shows that people who have lived in their own house for a longer period of residence in the community are more aware of the community character that is filled with 'free riders' in the waste management system. This explains why the respondents who owned their current houses for a longer period were more reluctant to participate in waste sorting.

The level of participation in community activities had a significant influence with a positive correlation on waste sorting behaviour. This is because people who participate in community activities more actively are generally more agreeable to be invited to participate in activities or programmes that support the environment. This is in accordance with previous research by Ekere et al. (2009), conducted in Uganda, which found that the behaviour of households in the utilisation and sorting of waste was partly influenced by participation in environmental organisations.

Membership in the trash bank had a positive and significant effect on waste sorting behaviour. Public participation in a recycling programme has a positive influence on a community's attitude to waste sorting (Khaliesah, Ho & Latifah, 2015; Ekere et al., 2009). By participating in the recycling programme provided by trash banks, people would be more motivated to conduct selective waste sorting, whether due to the monetary incentive or environmental awareness.

Age of trash bank in a community significantly influenced the behaviour of garbage sorting with a positive correlation. Similar to the residence period factor in a community that influences a person's decision to sort his or her household garbage, the longer the existence of such a programme as a trash bank in a community, the higher the expected awareness among the community members of the importance of waste recycling and current environmental issues and conditions. Benefits from the trash bank in the form of monetary compensation for waste sorting can influence the community around the trash bank to participate in the activities planned by the organisation.

Education level of trash bank head was also proven to have a significant effect on waste sorting behaviour with a negative correlation. This means that trash bank heads with higher education such as a Bachelor degree (S1) had a lesser influence on increasing waste sorting behaviour in the community compared to those who only completed high school education. Shawn (1991) previously found that a leader in a community is an important factor to assist community members in making interventions that can be accepted by the society. Therefore, the level of education is commonly used as a parameter of a person's ability to create interventions within the community to make the community members behave as expected. The negative correlation found in this result might be due to the better social approach possessed by the heads of the trash banks with only high school education. As the kind of monetary incentive offered by trash banks is more likely to be needed by people from a lower income group, trash bank heads with high school education may have better advantages in both previous relationship ties and ability to effectively communicate with this level of income group. These advantages may help create a higher participation rate in waste sorting behaviour in the community. According to Holil (1980), intensive communication between fellow citizens, and between citizens and their leader, as well as the existing social system in the community and the system outside of the community can affect the level of community participation in environmental activities. Additionally, the heads of the trash bank with a Bachelor degree (S1) may have a higher opportunity cost (e.g. time and resources) to developing trash banks for the community.

3.4 *Analysis according to respondents' membership in trash banks*

To determine whether there were significant differences regarding waste sorting behaviour between members and non-members of trash banks, both a descriptive analysis and a logistic regression on the survey data that separated the respondents based on the membership status in trash banks were summarised in Tables 4 and 5.

As can be seen in Table 4, respondents who were members of trash banks were more likely to perform garbage sorting with a percentage of 92%, while non-member respondents were less likely to sort their garbage with a percentage of only 50%. This result means that being a member of a trash bank tends to improve one's behaviour regarding sorting garbage. Similarly, for the frequency of waste sorting, respondents who were members of trash banks had a higher frequency of sorting garbage. As for the reason for sorting their garbage, both groups had the same tendency, answering that it was for environmental reasons. From these

Table 4. Factors affecting waste sorting behaviour based on trash bank membership status.

		Number of Respondents: 335							
		Trash Bank Members: 132				Non-Members 203			
		Sorting: 122		No-Sorting: 10		Sorting: 101		No-Sorting: 102	
Influencing factor		Freq	%	Freq	%	Freq	%	Freq	%
Knowledge level	Low	4	3.3	0	0.0	2	2.0	4	3.9
	Moderate	45	36.9	3	30.0	34	33.7	44	43.1
	High	73	59.8	7	70.0	63	62.4	54	52.9
Knowledge of existing waste sorting regulations	Do not know	46	37.7	2	20.0	44	43.6	55	53.9
	Know	76	62.3	8	80.0	57	56.4	47	46.1
Having the means/facilities to sort the trash	Yes	97	79.5	5	50.0	79	78.2	29	28.4
	No	25	20.5	5	50.0	22	21.8	73	71.6
Perceive environmental issues as a priority	Yes	55	45.1	4	40.0	48	47.5	33	32.4
	No	67	54.9	6	60.0	53	52.5	69	67.6
Participated in advocacy and communication programme regarding waste sorting	Yes	101	82.8	9	90.0	57	56.4	29	28.4
	No	21	17.2	1	10.0	44	43.6	73	71.6
Advocacy and communication programme organiser	Cleaning agency (*Dinas Kebersihan*)	4	4.0	–	–	–	–	–	–
	RT/RW	39	38.6	8	88.9	33	57.9	13	44.8
	Trash bank management	55	54.5	–	–	19	33.3	12	41.4
	Village/districts	2	2.0	–	–	3	5.3	–	–
	More	2	2.0	1	11.1	2	3.5	4	13.8
Waste sorting frequency	Always	59	48.4	–	–	33	32.7	–	–
	Often	32	26.2	–	–	33	32.7	–	–
	Sometimes	25	20.5	–	–	24	23.8	–	–
	Rarely	6	4.9	–	–	11	10.9	–	–
Reasons for waste sorting	Monetary incentive	40	32.8	–	–	23	22.8	–	–
	Environment	65	53.3	–	–	60	59.4	–	–
	Rules of RT/RW	14	11.5	–	–	12	11.9	–	–
	Local government regulation	1	0.8	–	–	0	0.0	–	–
	Other	2	1.6	–	–	6	5.9	–	–
What the respondents do with inorganic waste	Used privately	1	0.8	–	–	3	3.0	–	–
	Gave it for free	8	6.6	–	–	34	33.7	–	–
	Sold to ragpickers/flea market	7	5.7	–	–	28	27.7	–	–
	Deposited into trash bank	106	86.9	–	–	31	30.7	–	–
	Other	0	0.0	–	–	4	4.0	–	–

Table 5. Regression test results based on trash bank membership.

	Trash bank members		Non-trash bank members	
Variable	B	Odd ratio	B	Odd ratio
Knowledge	–	–	0.175**	1.191
Availability of time (1)	1.962**	7.114	0.791*	2.205
Availability of sorting facilities (1)	0.990	2.691	2.397***	10.986
View of environment as a priority (1)	–	–	1.412***	4.102
Education (1)	0.646	1.907	0.934*	2.545
Home ownership (1)	–2.531*	.080	.326	1.385
Number of family members	0.408	1.504	.213*	1.237
Length of stay	–	–	0.105***	1.111
Advocacy and communication programme (1)	–	–	0.704*	2.021
Interaction of home ownership with the length of stay (1)	–	–	–0.079**	.924
Participation in community	0.271**	1.311	0.135**	1.144
Distance between home and the trash bank	–0.584	.557	–0.832*	.435
Age of trash bank	.514*	1.673	0.288**	1.333
Education level of head of the trash bank (1)	–2.319**	.098	–0.849	.428
Constant	–4.859	0.008	–8.985	0.000
Pseudo R-squares	0.400		0.575	
LR chi-square	23.956		114.487	
Prob (chi-square)	0.004		0.000	
Total Observation	132		203	

* Significant at the 10% level;
** Significant at the 5% level;
*** Significant at the 1% level.

results, it can be said that membership in the trash bank is an important factor to form selective waste sorting behaviour. It may be caused by special attachment and responsibilities that are perceived as part of becoming a member of a trash bank.

Based on *perception of the sorting facilities ownership*, of the respondents who were not members of trash banks and did not sort their garbage into categories of trash, 71% perceived that they did not have the means to sort their garbage, this being the reason for them not doing so. This perspective is expected to change with the provision of waste sorting facilities. After facilities are made available, waste sorting behaviour should be further improved.

Regarding the management of inorganic waste that had been sorted, 87% of the members of trash banks deposited their sorted waste to the trash banks, while only 30% of those who were not members deposited their trash to the banks. The rest of the non-members sold their inorganic trash to ragpickers or in the flea market (27%), or gave away their trash to others (33%) for free to their low-income neighbour as their additional income. This finding showed that the bank's role in providing a place to deposit sorted trash is quite important, both for members and non-members of the bank. Despite the availability of *peloak* (ragpickers) around respondents' neighbourhoods that allows them to just sell their trash without having to transport it to another place, many respondents still chose to go to the trash bank due to the relatively higher price offered by the bank.

Based on whether or not respondents had participated in an advocacy and communication programme related to waste sorting activity, almost all the respondents who were members of the trash bank had participated in such a programme. Meanwhile, for the respondents who were not members of the trash bank but already sorted their garbage, 56% of them had participated in an advocacy and communication programme for selective waste sorting. As for the respondents who were neither members of a trash bank nor sorted their trash, only a small number of respondents had engaged in such a programme (28%).

The lack of advocacy and communication regarding selective waste sorting could be the reason why most of the respondents who were not members of the bank did not sort their garbage as the information regarding its importance, the facilities available, and mechanism of the waste sorting process were not known to them. Advocacy and communication is one means to convey information about the importance of waste sorting which also creates an emphasis on the responsibility of each individual to contribute and manage his or her garbage at least by sorting it according to type. Improvement of the current advocacy and communication programme from the range of its beneficiaries, the scale of the programme, and also the relevancy of the information given is expected to increase the tendency to sort household garbage in a community. The local government, for example, could use various types of media to disseminate information about waste sorting so that the information flow is not limited to the direct face-to-face method that requires time and place, which might be ineffective for citizens of Jakarta who have very busy schedules. The use of digital media such as social media and the Internet, or passive media such as banners, could be alternatives to delivery of information to a wider scope of society.

Socialization on the importance of sorting waste and the benefit of waste bank usually carried out by the waste bank activist. For non-waste bank member, the socialization is usually carried out by neighbourhood association (RT/RW). The data showed that the roles of trash bank management and RT/RW in disseminating information related to waste sorting were more influential compared to local government's role. To expand sorting behaviour in community, the local government need to facilitate and improve supporting facilities (waste-related infrastructure) such as separate trash mull and involve more in disseminating the importance of waste separation. The involvement of government officials through the Local Government Cleaning Agency (*Dinas Kebersihan*) and villages/districts still needs to be improved in order to change public attitude and behaviour which are not only driven by the community but also by the government whose roles are not only limited to regulation makers but act also as enforcers of regulation. It can be concluded that the current role of the trash bank is crucial to form waste sorting behaviour in the community. This is further indicated by the dominant role of trash bank management boards in providing advocacy and information on waste sorting.

Based on the regression test results in Table 5, it can be observed that the model is better applied to those respondents who were not members of trash banks. This conclusion is reached from the result of the model significance test in which the model for non-members has smaller prob (chi-square) and also Negelke R-square that is greater than the result of the model applied for members of trash banks. In addition, there are several variables that contribute more significantly to the model for non-trash bank members.

In the model for the trash bank members, the variables that significantly affected the behaviour of waste sorting are much fewer than in the model for non-members. These variables include the perception of time availability, home ownership status, level of participation in the community, age of trash bank since establishment, and education level of the trash bank heads. These variables tended to change the behaviour of trash bank members, while other variables such as knowledge, availability of facilities, number of family members, and distance between home and trash bank were not proven to create a significant influence on waste sorting behaviour. The insignificance of these factors that supposedly affect waste sorting behaviour may indicate that the behaviour of the trash bank members had been more consistent compared to that of the non-trash bank members.

In the model for non-trash bank members, all the independent variables significantly influenced the behaviour of waste sorting except for home ownership. It was also found that there were several variables that did not significantly affect the general model but had significant effects on the model for non-trash bank members; these variable were the advocacy and communication programme as well as distance between home and trash bank. As described earlier, an advocacy and communication programme tended to affect non-trash bank members more than members. This was proven by observing that only 28% of the respondents who were non-trash bank members and did not sort their waste had previously received advocacy on waste sorting. Thus, this essentially means that advocacy is needed specifically for the community members who have not become members of trash banks.

Other variables that showed different results with the general model are the distance between home and the trash bank. For respondents who were not members of the trash bank, distance from the house to the trash bank significantly influenced the waste sorting behaviour with a negative correlation. This means that the greater the distance between the respondent's house and the trash bank, the smaller the probability of the respondent to sort their waste. The distance between house and trash bank is a constraint in waste sorting that can also be seen in Table 4, where the majority of respondents sold their sorted inorganic waste to *peloak* (ragpickers/flea market) or handed them over to a neighbour who was a member of trash bank. Therefore, it can be concluded that on average trash banks should have a smaller coverage area with closer distance between the trash bank and the trash bank members' houses, which would incentivise more people to contribute to the trash bank. The current situation of trash banks is that they have a large coverage area (one RW area) as it can covers up to 18 RTs. As a result, members of the trash banks were usually dominated by people living in the closest RTs to the location of the trash banks. For future reference, trash banks would function more effectively and efficiently if they are established based on RT area (smaller coverage) for example, provide one trash bank every two or three RTs.

Several differences between the model results of the members and non-members of the trash banks showed that the behaviour of the trash bank members in waste sorting is relatively more consistent, while the waste sorting behaviour of non-members is significantly influenced by a number of different factors. Some factors that could be used to increase the probability of non-members to perform waste sorting are through the provision of facilities for waste sorting, the organisation of advocacy and communication programmes for a wider scope of audience, and the replication of trash banks for a smaller area.

4 CONCLUSION

To reduce the waste problem in Jakarta, which can be observed from the waste volume that is not evenly proportioned to the capacity of TPA (final landfill), Jakarta Provincial Government issued Regional Regulation No. 3 of 2013 on Waste Management, which requires communities to sort their trash. In some areas of Jakarta, several trash banks were already established. These banks provide incentives to the public for sorting their household garbage and depositing inorganic trash into these trash banks. However, not all households have become members and not all people participate in trash bank activities. This study aims to identify the factors that influence waste sorting behaviour of households in the trash bank communities. These factors include the respondents' knowledge and perceptions about the 3Rs, household characteristics, community characteristics and characteristics of the trash banks. The survey was conducted by taking respondents from households in the trash bank communities, both trash bank members and non-members, as well as the heads of the trash banks. The data obtained were then processed using the binary logistic regression method.

The results of this study showed that knowledge as well as advocacy and communication programmes significantly influenced the behaviour of waste sorting with positive correlations especially for non-trash bank members. Thus, collaboration between government and non-government parties through local government officers, trash bank staff and RT/RW officers is needed to provide correct information about the impact of waste on the environment and the importance of waste sorting. This advocacy and communication programme should include information which can change people's perception about the urgency of waste management, especially waste sorting in our own houses, the ease of performing waste sorting and the little time required to do so.

Availability of facilities also positively influenced the waste sorting behaviour. The government is expected to provide adequate facilities to the public both individually and collectively so that the public is increasingly attracted to participate in waste sorting activities. The required facilities include garbage transport systems with separate slots for each garbage type in order to support waste sorting activities performed by households, as each type of waste can be processed accordingly.

The role of trash banks in regard to waste sorting behaviour is decisive in shaping public attitude and perception through the management board's role in disseminating information related to waste sorting and incentives that can be gained by the community. Local government needs to conduct intensive developmental guidance and advocacy programmes on trash banks in order to keep existing trash banks active and be able to further reduce waste problems. Developmental guidance programmes for the trash banks might include the provision of operational facilities for trash banks, provision of media for dissemination of information to invite more participants, provision of information on *pelapak*, and training for trash bank management on organisational management. In addition, local governments should also foster initiatives to establish new trash banks up to their actual formation as well as monitor their existence so that the replication of trash banks can provide larger coverage by establishing an active trash bank in each community area. The sooner a trash bank can be established, the sooner the behavioural change in waste sorting can be expected in society.

One limitation of this study is the sampling method where all the respondents selected in this study had to reside in the community where the trash banks already existed. Further research may be expanded by including areas outside trash bank communities in order to understand the differences in people, and location characteristics between areas without any recycling facilities and those with recycling facilities.

REFERENCES

Arcury, Thomas, and Eric Howard Christianson. (1993). Rural-urban differences in environmental knowledge and actions. *The Journal of Environmental Education, 25*(1), 19–25. doi:10.1080/00958964.1993.9941940.

Berger, Ida E. (1997). The demographics of recycling and structure of environmental behaviour. *Environment and Behaviour, 29*(4), 515–531.

Badan Pusat Statistik (BPS). (2014). *Jumlah Penduduk dan Rasio Jenis Kelamin Menurut Kabupaten/Kota Administrasi, 2014.* Jakarta: Indonesian Central Bureau of Statistics.

Chanda, Raban. (1999). Correlates and dimensions of environmental quality concern among residents of an African subtropical city: Gaborone, Botswana. *Journal of Environmental Education, 30*(2), 31–39.

Cointreau-Levine, Sandra. (1994). Private sector participation in municipal solid waste services in developing countries, volume 1: The formal sector. *Journal of Urban Management and the Environment, 1.* doi:10.1016/S0034-3617(03)80041-3.

Cornes, Richard & Todd Sandler. (1986). *The theory of externalities, public goods, and club goods*. Cambridge: Cambridge University Press.

D'Elia, Jose L.I. (2008). Determinants of household waste recycling in Northern Ireland. *Economic Research Institute of Northern Ireland. http://eservices.afbini.gov.uk/erini/pdf/ERINIMon23.pdf.*

Dhokhikah, Yeny, & Yulinah Trihadiningrum. (2012). Solid waste management in Asian developing countries: Challenges and opportunities. *J. Appl. Environ. Biol. Sci. Journal of Applied Environmental and Biological Sciences, 2*(7), 329–35. www.textroad.com.

Dhokhikah, Yeny, Yulinah Trihadiningrum, & Sony Sunaryo. (2015). Community participation in household solid waste reduction in Surabaya, Indonesia. *Resources, Conservation and Recycling, 102*: 153–62. doi:10.1016/j.resconrec.2015.06.013.

Ekere W., Mugisha J., Drake L. (2009). Factors influencing waste separation and utilization among households in the Lake Victoria crescent, Uganda. *Waste Management, 29*: 3047–3051.

Gamal, Yuliman. (2009). *Faktor-faktor yang Mempengaruhi Perilaku Prolingkungan pada Masyarakat Perkotaan* (Survey terhadap masyarakat Kota Jakarta Selatan sebagai Peraih Adipura). PhD Diss., Fakultas Ilmu Sosial dan Ilmu Politik, Universitas Indonesia.

Gifford, Robert, & Andreas Nilsson. (2014). Personal and social factors that influence pro-environmental concern and behaviour: A Review. *International Journal of Psychology, 49*(3), 141–57. doi:10.1002/ijop.12034.

Grazhdani, Dorina. (2015). Assessing the variables affecting on the rate of solid waste generation and recycling: An empirical analysis in Prespa Park. *Waste Management, 48*: 3–13.

Hosmer Jr., D.W. & Lemeshow, S. (2000). *Applied logistic regression* (2nd ed.). New York: John Wiley & Sons.

Howenstine, E. (1993). Market segmentation for recycling. *Environment and Behavior, 25* (March), 86–102. doi:10.1177/0013916593251004.

Hsu, Shih J., & Robert E. Rothe. (1996). An assessment of environmental knowledge and attitudes held by community leaders in the Hualien area of Taiwan. *Journal of Environmental Education, 28*(1), 24–31.

Jan Krajhanzl. (2010). Environmental and proenvironmental behaviour. School and health 21. *Health Education: International Experience*.

Kirakozian, Ankinee. (2014). *Selective sorting of waste: A study of individual behaviour*. France: Universite de Nice-Sophia Antipolis.

Levine, Debra Siegel, & Michael J. Strube. (2012). Environmental attitudes, knowledge, intentions and behaviors among college students. *The Journal of Social Psychology, 152*(3), 308–26. doi:10.1080/00224545.2011.604363

Malhotra, Naresh K. (2010). *Marketing Research: An Applied Orientation* (6th ed.). New Jersey: Prentice Hall Inc.

Nur Khaliesah, Abdul Malik, Abdullah Sabrina Ho, & Abd Manaf Latifah. (2015). Community participation on solid waste segregation through recycling programmes in Putrajaya. *Procedia Environmental Sciences, 30*: 10–14. doi:10.1016/j.proenv.2015.10.002

Omran, A., Mahmood, A., Abdul Aziz, H. & Robinson, G.M. (2009). Investigating households' attitude towards recycling of solid waste in Malaysia: A case study. *International Journal of Environmental Research, 3*(2), 275–288.

Oskamp, Stuart, Maura Harrington, Todd Edwards, Deborah Sherwood, Shawn Okuda, & Deborah Swanson. (1991). Factors influencing household recycling behavior. *Environment and Behavior, 23*(4), 494–519. doi:0803973233

Ross, Murray G., & B.W. Lappin. (1967). *Community Organization: Theory, Principles and Practice Second Edition*. New York: Harper and Row Publishers.

Shawn M. Burn. (1991). Social psychology and the stimulation of recycling behaviors: The block leader approach. *Journal of Applied Social Psychology, 21*: 611–629.

Starr, Jared, & Craig Nicolson. (2015). Patterns in trash: Factors driving municipal recycling in Massachusetts. *Resources, Conservation and Recycling, 99*, 7–18. doi:10.1016/j.resconrec.2015.03.009

Tonglet, Michele, Paul S. Phillips, & Adam D. Read. (2004). Using the theory of planned behaviour to investigate the determinants of recycling behaviour: A case study from Brixworth, UK. *Resources, Conservation and Recycling, 41*(3), 191–214. doi:10.1016/j.resconrec.2003.11.001

Competition and Cooperation in Economics and Business – Gani et al. (Eds)
© 2018 Taylor & Francis Group, London, ISBN 978-1-138-62666-9

Effectiveness analysis of machinery restructuring program in Indonesia's Textile and Clothing Industry (ITPT) 2002–2011

Lailatus Shofiyah & T.M. Zakir Machmud
Department of Economics, Faculty Economics and Bisnis, Universitas Indonesia, Depok, Indonesia

ABSTRACT: This study aims to examine the effect of the machinery restructuring program on the technology use using electricity consumption as the proxy, efficiency, productivity, and competitiveness using export as the proxy. This study also measured the competitiveness of textile industry by using Net Export Index (NEI). The sample data used were 50 observations made up of the five textile industry sub-sectors and textile industry export import data for 10 years (2002–2011). By using panel data estimation, it was found that the machinery restructuring program in Indonesia's textile industry positively and significantly affected efficiency and productivity and also reduced electricity consumption. However, this program was not proven to improve competitiveness. The calculation of NEI also found that the competitiveness of Indonesia's textile industry after the machinery restructuring program decreased.

1 INTRODUCTION

Textile and Clothing industry (TPT) is one of the vital industries for Indonesian economy due to its advantage in gaining foreign exchange reserve from export, massive number of labor absorption, and as a supplier for the domestic market especially for clothing needs.[1] In addition, the upstream and downstream subsectors of Indonesia's TPT industry have been integrated and it has a tight linkage with each other.

One of the problems that obstructs the performance of TPT industry this time is the use of outdated machinery for the production process. More than 60 percent of the machinery/tools for production in TPT industry from all types of industry are more than 20 years old, outdated by three to four generations from the modern technology. The recent condition of TPT industry's machinery is presented in Table 1.

Responding to this condition, in 2007, the government released a program to increase the technology through TPT industry machinery restructuring program. The machinery/tools in TPT industry that were already over 20 years old had to be renewed or replaced with the modern machinery/tools to allow the production of more competitive products and to improve the overall performance. This program is expected to help TPT industry to renew the machinery or to invest in the increasing use of technology, efficiency, productivity, and competitiveness of the national TPT industry.

---

1. TPT industry is a labour intensive industry that absorbs many workers. The export value of TPT industry from 2003 to 2007 reached \$ 7 billion to \$ 10 billion. Its growth relatively increased. The trade surplus of TPT industry also continuously increased. From 2001 to 2006, textile and footwear industry contributed about 3.0% to 3.25% to national GDP. TPT industry contribution to GDP in 2007 through 2011 was 1.9% to 2.3%. In 2012, TPT industry contributed nearly 1.56% to the total GDP of Indonesia. In 2011, the number of firms in TPT industry was 2880 units with a total investment of \$ 15.6 million. The growth of TPT industry increased by 5% in 2012. TPT industry is one of the leading industrial clusters driving economic growth in Indonesia. Sales value of TPT industry increased from 2004 to 2008, both for sales in the domestic market and in the global market.

Table 1. Number of old machinery in TPT industry.

Industry type	Units	Number of machines	>20-year-old machines	
			Number	Percentage
Spinning	MP	7,803,241	5,025,287	64.4
Weaving	ATM	248,957	204,393	82.1
Knitting	MR	41,312	34,743	84.1
Finishing	Unit	349	325	93.2
Confection	MSJ	290,838	226,854	78.0

Source: Technical guidelines for industrial technology program improvement TPT for 2007 and 2009 Ditjen ILMTA.

In the technical manual of textile industry machinery restructuring program, it is mentioned that the government released the program to increase the use of technology, competitiveness, efficiency, and productivity of the national textile industry.[2] In relation to that, a question rises whether the assistance for machinery purchases through the restructuring program is really capable of increasing the use of technology, efficiency, productivity, and competitiveness of the national TPT industry as targeted by the government. Thus, this research focuses on the effectiveness analysis of TPT industry machinery restructuring program.

The question investigated in this research is whether the machinery restructuring program is capable of increasing the use of technology, efficiency, productivity, and competitiveness of the TPT industry. It also examines the development of TPT Industry's competitiveness in 2002–2011 and the comparison of TPT industry before and after the machinery restructuring program. The measurement of the impact of the assistance program in machinery purchases on the program's objective indicators would show whether the program was effective in affecting TPT industry's performance in general. Furthermore, this research also aims to measure the competitiveness of TPT industry before and after the machinery restructuring program from 2002 until 2011 by using Net Export Index (NEI).

1.1 *Literature review*

A number of studies related to the machinery restructuring program of TPT industry in Indonesia include the analysis of restructuring program in textile industry against the export of textile and clothing (Setiawan, 2013) and analysis on machinery restructuring program or equipment (enhanced technology) of ITPT and its relation with the agreement on subsidies and countervailing measures (Sudrajat, 2007). The machinery restructuring program is closely related to subsidy, as this program provides assistance for the purchase of machinery in the company; therefore, in this study, a literature review on the impact of subsidies on the performance of the industry is also included (e.g. Emvalomatis *et al.* 2008; Zhu, Demeter, and Lansink, 2012; Bavorova, 2003; Lachaal, 1994).[3]

2. It shows that the Government through the Ministry of Industry wants to help the textile industry not only to big companies, but also small companies that exist in the TPT industry. Based on machine restructuring program data of purchasing way of the Industry Ministry, more than 50% participants were able to buy a machine/equipment by using their own funds. Based on that data, it indicated that companies which enjoy a rebate of the aid program is the big companies in TPT Industry.

3. Studies on electrical energy demand, efficiency, productivity, and exports have been conducted. The purpose of reviewing these studies is to determine the control variables that would be used in this study and determine the appropriate model equations. The summary of the related variables and the justification are attached in Appendix A.

1.2 *Gap analysis*

Several studies on the machinery restructuring program of TPT industry have been done, but these studies primarily focus on the machinery restructuring program's impact on textile industry export performance (Setiawan, 2011) and on the analysis of machinery restructuring program in terms of international trade law (Sudrajat, 2007). Studies analyzing the effectiveness of the textile industry machinery restructuring program remain non-existent. Therefore, this study was conducted to analyze the effectiveness of the machinery restructuring program in the TPT industry to measure the impact of the program based on some objectives.

1.3 *Machinery restructuring program in Indonesia*

This program consists of two SKIMs. SKIM 1 is a discounted price from the investment value. SKIM 2 is a credit allocation with a maximum period of five years, but SKIM 2 program already ended in 2009. USAID (2008) data collected from firms taking SKIM 1 program show that firms obtaining the rebates have approximately 1,725 workers. USAID (2008) forecast the average amount of sales of firms in that size is Rp250 billion (US$ 26 million). The assistance through SKIM 1 is channeled by the government in the form of discounted price of the machinery as much as 11% of the purchasing value in 2007 and 10% in 2008 until now. The 25% discounted price is for the program participants that use minimum 25% machinery/tools with Domestic Level of Component (TKDN). The amount of the maximum discounted price depends on the availability of the remaining DIPA fund in the current budget year and the number of program participants registered. SKIM 2 program is a credit allocation with a low interest rate from 2007 until 2009, with a maximum loan period of five years and credit value between IDR 100 million and IDR 5 billion. Since the launch of this program, the number of participants has been increasing. This shows that interest in this program among firms in TPT industry is relatively high. Most of the machinery investment is in the spinning industry subsector, shown by the highest number of total addition of machines in this subsector compared to that in the other industry subsectors.

2 METHOD

2.1 *Model 1: The impact of machinery restructuring program on the use of technology*

$$\ln\_kWh_{it} = \beta_0 + \beta_1 \ln\_Dummy\_Program_{it} + \beta_2 \ln B\_listrik_{it} + \beta_3 \ln B\_Bakar_{it} + e_{it}$$ [4]

where kWh_{it} is the amount of electricity consumed by the i-th sub-industry in t-th year, $Dummy\_Program_{it}$ is the *dummy* of the machinery restructuring program in the i-th subindustry in t-th year, $B\_listrik_{it}$ is the electricity cost of the i-th sub-industry in t-th year, B_Bakar_{it} is the fuel used by the i-th sub-industry in t-th year, and e_{it} is the *random error.*[5]

2.2 *Model 2: The impact of machinery restructuring program on efficiency*

$$\ln\_Efisiensi_{it} = \beta_0 + \beta_1 Dummy\_Program_{it} + \beta_2 \ln Size_{it} + \beta_3 \ln C\_I_{it} + e_{it}$$

4. We use natural log beause we want to know the change of independent variables in percentage (growth) when the dependent variable change.

5. Dummy Program: 0 for 2002–2006 (when there is no machinery restructuring program). Dummy Program: 1 for 2007–2011 (when there is machinery restructuring program). We use kWh as proxy of Technology Using because we assume that old machine will consume more electricity. Once, the old machine is changed to the new machine, its electricity consumption is decreasing.

where $Efisiensi_{it}$ is the efficiency in the i-th subindustry in t-th year, $Dummy\_Program_{it}$ is the *dummy* of the machinery restructuring program in the i-th sub-industry in t-th year, $Size_{it}$ is the total number of workers in the i-th subindustry n t-th year, C_I_{it} is the capital intensity of the i-th sub-industry n t-th year, and e_{it} is the *random error.*

2.3 *Model 3: The impact of machinery restructuring program on productivity*

$$\ln\_Prod_{it} = \beta_0 + \beta_1 Dummy\_Program_{it} + \beta_2 \ln\_Upah_{it} + \beta_3 H\_Cap_{it} + e_{it}$$

where $Prod_{it}$ is the productivity of the i-th sub-industry in t-th year, $Dummy\_Program_{it}$ is the *dummy* of the machinery restructuring program in the i-th sub-industry in t-th year, $Upah_{it}$ is the wage paid to the workers in the i-th sub-industry in t-th year, H_Cap_{it} is the raw material of the i-th sub-industry n t-th year, and e_{it} is the *random error.*

2.4 *Model 4: The impact of machinery restructuring program on competitiveness*

$$\ln Ekspor_{it} = \beta_0 + \beta_1 \ln Dummy\_Program_{it} + \beta_2 \ln Produksi_{it} \\ + \beta_3 \ln W\_GDP_t + \beta_4 \ln REER_t + e_{it}$$

where, $Ekspor_{it}$ is the real export value of the i-th subindustry in t-th year, $Dummy\_Program_{it}$ is the *dummy* of the machinery restructuring program in the i-th subindustry in t-th year, $Produksi_{it}$ is the real production of the i-th subindustry in t-th year, W_GDP_t is the world real GDP per capita in a certain year (t), $REER_t$ is the Real Effective Exchange Rate sub in a certain year (t), and e_{it} is the *random error*.

This research used panel regression model since the data used are panel data. Panel data can explain two kinds of information, i.e. cross-section information in the differences between the individuals/subjects and the time-series information reflecting the changes in a certain time period. In panel data regression, there are three estimation methods that can be used, i.e. Panel Least Square (PLS), Fixed Effect Model (FEM), and Random Effect Model (REM).

2.5 *Competitiveness measurement*

Besides the impact of the machinery restructuring assistance program on competitiveness, with export as the proxy, the competitiveness of TPT industry before and after the assistance program was also measured. A simple measurement method of competitiveness index, namely Net Export Index (NEI) used by Latruffe (2010) was employed. NEI is the spread between the export in a country or the export in an industry sector with its import divided by the total value of trade (Banterle and Carraresi, 2007). NEI is defined as follows:

$$NEI = X_{ij} - M_{ij} / X_{ij} M_{ij}$$

X is the export, M is the import, j shows the sector/product, and i shows the country or sector. Export and import data of the textile industry from 2002 to 2011 were used to measure NEI.

3 RESULTS

The regression results are explained in this section. Tables containing the econometric test results are attached in the Appendix section.[6]

6. All regression results presented is already passed the Robust Test.

3.1 *The impact of machinery restructuring program on the use of technology*

Based on the estimation, the machinery restructuring program was proved to save electricity consumption. This finding is supported by the program evaluation result that has been done by the government. It is stated that, after the program, energy saving achieved 6–18% (*Kajian Pengembangan Industri Tekstil dan Produk Tekstil* 2011). The estimation result of the impact of the program on electricity consumption is supported by the descriptive graphic evidence of the amount of electricity consumption in TPT industry in Figure 1 in Appendix C.

3.2 *The impact of machinery restructuring program on efficiency*

Based on the examination, it is proved that the restructuring program had a vital role in increasing the production efficiency of the TPT industry. With the use of modern technology machinery, the industry's ability to maximize its production with several given inputs increases. This is aligned with the theory that firm's efficiency may increase by changing and renewing the technology. The findings showing that the machinery restructuring program has a positive impact on TPT industry's efficiency can be justified by the description in Figure 2.

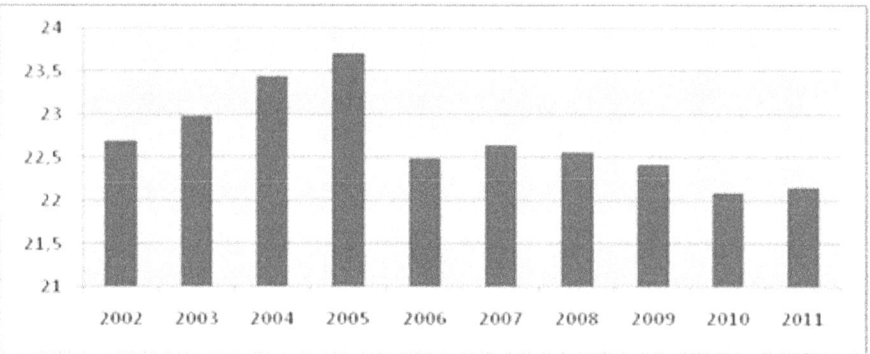

Figure 1. Electricity consumption.
Source: Badan Pusat Statistik (BPS). The restructuring program allows revitalization of machinery, bringing in new machines which are more energy efficient and more productive. The figure above shows that electricity consumption from 2007 to 2011 decreased compared to the amount of electricity consumption in the previous years.

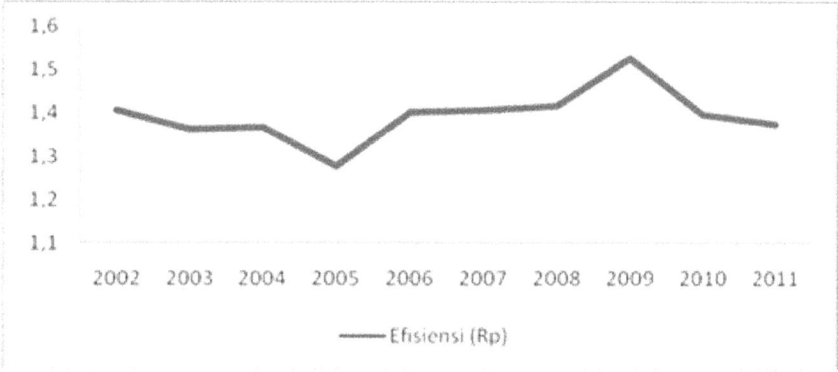

Figure 2. Efficiency of TPT industry in Indonesia.
Source: Badan Pusat Statistik (BPS). The figure above indicates that the efficiency of textile industry increased relatively from 2002 to 2011. Although it fell sharply in 2005 and 2010, it began to rise again in 2006 and the subsequent years. It increased continuously from the first year of the program in 2007 to 2009.

3.3 *The impact of machinery restructuring program on productivity*

The estimation results show that the machinery restructuring program could increase labor productivity in Indonesia TPT industry. This is aligned with the theory of technological progress stating that technological improvement may increase productivity. Productivity in general is the ability of the production factors to produce output (Latruffe, 2010). Increasing productivity means getting more output with the same level of input or getting the same level of output with less input. Changing old machinery with the new one shows that there is an increase in the production factor's ability to produce output so that the productivity level becomes higher. The findings on the correlation between the machinery restructuring program and TPT industry's productivity are justified by the graphic description in Figure 3.

3.4 *The impact of machinery restructuring program on competitiveness (export)*

Based on the estimation, it was found that the machinery restructuring program in TPT industry was not effective in increasing competitiveness. This result is not aligned with the hypothesis proposed. Figure 4 shows that the export performance of TPT industry in 2007 was higher than that in 2006. However, in 2009, TPT industry's export declined. This is assumed to be the effect of the global crisis that affected the world demand for textile products and clothing. In 2010, export was relatively high until 2011. The estimation results from Model 4 show that the export of Indonesia's TPT industry do not depend only on the machinery used in the production process but also on other factors that are more vital in affecting the export performance of TPT industry in Indonesia (see Figure 4).

3.5 *Competitiveness measurement results*

Net Export Index (NEI) measurement of the TPT industry per subsector from 2002 until 2011 shows declining competitiveness, as can be seen in the declining value of NEI. NEI in 2007 started declining and a sharp decline occurred in 2008. Later in 2009, TPT industry's competitiveness started to increase, but back on declining trend in the following years. NEI measurement results are presented in Table 2.

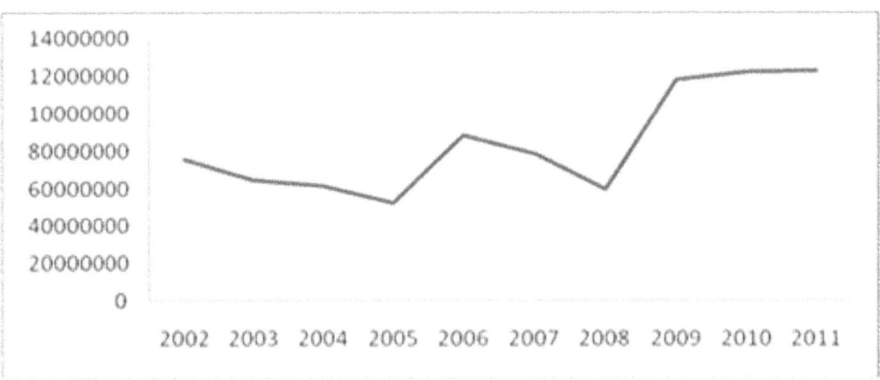

Figure 3. Productivity of TPT industry in Indonesia.
Source: Badan Pusat Statistik (BPS). The findings about the relation between the machinery restructuring program and the productivity of textile industry is justified by the description of the above figure. There was a relative increase of the productivity of textile industry in 2007 to 2011 compared to that in the years before the program despite the decline in 2008. It might have been due to the global crisis (external shock) causing a decline in value added and employment. However, as seen in the figure above, the impact of the crisis did not cause a prolonged effect on the productivity of the textile industry because after 2009, the industry began to increase its productivity.

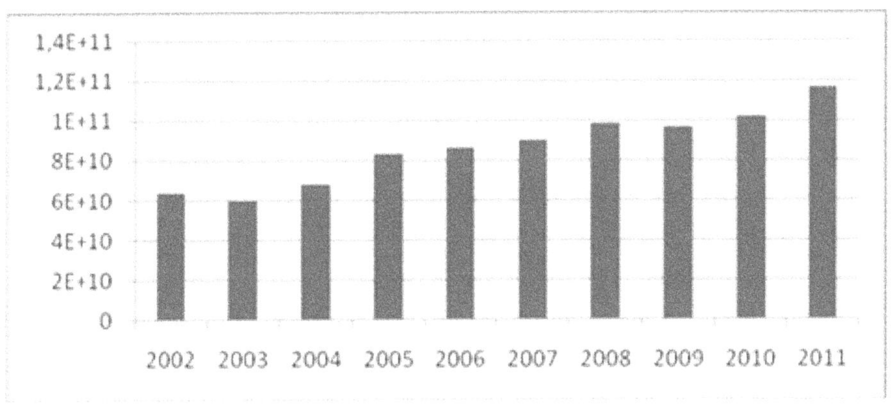

Figure 4. Export of TPT industry in Indonesia.
Source: Badan Pusat Statistik (BPS). The results obtained are justified by the model specification and indicate that the effect of the restructuring program was consistent engine is negatively related to export, but not significant. The relationship between the restructuring program and its own export engine was actually positive although production variables were entered. However, once they were entered, the relationship between the world per capita income variable and the real exchange rate became negative.

Table 2. NEI measurement results.

Year	NEI
2002	0.58109375
2003	0.646614726
2004	0.633023997
2005	0.685487437
2006	0.692815657
2007	0.661571607
2008	0.330610045
2009	0.379066613
2010	0.289346832
2011	0.216930716

Annotation: NEI value = −1 (when a country or an industry only imports) < NEI < 1 (when a country or an industry only exports). If the value is zero, it means that the size of export and import is equal. If NEI value > 0, then export > import.

The decline in TPT industry's competitiveness is caused by the high growth of import compared to the export growth. Based on World Integrated Trade Solution, TPT industry's export from 2002 to 2011 was relatively increasing.

The sales of Indonesia's TPT in the last few years have been absorbed more by the domestic market, as the international market is highly affected by the international demand and global economic stability. Moreover, some export destination countries have certain special criteria of imported goods. For example, the European Union market tends to choose to import natural and environmentally friendly products. This can be challenging for the national TPT industry export. Even when the production machinery in TPT industry has been renewed, it does not guarantee that the competitiveness or export performance of the national TPT industry will increase because there are also external factors affecting the demand for Indonesia's TPT industry export.

4 CONCLUSION

The problem with the outdated machinery in TPT industry has become an obstacle to the industry to improve its performance and competitiveness. More than 50% of the production machinery used by every subsector in TPT industry was more than 20 years old. This problem obstructs TPT industry from seizing the existing opportunity, including the chance to expand TPT products in the global market as a result of ATC (Agreement on Textile and Clothes) in 2005. This condition encouraged the government to impose a strategic policy as an act to save the national TPT industry. In 2007, the government through the Ministry of Industry imposed the machinery restructuring program for TPT industry used as a stimulus for firms in the industry to renew their production machinery to modern technology.

The goal of the machinery restructuring program is to increase the use of technology, efficiency, productivity, and competitiveness. Since this program uses public resources in a large amount, it is necessary to analyze its effectiveness so that misallocation of resources will not occur. An examination using four simple models estimated by panel data estimation method found that the machinery restructuring program in TPT industry was proved to be effective to save electricity (as a proxy of the use of technology), efficient (ratio of production to total input), and productive (ratio of value added to number of workers). However, the program did not appear effective to increase export (as a proxy of competitiveness).

The negative correlation between the machinery restructuring program and export was also supported by the measurement result of Net Export Index (NEI). It was proved that TPT industry's competitiveness declined after the machinery restructuring program although there are in fact other factors causing the decline, including massive import. However, Indonesia's TPT trade remains still surplus. It shows that so far Indonesia's TPT export is still larger than the import and still has good competitiveness.

REFERENCES

Banterle, Alessandro, and Laura Carraresi. 2007. "Competitive Performance Analysis and European Union Trade: The Case of the Prepared Swine Meat Sector." *Food Economics ? Acta Agricult Scand C* 4: 159–72. doi:10.1080/16507540701597048.

Bavorova, Miroslava. 2003. Influence of Policy Measures on the Competitiveness of the Sugar Industry in the Czech Repubic. *Agricultural Economics (AGRICECON)* 49 (6): 266–274.

Emvalomatis, Grigorios. *et al.* 2008. Paper presented at the 107th EAAE Seminar *"Modelling of Agricultural and Rural Development Policies"*, Sevilla, Spain. (ini gaada judulnya, namanya juga, soalnya dicari di google jg gaada ttg ini).

Lachaal, Lassaad. 1994. "Subsidies, Endogenous Technical Efficiency and the Measurement of Productivity Growth." *Journal Agricultural and Applied Economies* 26 (1): 299–310.

Latruffe, Laure. 2010. "Competitiveness, Productivity and Efficiency in the Agricultural and Agri-Food Sectors." *OECD Food, Agriculture and Fisheries Papers* 30 (30): 1–63. doi:10.1787/5 km91nkdt6d6-en.

Setiawan, Didik. 2013. Dampak Program restrukturisasi industri Tekstil terhadap Kinerja Ekspor Industri TPT Indonesia. *Tesis: FEUI.*

Sudrajat, Eko Prilianto. 2007. Analisis Tentang Program Restrukturisasi Mesin atau Peralatan (Peningkatan Teknologi) ITPT dan Kaitannya Dengan Agreement on Subsidies and Countervailing Measures. *Tesis: FHUI.*

Zhu, Xueqin, Róbert Milán Demeter, and Alfons Oude Lansink. 2012. "Technical Efficiency and Productivity Differentials of Dairy Farms in Three EU Countries: The Role of CAP Subsidies." *Agricultural Economics Review.*

APPENDIX A

Description of variables

Variable	Explanation	Literature	Relationship
kWh	Electricity consumption	Griliches and Jorgenson (1967), Costello (1993), Burnside, *et al.* (1995)	
Dummy program	Before program (2002–2006) = 0 After program (2007–2011) = 1		–
Electricity cost	Value of spending on electricity/kWh	Putra (1993), Ayu (2010), Surbakti and Kodoatie (2013)	–
Fuel	Value of spending on fuel	Asmara, Purnamadewi, Mulatsih, and Novianti (2013)	+
Efficiency	Value of production/ Value of total input	Latruffe (2010)	
Dummy program	Before program (2002–2006) = 0 After program (2007–2011) = 1		+
Size of firm	Number of workers	Samad and Patwary (2003), Caparas and Teresa (2006), Bhandari and Maiti (2007), Gani (2013).	+
Capital intensity	Total of capital/total of labor	Gani (2013), Fallahi, Sajoodi, and Aslaninia (2011), Greenaway and Zhihong (2004)	+
Productivity	Value added/number of workers	Fallahi, Sajoodi, and Aslaninia (2011), Adhadika (2013), Choudhry (2009), Bavorova (2003), Greenaway and Yu (2004)	
Dummy program	Before program (2002–2006) = 0 After program (2007–2011) = 1	Asmara, Purnamadewi, Mulatsih, and Novianti (2013), Fallahi, Sajoodi, and Aslaninia (2011), Adhadika (2013), Putra (1993)	+
Wage	Total wage		+
Human capital	*skilled labor* ratio to the total workforce of a firm	Fallahi, Sajoodi, and Aslaninia (2011), Gani (2013), Choudhry (2009)	–
Export	Export value	Latruffe (2010), Setiawan (2013), Siddiq, *et al.* (2012), Supriyati (2014), Chintia (2008)	

(*Continued*)

APPENDIX A (*Continued*)

Variable	Explanation	Literature	Relationship
Dummy program	Before program (2002–2006) = 0 After program (2007–2011) = 1		+
Production	Value of production	Latruffe (2010)	+
World GDP per capita	World real GDP per capita	Supriyati (2014), Fallahi, Sajoodi, and Aslaninia (2011)	+
REER	The real exchange rate of the Indonesian rupiah against the US dollar	Setiawan (2013), Supriyati (2014)	+

APPENDIX B

Test results of Model 1: The impact of machinery restructuring program on the use of technology

Uji Chow	0.0001	
Uji LM	0.0015	
Uji Hausman	0.1175	
	lnkWh (RE)	
Model 1	Coef	St. Error
D_Program	–0.5716374***	0.145275
lnB_Listrik	–0.2219856**	0.0975733
lnB_Bakar	0.6103696***	0.0966239
Cons	10.31183***	1.777841
Number of Obs	50	
Prob > Chi^2	0.0000	
R-squared	0.6737	

Note: ***) Significant at level α = 1%, **) Significant at level α = 5%, *) Significant at level α = 10%.

Description: kWh = total electricity consumption, a proxy for the use of technology/capital machinery; D_Program = dummy program, 2002–2006 (no program) = 0; 2007–2011 (no program) = 1; B_Listrik = cost of electricity measured by dividing the nominal value of the expenditure for electricity by the amount of electricity used (kWh); B_Bakar = fuel expenditures for fuel value.

The regression results indicate that the best model for model 1 is the Random Effect Model (REM). REM model already uses GLS (General Least Square) in its estimation, so a test to check the presence of multicollinearity, heteroscedasticity, or autocorrelation problems is not necessary. The result of Haussmann test is greater than alpha level of 1%, 5% and 10%. Dummy variable program (D_Program) was associated negative and significant at α = 1% against the use of technology, described by a variable number of the electricity consumption (kWh)—with a coefficient of –0.5716374.1. This means when there is the machine restructuring program and the other independent variables are fixed/constant, savings of electricity consumption amounting to 77.116% in average can take place.

Test results of Model 2: The impact of machinery restructuring program on efficiency

Uji Chow	0.0000	
Uji LM	0.0000	
Uji Hausman	0.0022	
	Ln_Efisiensi (FE)	
Model 2	Coef	Robust St. Error
Dummy Program	0.1942957**	0.0656399
ln_Size	0.8234845***	0.0264762
lnC_I	-.0140411	0.0391816
Cons	-3.072239**	0.8477449
Number of Obs	50	
Prob > F	0.0000	
R-squared	0.5389	

Note: ***) Significant at level $\alpha = 1\%$, **) Significant at level $\alpha = 5\%$, *) Significant at level $\alpha = 10\%$.

Description: Efficiency = is obtained by dividing the value of the production sub-sectors of the textile industry with a total input used; D_Program = Dummy Program, 2002–2006 (no program) = 0; 2007–2011 (no program) = 1; Size = size of the company, in-proxy using the total workforce in a company; C_I = Capital Intensity, capital intensity as measured by capital ratio (total capital) to total employment.

The results of the regression estimation for the two models show that the best model is the Fixed Effects Model (FEM). The estimation results indicate that there was a multicollinearity problem in variable size and capital intensity (C_I). To overcome this, a robust test was used. The dummy variable of machine restructuring program had a significantly positive effect on the level of $\alpha = 5\%$ of the production efficiency coefficient of 0.1942957. This means that when the restructuring program and other independent variables are assumed constant, then the production efficiency variable will increase by 21.446%. The table below shows the efficiency average difference among subsectors.

	lnEfisiensi	
Model 2	Coef	St. Error
D_Program	0.1942957**	0.0785313
lnSize	0.8234845***	0.1185446
lnC_I	-0.0140411	0.031031
Subsektor 2	-0.3706466***	0.1289446
Subsektor 3	-1.57114***	0.2748787
Subsektor 4	-1.784302***	0.3569349
Subsektor 5	2.84E-01	0.2674997
Cons	-2.383915	1.679593
Number of obs	50	
Prob > F	0.0000	
R squared	0.9759	

Based on the estimates above, the average value of the sub-sector 1 efficiency can be seen from a constant value of −2384, which means an annual average efficiency in the sub-sector 1 (Fibre/Filament) was −2384. Meanwhile, the average efficiency of the sub-sector 2 (Spinning) was −2755. The average efficiency of the sub-sector 3 (Cain Making) was −3955. The average efficiency for sub-sector 4 (Garment) was −4168. Sub-sector 5 (TPT Other) had an average efficiency of −2668. Based on these results, the aid of the most effective machine restructuring program is intended for other textile sub-sectors related to the efficiency indicators. These intercept differences can be obtained if the best model, FEM is used. REM models can not be used to obtain these differences.

Test results of Model 3: The impact of machinery restructuring program on productivity

Uji Chow	0.0000	
Uji LM	0.0000	
Uji Hausman	0.1376	
	ln_Produktivitas (RE)	
Model 3	Coef	St. Error
Dummy Program	0.3323661***	0.0813485
ln_Upah	0.6906376***	0.0847844
H_Cap	0.0000226	0.0001703
Cons	1.704575	1.757864
Number of Obs	50	
Prob > Chi^2	0.0000	
R-squared	0.8626	

Note: ***) Significant at level $\alpha = 1\%$, **) Significant at level $\alpha = 5\%$, *) Significant at level $\alpha = 10\%$.
Description: Productivity = labor productivity, which is measured by dividing the value added (value added) by the total number of workers; D_Program = Dummy Program, 2002–2006 (no program) = 0; 2007–2011 (no program) = 1; Wages = total value of wages paid by the company to its workforce; H_Cap = Human Capital (human capital), as measured by the ratio of skilled labor to total employment, skilled labor in the proxy uses the total number of non-production workers.

The regression estimates indicate that the best model for model 3 is REM (Random Effect Model). REM model already uses GLS (General Least Square) in its estimation, so a test to check the presence of multicollinearity, heterokedastisitas, or autocorrelation problems is not necessary. The dummy variable of machine restructuring program had a significantly positive effect on the level of $\alpha = 1\%$ on labor productivity with a coefficient of 0.3323661. This means the program increases productivity by an average of 39.426%.

Test results of Model 4: The impact of machinery restructuring program on competitiveness

Uji Chow	0.0005	
Uji LM	0.0000	
Uji Hausman	1.0000	
	lnEkspor (RE)	
Model 4	Coef	St. Error
Dummy Program	−1.509269	2.945824
lnProduksi	10.61445***	1.623227
lnW_GDP	20.10352	39.35418
lnREER	−16.25348	52.94337
Cons	−418.9086	309.9112
Number of Obs	50	
Prob > Chi²	0.0000	
R-squared	0.6488	

Note: ***) Significant at level $\alpha = 1\%$, **) Significant at level $\alpha = 5\%$, *) Significant at level $\alpha = 10\%$.

Description: Exports = exports, a proxy of competitiveness; D_Program = Dummy Program, 2002–2006 (no program) = 0; 2007–2011 (no program) = 1; Production = Value of production per sub-sector; W_GDP = Value of real world GDP per capita at constant 2000 prices that have been adapted into rupiah currency; REER = real exchange rate of the rupiah against the US dollar.

Model 4 estimation results indicate that the best model used is REM (Random Effect Model). REM model already uses GLS (General Least Square) in its estimation, so a test to check the presence of multicollinearity, heterokedastisitas, or autocorrelation problems is not necessary. The results suggest that the machine restructuring program reduces export.

Competition and Cooperation in Economics and Business – Gani et al. (Eds)
 ISBN 978-1-138-62666-9

The fragility of competition and cooperation between Indonesia, China and the EU from an SME and cooperative point of view

N.I. Soesilo
Department of Economics, Faculty of Economics and Business, Universitas Indonesia, Depok, Indonesia

ABSTRACT: The increased maritime cooperation under the revival of the Silk Road theme has been initiated since President Jokowi's leadership of Indonesia. Many controversial efforts have been implemented since then to materialise this cooperation with China. Yet Indonesian Small Medium Enterprises (SMEs) and cooperatives have been worried by the influx of Chinese products due to their low price. This is not cooperation but the opposite, they compete with one another. However, both Indonesia and Chinese SMEs and cooperatives are donor recipients of European Union (EU) funding, some of which are disbursed through IFIs (International Financial Institutions) in cooperation with both governments. As stated in the Fragile States Index these recipient countries are considered by donor countries to be fragile States. The Triangular Cooperation (TrC) of Indonesia, China and the EU has existed as additional to the South-South cooperation between Indonesia and China. This paper will evaluate the relevance of the EU's point of view as the donor country using their ten principles to cope with Fragile States, especially in assisting the TheySMEs and cooperatives in both Indonesia and China. Despite the upper position of the EU as the donor country in the lending scheme to China and Indonesia, this higher position of EU has been labelled only 'normative power' not the real power, especially when dealing with China as the world's new superpower.

1 INTRODUCTION

China and Indonesia were categorised as fragile States with the same rank in 2016 by the Fragile State Index sponsored by the European Union (EU) within the Triangular Cooperation (TrC). Aid to fragile States represents 30% of all ODA (Official Development Assistance) missions along with 116,000 UN peacekeeping personnel (Blue Helmets) in 2010 which was eight times more than in 1999. Both Indonesia and China encouraged inclusive growth through the development of Small Medium Enterprises (SMEs) as well as cooperatives, despite their objective in a rigorous national economic development that prioritized growth that often create trade-off with equality. Fragile States are considered as very important from the point of view of donor countries such as the EU, because Fragile States are mostly off track for most Millenium Development Goals (Ramalingam, 2012), although many agendas of development, security, diplomacy, trade, migration, and beyond are actively deployed by international actors. However, EU influence has been labelled as normative power (Manners, 2000) rather than the doer. Since 2005, the Fragile State Index (FSI) has been calculated annually in many countries worldwide, supporting the EU international assistantship programme, and is available to be used for international economic, political and social research. However, beyond traditional conceptions of the EU's international role (Hedley Bull in Manners, 2002), this role dismisses the suggestion that the European Community (EC) represents a 'civilian power' in international relations, maintaining that in fact it is only a normative power. Kaplan states that EU as donor countries have often been confused by the definition and categorization of Fragile States, with some even calling them a 'wicked problem' because of its obscured causality. Both Indonesia and China are praised by Kaplan because of the social glue that lubricates business and has encouraged leaders to focus on inclusive development, even though sometimes they are not compatible with the rule of thumb of good governance, or progressive taxes effort, as often encouraged by common Western policy playbook.

Die Seidenstrasse (the Silk Road) was named by the German geologist, Baron Ferdinand von Richthofen, as the trade and communication network, enriched by the intellectual (science and technologies) and cultural (language, art, literature and craft) exchange and religion (Silk Road Foundation, 2007) that reflects cooperation, which is embedded in fragile state support. However, this does not mean that as fragile states, both China and Indonesia as EU's assisted countries do not have their own cooperation or competition. The SSC and TrC partnerships are vital forovercoming the current development challenges and reaching internationally agreed development goals, including the Millennium Development Goals as elaborated by the UNDP.

The increased maritime cooperation between Indonesia and China under the revival of the Silk Road theme has been implemented since President Jokowi's leadership of Indonesia. Many controversial efforts have been conducted since then to materialise this cooperation with China. Yet the Indonesian SMEs and cooperatives have been worrying about the influx of Chinese products because of their low price, and have therefore been competing with one another. However, both Indonesian and Chinese SMEs and cooperatives are donor recipients of EU funding, eventhough some of the funding are disbursed indirectly because the International Financial Institutions give the money to the governments first before goes to them. It is because both countries are considered as Fragile States due to their high fragility magnitude as stated by Fail States or Fragile State Index.

Cooperation is the opposite of competition. IPorter's national diamond 'Competitive Advantage of Nations (1990)' he proclaims a bridge between the theoretical literatures in strategic management and international economics and provides the basis for improved national policies on competitiveness (Porter in Davies & Ellis, 2000). The 'competitive advantage' of a nation's industries is determined by the configuration of four broad attributes of the national location, referred to as the 'home base'. These are the now-familiar corners of the 'national diamond', namely, (1) factor conditions, (2) demand conditions, related and supporting industries, and firm strategy, structure, and rivalry (Davies & Ellis, 2000). On the other hand, a distinguished commercial and social entrepreneurship cooperation is established (Austin et al. 2006), as both built alliances are created by mutuality of interests, similar to incongruity that ends up with an unexpected strategic fit. There is no difference between business and non-profit partnerships, and those between business and business. However, commercial entrepreneurship is different from social entrepreneurship in the identification, evaluation and exploitation of opportunities leading to profits because of social value. Fragility is the opposite of competitiveness, as it represents a weakness as well as a critical aspect that should be prevented immediately to mitigate further risk.

1.1 *Review of literature*

Unlike Porter's work of improved national policies based on competitiveness, the fragility being used in this research is based on Fund for Peace's Fail State Index (FSI) or Fragile State Index, which has been creating some calculation about country fragility since 2005 focusing on the indicators of risks using the CAST Software in its computation and based on thousands of articles and reports that are processed from electronically available sources. This index is produced annually and usage is encouraged to develop ideas for promoting greater stability worldwide. Using the FSI data for calculating total annual fragility growth from 2005 to 2016, it was found that fragility in Indonesia has decreased by 1.4 per cent, while in China it has increased by 0.7 per cent, and in the EU decreased 0.1 per cent (using annual growth of 2006–2016). In this case, Indonesia shows a better picture compared to the other two countries, as seen in Tables 1, 2 and 3.

According to the UN, the Triangular Cooperation (TrC) is a common Southern-driven partnership between two or more developing countries, supported by a developed country(ies)

Table 1. The growth of fragility in China and Indonesia based on the FSI data.

Year	2005		2016		2017		Fragility growth	
Country	IDN	China	IDN	China	IDN	China	IDN	China
Fragility Rank	47	75	86	86	Projection			
							Decrease	Increase
Total Fragility	87	72.3	74.9	74.9	73.5	78.1	–1.4%	0.7%
Demographic Pressures	8.6	6.8	6.8	6.9	7	7.5	–1.7%	0.9%
Refugees and IDPs	7	5	5.6	5.1	5.4	6.0	–2.1%	1.7%
Group Grievance	6.3	7.4	7.3	8.1	7.6	8.4	1.9%	1.1%
Human Flight	8.9	6	6.6	4.6	5.7	4.4	–3.3%	–2.3%
Uneven Development	9	9	6	7.5	5.9	7.4	–3.1%	–1.5%
Poverty and Economic Decline	4	0.5	5.5	4.1	5.7	4.3	3.8%	63.9%
Legitimacy of the State	9.2	8.6	5.3	8.3	5.4	7.9	–3.7%	–0.7%
Public Services	4	2.9	6.1	5.9	6.3	6.8	5.2%	11.2%
Human Rights	8.6	8.9	7.4	8.7	6.3	9.0	–2.4%	0.1%
Security Apparatus	7.6	7	5.9	5.6	6.1	5.9	–1.8%	–1.3%
Factionalised Elites	8.8	8.4	7	7.2	6.6	6.8	–2.3%	–1.6%
External Intervention	5	1.8	5.4	2.9	5.5	3.7	1.0%	8.7%

IDP = Internally Displaced Person.

Table 2. International migrant stock is the number of people born in a country other than that in which they live. It also includes refugees.

Country code	1965	1975	1985	1995	2005	2015
CHN	260479	298474	323937	442198	678947	978046
IDN	1475121	928340	593098	378960	289568	328846
CHN/IDN	0.176581	0.321514	0.546178	1.166872	2.344689	2.974176

or multilateral organisation(s), to implement development cooperation programmes and projects. In this case, the TrC comprises Indonesia, China and the EU. Funding from EU has been dominating IFIs. One of them is the UN IFAD (International Fund for Agricultural Development). On 31 December 2012, EU dominated 31% from the majority share of non-member countries, whose number reached 50.2% in the UN IFAD, while the World Bank were only 7.85%). It is based on consolidated financial statements audited by Deloitte (Soesilo, 2014). The number were calculated accumulatively from 1978–2012, including Supplementary Contribution and project co-financing to UN IFAD (IFAD, 2013) as well as the contribution of Belgium BFFSJP (Belgian Fund for Food Security Joint Programme) that had the biggest share of 16.75%, followed by the Netherlands (7.9%) and the UK (2.95%), while Indonesia was only 0.004%. In 2014, from member state shares, China's share was 1.3% and Indonesia was 0.8%. For the purpose of providing aid, the EU utilized the Fragile State Index (FSI) report as a modification of Fail State Indeks report, even though it was not publicly announced.

At the initial formation, the accumulation score of countries shown by the FSI annual index was categorised into the following: 'alert', 'warning', 'moderate' and 'sustainable'. Both the Fail State Index or the Fragile State Index use the categorisations of: 'very high alert', 'high alert', 'alert', 'warning', 'high warning', 'elevated warning', 'stable', 'more stable', 'very stable', 'sustainable', and 'very sustainable'. By looking at the trend of time series scores, the growth of FSI components can be used to calculate the most critical aspect that should be prevented immediately. It is a tool for highlighting the most critical warning pressures that a country experiences. Among the social, economic and political aspects, which is the most fragile can be easily found and then underlined to avoid entering the brink of failure.

Table 3. EU countries and their fragilities based on FSI's data from 2006–2016 and its projections.

EU countries	Growth	2017	2016	2015	2014	2013	2012	2011	2010	2009	2008	2007	2006
Austria	0.6%	27.7	27.5	26.0	28.5	26.9	27.5	27.3	27.2	27.6	25.9	26.0	26.1
Belgium	3.1%	32.3	29	30.4	32.0	30.9	33.5	34.1	32.0	33.5	29.0	25.5	24.0
Bulgaria	–0.7%	57.5	53.7	55.4	54.4	55.0	56.3	59.0	61.2	61.5	58.5	60.3	62.1
Croatia	–0.8%	56.3	52.4	51.0	52.9	54.1	56.3	57.3	59.0	60.1	59.4	60.5	61.9
Cyprus	–0.4%	67.7	64	66.2	67.9	67.0	66.8	67.6	68.0	68.9	69.7	70.2	70.5
Czechia	–0.2%	40.7	40.8	37.4	39.4	39.9	39.5	42.4	41.5	42.6	42.1	42.1	41.8
Denmark	–0.8%	22.6	21.5	21.5	22.8	21.9	23.0	23.8	22.9	23.2	21.5	22.2	24.8
Estonia	–0.5%	48.0	43.4	43.8	45.2	45.3	47.5	49.3	50.7	51.2	51.0	50.5	51.0
Finland	0.3%	18.8	18.8	17.8	18.7	18.0	20.0	19.7	19.3	19.2	18.4	18.5	18.2
France	0.0%	34.2	34.5	33.7	34.8	32.6	33.6	34.0	34.9	35.3	34.8	34.1	34.3
Germany	–1.4%	33.6	28.6	28.1	30.6	29.7	31.7	33.9	35.4	36.2	37.3	38.4	39.7
Greece	1.6%	48.3	55.9	52.6	52.1	50.6	50.4	47.4	45.9	46.1	45.4	43.5	41.1
Hungary	0.5%	49.5	52.7	49.1	48.3	47.6	48.3	48.7	50.1	50.7	50.9	51.2	46.7
Ireland	2.1%	22.9	22.5	24.7	26.1	24.8	26.5	25.3	22.4	21.6	19.9	19.5	18.6
Italy	1.9%	42.5	43.1	43.2	43.4	44.6	45.8	45.8	45.7	43.9	39.9	37.1	35.1
Latvia	–0.6%	52.3	47.4	48.6	48.0	47.9	51.9	54.2	55.4	54.6	54.5	56.7	56.2
Lithuania	–0.7%	45.8	42.4	43.0	43.2	43.0	44.2	45.3	47.8	48.0	48.7	49.0	49.7
Luxembourg	–0.8%	25.9	24.1	22.2	24.6	23.3	25.5	26.1	27.3	27.6	27.9	28.1	
Malta	–0.7%	45.3	39.6	40.9	43.0	42.4	43.8	45.4	48.2	48.8	48.3	48.5	
Netherlands	–0.1%	27.8	28.2	26.8	28.6	26.9	28.1	28.3	27.9	27.0	27.3	28.6	28.1
Poland	–0.5%	45.1	40.7	39.8	42.1	40.9	44.3	46.8	49.0	49.6	47.6	47.6	47.9
Portugal	–0.2%	32.2	29.2	29.7	33.1	32.6	34.2	32.3	33.1	32.7	31.8	32.4	32.7
Romania	–0.6%	58.7	52.9	54.2	56.9	57.4	59.5	59.8	60.2	61.3	59.9	60.9	62.6
Slovakia	–0.5%	47.1	44.9	42.6	45.3	45.3	47.4	47.1	48.8	48.6	48.8	49.3	49.9
Slovenia	–0.5%	34.9	33.9	31.6	32.6	32.3	34.0	35.5	36.0	36.3	37.1	37.5	36.8
Spain	1.1%	41.7	39.8	40.9	43.1	44.4	42.8	43.1	43.5	43.3	41.6	39.2	37.4
Sweden	1.2%	20.6	22.6	20.2	21.4	19.7	21.3	22.8	20.9	20.6	19.8	19.3	18.2
UK	–0.1%	33.8	32.4	33.4	34.3	33.2	35.3	34.1	33.9	33.6	32.9	34.1	34.2
Average	–0.1%	39.7	38.09	37.7	39.0	38.5	40.0	40.6	41.0	41.2	40.4	40.4	40.4
Max			64	66.2	67.9	67.0	66.8	67.6	68.0	68.9	69.7	70.2	70.5
Min			18.8	17.8	18.7	18.0	20.0	19.7	19.3	19.2	18.4	18.5	18.2

By underlining pertinent issues in Fragile State Indexes over time, the risk assessment can be used as an early warning of conflict that is accessible to policymakers and the public at large.

Both Indonesia and China are categorised as elevated warning countries in 2016 and positioned at the same 86th rank. However, based on previous record, in 2005 Indonesia was ranked 47th in the Failed State Index, while China was in the 75th. It means China has worsening trends compared to Indonesia. To see the more detail performance, from the similar worsening trends in Indonesia and China, they are grouped in the three following components with their increasing annual fragility growths in brackets: group grievance (Indonesia and China deteriorated by 1.90% and 1.1% consecutively), poverty and economic decline (Indonesia and China deteriorated by 3.80% and 63.9% sequentially), public services (deteriorating by 5.18% and 11.2% for Indonesia and China consecutively) and worsening fragility due to external intervention at a rate of 0.97 per cent for Indonesia, and 8.7 per cent for China. However, China has three additional failing trends, while Indonesia is better off due to the opposite inclination. Pairing them together with their annual growth in brackets, the findings are as follows: the Demographic Pressures fragility (China increased by 0.9%, while Indonesia decreased by 1.7%), fragility due to Refugees and Internally Displaced Persons (China increased by 1.7%, while Indonesia decreased by 2.1%), and Human Rights fragility (China increased by 0.1%, while Indonesia decreased by 2.4%). A similar annual decreasing fragility happened for both countries in increasing state legitimacy; Indonesia and China are both better off with 3.7 per cent and 1.7 per cent consecutively. In total, the number of

decreasing risk indicators in Indonesia is eight, while in China it is five, which indicates that Indonesia has better trends compared to China as can be seen in Table 1.

Meanwhile, the EU is a politico-economic union of 28 member states that are located primarily in Europe with an estimated population of over 510 million, about two times of that of Indonesia and one third of China. Brexit is the referendum result which urged that the UK should leave the EU, based on more than 30 million people voting on Thursday 23 June 2016 (Wheeler and Hunt, 2016).

Historically, the Silk Road, *die Seidenstrasse* was named by the German geologist, Baron Ferdinand von Richthofen (UNESCO, 2000), as the trade and communication network enriched by intellectual (science and technologies) and cultural (language, art, literature and craft) exchange and religion. In recent developments, as Jokowi put it, Indonesia has rich natural resources and China has knowledge and experience in the field of infrastructure. President Xi Jinping from China promised to support Indonesia's plans to develop its maritime infrastructure (the diplomat.com, 2015).

The increased maritime cooperation under the revival of the Silk Road theme has been conducted since President Jokowi has been leading Indonesia. Many controversial efforts have been made since then to materialise this cooperation, such as the development of infrastructure in Indonesia by employing Chinese workers and inviting the same number of Indonesians to China. A lot of resentment spurred by social media that might ignite conflicts have been setoff. This paper will elaborate ten ways on how to cope with fragile States Principles especially regarding the involvement and participation of micro small enterprises and cooperatives.

In the eyes of the EU (Farnell and Crookes, 2016), China is often characterised as operating in ways that accord with a broadly realist world view of international relations, premised on competing power dynamics between sovereign states, building alliances and mitigating threat perceptions in a modern international system whose anarchy is only partially offset by the evolution of issue-based institutions and regulatory regimes. The EU, by contrast, is typically characterised by a distinctive post-Westphalian organisation whose members have willingly agreed to bring together sovereignty and accepted a hint of supranational authority over major areas of domestic and international policymaking.

The Fragile State Index, however, is calculated from the point of view of the EU (representing the West). No wonder that both Indonesia and China are considered as at 'elevated warning' levels in Fragile State Indexes. Both China and Indonesia are non-Western Countries and recipients of the funding. As a whole, according to the FSIs, the EU is considered as stable, despite the fact that Brexit has been agreed upon. From 28 member countries in EU, there are ten states that have shown an increase in their fragility in the period 2006–2016, that is, Austria, Belgium, Finland, France, Greece, Hungary, Ireland, Italy, Spain and Sweden, as seen in Table 3. However, the following states show more stability: Bulgaria, Croatia, Cyprus, Czechia, Denmark, Estonia, Germany, Latvia, Lithuania, Luxembourg, Malta, the Netherlands, Poland, Portugal, Slovakia, Slovenia, the United Kingdom. The average growth of fragility from 2006 to 2016 is minus 0.1%.

Using Hellsten, Sirkku's dual analytical tool, first the polarisation of Western and non-Western ways of thinking, and second, the gap analysis about the desired and existing outcome in a multicultural environment and/or global context, the similarity of conflicts between two different theoretical and ethical frameworks exists. In international human rights discourse, two theoretically incompatible views exist. Third World communitarians claim that a form of cultural imperialism smuggles the undesirable Western egoist lifestyle (like the EU) into non-Western traditional communalist cultures as an attempt by Western liberal individualists to universalise their value orientation. On the other hand, the inability to provide moral impartiality in the communitarian approach, whose cultural attachment makes its arguments so culturally embedded, is driven by individualist liberalism. The collective values and consensus trampling on the rights of individual members of these (as well as other) communities is inevitable. In other words, the communitarian position seems to equate a liberal, individualistic promotion of universal values with cultural assimilation that results in the fragmentation and demise of various traditional cultures.

Even though considered to be fragile States, both China and Indonesia are praised by Kaplan (2012). He states,

> *The countries most successful at promoting development and reducing poverty have been horizontally social cohesive, not vertically so, because of how important this is when governments do not work well. Places like China, Vietnam, Indonesia, and the rest of East Asia have improved the lives of over a billion poor people over the past two generations not because they have had good governance, progressive taxes, or followed a certain policy playbook, but because they have had the social glue that lubricated business and encouraged leaders to focus on inclusive development. Donors have often been befuddled by fragile States, with some even calling them a wicked problem. These difficulties, however, say more about donors than they do about fragile States.*

From 1998 to 2003, based on statistics from China's Information Office of the State Council, approximately 19 million workers laid off by State Owned Enterprises (SOEs), were re-employed, most of them by SMEs (Li, 2012). In 2010, the tax revenue collected from private enterprises in China accounted for almost 11 per cent of the total tax intake (Li, 2012). It is possible that the data of private enterprises in China includes larger enterprises, because in comparison, the contribution of tax revenue from micro small and medium enterprises (MSMEs) in Indonesia is still minimal, only three per cent of the total tax revenue (Financial Business.com, 2015). In Indonesia Small Medium Enterprises face a heavy burden of one per cent of tax from business turnover. However, the modification of the regulation is still being processed. SMEs with a turnover range of IDR300 million per month will be charged at a lower rate layer, that is 0.25 per cent, while SMEs with an annual turnover range of between IDR300 million to IDR 4.8 billion will be subject to a higher tariff layer, that is, 0.5 per cent (Adelia, 2016). SMEs account for around 80 per cent of China's manufacturing employment and around 80 per cent of new urban employment; they employ more low income workers and socially vulnerable groups and sometimes are the only source of employment in poorer regions (Li, 2012). SMEs are estimated to contribute more than 60 per cent of China's GDP (Li, 2012) while in Indonesia they contribute 58 per cent of total GDP.

Cooperatives are civil society actors (Coopeurope.coop, 2012). A cooperative is categorized as social enterprise. This classification is based on a discussion about social entrepreneurship with the British Council in Jakarta in July 2016 at the University of Atmajaya, because cooperative's goal is to produce surplus which is comparable to profits in private enterprise (Skurnik, 2002); however, both strive to be economically efficient. A cooperative is an entrepreneurial business entity. The difference between a cooperative and a limited company is that the production of profit is for its owners while the cooperative is for its members (Skurnik, 2002). The surplus is returned to the members in proportion to their use of the company's services, after keeping the reserves necessary to develop the company. A cooperative is just a different variety of company, and hence only one form of enterprise among many. The average cooperative business size in Indonesia is IDR872.51 million, thus it fits within the category of small businesses, based on the SME law or the Act No.20/2008 classification (bi.go.id, 2017). All 36.4 million cooperative members contribute 30 per cent of the total labour force (Meisari, 2016). In China, cooperatives are the most common form of association that are supported by the ICCIC (International Committee for the Promotion of Chinese Industrial Cooperatives) in the enhancement of women's position, poverty alleviation, etc. which can be found in Gansu, Sichuan, Hebei, Henan, Shanghai, Yunnan, Guizhou, Anhui, and Jilin.

Gung Ho is the slogan of a project initially founded by Rewi Alley in 1938 to foster genuine democratic cooperatives in various parts of China. Unfortunately, the farmers' associations have a weak link in overcoming legal difficulties since the legal status of these associations was only clarified with the new Cooperatives Law in 2006 (Soesilo, 2012).

Silk Road (Unesco, 2000) is, in fact, a relatively recent term introduced by President Joko Jokowi Widodo when he mentioned the Silk Road Fund through the Asian Infrastructure Investment Bank. At that time where he invited Chinese companies and others to help build a maritime highway from the west to the east of Indonesia and 24 seaports and deep seaports, to reduce commodity price disparities. This was mentioned during the CEO Summit of the Asia-Pacific Economic Cooperation forum in Beijing on 10 November 2014. Historically, as Jokowi and President Xi Jinping put it, Indonesia provides natural resources and China supports Indonesia's maritime infrastructure (Luhulima, 2016).

If Porter's work underlined the constructive side, the FSI emphasized the opposite. Porter's competitiveness is considered as a bridge between strategic business management theory and international economics. The country competitiveness index has been used worldwide to calculate comparable positive side, while the FSI calculates the negative side. The FSI uses four social indicators, two economic indicators and six political indicators. The social indicators are demographic pressures, refugees and IDPs, group grievances, and human flight. The economic indicators are uneven development, poverty and economic decline. The political indicators are legitimacy of the state, public services, human rights, security apparatus, factionalised elites and external intervention.

2 METHOD

This research has three steps because of the existence of three actors, Indonesia, China and the EU, which creates the Triangular Cooperation (TrC). The first research step is the power of the EU (normative power) by using the Fragility State Index as a yardstick, the second step is the loan analysis using the ten principles of the EU as a benchmark, and the third step is analysing whether the cooperation or competition exists using the principle of the SSC (see Figure 1).

The Triangular Cooperation (TrC) sees the connection among Indonesia, China, and the EU, as 'Southern-driven partnerships between two or more developing countries, supported by a developed country(ies) or multilateral organisation(s), to implement development cooperation programs and projects' (UNDP, 2009).

Between Indonesia and China, the analysis uses the SSC principle which is

> *a process whereby two or more developing countries pursue their individual and/or shared national capacity development objectives through exchanges of knowledge, skills, resources, and technical know-how, and through regional and interregional collective actions, including partnerships involving Governments, regional organizations, civil society, academia and the private sector, for their individual and/or mutual benefit within and across regions. South-South cooperation is not a substitute for, but rather a complement to, North-South cooperation* (UNDP, 2009).

In elaborating the position of Indonesia, China and the EU, the UNDP standpoint can be used as the basis of the analysis in three activities. In elaborating the position of Indonesia,

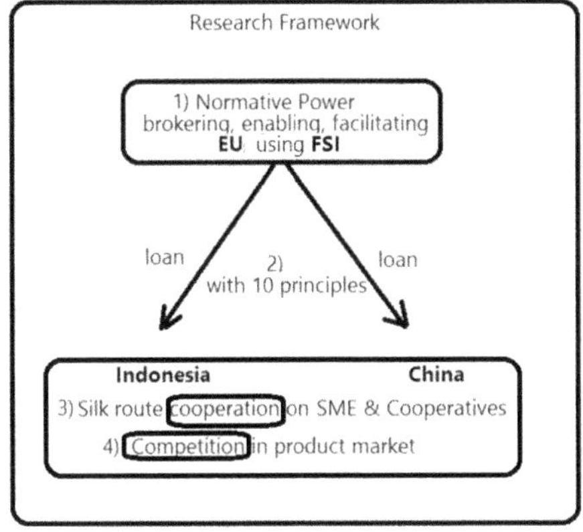

Figure 1. Research framework.

Table 4. Normative power of EU to be imposed by China and Indonesia.

	Unit Cost for Manajer per day					
	EU group					
Countries	1	2	3	4	China	Indonesia
EURO	294	280	164	88	77	47
Ratio to Indonesia	6.26	5.96	3.49	1.87	1.64	1.00

EU 1: Austria, Denmark, Ireland, Luxembourg, Netherlands, Sweden.
EU 2: Belgium, Finland, France, Germany, Italy, UK.
EU 3: Cyprus, Czechia, Greece, Malta, Portugal, Slovenia, Spain.
EU 4: Bulgaria, Estonia, Croatia, Latvia, Lithuania, Hungary, Poland, Romania, Slovakia.
Source: Guidelines for the use of the grant, Erasmus+Programme, 2017.

China, and the EU, the UNDP standpoint (UNDP, 2009) has a strong role to play as a knowledge broker, capacity development supporter, and partnership facilitator using EU guidelines as normative power, as seen in Table 4.

The UNDP will focus on: (1) brokering knowledge on scalable development solutions and analyse what has worked and what has not, with systematic information of who, where, and what is happening in South-South and Triangular Cooperation; (2) enabling harmonisation of policies, legal frameworks, and regulations to increase opportunities and maximise mutual benefits of South-South exchanges, while supporting capacity development of Southern partners to better implement SSC and TrC initiatives; and (3) facilitating partnerships, fostering innovations, and promoting the scaling-up of promising ideas. In its South-South and Triangular Cooperation efforts, the UNDP connects and works together with a wide variety of interested stakeholders from governments to private sectors and civil societies.

From the donor side, this analysis uses the EU benchmark (Morcos & Roder, 2007) in giving aid to fragile States that should be as follows: (1) Context is the starting point; (2) Do no harm; (3) State building is the central objective; (4) Prioritise prevention; (5) Political, security and development objectives are linked; (6) Promote non-discrimination as a basis for inclusive and stable societies; (7) Align with local priorities in different ways and in different contexts; (8) Agree on practical coordination mechanisms between international actors; (9) Act fast… but stay engaged long enough to give success a chance; (10) Avoid pockets of exclusion (aid orphans).

3 RESULTS

From the point of view of SSCs regarding the ten principles of aid, only four elements can be used for cooperation between Indonesia and China. The four elements are the first, sixth, seventh, eighth principles, and these are elaborated on as follows:

***First Principle—Context is the starting point*:** It linked Indonesia and China in entrepreneurship both in positive and negative side. In 2002, the Central Government of China declared the SME Promotion Law, showing a major change in attitude from the central government towards SMEs after years of market reform, even though they were not permitted to exist until 1988, ten years after economic reform started in 1978 (Li, 2012). In Indonesia the SME Law Number 20, 2008, was launched with the aim to foster and expand its business in order to build a national economy based on equitable economic democracy. Since reaching it highest annual economic growth of 13.6 per cent in 2007, China has been showing a slowdown. China's growing middle class meant it could not constantly be a low-wage and low-cost producer, but would be able to shift to a US-type consumption model within a decade (Whittaker 2016). Some migration of apparel manufacturing from China to Indonesia is an inevitable consequence, and even encouraged by the Chinese government because China

must embrace research and high technology production to transform its economy as South Korea and Japan once did more efficiently at mass manufacturing. However, vigorous economic growth demands that China increases its service sector and prioritize higher-skilled manufacturing jobs with greater compensation (Chu 2013). While China's economy has been diminishing, Indonesia has been increasing after viewing the potential in its huge consumer market. For example, the teak furniture maker Broxo Indonesia, which has been exhibiting at the China-ASEAN Expo since 2013, had an annual sales growth to China of 30 per cent until 2015 (Tan, 2015).

As the second, third, fourth, and fifth principles are not relevant, the elaboration goes to the ***Sixth Principle—Promote non-discrimination as a basis for inclusive and stable societies***. Promoting MSEs, the majority of which are privately owned, is a government priority in the drive to build a well-off and harmonious society. While MSMEs as a whole remain credit constrained, it is the MSEs that have over time become the forgotten segment. To overcome this, the Chinese government created The Programme for the Development of Chinese Women, a new legal and organisational framework for the period 2011–2020. Cooperatives are the most common form of association supported by the ICCIC. Unfortunately, the farmers' associations in China have weak links in overcoming legal difficulties since the legal status of these associations was only clarified with the new Cooperatives Law in 2006. The inclusive growth has to face the annual growth of group grievance in Indonesia which is 1.90 per cent with China at 1.1 per cent. In this case China is better in comparison to Indonesia because it has become the first developing country to realise the UN's MDGs. China also follows the Beijing Platform for Action, the Outcome Document, Convention on the Elimination of All Forms of Discrimination against Women and the MDGs. Three decades of China economic growth have focused more on infrastructure building than on a functioning society. However, within the MSME sector, micro and small enterprises (MSEs) will play an indispensable role in meeting these challenges. Though, in the case of Hubei in China, entrepreneurship movement went too far, as there was also a shift to commercial farming under Dragonhead Enterprises that increased economic gains to participating countries but which also raised concerns about targeting the poor. While China is able to meet its MDGs, Indonesia is not, because its target is off track, although many agendas of development, security, diplomacy, trade, migration and beyond are actively deployed by international actors. In Indonesia, cooperative discrimination policies exist, such as Article 21 from Law 44, 2009 about hospitals (hukumonline.com, 2009). Private hospital management in Indonesia excludes cooperatives as a formal legal entity, while in other countries, health cooperatives are very common in managing hospitals (IHCO Coop 2016). No wonder that cooperatives as an inclusive growth actor contribute only 1.9 per cent of total GDP, lagging behind SOEs and private companies. While the annual growth of Competition Index in Indonesia has been decreasing at a rate of –1.2 per cent, at the same time the rate of Indonesia Gini coefficient rose 2 per cent from 2007 to 2016 as seen in Table 5. No wonder that Robison doubted the political and economic intentions and capacities of Indonesia to envisage the power of the state (Kompas.com, 2016 and Robison, 2016).

Seventh Principle—Align with local priorities in different ways and in different contexts: This principle is relevant to be discussed, but with several notes. A lot of resentment has been ignited, spurred by social media that might ignite conflicts, because the Silk Road Programme contradicts the local priorities to give employment to Indonesian people. The revocation of Article 26, Paragraph 1, the Manpower and Transportation Minister Regulation No 12/2013 deletes the regulation that requires that foreign workers must be able to speak local Indonesian. It is against Article 3 Paragraph 1 of the Manpower Minister Regulation No 16/2015 (Riyadi 2016). On the other hand in China since the mid-1990s, in order to align with local priorities, UN IFAD's strategic partners are the Government of China, and the provincial Association of Woman Federation and Rural Credit Cooperatives (RCC) network.

Eighth Principle—Agree on practical coordination mechanisms between international actors: Practical coordination is needed on several topics. However, the most urgent issue is drug trafficking between China and Indonesia. For this cooperation, the Indonesia Global Maritime Fulcrum (GMF) is more inward looking to Indonesia; it focuses on the internal waters

Table 5. Indonesia annual growth of regulation index and Gini coefficient compared to China.

Year	Indonesia competitiveness index					Gini	Coef
	Agriculture	Manufacture	Services	Trade	Total	IDN	China
2007	0.7	0.65	0.75	0.85	0.74	0.35	
2008	0.7	0.6	0.75	0.85	0.73	0.428	0.428
2009	0.7	0.6	0.8	0.85	0.74	0.37	
2010	0.7	0.55	0.8	0.85	0.73	0.38	
2011	0.7	0.55	0.85	0.85	0.74	0.41	
2012	0.65	0.55	0.8	0.8	0.7	0.41	0.422
2013	0.6	0.6	0.8	0.8	0.7	0.41	
2014	0.6	0.6	0.75	0.8	0.69	0.41	
2015	0.65	0.7	0.75	0.8	0.73	0.4	
2016	0.61	0.56	0.77	0.75	0.67	0.39	
Annual growth	–1.48%	–1.56%	–0.18%	–1.54%	–1.21%	2.04%	

Source: Indonesia Competitive Index is calculated from Business Competition Supervisory Committee data, 2016. The news also appears in Kontan, 2016. Gini coefficient is calculated from *www.bps.go.id* as well as data in Indonesia-investment.com (2016) in *www.indonesia-investments.com/id/news/todays-headlines/gini-ratio-indonesia-declines-inequality-narrows/item7113*; In China, the data above is from World Bank, while other sources can be found in: *https://www.theguardian.com/world/2014/jul/28/china-more-unequal-richer*; the Gini coefficient in China in 2012 was 0.78.

of Indonesia. From a micro enterprise and cooperative point of view, this is seen as beneficial due to the banning of fishing nets used by big companies to harvest all sizes and varieties of fish to protect the environment. However, it is actually ruining the fishing industry in Indonesia due to the short supply of tuna fish from the internal seas. In a maritime country like Indonesia, it is an irony that it has to import tuna from overseas in order to supply the domestic industry (Bisnispos 2016).

4 CONCLUSION

The increased maritime cooperation under the revival of the Silk Road has made some progress in economic development in Indonesia; however, at the initial process, it is dominated by competition rather than cooperation. Indonesian SMEs and cooperatives have been worried by the influx of Chinese products because of their low prices. Later on, China started showing an economic slowdown and some migration of low-tech industry to Indonesia happened. For example the apparel manufacturing has been moving to Indonesia because China must embrace research and high technology production to transform its economy to be more efficient at mass manufacturing. For Indonesia and China, growth and comfort never coexist as noted by Ginni Romety (Business Insider, 2015). However, the process is already happening if analyzed using the international relations framework in the form of both competition as well as cooperation (the SSC).

When dealing with Triangular Cooperation (TrC), even though considered to be fragile States by the EU, both China and Indonesia possess their own footpath of progress. Both countries are praised by Kaplan not because he is impressed by good governance, progressive taxes, or followed a certain policy playbook, but because they have the social glue that lubricates business and encourages leaders to focus on inclusive development. Kaplan states that donors have often been confused by fragile States, with some even calling them a 'wicked problem'. These difficulties, however, say more about donors than they do about fragile States. Compare to Indonesia and China, fragility problems within the EU member states are already increasing (one big example of which is Brexit), even though some part of the FSI calculation for EU's fragility are decreasing (as seen in Table 3). Therefore, it would be wiser, especially for China, to put the FSI implications aside and insist on their own path of development (Xie, 2016).

If using the following three basis analysis, EU mostly possesses comparative advantage in; (1) brokering knowledge, (2) enabling harmonisation of policies and, (3) facilitating partnerships. In giving assistance to Indonesia and China, as the supply side of aid in Triangular Cooperation, it is beneficial for EU if fragility consideration is used for evaluating the recipient countries. However, based on the demand side of aid, both Indonesia and China have their own dynamics of cooperation and competition which is completely different. They are fundamentally distinctive. Many would group these types of initiative together as 'aid' as opposed to a 'trade' type activity. China, due to its impressive progress, is often considered as a doer country. It is completely the opposite compared to EU, that typically characterised as a distinctive post-Westphalian organisation with ten principles of assistance. When using all ten principles of fragile state assistance from the EU in the case of SMEs and cooperatives, only the first, sixth, seventh, eighth, ninth and tenth principles can be used.

When analysing the competing power dynamics between China and Indonesia as sovereign states in the hope of alliance building with the revival of the Silk Road since President Jokowi's administration, some precaution have to be taken into account. It is important to mitigate threat perceptions, especially from the Indonesia side. In modern international system, whose lawlessness is only partially offset by the evolution of issue-based institutions and regulatory regimes, some elements of cooperation only relevant if the win-win solution is achieved. This is unlike the EU whose members have voluntarily agreed to pool sovereignty and accept the appearance of supranational authority over key areas of domestic and international policymaking.

From the EU ten principles of giving aid, there are five principles that are not relevant in the South-South Cooperation of Indonesia and China, especially the second principle (do no harm), the third principle (state building is the central objective), the fourth principle (prioritise prevention), the fifth principle (political security), the ninth principle (act fast), and finally the tenth principle (aid orphans). This is elaborated on as follows:

The second ***'Do no harm' principle*** cannot create cooperation because first, it is more about competition, such as the influx of Chinese products in Indonesia that has caused bankruptcy for some Indonesians SME, and second, due to lawlessness. Chinese ships seized by the Indonesians six years ago were scuttled and detonated and then in retaliation, an Indonesian ship was threatened at gunpoint by Chinese vessels off the Natuna islands for having arrested Chinese trawlers. The third principle, ***'State building as the central objective'*** is not relevant for SMEs and cooperatives due to their frantic schedule in earning their own living, which does not allow time to think about big issues like state building. The fourth principle ***'Prioritise prevention'*** is more about government policy rather than SME and cooperative daily activities; while the fifth principle, ***'Political, security and development objectives are linked'*** is also not relevant for SMEs and cooperatives. In the ninth principle, ***'Act fast... but stay engaged long enough to give success a chance'***, Indonesian and Chinese cooperation has happened very fast; especially in infrastructure for example the development of fast trains connecting Jakarta and Bandung (Afrianto, 2016). This has created potential conflict due to a lack of proper planning documents and bias towards big companies, leaving no ample benefits for SMEs and cooperatives. In the bilateral comprehensive strategic partnership between two countries, Xi called for strengthened high-level coordination, connecting China's twenty-first century Maritime Silk Road initiative with Indonesia's Global Maritime Axis vision, in which Indonesia also supports and is willing to strengthen cooperation with China in trade, investment, finance, infrastructure and other areas (China.org 2016). Finally the tenth principle, ***'Avoid pockets of exclusion (aid orphans)'*** means that each project should not be in the form of enclave, but should have good impact for the environment and community such as mentioned by the spirit of SSC. For example, when Chinese workers came to Indonesia along with the Bilateral Agreements of Foreign Debt between Indonesia and China the sociological impact should be calculated, because job seekers in Indonesia are also still abundant. In general, this agreement requires that infrastructure development undertaken by Chinese companies consists of one package using Chinese workers brought from China. With increasing demographic pressure in China, the perceived Chinese influx to Indonesia is inevitable from the point of view of China. But, based on reciprocity principle of the Silk Road programme, some resentment have to be reduced. As planned, by employing Chinese workers in Indonesia, China is also able to invite the same number of Indonesians to China. Chinese investment in Indonesia accounts for six per cent of the combined investment realisation of $7.29 billion in the first quarter, placing China in fourth position after Singapore, Japan and Hong Kong. The pocket of exclusion of Chinese investment is perceived as dangerous from an Indonesian point of view, and it is not in line with the spirit of SSC.

The challenges faced by fragile States are multidimensional. Third World communitarians claim that a form of cultural imperialism that smuggles the undesirable Western egoist lifestyle (like the EU) into non-Western traditional communalist cultures is an attempt by Western liberal individualists to universalise their value orientation. The political, security, economic and social spheres are interdependent factors, such as the OECD principles in the ambitious goal of assisting in the building of effective, legitimate and resilient states. However, it seems too normative for the SME and cooperative case, since State effectiveness is a complex and multidimensional phenomenon. However, SMEs and cooperatives cannot be effective if there is no power. Despite the upper-hand position of the EU as the lender, or principal donor entity in the lending scheme, the EU has been labelled as possessing only normative power, especially when dealing with China as the new world superpower. There are six utmost critical functions of capacities to perform effectiveness, these are to: (1) monopolise the legitimate use of violence (coercive capacity), (2) extract resources (extractive capacity), (3) shape national identity (assimilative capacity), (4) regulate the society and economy (regulatory capacity), (5) maintain internal coherence of state institutions (amalgamating capacity), and (6) distribute resources (redistributive capacity) (Wang, 2014; Pye, 1966; Binder et al., 1971; Grew, 1978 in Metreveli, 2016). The EU is an ideational, value driven power in the TrC framework. It is dominated by supply driven assistance's motive on the international arena rather than partnership and demand side. It is not based on the analysis of international identity in traditional nationstate discourse (Manners, 2002; Youngs, 2004 in Metreveli 2016). In this case, it is good to become a reminder that both the TrC and SSC are aimed mostly to implement development cooperation rather than competition, and there is no exception for the case of relationship between Indonesia, China and the EU. In a nutshell, not all ten principles of EU's point of view as the donor country are relevant to cope with Fragile States aid, especially when assisting the SMEs and cooperatives in both Indonesia and China.

REFERENCES

Adelia, Melly Kartika. (2016). Tarif Pajak UMKM akan Diturunkan 0.25–0.50 Persen. Bengkel UKM. http://www.bengkelumkm.com/id-288-post-tarif-pajak-untuk-umkm-turun-menjadi-02505-persen.html

Afrianto, Dedy. (2016). Top Speed 350 Km/Jam, Begini Penampakan Kereta Cepat Jakarta-Bandung. Okezone Finance. http://economy.okezone.com/read/2016/09/14/320/1489133/top-speed-350-km-jam-begini-penampakan-kereta-cepat-jakarta-bandung. Retrieved on September, 10 2016.

Austin, James. (2000). The collaboration challenge: How nonprofits and businesses succeed through strategic alliances. *Strategic Marketing for Nonprofit Organizations, page 224.*

Bi.go.id. (2017). Data Kredit UMKM Desember 2016, http://www.bi.go.id/id/umkm/kredit/data

Binder, Leonard, James S. Coleman, Joseph LaPalombara, Lucian W. Pye, Sidney Verba, & Myron Weiner. (1971). *Crisis and sequences in political development*. Princeton: Princeton University Press.

Chu, Kathy. (2013). *China manufacturers survive by moving to Asian neighbors. http://www.wsj.com/articles/SB10001424127887323798104578453073103566416*, retreived on August, 26 2016.

Davies, Howard, & Ellis Paul. (2000). Porter's competitive advantage of nations: Time for the final judgement? *Journal of Management Studies, 37*(8), 1189–1213. doi:10.1111/1467-6486.00221

Erasmus+Programme. (2017). *Guidelines for the use of the grant, capacity building in the field of higher education*, for grants awarded in 2016 under Call EAC/A04/2015

Farnell, John & Paul Irwin Crookes. (2016). *The politics of EU-China economic relations: An uneasy partnership*. United Kingdom: Palgrave Macmillan.

Financial Business.com. (2015). Kontribusi Penerimaan Pajak dari UMKM Hanya 3%. http://finansial.bisnis.com/read/20150903/10/468689/kontribusi-penerimaan-pajak-dari-umkm-hanya-3. Retreived on September 1 2016.

Meisari, Dewi. (2016). Koperasi: Solusi Pembangunan dengan Pemerataan? Seminar Koperasi dan Pembangunan Berkelanjutan, MM FEB UI, 27 July 2016, Kampus UI Salemba.

Hellsten, Sirkku K. (2010). Empowering the invisible: Women, local culture and global human rights protection. *Thought and Practice: A Journal of the Philosophical Association of Kenya (PAK) New Series, 2*(1), 37–57.

IFAD. (2013). *Consolidated financial statements of IFAD as at 31 December 2012*. https://webapps.ifad.org/members/eb/108/docs/EB-2013-108-R-13-Rev-1.pdf. Retreived on September 1 2013.

Kaplan, Seth, (2012). Horizontal versus vertical social cohesion: Why the differences matter. *Fragile-states*. http://www.fragilestates.org/2012/03/12/horizontal-versus-vertical-social-cohesion-why-the-differences-matter/. Retreived on August, 26 2016.

Kompas.com. (2016). Profesor Australia: Indonesia Tak Punya Kapasitas untuk Jadi Kekuatan Baru di Dunia http://internasional.kompas.com/read/2016/07/08/13300091/profesor.australia.indonesia.tak.punya.kapasitas.untuk.jadi.kekuatan.baru.di.dunia

Li, Wei. (2012). Small and medium enterprises—the source of China's economic miracle—and their financing challenges. *CSC Academic Group: Enterprise Development*. http://sydney.edu.au/china_studies_centre/china_express/issue_3/features/small-and-medium-enterprises.shtml. Retreived on August, 26 2016.

Luhulima, C.P.F. (2016). Superimposing China's 'maritime Silk Road' on Indonesia. *The Jakarta Post*. http://www.thejakartapost.com/news/2016/06/10/superimposing-china-s-maritime-silk-road-indonesia.html

Manners, Ian. (2000). *Normative power Europe: A contradiction in terms*? http://rudar.ruc.dk:8080/bitstream/1800/8930/1/Ian_Manners_Normative_Power_Europe_A_Contradiction_in_Terms_COPRI_38_2000.pdf.on September 10 2016.

Metreveli, Tornike. (2016). *The EU's normative power—its greatest strength or its greatest weakness?* http://www.atlantic-community.org/app/webroot/files/articlepdf/EUNormativePower.pdf. Retreived on September 1 2016.

Morcos, K. & Roder, Anne Friederike. (2007). The OECD's work on fragile states and situations. *Transforming Fragile States-Examples of Practical Experience*. Nomos Verlagsgesellschaft mbH & Co. KG.

Pye, Lucian W. (1966). *Aspects of political development*. Boston: Little, Brown and Company.

Ramalingam, Ben. (2012). *State fragility as a wicked problem?* http://aidontheedge.info/2012/02/15/state-fragility-as-a-wicked-problem/. Retreived on October 10, 2012.

Riyadi, Muhammad A. (2016). *Illegal Chinese workers invade Indonesia*. Gres.news. http://gres.news/news/politics/108616-illegal-chinese-workers-invade-indonesia/0/. Retreived on August, 26 2016.

Robison, Richard. (2016). *Why Indonesia will not be Asia's next giant*. The University of Melbourne, 14 July, 2016 http://indonesiaatmelbourne.unimelb.edu.au/why-indonesia-will-not-be-asias-next-giant/. Retreived on August, 26 2016.

Silk Road Foundation. (2007). Richthofen's silk roads: Toward the archaeology of a concept. *The Bridge between Eastern and Western Cultures, 5*(1).

Skurnik, Samuli. (2002). The role of cooperative entrepreneurship and firms in organizing economic activities—past, present and future. http://lta.hse.fi/2002/1/lta_2002_01_d6.pdf. Retreived on December 1, 2012.

Soesilo, Nining I. (2014). *Corporate level evaluation on fragile states and conflict situations*. Case Study in Indonesia and the Philippines. UN IFAD.

Tan, Valarie. (2015). *Indonesia struggles to stay competitive in Chinese market*. Channel NewsAsia. http://www.channelnewsasia.com/news/business/indonesia-struggles-to/2144352.html. Retreived on September 2 2016.

The Diplomat.com. (2015). *Indonesia, China seal 'maritime partnership,'* http://thediplomat.com/2015/03/indonesia-china-seal-maritime-partnership/. on October 22 2016.

UNESCO. (2000). *About the Silk Road*. http://en.unesco.org/silkroad/about-silk-road. Retreived on August, 26 2016.

UNDP. (2009). *Frequently asked question, the south-south and triangular cooperation*. http://www.undp.org/content/dam/undp/library/Poverty%20Reduction/Development%20Cooperation%20and%20Finance/SSC_FAQ%20v1.pdf. Retreived on February 2 2016.

Wang, Shaoguang. (2014). *Democracy and state effectiveness*. www.cuhk.edu.hk/gpa/wang_files/DemocracyandState.doc. Retreived on September 19 2016.

Wheeler, Brian Wheeler & Alex Hunt. (2016). *Brexit: All you need to know about the UK leaving the EU*. http://www.bbc.com/news/uk-politics-32810887. Retreived on September 20 2016.

Whittaker, Matt. (2016). *How to invest in China's changing economy*. http://money.usnews.com/investing/articles/2016-04-01/how-to-invest-in-chinas-changing-economy. Retreived on September 5 2016.

Xie, Chao. (2016). Fragile States Index 2016, China and the FSI: Decennary Trends, 2007–2016. IPCS. http://www.ipcs.org/article/india/china-and-the-fsi-decennary-trends-2007-2016-5123.html. Retreived on October 21 2016.

Youngs, R. (2004). Normative dynamics and strategic interests in the EU's external identity. *Journal of Common Market Studies 42*(2), 415–35. http://media.library.ku.edu.tr/reserve/resfall08_09/INTL533_BRumelili/22ndDecember/Normative_dynamics_wiley.pdf. Retrieved on October 22 2016.

Competition and Cooperation in Economics and Business – Gani et al. (Eds)
© 2018 Taylor & Francis Group, London, ISBN 978-1-138-62666-9

Relationship between living arrangements and marital status and the obesity of productive age women in Indonesia

U. Naviandi, T. Wongkaren & L.H.M. Cicih
Department of Economics, Faculty of Economics and Business, Universitas Indonesia, Depok, Indonesia

ABSTRACT: Obesity has been defined as a global epidemic with various impacts on social, economic, psychological, and health conditions not only in individuals but also in families. It affects life cycle. The prevalence of overweight and obese adults in Indonesia was quite high, especially among women (32.9 per cent) in 2013. This poses a threat for the Indonesian demographic dividend, considering the female population will reach 152.6 million by 2035. This paper examines the effects of living arrangements and marital status on women's obesity by analysing the data of 301,119 women aged 15–54 included in Indonesia's Basic Health Research of 2013. It uses a descriptive analysis and multinomial logistic regression. The regression is controlled for socio-demography and economic characteristics, such as smoking, consumption risk and physical activities. The preliminary results suggest that women living with a spouse plus others are more likely to be obese than women living alone or in other living arrangements. In addition, 'marital status' is a significant predictor of obesity. Married women are more likely to be obese than divorced or single women.

1 INTRODUCTION

Obesity has become a serious problem as the leading cause of death in the world. Borrell and Samuel (2014) found that adult obesity is not only associated with a high mortality rate but also with premature mortality, dependent on behavioural factors and socio-demographics. This is because obesity is closely related to a person's risk of complications of degenerative diseases (Handajani et al., 2009). The prevalence of obesity throughout the world, including Indonesia, continues to increase every year. Murray and Marie (2013) found that no country in the world has succeeded in reducing the rate of obesity in the last 33 years. In fact, Indonesia was the tenth ranked country with the largest population of adult obesity in the world after Germany and Pakistan (Institute for Health Metrics and Evaluation, 2013).

Indonesia's Basic Health Research (2013) shows that the prevalence of overweight and obese adults in Indonesia was 26.6 per cent, an increase compared to the year 2010 which amounted to 21.7 per cent. Women have a higher prevalence (32.9 per cent) than men (19.6 per cent). This condition is different from developed countries, where men have a higher rate of obesity prevalence than women (Murray & Marie, 2013). This is certainly the challenge for the future, given that the female population in Indonesia is estimated to reach 152.6 million in 2035, with the women of productive age (15–54 years) estimated at 85.9 million (56.2 per cent of the total female population).

The high prevalence of obese women is a big question that needs to be investigated, particularly that the increase in obesity tends to occur in productive age women. Indonesia's Basic Health Research in 2013 shows that the prevalence of overweight and obese women increases in the 16–18 year-old category and reaches its peak between 40–49 years. However, in infants and children, obesity prevalence in girls is lower than that of boys. This indicates that there are certain factors that affect the increase of obesity in productive age women.

Specialised in women, obesity needs more concern. In addition to the impact on disease risks, obesity also affects fertility rates (Jokela et al., 2007) and menopause (Akahoshi et al.,

2002). Moreover, obesity is associated with aspects of social psychology, as most women hope to have an 'ideal' body, especially women who are in conflict with their body shape (Coward, 1985). Obesity is negatively associated with beauty, and even the issue of 'not liking fat people' has become an important issue in America with regard to discrimination and human rights (Gilman, 2008). Other negative consequences of obesity in women affect the life cycle with psychosocial, economic and biological implications (Ryan, 2010). Obesity has an impact on lower opportunities to work or be married.

Creating a healthy productive age population is the first step to realising that the ageing population is autonomous and productive anyway. Moreover, the female population in Indonesia still has a statistically higher life expectancy than men. On the other hand, healthy productive age women are expected to give birth and create a healthy and quality generation. Economically, obesity also lowers the productivity of labour (Gates et al., 2008). Hence, an increasing number of people with obesity allegedly reduces public productivity, especially if obesity occurs in the productive population, which is the main asset in the challenge of the demographic dividend.

So far, research related to obesity has been conducted quite often in Indonesia, especially concerning aspects of nutrition and public health. However, further research related to obesity, especially in productive age women, is still very limited and very much needed, primarily associated with the living arrangements and marital status aspects that are an important factor for obesity in productive age women. Living arrangements and marriage are normally associated with potential support both economically and socially, and have an important role in the status of women's health, including obesity. This study focuses on viewing the relationship between living arrangements and marital status and the obesity of productive age women with regard to socioeconomic characteristics, demographics and lifestyles.

This research is expected to be useful in providing an understanding of the relationship between socioeconomic variables and demographics, especially living arrangements and marital status on the incidence of obesity in productive age women. Thus, it will provide input and recommendations for everyone, especially policymakers, in order to make the right decision in addressing this issue. In addition, it could create awareness for the public about a causal factor of obesity and its impact on health in later life.

2 THEORETICAL BACKGROUND

Obesity and overweight are defined as the abnormal or excessive fat accumulation that indicates a risk to health (WHO, 2016). According to Harahap, Widodo and Mulyati (2005), obesity is a condition where excessive body fat accumulates, to the point that a person's weight is far above normal and can endanger health; overweight is a condition where a person's weight exceeds normal weight. Measurements to determine obesity can be done in various ways: anthropometry, bioelectrical impedance or regional fat distribution. The most common method used to measure the rate of obesity is the Body Mass Index (BMI), which is obtained by dividing body weight (kg) by the square of height (metres). BMI can be used to determine how much a person can be exposed to a disease or even the risk of death due to his/her weight. Curve-U (or curve-J) in Figure 1 illustrates the relationship between BMI (illustrated by weight) and the high number (height) of deaths (mortality) which is the result of replication of various studies in multiple levels of the population (Gordon & Doyle in Miles & Himes, 1995).

Regarding the influence of age, Lee and Manson (1998) found a strong association between BMI and mortality in middle-aged women more than among older women. BMI may reflect less adiposity in older women because they tend to have more fat, which is concentrated in the middle and upper parts of the body, so the increased waist circumference is a strong predictor of death from obesity. In contrast, for younger women BMI is considered as one of the best predictors to describe death from obesity due to the distribution of fat and muscle which is considered more evenly, and which is different from older women.

The relationship between living arrangements and health according to causation-social theory (Joutsenniemi, 2007) is that marriage or living together has the effect of mutual

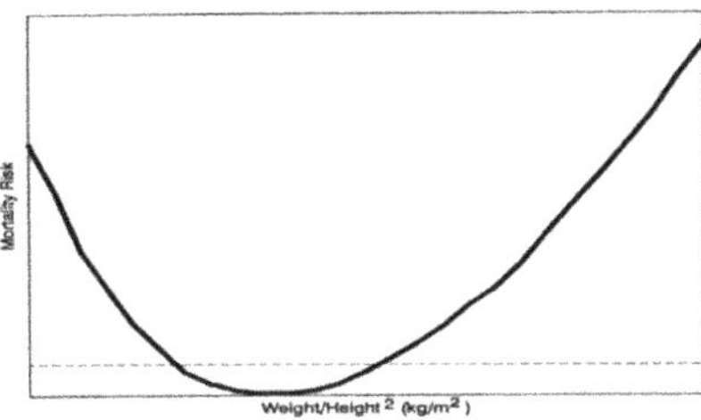

Figure 1. Mortality according to height/weight.
Source: Gordonand Doyle (1988).

support and protection against health risks, whereas living alone or with someone other than a spouse may have a detrimental effect on health. Living arrangements are not assumed to form one's health, but their effect is mediated by a number of factors.

Michael et al. (2001) found that women who live with others have a higher prevalence of obesity than those who live alone or with a partner. Rimmer and Hsieh (2010) found that the prevalence of overweight and obesity in American adults who live alone tends to be higher than that of those who stay with family. Lancaster's research (2009) in Australia found a relationship between living arrangements and obesity in both men and women, but found it to be stronger in women. Its influence is reflected through social support, the frequency of gathering with family members, and marital status.

According to Davis et al. (2000), women who live only with their spouses have lower BMI than those who live with others besides their spouses. They have less potential to smoke, are more likely to exercise regularly, and have better health reports. On the other hand, women who live with their spouses plus other household members have a higher BMI than women who live with others than their spouses because they are more likely to work and have a higher energy intake. Conversely, women who live with other non-spouses are more likely to have low levels of education, less likely to exercise, and have a lower energy intake. In addition, they are also more likely to smoke and have a poor health report.

Duvall and Miller (1985) define marriage as a relationship between men and women which provides sexual relations, offspring and division of roles between husband and wife. There is an association between marital status and health. According to Kim (2000), those who are married are healthier than non-married people. Similarly, widowed or divorced people also have worse health status than married people. Furthermore, marital status also has an influence on the tendency of women to be obese. Soriguer et al. (2003) found that marital status and education have a relationship to obesity. According to Borrell and Samuel (2014), the prevalence of obesity in American adults who are married is higher than that of single ones. Jeffery and Rick (2002) indicate the association of marriage with body weight; a significant weight gain occurs after a duration of two years of marriage. In contrast, those who are divorced will lose weight significantly in the first two years of separation.

3 METHODS

The study uses a cross-sectional design with secondary data from Indonesia's Basic Health Research conducted by the Ministry of Health in 2013. The population was all households representing 33 provinces in Indonesia. By the 11,986 census blocks that are visited, there are 294,959 households interviewed, with a sample of 301,119 women of productive age (15–54 years old) who reported their weight and height.

This research uses a descriptive analysis to gain an overview of living arrangements, marital status, and the obesity status of productive age women (15–54 years old) with a few control variables of socioeconomic characteristics, demographics, and lifestyles. The obesity status is categorised into non-obese ($BMI \leq 25.0$), overweight ($25.0 < BMI \leq 27.0$), and obesity ($BMI >$

27.0). Moreover, the living arrangements are grouped into: 'alone' for women who live alone at home, 'spouse' for women who live with only their spouse in the household, 'both non-spouse' for women who live with someone else but not their spouse in the household, 'spouse plus other' for women who live with their spouses plus other household members, and 'the other' for women who live together with other household members (>3 persons), but not with their spouses. Furthermore, the inferential analysis was used to test the hypothesis based on sample data. Multinomial logistic regression was used.

4 RESULTS AND DISCUSSION

The obesity rate of productive age women varies according to living arrangements. Women who live with their spouses only or with their spouses plus other household members have the highest percentage of obesity with 22.96 per cent and 24.04 per cent respectively. On the contrary, women who live with others have the lowest percentage of obesity, at 8.21 per cent. This condition is contrary to the study of Michael et al. (2001) in America where women who live with others actually have a higher prevalence of obesity due to lack of social communication and physical activity. This result is also different from Rimmer and Hsieh (2010), where the prevalence of adults with obesity who live alone tends to be higher than that of those who stay with family. Meanwhile based on Indonesia's Basic Health Research (2013), the percentage of women with obesity who live alone is 16.29 per cent, which is lower than that of women living with family, especially the spouse.

In general, the average BMI of women who live alone is 23.0 which is slightly higher than that of those living with a non-spouse (22.92). This means that women who live alone by average are more overweight than women who live with a non-spouse. This happens because women who live alone do physical activities less frequently than women who live with a non-spouse. Based on Indonesia's Basic Health Research in 2013, the percentage of women living alone who routinely perform strenuous activities (such as sports) is 25.46 per cent, lower than women who live with a non-spouse (26.34 per cent). This result contrasts with the research of Davis et al. (2000) which found the women who live with a non-spouse exercise less than those living alone (Table 1).

Based on data processing using the multinomial logistic regression model (Table 2), there are statistically significant differences in opportunities for obesity among women in the categories of 'alone' living arrangements, living with their spouses only, living with someone who is non-spouse, living with their spouses plus others, and living arrangements with others ($\alpha = 0.05$). This produces all positive coefficient of β, showing that women with 'alone' living

Table 1. Obesity distribution of productive age women according to each variable.

Variable	Obesity class			Total
	Normal	Overweight	Obese	N = 301.119
Living arrangements				
Alone	72.08	11.63	16.29	100.00
Spouse only	62.60	14.44	22.96	100.00
Non-spouse	72.81	10.78	16.42	100.00
Spouse plus others	60.78	15.18	24.04	100.00
Others	85.61	6.17	8.21	100.00
Marital status				
Divorced	66.97	12.89	20.14	100.00
Married	61.21	15.05	23.73	100.00
Single	90.57	4.30	5.13	100.00

Note: Percentages were calculated using Indonesia's Basic Health Research from 2013.

Table 2. Multinomial logit regression of productive aged women opportunities to obese, 2013.

	Model 1		Model 2		Model 3		Model 4	
Parameter	B	S.E	B	S.E	S.E	B	S.E	Odd ratio
Living Arrangements								
Alone	0.857***	0.048			0.050	0.153***	0.051	1.166
Spouse only	1.341***	0.023			0.031	0.214***	0.032	1.238
Non-spouse	0.855***	0.033			0.035	0.243***	0.036	1.275
Spouse plus others	1.416***	0.014			0.026	0.277***	0.026	1.319
Others	0b					0b		
Marital Status								
Divorced			1.670***	0.027	0.028	0.976***	0.031	2.655
Married			1.924***	0.018	0.030	1.045***	0.033	2.843
Single			0b			0b		
Age Group								
15–25						–1.116***	0.022	0.328
26–45						–0.138***	0.012	0.871
46–55						0b		
Urban/Rural								
Urban						0.246***	0.011	1.278
Rural						0b		
Education								
≤ Primary school						0.037*	0.020	1.038
Junior/Senior high school							0.030	0.019
> Senior high school							0b	
Economy								
40% lower						–0.704***	0.016	0.494
40% middle						–0.186***	0.013	0.831
20% higher						0b		
Working								
Yes						–0.111***	0.010	0.895
No						0b		
Mental Health								
Mental Disorders						0.008	0.024	1.008
Healthy						0b		
Physical Activity								
Very active						–0.132***	0.012	0.877
Moderate and sedentary							0b	
Smoking								
Yes						–0.003	0.034	0.997
No (but ever)						0.354***	0.056	1.424
Never						0b		
Consumption Risk Food								
Rarely						–0.085***	0.014	0.919
Often						0b		
Intercept	–2.344***	0.013	–2.871***	0.017	0.018	–1.807***	0.032	
Pseudo R-Square	0.73		0.95		0.97			0.137

Note: The outcome of each regression model from Indonesia's Basic Health Research (2013); Model 1 = only includes living arrangements as an independent variable, Model 2 = only includes marital status as an independent variable, Model 3 = living arrangements and marital status collectively as independent variables, Model 4 = living arrangements and marital status controlled by socioeconomic, demographic, and lifestyle variables; *** = significant at p ≤ 0,01, ** = significant at p ≤ 0,05, * = significant at p ≤ 0,1.

arrangements, living with their spouses only, living with someone who is non-spouse, and living with their spouses plus others have a higher risk of becoming obese than women with living arrangements with other household members (Model 1).

Furthermore, after controlled by the marital status variable (Model 3), socioeconomic demographic and lifestyle (Model 4), living arrangements still have a statistically significant effect on the tendency of productive age women to become obese ($\alpha = 0.05$), with a β coefficient direction that is positive. This result suggests that there is a strong (robust) relationship between the effects of living arrangements and the tendency of productive age women to become obese.

However, the effect of each category tends to weaken after tested by control variables. This is seen from the coefficient of β which falls into each category of living arrangements. The largest decline of the β coefficient is after being controlled by the marital status variable (between Model 1 and Model 3), mainly in the category of living arrangements with spouse only (–1.045) and spouse plus others (–1.045). The category of 'alone' living arrangements and with non-spouse is –0.609 and –0.530 each, respectively. It reflects that partly the effect of the living arrangements variable against the tendency to obesity has been described through the marital status variable. As for Model 3 and Model 4, the reduction of β coefficient for the four categories of living arrangements is much smaller ($\beta1 = -0.094$; $\beta2 = -0.082$; $\beta3 = -0.081$; $\beta4 = -0.094$). It means that the influence of the controlled socioeconomic and demographic variables is able to explain the effect of living arrangements on obesity in only a small percentage.

The value ($\beta1 = 0.153$, $p = 0.002$) shows that women with 'alone' living arrangements have a tendency to become obese 1.17 times more than women with living arrangements with others. This is because women who live alone are more likely to have higher education (12.62 per cent) and to work (65.77 per cent) than those living with others (7.89 per cent; 33.82 per cent), and thus they require a higher energy intake (Davis et al., 2000) (Figure 2).

Women who live alone tend to have lower social support. Lancaster (2009) found that those who have low social support and are rarely with other people are more predisposed to obesity, especially women. Other factors are similar to the studies of Brunt and Rhee (2008) and Davis et al. (2000) that found that women with 'alone' living arrangements are more likely to consume alcohol or smoke and are at risk for obesity, especially in adolescence. Based on data from Indonesia's Basic Health Research in 2013, the percentage of women living alone that smoke is 5.23 per cent and that ever smoked 1.36 per cent, which is more than the percentage of women with living arrangements with other household members (respectively 1.45 per cent and 0.51 per cent).

Furthermore, women in living arrangements with their spouses only ($\beta2 = 0.214$, $p = 0.000$) have a 1.24 times greater tendency to become obese than women in living arrangements with others. Similarly, women in living arrangements with their spouses plus others ($\beta4 = 0.277$,

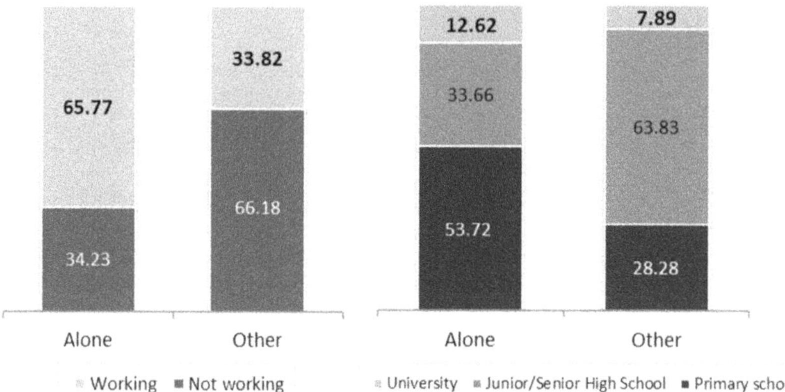

Figure 2. Distribution of 'alone' living arrangements and others according to education and employment status of women at productive age.
Source: Indonesia's Basic Health Research (2013).

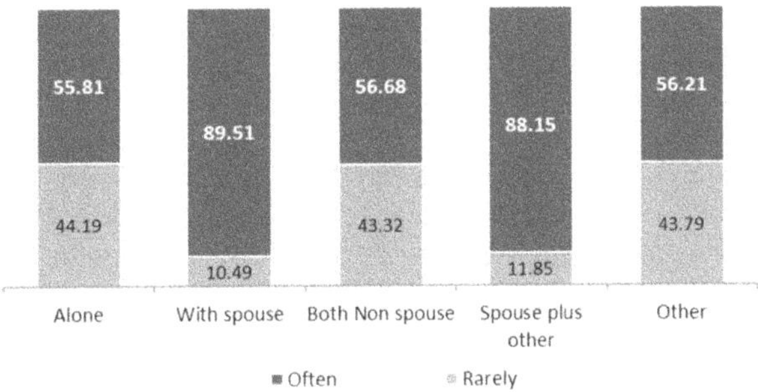

Figure 3. Percentage of productive aged women according to living arrangements and consumption of risk foods.
Source: Indonesia's Basic Health Research (2013).

p = 0.000) have the trend of obesity of 1.32 times greater than women in living arrangements with others. Based on Model 1 and Model 3, the effects of living arrangements on the tendency for obesity can be seen, and this is described through the marital status of women, especially for those who live with their spouses only (Category 2) and spouses plus others (Category 4).

The higher percentage of women with obesity living their spouse is not caused by lack of physical activities. The percentage of women who routinely perform strenuous physical activities (including sports) is higher in those with living arrangements with their spouses only (29.56 per cent) or with their spouses plus others (28.15 per cent) than in those with living arrangements with others (19.25 per cent). The effect is due rather to the high consumption of risky foods. Based on Indonesia's Basic Health Research of 2013, the percentage of women with living arrangements with their spouses only and with their spouses plus others who frequently consume risk foods is 89.51 per cent and 88.15 per cent respectively, more than the percentage of women in living arrangements with others (56.21 per cent) (Figure 3).

Value ($\beta 5 = 0.976$, $p = 0.000$) in the model indicates that divorced women have a tendency to become obese 2.66 times greater than unmarried women. Meanwhile, married woman ($\beta 6 = 1.045$, $p = 0.000$) have the tendency to become obese 2.84 times greater than unmarried women. The mechanism of married women for becoming obese is through pregnancy and bearing children, duration of marriage, contraceptive use, husband's employment status and education. The probability of women using hormonal contraception for becoming obese is 1.35 times higher than that of the women who never use contraceptives. Moreover, the probability of women using non-hormonal contraception of becoming obese is 1.68 times higher than that of women who never use contraceptives.

5 CONCLUSION

Living arrangements have an effect on the obesity status of productive age women. Women with 'alone' living arrangements, living with their spouses only, living with non-spouse, and living with their spouses plus others, have a higher tendency to become overweight and suffer from obesity than those who live with others. Especially for women who live alone, the effect is explained by the higher level of education, employment status, low social support and the lifestyle of smoking. Moreover, for women in living arrangements with their spouses only and spouses plus others, the effect is explained by marital status and consumption of riskier foods. For women in living arrangements with non-spouses, the effect is explained by their propensity to consume risky food and to smoke.

The marital status variable also has a significant influence on the trend of productive age women to be overweight and obese, and it is even greater than the living arrangements variable. Married and divorced women are more likely to be obese than unmarried women.

Pregnancy and having children are the causes in this case. Women who have children, giving birth to more than two children, will have a tendency to be overweight or obese; the tendency is higher than women who do not have children or have 1–2 children. Furthermore, women who have been married more than 17 years will have a higher probability to become obese than newly married women (≤17 years). The use of contraceptives also has a positive effect on the tendency of women to become overweight or obese. Similarly husband's education has positive effects on obesity in married women. The husband who has higher education tends to have a more overweight wife than he who has lower education.

However, this research is a cross-sectional study, so it cannot see the effects of time, such as a woman's health history and nutrition when they were in the womb, in infancy, and in early childhood. The time accumulation related to the process of women becoming obese cannot be explained in the study so this is a limitation. In addition, the variables related to social support are also not available in the data set. Further research is recommended to see a variety of forms of social support from family or environment. In addition, the effects of spousal obesity or other household members on the probability of women having obesity should be examined.

REFERENCES

Akahoshi, M., Soda, M., Nakashima, E., Tsuruta, M., Ichimaru, S., Seto, S. & Yano, K. (2002). The effect of body mass index on age at menopause. *International Journal of Obesity, 26*(7), 961–968.

Indonesia Ministry of Health. (2013). *Indonesia basic health research*. Jakarta. Ministry of Health.

Institute for Health Metrics and Evaluation. 2013. *Global burden of disease study*. Washington. University of Washington. Retrieved from http://www.healthdata.org/news-release/nearly-one-third-world%E2%80%99 s-population-obese-or-overweight-new-data-show

Borrell, L.N. & Samuel, L. (2014). Body mass index categories and mortality risk in US adults: The effect of overweight and obesity on advancing death. *American Journal of Public Health, 104*(3), 512–19. doi:10.2105/AJPH.2013.301597

Coward, Rosalind. (2002). The body beautiful. In Grewal, Inderpal, and K. Caren. *An Introduction to Woman Studies: Gender in Transnational Word. (pp. 366–375)*. New York: McGraw-Hill Companies.

Davis, M.A., Murphy, S.P., Neuhaus, J.M., Gee, L., and Quiroga, S.S. (2000). Living arrangements affect dietary quality for U.S. adults aged 50 years and older: NHANES III 1988–1994. *The Journal of Nutrition, 130*(9), 2256–64.

Duvall, M. & Miller, B.C. 1985. *Marriage and family development* (6th ed.). New York: Harper and Row Publisher Inc.

Gates, D.M., Succop, P., Brehm, B.J., Gillespie, G.L., & Sommers, B.D. (2008). Obesity and presenteeism: The impact of body mass index on workplace productivity. *Journal of Occupational Environmental Medicine, 50*(1), 39–45. doi: 10.1097/JOM.0b013e31815d8db2

Gilman, Sander L. (2008). *Fat; a cultural history of obesity*. Cambridge, UK: Polity Press.

Handajani, A., Roosihermiatie, B., & Maryani, H. (2009). Faktor-faktor yang berhubungan dengan pola kematian pada penyakit degeneratif di Indonesia. *Buletin Penelitian Sistem Kesehatan, 13*, 42–53.

Harahap, H., Widodo, Y., & Mulyati, S. (2005). Penggunaan berbagai *Cut-Off* Indeks Massa Tubuh Sebagai Indikator Obesitas Terkait Penyakit Degeneratif di Indonesia. *Gizi Indonesia, 2*(28).

IOTF/WHO. (2000). *The Asia Pacific perspective: Redefining obesity and its treatment*. Melbourne: Health Communication Australia.

Jeffery, R.W. & Rick, A.M. (2002). Cross sectional and longitudinal associations between body mass index and marriage-related factors. *Obesity Research, 10*, 809–815.

Jokela, M., Kivimäki, M., Elovainio, M., Viikari, J., Raitakari, O.T., & Järvinen, L.K. (2007). Body mass index in adolescence and number of children in adulthood. *Epidemology, 18*(5), 599–606.

Joutsenniemi, Kaisla. (2007). *Living arrangements and health*. Helsinki: Department of Public Health, University of Helsinki.

Kim, Jung Ki. (2000). *Marriage and health: The effect of marital status on health and the mechanisms by which marital status affects health among elderly people*. Dissertation, Faculty of Graduate School, University of Southern California.

Lancaster, Kate Germaine. (2009). *Overweight and obesity among Australian youth: Associations with family background and social networks*. Dissertation, School of Political and Social Inquiry, Monash University.

Lee, I-Min, & Manson, Joann E. (1998). Body weight and mortality: What is the shape of the curve? *Epidemiology, 9*(3).

Michael, Y.L., Berkman, L.F., Colditz, G.A., & Kawachi, I. (2001). Living arrangements, social integration, and change in functional health status. *American Journal of Epidemiology, 153*(2), 123–31. doi:10.1093/aje/153.2.123

Miles, T.P., & Himes, C. (1995). Biological and social determinants of body size across the life span—a model for the integration of population genetics and demography. *Population Research and Policy Review, 14*(3), 327–46. doi:10.1007/BF01074396

Murray, Christopher J.L., & Marie Ng. (2013). Global, regional, and national prevalence of overweight and obesity in children and adults during 1980–2013: Asystematic analysis for the Global Burden of Disease Study 2013. *Published in the Lancet, 384,* 766–781.

Rimmer, James, & Hsieh Kelly. (2010) Longitudinal health and intellectual disability study (LHIDS) on obesity and health risk behaviour. *Working paper:* University of Illinois at Chicago.

Ryan, Donna. (2007). Obesity in women: A life cycle of medical risk. *International Journal of Obesity*, 53–57. doi:10.1038/sj.ijo.0803729

Soriguer, F., G. Rojo-Martínez, I. Esteva De Antonio, M.S. S Ruiz De Adana, M. Catalá, M.J. J Merelo, M. Beltrán, et al. (2004). Prevalence of obesity in South-East Spain and its relation with social and health factors. *Eur J Epidemiol, 19*(1), 33–40. doi:10.1023/B:EJEP.0000013254.93980.97

World Health Organization. (1995). Physical status: The use and interpretation of anthropometry. *Report of WHO Expert Committee*. Geneva: WHO.

World Health Organization, Media Centre. (2013). Noncommunicable diseases. Accessed on 15 Feb 2016. Retrieved from http://who.int/mediacentre/factsheets/fs355/en/

Competition and Cooperation in Economics and Business – Gani et al. (Eds)
© 2018 Taylor & Francis Group, London, ISBN 978-1-138-62666-9

Author index

Milton Keynes UK
Ingram Content Group UK Ltd.
UKHW020843141024
449569UK00009B/594

9 781138 626669